Handbook of Geometry and Topology of Singularities II

José Luis Cisneros-Molina · Dũng Tráng Lê ·
José Seade
Editors

Handbook of Geometry and Topology of Singularities II

Springer

Editors
José Luis Cisneros-Molina
Instituto de Matemáticas
Unidad Cuernavaca
Universidad Nacional Autónoma de México
Cuernavaca, Mexico

Dũng Tráng Lê
Centre de Mathématiques et Informatique
Université d'Aix-Marseille
Marseille, France

José Seade
Instituto de Matemáticas
Universidad Nacional Autónoma de México
Mexico City, Mexico

ISBN 978-3-030-78023-4 ISBN 978-3-030-78024-1 (eBook)
https://doi.org/10.1007/978-3-030-78024-1

This Springer imprint is published by the registered company Springer Nature Switzerland AG
The registered company address is: Gewerbestrasse 11, 6330 Cham, Switzerland

Preface

This is the second volume of the Handbook of the Geometry and Topology of Singularities, a subject which is ubiquitous in mathematics, appearing naturally in a wide range of different areas of knowledge. The scope of singularity theory is vast; its purpose is multifold. This is a meeting point where many areas of mathematics, and science in general, come together.

Let us recall Bernard Teissier's words in his foreword to the Handbook in Volume I:

> I claim that Singularity Theory sits inside Mathematics much as Mathematics sits inside the general scientific culture. The general mathematical culture knows about the existence of Morse theory, parametrizations of curves, Bézout's theorem for plane projective curves, zeroes of vector fields and the Poincaré-Hopf theorem, catastrophe theory, sometimes a version of resolution of singularities, the existence of an entire world of commutative algebra, etc. But again, for the singularist, these and many others are lineaments of a single landscape and she or he is aware of its connectedness. Moreover, just as Mathematics does with science in general, singularity theory interacts energetically with the rest of Mathematics, if only because the closures of non-singular varieties in some ambient space or their projections to smaller dimensional spaces tend to present singularities, smooth functions on a compact manifold must have critical points, etc. But singularity theory is also, again in a role played by Mathematics in general science, a crucible where different types of mathematical problems interact and surprising connections are born.

The Handbook has the intention of covering a wide scope of singularity theory, presenting articles on various aspects of the theory and its interactions with other areas of mathematics in a reader-friendly way. The authors are world experts; the various articles deal with both classical material and modern developments.

Volume I of this collection gathered together ten articles with foundational aspects of the theory. These include

- The combinatorics and topology of plane curves and surface singularities.
- An introduction to four of the classical methods for studying the topology and geometry of singular spaces, namely resolution of singularities, deformation theory, stratifications, and slicing the spaces à la Lefschetz.

- Milnor fibrations and their monodromy.
- Morse theory for stratified spaces and constructible sheaves.
- Simple Lie algebras and simple singularities.

Volume II also consists of ten articles. These cover foundational aspects of the theory and related topics, including

- The analytic classification of plane curve singularities and the existence of complex and real algebraic curves in the plane with prescribed singularities.
- An introduction to the limits of tangents to a complex analytic surface, a subject that originates in Whitney's work on understanding the set of limits of tangents at smooth points as one approaches the singular set.
- Introductions to Zariski's equisingularity and Intersection homology, which are two of the main current viewpoints for studying singularities. Equisingularity means equivalent or similar singularity in some sense that has to be made precise and it is a vast field of current research. Intersection homology was introduced by Mark Goresky and Robert MacPherson and is a brilliant way of making the famous duality theorems for compact oriented manifolds work for singular varieties.
- An overview of Milnor's fibration theorem for real and complex singularities, as well as an introduction to Massey's theory of Lê cycles, which encode deep information about the geometry and topology of the Milnor fibers of complex hypersurface singularities.
- A discussion of mixed singularities, which are real analytic singularities with a rich structure that allows their study via complex geometry. This uses the method of the non-degenerate Newton boundary and toric modifications, which are powerful tools for the study of complex analytic singularity theory.
- The study of intersections of concentric ellipsoids in $\mathbb{R}^n$ and its relation with several areas of mathematics, from holomorphic vector fields to singularity theory, toric varieties, and moment-angle manifolds.
- A review of the topology of quasi-projective varieties and generalizations of results about the topology of the complements of singular plane curves and hypersurfaces in projective space.

Each chapter in Volume II has its own introduction and a large bibliography for further reading, and there is a global index of terms at the end.

This collection, the Handbook of Geometry and Topology of Singularities, will continue with three more volumes. Volumes III and IV will include contributions on Zariski equisingularity; the basic theory of $\mathcal{A}$-equivalence and density of stable maps due to John Mather, Terry Wall, and others; various aspects of the theory of Chern classes for singular varieties; indices of vector fields, 1-forms and foliations, extending the classical local index of Poincaré-Hopf; Lipschitz geometry in singularity theory; an introduction to mixed Hodge structures; limits of tangent spaces in high dimensions; tropical geometry; determinantal varieties; constructible sheaves and other important aspects of singularity theory. Volume V will be devoted to singular holomorphic foliations in complex manifolds.

Re-phrasing Bernard Teissier's words above, these topics, among many others, are lineaments of the single landscape that goes under the name "singularity theory".

There is a lot more that ought to be included in this collection but, happily, the vastness of this rich area of mathematics makes impossible the task of gathering together in five volumes so many important aspects. Yet, these five volumes together will cover a wide spectrum of singularity theory and its interactions with other related areas of mathematics.

This book is addressed to graduate students and newcomers to the theory, as well as to specialists who can use it as a guidebook, and it provides an accessible account of the state of the art in several aspects of the subject, its frontiers, and its interactions with other areas of research.

Cuernavaca, Mexico
Marseille, France
Mexico City, Mexico
March 2021

José Luis Cisneros-Molina
Dũng Tráng Lê
José Seade

Contents

Contributors

Jean-Paul Brasselet I2M, CNRS and Université d'Aix-Marseille, Marseille, France

José Luis Cisneros-Molina Instituto de Matemáticas, Unidad Cuernavaca, Universidad Nacional Autónoma de México, Cuernavaca, Mexico;
Laboratorio Solomon Lefschetz, International Research Laboratry CNRS, Cuernavaca, Mexico

Gert-Martin Greuel Universität Kaiserslautern, Kaiserslautern, Germany

Abramo Hefez Universidade Federal Fluminense, IME, R. Prof. Marcos Waldemar de Freitas Reis, Niterói-RJ, Brazil

Marcelo Escudeiro Hernandes Universidade Estadual de Maringá, Maringá-Pr, Brazil

Tráng Dũng Lê University of Aix-Marseille, Marseille, France

Anatoly Libgober Department of Mathematics, University of Illinois, Chicago, IL, USA

Santiago López de Medrano Instituto de Matemáticas, Universidad Nacional Autónoma de México, Mexico City, Mexico

David B. Massey Department of Mathematics, Northeastern University, Boston, MA, USA

Mutsuo Oka Tokyo University of Science, Shinjuku-ku, Tokyo, Japan

Adam Parusiński Université Côte d'Azur, CNRS, LJAD, UMR 7351, Nice, France

José Seade Laboratorio Solomon Lefschetz, International Research Laboratry CNRS, Cuernavaca, Mexico;
Instituto de Matemáticas, Universidad Nacional Autónoma de México, Mexico City, Mexico

Eugenii Shustin Tel Aviv University, Tel Aviv, Israel

Jawad Snoussi Universidad Nacional Auta de México, Instituto de Matemáticas, Unidad Cuernavaca, Avenida Universidad S/n, Lomas de Chamilpa, Cuernvaca, México;
Unidad Mixta Internacional CNRS, Laboratorio Solomon Lefschetz, Cuernavaca, Mexico

Chapter 1
The Analytic Classification of Irreducible Plane Curve Singularities

Abramo Hefez and Marcelo Escudeiro Hernandes

Abstract In 1973, Oscar Zariski gave a course at the *École Polytechnique* (cf. the lecture notes [23] or the monograph [24]), where he discussed in the local case the analogous problem to the construction of the moduli space of algebraic curves of genus g; that is, he proposed to search for moduli spaces with respect to analytic equivalence for germs of irreducible complex analytic plane curves having the same topological type. Zariski recognized that this problem was at that stage very difficult and concentrated his efforts on explicit calculations in some particular cases. Since then, the subject has substantially advanced and our purpose here is to further detail the solution of the moduli problem for plane branches given in [12], using the same framework as Zariski's, adding to his methods singularity tools that were starting to blossom at this time. The results we present improve several ones found in the literature and shed light over many questions asked by Zariski, using techniques that may be useful to solve other relevant related questions. The exposition is kept as elementary as possible to highlight the beauty and simplicity of the solution of Zariski's problem and since it is not intended to be a compendium on the subject, several results from algebra and singularity theory will be invoked, giving precise statements and references, where they may be found, without concern about quoting primary sources.

Contents

A. Hefez (✉)
Universidade Federal Fluminense, IME, R. Prof. Marcos Waldemar de Freitas Reis, s/n, Bloco H, Campus do Gragoatá, São Domingos, Niterói-RJ 24210-201, Brazil
e-mail: ahefez@id.uff.br

M. E. Hernandes
Universidade Estadual de Maringá, Av. Colombo 5790, Maringá-Pr 87020-900, Brazil
e-mail: mehernandes@uem.br

J. L. Cisneros-Molina et al. (eds.), *Handbook of Geometry and Topology of Singularities II*, https://doi.org/10.1007/978-3-030-78024-1_1

1.1 Background

In this section we will introduce our objects of study: germs of complex analytic plane curves and their singularities. Although, apparently, it may seem to be a shallow subject, surprisingly, this is not the case, since it relates important branches of mathematics as algebra, topology, analytic geometry and theory of singularities.

Since our major problem is the analytic classification of plane curve singularities, we will start by setting notation, defining our basic objects and studying the relations among them. This part may be used as an introduction to the study of germs of plane analytic curves.

1.1.1 Plane Curve Singularities

A complex *plane affine curve* is a set of the form

$$C_f = \{(x, y) \in \mathbb{C}^2;\ f(x, y) = 0\},$$

where $f \in \mathbb{C}[X, Y]$ is a non-constant polynomial without multiple factors.

A point $P = (a, b) \in \mathbb{C}^2$ will be called a *singular point* of C_f, if

$$f(a, b) = \frac{\partial f}{\partial X}(a, b) = \frac{\partial f}{\partial Y}(a, b) = 0.$$

There are at most finitely many singular points on C_f, and our objective here is to study such exceptional points from different points of view: algebraic, topologic and analytic, as we will explain in the course of these notes.

To C_f there is associated its coordinate ring $R_f = \mathbb{C}[X, Y]/\langle f \rangle$, where $\langle f \rangle$ denotes the ideal of $\mathbb{C}[X, Y]$ generated by f. This ring characterizes algebraically, in a categorical sense, the geometric object C_f, but the behavior of C_f at a particular point $P = (a, b)$ could be better understood if one considers, instead, the localization of R_f at P, that is, the local ring

$$R_{f,P} = \left\{ \frac{g}{h};\ g, h \in R_f,\ h \notin \mathcal{M}_P \right\} \subset \mathcal{K}_f,$$

where $\mathcal{M}_P$ is the maximal ideal of R_f corresponding to the point $P = (a, b)$ and $\mathcal{K}_f$ is the total ring of fractions of R_f.

Still, this ring is not "sufficiently local" to describe properly the properties of C_f at P. For example, the real trace of the irreducible nodal curve C_f defined by $f = Y^2 - X^2(X + 1)$ looks like Fig. 1.1, suggesting that this curve is locally reducible in a neighborhood of the origin in the classical topology of $\mathbb{C}^2$.

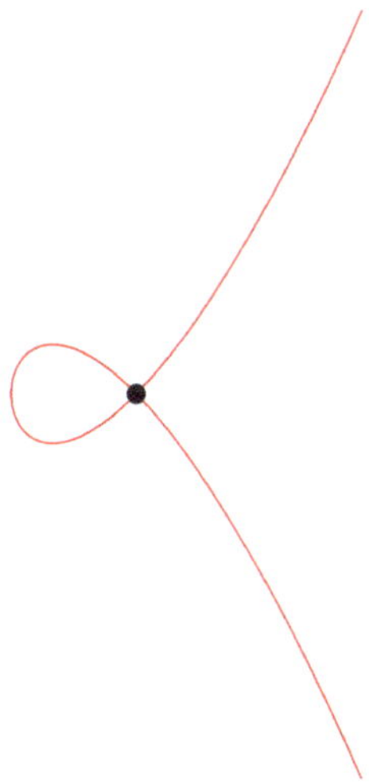

Fig. 1.1 The nodal curve

This fact is not reflected by the ring $R_{f,P}$ which, in this example, is a domain, since it is a subring of the field of fractions $\mathcal{K}_f$ of the domain R_f. The trouble is that the local character of $R_{f,P}$ refers to the Zariski topology of $\mathbb{C}^2$ which is coarser than the classical topology.

To remedy this situation, we may think of the polynomial f as an element of $\mathcal{O} = \mathbb{C}\{X, Y\}$, the ring of complex convergent power series. In this ring, f decomposes as $f = (Y - X\sqrt{X + 1})(Y + X\sqrt{X + 1})$, therefore, there exists a neighborhood U

of the origin in $\mathbb{C}^2$ such that

$$C_f \cap U = \{(x, y) \in U;\ y = x\sqrt{x+1}\} \cup \{(x, y) \in U;\ y = -x\sqrt{x+1}\}.$$

This point of view is very promising since the ring $\mathcal{O}$ carries only information around the origin in the classical topology of $\mathbb{C}^2$ and has very good algebraic properties: it is a Noetherian local regular and factorial domain (cf. [17, Chap. 3]). The group $\mathcal{O}^*$ of units of $\mathcal{O}$ consists of the series u such that $u(0,0) \neq 0$, hence $\mathcal{M} = \langle X, Y \rangle$ is the maximal ideal of $\mathcal{O}$.

So, we may associate to $f \in \mathcal{M} \setminus \{0\}$ a germ of a complex analytic plane curve through the origin of $\mathbb{C}^2$, which we still denote by C_f. The series f is called a *Cartesian equation* of C_f, and determines an analytic function in a non specified open neighborhood of the origin $0 := (0, 0)$ of $\mathbb{C}^2$, which we denote by $(\mathbb{C}^2, 0)$.

From the Hilbert-Rückert Nullstellensatz [17, Theorem 3.4.4], to have a good correspondence between germs of plane curves and their equations, we will always assume that these are reduced. This result implies that for $f, g \in \mathcal{M} \setminus \{0\} \subset \mathcal{O}$, reduced, one has

$$C_f = C_g \iff f = ug, \quad \text{for some } u \in \mathcal{O}^*.$$

The above equivalence asserts that a germ of analytic plane curve determines its equation modulo multiplication by a unit. Hence, if we denote by $\mathrm{Red}(\mathcal{M})$ the set of reduced series in $\mathcal{M}$, then we may identify the set of germs of analytic plane curves with the quotient $\mathrm{Red}(\mathcal{M})/\mathcal{O}^*$.

Let us write f in the form $f = f_n + f_{n+1} + \cdots$, where each f_i is a homogeneous polynomial of degree i or the zero polynomial and $f_n \neq 0$. The polynomial f_n is called the initial form of f and n the *multiplicity* of f or of C_f. We will write $n = \mathrm{mult}(f) = \mathrm{mult}(C_f)$. Since f_n is a homogeneous polynomial in two variables and $\mathbb{C}$ is algebraically closed, it follows that f_n splits into a product of linear homogeneous polynomials:

$$f_n = \prod_{i=1}^{l} (a_i X + b_i Y)^{r_i}, \quad \text{with } r_1 + \cdots + r_l = n.$$

The union of lines defined by the equations $a_i X + b_i Y = 0, i = 1, \ldots, l$, is called the tangent cone of C_f, denoted by TC_f. Using the same definition for a singular point as for plane affine curves in $\mathbb{C}^2$, it follows that C_f is singular at the origin if and only if $n = \mathrm{mult}(f) > 1$; that is, $f \in \mathcal{M}^2$. If C_f is not singular at the origin, it will be called *nonsingular*.

When $f \in \mathcal{M} \setminus \{0\}$ is irreducible, by Hensel's Lemma (cf. [17, Corollary 3.3.21]), one has that $f_n = (aX + bY)^n$, so TC_f consists of only one line.

From the Inverse Function Theorem (cf. [17, Corollary 3.3.7]), an analytic map germ $\varphi\colon (\mathbb{C}^2, 0) \to (\mathbb{C}^2, 0)$ is an isomorphism if and only if

$$\varphi(X, Y) = (aX + bY + g, cX + dY + h),\ \ g, h \in \mathcal{M}^2\ a, b, c, d \in \mathbb{C},\ ad - bc \neq 0.$$

If this is the case, φ is called an automorphism of $(\mathbb{C}^2, 0)$. The set of all such automorphisms form a group under composition, denoted by $\mathrm{Aut}(\mathbb{C}^2, 0)$.

Definition 1.1.1 We say that the curves C_f and C_g are *analytically equivalent*, writing $C_f \simeq C_g$, if there exists an *automorphism* φ of $(\mathbb{C}^2, 0)$ such that $\varphi(C_f) = C_g$.

In terms of equations, one has that $C_f \simeq C_g$ if and only if there exist an automorphism φ of $(\mathbb{C}^2, 0)$ and a unit u of $\mathcal{O}$ such that $f = u(g \circ \varphi)$. The above condition is an equivalence relation in the set of reduced elements in $\mathcal{M}$, called analytic equivalence and denoted by $f \sim g$.

Given $f, g \in \mathcal{O}$, we define the *intersection multiplicity* at the origin of f and g as being

$$\mathrm{I}(f, g) = \dim_{\mathbb{C}} \mathcal{O}/\langle f, g\rangle.$$

We reproduce below, for the reader's convenience, the well known facts about intersection multiplicities (cf. [11, Theorems 4.14 and 4.18]).

Property 1.1.2 Properties of intersection multiplicities:

(a) $\mathrm{I}(f, g) = \infty$ (respectively, $\mathrm{I}(f, g) = 0$) if and only if f and g have a nontrivial common factor (respectively, f or g is a unit);
(b) $\mathrm{I}(f, g) = \mathrm{I}(g, f)$;
(c) $\mathrm{I}(f, g) = \mathrm{I}(f \circ \varphi, g \circ \varphi) = \mathrm{I}(uf, vg)$, for all $\varphi \in \mathrm{Aut}(\mathbb{C}^2, 0)$ and all $u, v \in \mathcal{O}^*$;
(d) $\mathrm{I}(f, gh) = \mathrm{I}(f, g) + \mathrm{I}(f, h)$, for all $h \in \mathcal{O}$;
(e) $\mathrm{I}(f, g) = 1$ if and only i f$f, g \in \mathcal{M} \setminus \mathcal{M}^2$ and $TC_f \cap TC_g = \{0\}$;
(f) $\mathrm{I}(f, g) = \mathrm{I}(f, g - hf)$, for all $h \in \mathcal{O}$.
(g) $\mathrm{I}(f, g) \geq \mathrm{mult}(f)\,\mathrm{mult}(g)$ with equality if and only if $TC_f \cap TC_g = \{0\}$.

Given two germs of curves C_f and C_g, because of Property 1.1.2 (c), we may define their intersection multiplicity as $C_f \cdot C_g = \mathrm{I}(f, g)$, which is invariant under automorphisms of $(\mathbb{C}^2, 0)$.

For irreducible $f \in \mathcal{M} \setminus \{0\}$, we define

$$\Gamma(f) = \{\mathrm{I}(f, g);\ g \in \mathcal{O}\} \subset \mathbb{N} \cup \{\infty\}.$$

From Properties 1.1.2 (a) and (d), this set has the structure of a semigroup,[1] called the *semigroup of values* of C_f. Since $\Gamma(f)$ describes the way C_f intersects all germs of curves, not necessarily reduced, at the origin of $\mathbb{C}^2$, it is conceivable that it will play an important role, as will be confirmed later.

From Property 1.1.2 (c), it follows that

$$C_f \simeq C_g \implies \Gamma(f) = \Gamma(g). \tag{1.1}$$

[1] By a semigroup we mean a subset of $\mathbb{N} \cup \{\infty\}$ that contains $\{0, \infty\}$ and is closed under addition.

Any property or numerical function, either on f or on C_f, invariant by analytic equivalence, will be said to be an *analytic invariant*.

For instance, from (1.1), it follows that $\Gamma(f)$ is an analytic invariant attached to the curve C_f. On the other hand, since the multiplicity of f is invariant by composition with an automorphisms of $(\mathbb{C}^2, 0)$ and multiplication by units, this is also an analytic invariant. These two invariants are related because from Property 1.1.2 (g) it follows that $\text{mult}(f) = \min(\Gamma(f) \setminus \{0\})$, hence, also for this reason, the multiplicity is an analytic invariant.

Remark 1.1.3 If C_f is nonsingular, then $f = aX + bY + hot$, with $a \neq 0$ or $b \neq 0$, where hot stands for an element in $\mathcal{M}^2$. By taking $\varphi(X, Y) := (cX + dY, f)$ with $ad - bc \neq 0$ and $g(X, Y) := Y$, we have that $f = 1 \cdot g \circ \varphi$, which shows that $C_f \simeq C_Y$. This implies that all nonsingular curves are analytically equivalent to the germ $(\mathbb{C}, 0)$.

The above equivalence $f \sim Y$ for f of multiplicity one is generalized by the Weierstrass Preparation Theorem (cf. [17, Lemma 3.2.2 and Theorem 3.2.4]), that asserts that any curve C_f is analytically equivalent to one with an equation of a special form, called a *Weierstrass polynomial*, as described next.

Given $f \in \mathcal{M}$ of multiplicity n, then $f \sim P$, where P is a Weierstrass polynomial, that is,

$$P(X, Y) = Y^n + a_1(X)Y^{n-1} + \cdots + a_n(X), \quad \text{with } a_1(X), \ldots, a_n(X) \in X\mathbb{C}\{X\},$$

with $\text{mult}(a_i(X)) \geq i$, for $i = 1, \ldots, n$, because $\text{mult}(P) = \text{mult}(f) = n$.

It is this representation of elements of $\mathcal{M}$ that allows one to prove the good algebraic properties of $\mathcal{O}$ such as noetherianity and factoriality.

From Property 1.1.2 (e), nonsingularity may be characterized as follows:

$$C_f \text{ is nonsingular} \iff \Gamma(f) = \mathbb{N} \cup \{\infty\}.$$

The *ring* of a plane curve germ C_f, defined as being the quotient $\mathbb{C}$-algebra

$$\mathcal{O}(f) = \frac{\mathcal{O}}{\langle f \rangle},$$

is an algebraic object that encodes all the analytic properties of C_f, in the following sense:

Proposition 1.1.4 *Let $f, g \in \mathcal{M} \setminus \{0\}$ reduced. The following are equivalent:*

(i) C_f and C_g are analytically equivalent;
(ii) C_f and C_g are analytically isomorphic as germs of analytic spaces;
(iii) $\mathcal{O}(f) \simeq \mathcal{O}(g)$ as $\mathbb{C}$-algebras.

Proof (i) $\Rightarrow$ (ii) is obvious.

(ii) $\Rightarrow$ (iii) This follows immediately from the fact that to a morphism of germs of analytic spaces $C_f \to C_g$ it corresponds functorially a homomorphism of $\mathbb{C}$ algebras $\mathcal{O}(g) \to \mathcal{O}(f)$ (cf. [17, Lemma 3.4.20 (3)]).
(iii) $\Rightarrow$ (i) This proof is contained in [11, Theorem 4.1]. □

One of the main problems in this theory is to classify the reduced elements in $\mathcal{M}$ with respect to analytic equivalence. This problem as stated until now is intractable because the group action $(u, \varphi)f = u(f \circ \varphi)$ of the semidirect product $\mathcal{G} = \mathcal{O}^* \rtimes \mathrm{Aut}(\mathbb{C}^2, 0)$ on the set $\mathrm{Red}(\mathcal{M})$ of reduced elements of $\mathcal{M}$ representing analytic equivalence is too much intricate to be successfully understood.

Example 1.1.5 In fact, given

$$\begin{aligned} f = {} & Y^8 + (1 - 13X)Y^7 + (-14X + 70X^2)Y^6 + (84X^2 - 196X^3)Y^5 + \\ & + (-1 - 280X^3 + 280X^4)Y^4 + (-1 + 2X + 560X^4 - 112X^5)Y^3 + \\ & + (3X - 672X^5 - 224X^6)Y^2 + (-3X^2 - 2X^3 + 448X^6 + 320X^7)Y + \\ & + X^3 + X^4 - 128X^7 - 128X^8, \end{aligned}$$

how one could suspect that the corresponding curve C_f is analytically equivalent to the curve C_g with the much simpler equation $g = Y^3 - X^7$?

This is so, because f was produced by taking $f = u(g \circ \varphi)$, with $u = 1 + X + Y \in \mathcal{O}^*$ and $\varphi(X, Y) = (X + Y, X - Y)$.

The strategy, then, will be to partition $\mathrm{Red}(\mathcal{M})$ in such a way that the action of $\mathcal{G}$ could be better understood on each set of the partition. We will start by separating the elements of $\mathrm{Red}(\mathcal{M})$ by their number of irreducible components. Then separate the members of each subset of $\mathrm{Red}(\mathcal{M})$ with a fixed number of irreducible components by topological type, as defined below.

Definition 1.1.6 We will say that two germs of curves C_f and C_g are *topologically equivalent*, or *equisingular*, writing $C_f \equiv C_g$, if there exists a germ of *homeomorphism* φ of $(\mathbb{C}^2, 0)$ such that $\varphi(C_f) = C_g$.

Notice the similarity with analytic equivalence, where φ was taken as an analytic isomorphism, instead of simply a homeomorphism. In terms of equations, this equivalence is expressed by the existence of a germ of homeomorphism φ of $(\mathbb{C}^2, 0)$ and a unit u in $\mathcal{O}$ such that $f = u(g \circ \varphi)$.

At a first glance it looks intractable to decide whether two given germs of plane curves C_f and C_g are topologically equivalent, or not, because there is no explicit description, as in the case of analytic isomorphisms, for the homeomorphisms of $(\mathbb{C}^2, 0)$. Fortunately, as we will see in Sect. 1.1.3, there are deep results that lead to easy tests to verify if two germs of plane curves are topologically equivalent or not, which is far for being the case for analytic equivalence.

Obviously, analytically equivalent plane curve germs are topologically equivalent, so the group $\mathcal{G}$ acts on any equisingularity class $\mathcal{L}$ of germs of curves. But, as one may naturally suspect, it is not possible to give a reasonable geometric model for $\mathcal{L}$ and for the action of $\mathcal{G}$ on it.

So, to simplify our problem, we will focus on the analytic classification of equisingular plane irreducible curves, called *plane branches*, and represent them by means of parametrization. This is what we are going to do in the next subsection.

1.1.2 Irreducible Plane Curve Singularities

As said before, the problem we are interested in is the analytic classification of equisingular plane branches. This problem was first addressed by Ebey [9] and then by Zariski in [23]. In general lines, given an equisingularity class $\mathcal{L}$, this problem consists in determining a constructible set Σ in some finite dimensional affine space, called a *parameter space*, such that its points represent at least one branch of each analytic class in $\mathcal{L}$, with the property that its quotient by the analytic equivalence has a good geometric structure.

In this subsection, we will make the first steps toward the construction of such a parameter space Σ.

We will assume, from now on, that C_f is a plane branch. This corresponds to irreducible equations f in $\mathcal{O}$, so $\mathcal{O}(f)$ is a domain and $\mathcal{K}_f$ is its field of fractions.

A branch may be defined alternatively through parametrization instead of Cartesian equation, as will be described below. Before, we make some considerations.

We denote by $\mathcal{O}_1$ the $\mathbb{C}$-algebra $\mathbb{C}\{t\}$ of convergent power series in one indeterminate with complex coefficients and denote by $\mathcal{M}_1$ its maximal ideal $t\mathbb{C}\{t\}$. An analytic map germ $\rho\colon (\mathbb{C}, 0) \to (\mathbb{C}, 0)$ is determined by an element $\rho(t) = \sum_{i\geq 1} a_i t^i$ of $\mathcal{M}_1$. The least i for which $a_i \neq 0$ is called *the order* of $\rho(t)$, or of ρ, and is denoted by $\mathrm{ord}_t(\rho(t))$ or by $\mathrm{ord}_t(\rho)$. Notice that ρ is an isomorphism if and only if $\mathrm{ord}_t(\rho) = 1$. The set of these isomorphisms form a group under composition which we denote by $\mathrm{Aut}(\mathbb{C}, 0)$.

Any element $\xi(t) = (x(t), y(t)) \in \mathcal{M}_1 \times \mathcal{M}_1 \setminus \{(0, 0)\}$ will be called a *parametrization*. To a parametrization $\xi(t)$ there is associated naturally the germ of analytic morphism $\xi\colon (\mathbb{C}, 0) \to (\mathbb{C}^2, 0)$, defined by $z \mapsto (x(z), y(z))$, that parametrizes $\mathrm{Im}(\xi)$.

In the set of all parametrizations one has a partial order defined as follows: we say that a parametrization $\xi_1(t)$ *precedes* a parametrization $\xi_2(t)$, writing $\xi_1(t) \prec \xi_2(t)$, if there is an element $\rho(t) \in \mathcal{M}_1$ such that $\xi_2(t) = \xi_1(\rho(t))$. If $\mathrm{ord}_t(\rho(t)) = 1$, both parametrizations will be considered *equivalent*.

A parametrization $\xi(t)$ will be called a *primitive parametrization* if for every parametrization $\xi_1(t)$ with $\xi_1(t) \prec \xi(t)$, one has that $\xi_1(t)$ is equivalent to $\xi(t)$. In this sense, a primitive parametrization is minimal with respect to the partial order $\prec$.

Example 1.1.7 Just to illustrate the above definition, consider the family of parametrizations

$$\xi_k(t) = (t^{kn}, t^{km}) \in \mathcal{M}_1 \times \mathcal{M}_1, \quad \text{with } k, n, m \in \mathbb{N} \setminus \{0\}, \text{ and } \mathrm{GCD}(n, m) = 1.$$

One has that $\xi_k(t)$ is primitive if and only if $k = 1$.

Indeed, if $k > 1$ then $\xi_1(t) \prec \xi_k(t) = \xi_1(\rho(t))$ where $\rho(t) = t^k$ with $\text{ord}_t(\rho(t)) = k > 1$; hence $\xi_k(t)$ is a non-primitive parametrization.

Observe that any element $\zeta(t) \in \mathcal{O}_1 \setminus \{0\}$ may be written as $\zeta(t) = t^n u(t)$, where $u(0) \neq 0$, that is, $u(t)$ is a unit in $\mathcal{O}_1$. So, any $\xi(t) \in \mathcal{M}_1 \times \mathcal{M}_1 \setminus (\mathcal{M}_1 \times \{0\} \cup \{0\} \times \mathcal{M}_1)$ may be written as $\xi(t) = (t^{n_1}u_1(t), t^{n_2}u_2(t))$, where $n_1, n_2 \in \mathbb{N} \setminus \{0\}$ and $u_1(t), u_2(t)$ are units in $\mathcal{O}_1$.

Suppose that $\xi \prec \xi_1$ then there exists $\rho(t) = t^j v(t) \in \mathcal{M}_1$, with $j \in \mathbb{N} \setminus \{0\}$ and $v(t)$ a unit in $\mathcal{O}_1$, such that $\xi_1(t) = \xi(\rho(t))$; that is,

$$t^n = t^{jn_1} v^{n_1}(t) u_1(t^j v(t)), \quad t^m = t^{jn_2} v^{n_2}(t) u_2(t^j v(t)).$$

Since $u_1(t), u_2(t)$ and $v(t)$ are units in $\mathcal{O}_1$, we must have $n = jn_1$ and $m = jn_2$. As $\text{GCD}(n, m) = 1$, it follows that $\text{ord}_t(\rho(t)) = j = 1$ and, therefore, $\xi_1(t)$ is a primitive parametrization.

In general, it is a hard problem to determine if a given parametrization is primitive or not. Fortunately, in Proposition 1.1.17, we will present an easy criterion that will apply to a special kind of parametrizations, the so called *Puiseux parametrizations*, which are extensively used in this context.

Given a parametrization $\xi(t) = (x(t), y(t))$, we define the ring

$$R_\xi : = \mathbb{C}\{\xi(t)\} = \{g(x(t), y(t));\ g \in \mathcal{O}\},$$

and set $\text{ord}_t(\xi(t)) = \min\{\text{ord}_t\, x(t), \text{ord}_t\, y(t)\}$. Note that if $\xi(t)$ is a parametrization such that $\text{ord}_t(\xi(t)) = 1$, then $\xi(t)$ is primitive. The ring R_ξ is a domain, since it is a subring of $\mathcal{O}_1$. We denote its field of fractions by $\mathcal{K}_\xi$.

The ring R_ξ will be said to have a *conductor*, if there exists $\alpha \in \mathbb{N}$ such that $t^\alpha \mathcal{O}_1 \subset R_\xi$.

A parametrization $\xi(t)$ will be called a parametrization of a branch C_f, if $f(\xi(t)) = 0$, as an element of $\mathcal{O}_1$.

If $\xi(t)$ is a parametrization of C_f, then for any $\rho(t) \in \mathcal{M}_1 \setminus \{0\}$ one has that $\xi(\rho(t))$ is a parametrization of C_f, since

$$f(\xi(\rho(t)) = \rho(f(\xi(t))) = \rho(0) = 0.$$

At this point, it is not obvious that a plane branch admits non trivial parametrizations. The theorem below will guarantee their existence, but first we set a definition.

The integral closure of a domain R in its field of fractions $\mathcal{K}_R$, denoted by $\overline{R}$, is the set of elements in $\mathcal{K}_R$ that satisfy an equation of the form $Z^n + a_1 Z^{n-1} + \cdots + a_n = 0$, where $a_1, \ldots, a_n \in R$.

Theorem 1.1.8 *Let $f \in \mathcal{M}$ be irreducible. Then there exists $\xi(t) \in \mathcal{M}_1 \times \mathcal{M}_1$ such that, if we consider $\mathcal{O}_1/R_\xi$ as quotient of $\mathbb{C}$-vector spaces, one has*

(i) $\dim_{\mathbb{C}} \mathcal{O}_1/R_\xi < \infty$;

(ii) $f(\xi(t)) = 0$;
(iii) $\mathcal{O}(f) \simeq R_\xi$.

Conversely if $\xi(t) \in \mathcal{M}_1 \times \mathcal{M}_1$ *is such that Condition (i) above is satisfied, then there exists an irreducible* $f \in \mathcal{M}$ *for which* $f(\xi(t)) = 0$*. Moreover,* $\mathcal{O}(f) \simeq R_\xi$ *and* $\mathcal{O}_1$ *is the integral closure of* R_ξ *in its field of fractions* $\mathcal{K}_\xi$*.*

Proof The proof of the existence of $\xi(t)$ satisfying (i) and (ii) is given in [17, Theorem 5.1.1]. To complete the proof of the first part of the theorem, we will show that any $\xi(t)$ satisfying (i) and (ii) also satisfies (iii). Indeed, consider the $\mathbb{C}$-algebras homomorphism $\xi^*\colon \mathcal{O} \to \mathcal{O}_1$, defined by $\xi^*(g) = g(\xi(t))$. Then by (ii) we have that $f \in \text{Ker}(\xi^*)$. Suppose that there exists $g \in \text{Ker}(\xi^*)$ which is not a multiple of f, then because f is irreducible, f and g have no nontrivial common factor, hence from Property 1.1.2 (a) it follows that

$$\dim_{\mathbb{C}} R_\xi = \dim_{\mathbb{C}} \text{Im}(\xi^*) = \dim_{\mathbb{C}} \mathcal{O}/\text{Ker}(\xi^*) \leq \dim_{\mathbb{C}} \mathcal{O}/\langle f, g\rangle < \infty,$$

which contradicts (i), since $\dim_{\mathbb{C}} \mathcal{O}_1 = \infty$. Therefore, $\text{Ker}(\xi^*) = \langle f\rangle$, showing that (iii) holds.

The proof of the converse may be found in [17, Theorem 5.1.3]. □

The fact that $\mathcal{O}_1$ is the integral closure $\overline{R_\xi}$ of R_ξ in $\mathcal{K}_\xi$ implies that $\mathcal{K}_\xi$ is isomorphic to the field of fractions $\mathbb{C}((t))$ of $\mathcal{O}_1$, called the *field of Laurent series*. Hence one has that

$$\mathcal{O}(f) \simeq R_\xi, \quad \overline{\mathcal{O}(f)} \simeq \overline{R_\xi} = \mathcal{O}_1 \quad \text{and} \quad \mathcal{K}_f \simeq \mathcal{K}_\xi = \mathbb{C}((t)).$$

Remark 1.1.9 Let $\xi(t) = (x(t), y(t))$ be a parametrization of order 1 and let C_f be the associated branch. Without loss of generality, we may assume that $\text{ord}_t\, x(t) = 1$. Then defining $\rho(t) := x(t)$, we have that $\rho \in \text{Aut}(\mathbb{C}, 0)$, hence we may consider ρ^{-1}. Suppose that $\rho^{-1}(t) = \sum_{i\geq 1} a_i t^i$, then defining $g(X, Y) = \sum_{i\geq 1} a_i X^i \in \mathcal{O}$, we have that $t = g(\xi(t)) \in R_\xi$, which shows that $\mathcal{O}(f) \simeq R_\xi = \mathcal{O}_1 \simeq \mathbb{C}\{X\} = \mathcal{O}(Y)$. This, in view of Proposition 1.1.4, implies that $C_f \simeq C_Y$, hence $\text{mult}(f) = \text{mult}(Y) = 1$, which shows that C_f is nonsingular.

Since any nonzero element $\zeta(t) \in \mathbb{C}((t))$ may be written uniquely as $\zeta(t) = t^\alpha z(t)$, where $\alpha \in \mathbb{Z}$, $z(t) \in \mathcal{O}_1$ and $z(0) \neq 0$, we define the order of $\zeta(t)$ as being the number $\text{ord}_t(\zeta(t)) = \alpha$, and set by convention $\text{ord}_t(0) = \infty$. This extends the order function we defined on $\mathcal{O}_1$, determining a valuation v on $\mathbb{C}((t))$, in such a way that $v(\mathcal{O}_1) = \mathbb{N} \cup \{\infty\}$.

Remark 1.1.10 Let S be a subset of $\mathcal{O}_1$ such that its elements have distinct values and $v(S) \subset v(\mathcal{O}_1) \setminus v(R_\xi) = \mathbb{N} \setminus v(R_\xi)$, then the elements of the set $\overline{S} = \{\overline{s};\ s \in S\} \subset \mathcal{O}_1/R_\xi$ are linearly independent over $\mathbb{C}$. This implies that $\dim_{\mathbb{C}} \mathcal{O}_1/R_\xi \geq \#(\mathbb{N} \setminus v(R_\xi))$.

Suppose now that R_ξ has a conductor. Then for some $\alpha \in \mathbb{N}$ one has $t^\alpha \mathcal{O}_1 \subset R_\xi$, which implies that $\alpha + \mathbb{N} \cup \{\infty\} = v(t^\alpha \mathcal{O}_1) \subset v(R_\xi)$, hence $\mathbb{N} \setminus v(R_\xi) \subset \mathbb{N} \setminus \alpha +$

$\mathbb{N}$, where the last set is finite. Take any subset $S = \{h_1, \ldots, h_r\}$ of $\mathcal{O}_1$ such that $v(S) = \mathbb{N} \setminus v(R_\xi)$. We will show that the set $\overline{S}$ is a set of generators of the $\mathbb{C}$-vector space $\mathcal{O}_1/R_\xi$. Indeed, let $h \in \mathcal{O}_1$. Then either $v(h) \in v(R_\xi)$, in which case there exists $g_1 \in R_\xi$ such that $v(h - g_1) > v(h)$, or $v(h) \in \mathbb{N} \setminus v(R_\xi)$, in which case there exist $h_{i_1} \in S$ and $a_{i_1} \in \mathbb{C}$ such that $v(h - a_{i_1}h_{i_1}) > v(h)$. Now, repeat recursively this procedure to the resulting element whose order is greater than the previous one, until one reaches an element with value greater or equal to $\max(\mathbb{N} \setminus v(R_\xi))$, which belongs to R_ξ.

Combining together the above considerations we get, in any case, that

$$\dim_{\mathbb{C}} \frac{\mathcal{O}_1}{R_\xi} = \#(\mathbb{N} \setminus v(R_\xi)),$$

and $\dim_{\mathbb{C}} \mathcal{O}_1/R_\xi$ is finite if R_ξ has a conductor.

The following result will relate condition (i) in Theorem 1.1.8, the existence of a conductor for R_ξ and the notion of primitive parametrization.

Proposition 1.1.11 *Let* $\xi(t) = (x(t), y(t)) \in \mathcal{M}_1 \times \mathcal{M}_1 \setminus \{(0, 0)\}$ *be a parametrization. If* $\mathcal{K}_\xi$ *is the field of fractions of* R_ξ*, then the following conditions are equivalent:*

(i) R_ξ *has a conductor;*
(ii) $\dim_{\mathbb{C}} \mathcal{O}_1/R_\xi < \infty$*;*
(iii) $\xi(t)$ *is a primitive parametrization;*
(iv) $\mathcal{K}_\xi$ *contains a n dement of order one.*

Proof (i) $\Rightarrow$ (ii) This follows from Remark 1.1.10.
(ii) $\Rightarrow$ (iii) Suppose that $\xi(t)$ is not a primitive parametrization, then there is a parametrization $\xi_1(t)$ and an element $\rho(t) \in \mathcal{M}_1$ with $\mathrm{ord}_t(\rho(t)) = p > 1$ such that $\xi(t) = \xi_1(\rho(t))$. Therefore $R_\xi \subset \mathbb{C}\{\rho(t)\}$, hence $v(R_\xi) \subset p\mathbb{N}$. This implies that $\mathbb{N} \setminus v(R_\xi)$ is not finite, hence from Remark 1.1.10, one has $\dim_{\mathbb{C}} \mathcal{O}_1/R_\xi = \infty$.
(iii) $\Leftrightarrow$ (iv) If $s \in \mathcal{K}_\xi$ is an element of minimal positive order p, then any element in $\mathcal{K}_\xi$ has order divisible by p. Indeed, if $s = t^p u$, and if $h = t^m v$ is any element in $\mathcal{K}_\xi$, where u and v are units in $\mathcal{O}_1$, then writing $m = pq + r$, where $q, r \in \mathbb{Z}$ and $0 \leq r < p$, we have $h/s^q = t^r v/u^q \in K_\xi$, implying that $v(h/s^q) = r$, which is a contradiction unless $r = 0$. Now, we will show that $\mathcal{K}_\xi \subset \mathbb{C}((s))$. For, if $\zeta \in \mathcal{K}_\xi$ is of order pn_1, then there exists a complex number c_1 and an element $\zeta_1 \in \mathcal{K}_\xi$ such that $\zeta - c_1 s^{n_1} = \zeta_1$, such that $\mathrm{ord}_t(\zeta_1) = pn_2$ with $n_2 > n_1$. Repeating this argument for ζ_1 and so on, we see that there is a formal Laurent series $h(t)$ such that $\zeta = h(s)$. Since ζ and s are convergent, we have that h is convergent. So, $\zeta \in \mathbb{C}((s))$.

Since $x(t), y(t) \in K_\xi \subset \mathbb{C}((s))$, and $\xi(t) = (x(t), y(t))$ is primitive, it follows that $\mathrm{ord}_t(s) = 1$.

Conversely, suppose that ξ is not primitive, hence $x(t), y(t) \in \mathbb{C}\{s\}$ for some s in $\mathbb{C}\{t\}$ of order $p > 1$. Since $\mathcal{K}_\xi \subset \mathbb{C}((s))$, it follows that the elements of $\mathcal{K}_\xi$ have orders multiple of p, therefore $\mathcal{K}_\xi$ does not contain any element o f order one.
(iv) $\Rightarrow$ (i) It will be proved later (see Corollary 1.1.15). □

Notice that we proved that (iii) and (iv), above, are equivalent. The proof that (iv) implies (i) will be done in Corollary 1.1.15, by showing that (iii) implies (i). To prove this result, we need to show that we may, by means of a change of parameter, transform any parametrization into a special very useful form.

If $\xi(t) = (z(t), w(t))$ is a parametrization of the branch C_f, then $\xi^\dagger(t) = (w(t), z(t))$ is a parametrization of $C_{f^\dagger}$, where $f^\dagger(X, Y) = f(Y, X)$, which is analytically equivalent to C_f. Moreover, $R_\xi = R_{\xi^\dagger}$. So, at the cost of changing our branch by an analytically equivalent one, we may assume that $n: = \mathrm{ord}_t(z(t)) \leq \mathrm{ord}_t(w(t))$. Since $z(t) = t^n u(t)$, where $u(t)$ is a unit in $\mathcal{O}_1$, hence $u(0) \neq 0$, we may take a branch $v(t) = \sqrt[n]{u(t)}$ of the n-th root function of $u(t)$ around the origin, which is also a unit in $\mathcal{O}_1$. Now, by defining $\rho(t) = tv(t)$, one has $\rho \in \mathrm{Aut}(\mathbb{C}, 0)$, so if we consider its inverse ρ^{-1}, one gets the equivalent parametrization of $\xi(t)$:

$$\xi(\rho^{-1}(t)) = (t^n,\ y(t)),\ \text{ where } y(t) = w(\rho^{-1}(t)),\ \text{ and } \mathrm{ord}_t\, y(t) \geq n.$$

A parametrization of the form $(t^n,\ y(t)) \in \mathbb{C}\{t\}^2$, with $\mathrm{ord}_t\, y(t) \geq n$, is called a *Newton-Puiseux parametrization*, or shortly a *Puiseux parametrization*. Notice that because we made a particular choice of the n-th root of $v(t)$, other choices would produce the following equivalent parametrizations to $\xi(t)$:

$$\xi_i(t) = (t^n,\ y(\epsilon^i t)),\ \text{ where } \epsilon \text{ is a primitive } n\text{-th root of 1 and } i = 1, \ldots, n,$$

called the *associated Puiseux parametrizations* to the parametrization $\xi(t)$, which are the only parametrizations of C_f of the form $(t^n, y(t))$. Notice that $\xi(t)$ is primitive if and only if any one of its associated Puiseux parametrizations is primitive, since primitivity is invariant under equivalence of parametrizations.

The following result will show the interplay between Puiseux parametrizations and Cartesian equations of branches to which they correspond.

Theorem 1.1.12 *Let ϵ be a primitive n-th root of unity.*

(i) Let $g = Y^n + a_1(X)Y^{n-1} + \cdots + a_n(X)$ be an irreducible Weierstrass polynomial. Then there exists $y(t) \in \mathcal{O}_1$ such that

$$g(X, Y) = \prod_{i=1}^{n}(Y - y(\epsilon^i t)),\quad \text{where } t^n \text{ is replaced by } X. \tag{1.2}$$

(ii) Let $y(t) \in \mathcal{M}_1$ be such that $\dim_{\mathbb{C}} \mathcal{O}_1/R_\xi < \infty$, where $\xi(t) = (t^n, y(t))$. Then the polynomial $\prod_{i=1}^{n}(Y - y(\epsilon^i t))$, where t^n is replaced by X, is an irreducible Weierstrass polynomial in $\mathcal{O}$.

Proof See [17, Theorem 5.1.7]. □

The following result connects intersection multiplicities to parametrizations:

Proposition 1.1.13 *Let $\xi(t)$ be a primitive Puiseux parametrization and let g be the irreducible Weierstrass polynomial associated to it as in Theorem 1.1.12 (ii). Let $h \in \mathcal{O}$, then one has that*

$$\mathrm{I}(g, h) = \mathrm{ord}_t(h(\xi(t))).$$

Proof See [11, Theorem 4.17], or [17, Lemma 5.1.5]. □

Observe that it is not necessary to suppose that $\xi(t)$ is a Puiseux parametrization, but it is enough to require that it is primitive, since for any $\rho \in \mathrm{Aut}(\mathbb{C}, 0)$ one has that

$$\mathrm{ord}_t(h(\xi(\rho((t)))) = \mathrm{ord}_t(\rho(h(\xi(t)))) = \mathrm{ord}_t(h(\xi(t))).$$

So, for any primitive parametrization $\xi(t)$ of C_f, we have that

$$\Gamma(f) = \{\mathrm{ord}_t(g(\xi(t))),\ g \in \mathcal{O}\} = \Gamma(\xi),$$

where

$$\Gamma(\xi) : = \{\mathrm{ord}_t(h(t)),\ h \in R_\xi\}.$$

The following result is crucial.

Proposition 1.1.14 *Let $\xi(t) = (t^n,\ y(t))$ be a primitive Puiseux parametrization and let $g(X, Y)$ be the associated Weierstrass polynomial as given in Theorem 1.1.12 (ii). If $D_Y g(X) \in \mathbb{C}\{X\}$ denotes the discriminant of $g(X, Y)$ as a polynomial in Y, then*

$$D_Y g(t^n)\mathbb{C}\{t\} \subset R_\xi.$$

Proof See [24, Theorem on p. 6], or [11, Theorem 4.4] for an elementary proof. □

Let us write $D_Y g(t^n) = t^\alpha u(t)$, where $\alpha \in \mathbb{N}$ and $u(t)$ is a unit in $\mathcal{O}_1$, then one has that $t^\alpha \mathcal{O}_1 \subset R_\xi$, hence R_ξ has a conductor. Let γ denote the smallest natural number such that $t^\gamma \mathcal{O}_1 \subset R_\xi$, so $t^\gamma \mathcal{O}_1$ is the largest common ideal of $\mathcal{O}_1$ and R_ξ, called the *conductor ideal* of R_ξ. We call γ the *conductor exponent* of R_ξ.

In general, if $\xi(t)$ is a primitive parametrization, not necessarily in Puiseux form, of an irreducible $f \in \mathcal{O}$, as we may suppose that $n = \mathrm{I}(f, X) \leq \mathrm{I}(f, Y)$ (at possibly the cost of interchanging X and Y), there exists a primitive Puiseux parametrization and an automorphism ρ of $(\mathbb{C}, 0)$ such that $(\rho(t^n), y(\rho(t)) = \xi(t)$. Since there exists $\gamma \in \mathbb{N}$ such that $t^\gamma \mathcal{O}_1 \subset \mathbb{C}\{t^n,\ y(t)\}$, it follows that $\rho(t^\gamma)\mathbb{C}\{\rho(t)\} \subset \mathbb{C}\{\xi(t)\} = R_\xi$. But, $\mathbb{C}\{\rho(t)\} = \mathbb{C}\{t\} = \mathcal{O}_1$ and $\rho(t^\gamma) = t^\gamma u$, where u is a unit in $\mathcal{O}_1$, then $t^\gamma \mathcal{O}_1$ is also the conductor ideal of R_ξ.

Corollary 1.1.15 *If $\xi(t)$ is a primitive parametrization, then the ring R_ξ has a conductor.*

From the existence of the conductor ideal $t^\gamma \mathcal{O}_1$ of the ring R_ξ associated to a primitive parametrization $\xi(t)$, it follows that for any $h \in \mathcal{O}_1$ such that $\mathrm{ord}_t(h) \geq \gamma$ we have that $h \in R_\xi$, and therefore $\mathrm{ord}_t(h) \in \Gamma(\xi)$. This implies that

$$\gamma + \mathbb{N} \subset \Gamma(\xi).$$

Let c be the smallest natural number such that $c + \mathbb{N} \subset \Gamma(\xi)$. This number c is called the conductor of $\Gamma(\xi)$. We have the following result:

Lemma 1.1.16 *Let $\xi(t)$ be a primitive parametrization, The conductor c of $\Gamma(\xi)$ coincides with the conductor exponent γ of R_ξ; that is, $\gamma = c$.*

Proof We have obviously that $c \le \gamma$. Suppose by reductio ad absurdum that $c < \gamma$. Let h be an arbitrary element of $\mathcal{O}_1$, then $\mathrm{ord}_t\, t^c h \ge c$, so there exist $h_1, \ldots, h_{\gamma-c} \in R_\xi$ such that

$$\mathrm{ord}_t(t^c h - h_1 - h_2 - \cdots - h_{\gamma-c}) > \gamma,$$

hence $t^c h - h_1 - h_2 - \cdots - h_{\gamma-c} \in R_\xi$, which implies that $t^c h \in R_\xi$, therefore, $t^c\mathcal{O}_1 \subset R_\xi$, a contradiction with the minimality of γ. □

For a branch C_f, we know from Theorem 1.1.8 that a parametrization $\xi(t)$ of C_f induces an isomorphism from $\mathcal{K}_f$ onto $\mathbb{C}((t))$, in such a way that $\overline{\mathcal{O}(f)}$ corresponds to $\mathcal{O}_1$ and $\mathcal{O}(f)$ corresponds to R_ξ. So, the inverse image of the conductor of R_ξ is the largest common ideal of $\overline{\mathcal{O}(f)}$ and $\mathcal{O}(f)$, which we denote by $\mathcal{C}$. Therefore, one has that

$$\dim_{\mathbb{C}} \frac{\overline{\mathcal{O}(f)}}{\mathcal{C}} = \dim_{\mathbb{C}} \frac{\mathcal{O}_1}{t^c\mathcal{O}_1} = c. \tag{1.3}$$

Although in general, given a parametrization, without the help of Computer Algebra it is difficult to verify if it is primitive or not, we will see below that there is an easy criterion for doing this when the parametrization is in Puiseux form.

Given a Puiseux parametrization $\xi(t) = (t^n,\ y(t))$, where $y(t) = \sum_{i=n}^{\infty} a_i t^i$, we will define two associated fundamental sequences of integers as follows:

Put $\beta_0 = n$. Let β_1 the first exponent i in $y(t) = \sum_{i=n}^{\infty} a_i t^i$ such that $a_i \neq 0$ and $n \nmid i$, if it exists, otherwise we stop the process. Define $e_0 = n$ and $e_1 = \mathrm{GCD}(n, \beta_1)$, and let β_2 be the first exponent in $y(t)$ greater than β_1 which is not divisible by e_1, if it exists, otherwise we stop the process. Define $e_2 = \mathrm{GCD}(e_1, \beta_2) = \mathrm{GCD}(n, \beta_1, \beta_2)$, and continue this process. In this way, we obtain two sets of integers $n, \beta_1, \beta_2, \ldots$, and $e_i = \mathrm{GCD}(n, \beta_1, \ldots, \beta_i)$, $i \ge 1$. Since the e_i are not increasing, the sequence must stabilize, that is, $e_i = d$ for $i \ge g$, for some g.

Proposition 1.1.17 *A Puiseux parametrization $\xi(t) = (t^n,\ y(t))$, where $y(t) = \sum_{i=n}^{\infty} a_i t^i$, is primitive if and only if $d := \mathrm{GCD}\{n, j;\ a_j \neq 0\} = 1$.*

Proof Suppose that $d > 1$, then $t^n, y(t) \in \mathbb{C}\{t^d\}$, hence $\xi(t)$ is not primitive. So, we have proved that if $\xi(t)$ is primitive, then $d = 1$.

Conversely, suppose that $d = 1$, then one must have $e_g = 1$ for some g. If we subtract from $y(t)$ a linear combination of powers of t^n, we get $y_1 = \sum_{i \ge \beta_1} a_i t^i$ as an element in $\mathcal{K}_\xi$, the field of fractions of $R_\xi = \mathbb{C}\{t^n, y(t)\}$. Now if $e_1 = \lambda_1 n + \mu_1\beta_1$, we have that $z_1(t) := t^{\lambda_1 n} y_1(t)^{\mu_1}$ is in $\mathcal{K}_\xi$ and has order e_1. Now, subtracting a linear combination of powers of $z_1(t)$ from $y_1(t)$, we get an element in $\mathcal{K}_\xi$ of with order β_2. We continue this process until we get $y_1(t), \dots, y_g(t)$ of orders $\beta_1, \dots, \beta_g$, with $\mathrm{GCD}(n, \beta_1, \dots, \beta_g) = 1$. So, there exist integers η_i such that $\eta_0 n + \eta_1\beta_1 + \cdots + \eta_g\beta_g = 1$, therefore $t^{\eta_0 n} y_1(t)^{\eta_1} \cdots y_g(t)^{\eta_g}$ is in $\mathcal{K}_\xi$ and has order 1. This, in view o f Poposition 1.1.11 (iv) $\Rightarrow$ (iii), implies that $(t^n, y(t))$ is a primitive parametrization. □

The two sets of integers $\beta_0 = n, \beta_1, \dots, \beta_g$ and $e_0 = n, e_1, \dots, e_g$, with $e_g = 1$, obtained from a primitive Puiseux parametrization, play a very important role in this theory.

Example 1.1.18 Given the primitive Puiseux parametrization

$$\xi(t) = (t^{36}, 2t^{36} + 5t^{54} + t^{72} - t^{78} + 3t^{84} + 2t^{90} + t^{92} + 6t^{94} - 2t^{95} + 5t^{96}),$$

one has

$$\begin{array}{ll} \beta_0 = n = 36, & e_0 = \beta_0 = n = 36; \\ \beta_1 = 54, & e_1 = GCD(e_0, \beta_1) = GCD(36, 54) = 18; \\ \beta_2 = 78, & e_2 = GCD(e_1, \beta_2) = GCD(18, 78) = 6; \\ \beta_3 = 92, & e_3 = GCD(e_2, \beta_3) = GCD(6, 92) = 2; \\ \beta_4 = 95, & e_4 = GCD(e_3, \beta_4) = GCD(2, 95) = 1. \end{array}$$

Let us now interpret geometrically the above results about parametrized plane branches.

In Theorem 1.1.8 (and in its proof) we saw that to a given primitive parametrization $\xi(t)$ there is associated a unique branch C_f and a morphism $\xi : (\mathbb{C}, 0) \to (\mathbb{C}^2, 0)$ such that $\mathrm{Im}(\xi) \subset C_f$. From the same quoted theorem, we have that the map $\xi : (\mathbb{C}, 0) \to C_f$ is the normalization map, hence it is a germ of homeomorphism between the two germs of analytic curves, so $\mathrm{Im}(\xi) = C_f$. This implies that any plane branch is intrinsically homeomorphic to the germ of disc $(\mathbb{C}, 0)$, hence they are all topologically equivalent to each other. This is why we needed to define equisingularity of germs of curves as we did in Definition 1.1.6, taking into account the embedding of the germ in $\mathbb{C}^2$.

From what we saw until now, we have that a branch C_f determines a unique class of equivalent primitive parametrizations and, conversely, any primitive parametrization determines a unique germ of plane branch C_f. Hence, we may indistinctly work with plane branches or with primitive parametrizations.

Since we are going to work with branches represented by parametrizations, it will be important to characterize analytic equivalence of branches in terms of their parametrizations. Before we do that, we will need the following result.

Lemma 1.1.19 *If* $\xi_i : (\mathbb{C}, 0) \to (\mathbb{C}^2, 0)$, $i = 1, 2$, *are two non-constant morphisms corresponding to primitive parametrizations* $\xi_i(t)$, *then* $\mathrm{Im}(\xi_1) = \mathrm{Im}(\xi_2)$ *if and only if* $\xi_1(t)$ *and* $\xi_2(t)$ *are equivalent parametrizations; that is, there is an automorphism* ρ *of* $(\mathbb{C}, 0)$ *such that* $\xi_2(t) = \xi_1(\rho^{-1}(t))$.

Proof If there exists an automorphism ρ of $(\mathbb{C}, 0)$ such that $\xi_2(t) = \xi_1(\rho^{-1}(t))$, then clearly $\mathrm{Im}(\xi_1) = \mathrm{Im}(\xi_2)$. On the other hand, if $\mathrm{Im}(\xi_1) = \mathrm{Im}(\xi_2)$, then ξ_1 and ξ_2 are two normalization maps of their common image, therefore, by the uniqueness of normalizations (cf. [17, Remark 4.4.6]), it follows that there exists an analytic isomorphism $\rho\colon (\mathbb{C}, 0) \to (\mathbb{C}, 0)$ such that $\xi_2 = \xi_1 \circ \rho^{-1}$, hence $\xi_2(t) = \xi_1(\rho^{-1}(t))$. □

The next result will characterize primitive parametrizations that represent analytic equivalent branches.

Proposition 1.1.20 *Two primitive parametrizations $\xi_1(t)$ and $\xi_2(t)$ define analytically equivalent branches if and only if there exist $\varphi \in \mathrm{Aut}(\mathbb{C}^2, 0)$ and $\rho \in \mathrm{Aut}(\mathbb{C}, 0)$ such that $\xi_2(t) = (\varphi \circ \xi_1 \circ \rho^{-1})(t)$.*

Proof Let C_{f_i} be the images of the morphisms $\xi_i\colon (\mathbb{C}, 0) \to (\mathbb{C}^2, 0)$ associated to the parametrizations $\xi_i(t)$, for $i = 1, 2$. If C_{f_1} and C_{f_2} are analytically equivalent, then, by definition, there is a $\varphi \in \mathrm{Aut}(\mathbb{C}^2, 0)$ such that $\varphi(C_{f_1}) = C_{f_2}$. This implies that ξ_2 and $\varphi \circ \xi_1$ have the same image, hence from Lemma 1.1.19, there exits $\rho \in \mathrm{Aut}(\mathbb{C}, 0)$ such that $\xi_2 = (\varphi \circ \xi_1) \circ \rho^{-1}$, proving one direction of the proposition.

Conversely, if $\xi_2 = \varphi \circ \xi_1 \circ \rho^{-1}$, then the isomorphism φ of $(\mathbb{C}^2, 0)$ induces an isomorphism between the images C_{f_1} of ξ_1 and C_{f_2} of ξ_2, showing that they are analytically equivalent. □

Remark 1.1.21 If we consider the group $\mathcal{A} = \mathrm{Aut}(\mathbb{C}^2, 0) \times \mathrm{Aut}(\mathbb{C}, 0)$ acting on the set $\mathcal{M}_1 \times \mathcal{M}_1$ by the rule

$$(\varphi, \rho) \cdot \xi(t) \colon = (\varphi \circ \xi \circ \rho^{-1})(t) = \varphi(\xi(\rho^{-1}(t))),$$

then the criterion in Proposition 1.1.20 may be rephrased as follows:

The primitive parametrizations $\xi_1(t)$ and $\xi_2(t)$ define two analytically equivalent branches if and only if they are $\mathcal{A}$-equivalent; that is, they belong to the same orbit under the action of $\mathcal{A}$.

In particular, if $\xi_1(t)$ and $\xi_2(t)$ are $\mathcal{A}$-equivalent, then $\Gamma(\xi_1) = \Gamma(\xi_2)$.

Example 1.1.22 The primitive Puiseux parametrizations $\xi_1(t) = (t^2, t^3)$ and $\xi_2(t) = (t^2, \frac{1}{4}t^2 - \frac{1}{8}t^3 + \frac{1}{16}t^4 - \frac{1}{32}t^5)$ are $\mathcal{A}$-equivalent. Indeed, if we take $\varphi(X, Y) = (4X, X - Y + X^2 - XY)$ and $\rho(t) = 2t$, we have that $\xi_2(t) = (\varphi \circ \xi_1 \circ \rho^{-1})(t)$.

By Property 1.1.2 (g) we have $2 = n = \min \Gamma(\xi_1) \setminus \{0\} = \min \Gamma(\xi_2) \setminus \{0\}$; and since any $m \geq 2$ may be written as $m = 2a + 3b$ for $a, b \in \mathbb{N}$, we have that $m = \mathrm{ord}_t(h(\xi_1(t))) = \mathrm{ord}_t(g(\xi_2(t)))$ where $h(X, Y) = X^a Y^b$ and $g(X, Y) = X^a(Y - \frac{1}{4}X)^b$. Hence, $\Gamma(\xi_1) = \Gamma(\xi_2) = \mathbb{N} \setminus \{1\}$.

Notice that in the case of parametrizations, the group action that characterizes analytic equivalence is the product of two groups, while in the case of equations, it is the direct product of two groups, hence a more complicated action.

Up to now, we know that plane branches may be defined through primitive Puiseux parametrizations. We will use automorphisms φ of $(\mathbb{C}^2, 0)$ and ρ of $(\mathbb{C}, 0)$, as above, to transform a Puiseux parametrization into another simpler Puiseux parametrization. This strategy was initiated, to the best of our knowledge, by Ebey in [9] and developed further by Zariski in [23].

To simplify notation, we will always denote by n the number β_0 and by m the number β_1.

Remark 1.1.23 If C_f is nonsingular, then from Remark 1.1.3 one knows that $f \sim Y$. It then follows that any nonsingular plane branch is equivalent to one with Puiseux parametrization $\xi(t) = (t, 0)$. On the other hand, if $\xi(t) = (t^n, \sum_{i\geq n} a_i t^i)$ is a primitive Puiseux parametrization defining a singular branch, then, from Proposition 1.1.17, there exists $a_j \neq 0$ such that $n \nmid j$ and the least such j is what we denoted by β_1 just before Proposition 1.1.17 and we are denoting now by m.

1.1.3 Equisingularity of Branches

Thanks to the works of Brauner, Kähler, Burau and Zariski in the first three decades of last century, one knows that the topological type of an immersed plane branch is encoded by its Puiseux exponents, which we define below.

Let $(t^n, \sum_{i\geq n} a_i t^i)$ be a Puiseux primitive parametrization of a plane branch given by a Weierstrass polynomial P of multiplicity n. The numbers $n = \beta_0, m = \beta_1, \ldots, \beta_g$ defined just before Proposition 1.1.17 are the so called *Puiseux exponents* or *characteristic exponents* of C_P. The set of characteristic exponents associated to a Puiseux primitive parametrization $\xi(t)$ will be denoted by $\mathcal{E}xp(\xi(t))$.

The following is a deep classical theorem which is comprehensively discussed in [5, Theorem 12 pp. 438-439] and gives a manageable criterion for equisingularity of branches.

Theorem 1.1.24 *Let two plane branches C_1 and C_2 be given by primitive Puiseux parametrizations $\xi_1(t)$ and $\xi_2(t)$, respectively. Then*

$$C_1 \equiv C_2 \iff \mathcal{E}xp(\xi_1(t)) = \mathcal{E}xp(\xi_2(t)).$$

This means that the characteristic exponents of Puiseux parametrizations are a complete topological invariant for the branches they represent. In particular, an immediate consequence of this theorem is the nontrivial fact that the multiplicity of a plane branch is a topological invariant. This led Zariski to formulate the famous conjecture that the same result holds more generally for hypersurface germs, instead of plane curves (cf. [26]).

When the plane curve germs C_1 and C_2 are not irreducible, Zariski has shown in [27, Lemma 7.1] that $C_1 \equiv C_2$ if and only if C_1 and C_2 have the same number r of irreducible components and there is an ordering of their components $C_{1,i}$ and $C_{2,i}$ such that $C_{1,i} \equiv C_{2,i}$ and $C_{1,i} \cdot C_{1,j} = C_{2,i} \cdot C_{2,j}$, for all $i, j = 1, \ldots, r$.

The only restriction for a set $\beta_0, \ldots, \beta_g$, whose GCD is 1, to be a set of characteristic exponents of a Puiseux parametrization is that they form an increasing sequence such that $\text{GCD}(\beta_0, \ldots, \beta_{i-1}) \nmid \beta_i$, $i = 1, \ldots, g$.

To the characteristic exponents there are attached the important invariants,

$$e_0 = \beta_0; \ \ e_i = \text{GCD}(\beta_0, \ldots, \beta_i), \ \ i = 1, \ldots, g,$$

with $e_g = 1$, which already have been defined just before Proposition 1.1.17, and

$$n_0 = 1; \ \ n_i = \frac{e_{i-1}}{e_i}, \ \ i = 1, \ldots, g.$$

Let us stress the fact that not every Puiseux parametrization $\xi(t) = (t^n, y(t))$, where $y(t) = \sum_{i \geq n} a_i t^i$ with $a_{\beta_i} \neq 0$, $i = 1, \ldots, g$, represents a plane branch such that $\mathcal{E}xp(\xi(t)) = \{\beta_0, \ldots, \beta_g\}$. This is the case if, and only if, for all $j = 0, \ldots, g - 1$,

$$\forall i \text{ with } \beta_j \leq i < \beta_{j+1} \text{ and } e_j \nmid i \Longrightarrow a_i = 0. \tag{1.4}$$

In the literature, very often, instead of giving the characteristic exponents of a Puiseux parametrization, one gives the *Puiseux pairs* (n_i, m_i), $i = 1, \ldots, g$, where n_i is as above and $m_i = \frac{\beta_i}{e_i}$. Since $e_i = \text{GCD}(e_{i-1}, \beta_i)$, it follows that $\text{GCD}(n_i, m_i) = 1$, for all $i = 1, \ldots, g$.

It is an exercise to verify that Puiseux pairs and the characteristic exponents of a primitive Puiseux parametrization determine each other.

We now retake the study of the semigroup of values of a plane branch C_f, which is by itself a rich subject.

Given $v_0, \ldots, v_r \in \mathbb{N}$ we define

$$\langle v_0, \ldots, v_r \rangle = \{x_0 v_0 + \cdots + x_r v_r; \ x_0, \ldots, x_r \in \mathbb{N}\} \cup \{\infty\},$$

called the *semigroup generated* by $v_0, \ldots, v_r$.

Lemma 1.1.25 *Given a semigroup Γ of $\mathbb{N} \cup \{\infty\}$, there exists a unique finite set of integers $v_0, \ldots, v_r \in \Gamma$ such that*

(i) $v_0 < v_1 < \cdots < v_r$, *and* $v_i \notin \langle v_0, \ldots, v_{i-1} \rangle$, *for all* $i = 1, \ldots, r$;
(ii) $\Gamma = \langle v_0, \ldots, v_r \rangle$;
(iii) $\{v_0, \ldots, v_r\}$ *is contained in any set of generators of* Γ.

Proof See [11, Proposition 6.1]. □

The set $\{v_0, \ldots, v_r\}$ is called the *minimal set of generators* of Γ and the integer r the *genus* of Γ.

Lemma 1.1.26 *Given a semigroup Γ of $\mathbb{N} \cup \{\infty\}$, the following assertions are equivalent:*

(i) Γ *has a conductor;*
(ii) The GCD of all elements of $\Gamma \setminus \{\infty\}$ *is* 1.

Proof See [11, Proposition 6.2]. □

For a primitive parametrization $\xi(t)$, we know, from Corollary 1.1.15 and Lemma 1.1.16, that $\Gamma(\xi)$ has a conductor, it follows that its minimal set of generators $v_0, \ldots, v_r$ has GCD equal to one.

The interesting fact is that there are remarkable relations among the characteristic exponents $\beta_0, \ldots, \beta_g$ of a primitive Puiseux parametrization $\xi(t)$ and the minimal set of generators $\{v_0, \ldots, v_r\}$ of the semigroup $\Gamma(\xi)$.

Theorem 1.1.27 *Let* $\xi(t)$ *be a primitive Puiseux parametrization with characteristic exponents* $n = \beta_0, \beta_1, \ldots, \beta_g$ *and associated integers* $e_0, \ldots, e_g = 1$. *Let* $v_0, \ldots, v_r$ *be the minimal set of generators o f* $\Gamma(\xi)$. *Then*

(i) $r = g$;
(ii) $v_0 = \beta_0 = n$, *and*

$$v_j = \sum_{k=1}^{j-1} \frac{e_{k-1} - e_k}{e_{j-1}} \beta_k + \beta_j, \quad j = 1, \ldots, g;$$

(iii) $e_i = \mathrm{GCD}(v_0, \ldots, v_i)$, *and* v_j *is the smallest element in* $\Gamma(\xi)$ *which is not divisible by* e_{j-1}.

Proof See [11, Theorem 6.12]. □

From this we get easily the following formulas (cf. [11, (6.4)]):

$$v_i = n_{i-1} v_{i-1} - \beta_{i-1} + \beta_i, \; i = 1, \ldots, g. \tag{1.5}$$

In view of Theorem 1.1.27 (ii), equalities (1.5) imply that the characteristic exponents of a Puiseux parametrization $\xi(t)$ determine $\Gamma(f)$, where C_f is the branch determined by $\xi(t)$ and conversely. Hence, one has that

$$C_{f_1} \equiv C_{f_2} \iff \Gamma(f_1) = \Gamma(f_2).$$

Formulas (1.5) also imply that

$$v_i > n_{i-1} v_{i-1}, \quad i = 1, \ldots, g. \tag{1.6}$$

A semigroup with the above relations (1.6) among the elements of its minimal set of generators will be called *strongly increasing*. These necessary conditions for a semigroup $\langle v_0, \ldots, v_g \rangle$ to be the semigroup of a branch are also sufficient (cf. [11, Theorem 6.14]).

Formulas (1.5), when inverted, give the following ones:

$$\beta_i = v_i - (n_{i-1} - 1)v_{i-1} - \cdots - (n_1 - 1)v_1, \ i = 1, \ldots, g. \tag{1.7}$$

A semigroup Γ with conductor c will be called symmetric if the following condition is satisfied:

$$\forall z \in \mathbb{N}, \ \ z \in \Gamma \iff c - 1 - z \notin \Gamma.$$

The *gaps* of Γ are the elements in $\mathbb{N} \setminus \Gamma$, so the symmetry of Γ means that in the interval $[0, c-1]$ there are as many gaps as elements in Γ.

One has the following result:

Theorem 1.1.28 *Let $\Gamma(f)$ be the semigroup of values of a plane branch C_f, then $\Gamma(f)$ is symmetric and the conductor c of $\Gamma(f)$ satisfies*

$$c = \sum_{i=1}^{g} (n_i - 1)v_i - v_0 + 1. \tag{1.8}$$

Proof See [11, Propositions 7.9 and 7.5]. □

A consequence of the symmetry of $\Gamma(f)$ is that the number of its gaps is precisely $\frac{c}{2}$, hence c is always an even number. This has the following notable consequence. Let $\xi(t)$ be a primitive parametrization for C_f, since, from Theorem 1.1.8 and Remark 1.1.10, one has

$$\dim_{\mathbb{C}} \frac{\overline{\mathcal{O}(f)}}{\mathcal{O}(f)} = \dim_{\mathbb{C}} \frac{\mathcal{O}_1}{R_\xi} = \#(\mathbb{N} \setminus \Gamma(f)) = \frac{c}{2},$$

it follows from equalities (1.3) and the fact that $\gamma = c$, that

$$\dim_{\mathbb{C}} \frac{\mathcal{O}(f)}{\mathcal{C}} = \frac{c}{2} = \dim_{\mathbb{C}} \frac{\overline{\mathcal{O}(f)}}{\mathcal{O}(f)}.$$

The above equality is what characterizes the so called *Gorenstein rings*.

Since $\Gamma(f)$ is determined by its gaps, a finite set, we have that this set of integers is also a complete topological invariant for the branch C_f.

In general, a given integer s in a semigroup $\Gamma = \langle v_0, \ldots, v_g \rangle$ may be written in several different ways as a combination of the v_i, but there is one privileged way to write it when these generators satisfy relations (1.6), as described in the result below.

Proposition 1.1.29 *Let $\Gamma = \langle v_0, \ldots, v_g \rangle$ be a semigroup whose generators satisfy relations (1.6), then any natural number s is uniquely representable in the form*

$$s = \sum_{i=0}^{g} s_i v_i, \ \ 0 \le s_i < n_i, \ \ i = 1, \ldots, g, \ \ s_0 \in \mathbb{Z},$$

in such a way that $s \in \Gamma$ if and only if $s_0 \ge 0$.

Proof See [11, Proposition 7.5]. □

In the sequel we discuss another set of invariants that characterize an equisingularity class of plane branches with semigroup of values Γ introduced by R. Apéry in [2].

Let $n = \min(\Gamma \setminus \{0\})$. The Apéry sequence $\mathfrak{a}_0 = 0, \mathfrak{a}_1, \ldots, \mathfrak{a}_{n-1}$, associated to Γ is defined as follows:

$$\mathfrak{a}_i = \min\{\alpha \in \Gamma;\ \alpha \not\equiv 0, \mathfrak{a}_1, \ldots, \mathfrak{a}_{i-1} \bmod n\},\quad i = 1, \ldots, n-1.$$

Notice that Γ determines the Apéry sequence.

Proposition 1.1.30 *The Apéry sequence has the following properties:*

(i) $0 = \mathfrak{a}_0 < \mathfrak{a}_1 < \cdots < \mathfrak{a}_{n-1}$;
(ii) $\mathfrak{a}_i \not\equiv \mathfrak{a}_j \bmod n$, *if* $0 \le i < j \le n-1$;
(iii) $\Gamma \setminus \{\infty\} = \bigcup_{i=0}^{n-1} \left(\mathfrak{a}_i + n\mathbb{N}\right)$;
(iv) $c = \mathfrak{a}_{n-1} - (n-1)$.

Proof See [11, p. 92]. □

Item (iii) above shows that the Apéry sequence determines Γ. On the other hand, it is clear that the gaps of Γ, that is the elements in $\mathbb{N} \setminus \Gamma$, are the integers of the form

$$\mathfrak{a}_i - n, \mathfrak{a}_i - 2n, \ldots, \mathfrak{a}_i - \left[\frac{\mathfrak{a}_i}{n}\right]n,\ i = 1, \ldots, n-1.$$

Example 1.1.31 Consider the semigroup

$$\begin{aligned}\Gamma = \langle 6, 9, 19\rangle = \{&0, 6, 9, 12, 15, 18, 19, 21, 24, 25, 27, 28,\\ &30, 31, 33, 34, 36, 37, 38, 39, 40, 42, \cdots\}.\end{aligned}$$

Then $n = \min(\Gamma \setminus \{0\}) = 6$ and its conductor is $c = 42$. The Apéry sequence of Γ is

$$\begin{aligned}
\mathfrak{a}_0 &= 0,\\
\mathfrak{a}_1 &= \min\{\alpha \in \Gamma; \alpha \not\equiv 0 \mod 6\} = 9,\\
\mathfrak{a}_2 &= \min\{\alpha \in \Gamma; \alpha \not\equiv 0, 9 \mod 6\} = 19,\\
\mathfrak{a}_3 &= \min\{\alpha \in \Gamma; \alpha \not\equiv 0, 9, 19 \mod 6\} = 28,\\
\mathfrak{a}_4 &= \min\{\alpha \in \Gamma; \alpha \not\equiv 0, 9, 19, 28 \mod 6\} = 38,\\
\mathfrak{a}_5 &= \min\{\alpha \in \Gamma; \alpha \not\equiv 0, 9, 19, 28, 38 \mod 6\} = 47.
\end{aligned}$$

Notice that $c = \mathfrak{a}_5 - (n-1) = 42$.

1.1.4 Semiroots of a Branch

In this section, we will study special elements of $\mathcal{O}$ whose intersection multiplicity realize the generators of the semigroup $\Gamma(f)$, of a branch C_f. These elements are important in several contexts, since the branches associated to them are simpler than the original branch and contain a great amount of information about it.

Let $f \in \mathbb{C}\{X\}[Y]$ be an irreducible Weierstrass polynomial and let $\xi(t)$ be a primitive Puiseux parametrization of C_f. As mentioned in the previous section, $\Gamma(f) = \langle v_0, \ldots, v_g \rangle$ is a complete topological invariant for the plane branch C_f.

Notice that if $h \in \mathcal{O}$ is such that $I(f, h) = v_i$ for some $0 \leq i \leq g$, then h must be irreducible. In fact, if $h = h_1 \cdot h_2$ with $h_1, h_2 \in \mathcal{M}$, then

$$v_i = I(f, h) = I(f, h_1 \cdot h_2) = I(f, h_1) + I(f, h_2),$$

with $I(f, h_1), I(f, h_2) \in \Gamma(f) \setminus \{0\}$, which contradicts the fact that the v_i, $i = 0, \ldots, g$, form a minimal set of generators.

As remarked before, we may suppose that $\xi(t) = (t^n, \sum_{i \geq n} a_i t^i)$, where $n = \deg_Y(f)$.

Among all the elements in $\mathcal{O}$ whose intersection multiplicity with C_f belong to the minimal set of generators of $\Gamma(f)$ we will consider a particular subset.

Definition 1.1.32 For any $k = 0, \ldots, g-1$, a k-semiroot of f is a monic polynomial $f_k \in \mathbb{C}\{X\}[Y]$ such that $\deg_Y(f_k) = \frac{n}{e_k}$ and $\mathrm{I}(f, f_k) = v_{k+1}$. We define $f_g = f$ and call $\{f_0, f_1, \ldots, f_g\}$ a complete system of semiroots for f.

The next result shows that complete systems of semiroots for f exist.

Theorem 1.1.33 *Let $\xi(t) = (t^n, \sum_{i \geq n} a_i t^i)$ be a primitive Puiseux parametrization for a plane branch C_f and let $f_k \in \mathbb{C}\{X\}[Y]$, for $0 \leq k < g$, be the minimal polynomial of $\sum_{n \leq i < \beta_{k+1}} a_i X^{\frac{i}{n}}$ over $\mathbb{C}((X))$. We have*

(i) $\xi_k(t) = (t^{\frac{n}{e_k}}, \sum_{n \leq i < \beta_{k+1}} a_i t^{\frac{i}{e_k}})$ *is a primitive Puiseux parametrization for* C_{f_k}*;*
(ii) $\deg_Y(f_k) = \frac{n}{e_k}$ *and* $\mathrm{I}(f, f_k) = \mathrm{ord}_t(f_k(\xi(t))) = \mathrm{ord}_t(f(\xi_k(t))) = v_{k+1}$*;*
(iii) $\sum_{i=1}^{j}(n_i - 1)v_i + \beta_{j+1}$ *is the minimal order of a term in* $f_k(\xi(t))$ *not divisible by* e_j*, for any fixed* j *such that* $k \leq j < g$*.*

In particular, f_k *is a* k*-semiroot of* f*.*

Proof The first two items follow from the computations in [24, Theorem 3.9 on p. 11], or [11, Theorem 6.12].

We now proceed to the proof of (iii). Denoting $G_j = \{\eta \in \mathbb{C};\ \eta^{e_j} = 1\}$, then Theorem 1.1.12 gives

$$f_k = \prod_{\overline{\epsilon} \in \frac{G_0}{G_k}} \left(Y - \sum_{n \leq i < \beta_{k+1}} a_i \epsilon^i X^{\frac{i}{n}} \right),$$

and, consequently,

$$f_k(\xi(t)) = \prod_{\overline{\epsilon}\in\frac{G_0}{G_k}} \Big(\sum_{n\le i<\beta_{k+1}} a_i(1-\epsilon^i)t^i + \sum_{i\ge\beta_{k+1}} a_i t^i \Big).$$

From [11, Lemma 6.10] we have that

$$\mathrm{ord}_t \Big(\sum_{n\le i<\beta_{k+1}} a_i(1-\epsilon^i)t^i + \sum_{i\ge\beta_{k+1}} a_i t^i \Big) = \begin{cases} \beta_j \text{ if } \epsilon \in G_{j-1}\setminus G_j;\ j<k \\ \beta_k \text{ if } \epsilon \in G_k. \end{cases}$$

As $G_{j-1}\setminus G_j$ is the union of $\frac{e_{j-1}-e_j}{e_k}$ cosets of G_k, then, for any j such that $k \le j < g$, the minimal order of a term in $f_k(\xi(t))$ not divisible by e_j is

$$\sum_{i=1}^{j} \frac{e_{j-1}-e_j}{e_k}\beta_i + \beta_{j+1} = \sum_{i=1}^{j}(n_i-1)v_i + \beta_{j+1},$$

as we wanted to show. □

A k-semiroot of f, as described in the previous theorem, will be called a *characteristic k-semiroot.*

Semiroots are not uniquely determined. In fact, S. Abhyankar and T. Moh in [1] define a special type of semiroots, called approximate roots, which have some particularities that characterize them uniquely. On the other hand, in [13] we present an algorithm that allows one to compute *Standard Basis* for the coordinate ring of any irreducible curve, not necessarily plane. For a plane branch, elements in a minimal Standard Basis are semiroots. So, characteristic semiroots, approximate roots, or elements in a minimal Standard Basis are all semiroots, but may differ from each other.

As before, we denote $\mathcal{E}xp(\xi) = \{\beta_0,\dots,\beta_g\}$ and $\Gamma(f) = \langle v_0,\dots,v_g\rangle$. Recall that we defined $n=\beta_0=v_0$ and $m=\beta_1=v_1$. We have the following immediate corollary.

Corollary 1.1.34 *For f_k and $\xi_k(t)$, with $0\le k\le g$, as in Theorem 1.1.33, one has*

$$\mathcal{E}xp(\xi_k) = \left\{\frac{\beta_0}{e_k},\dots,\frac{\beta_k}{e_k}\right\} \ \textit{and}\ \Gamma(f_k) = \left\langle \frac{v_0}{e_k},\dots,\frac{v_k}{e_k}\right\rangle.$$

Observe that, for $j=1,\dots,k$, we have

$$n_j = \frac{e_{j-1}}{e_j} = \frac{GCD(\beta_0,\dots,\beta_{j-1})}{GCD(\beta_0,\dots,\beta_j)} = \frac{GCD\Big(\frac{\beta_0}{e_k},\dots,\frac{\beta_{j-1}}{e_k}\Big)}{GCD\Big(\frac{\beta_0}{e_k},\dots,\frac{\beta_j}{e_k}\Big)}.$$

As a consequence, we have that the conductor c_k of $\Gamma(f_k)$ is given by

$$c_k = \sum_{j=1}^{k}(n_j - 1)\frac{v_j}{e_k} - \frac{v_0}{e_k} + 1. \tag{1.9}$$

If f_k is a k-semiroot of f and h_j is a j-semiroot of f_k, then it follows immediately that h_j is a j-semiroot of f.

Applying successive Euclidean division, and by Proposition 1.1.29 we obtain the following consequence:

Corollary 1.1.35 *Let $\{f_0, \ldots, f_g\}$ be a complete system of semiroots. Every $h \in \mathcal{O}$ can be uniquely written in the form*

$$h = \sum_{i_0,\ldots,i_g} a_{i_0,\ldots,i_g} f_0^{i_0} \cdots f_g^{i_g},$$

with $i_g \geq 0$, $0 \leq i_k < n_k$, for all $0 \leq k < g$, and $a_{i_0,\ldots,i_g} \in \mathbb{C}\{X\}$. Moreover,

(i) $\deg_Y(f_0^{i_0} \cdots f_g^{i_g}) \neq \deg_Y(f_0^{j_0} \cdots f_g^{j_g})$, *if* $(i_0, \ldots, i_g) \neq (j_0, \ldots, j_g)$*;*

(ii) If $(i_0, \ldots, i_{g-1}) \neq (j_0, \ldots, j_{g-1})$, *then*

$$\mathrm{I}(f, a_{i_0,\ldots,i_{g-1},0} f_0^{i_0} \cdots f_{g-1}^{i_{g-1}}) \neq \mathrm{I}(f, a_{j_0,\ldots,j_{g-1},0} f_0^{j_0} \cdots f_{g-1}^{j_{g-1}});$$

(iii) If $h \in \mathbb{C}\{X\}[Y]$ *with* $\deg_Y(h) < \deg_Y(f_k) = \frac{n}{e_k}$, *then* $\mathrm{I}(f, h) = e_k \mathrm{I}(f_k, h)$.

Proof For (i) and (ii) see [21, Corollary 5.4].
Now, we prove (iii). If $h \in \mathbb{C}\{X\}[Y]$ with $deg_Y(h) < deg_Y(f_k)$, then by (i) we may write $h = \sum_{i_0,\ldots,i_{k-1}} a_{i_0,\ldots,i_{k-1}} f_0^{i_0} \cdots f_{k-1}^{i_{k-1}}$, and by (ii), for some $(j_0, \ldots, j_{k-1})$,

$$\begin{aligned}\mathrm{I}(f, h) &= \min_{(i_0,\ldots,i_{k-1})}\{\mathrm{I}(f, a_{i_0,\ldots,i_{k-1}} f_0^{i_0} \cdots f_{k-1}^{i_{k-1}})\} \\ &= \mathrm{I}(f, a_{j_0,\ldots,j_{k-1}} f_0^{j_0} \cdots f_{k-1}^{j_{k-1}}).\end{aligned}$$

But, since $\{f_0, \ldots, f_{k-1}\}$ is a complete system of semiroots of f_k, it follows that $\mathrm{I}(f_k, h) = \mathrm{I}(f_k, a_{j_0,\ldots,j_{k-1}} f_0^{j_0} \cdots f_{k-1}^{j_{k-1}})$. As $\mathrm{I}(f, X) = n = e_k \mathrm{I}(f_k, X)$ and $\mathrm{I}(f, f_j) = v_{j+1} = e_k \mathrm{I}(f_k, f_j)$, for $0 \leq j < k$, we get the result. □

Example 1.1.36 Let $\xi(t) = \xi_3(t) = (t^8, t^{12} + t^{14} + t^{15})$ be a primitive Puiseux parametrization for a plane branch C_f. The characteristic exponents of C_f are $\mathcal{E}xp(\xi_3) = \{8, 12, 14, 15\}$, $e_0 = 8$, $e_1 = 4$, $e_2 = 2$, $e_3 = 1$ and according to Theorem 1.1.27 we have $\Gamma(f) = \langle 8, 12, 26, 53 \rangle$.

By Theorem 1.1.33,

$$\xi_0(t) = (t, 0), \quad \xi_1(t) = (t^2, t^3), \quad \xi_2(t) = (t^4, t^6 + t^7)$$

are primitive Puiseux parametrizations for C_{f_k} with f_k a k-semiroot of f for $0 \leq k \leq 2$.

Using Theorem 1.1.12 we obtain

$$f_0 = Y, \quad f_1 = Y^2 - X^3, \quad f_2 = Y^4 - 2X^3Y^2 - 4X^5Y + X^6 - X^7,$$

$$\begin{aligned} f = f_3 = Y^8 - 4X^3Y^6 - 8X^5Y^5 + (6 - 26X)X^6Y^4 + \\ +(16 - 24X)X^8Y^3 + (-4 + 36X - 20X^2)X^9Y^2 + \\ +(-8 + 16X - 8X^2)X^{11}Y + (1 + 6X + 21X^2 - X^3)X^{12}. \end{aligned}$$

Recall that

$$\begin{aligned} I(f, f_0) &= \mathrm{ord}_t(f_0(\xi(t))) = 12 = v_1, \\ I(f, f_1) &= \mathrm{ord}_t(f_1(\xi(t))) = 24 = v_2, \\ I(f, f_2) &= \mathrm{ord}_t(f_2(\xi(t))) = 53 = v_3. \end{aligned}$$

1.2 Zariski's Approach

Zariski's program to attack the problem of analytic classification of plane branches belonging to an equisingularity class $\mathcal{L}$, consisted in constructing an appropriate parameter space Σ that represents all the analytic types of the members of $\mathcal{L}$, and then to quotient it modulo the relation that identifies two point that represent analytically equivalent branches, to get the corresponding moduli space, as specified below.

One wants Σ to be a constructible nonempty set in some finite dimensional affine space $\mathbb{C}^N$, together with a function $F\colon \Sigma \to \mathcal{L}$, in such a way that $F(\Sigma)$ contains at least one representative of each class of analytic equivalence in $\mathcal{L}$. To realize this program it will be necessary to interpret the equivalence relation $\sim$ on Σ induced by the action of $\mathcal{G} = \mathcal{O}^* \rtimes \mathrm{Aut}(\mathbb{C}^2, 0)$ on $\mathcal{L}$, namely,

$$\sigma_1 \sim \sigma_2 \iff F(\sigma_1) \in \mathcal{G}(F(\sigma_2)),$$

where $\mathcal{G}(F(\sigma_2))$ denotes the orbit of $F(\sigma_2)$ under the action of $\mathcal{G}$, and to study the space $\Sigma/\sim$. It is intuitive that the smaller is the integer N, the simpler the equivalence relation $\sim$ on Σ will be, since there will be less parameters to deal with.

1.2.1 A Parameter Space

Zariski's strategy, sketched in [23], consisted in constructing a parameter space by means of the coefficients of $y(t)$ in primitive Puiseux parametrizations of the form $\xi(t) = (t^n, y(t))$ that correspond to branches in a given equisingularity class $\mathcal{L}$. The trouble is that this would yield to an infinite dimensional space, but he overcomes this inconvenient by observing that there is a uniform bound for $\mathcal{L}$ such that each branch in $\mathcal{L}$ is analytically equivalent to a branch with a Puiseux parametrization $(t^n, y(t))$, where $y(t)$ is a polynomial of degree less or equal than this bound. In this way, the coefficients of such $y(t)$ determine a point in a finite dimensional affine space. The method proceeds by introducing some uniform elimination criteria of coefficients in $y(t)$, depending only on $\mathcal{L}$, without changing the analytic type of the branch it represents, in order to lower the dimension of this affine space.

So, we start with the reduction that will tell us that our parametrizations are finitely determined in the sense we make precise below.

Given an element $h \in \mathbb{C}\{X_1, \ldots, X_r\}$, written as $h = \sum_{i=0}^{\infty} h_i$, where each h_i is a homogeneous polynomial of degree i in $\mathbb{C}[X_1, \ldots, X_r]$ or the zero polynomial, we denote by $j^k h$ the truncation $\sum_{i=0}^{k} h_i$ of h and call it *the k-th jet* of h.

First Reduction: Let $\xi(t) = (x(t), y(t))$ be a primitive parametrization such that $n\colon = \text{ord}_t(\xi(t)) > 2$ and let c be the conductor of $\Gamma(\xi)$. Then one has that $\xi(t)$ and $j^{c-1}\xi(t)$ are $\mathcal{A}$-equivalent. The case $n = 2$ will be treated separately in Proposition 1.2.1.

Indeed, since $(x(t), y(t))$ and $(y(t), x(t))$ are $\mathcal{A}$-equivalent, we may assume that $x(t) = \sum_{i\geq n} a_i t^i$ and $y(t) = \sum_{i\geq r} b_i t^i$, where $a_n b_r \neq 0$ and $n \leq r$. Suppose that $\Gamma(\xi) = \langle v_0, \ldots, v_g\rangle$, then $v_0 = n$ and $v_1 \geq r$. It follows, from Formula (1.8) that $c \geq r + 1 \geq n + 1$ (here we use $n > 2$), hence we may write $x(t) = \sum_{i=n}^{c-1} a_i t^i + \sum_{i\geq c} a_i t^i$ and $y(t) = \sum_{i=r}^{c-1} b_i t^i + \sum_{i\geq c} b_i t^i$. As $\sum_{i\geq c} a_i t^i, \sum_{i\geq c} b_i t^i \in R_\xi$, i t fol-lows that there exist $p, q \in \langle X, Y\rangle^2$ such that $p(\xi(t)) = \sum_{i\geq c} a_i t^i$ and $q(\xi(t)) = \sum_{i\geq c} b_i t^i$. So, if we take $\rho(t) = t$ and $\varphi(X, Y) = (X - p(X, Y), Y - q(X, Y))$, one has $\varphi(\xi(\rho^{-1}(t))) = j^{c-1}\xi(t)$.

In particular, if $\xi(t) = (t^n, y(t))$ is a primitive Puiseux parametrization, with $n > 2$, then $\xi(t)$ is $\mathcal{A}$-equivalent to $(t^n, j^{c-1}y(t))$.

The following reduction will give us a uniform way, depending on $\mathcal{L}$, to start the series $y(t)$.

Second Reduction: Suppose that $\xi(t) = (t^n\ y(t))$ is a given Puiseux parametrization such that $y(t) = Q(t^n) + a_m t^m + \cdots$, where $a_m \neq 0$, m is not a multiple of n and Q is a univariate polynomial with $\deg_t Q(t^n) < m$. Let us define

$$\rho(t) = t \quad \text{and} \quad \varphi(X, Y) = \left(X, \frac{Y - Q(X)}{a_m}\right),$$

then ρ and φ are, respectively, automorphisms of $(\mathbb{C}, 0)$ and of $(\mathbb{C}^2, 0)$ such that

$$\varphi(\xi(\rho^{-1}(t))) = (t^n,\ t^m + \cdots).$$

So, we may eliminate all initial terms of order multiple of n. Notice that besides the elimination of the terms in $y(t)$ of degrees divisible by n and lower that m, one may transform any term $a_i t^i$ with $a_i \neq 0$ in $y(t)$, instead of $a_m t^m$, into t^i.

The following result is an example of how these reductions may help us in our intent.

Proposition 1.2.1 *Any branch of multiplicity 2 is analytically equivalent to a branch parametrized by (t^2, t^m), where m is odd.*

Proof Let C_f be a branch of multiplicity 2, then there exists a primitive parametrization $\xi(t) = (x(t), y(t))$ such that $\mathrm{ord}_t\, \xi(t) = 2$, hence at the cost of permuting X and Y, we may assume that $2 = \mathrm{ord}_t\, x(t) \leq \mathrm{ord}_t\, y(t)$. So, we may reparametrize $\xi(t)$ is such a way that, after applying the Second Reduction, we get a primitive Puiseux parametrization $\xi_1(t) = (t^2, \sum_{i \geq m} a_i t^i)$, where m is odd. This implies that $\Gamma(\xi) = \langle 2, m\rangle$, hence, by Theorem 1.1.28, its conductor is $c = m - 1$. So, $\sum_{i>m} a_i t^i \in R_{\xi_1}$, which means that there exists $q \in \langle X, Y\rangle^2$ such that $q(\xi_1(t)) = \sum_{i>m} a_i t^i$. Now, if we take

$$\rho(t) = t \text{ and } \varphi(X, Y) = \left(X, \frac{Y - q(X, Y)}{a_m}\right),$$

we get $\varphi(\xi_1(\rho^{-1}(t))) = (t^2, t^m)$. □

Remark 1.2.2 Given a branch defined by a Weierstrass polynomial $f = Y^n + \sum_{i=1}^{n} c_i(X) Y^{n-i}$ in $\mathbb{C}\{X\}[Y]$ with a primitive Puiseux parametrization $\xi(t)$, in the Second Reduction we can take $Q(t^n) \in \mathbb{C}\{t^n\}$ and obtain $\xi_1(t) = \varphi(\xi(\rho^{-1}(t))) = (t^n, y(t))$ with $y(t)$ without any term of the form $a_{kn} t^{kn}$. Consequently, by Theorem 1.1.12, the branch with parametrization $\xi_1(t)$ admits a Weierstrass polynomial $Y^n + \sum_{i=2}^{n} b_i(X) Y^{n-i}$, with $b_i(X) \in \mathbb{C}\{X\}$, as Cartesian equation. Notice that this corresponds to apply the Tschirnhausen operator on f, that is, the automorphism $\varphi \in \mathrm{Aut}(\mathbb{C}^2, 0)$ defined by $\varphi(X, Y) = (X, Y - \frac{c_1(X)}{n})$.

Note that in the First and Second Reductions, applied to a Puiseux parametrization, we only played with the automorphism φ, taking in both cases $\rho(t) = t$. Now, playing with both automorphisms φ and ρ, we may characterize the transformations that preserve the form of a Puiseux parametrization.

Proposition 1.2.3 *Let $\xi(t) = (t^n, y(t)) \in \mathcal{M}_1 \times \mathcal{M}_1$, where $y(t) = t^m + hot$, with $n < m$, and hot means an element in $\mathcal{M}_1^{m+1}$. Then the automorphisms φ and ρ of $(\mathbb{C}^2, 0)$ and $(\mathbb{C}, 0)$, respectively, are such that $\varphi(\xi(\rho^{-1}(t))) = (t^n, t^{m_1} + hot)$, with $n < m_1$, and hot means an element in $\mathcal{M}_1^{m_1+1}$, if, and only if, $m = m_1$ and there exist $\epsilon \in \mathbb{C}^*$, $p(X, Y) \in \langle X^2, Y\rangle$ and $q(X, Y) \in \langle X, Y\rangle^2$ such that*

$$\rho(t) = t\left(\epsilon^n + \frac{p(\xi(t))}{t^n}\right)^{\frac{1}{n}}, \quad \varphi(X, Y) = (\epsilon^n X + p(X, Y), \epsilon^m Y + q(X, Y)).$$

Proof Let us write

$$\varphi(X, Y) = (\alpha X + \beta Y + p_1(X, Y), \gamma X + \delta Y + q(X, Y)),$$

where $p_1(X, Y), q(X, Y) \in \langle X, Y\rangle^2$.

From the equality $\varphi(\xi(\rho^{-1}(t))) = (t^n, t^{m_1} + hot)$, we get

$$\alpha(\rho^{-1}(t))^n + \beta y(\rho^{-1}(t)) + p_1(\xi(\rho^{-1}(t))) = t^n,$$

$$\gamma(\rho^{-1}(t))^n + \delta y(\rho^{-1}(t)) + q(\xi(\rho^{-1}(t))) \quad = t^{m_1} + hot.$$

Evaluating orders in the above equalities, it follows that $\alpha \neq 0$, $\gamma = 0$ and $m_1 = m$.

If we write $\beta Y + p_1(X, Y) =: p(X, Y)$, w e have that $p(X, Y) \in \langle X^2, Y\rangle$, and

$$\alpha(\rho^{-1}(t))^n + p(\xi(\rho^{-1}(t))) = t^n, \text{ and } \varphi(X, Y) = (\alpha X + p(X, Y), \delta Y + q(X, Y)).$$

Hence, $(\rho(t))^n = \alpha t^n + p(\xi(t))$, and from the fact that

$$\delta y(\rho^{-1}(t)) + q(\xi(\rho^{-1}(t))) = t^m + hot,$$

putting $\epsilon = \sqrt[n]{\alpha}$, it follows that $\delta = \epsilon^m$ and the proposition is proved. □

The next result will give us a way to compare the parametrization $\varphi(\xi(\rho^{-1}(t)))$, where φ and ρ are automorphisms as in the above proposition with $\epsilon = 1$, with the Puiseux parametrization $\xi(t)$ such that $a_m = 1$. We will denote by $y'(t)$ the derivative of $y(t)$ with respect to t.

Corollary 1.2.4 *For the automorphisms φ and ρ, with $\epsilon = 1$, and for p and q as described in Proposition 1.2.3, if $\xi(t) = (t^n, y(t))$ is such that $y(t) = \sum_{i\geq m} a_i t^i$, with $a_m = 1$, then $\varphi(\xi(\rho^{-1}(t))) = \xi(t) + (0, \theta(\rho^{-1}(t)))$, where*

$$\theta(t) = q(\xi(t)) - \frac{y'(t)}{nt^{n-1}} p(\xi(t)) - \sum_{i\geq m} a_i t^i \sum_{l=2}^{\infty} \binom{\frac{i}{n}}{l} \left(\frac{p(\xi(t))}{t^n}\right)^l, \tag{1.10}$$

and is such that the initial term of $\theta(\rho^{-1}(t))$ is equal to that of $\theta(t)$.

Proof From Proposition 1.2.3 we have

$$(\rho(t))^i = t^i \left(1 + \frac{p(\xi(t))}{t^n}\right)^{\frac{i}{n}} = t^i + t^i \sum_{l=1}^{\infty} \binom{\frac{i}{n}}{l} \left(\frac{p(\xi(t))}{t^n}\right)^l,$$

and consequently,

$$(\rho^{-1}(t))^i = t^i - \rho^{-1}(t)^i \sum_{l=1}^{\infty} \binom{\frac{i}{n}}{l} \left(\frac{p(\xi(\rho^{-1}(t)))}{\rho^{-1}(t)^n}\right)^l,$$

hence

$$\varphi(\xi(\rho^{-1}(t))) = \xi(t) + (0, \theta(\rho^{-1}(t))),$$

where,

$$\theta(t) = q(\xi(t)) - \sum_{i\geq m} a_i t^i \sum_{l=1}^{\infty} \binom{\frac{i}{n}}{l} \left(\frac{p(\xi(t))}{t^n}\right)^l$$

$$= q(\xi(t)) - \sum_{i\geq m} a_i t^i \frac{ip(\xi(t))}{nt^n} - \sum_{i\geq m} a_i t^i \sum_{l=2}^{\infty} \binom{\frac{i}{n}}{l} \left(\frac{p(\xi(t))}{t^n}\right)^l$$

$$= q(\xi(t)) - \frac{y'(t)}{nt^{n-1}} p(\xi(t)) - \sum_{i\geq m} a_i t^i \sum_{l=2}^{\infty} \binom{\frac{i}{n}}{l} \left(\frac{p(\xi(t))}{t^n}\right)^l.$$

Finally, since

$$\theta(\rho^{-1}(t)) = \theta(t + hot) = \theta(t) + \epsilon(t), \text{ where } \mathrm{ord}_t(\epsilon(t)) > \mathrm{ord}_t(\theta(t)),$$

where hot means an element in $\mathcal{M}_1^2$, it follows that $\theta(\rho^{-1}(t))$ and $\theta(t)$ have same initial term. □

The following is a slight generalization of [24, Proposition 1.2, Chap. 3, p. 19] and [24, (2.5) and (2.6) on p. 23].

Proposition 1.2.5 *Let $\xi(t) = (t^n, t^m + \sum_{i>m} a_i t^i)$, with $n \nmid m$, be a primitive Puiseux parametrization of a branch C_f. If $k > m$ is such that $k \in \Gamma(f)$, or $k \in \Gamma(f) + m - n$, then C_f is analytically equivalent to a branch with a Puiseux parametrization $(t^n, t^m + \sum_{m<i<k} a_i t^i + \sum_{i>k} a_i' t^i)$, where $a_i' \in \mathbb{C}$, for $i > k$.*

Proof If $k > m$ and $k \in \Gamma(f)$, then there exists $q \in \langle X, Y\rangle^2$ such that $j^k(q(\xi(t))) = -a_k t^k$. Considering $\epsilon = 1$ and $p = 0$ in Proposition 1.2.3, we obtain the result.

On the other hand, if $k > m$ and $k \in \Gamma(f) + m - n$, there exists $p \in \langle X^2, Y\rangle$ with $j^{k-m+n}(p(\xi(t)) = -\frac{n}{m} a_k t^{k-m+n}$. We obtain the result taking $\epsilon = 1$ and $q = 0$ in Proposition 1.2.3. □

The next result is a substantial generalization of the above proposition.

Theorem 1.2.6 *Let $\xi(t) = (t^n, t^m + \sum_{i>m} a_i t^i)$ be a primitive Puiseux parametrization representing a plane branch C_f with semigroup of values $\Gamma(f) = \langle v_0, v_1, \ldots, v_g\rangle$, where $v_0 = n$ and $v_1 = m$. If $j > m$ and $j \in \Gamma(f) + v_k - n$, for some $0 \leq k \leq g$, then C_f is analytically equivalent to a branch with parametrization $(t^n, t^m + \sum_{i>m} a_i' t^i)$ with $a_i' = a_i$ for $i < j$, and $a_j' = 0$.*

Proof For $k \in \{0, 1\}$ the result corresponds to Proposition 1.2.5.

Let us suppose that $j > m$ and $j \in \Gamma(f) + v_k - n$ for some k with $2 \leq k \leq g$. Consider the characteristic $(k-1)$-semiroot f_{k-1} as described in Theorem 1.1.33.

Since $f_{k-1} \in \mathbb{C}\{X\}[Y]$ is an irreducible Weierstrass polynomial of degree $\frac{n}{e_{k-1}} = n_1 \cdots n_{k-1}$, and $n_i \geq 2$ for $i = 1, \ldots, g$, we have that $(f_{k-1})_Y \in \langle X^2, Y\rangle$ and $(f_{k-1})_X \in \langle X, Y\rangle^2$. Moreover,

$$\begin{aligned} \mathrm{I}(f, (f_{k-1})_Y) &= e_{k-1}\mathrm{I}(f_{k-1}, (f_{k-1})_Y) = e_{k-1}(c_{k-1} + \tfrac{n}{e_{k-1}} - 1) \\ &= \textstyle\sum_{i=1}^{k-1} (n_i - 1) v_i \geq m > n, \end{aligned} \tag{1.11}$$

where the first equality, above, follows from Corollary 1.1.35 (iii), the second equality follows from the formula $\mathrm{I}(f, f_Y) = c + n - 1$ (cf. [11, Corollary 7.16]) applied to f_{k-1} instead of f and, finally, the third equality follows from (1.9).

Considering $j > m$ and $j = s + v_k - n$, where $s = \sum_{i=0}^{g} s_i v_i \in \Gamma(f)$ with $0 \leq s_i < n_i$ and $s_0 \geq 0$ as described in Proposition 1.1.29, we have that $h = X^{s_0} f_0^{s_1} \cdot \ldots \cdot f_{g-1}^{s_g}$ is such that $\mathrm{I}(f, h) = s$.

Taking, in Proposition 1.2.3, $\epsilon = 1$, $p = -\alpha h (f_{k-1})_Y$ and $q = \alpha h (f_{k-1})_X$, where α is an arbitrary complex number, we get $\varphi \in \mathrm{Aut}(\mathbb{C}^2, 0)$ and $\rho \in \mathrm{Aut}(\mathbb{C}, 0)$, which in view of Corollary 1.2.4, are such that

$$\varphi(\xi(\rho^{-1}(t))) = \xi(t) + (0, \theta(\rho^{-1}(t))),$$

with

$$\theta(t) = \alpha h(\xi(t)) \frac{(f_{k-1}(\xi(t)))'}{n t^{n-1}} - \sum_{i \geq m} a_i t^i \sum_{l=2}^{\infty} \binom{\frac{i}{n}}{l} \left(\frac{-\alpha (h (f_{k-1})_Y)(\xi(t))}{t^n} \right)^l,$$

and the initial terms of $\theta(\rho^{-1}(t))$ and of $\theta(t)$ are equal.

Notice that since $\mathrm{ord}_t \left((f_{k-1}(\xi(t)))' \right) = v_k - 1$, then

$$\mathrm{ord}_t \left(h(\xi(t)) \frac{(f_{k-1}(\xi(t)))'}{n t^{n-1}} \right) = s + v_k - n = j.$$

On the other hand, the order in t of the second term in the expression of $\theta(t)$ is $2(s + \mathrm{I}(f, (f_{k-1})_Y) - n) + m$.

Case 1. $j = s + v_k - n \leq 2(s + \mathrm{I}(f, (f_{k-1})_Y) - n) + m$.

By choosing a convenient value for α we can eliminate the term that assume order j in $\varphi(\xi(\rho^{-1}(t)))$.

Case 2. $j = s + v_k - n > 2(s + \mathrm{I}(f, (f_{k-1})_Y) - n) + m$.

In this case, from (1.11), it follows that $s < v_k$ and, also,

$$\begin{aligned} & 2(s + \mathrm{I}(f, (f_{k-1})_Y) - n) + m > \mathrm{I}(f, (f_{k-1})_Y) - n + m = \\ & e_{k-1} \mathrm{I}(f_{k-1}, (f_{k-1})_Y) - n + m = \\ & e_{k-1} \left(c_{k-1} + \frac{n}{e_{k-1}} - 1 \right) - n + m > e_{k-1} c_{k-1}. \end{aligned}$$

Since $s < v_k$, w e may take $h = X^{s_0} f_0^{s_1} \cdot \ldots \cdot f_{k-2}^{s_{k-1}}$. By Theorem 1.1.33 (iii) and some elementary computations, it follows that the lowest order not multiple of e_k of a term in $t^i \binom{\frac{i}{n}}{l} \left(\frac{(h (f_{k-1})_Y)(\xi(t))}{t^n} \right)^l$, for $i \geq m$ and $l \geq 2$, is

$$m+2(s+\sum_{i=1}^{k-1}(n_i-1)v_i-v_{k-1}+\beta_k-2n)>s+v_k-n=j.$$

The above analysis shows that in $\sum_{i\geq m}a_it^i\sum_{l=2}^{\infty}\binom{\frac{i}{n}}{l}\left(\frac{-\alpha(h(f_{k-1})_Y)(\xi(t))}{t^n}\right)^l$ all terms have order $r>e_{k-1}c_{k-1}$ and e_{k-1} divides r for all $r<s+v_k-n=j$.

Since c_{k-1} is the conductor of $\Gamma(f_{k-1})$, there exists

$$h_1=\sum_{e_{k-1}c_{k-1}<r<s+v_k-n}a'_rX^{r_0}f_0^{r_1}\cdots f_{k-1}^{r_{k-1}}\in\langle X,Y\rangle^2,$$

where $r=r_0v_0+r_1v_1+\cdots+r_{k-1}v_{k-1}$, with $\mathrm{I}(f,X^{r_0}f_0^{r_1}\cdots f_{k-2}^{r_{k-1}})=r$, such that

$$\mathrm{ord}_t\left(h_1(\xi(t))-\sum_{i\geq m}a_it^i\sum_{l=2}^{\infty}\binom{\frac{i}{n}}{l}\left(\frac{-\alpha(h(f_{k-1})_Y)(\xi(t))}{t^n}\right)^l\right)>s+v_k-n=j.$$

Redefining $p:=-\alpha h(f_{k-1})_Y$ and $q:=\alpha h(f_{k-1})_X+h_1$ we obtain

$$\varphi(\xi(\rho^{-1}(t)))=\xi(t)+(0,\theta_1(\rho^{-1}(t))),$$

with

$$\theta_1(t)=\alpha h\frac{(f_k(\xi(t)))'}{nt^{n-1}}+h_1(\xi(t))-\sum_{i\geq m}a_it^i\sum_{l=2}^{\infty}\binom{\frac{i}{n}}{l}\left(\frac{-\alpha(h(f_{k-1})_Y)(\xi(t))}{t^n}\right)^l,$$

where $\mathrm{ord}_t(\theta_1(\rho^{-1}(t))-\theta_1(t))>\mathrm{ord}_t(\theta_1(\rho^{-1}(t)))=\mathrm{ord}_t(\theta_1(t))=s+v_k-n=j$, concluding our proof, after a suitable choice of α. □

The above theorem gives us the following reduction:

Third Reduction: Let $\xi(t)=(t^n,t^m+\sum_{i>m}a_it^i)$ be a primitive Puiseux parametrization of C_f with semigroup $\Gamma(f)=\langle v_0,v_1,\ldots,v_g\rangle$, where $v_0=n$ and $v_1=m$. If $s>m$ and $s\in\Gamma(f)-n$, there exists a branch analytically equivalent to C_f with parametrization $(t^n,t^m+\sum_{i>m}a'_it^i)$ such that $a'_i=a_i$ for $i<s$ and $a'_s=0$. Indeed, as $m<s=j-n$ with $j\in\Gamma(f)\setminus\{0\}$, we have $j=\alpha+v_k$, for some $\alpha\in\Gamma(f)$ and $0\leq k\leq g$, then the previous theorem allows us to perform this reduction.

We summarize the three above reductions in the following theorem which generalizes the results of Ebey in [9] and Zariski in [23].

Theorem 1.2.7 *Let $\xi_1(t)=(t^n,y(t))$ be a primitive Puiseux parametrization of a branch C with semigroup of values $\Gamma=\langle v_0,\ldots,v_g\rangle$, where $n=v_0$. If $n>2$, there exists a primitive Puiseux parametrization*

$$\xi(t) = \left(t^n,\ t^m + \sum_{\substack{i>m \\ i\notin\Gamma-n}} a_i t^i\right), \tag{1.12}$$

where $m = v_1$ and the coefficients a_i satisfy condition (1.4), such that the branch determined by $\xi(t)$ is $\mathcal{A}$-equivalent to C. If $n = 2$, $\xi(t)$ is $\mathcal{A}$-equivalent to (t^2, t^m), where m is odd.

Proof If $n > 2$, by the Second Reduction, we have that $\xi_1(t)$ and $\widetilde{\xi_1}(t) = (t^n,\ t^m + \sum_{i>m} a_i' t^i)$ determine analytically equivalent branches.

Applying the Third Reduction to $\widetilde{\xi_1}(t)$, for every $i \in \Gamma - n$, with $m < i < c$, we obtain

$$\widetilde{\xi}(t) = \left(t^n,\ t^m + \sum_{\substack{i>m \\ i\notin\Gamma-n}}^{c-1} a_i t^i + \sum_{i\geq c} a_i t^i\right).$$

Now, we apply the First Reduction to get the result in this case. If $n = 2$, the the result follows from Proposition 1.2.1. □

If $c - 1 - n > m$, since $\max\{i;\ i \notin \Gamma(f) - n\} = c - 1 - n$, the above result implies that a primitive parametrization $\xi_1(t) = (t^n,\ y(t))$ and the associated parametrization $\xi(t) = (t^n,\ j^{c-1-n}(y(t)))$ determine two analytically equivalent branches. This is the case when $n > 3$ or when $n = 3$ and $m > 6$. On the other hand, if $n = 2$, or $n = 3$ and $4 \leq m \leq 5$, then the curve is analytically equivalent to one with Puiseux parametrization (t^n, t^m).

Remark 1.2.8 The above theorem gives the best possible reduction if one considers the whole equisingularity class determined by an arbitrary semigroup of values Γ. But, for some special semigroups Γ, it could happen that possibly shorter parametrizations represent all analytic classes in the equisingularity class determined by Γ. What we are claiming is that if we suppress a term of degree greater than m and that does not belong to $\Gamma - n$ in the expression of $y(t) = t^m + \sum_{\substack{i>m \\ i\notin\Gamma-n}} a_i t^i$, then possibly some analytic class belonging to the equisingularity class determined by Γ would not be represented.

Now, we are in position to define a parameter space for the equisingularity class of branches determined by a semigroup of values Γ, better than the one obtained by Zariski by means of what he called *short parametrizations*.

Definition 1.2.9 Let $\Gamma = \langle v_0, \ldots, v_g\rangle$ be a semigroup of values of a branch. We define the parameter space $\Sigma(\Gamma)$ as being the Zariski open set in $\mathbb{C}^N$, with $N = \#[(m, +\infty) \setminus (\Gamma - n)]$, where $n = v_0$ and $m = v_1$, whose points have as components the coefficients a_i in the Puiseux parametrization

$$\left(t^n,\ t^m + \sum_{\substack{i>m \\ i\notin\Gamma-n}} a_i t^i\right),$$

under the restriction that $a_{\beta_2}\cdots a_{\beta_g}\neq 0$ and the coefficients a_i satisfy condition (1.4).

Next, we compute the dimension N of $\Sigma(\Gamma)$.

The condition $i>m$ and $i\notin\Gamma-n$, means that i is a gap of Γ greater than m and such that $i+n$ is still a gap. These elements are exactly those of the form

$$\mathfrak{a}_i-2n,\ldots,\mathfrak{a}_i-\left[\frac{\mathfrak{a}_i-m}{n}\right]n,\ i=2,\ldots n-1,$$

where $\mathfrak{a}_i$, $i=0,\ldots,n-1$, is the Apéry sequence attached to Γ. So, we get the estimate

$$N\le\sum_{i=2}^{n-1}\left(\left[\frac{\mathfrak{a}_i-m}{n}\right]-1\right)=\sum_{i=2}^{n-1}\left[\frac{\mathfrak{a}_i-m}{n}\right]+n-2.$$

Let now γ_i, $i=1,\ldots,g$, be the numbers of gaps of Γ in the interval $[\beta_i,\beta_{i+1})$, where $\beta_{g+1}=\infty$, which are divisible by e_i and do not belong to $\Gamma-n$, then

$$N=\gamma_1+\cdots+\gamma_g.$$

Suppose $g\ge 2$. Because (cf. Formulas (1.7))

$$\beta_i=v_i-(n_{i-1}-1)v_{i-1}-\cdots-(n_1-1)v_1,\ i=1,\ldots,g,$$

and since $e_i\mid\beta_i$, and $n_i\ge 2$, for $i\ge 1$, it follows that

$$\beta_i+v_0=v_i-(n_{i-1}-1)v_{i-1}+\cdots+(n_1-1)v_1+v_0\notin\Gamma.$$

This, because otherwise, since $\beta_i+n<v_i$, we would have

$$\beta_i+v_0\in\langle v_0,\ldots,v_{i-1}\rangle,$$

which is impossible since e_{i-1} divides every element in $\langle v_0,\ldots,v_{i-1}\rangle$, but does not divide β_i.

This implies that if $g\ge 2$, then $\gamma_i\ge 1$, for $i=2,\ldots,g$, and consequently, $N\ge g-1$.

Remark 1.2.10 One has $N=0$ if and only if $v_0=2$, or $v_0=3$ and $v_1=4$ or $v_1=5$.

Indeed, $N=0$ implies $g=1$. So, we have $\Gamma=\langle v_0,v_1\rangle$, with $\mathrm{GCD}(v_0,v_1)=1$. In this case, $\mathfrak{a}_{v_0-1}=v_1(v_0-1)$. On the other hand, $N=0$ if, and only if,

$$\mathfrak{a}_{v_0-1}-2v_0<v_1,$$

which is equivalent to $v_1(v_0-2)<2v_0$, implying that $v_0<4$.

For $v_0=2$, v_1 is any odd natural number greater than 2. For $v_0=3$, $v_1=4$ or $v_1=5$.

In conclusion, $N = 0$ if, and only if, $\Gamma = \langle 2, m\rangle$, where m is odd, or $\Gamma = \langle 3, m\rangle$, where $m = 4$, or $m = 5$.

Example 1.2.11 Let us consider the equisingularity class of branches with semigroup $\Gamma = \langle 3, 7\rangle$.

As $n = v_0 = 3$ and $m = v_1 = \beta_1 = 7$, then the conductor of Γ is $c = 12$ (cf. Theorem 1.1.28) and any branch with semigroup Γ admits a primitive Puiseux parametrization as $\xi_3(t) = (t^3, b_3t^3 + \sum_{i\geq 6} b_it^i)$ with $b_7 \neq 0$.

Following the steps in the proof of Theorem 1.2.7, we apply first the Second Reduction to $\xi_3(t)$, by considering

$$\rho(t) = t \text{ and } \varphi(X, Y) = \left(X, \frac{Y - b_3X - b_6X^2}{b_7}\right),$$

obtaining

$$\xi_2(t) = \varphi(\xi_3(\rho^{-1}(t))) = (t^3, t^7 + \sum_{i\geq 8} a_it^i)$$

with $a_i = \frac{b_i}{b_7}$.

Now, we apply the Third Reduction to $\xi_2(t)$ for every $i \in \Gamma - 3$ with $7 < i < c = 12$, that is, for $i \in \{9, 10, 11\}$. As $9, 10 \in \Gamma$ and $11 = v_1 + v_1 - n$, the reduction can be performed using Proposition 1.2.5 that indicates elements ϵ, p and q in order to obtain automorphisms ρ and φ as described in Proposition 1.2.3.

For $i = 9$, we take $\epsilon = 1$, $p(X, Y) = 0$ and $q(X, Y) = -a_9X^2$ in Proposition 1.2.3, getting $\rho(t) = t$ and $\varphi(X, Y) = (X, Y - a_9X^2)$. In this way,

$$\xi_{2,1}(t) = \varphi(\xi_2(\rho^{-1}(t))) = (t^3, t^7 + a_8t^8 + \sum_{i\geq 10} a_it^i).$$

For $i = 10$, we take $\epsilon = 1$, $p(X, Y) = 0$ and $q(X, Y) = -a_{10}XY$ in Proposition 1.2.3. So, $\rho(t) = t$ and $\varphi(X, Y) = (X, Y - a_{10}XY)$ and therefore

$$\xi_{2,2}(t) = \varphi(\xi_{2,1}(\rho^{-1}(t))) = (t^3, t^7 + a_8t^8 + \sum_{i\geq 11} a'_it^i).$$

For $i = 11$, we take $\epsilon = 1$, $q(X, Y) = 0$ and $p(X, Y) = \frac{3a'_{11}}{7}Y$ in Proposition 1.2.3, consequently $\rho(t) = t\left(1 + \frac{3a'_{11}}{7}(t^4 + a_8t^5 + \sum_{i\geq 11} a'_it^{i-3})\right)^{\frac{1}{3}}$ and $\varphi(X, Y) = (X - \frac{3a'_{11}}{7}Y, Y)$. By Corollary 1.2.4, we obtain

$$\xi_{2,3}(t) = \varphi(\xi_{2,2}(\rho^{-1}(t))) = (t^3, t^7 + a_8t^8 + r(t))$$

with $r(t) \in \mathcal{M}_1^{12}$.

Now, as the conductor of Γ is $c = 12$ it follows that $r(t) \in \mathcal{R}_{\xi_{2,3}}$ and there exists $h(X, Y) \in \langle X, Y\rangle^2$ with $h(\xi_{2,3}) = r(t)$. In this way, taking $\rho(t) = t$ and $\varphi(X, Y) =$

$(X, Y - h(X, Y))$ we get the First Reduction on $\xi_{2,3}(t)$:

$$\xi(t) = \varphi(\xi_{2,3}(\rho^{-1}(t))) = (t^3, t^7 + a_8 t^8).$$

In this case, the dimension of $\Sigma(\Gamma)$ is $N = 1$.

Notice that if we suppress the term $a_8 t^8$ in $\xi(t)$ we do not have the analytic class of $(t^3, t^7 + t^8)$ that is distinct from the analytic class of (t^3, t^7), because these parametrizations are not $\mathcal{A}$-equivalent, as we verify in the sequel.

In fact, if $\xi_1 = (t^3, t^7 + t^8)$ and $\xi_0 = (t^3, t^7)$ were $\mathcal{A}$-equivalent then by Corollary 1.2.4 there would exist $p(X, Y) \in \langle X^2, Y\rangle$ and $q(X, Y) \in \langle X, Y\rangle^2$ that give us automorphisms $\rho(t)$ and $\varphi(X, Y)$ such that

$$\xi_0(t) = (t^3, t^7) = \varphi(\xi_1(\rho^{-1}(t)) = (t^3, t^7 + t^8) + (0, \theta(\rho^{-1}(t))), \qquad (1.13)$$

with

$$\theta(t) = q(\xi_1(t)) - \frac{7t^6 + 8t^7}{3t^2} p(\xi_1(t)) - \sum_{i \in \{7,8\}} t^i \sum_{l=2}^{\infty} \binom{\frac{i}{3}}{l} \left(\frac{p(\xi_1(t))}{t^3} \right)^l .$$

Equality (1.13) implies that $\operatorname{ord}_t(\theta(t)) = \operatorname{ord}_(\theta(\rho^{-1}(t))) = 8$. As $p(X, Y) \in \langle X^2, Y\rangle$ we have $\operatorname{ord}_t(p(\xi_1(t))) \geq 6$ and

$$\operatorname{ord}_t \left(-\frac{7t^6 + 8t^7}{3t^2} p(\xi_1(t)) - \sum_{i \in \{7,8\}} t^i \sum_{l=2}^{\infty} \binom{\frac{i}{3}}{l} \left(\frac{p(\xi_1(t))}{t^3} \right)^l \right) \geq 10.$$

On the other hand, we have $q(X, Y) = dX^2 + s(X, Y)$, that is, $q(\xi_1(t)) = dt^6 + s(\xi_1(t))$ with $d \in \mathbb{C}$ and $\operatorname{ord}_t(s(\xi_1(t))) \geq 9$.

So, Equality (1.13) is impossible, i.e., $\xi_0(t)$ and $\xi_1(t)$ represent branches in distinct analytic classes.

We will show (see Proposition 1.4.5) that the dimension of the moduli space for branches with semigroup $\Gamma = \langle 3, 7\rangle$ is zero.

Notice that given $(a_i)_i \in \Sigma(\Gamma)$, then $(\zeta^{ji} a_i)_i$, for $j = 1, \ldots, n$, where ζ is a primitive n-th root of unity, also belong to $\Sigma(\Gamma)$ and the associated parametrizations $\xi_j(t) = (t^n, y(\zeta^j t))$ define the same branch. This is not the whole equivalence class of $(a_i)_i$ with respect to the equivalence relation $\sim$ induced by $\mathcal{G}$ on $\Sigma(\Gamma)$, but only part of it. The equivalence classes with respect to the relation $\sim$ are not well behaved enough to produce a good quotient $\Sigma(\Gamma)/\sim$. So, the strategy adopted by Zariski was to stratify further his parameter space analogous to our $\Sigma(\Gamma)$ by means of some analytic invariant. The invariant adopted by Zariski was what he called the λ-invariant, which we discuss in Sect. 1.2.3.

Several analytic invariants were considered in the literature: Tjurina numbers, Zariski λ-invariant and many others. The invariant we will use in Sect. 1.3, for the solution of Zariski's problem, is the set of values of Kähler differentials for being a more complete invariant. This invariant was partially considered by Zariski and we will introduce it in the next subsection.

1.2.2 *Kähler Differentials*

Let C_f be a plane branch, we define the $\mathcal{O}(f)$-module of Kähler differentials as being[2]

$$\Omega(f) = \frac{\mathcal{O}(f) \oplus \mathcal{O}(f)}{(f_x, f_y)\mathcal{O}(f)},$$

where f_x and f_y are respectively the images of f_X and f_Y in $\mathcal{O}(f)$.

If we denote by dx and dy the images of $e_1 = (1, 0)$ and $e_2 = (0, 1)$ in $\Omega(f)$, respectively, it follows that dx and dy are generators of $\Omega(f)$ as an $\mathcal{O}(f)$-module, which is not free, since one has the relation $f_x dx + f_y dy = 0$.

For $g \in \mathcal{O}(f)$, let us define

$$dg = g_x dx + g_y dy \in \Omega(f).$$

We will say that $\omega \in \Omega(f)$ is an *exact differential* if there exists $g \in \mathcal{O}(f)$ such that $\omega = dg$. The set of exact differentials will be denoted by $d\mathcal{O}(f)$.

An important invariant of C_f related to Kähler differentials is the *Tjurina number*, defined as

$$\tau(f) = \dim_{\mathbb{C}} \frac{\mathcal{O}}{\langle f, f_X, f_Y \rangle} = \dim_{\mathbb{C}} \frac{\mathcal{O}(f)}{J(f)},$$

where $J(f)$ is the ideal $\langle f_x, f_y \rangle$ in $\mathcal{O}(f)$.

One has that $\tau(f) < \infty$, because otherwise f would have a common factor with f_X and f_Y, since f is irreducible, this common factor would be f, which is not possible.

The Tjurina number of C_f is an analytic invariant. Indeed, if $g = uf$, where $u \in \mathcal{O}^*$, then $g_X = u_X f + u f_X$ and $g_Y = u_Y f + u f_Y$. So,

$$\langle g, g_X, g_Y \rangle = \langle uf, u_X f + u f_X, u_Y f + u f_Y \rangle = \langle uf, u f_X, u f_Y \rangle = \langle f, f_X, f_Y \rangle,$$

showing that $\tau(uf) = \tau(f)$. Also, if $g = f \circ \varphi$, where $\varphi = (\varphi_1, \varphi_2)$ is an automorphism of $(\mathbb{C}^2, 0)$, then by the chain rule

$$\begin{pmatrix} g_X \\ g_Y \end{pmatrix} = \begin{pmatrix} (\varphi_1)_X & (\varphi_2)_X \\ (\varphi_1)_Y & (\varphi_2)_Y \end{pmatrix} \begin{pmatrix} f_X \circ \varphi \\ f_Y \circ \varphi \end{pmatrix}, \quad \text{with} \quad \begin{pmatrix} (\varphi_1)_X & (\varphi_2)_X \\ (\varphi_1)_Y & (\varphi_2)_Y \end{pmatrix} \text{ invertible.}$$

This implies that

---

[2] This module is denoted by Zariski in [23] by $\mathcal{O}d\mathcal{O}$.

$$\langle g, g_X, g_Y \rangle = \langle f \circ \varphi, f_X \circ \varphi, f_Y \circ \varphi \rangle,$$

showing that $\tau(f \circ \varphi) = \tau(f)$.

Lemma 1.2.12 $\tau(f) = 0$ *if and only if* C_f *is nonsingular.*

Proof We have that $\tau(f) = 0$ if and only if $\langle f, f_X, f_Y \rangle = \mathcal{O}$. This is so, if and only if one of the three generators is a unit. Since $f \in \mathcal{M}$, this is equivalent to either f_X or f_Y is a unit, which is equivalent to $\mathrm{mult}(f) = 1$. □

Let us denote by $\mathcal{T}(f)$ the torsion submodule of $\Omega(f)$ and by $\ell(M)$ the length of a module M. Concerning the torsion module $\mathcal{T}(f)$, one has the following result:

Proposition 1.2.13 *Let* $\mathcal{T}(f)$ *be the torsion submodule of* $\Omega(f)$, *then*

$$\ell(\mathcal{T}(f)) = \tau(f).$$

Proof See [25, Theorem 1]. □

This result, together with Lemma 1.2.12, yield immediately the following:

Corollary 1.2.14 $\Omega(f)$ *is torsion free if and only if* C_f *is nonsingular.*

Another remarkable number attached $f \in \mathcal{O}$ is the *Milnor number* defined as

$$\mu(f) = \dim_{\mathbb{C}} \frac{\mathcal{O}}{\langle f_X, f_Y \rangle}.$$

Since $\mu(f) = \infty$ if, and only if, f_X and f_Y have a common factor, this number is finite for irreducible f.

It is not evident at all that $\mu(f)$ is an analytic invariant. It is clearly invariant by composition of f with automorphisms of $(\mathbb{C}^2, 0)$, but it is not clear that it is invariant by multiplication of f by units in $\mathcal{O}$. Indeed, the following result will give us more: $\mu(f)$ is a topological invariant!

Theorem 1.2.15 *For irreducible* $f \in \mathcal{M}$ *one has that* $\mu(f) = c$, *where* c *is the conductor of* $\Gamma(f)$.

Proof See [11, Theorem 7.18]. □

From the definitions it follows readily that

$$\ell(\mathcal{T}(f)) = \tau(f) \leq \mu(f) = c.$$

The equality in the above inequality was treated by Zariski and is the content of the following result:

Theorem 1.2.16 *Let* C_f *be a branch. Then one has that* $\mu(f) = \tau(f)$ *if, and only if,* $f \sim Y^n - X^m$, *where* $n, m \in \mathbb{N}$, *with* $\mathrm{GCD}(n, m) = 1$.

Proof See [25, Theorem 1]. □

Due to the inequality $\ell(\mathcal{T}(f)) \leq c$, the curves described in the above theorem are called curves with maximal torsion, that is, those for which $\ell(\mathcal{T}(f)) = c$. These curves are the most special ones, curves for which $\mu(f) - \tau(f) \leq 2$ were classified in [3] and will be revisited later.

Given a primitive parametrization $\xi(t) = (x(t), y(t))$ of C_f, we may consider the $\mathcal{O}(f) \simeq \mathbb{C}\{\xi(t)\}$-modules homomorphism

$$\begin{array}{rrl} \Psi: & \Omega(f) & \to \mathbb{C}\{t\} \\ & gdx + hdy & \mapsto [g(\xi(t))x'(t) + h(\xi(t))y'(t)]t, \end{array} \tag{1.14}$$

where $x'(t)$ and $y'(t)$ denote the derivatives of $x(t)$ and $y(t)$ with respect to t. We will denote by ψ the monomorphism $\mathcal{O}(f) \to \mathbb{C}\{t\}$, $g \mapsto g(\xi(t))$.

Proposition 1.2.17 *One has that* $\ker \Psi = \mathcal{T}(f)$.

Proof Let $\omega \in \mathcal{T}(f)$, then there exists an element $0 \neq p \in \mathcal{O}(f)$ such that $p\omega = 0$. Now,

$$0 = \Psi(p\omega) = \psi(p)\Psi(\omega),$$

and since $\psi(p) \neq 0$ and $\mathbb{C}\{t\}$ is a domain, it follows that $\Psi(\omega) = 0$.

Conversely, suppose that $0 \neq \omega = gdx + hdy \in \Omega(f)$ is such that

$$\Psi(\omega) = [g(\xi(t))x'(t) + h(\xi(t))y'(t)]t = 0.$$

On the other hand, from the relation $f_x dx + f_y dy = 0$, we have

$$f_x(\xi(t))x'(t) + f_y(\xi(t))y'(t) = 0.$$

This shows that the system of linear equation over $\mathbb{C}((t))$

$$\begin{cases} g(\xi(t))Z + h(\xi(t))W = 0 \\ f_x(\xi(t))Z + f_y(\xi(t))W = 0 \end{cases}$$

has a nontrivial solution $Z = x'(t)$, $W = y'(t)$, which implies that

$$0 = \det \begin{bmatrix} g(\xi(t)) & h(\xi(t)) \\ f_x(\xi(t)) & f_y(\xi(t)) \end{bmatrix} = g(\xi(t)) f_Y(\xi(t)) - h(\xi(t)) f_X(\xi(t)).$$

In this way, one has the relation $f_y g - h f_x = 0$ in $\mathcal{O}(f)$ and the relation $f_x dx + f_y dy = 0$ in $\Omega(f)$. Hence,

$$f_y\omega = f_y(gdx + hdy) = f_y gdx - hf_x dx = (f_y g - hf_x)dx = 0,$$

which shows that $\omega \in \mathcal{T}(f)$. □

The above theorem allows us to view, through the homomorphism Ψ, the $\mathcal{O}(f)$-module $\Omega(f)/\mathcal{T}(f)$ as a submodule of $\mathbb{C}((t))$. In view of this, we may define, for $\overline{\omega} \in \Omega(f)/\mathcal{T}(f) \setminus \{0\}$,

$$v(\overline{\omega}) = \text{ord}_t(\Psi(\overline{\omega})).$$

Notice that given a non unit $g \in \mathcal{O}(f)$, then

$$\begin{aligned} v(dg) &= v(g_x dx + g_y dy) = \text{ord}_t([g_x(\xi(t))x'(t) + g_y(\xi(t))y'(t)]t) \\ &= \text{ord}_t(g(\xi(t))') + 1 = \text{ord}_t(g(\xi(t)) = v(g). \end{aligned}$$

Let us define the *set of values of differentials* as

$$\Lambda(f) = v\left(\Omega(f)/\mathcal{T}(f)\right) \subset \mathbb{N} \cup \{\infty\}.$$

Then we have that

$$\Gamma(f) \setminus \{0\} \subset \Lambda(f), \quad \text{and} \quad \Gamma(f) + \Lambda(f) \subset \Lambda(f),$$

where the first inclusion follows from the equality $v(dg) = v(g)$, for all non invertible $g \in \mathcal{O}(f)$, and the second, because $v(g\omega) = v(g) + v(\omega)$, for all $g \in \mathcal{O}(f)$ and $\omega \in \Omega(f)$.

Because of the above second inclusion, one has that $c + \min(\Lambda(f)) + \mathbb{N} \subset \Lambda(f)$. The smallest natural number d such that $d + \mathbb{N} \subset \Lambda(f)$ will be called the conductor of $\Lambda(f)$.

The sets $\Lambda(f)$ will play a central role in our treatment of analytic classification of plane branches. To begin with, we show that they are analytic invariants. To see this, we will give another description of the $\mathcal{O}(f)$-module $\Omega(f)/\mathcal{T}(f)$.

Let us recall (cf. [11, Corollary 7.16]) that, for an irreducible Weierstrass polynomial $f \in \mathbb{C}\{X\}[Y]$, one has $\mathrm{I}(f, f_Y) = n + c - 1$, where $n = \mathrm{I}(f, X)$ and c is the conductor of $\Gamma(f)$. Consider now $\xi(t) = (x(t), y(t))$ a primitive parametrization of C_f and let

$$\zeta(t) = -\frac{tx'(t)}{f_y(\xi(t))} \in \mathbb{C}((t)),$$

hence $\text{ord}_t(\zeta(t)) = 1 + n - 1 - (n + c - 1) = -c + 1$.

Consider now the $\mathcal{O}(f)$-module homomorphism multiplication by $\zeta(t)$:

$$\begin{aligned} m_\zeta : \mathcal{O}(f) &\to \quad \mathbb{C}((t)) \\ g \quad &\mapsto \zeta(t) \cdot g(\xi(t)). \end{aligned}$$

We have the following result:

Proposition 1.2.18 *The map m_ζ is injective and $m_\zeta(J(f)) = \Psi(\Omega(f))$.*

Proof The map m_ζ is injective, because the map $\mathcal{O}(f) \to \mathbb{C}\{t\}$ is injective and $\zeta(t) \neq 0$. Notice that any element of $h \in J(f)$ may be written as $h = g_2 f_x - g_1 f_y$, so

$$\begin{aligned} m_\zeta(h) &= -\frac{tx'(t)}{f_y(\xi(t))}(g_2(\xi(t))f_x(\xi(t)) - g_1(\xi(t))f_y(\xi(t)) \\ &= t(g_1(\xi(t))x'(t) + g_2(\xi(t))y'(t)) \\ &= \Psi(g_1dx + g_2dy) \in \Psi(\Omega(f)), \end{aligned}$$

where the second equality follows from $f_x(\xi(t))x'(t) + f_y(\xi(t))y'(t) = 0$.

Conversely, let $\Psi(g_1dx + g_2dy) \in \Psi(\Omega(f))$, then

$$\begin{aligned} \Psi(g_1dx + g_2dy) &= t(g_1(\xi(t))x'(t) + g_2(\xi(t))y'(t)) \\ &= t(g_1(\xi(t))x'(t) - g_2(\xi(t))\frac{f_x(\xi(t))}{f_y(\xi(t))}x'(t)) \\ &= -\frac{tx'(t)}{f_y(\xi(t))}(g_2(\xi(t))f_x(\xi(t)) - g_1(\xi(t))f_y(\xi(t))) \\ &= m_\zeta(g_2f_x - g_1f_y) \in m_\zeta(J(f)), \end{aligned}$$

concluding our proof. □

Since $\Psi(\Omega(f)) \simeq \Omega(f)/\mathcal{T}(f)$, it follows that we have an $\mathcal{O}(f)$-modules isomorphism $\Omega(f)/\mathcal{T}(f) \simeq J(f)$, in such a way that

$$\Lambda(f) = v(J(f)) + v(\zeta) = v(J(f)) - c + 1. \tag{1.15}$$

From this, and the discussion on Tjurina's number done just before Lemma 1.2.12 one has that $\Lambda(f)$ is an analytic invariant.

The analytic invariant $\mu(f) - \tau(f)$ will be described in the following result.

Proposition 1.2.19 *We have that* $\sharp(\Lambda(f) \setminus \Gamma(f)) = \mu(f) - \tau(f)$.

Proof Let $\xi(t) = (x(t), y(t))$ be a primitive parametrization of C_f. Since $\mathcal{O}(f) \simeq R_\xi = \mathbb{C}\{x(t), y(t)\}$ and

$$J(f) \simeq S := \{h(\xi(t)),\ h \in \langle f, f_X, f_Y\rangle\mathbb{C}\{X, Y\}\}$$

are $\mathbb{C}$-vector subspaces of finite codimension in $\mathbb{C}\{t\}$, it follows that

$$\begin{aligned} \tau(f) &= \dim_{\mathbb{C}}\left(\frac{\mathcal{O}(f)}{J(f)}\right) \\ &= \sharp\left[\{\mathrm{ord}_t(q);\ q \in R_\xi\} \setminus \{\mathrm{ord}_t(q);\ q \in S\}\right]. \end{aligned}$$

Since $\{\mathrm{ord}_t(q);\ q \in R_\xi\} = \Gamma(f)$, from (1.15) we get

$$\begin{aligned} \tau(f) &= \sharp[\Gamma(f) \setminus \{l + c - 1;\ l \in \Lambda(f)\}] \\ &= \tfrac{c}{2} + \sharp[\{l + c - 1;\ l \in \mathbb{N}^* \setminus \Lambda(f)\}] \\ &= c - \sharp(\Lambda(f) \setminus \Gamma(f)). \end{aligned}$$

Now, we conclude the proof by using Theorem 1.2.15. □

As a consequence we get the following result.

Corollary 1.2.20 *Let C_f be a branch. The following statements are equivalent:*

(i) $f \sim Y^n - X^m$ *with* $GCD(n, m) = 1$*;*
(ii) $\mu(f) = \tau(f)$*;*
(iii) $\Lambda(f) = \Gamma(f) \setminus \{0\}$.

Proof This follows from Theorem 1.2.16 and the previous result. □

Example 1.2.21 Let $\xi(t) = (t^3, t^{10} + at^{11} + bt^{14})$ be a primitive Puiseux parametrization of a plane curve C_f with $a \neq 0$.

Remark that $n = v_0 = 3$, $m = v_1 = \beta_1 = 10$, that is, $\Gamma(f) = \langle 3, 10\rangle$ and $\xi(t)$ is expressed as (1.12).

The conductor of $\Gamma(f)$ is $c = 18$ (see (1.8)) and its gaps are the elements of the set $\{1, 2, 4, 5, 7, 8, 11, 14, 17\}$.

Considering $\omega_1 = 3xdy - 10ydx \in \Omega(f)$ we get

$$v(\omega_1) = \text{ord}_t(\Psi(\omega_1)) = 14 \quad \text{and} \quad v(x\omega_1) = \text{ord}_t(\Psi(x\omega_1)) = 17.$$

In this way, as $\Gamma(f) \setminus \{0\} \subseteq \Lambda(f)$ we have

$$D := \{3, 6, 9, 10, 12 + i;\ i \in \mathbb{N}\} \subseteq \Lambda(f).$$

We claim that the above inclusion is an equality.

Indeed, given $\gamma \in \Lambda(f)$, by definition of $\Lambda(f)$, there exists $\omega = gdx + hdy$ in $\Omega(f)$ such that $\gamma = v(\omega)$.

If $v(gdx) \neq v(hdy)$ then $\gamma = \min\{v(gdx), v(hdy)\} \in \Gamma(f) \setminus \{0\} \subset D$.

If $v(gdx) = v(hdy)$ then we must have $v(h) \geq 3$, $v(g) \geq 10$ and in this way $\gamma = v(\omega) \geq \min\{v(gdx), v(hdy)\} \geq 13$ and, consequently, $\gamma \in D$.

In conclusion, $\Lambda(f) = \{3, 6, 9, 10, 12 + i;\ i \in \mathbb{N}\}$.

In [13] we described algorithms to compute all possible sets $\Lambda(f)$ for f varying in a fixed equisingularity class.

1.2.3 The Zariski Invariant

For not being able to realize the analytic classification in the parameter space of short parametrizations, Zariski tried to stratify it by means of an analytic invariant that he denoted by λ and we call the *Zariski invariant*. Since this is not a topological invariant, equisingular branches may have distinct λ-invariants (see Example 1.2.23 below), but for a fixed value of λ, branches that share this common λ-invariant may admit further elimination of parameters.

Let C be a plane branch with semigroup of values $\Gamma = \langle n, m, v_2, \ldots, v_g \rangle$, with $n > 3$, or $n = 3$ and $m > 6$, and admitting a primitive Puiseux parametrization $(t^n, \sum_{i \geq n}^{\infty} a_i t^i)$. Let us define the Zariski λ-invariant as

$$\lambda = \min\{i \in \mathbb{N};\ i + n \notin \Gamma \text{ and } a_i \neq 0\}.$$

Then from Theorem 1.2.7, there exists an analytically equivalent branch to C with a primitive Puiseux parametrization of the form

$$\xi(t) = \left(t^n, t^m + b_\lambda t^\lambda + \sum_{\substack{i>\lambda \\ i \notin \Gamma - n}}^{c-1-n} a_i t^i \right), \tag{1.16}$$

where $b_\lambda \neq 0$ and c is the conductor of Γ.

We have the following result:

Proposition 1.2.22 *Let C_f be a plane branch with a parametrization of the form* (1.16), *with $\lambda + n \notin \Gamma(f)$ and $b_\lambda \neq 0$. Let $\omega_1 = mydx - nxdy \in \Omega(f)$, then $v(\omega_1) = \lambda + n = \min(\Lambda(f) \setminus \Gamma(f))$. For any differential $\omega \in \Omega(f)$ such that $v(\omega) \notin \Gamma(f)$ and $v(\omega) > v(\omega_1)$, one has $v(\omega) \geq v(\omega_1) + n$.*

Proof See [24, Lemma 4.2, p. 50]. □

It follows readily from Proposition 1.2.22 that $\lambda = \lambda(f)$ is an analytic invariant, since $\Lambda(f)$ and $\Gamma(f)$ are so.

The following example will show that λ is not a topological invariant, in the sense that it may vary in a given equisingularity class of branches.

Example 1.2.23 Let $\Gamma = \langle 3, 10 \rangle$, then $c = 18$. Since $11 + 3 \notin \Gamma$ and $14 + 3 \notin \Gamma$, it follows that $(t^3, t^{10} + t^{11})$ and $(t^3, t^{10} + t^{14})$ represent two equisingular branches with $\lambda = 11$ and $\lambda = 14$, respectively.

Observe that if $\xi(t) = (t^{v_0}, t^{v_1} + \sum a_i t^i)$ is a parametrization of C_f as (1.12), then the Zariski λ-invariant is given by $\lambda(f) = \min\{i;\ a_i \neq 0\}$. In particular, if $g \geq 2$ then $\lambda(f) \leq \beta_2$.

Notice that $\Gamma(f)$ and $\Lambda(f)$ determine $\tau(f)$ and $\lambda(f)$, but there are no relation between the last two invariants, as one may check in the following example.

Example 1.2.24 Consider the following primitive parametrizations with semigroup $\Gamma = \langle 5, 11 \rangle$:

$$\xi_1 = (t^5, t^{11} + t^{12}), \qquad \xi_2 = (t^5, t^{11} + t^{12} - \tfrac{1}{3} t^{14}),$$
$$\xi_3 = (t^5, t^{11} + t^{14}), \qquad \xi_4 = (t^5, t^{11} + t^{12} + \tfrac{22}{23} t^{13} + \tfrac{136}{121} t^{14}).$$

Denoting by $\lambda_i := \lambda(f_i)$, $\tau_i := \tau(f_i)$ and $\Lambda_i := \Lambda(f_i)$, where $f_i \in \mathcal{O}$ determines a branch with parametrization ξ_i, $i = 1, 2, 3, 4$, one has

$$\begin{aligned}
\lambda_1 &= 12, & \tau_1 &= 34, & \Lambda_1 \setminus \Gamma &= \{17, 23, 28, 29, 34, 39\};\\
\lambda_2 &= 12, & \tau_2 &= 35, & \Lambda_2 \setminus \Gamma &= \{17, 23, 28, 34, 39\};\\
\lambda_3 &= 14, & \tau_3 &= 35, & \Lambda_3 \setminus \Gamma &= \{19, 24, 29, 34, 39\};\\
\lambda_4 &= 12, & \tau_4 &= 35, & \Lambda_4 \setminus \Gamma &= \{17, 28, 29, 34, 39\},
\end{aligned}$$

where these numbers were calculated by means of the algorithms in [15].

For equisingular branches that share the same Zariski λ-invariant, there is an additional reduction process. This reduction was considered by Zariski in [23] and [25] in his study of branches with semigroup $\langle n, m\rangle$. However, we will extend the result for branches with an arbitrary semigroup of values.

Theorem 1.2.25 *Let $\xi(t) = (t^n, t^m + a_\lambda t^\lambda + \sum_{i>\lambda} a_i t^i)$ be a primitive Puiseux parametrization of a branch C with semigroup of values $\Gamma = \langle v_0, \ldots, v_g\rangle$, where $n = v_0$, $m = v_1$ and $\lambda = \min\{i;\ i + n \notin \Gamma \text{ and } a_i \neq 0\}$ its Zariski invariant. If $j > \lambda$ and $j \in \Gamma + \lambda$, then C is analytically equivalent to a branch parametrized by $(t^n, t^m + a_\lambda t^\lambda + \sum_{i>\lambda} a'_i t^i)$ with $a'_i = a_i$ for $i < j$, and $a'_j = 0$.*

Proof Initially, we will consider the case $g = 1$ or $0 < j - \lambda < v_2$. In this case, we have that $j = \alpha n + \beta m + \lambda$ with $\alpha \neq 0$ or $\beta \neq 0$.

Let us define $\varphi \in \mathrm{Aut}(\mathbb{C}^2, 0)$ and $\rho \in \mathrm{Aut}(\mathbb{C}, 0)$ as in Proposition 1.2.3, putting $\epsilon = 1$,

$$\begin{aligned}
p &= aX^{\alpha+1}Y^\beta \in \langle X^2, Y\rangle, \quad \text{where } a = \tfrac{na_j}{(\lambda-m)a_\lambda},\\
q &= \textstyle\sum_{k=1}^{\left[\frac{\lambda-m}{\alpha n+\beta m}\right]+1} \binom{\frac{m}{n}}{k} a^k X^{k\alpha} Y^{k\beta+1} \in \langle X, Y\rangle^2.
\end{aligned} \tag{1.17}$$

From Corollary 1.2.4, we have $\varphi(\xi(\rho^{-1}(t))) = \xi(t) + (0, \theta(\rho^{-1}(t)))$, where $\theta(t)$ is as defined there, having $\theta(\rho^{-1}(t))$ and $\theta(t)$ same initial terms.

Since

$$q(\xi(t)) = \sum_{k=1}^{\left[\frac{\lambda-m}{\alpha n+\beta m}\right]+1} \binom{\frac{m}{n}}{k} a^k t^{k\alpha n + m(\beta+1)} + \frac{a a_\lambda (1+\beta) m}{n} t^j + hot,$$

where here, and in the rest of the proof, hot means elements in $\mathcal{M}_1^{j+1}$, and

$$\frac{y'(t)}{nt^{n-1}} p(\xi(t)) + \textstyle\sum_{i\geq m} a_i t^i \sum_{k=2}^{\infty} \binom{\frac{i}{n}}{k} \left(\frac{p(\xi(t))}{t^n}\right)^k =$$

$$\textstyle\sum_{k=1}^{\left[\frac{\lambda-m}{\alpha n+\beta m}\right]+1} \binom{\frac{m}{n}}{k} a^k t^{k\alpha n + m(\beta+1)} + \frac{a a_\lambda (\lambda + \beta m)}{n} t^j + hot,$$

we have that

$$\theta(t) = \frac{-a a_\lambda (\lambda - m)}{n} t^j + hot = -a_j t^j + hot.$$

Now, because $\theta(\rho^{-1}(t))$ and $\theta(t)$ have same initial terms, the result follows.

Let us now suppose that $j - \lambda \geq v_2$. In this case, there exists some $h \in \langle X, Y\rangle^2$ such that $v(h) = j - \lambda$.

We define φ and ρ as in Proposition 1.2.3, with $p = aXh, q = \frac{m}{n}aYh$ and $\epsilon = 1$, where a is a complex number, and then we apply Corollary 1.2.4.

With our choices of p, q and ϵ, we have

$$q(\xi(t)) - \frac{y'(t)}{nt^{n-1}} p(\xi(t)) = \frac{\Psi(h\omega_1)}{nt^n},$$

where Ψ is the homomorphism defined in (1.14), and $\omega_1 = mydx - nxdy$, as defined in Proposition 1.2.22. In this way we get

$$v(q(\xi(t)) - \frac{y'(t)}{nt^{n-1}} p(\xi(t))) = j \leq \beta_2 + v(h) < m + 2v(h)$$

$$= v\left(\sum_{i\geq m} a_i t^i \sum_{k=2}^{\infty} \binom{\frac{i}{n}}{k} \left(\frac{p(\xi(t))}{t^n}\right)^k\right).$$

Consequently, $v(\theta(\rho^{-1}(t))) = v(\theta(t)) = j$; and by choosing conveniently the value of a, we may conclude the proof. □

The above theorem allows us to formulate one additional reduction process for equisingular branches with a given λ-invariant.

Fourth Reduction: Let $\xi(t) = (t^n, t^m + a_\lambda t^\lambda + \sum_{i>\lambda} a_i t^i)$ be a primitive Puiseux parametrization of C_f with semigroup $\Gamma(f)$ and Zariski invariant λ. If $s > \lambda$ and $s - \lambda \in \Gamma(f)$, there exists a branch analytically equivalent to C_f parametrized by $(t^n, t^m + a_\lambda t^\lambda + \sum_{i>\lambda} a'_i t^i))$ such that $a'_i = a_i$ for $i < s$ and $a'_s = 0$.

Notice that if we apply the four reductions to a primitive Puiseux parametrization of a branch with a semigroup Γ and Zariski invariant λ, we obtain a primitive parametrization $\xi_1(t) = (t^n, t^m + at^\lambda + \sum_{\substack{i\notin\Gamma-n \\ i\notin\Gamma+\lambda}} a_i t^i)$, where $a \neq 0$ and the parameters a_i respecting the necessary conditions to preserve the semigroup. Moreover, if we consider $\rho^{-1}(t) = a^{-\frac{1}{\lambda-m}} t$ and $\varphi(X, Y) = (a^{\frac{n}{\lambda-m}} X, a^{\frac{m}{\lambda-m}} Y)$ we have

$$\xi(t) = \varphi(\xi_1(\rho^{-1}(t))) = \left(t^n, t^m + t^\lambda + \sum_{\substack{i\notin\Gamma-n \\ i\notin\Gamma+\lambda}} a'_i t^i\right), \quad \text{with } a'_i = a_i a^{-\frac{i}{\lambda-m}}. \tag{1.18}$$

The above changes of parameter and coordinates show that in the analytic class of a branch given by a primitive Puiseux parametrization $(t^n, \sum_{i\geq m} a_i t^i)$, where $m > n$, the coefficients a_m and a_λ may be chosen arbitrarily in $\mathbb{C}^*$. The choice of $a_m = a_\lambda = 1$ represents a canonical normalization.

Example 1.2.26 Let $\xi(t) = (t^3, t^{10} + at^{11} + bt^{14})$, with $a \neq 0$, be a primitive Puiseux parametrization of a plane curve C_f, as considered in Example 1.2.21, that admits $\Gamma(f) = \langle 3, 10\rangle$ with conductor $c = 18$ and Zariski invariant $\lambda = 11$.

Since $14 - \lambda = 3 \in \Gamma(f)$, we may apply the Fourth Reduction.

Following the instructions in Theorem 1.2.25, we consider $\varphi \in \mathrm{Aut}(\mathbb{C}^2, 0)$ and $\rho \in \mathrm{Aut}(\mathbb{C}, 0)$ as in Proposition 1.2.3, putting

$$\epsilon = 1, \quad p = \frac{3b}{a}X^2 \text{ and } q = \frac{10b}{a}XY.$$

From Corollary 1.2.4, we have $\xi_1(t) = \varphi(\xi(\rho^{-1}(t))) = \xi(t) + (0, \theta(\rho^{-1}(t)))$, where $\theta(t)$ is as (1.10) with $\theta(\rho^{-1}(t))$ and $\theta(t)$ having the same initial term.

Since $\theta(t) = -bt^{14} + r(t)$ with $r(t) \in \mathcal{M}_1^{16}$ we get $\xi_1(t) = (t^3, t^{10} + at^{11} + s(t))$ with $s(t) \in \mathcal{M}_1^{15}$.

Now, as $i \in \Gamma(f) - 3$ for $i \geq 15$, applying the Third and First Reductions, it follows that $\xi_1(t)$ is $\mathcal{A}$-equivalent to $\xi_2(t) = (t^3, t^{10} + at^{11})$. In addition, considering $\rho^{-1}(t) = a^{-1}t$ and $\varphi(X, Y) = (a^3X, a^{10}Y)$ we get

$$\varphi(\xi_2(\rho^{-1}(t))) = (t^3, t^{10} + t^{11}).$$

So, $\xi(t)$ is $\mathcal{A}$-equivalent to $(t^3, t^{10} + t^{11})$.

From the discussion above Example 1.2.26, we have two distinct situations; namely,

(i) $\Lambda(f) \setminus \Gamma(f) = \emptyset$. Then C_f is analytically equivalent to a branch with a parametrization $\xi(t) = (t^n, t^m)$, with $\mathrm{GCD}(n, m) = 1$, or
(ii) $\Lambda(f) \setminus \Gamma(f) \neq \emptyset$. Then C_f is analytically equivalent to a branch with a parametrization $\xi(t)$ as in (1.18).

Till now, we have followed the strategy proposed by Zariski in [23], in which one performs a succession of reductions with growing complexity level. Zariski recognized that this problem was at that stage very difficult, and concentrated his efforts on explicit calculations in some particular cases. At the end of the introduction of his lecture notes, Zariski expresses his hope on the subject:

Nous espérons que ce cours stimulera de nouveaux travaux sur le sujet. Un des problèmes que nous recommandons au lecteur est celui-ci de l'identification de l'espace des modules avec un constructible d'une variété V.

So, to proceed successfully with Zariski's program, we are going from now on to adopt a new strategy that will allow us to obtain reductions without necessarily exhibiting explicitly the changes of parameter and of coordinates involved. This will be done by using techniques from Singularity Theory, which we introduce in the next section.

1.3 Singularity Theory Approach

Up to this point, we converted the classification problem of branches in an equisingularity class determined by a semigroup of values Γ to the study of an equivalence relation $\sim$ on the set of parameters $\Sigma(\Gamma)$ representing Puiseux parametrizations of such branches. Although this set has the good geometric structure of a principal open Zariski set in a finite dimensional affine space, hence itself a smooth affine variety, the trouble is that the equivalence relation $\sim$, induced by automorphisms φ and ρ as in Proposition 1.2.3, is not given by a group action, since, in this case, the change of parameters ρ is tailored by the specific parametrization on which it acts.

To apply techniques from Singularity Theory, one has to work with a geometric group acting on a geometric space. So, to put ourselves in this framework, we are going to relax the assumption that branches are represented by primitive Puiseux parametrizations, and go back to parametrizations in general as elements in $\Xi = \mathcal{M}_1 \times \mathcal{M}_1$, under the action of the group $\mathcal{A} = \mathrm{Aut}(\mathbb{C}^2, 0) \times \mathrm{Aut}(\mathbb{C}, 0)$, defined by

$$(\varphi, \rho) \cdot \xi(t) = (\varphi \circ \xi \circ \rho^{-1})(t).$$

The elimination criteria of parameters, we are looking for, in a given $\xi(t) = (x(t), y(t)) \in \Xi$, where $x(t) = \sum_{i=1}^{\infty} a_i t^i$ and $y(t) = \sum_{i=1}^{\infty} b_i t^i$, keeping $\mathcal{A}$-equivalence, may be summarized as follows.

To verify if a term $(a_k t^k, b_k t^k)$ of order k in $\xi(t)$ is eliminable, one has to show that there is an element $\xi_1(t) \in \Xi$ which is $\mathcal{A}$-equivalent to $\xi(t)$ and such that

$$j^{k-1}\xi(t) = j^{k-1}\xi_1(t) = j^k \xi_1(t).$$

1.3.1 The Complete Transversal Theorem

At this point Singularity Theory comes into the scene through the following fundamental tool:

Theorem 1.3.1 (Mather's Lemma) *Let G be a Lie group acting on a smooth manifold M and let $T_m M$ be the tangent space of M at a point m of M. A connected smooth submanifold W of M is contained in a single orbit of G if, and only if, the following conditions are fulfilled:*

(i) $T_w W \subseteq T_w G(w)$ *for all* $w \in W$;
(ii) $\dim T_w G(w)$ *is constant for* $w \in W$.

Proof See Lemma 3.1 in [19]. □

The following is a simple, but important, consequence of Mather's Lemma that appeared in [7, Proposition 1.3] and will be of our direct interest.

Corollary 1.3.2 (The Complete Transversal Theorem) *Let G be a Lie group acting on an affine space $\mathbb{A}$ with underlying vector space V and $W \subseteq V$ a subspace. If for $a \in \mathbb{A}$ we have*

(i) $W \subseteq T_a G(a)$;

(ii) $T_{a+w}G(a+w) = T_aG(a)$ *for all* $w \in W$,

then $a + W \subseteq G(a)$.

Proof Obviously, $a + W \subseteq \mathbb{A}$ is a connected submanifold of the affine space $\mathbb{A}$. Since W is a subspace of V we have

$$T_{a+w}(a+W) = W \subseteq T_aG(a) = T_{a+w}G(a+w), \; \forall w \in W.$$

In this way, $a + W$ satisfies the hypothesis of Mather's Lemma, hence $a + W$ is contained in a single orbit. More precisely, $a + W \subseteq G(a)$. □

The crucial fact is that under the conclusion that $a + W \subseteq G(a)$ of the Complete Transversal Theorem, we may choose an element $w \in W$ such that $a + w$ is G-equivalent to a. This gives a glimpse on the enormous potential of this result in obtaining new elimination criteria of coefficients in parametrizations, keeping unchanged the analytic type of the branches they represent.

Let us now outline the strategy we will follow to attack the problem of analytic classification of branches in which we make use of what observed above.

Let $n > 1$ be an integer and let us define

$$\Xi_n = \{\xi(t) \in \mathcal{M}_1 \times \mathcal{M}_1;\; j^n\xi(t) = (t^n, 0)\}.$$

Notice that every branch C of multiplicity n is equivalent to a branch with parametrization in Ξ_n, since this is equivalent to choose coordinates in such a way that the line $Y = 0$ is the tangent line to the branch and then normalize the coefficient of t^n in $x(t)$.

The largest subgroup of $\mathcal{A}$ that acts on this set is

$$\tilde{\mathcal{A}} = \{(\varphi, \rho) \in \mathcal{A};\; j^1(\rho(t)) = \alpha t,\; j^1(\varphi(X,Y)) = (\alpha^n X + bY, dY).$$

This is so, because any other (φ, ρ) acting on an element $\xi(t)$ in Ξ_n will affect its first jet.

We highlight the following subgroups of the group $\tilde{\mathcal{A}}$:

$$\mathcal{H} = \{(\varphi, \rho) \in \tilde{\mathcal{A}};\; \rho(t) = \alpha t, \text{ and } \varphi(X,Y) = (\alpha^n X, dY),\; \alpha d \neq 0\}, \; \textit{and}$$
$$\tilde{\mathcal{A}}_1 = \{(\varphi, \rho) \in \tilde{\mathcal{A}};\; j^1(\rho) = t \text{ and } j^1(\varphi) = (X + bY, Y)\}.$$

Although the elements in $\tilde{\mathcal{A}}_1$ and $\mathcal{H}$ do not commute, every element in $\tilde{\mathcal{A}}$ can be written as a composition of an element of each subgroup.

The action of the group $\mathcal{H}$, called the *Homothety group*, on a parametrization $\xi(t)$ does not introduce nor eliminate terms, hence we will initially focus on the group $\tilde{\mathcal{A}}_1$ that may allow elimination of terms. Observe that the reductions presented in Sect. 1.2 were obtained only with the action of $\tilde{\mathcal{A}}_1$.

To make use of the Complete Transversal Theorem (CTT), we need a Lie group G acting on a finite dimensional affine space $\mathbb{A}$ and identify which subspace W of the underlying vector space V of $\mathbb{A}$ we will consider.

For given $k > n$, we will consider the affine spaces

$$\mathbb{A} = \Xi_n^k = j^k \Xi_n,$$

with underlying vector space $V = j^k(\mathcal{M}_1^{n+1} \times \mathcal{M}_1^{n+1})$. The groups under consideration are the algebraic affine groups $G = \tilde{\mathcal{A}}_1^k$, with operation defined by

$$(j^k\varphi_1, j^k\rho_1) \cdot (j^k\varphi_2, j^k\rho_2) = j^k(j^k\varphi_1 \circ j^k\varphi_2, j^k\rho_1 \circ j^k\rho_2),$$

acting on $\mathbb{A}$ as follows:

$$(j^k\varphi, j^k\rho) \cdot j^k\xi(t) = j^k(j^k\varphi \circ j^k\xi \circ j^k\rho^{-1})(t),$$

which is consistent, since $j^k\rho^{-1}(t)$ depends only on the coefficients of $j^k\rho(t)$.

Remark 1.3.3 Because of the compatibility of the formation of jets with compositions, we have that the above definitions could be rephrased as follows:

$$(j^k\varphi_1, j^k\rho_1) \cdot (j^k\varphi_2, j^k\rho_2) = (j^k\varphi_1 \circ \varphi_2, j^k\rho_1 \circ \rho_2),$$

and

$$(j^k\varphi, j^k\rho) \cdot j^k\xi(t) = j^k(\varphi \circ \xi \circ \rho^{-1})(t).$$

Notice also that the groups $\tilde{\mathcal{A}}_1^k$ are unipotent, since $j^k\big((\mathrm{Id}, \mathrm{Id}) - (\varphi, \rho)\big)^{k+1} = 0$ for all $(\varphi, \rho) \in \tilde{\mathcal{A}}_1$.

On the other hand, the subspace W we consider will be one of the following sets: $W_{10}^k = (t^k, 0)\mathbb{C}$, $W_{01}^k = (0, t^k)\mathbb{C}$ or $W_{11}^k = (t^k, 0)\mathbb{C} + (0, t^k)\mathbb{C}$.

Remark 1.3.4 Notice that in our context one has

$$g(\xi(t) + w) = g(\xi(t)) + w, \ \forall g \in \tilde{\mathcal{A}}_1^k, \ \forall \xi \in \Xi_n^k, \text{ and } \forall w \in W_{10}^k,$$

implying that $\tilde{\mathcal{A}}_1^k(\xi(t) + w) = \tilde{\mathcal{A}}_1^k(\xi(t)) + w$, for all $\xi(t) \in \Xi_n^k$ and $w \in W_{10}^k$, which is stronger than condition (ii). We will show in Proposition 1.3.8 that condition (ii) in the statement of CTT is satisfied in general for $W = W_{11}^k$.

In conclusion, to apply CTT to our problem we have to characterize the tangent space to orbits under the action of $\tilde{\mathcal{A}}_1^k$, so, let us put our hands to work.

1.3.2 Tangent Spaces to Orbits

In what follows, we use the following notation:

$$\mathcal{L}_1 = \{\varphi \in \mathrm{Aut}(\mathbb{C}^2, 0);\ j^1(\phi) = (X + bY, Y),\ \text{ where } b \in \mathbb{C}\}, \text{ and}$$

$$\mathcal{R}_1 = \{\rho \in \mathrm{Aut}(\mathbb{C}, 0);\ j^1(\rho) = t\}.$$

So, $\tilde{\mathcal{A}}_1 = \mathcal{L}_1 \times \mathcal{R}_1$. We will denote by $\mathcal{G}$ any of the groups $\mathcal{L}_1$, $\mathcal{R}_1$ or $\tilde{\mathcal{A}}_1$; we will indicate by g and $\xi(t)$ elements in $\mathcal{G}^k$ and Ξ^k_n, respectively.

By definition, the orbit $\mathcal{G}^k(\xi(t))$ is the image of the map

$$\begin{array}{rcl} \psi_\xi : \mathcal{G}^k & \rightarrow & \Xi^k_n \\ g & \mapsto & \psi_\xi(g) := g \cdot \xi(t). \end{array}$$

Given $g_1 \in \mathcal{G}^k$ we have the commutative diagram:

$$\begin{array}{ccc} \mathcal{G}^k & \xrightarrow{\psi_\xi(t)} & \Xi^k_n \\ L_{g_1} \downarrow & \circlearrowright & \downarrow L'_{g_1} \\ \mathcal{G}^k & \xrightarrow{\psi_\xi(t)} & \Xi^k_n \end{array}$$

where $L_{g_1}(g) = j^k(g_1 \circ g)$ and $L'_{g_1}(\xi(t)) = j^k(g_1 \cdot \xi(t))$, for any $g \in \mathcal{G}^k$ and $\xi(t) \in \Xi^k_n$.

Denoting by $d_p F$ the differential of a map F at a point p, we have, for any $g \in \mathcal{G}^k$,

$$(d_{L_{g_1}(g)}\psi_\xi) \circ (d_g L_{g_1}) = (d_{\psi_\xi(g)} L'_{g_1}) \circ (d_g \psi_\xi).$$

Since L_{g_1} and L'_{g_1} are diffeomorphisms, taking $g = g_0$, the identity of $\mathcal{G}^k$, we conclude that $d_{g_1}(\psi_\xi)$ has the same rank for all $g_1 \in \mathcal{G}^k$ and that the tangent space to the orbit $\mathcal{G}^k(\xi)$ at $\xi(t)$, is the image of

$$d_{g_0}\psi_\xi : T_{g_0}\mathcal{G}^k \rightarrow T_\xi \Xi^k_n,$$

that is,

$$T_{\xi(t)}\mathcal{G}^k(\xi(t)) = d_{g_0}\psi_\xi(T_{g_0}\mathcal{G}^k). \tag{1.19}$$

Recall that the tangent space $T_{g_0}\mathcal{G}^k$ is the set of vectors v such that there exists a germ of complex curve $\gamma(u)$ in $\mathcal{G}^k$ with $\gamma(0) = g_0$ and $\gamma'(0) = v$.

If we write $g_0 = (e_1, e_2)$, where $e_1(X, Y) = (X, Y)$ and $e_2(t) = t$ are the identities of $\mathcal{L}_1$ and $\mathcal{R}_1$, respectively, then

$$
\begin{aligned}
T_{\xi(t)}\tilde{\mathcal{A}}_1^k(\xi(t)) &= d_{g_0}\psi_\xi(T_{g_0}\widetilde{\mathcal{A}}_1^k) \\
&= d_{g_0}\psi_\xi(T_{e_1}\mathcal{L}_1^k \times T_{e_2}\mathcal{R}_1^k) \\
&= d_{g_0}\psi_\xi\left((T_{e_1}\mathcal{L}_1^k \times \{e_2\}) \oplus (\{e_1\} \times T_{e_2}\mathcal{R}_1^k)\right) \\
&= d_{g_0}\psi_\xi\left(T_{e_1}\mathcal{L}_1^k \times \{e_2\}\right) + d_{g_0}\psi_\xi\left(\{e_1\} \times T_{e_2}\mathcal{R}_1^k\right).
\end{aligned}
\tag{1.20}
$$

Now, since

$$
\mathcal{L}_1^k = \{(X,Y) + j^k(p,q);\ p \in \langle X^2, Y\rangle,\ q \in \langle X, Y\rangle^2\},\ \text{and}
$$

$$
\mathcal{R}_1^k = \{t + j^k(r); r \in \langle t\rangle^2\},
$$

hence are affine spaces, it follows that

$$
\begin{aligned}
T_{e_1}\mathcal{L}_1^k &= j^k(\langle X^2, Y\rangle) \times j^k(\langle X, Y\rangle^2),\ \text{and} \\
T_{e_2}\mathcal{R}_1^k &= j^k(\langle t\rangle^2).
\end{aligned}
\tag{1.21}
$$

Proposition 1.3.5 *With the above notation we have*

$$
d_{g_0}\psi_\xi\left(T_{e_1}\mathcal{L}_1^k \times \{e_2\}\right) = \left\{j^k(p(\xi(t)), q(\xi(t)));\ p \in \langle X^2, Y\rangle,\ q \in \langle X, Y\rangle^2\right\}.
$$

Proof Considering the commutative diagram

$$
\begin{array}{ccc}
\mathcal{L}_1^k \times \{e_2\} & \stackrel{\psi_\xi}{\rightarrow} & \mathcal{L}_1^k(\xi(t)) \\
\pi_1 \downarrow & & \nearrow \psi_1 \\
\mathcal{L}_1^k & &
\end{array}
$$

Since, $d_{g_0}\psi_\xi = d_{e_1}(\psi_1) \circ d_{g_0}\pi_1$ and $d_{g_0}\pi_1$ is an isomorphism, we have

$$
d_{g_0}\psi_\xi\left(T_{e_1}\mathcal{L}_1^k \times \{e_2\}\right) = d_{e_1}\psi_1\left(T_{e_1}\mathcal{L}_1^k\right).
$$

Now, for $v \in T_{e_1}\mathcal{L}_1^k$, consider the curve $\gamma(u) = e_1 + uv$, $u \in \mathbb{C}$, in $\mathcal{L}_1^k$, hence we obtain the curve $\psi_1(\gamma(u)) = \gamma(u) \cdot \xi(t) = (e_1 + uv)(\xi(t))$ in the orbit $\mathcal{L}_1^k(\xi(t))$. So,

$$
d_{e_1}\psi_1(v) = \frac{\partial(e_1 + uv)(\xi)}{\partial u}(0) = v(\xi(t)).
$$

As $v \in T_{e_1}\mathcal{L}_1^k$, from (1.21), we have $v \in j^k(\langle X^2, Y\rangle) \times j^k(\langle X, Y\rangle^2)$, hence

$$
d_{e_1}(\psi_1)\left(T_{e_1}\mathcal{L}_1^k\right) = \left\{j^k(p(\xi(t)), q(\xi(t)));\ p \in \langle X^2, Y\rangle,\ q \in \langle X, Y\rangle^2\right\},
$$

concluding our proof. □

Similarly, one will describe $d_{g_0}\psi_\xi\left(\{e_1\}\times T_{e_2}\mathcal{R}_1^k\right)$ in the next proposition.

Proposition 1.3.6 *If $\xi(t)=(x(t),y(t))\in\Xi_n^k$, then*

$$d_{g_0}\psi_\xi\left(\{e_1\}\times T_{e_2}\mathcal{R}_1^k\right)=\left\{j^k(x'(t)r,\,y'(t)r);\ r\in\langle t\rangle^2\right\}.$$

Proof Consider the natural commutative diagram

$$\begin{array}{ccc}\{e_1\}\times\mathcal{R}_1^k & \stackrel{\psi_\xi}{\rightarrow} & \mathcal{R}_1^k(\xi)\\ \pi_2\downarrow & & \nearrow\psi_2\\ \mathcal{R}_1^k & & \end{array}$$

Since $\psi_\xi=\psi_2\circ\pi_2$, we obtain $d_{g_0}\psi_\xi=d_{e_2}(\psi_2)\circ d_{g_0}\pi_2$. But, $d_{g_0}\pi_2$ is an isomorphism and

$$d_{g_0}\psi_\xi\left(\{e_1\}\times T_{e_2}\mathcal{R}_1^k\right)=d_{e_2}(\psi_2)\left(T_{e_2}\mathcal{R}_1^k\right).$$

To describe $d_{e_2}(\psi_2)\left(T_{e_2}\mathcal{R}_1^k\right)$, let us consider the curve $\gamma(u)=e_2+uv\in\mathcal{R}_1^k$, where $v\in T_{e_2}\mathcal{R}_1^k$. Then $\psi_2(\gamma(u))=\xi(e_2+uv)$ is a curve in the orbit $\mathcal{R}_1^k(\xi)$. Hence,

$$d_{e_2}\psi_2(v)=\frac{\partial\xi(e_2+uv)}{\partial u}(0)=d_{e_2}\xi(e_2)v.$$

As $e_2(t)=t$ and $v\in T_{e_2}\mathcal{R}_1^k=j^k\left(\langle t\rangle^2\right)$, we obtain

$$d_{g_0}\psi_\xi\left(\{e_1\}\times T_{e_2}\mathcal{R}_1^k\}\right)=d_{e_2}(\psi_2)\left(T_{e_2}\mathcal{R}_1^k\right)=\left\{j^k(x'(t)r,\,y'(t)r);\ r\in\langle t^2\rangle\right\},$$

proving the result. □

From Propositions 1.3.6, 1.3.5 and (1.20), we finally obtain the following description of the tangent space of the orbit $\tilde{\mathcal{A}}_1^k(\xi(t))$ at $\xi(t)\in\Xi_n^k$.

Theorem 1.3.7 *For $\xi(t)\in\Xi_n^k$, one has*

$$T_\xi\tilde{\mathcal{A}}_1^k(\xi(t))=\left\{j^k(\xi'(t)r+(p(\xi(t)),q(\xi(t)))),\,p\in\langle X^2,Y\rangle,\,q\in\langle X,Y\rangle^2,\,r\in\langle t\rangle^2\right\},$$

where $\xi'(t)=(x'(t),y'(t))$.

Now, that we have the description of the tangent space of the orbit at a point, we are in position to prove that Condition (ii) in Theorem 1.3.2 is verified for $W=W_{1,1}^k$.

Proposition 1.3.8 *Let $\xi(t)=(x(t),y(t))\in\Xi_n^k$ and let $W=(t^k,0)\mathbb{C}+(0,t^k)\mathbb{C}$. We have, for all $k>n$, that*

$$T_{\xi(t)+w}\tilde{\mathcal{A}}_1^k(\xi(t)+w) = T_{\xi(t)}\tilde{\mathcal{A}}_1^k(\xi(t)), \ \forall w \in W.$$

Proof Initially we show that $(t^k, 0) \in T_{\xi(t)+w}\tilde{\mathcal{A}}_1^k(\xi(t)+w)$ for all $w = (a_1t^k, a_2t^k) \in W$. Indeed, putting $r = t^k/(x'(t)+a_1t^{k-1})$ we have that $\mathrm{ord}_t(r) = k-(n-1) \geq 2$, hence $r \in \langle t\rangle^2$. So, taking $p = q = 0$, Theorem 1.3.7 implies that

$$(t^k, 0) = j^k((x'(t)+ka_1t^{k-1})r, (y'(t)+ka_2t^{k-1})r) \in T_{\xi(t)+w}\tilde{\mathcal{A}}_1^k(\xi(t)+w). \tag{1.22}$$

Given $v \in T_{\xi(t)+w}\tilde{\mathcal{A}}_1^k(\xi(t)+w)$, where $w = (a_1t^k, a_2t^k)$, by Theorem 1.3.7, there exist $p = bY + p_1$, $p_1, q \in \langle X, Y\rangle^2$ and $r \in \langle t\rangle^2$ such that

$$\begin{aligned} v &= j^k((x'(t)+ka_1t^{k-1})r + p(\xi(t)+w), (y'(t)+ka_2t^{k-1})r + q(\xi(t)+w)) \\ &= j^k(x'(t)r + p(\xi(t)+w), y'(t)r + q(\xi(t)+w)) \\ &= j^k(x'(t)r + p(\xi(t)), y'(t)r + q(\xi(t))) + ba_1(t^k, 0). \end{aligned} \tag{1.23}$$

As $j^k(x'(t)r + p(\xi(t)), y'(t)r + q(\xi(t))) \in T_{\xi(t)}\tilde{\mathcal{A}}_1^k(\xi(t))$ and, by (1.22), $ba_1(t^k, 0) \in T_{\xi(t)}\tilde{\mathcal{A}}_1^k(\xi(t))$, we have that $v \in T_{\xi(t)}\tilde{\mathcal{A}}_1^k(\xi(t))$; and consequently, $T_{\xi(t)+w}\tilde{\mathcal{A}}_1^k(\xi(t)+w) \subseteq T_{\xi(t)}\tilde{\mathcal{A}}_1^k(\xi(t))$.

On the other hand, if $v = j^k(x'(t)r + p(\xi(t)), y'(t)r + q(\xi(t))) \in T_{\xi(t)}\tilde{\mathcal{A}}_1^k(\xi(t))$, then, by (1.22) and (1.23),

$$v = j^k((x'(t)+ka_1t^{k-1})r + p(\xi(t)+w), (y'(t)+ka_2t^{k-1})r + q(\xi(t)+w)) - ba_1(t^k, 0),$$

which belongs to $T_{\xi(t)+w}\tilde{\mathcal{A}}_1^k(\xi(t)+w)$, finishing the proof. □

It will be convenient to highlight from what was done in this subsection the result that follows.

Theorem 1.3.9 *Let $\xi(t) \in \Xi_n$. If, for some integer $k > n$, $(a_1t^k, a_2t^k) \in T_{j^k\xi(t)}\tilde{\mathcal{A}}_1^k(j^k\xi(t))$, then there is a parametrization $\xi_1(t) \in \Xi_n$ which is $\tilde{\mathcal{A}}_1$-equivalent to $\xi(t)$ and such that $j^k\xi_1(t) = j^k\xi(t) + (a_1t^k, a_2t^k)$.*

Remark 1.3.10 Observe that Theorems 1.3.7 and 1.3.9 allow us to recover part of the reductions of the previous section.

For example, given $\xi_0(t) \in \Xi_n$ with semigroup of values $\Gamma = \langle n, m, \ldots, v_g\rangle$, taking $r = 0$, $p = 0$ and $q \in \langle X, Y\rangle^2$ with $\mathrm{ord}_t q(\xi_0(t)) = k \in \Gamma$ in Theorem 1.3.7, we get $(0, t^k) \in T_{j^k\xi_0(t)}\tilde{\mathcal{A}}_1^k(j^k\xi_0(t))$, for every $k > n$. In this way, $\xi_0(t)$ and $\xi_1(t) = (t^n, t^m + \sum_{\substack{i>m \\ i\notin\Gamma}} b_it^i)$ determine analytically equivalent curves.

In the next subsection, we will connect the elements of the invariant Λ of a branch parametrized by $\xi(t)$ to the tangent spaces $T_{j^k\xi(t)}\tilde{\mathcal{A}}_1^k(j^k\xi(t))$.

1.3.3 The Analytic Classification

For a given a branch C with semigroup Γ and set of values of differentials Λ, we know that all terms of order in $(\Gamma - n) \cup (\Gamma \setminus \{0\} + \lambda)$ in a primitive Puiseux parametrization $\xi(t)$ of C are eliminable (cf. Theorem 1.2.7 and the Fourth Reduction). Now, since $\Gamma \setminus \{0\} \subseteq \Lambda$, $\lambda + n = v(mydx - nxdy) \in \Lambda$ (cf. Proposition 1.2.22) and $\Gamma + \Lambda \subset \Lambda$, it follows that

$$(\Gamma - n) \cup (\Gamma + \lambda) \subseteq \Lambda - n.$$

In this subsection, by using Theorem 1.3.9, we will unify all the reductions so far presented in a sole one that will allow us to discard all terms in a primitive Puiseux parametrization with exponents in $\Lambda - n \setminus \{m, \lambda\}$ where $m = \beta_1$ and λ is the Zariski invariant.

In the next result we will characterize the integers $k > m$ such that $(0, t^k) \in T_{j^k\xi(t)}\tilde{\mathcal{A}}_1^k(j^k\xi(t))$, which, in virtue of Theorem 1.3.9, is crucial for eliminating the term of order k in $\xi(t)$.

Proposition 1.3.11 *Let $\xi(t) = (t^n, y(t)) \in \Xi_n$ be a parametrization of a branch C_f with $\Gamma(f) = \langle v_0, v_1, \ldots, v_g\rangle$, where $v_0 = n$ and $v_1 = m$. For $k > m$, $(0, t^k) \in T_{j^k\xi(t)}\tilde{\mathcal{A}}_1^k(j^k\xi(t))$ if and only if $k = v(\omega) - n$, where $\omega = \overline{q}dx - \overline{p}dy \in \Omega(f)$ with $p \in \langle X^2, Y\rangle$, $q \in \langle X, Y\rangle^2$ and $\overline{h}$ stands for the image of $h \in \mathcal{O}$ in $\mathcal{O}(f)$.*

Proof By Theorem 1.3.7, $(0, t^k) \in T_{j^k\xi(t)}\tilde{\mathcal{A}}_1^k(j^k\xi(t))$ if and only if there exist $p \in \langle X^2, Y\rangle$, $q \in \langle X, Y\rangle^2$ and $r \in \langle t\rangle^2$ such that

$$0 = j^k(x'(t)r + p(\xi(t))) \text{ and } t^k = j^k(y'(t)r + q(\xi(t))). \tag{1.24}$$

A solution for the first equation is $r = -\frac{p(\xi(t))}{x'(t)} + t^{k-n+2}s$, for some $s \in \mathcal{O}_1$. Replacing this expression for r in the second equation, we get

$$\begin{aligned} t^k &= j^k\left(q(\xi(t)) - \tfrac{p(\xi(t))y'(t)}{x'(t)} + t^{k-n+2}y'(t)s\right) \\ &= j^k\left(\tfrac{q(\xi(t))x'(t) - p(\xi(t))y'(t)}{x'(t)}\right) = j^k\left(\tfrac{\Psi(qdx - pdy)}{nt^n}\right), \end{aligned}$$

where Ψ is as given in (1.14). Hence $k = v(\overline{q}dx - \overline{p}dy) - n$.

On the other hand, given $w = \overline{q}dx - \overline{p}dy \in \Omega(f)$, where $p \in \langle X^2, Y\rangle$ and $q \in \langle X, Y\rangle^2$; if we put $k = v(\omega) - n$, then $r = -\frac{p(\xi(t))}{x'(t)}$ solves the system (1.24), completing the proof of the proposition. □

Using Proposition 1.3.11, Theorem 1.3.9 and the next lemma, we will recover below the Third Reduction.

Lemma 1.3.12 *Let C_f be a branch. If $k \in \Gamma(f) - n$ and $k > m$, then there exists $\omega = \overline{q}dx - \overline{p}dy \in \Omega(f)$, with $p \in \langle X^2, Y\rangle$ and $q \in \langle X, Y\rangle^2$, such that $k = v(\omega) - n$.*

Proof In fact, let k be such that $m < k = \gamma - n$, with $\gamma \in \Gamma(f)$. Considering $\{f_0 = Y, \ldots, f_g\}$ a complete system of semiroots of f, by Corollary 1.1.35, there exists $h = X^{\alpha_0}Y^{\alpha_1}f_1^{\alpha_2}\cdot\cdots\cdot f_{g-1}^{\alpha_g}$ with $\alpha_0 \geq 0, 0 \leq \alpha_i < n_i$, for $i = 1, \ldots, g$, such that $v(\overline{h}) = \gamma$. Observe that for each $j = 0, \ldots, g$, one has that $f_j \in \mathbb{C}\{X\}[Y]$ is a Weierstrass polynomial of degree $\frac{n}{e_j} = n_0 \cdot \cdots \cdot n_j$ that coincides to its multiplicity.

If there exists an index $j > 1$ such that $\alpha_j > 0$, then considering $d\overline{h} = h_x dx + h_y dy$, we have that $k = v(d\overline{h}) - n$ with $h_X \in \langle X, Y\rangle^2$ and $h_Y \in \langle X^2, Y\rangle$.

On the other hand, if for all $j > 1$, one has $\alpha_j = 0$, put $h = X^{\alpha_0}Y^{\alpha_1}$. Since $v(\overline{h}) = \gamma = k + n > m + n$, one of the following three possibilities occurs: $\alpha_1 \geq 2$; $\alpha_1 = 1$ and $\alpha_0 \geq 2$; or $\alpha_1 = 0$ and $\alpha_0 > \left[\frac{m}{n}\right] + 1 > 2$. In any case, $d\overline{h} = h_x dx + h_y dy$ is such that $h_X \in \langle X, Y\rangle^2$ and $h_Y \in \langle X^2, Y\rangle$. □

In this way, by Proposition 1.3.11 and Theorem 1.3.9 we obtain, again, by another method, that C_f is analytically equivalent to a branch defined by a parametrization as in Theorem 1.2.7.

Moreover, if $a_i = 0$ for all $i \notin \Gamma - n$ then the corresponding parametrization is $\xi(t) = (t^n, t^m)$. Otherwise, we have that $\lambda = \min\{i;\ a_i \neq 0\}$ and we recover the parametrization

$$\xi(t) = \left(t^n, t^m + a_\lambda t^\lambda + \sum_{\substack{i>\lambda \\ i\notin\Gamma-n}} a_i t^i\right),$$

that will be considered in the sequel.

Proposition 1.3.13 *Let $\xi(t)$ be a parametrization of a branch C_f as above. If $\delta > \lambda + n$ and $\delta \in \Lambda(f)\setminus\Gamma(f)$, then $\delta = v(\overline{q}dx - \overline{p}dy)$, where $q \in \langle X, Y\rangle^2$ and $p \in \langle X^2, Y\rangle$.*

Proof Let $\omega = \overline{q}dx - \overline{p}dy$ be such that $v(\omega) = \delta \notin \Gamma(f)$. Then we must have $p, q \in \langle X, Y\rangle$ and $\delta = v(\omega) > v(\overline{q}dx) = v(\overline{p}dy)$.

Suppose that $p = \alpha X + p_1$, with $p_1 \in \langle X^2, Y\rangle$ and $\alpha \neq 0$. Hence, $v(\overline{p}) = n$ which implies, in view of the equality $v(\overline{q}dx) = v(\overline{p}dy)$, that $v(\overline{q}) = m$; that is, $q = \beta Y + q_1$, with $\beta \neq 0$ and $q_1 \in \langle X, Y\rangle^2$.

The inequality $v(\omega) > v(\overline{q}dx) = v(\overline{p}dy)$ imposes that $\alpha = n$ and $\beta = m$. In this way, $\omega = \omega_1 + \overline{q_1}dx - \overline{p_1}dy$, where $\omega_1 = mydx - nxdy$. Notice that, by hypothesis we do not have $v(\omega) = v(\omega_1) = \lambda + n$ nor $v(\omega) = v(\overline{q_1}dx - \overline{p_1}dy) < v(\omega_1) = \lambda + n$. So, we must have $\lambda + n = v(\omega_1) = v(\overline{q_1}dx - \overline{p_1}dy)$ with $\overline{q_1}dx - \overline{p_1}dy \in \Omega(f)$ satisfying the conditions in Proposition 1.3.11. Hence, $(0, t^\lambda) \in T_{j^\lambda\xi(t)}\tilde{\mathcal{A}}_1^\lambda(j^\lambda\xi(t))$ and consequently, by Theorem 1.3.9, there would exist a primitive parametrization defining a branch analytically equivalent to C_f with distinct (or without) Zariski invariant λ, a contradiction. Hence we must have $p = p_1 \in \langle X^2, Y\rangle$.

Now, suppose that $q \notin \langle X, Y\rangle^2$. As $v(\overline{q}) + n = v(\overline{p}) + m$, we must have $q = \beta Y + q_1$ with $\beta \neq 0$ and $q_1 \in \langle X, Y\rangle^2$. As $v(y) \neq v(\overline{q_1})$, we have that $v(\overline{p}) + m = v(\overline{q}) + n \leq m + n$, that is, $v(\overline{p}) \leq n$, a contradiction. Therefore, we have $q \in \langle X, Y\rangle^2$. This concludes the proof of the proposition. □

We are ready now to present the ultimate reduction with respect to the group $\tilde{\mathcal{A}}_1$.

Theorem 1.3.14 *Let $\xi_1(t)$ be a primitive parametrization of a plane branch C_f with semigroup of values Γ and value set of differentials Λ. Suppose that $\Lambda \setminus \Gamma \neq \emptyset$, then $\xi_1(t)$ is $\tilde{\mathcal{A}}_1$-equivalent to a parametrization of the form*

$$\xi(t) = \left(t^n, t^m + a_\lambda t^\lambda + \sum_{\substack{i>\lambda \\ i \notin \Lambda - n}} a_i t^i\right), \tag{1.25}$$

where $n = \min \Gamma \setminus \{0\}$, $m = \min \Gamma \setminus \langle n \rangle$ and $\lambda = \min(\Lambda \setminus \Gamma) - n$. Moreover, two such parametrizations with same λ-jet are $\tilde{\mathcal{A}}_1$-equivalent if and only if they are equal.

Proof **Existence.** For each k such that $\lambda < k \in \Lambda - n$ we have that $k + n \in \Gamma$ or $k + n \in \Lambda \setminus \Gamma$.

If $\lambda < k \in \Gamma - n$, then by what we said before Proposition 1.3.13 there exists a parametrization $\xi_2(t)$ which is $\tilde{\mathcal{A}}_1$-equivalent to a parametrization $\xi_1(t)$ satisfying $j^k\xi_2(t) = j^{k-1}\xi_2(t) = j^{k-1}\xi_1(t)$.

If $\lambda + n < k + n \in \Lambda \setminus \Gamma$, by Proposition 1.3.13, there exists $\omega \in \Omega(f)$ with $v(\omega) = k + n$ satisfying the hypothesis of Proposition 1.3.11; that is, $(0, t^k) \in T_{j^k\xi_1(t)}\tilde{\mathcal{A}}_1^k(j^k\xi_1(t))$. Hence, by Theorem 1.3.9, we get a parametrization $\xi_2(t)$ which is $\tilde{\mathcal{A}}_1$-equivalent to $\xi_1(t)$ with $j^k\xi_2(t) = j^{k-1}\xi_2(t) = j^{k-1}\xi_1(t)$.

Now, repeating this reduction for all k with $\lambda < k < c$, where c is the conductor of Γ, and apply an element of $\tilde{\mathcal{A}}_1$ that performs the Third Reduction, we obtain a parametrization $\xi(t)$ as in the statement of the proposition which is $\tilde{\mathcal{A}}_1$-equivalent to $\xi_1(t)$.

Unicity. Let $\xi(t)$ be as in (1.25) and let $k > \lambda$ be an integer such that $k + n \notin \Lambda$. Consider the one dimensional affine space

$$N_\xi^k = j^{k-1}\xi(t) + \mathbb{C}(0, t^k),$$

and let G_ξ^k be the algebraic subgroup of $\tilde{\mathcal{A}}_1^k$ that leaves N_ξ^k invariant. Since $\tilde{\mathcal{A}}_1^k$ is an algebraic unipotent group over $\mathbb{C}$, it follows that G_ξ^k is an algebraic connected group (cf. [4, p. 8]). We will show that the orbit $G_\xi^k(j^k\xi(t))$ is just $\{j^k(\xi(t))\}$. This is so, because $G_\xi^k(j^k\xi(t))$ is a connected closed set in $N_\xi^k \simeq \mathbb{C}$ (cf. [22, Theorem 2]), so it consists either of one point, or it is the whole N_ξ^k.

If $G_\xi^k(j^k\xi(t))$ does not consist of a single point, then since $G_\xi^k(j^k\xi(t)) \subset \tilde{\mathcal{A}}_1^k(j^k\xi(t))$, it would follow that

$$\mathbb{C}(0, t^k) = T_{j^k\xi(t)}G_\xi^k(j^k\xi(t)) \subset T_{j^k\xi(t)}\tilde{\mathcal{A}}_1^k(j^k\xi(t)),$$

which from Propositions 1.3.11 and 1.3.13 yields to a contradiction. □

Remark 1.3.15 As observed in Sect. 1.2.2, each admissible set Λ in an equisingularity class determined by the semigroup Γ with conductor c, admits a conductor

$d \leq c$. So, the above theorem guarantees that an element $\xi(t)$ in Ξ_n with semigroup of values Γ and set of values of differentials Λ is $\tilde{\mathcal{A}}_1$-equivalent (consequently $\mathcal{A}$-equivalent) to $j^{d-n-1}\xi(t)$.

Remark 1.3.16 Not every parametrization $\xi(t)$ of the form (1.25) has semigroup of values equal to Γ, because the conditions $a_{\beta_2} \cdots a_{\beta_g} \neq 0$ and those contained in (1.4), as well, should be satisfied. On the other hand, the set of values of differentials of the branch it represents is not necessarily equal to Λ. This is a more subtle issue, since one can show (cf. [13]) that this imposes conditions on the coefficients of $\xi(t)$, so that the subset $\Xi_\Gamma(\Lambda)$ of elements in Ξ_n^{c-1} that represent branches with semigroup of values Γ and admissible set of values of differentials Λ form a constructible subset invariant under the action of $\tilde{\mathcal{A}}_1^{c-1}$. If Λ_i, $i = 1, \ldots, s$, are the admissible set of values of differentials for branches in the equisingularity class determined by Γ, then the subsets $\Xi_\Gamma(\Lambda_i)$, $i = 1, \ldots, s$, form a partition by constructible subsets of the constructible set Ξ_Γ, the set of elements in Ξ_n^{c-1} which have semigroup of values equal to Γ.

Recall that we are interested in the classification of equisingular branches with respect to analytic equivalence, which was translated into the classification with respect to the action of the group $\tilde{\mathcal{A}}$ on primitive parametrizations determining a given semigroup of values. Since the action of this group on parametrizations is obtained by composition of the $\tilde{\mathcal{A}}_1$-action with the $\mathcal{H}$-action, to perform the analytic classification of plane branches, it is enough to analyze the $\mathcal{H}$-action on primitive parametrizations as in (1.25), getting the result [12, Theorem 2.1], below.

Theorem 1.3.17 (Normal Form Theorem) *Let C be a branch with semigroup of values Γ and set of values of differentials Λ. If $\Lambda \setminus \Gamma \neq \emptyset$, then C is analytically equivalent to a branch with parametrization*

$$\xi(t) = \left(t^n, t^m + a_\lambda t^\lambda + \sum_{\substack{i>\lambda \\ i \notin \Lambda - n}} a_i t^i\right),$$

where $n = \min \Gamma \setminus \{0\}$, $m = \min \Gamma \setminus \langle n \rangle$ and $\lambda = \min(\Lambda \setminus \Gamma) - n$. Otherwise C is analytically equivalent to a branch with parametrization $\xi(t) = (t^n, t^m)$.

Moreover, a branch parametrized by $\xi_1(t) = (t^n, t^m + b_\lambda t^\lambda + \sum_{\substack{i>\lambda \\ i \notin \Lambda - n}} b_i t^i)$ is analytically equivalent to C if, and only if, there exists $\alpha \in \mathbb{C}^*$ such that $b_i = \alpha^{i-m} a_i$, for all i.

Proof As observed, just before the statement of the theorem, this is a consequence of the composition of the actions of $\tilde{\mathcal{A}}_1$ and $\mathcal{H}$ on a parametrization $\xi(t)$ as in Theorem 1.3.14. □

1.4 Final Remarks

In this final section we will derive some consequences of the Normal Form Theorem and compare them with related results in the literature.

1.4.1 Comparison with Other Works

We believe that the conclusion of the Normal Form Theorem is what Zariski was expecting to achieve in general. To support our belief, we quote a result about the analytic classification of curves with semigroup $\Gamma = \langle n, n+1\rangle$ that Zariski proves in his course at the École Polytéchnique [23].

Theorem 1.4.1 ([24, Theorem 6.12 on p. 104]) *For $n \geq 5$ consider two primitive parametrizations with* generic *coefficients*

$$\xi_1(t) = \left(t^n, t^{n+1} + \sum_{i=n+3}^{2n-1} a_i t^i + \sum_{i\in\bigcup_{j=2}^{s} E_j} a_i t^i\right),$$

$$\xi_2(t) = \left(t^n, t^{n+1} + \sum_{i=n+3}^{2n-1} b_i t^i + \sum_{i\in\bigcup_{j=2}^{s} E_j} b_i t^i\right),$$

where $E_j = \{(j+1)n - (n-j-1), \ldots, (j+1)n - 1\}$ and $a_i = 0$ whenever i is one of the first $j+1$ element of E_j for all $2 \leq j \leq s$ with $s = \left[\frac{n-3}{2}\right]$. The parametrizations define branches analytically equivalent if and only if there exists $r \in \mathbb{C}^*$ such that $r^{i-(n+1)} a_i = b_i$ for all $i > n+1$.

Remark 1.4.2 We made a correction in the original statement by Zariski, where it is stated that $s = \left[\frac{n-4}{2}\right]$, but according to [23, Definition 6.10 and Remark 6.11], one should have $s = \left[\frac{n-3}{2}\right]$, as we stated above.

Zariski observed that Theorem 1.4.1 holds for $n = 5, 6$, without the hypothesis of genericity on the coefficients and asked the following question:

Does Theorem 6.12 remain true without assuming that $\xi_1(t)$ and $\xi_2(t)$ are generic?

The answer is no, as we show in the following example.

Example 1.4.3 Let us consider $\xi_1(t) = (t^7, t^8 + t^{10} + t^{11} + \frac{11}{4}t^{12} + at^{13} + b_1 t^{20})$ and $\xi_2(t) = (t^7, t^8 + t^{10} + t^{11} + \frac{11}{4}t^{12} + at^{13} + b_2 t^{20})$ with $a \neq \frac{21}{4}$ and $b_1 \neq b_2$, which are in the form of the above theorem. So, if Theorem 1.4.1 was true, without the genericity condition, the associated branches would not be analytically equivalent.

But, applying the algorithms in [13] with the restriction $a \neq \frac{21}{4}$, one finds $\Lambda \setminus \Gamma = \{17, 25, 27, 33, 34, 41\}$ and $d = 27$. In this way, any natural number greater or equal to $d - n = 20$ belongs to $\Lambda - n$. Consequently, the normal forms in Theorem 1.3.17 for $\xi_1(t)$ and $\xi_2(t)$ are the same, namely: $(t^7, t^8 + t^{10} + t^{11} + \frac{11}{4}t^{12} + at^{13})$. Hence, the associated branches are analytically equivalent, giving a negative answer to Zariski's question.

In [14] we give the analytic classification of all branches in the equisingularity class determined by the semigroup $\langle 7, 8\rangle$.

One possible consequence of our result is an easy way to characterize branches with $\tau = \mu - 1$, obtained in [3], as shown below.

Corollary 1.4.4 *A branch C_f satisfies $\tau(f) = \mu(f) - 1$ if, and only if, it is analytically equivalent to a branch with a parametrization $(t^n, t^m + t^{(n-1)m-2n})$ with* $\mathrm{GCD}(n, m) = 1$.

Proof Observe that, by Proposition 1.2.19, the result is equivalent to characterize the branches with $\Lambda \setminus \Gamma = \{\lambda + n\}$.

It is clear that if $\Gamma = \langle n, m\rangle$ and $\lambda = (n-1)m - 2n$, then $\Lambda \setminus \Gamma = \{\lambda + n\}$.

Conversely, recall that $v_1 < \lambda \leq \beta_2 = v_2 + m - m_1 n$, with $m_1 = \frac{m}{e_1} > n_1 \geq 2$, and $\lambda + n + \Gamma \subseteq \Lambda$.

If $\Gamma = \langle n, m, v_2, \ldots, v_g\rangle$ with $\lambda = \beta_2 = v_2 + m - m_1 n$, then, by Proposition 1.3.14, $\lambda + 2n = v_2 + m - (m_1 - 2)n \in \Lambda \setminus \Gamma$.

In this way, we must have $\lambda + n = s_1 m - s_0 n$ with $2 \leq s_1 \leq n_1 - 1$ and $1 \leq s_0$. If $g \geq 2$, then $v_2 + \lambda + n = v_2 + s_1 m - s_0 n \in \Lambda \setminus \Gamma$. It follows that the only possibility for the semigroup is $\Gamma = \langle n, m\rangle$.

If $s_1 < n_1 - 1$, then $\lambda + n + m = (s_1 - 1)m - s_0 n \in \Lambda \setminus \Gamma$. If $s_0 \geq 2$ then $\lambda + 2n = s_1 m - (s_0 - 1)n \in \Lambda \setminus \Gamma$.

So, if $\Lambda \setminus \Gamma = \{\lambda + n\}$, then $\Gamma = \langle n, m\rangle$ and $\lambda = (n-1)m - 2n$. By Proposition 1.3.14 and normalizing the coefficient a_λ as in (1.18), any natural number $k > (n-1)m - 2n$ is an element of $\Gamma - n \subseteq \Lambda - n$, then by the Normal Form Theorem we have the result. □

1.4.2 Computability

Let us observe that it is computationally manageable to obtain the normal form presented in (1.25) and to decide if two plane branches are analytically equivalent or not.

In fact, given a primitive Puiseux parametrization we can easily determine its Puiseux exponents $\beta_0, \ldots, \beta_g$ which, by Theorem 1.1.27, allows us to get the semigroup Γ. The four reductions we presented give us explicit elements in $\mathcal{A}$ that reduce our parametrization to the form $\xi(t) = (t^n, t^m + b_\lambda t^\lambda + \sum_{i>\lambda}^{c-n-1} b_i t^i)$.

Using the algorithm presented in [13], we may compute the set Λ and its conductor d. As we pointed out in Remark 1.3.15, it is sufficient to consider $j^{d-n-1}\xi(t)$. Moreover, with these data, we determine the elements less than $d - n$ in $\Lambda - n$ that correspond to terms that we can eliminate in $j^{d-n-1}\xi(t)$.

In order to do this, preserving the Puiseux form, we consider an element $(\varphi, \rho) \in \tilde{\mathcal{A}}_1$, as described in Corollary 1.2.4, that is, $q \in \langle X, Y\rangle^2$ and $p \in \langle X^2, Y\rangle$ with $\mathrm{ord}_t\, q(\xi(t)) < d - n$ and $\mathrm{ord}_t\, p(\xi(t)) < d - m$. In particular, it is sufficient to take $q = \sum_{in+jm<d-n} \alpha_{ij} X^i Y^j$ and $p = \sum_{in+jm<d-m} \beta_{ij} X^i Y^j$. As $j^{d-n-1}\xi(t)$, p and q are polynomials, the method to obtain the normal form is effective and computable. In [15] we give details about the computational implementation for the normal formal process and the analytic classification of plane branches.

1.4.3 A Solution for the Moduli Problem

As remarked in previous sections, fixing an equisingularity class $\mathcal{L}$, determined by a semigroup of values Γ, there exist a finite number of distinct possible value sets of Kähler differentials $\Lambda_1, \ldots, \Lambda_s$. In this way, we may stratify $\mathcal{L}$ by these analytic invariants, obtaining a disjoint union $\mathcal{L} = \bigcup_{j=1}^{s} \mathcal{L}_j$, where $\mathcal{L}_j$ indicates the set of all branches with semigroup Γ and value set of Kähler differentials Λ_j.

Let us consider the parameter space Σ_Γ consisting of all tuples formed with the coefficients of $j^{c-1}y(t)$ of Puiseux parametrizations $\xi(t) = (t^n, y(t))$, where c is the conductor of Γ and $\Gamma_\xi = \Gamma$. This set is a Zariski open set in a finite dimensional affine space. Let us denote by $\Sigma_\Gamma(\Lambda_j)$ the subset of Σ_Γ consisting of points representing elements in $\mathcal{L}_j$, hence we get a stratification

$$\Sigma_\Gamma = \bigcup_{j=1}^{s} \Sigma_\Gamma(\Lambda_j).$$

Each $\Sigma_\Gamma(\Lambda_j)$ is a constructible set in Σ_Γ that may be explicitly described via the algorithms developed in [13] and [15], which also show that one of these sets is Zariski open in Σ_Γ. The quotient of this set by the equivalence relation induced by analytic equivalence on branches will be called *the generic component of the moduli space*.

Inside each $\Sigma_\Gamma(\Lambda_j)$ there is a subset N_j representing elements in the normal form (1.25), which in turn is a constructible set in $\Sigma_\Gamma(\Lambda_j)$ and may be realized as a constructible set in the affine space $\mathbb{C}^{d_j}$, where $d_j = 1 + \sharp\{i;\ i > \lambda,\ i \notin \Lambda_j - n\}$.

So, if we define the function

$$\Phi : \bigcup_{j=1}^{s} N_j \to \mathcal{L}$$

that maps the coefficients of $y(t)$ to the branch corresponding to the Puiseux parametrization $(t^n, y(t))$, then, by the *Normal Form Theorem*, the image of Φ contains one representative of each analytic class in the equisingularity class $\mathcal{L}$ and analytic equivalence is translated into the $\mathbb{C}^*$-action on each N_j as described in

Theorem 1.3.17, offering in this way a solution to the moduli problem proposed by Zariski.

1.4.4 Dimensions of Components of the Moduli Space

The problem of determining the dimension of the generic component of the moduli space for branches in a given equisingularity class was addressed by several authors, starting with Zariski.

Zariski dedicated a substantial part of [23] to this problem in the particular case in which the curves have semigroup of values $\langle n, nk+1\rangle$. Some time later, C. Delorme studied in [8] the case of branches with semigroup $\langle n, m\rangle$, giving a closed formula for the dimension of the generic component of the moduli space in this case. In [20], R. Peraire presented an algorithm to compute the dimension of the generic component of the moduli space for curves with semigroup $\langle n, m\rangle$ and a fixed Zariski λ-invariant. In [10], Y. Genzmer gave a closed formula to compute the dimension of the generic component of the moduli space for branches in any fixed equisingularity class.

The Normal Form Theorem gives us the description of any stratum of the moduli space determined by an analytic invariant Λ, its structure and, in particular, its dimension, not only for the generic component.

In fact, to compute the dimension of each stratum it is sufficient to fix a semigroup Γ and to consider a generic primitive parametrization $\xi(t)$ as in (1.12). By the algorithms developed in [13], one can describe all possible N_j and determine all algebraic restrictions on their points in order to represent branches in $\mathcal{L}_j$. So, the dimension of $N_j/\mathbb{C}^*$, which is equal to the dimension of N_j minus one (due to the normalization $a_\lambda = 1$) is the number of elements in $\{i;\ i > \lambda,\ i \notin \Lambda_j - n\}$ corresponding to free parameters in the set of algebraic restrictions that determine Λ_j.

Now, using the above discussion and the *Normal Form Theorem*, we may characterize the equisingularity classes that have zero-dimensional moduli space.

Proposition 1.4.5 *The moduli space for branches in an equisingularity class determined by $\Gamma = \langle v_0, v_1, \ldots, v_g\rangle$, where $v_0 = n$ and $v_1 = m$, is zero-dimensional if and only if Γ is one of the following semigroups:*

$$\mathbb{N};\ \langle 2, m\rangle;\ \langle 3, m\rangle;\ \langle 4, m\rangle,\ m = 5, 7 \text{ or } \langle 4, 6, v_2\rangle.$$

Moreover, for $n \leq 2$ or for $\langle 4, 6, v_2\rangle$, the moduli space is a single point; for $n = 3$, it is a set with $\left[\frac{m}{3}\right]$ isolated points and for $n = 4$ and $m = 5, 7$ the moduli space consists of $\left[\frac{m}{2}\right]$ isolated points.

Proof If $g > 2$, we have necessarily $\beta_3 > \lambda$ and $\beta_3 \notin \Lambda - n$, because otherwise, by the Normal Form Theorem, β_3 would be eliminable, provoking a change in the equisingularity class. So, for these semigroups we have parametrizations of branches with at least one free parameter a_{β_3} and, consequently, the moduli space admits a stratum with dimension greater than 0. Observe that for $g = 2$ and $\beta_2 > \lambda$ the same argument works.

Suppose now that $g = 2$ and $\lambda = \beta_2$. Then from (1.7) one has $\lambda = v_2 + v_1 - n_1 v_1 = v_2 + v_1 - m_1 v_0$, with $m_1 = \frac{m}{e_1} > n_1 \geq 2$. Taking $k = \left[\frac{(n_2-2)v_2+(n_1-1)m}{n}\right]$, we have that $\lambda < (n_2 - 1)v_2 - kn < \lambda + n$. If either $n_2 > 2$ or $n_1 > 2$ or $m > 2n$, then $k \geq 2$ and $i = (n_2 - 1)v_2 - kn \notin \Gamma - n$. By Proposition 1.2.22, we conclude that $i = (n_2 - 1)v_2 - kn \notin \Lambda - n$. In this way, we have parametrizations with a free parameter a_i, so there is a stratum of the moduli space with positive dimension. So, for $g = 2$, to have a zero dimensional moduli space, we must have $n_1 = n_2 = 2$ and $m < 2n$; that is, $\Gamma = \langle 4, 6, v_2 \rangle$. In this case, $\Lambda \setminus \Gamma = \{\lambda + 4, \lambda + 8\}$, and any integer greater than $\lambda = v_2 - 6$ belongs to $\Lambda - 4$. By the Normal Form Theorem, any curve with semigroup of values $\Gamma = \langle 4, 6, v_2 \rangle$ is equivalent to a curve given by a parametrization $(t^4, t^6 + t^{v_2 - 6})$ and consequently the moduli space is a single point.

Now, as we observed in Remark 1.2.10, all equisingular branches with $n = 1$, $n = 2$, or $n = 3$ and $4 \leq m \leq 5$ are analytically equivalent, so for such semigroups the moduli space reduces to a single point.

It remains to consider the equisingularity classes determined by semigroups $\Gamma = \langle n, m \rangle$ with $n \geq 3$.

If $n>4$, then $\lambda = 3m - 2n$ is a Zariski invariant. Taking $i = 4m - ([m/n + 2])\, n$, we have that $i \notin \Gamma - n$ and $\lambda < i < \lambda + n$. As before, Proposition 1.2.22 give us that $i \notin \Lambda - n$ and there exist parametrizations with a free parameter a_i and, consequently, a stratum of the moduli space with positive dimension. We have the same conclusion for $n = 4$ and $m \geq 9$ by considering $\lambda = 2m - 8$ and $i = 3m - ([m/4 + 2])\, 4$.

If $n = 4$ and $m \in \{5, 7\}$, then the only possibilities for λ are $3m - 4k$ with $2 \leq k \leq \left[\frac{m}{2}\right]$. For each of these possibilities, any natural number greater than λ is an element of $\Lambda - 4$. By the Normal Form Theorem every curve with semigroup $\Gamma = \langle 4, m \rangle$ with $m = 4, 7$ is analytically equivalent to a branch with parametrization (t^4, t^m) or $(t^4, t^m + t^\lambda)$, that is, we have $\left[\frac{m}{2}\right]$ possible analytic classes.

Finally, for $n = 3$ and $m > 6$, the only possible Zariski invariants are $2m - 3k$ with $2 \leq k \leq \left[\frac{m}{3}\right]$. As before, the normal forms are (t^3, t^m) or $(t^3, t^m + t^\lambda)$. So, we have $\left[\frac{m}{3}\right]$ possible analytic classes. □

The above result was obtained, by other methods, by J.W. Bruce and T.J. Gaffney in [6].

Given now a semigroup of values Γ, we denote by Λ_Γ the analytic invariant corresponding to the generic component of the moduli space of the equisingularity class $\mathcal{L}$ determined by Γ. The next result will show how to determine easily the dimension of the generic component of its moduli space.

Theorem 1.4.6 *The dimension of the generic component of the moduli space of the equisingularity class determined by a semigroup of values Γ is equal to*

$$\sharp\{i;\ i > \lambda,\ i \notin \Lambda_\Gamma - n \text{ and } e_j \mid i \text{ if } i \geq \beta_j,\ 1 \leq j \leq g\}, \tag{1.26}$$

where $\lambda = \min \Lambda_\Gamma - n$, $n = \min \Gamma \setminus \{0\}$, $n, \beta_1, \ldots, \beta_g$ *are the characteristic exponents of any Puiseux parametrization* ξ *such that* $\Gamma_\xi = \Gamma$ *and* $e_j = \mathrm{GCD}(n, \beta_1, \ldots, \beta_j)$.

Proof In fact, because $\Sigma_\Gamma(\Lambda_\Gamma)$ is an open Zariski set in Σ_Λ and in view of Theorem 1.3.17, the dimension of the generic component of the moduli space is precisely the number of coefficients a_i in a generic parametrization of the form $(t^n, t^m + t^\lambda + \sum_{\lambda<i\notin\Lambda_\Gamma-n} a_i t^i)$, provided $a_i = 0$ if $\beta_j < i$ and $e_j \nmid i$ for $1 \leq j \leq g$. From this, the result follows. □

In a similar way, if we choose an admissible Zariski λ-invariant, that is, any element in $\{i;\ i > m \text{ and } i \notin \Gamma - n\}$ and consider a primitive parametrization as in (1.18), with generic coefficients, we can compute the generic Λ for branches with such Zariski invariant and get the same expression as in (1.26) for the dimension of the generic component with fixed λ. This is a generalization of the result obtained by Peraire in [20], where she considers the case $g = 1$.

Notice that in Theorem 1.4.6, for branches with semigroup of genus one, that is, $\Gamma = \langle n, m\rangle$, we do not have the restrictions described in (1.4), so, the expression (1.26) reduces to $\sharp\{i,\ \lambda < i \notin \Lambda_\Gamma - n\}$. In this situation we recover the result [18, Theorem 5.1].

Corollary 1.4.7 *The dimension of the generic component of the moduli space for branches with semigroup* $\Gamma = \langle n, m\rangle$ *is* $l(n, m) - c + \tau_{min}$, *where* $c = (n-1)(m-1)$ *is the conductor of* Γ, τ_{min} *is the least Tjurina number for branches with semigroup* Γ *and* $l(n, m) = \sharp\{i,\ m < i \notin \Gamma - n\}$.

Proof The dimension of the generic component of the moduli space is the largest dimension of all strata, in which case $\sharp\Lambda_\Gamma \setminus \Gamma$ is maximum in the equisingularity class and, consequently, by Proposition 1.2.19 and Theorem 1.2.15, the dimension of the stratum with minimum Tjurina number τ_{min} is $\sharp\Lambda_\Gamma \setminus \Gamma = c - \tau_{min}$.

Now, since $\Gamma = \langle n, m\rangle$, the dimension of the generic component of the moduli space is the cardinality of the set

$$\{i,\ \lambda < i \notin \Lambda_\Gamma - n\} = \{i,\ m < i \notin \Gamma - n\} \setminus \{i,\ m < i \notin \Lambda_\Gamma - n \setminus \Gamma - n\}.$$

But, since

$$\sharp\{i,\ m < i \notin \Lambda_\Gamma - n \setminus \Gamma - n\} = \sharp\Lambda_\Gamma \setminus \Gamma,$$

the result follows. □

In the same way, if we consider the least Tjurina number for branches with semigroup $\Gamma = \langle n, m\rangle$ and Zariski λ-invariant fixed, we obtain a similar formula.

1.4.5 An Example

To illustrate our results, we will describe here all normal forms for branches in the equisingularity class determined by the semigroup $\Gamma = \langle 4, 11 \rangle$. This example is a particular case of the entire classification of singularities of multiplicity 4 to be found in [16].

The table below exhibits the normal forms with the respective algebraic restrictions on their coefficients to ensure that they possess the given $\Lambda \setminus \Gamma$ invariant.

Restrictions	Normalized normal form	$\Lambda \setminus \Gamma$
	$(t^4, t^{11} + t^{13} + a_{14}t^{14})$	$\{17, 21, 25, 29\}$
$a_{17} \neq \frac{25}{22}$	$(t^4, t^{11} + t^{14} + a_{17}t^{17})$	$\{18, 25, 29\}$
$a_{17} = \frac{25}{22}$	$(t^4, t^{11} + t^{14} + a_{17}t^{17} + a_{21}t^{21})$	$\{18, 29\}$
	$(t^4, t^{11} + t^{17})$	$\{21, 25, 29\}$
	$(t^4, t^{11} + t^{21})$	$\{25, 29\}$
	$(t^4, t^{11} + t^{25})$	$\{29\}$
	(t^4, t^{11})	$\emptyset$

The above table shows that the moduli space of the equisingularity class determined by $\Gamma = \langle 4, 11 \rangle$ has seven strata, the first row corresponds to the generic component of the moduli that has dimension one. The next two rows give also one-dimensional strata but non-generic, and the last four correspond to zero-dimensional strata.

1.4.6 Analytic Versus Formal

It is a fact, as we will see soon, that many of the properties of C_f may be studied algebraically, without any reference to convergence problems by considering f as an element of $\mathbb{C}[[X, Y]]$, the ring of formal power series in two indeterminates with coefficients in $\mathbb{C}$, viewed only as an algebraically closed field without any reference to its topology. Although we loose the geometric interpretation for C_f, which is useless for our purpose, we gain a remarkable simplification of the theory and at the same time the possibility to extend the related problems to a wider context, for instance, for curves over any algebraically closed ground field.

The ring $\mathbb{C}[[X, Y]]$ shares most of the algebraic properties of the ring $\mathbb{C}\{X, Y\}$ as, for instance, the fact that it is a local ring with maximal ideal $\mathcal{M} = \langle X, Y \rangle$ and is a unique factorization domain.

By an *algebroid plane curve* we mean $C_f = \text{Spec}\, R(f)$, where $R(f) = \mathbb{C}[[X, Y]]/\langle f \rangle$ for some $f \in \mathcal{M}$. Now, in contrast with the analytic case, C_f has only two points: the closed point 0, corresponding to the maximal ideal $\overline{\mathcal{M}}$ and the generic point η corresponding to the zero ideal $\overline{\langle 0 \rangle}$. Since in the analytic case, the ring $\mathcal{O}(f)$ carries all the analytic informations about C_f, it is natural to concentrate on the study of the ring $R(f)$ and to define the relation

$$C_f \sim C_g \iff R(f) \simeq R(g),$$

where $\simeq$ represents isomorphism as $\mathbb{C}$-algebras.

On the other hand, the topological equivalence of C_f and C_g in the analytic case may be replaced in the formal context by equisingularity relation as follows:

$$C_f \equiv C_g \iff \Gamma(f) = \Gamma(g),$$

where $\Gamma(f)$ and $\Gamma(g)$ are defined as in the analytic setting.

In this more general context, Puiseux parametrizations exist, but are given by formal series. Everything we did in these notes works as well in this formal setting, because all series involved are finitely determined with respect to the associated group actions, in such a way that we only have to work with polynomials.

At the contrary of what one could expect, the theory one gets in this new framework is not more general than in the analytic one, as remarked by Zariski in [24, p. 3].

Acknowledgements This work was partially supported by grants from CNPq. The authors wish to thank José Seade for inviting them to contribute to this volume and the anonymous referee for his careful reading of the manuscript and for his comments.

References

1. Abhyankar, S.S. and Moh, T., *Newton-Puiseux expansion and generalized Tschirnhausen transformation*, J. Reine Angew. Math. 260, 47–83 (1973) and 261, 29–54 (1973).
2. Apéry, R., *Sur les branches superlinéaires de courbes algébriques*, C. R. Math. Acad. Sci. Paris. 222, 1198–1200 (1946).
3. Bayer, V. and Hefez, A., *Algebroid plane curves whose Milnor and Tjurina numbers differ by one or two*, Bull. Braz. Math. Soc. (N.S) 32(1), 63–81 (2001).
4. Borel, A., *Linear algebraic groups* in ALGEBRAIC GROUPS AND DISCONTINUOUS SUBGROUPS, Proceedings of Symposia in Pure Mathematics Volume 9, Armand Borel and George D. Mostow Editors (1966).
5. Brieskorn, E. and Knörrer, H., *PLANE ALGEBRAIC CURVES*, Birkhäuser Verlag Basel (1986).
6. Bruce, J.W. and Gaffney, T.J., *Simple singularities of mappings* $\mathbb{C}, 0 \longrightarrow \mathbb{C}^2, 0$, J. Lond. Math. Soc. (2) 26, 465–474 (1982).
7. Bruce, J. W., Kirk, N. P. and du Plessis, A. A., *Complete transversals and the classification of singularities*, Nonlinearity, 10, N.1, 253–275 (1997).
8. Delorme, C., *Sur les modules des singularités de courbes planes*, Bull. Soc. Math. France 106, 417-446 (1978).
9. Ebey, S., *The classification of singular points of algebraic curves*, Trans. Amer. Math. Soc. 118, 454–471 (1965).
10. Genzmer, Y., *Dimension of the moduli space of a germ of curve in* $\mathbb{C}^2$, International Mathematics Research Notices, 2020, rnaa237, https://doi.org/10.1093/imrn/rnaa237.
11. Hefez, A., *Irreducible plane curve singularities* in REAL AND COMPLEX SINGULARITIES, Dekker Series of Lecture Notes in Pure and Applied Mathematics, Volume 232, D. Mond and M. J. Saia, Editors (2003).
12. Hefez, A.; Hernandes, M. E., *The analytic classification of plane branches*, Bull. Lond. Math. Soc. Vol. 43 (2), 289–298 (2011).

13. Hefez, A. and Hernandes M. E., *Standard bases for local rings of branches and their modules of differentials*, J. Symbolic Comput. **42**, 178–191 (2007).
14. Hefez, A. and Hernandes M. E., *Answer to a question of Zariski* in REAL AND COMPLEX SINGULARITIES, Cambridge University Press, M. Manoel, M. C. Romero Fuster and C. T. C. Wall, Editors (2010).
15. Hefez, A. and Hernandes, M. E., *Algorithms for the implementation of the analytic classification of plane branches*, J. Symbolic Comput. **50**, 308–313 (2013).
16. Hefez, A.; Hernandes, M. E., *The analytic classification of plane branches up to multiplicity 4*, J. Symbolic Comput. **44**, 626–634 (2009).
17. Jong, Th. de and Pfister, G., *LOCAL ANALYTIC GEOMETRY-Basic Theory and Applications*, Advanced Lectures in Mathematics, Springer (2000).
18. Laudal, O.A. and Pfister, G., *LOCAL MODULI AND SINGULARITIES*, Lecture Notes in Math., 1310 Springer-Verlag, Berlin (1988).
19. Mather, J. N., *Stability of C^∞ mappings, IV: Classification of stable germs by R-algebras*, Pub. Math. IHES, 37, 223–248 (1969).
20. Peraire, R., *Moduli of plane curve singularities with a single characteristic exponent*, Proc. Amer. Math. Soc. 126 (1), 25–34 (1998).
21. Popescu-Pampu, P. *Approximate roots*, Fields Institute Communications, vol 33 (2003).
22. Rosenlicht, M., *On quotient varieties and the afine embedding of certain homogeneous spaces*, Trans. Amer. Math. Soc. 101, 211–223 (1961).
23. Zariski, O., *LE PROBLÈME DES MODULES POUR LES BRANCHES PLANES*, Cours donné au Centre de Mathématiques de L'École Polytechnique (1973).
24. Zariski, O., *THE MODULI PROBLEM FOR PLANE BRANCHES*, University Lecture Series, Vol. 30, AMS (2006).
25. Zariski, O., *Characterization of plane algebroid curves whose module of differentials has maximum torsion*, Proc. Nat. Acad. of Sc. U.S.A. 56, 781–786 (1966).
26. Zariski O., *Some open questions in the theory of equisingularities*, Bull. Amer. Math. Soc. 77, 481–491 (1971).
27. Zariski O., *General theory of saturation and saturated local rings II: Saturated local rings of dimension 1*, Amer. J. Math. 93 (4), 872–964 (1971).

Chapter 2
Plane Algebraic Curves with Prescribed Singularities

Gert-Martin Greuel and Eugenii Shustin

Abstract We give a survey on the known results about the problem of the existence of complex and real algebraic curves in the plane with prescribed singularities up to analytic and topological equivalence. The question is whether, for a given positive integer d and a finite number of given analytic or topological singularity types, there exist a plane (irreducible) curve of degree d having singular points of the given type as its only singularities. The set of all such curves is a quasiprojective variety, which we call an equisingular family, denoted by ESF. We describe, in terms of numerical invariants of the curves and their singularities, the state of the art concerning necessary and sufficient conditions for the non-emptiness and T-smoothness (i.e., being smooth of expected dimension) of the corresponding ESF. The considered singularities can be arbitrary, but we pay special attention to plane curves with nodes and cusps, the most studied case, where still no complete answer is known in general. A n important result is, however, that the necessary and the sufficient conditions show the same asymptotics for T-smooth equisingular families if the degree goes to infinity.

Contents

G.-M. Greuel (✉)
Universität Kaiserslautern, Kaiserslautern, Germany

E. Shustin
Tel Aviv University, Tel Aviv, Israel
e-mail: shustin@tauex.tau.ac.il

J. L. Cisneros-Molina et al. (eds.), *Handbook of Geometry and Topology of Singularities II*, https://doi.org/10.1007/978-3-030-78024-1_2

2.1 Introduction

Singular algebraic curves, their existence, deformation, families (from the local and global point of view) attract continuous attention of algebraic geometers since the last century. The aim of this survey is to give an account of results, trends and bibliography related to the existence of curves with prescribed singularities with a focus on algebraic curves in the plane. We consider the existence problem for complex and real plane curves with given singularities up to analytic and topological equivalence. The general problem is: given an integer $d > 0$ and analytic or topological singularity types $S_1, \ldots, S_r$, does there exist a curve (resp. an irreducible curve) of degree d in $\mathbb{P}^2$ having r singular points of types $S_1, \ldots, S_r$, respectively, as its only singularities?

An important particular case is the same problem for one singularity. Namely, let S be an analytic or topological type. What is the minimal degree $d(S)$ of a curve in $\mathbb{P}^2$ having a singular point of type S? In other words, we ask about a polynomial normal form of minimal degree of the given singularity.

The space $|dH| = |H^0(\mathbb{P}^2, \mathcal{O}_{\mathbb{P}^2}(d))|$ of all curves of degree d in $\mathbb{P}^2$, H a hyperplane in $\mathbb{P}^2$, can be identified with the punctured vector space of homogeneous polynomials of degree d in 3 variables modulo multiplication with a non-zero constant. That is, $|dH| = \mathbb{C}[x_0, x_1, x_2]_d \setminus \{0\}/\mathbb{C}^*$ is a projective space of dimension $N = (d^2 + 3d)/2$. The subspace of this $\mathbb{P}^N$, consisting of (irreducible) curves of degree d in $\mathbb{P}^2$ having r singular points of types $S_1, \ldots, S_r$ (and maybe other unspecified singularities) is the *Equisingular Family* (ESF) which we denote by

$$V_d^{(irr)}(S_1, \ldots, S_r)$$

(it may be empty). This description of $V_d^{(irr)}(S_1, \ldots, S_r)$ is set-theoretically, but it is shown in [28] that these sets are quasi-projective subvarieties of $\mathbb{P}^N$ (see [28, Propositions I 1.61 and I 1.71] for a simple proof in the case of one singularity), which can be endowed with a unique (not necessarily reduced) scheme structure representing the functor of equianalytic resp. equisingular deformations (see [28, Theorem II 2.36]).

The following geometric problems concerning equisingular families of plane curves have been of interest to algebraic geometers since the early 20th century:

- *Existence Problem*: Is $V_d^{(irr)}(S_1, \ldots, S_r)$ non-empty?
- *T-Smoothness Problem*: If $V_d^{(irr)}(S_1, \ldots, S_r)$ is non-empty, is it T-smooth, i.e. smooth and of the "expected" dimension (see end of the Preliminaries)?
- *Irreducibility Problem*: Is $V_d^{(irr)}(S_1, \ldots, S_r)$ irreducible?
- *Deformation Problem*: What are the adjacencies of the singularities of a curve of degree d if it varies inside $|dH|$?

First of all, a *complete* answer to these questions is known only for the case of plane nodal curves (Severi [73], Harris [37]): the inequality $0 \le n \le \frac{(d-1)(d-2)}{2}$ is necessary and sufficient for the nonemptiness, T-smoothness, and irreducibility of the family $V_d^{irr}(nA_1)$ of irreducible plane curves of degree d with n nodes as their only singularities, and, additionally, for the independent smoothing of prescribed nodes while keeping the others, induced by the space of plane curves of degree d.

Already for plane curves with ordinary cusps a reasonable complete answer is hardly possible, due to a large gap between the known upper bounds of the number of cusps and the known examples of curves with many cusps. Due to the irregular behavior of such examples, it seems unrealistic to expect a sufficient condition for either non-emptiness, or T-smoothness, or irreducibility, which is at the same time necessary (as in the case of plane nodal curves).

This situation has motivated us to pursue the following goal: describe the *regular region* of $V_d^{(irr)}(S_1, \ldots, S_r)$ (i.e. the nonempty and T-smooth part), in a possibly precise form, which should be

(i) *universal*, i.e. applicable to arbitrary singularities,
(ii) *numerical*, i.e. expressed as relations (inequalities) for numerical invariants of the curves and their singularities,
(iii) *asymptotically optimal* or *asymptotically proper*, i.e. having either the same asymptotics or an asymptotics that coincides up to multiplication with a positive constant with the known examples of irregular (empty or non-T-smooth) equisingular families if d goes to infinity.

We like to emphasize that one can expect asymptotically optimal or asymptotically proper results (about nonemptiness, T-smoothness, irreducibility, ...) only for the regular region; we do not see any systematic behavior for the irregular region of $V_d^{(irr)}(S_1, \ldots, S_r)$ if $d \to \infty$.

In this survey we focus mainly on the existence problem and give only a short account on answers to the other problems. We give always precise references, including original sources and in addition hints to the methods whenever appropriate. We feature both complex and real singular curves. A special attention is paid to curves with nodes and cusps, curves with simple, ordinary, and semi-quasihomogeneous singularities, in which cases one can apply specific constructions and formulate general restrictions in a simpler form.

In general, there is only one universal approach which provides sufficient existence results for arbitrary topological and analytical singularity types and any degree, both over the complex and over the real fields, and which is asymptotically comparable with the necessary conditions. This approach combines two main ingredients: the *theory of zero-dimensional schemes* related to planar curve singularities coupled to the cohomology vanishing theory for their ideal sheaves, and the *patchworking construction*. While the cohomological approach, which builds a bridge between the local and global geometry of singular algebraic curves, is not treated in this survey, we explain the parchworking method in several interesting situations. Furthermore, we mention important results on the existence of curves with nodal singularities on other algebraic surfaces and in the projective space, and address several related problems.

For a comprehensive treatment of these problems and detailed proofs, and more generally of the theory of topologically and analytically equisingular families of curves on surfaces, see the monograph [31]. In the present survey, we basically follow the main thread of the monograph [31] providing more details in certain places, for instance, in Sect. 2.4.1 as well as in Sect. 2.4.2, where Theorems 2.4.5, 2.4.6 and 2.4.7 are new.

2.1.1 Preliminaries: Isolated Singularities

We work mainly with algebraic varieties (not necessarily reduced or irreducible) but use the Euclidean topology and analytic structure sheaf (unless otherwise stated). For this reason we call them *algebraic complex spaces* (see [31, Notations and Conventions] for a precise definition). An *algebraic curve* resp. *algebraic surface* means an algebraic complex space of pure dimension one resp. two. By a *real algebraic variety* resp. *real analytic variety* we mean an algebraic resp. analytic variety equipped with an anti-holomorphic involution. By a *hypersurface* we mean an effective Cartier divisor in a smooth variety Σ.

A *singularity* is by definition the germ (X, z) of a complex space, may be smooth. A singularity (X, z) is *isolated* if $X \setminus \{z\}$ is smooth for some representative X. Two hypersurface singularities $(X, z) \subset (\Sigma, z)$ and $(X', z') \subset (\Sigma, z')$ are called *analytically equivalent* (resp. *topologically equivalent*) if there exists an analytic isomorphism (resp. a homeomorphism) of neighborhoods of z resp. z' in Σ mapping (X, z) to (X', z').

The analytic equivalence can be expressed as an isomorphism of the analytic local rings: $\mathcal{O}_{X,z} \cong \mathcal{O}_{X',z'}$. The topological equivalence is used in this paper only for reduced plane curve singularities where it is completely characterized by discrete invariants (see [6, 28, 90, 97, 107]): Namely, two reduced plane curve singularities (C, z) and (C', z') are *topologically equivalent* iff there exists a bijection of local branches such that the Puiseux pairs of the corresponding branches coincide, as well as the pairwise intersection multiplicities of the corresponding branches; equivalently if they have embedded resolutions by blowing up points such that the systems of multiplicities of the reduced total transforms coincide. The second definition is the preferred one since it generalizes to deformations over non-reduced base spaces.

Analytic resp. topological equivalence classes of isolated singular points are called *(contact) analytic types* resp. *topological types* (or *analytic* resp. *topological singularities*). For *simple* or *ADE* singularities (cf. [28]) analytic and topological types coincide and we talk simply about their type. Of particular interest are the simple singularities of type A_1, called *nodes* , given in local analytic coordinates as $x^2 + y^2 = 0$ and of type A_2, called *(ordinary) cusps* , given as $x^2 + y^3 = 0$.

Important numerical invariants are the Milnor number, the delta invariant and the kappa-invariant. Let $(X, z) \subset (\Sigma, z) \cong (\mathbb{C}^n, 0)$ be an isolated hypersurface singularity and $f \in \mathbb{C}\{x_1, \ldots, x_n\} \cong \mathcal{O}_{\Sigma,z}$ a defining power series in local coordinates $x_1, \ldots, x_n$. Then

$$\mu(X, z) := \dim_{\mathbb{C}} \mathbb{C}\{x_1, \ldots, x_n\}/\langle \frac{\partial f}{\partial x_1}, \ldots, \frac{\partial f}{\partial x_n}\rangle$$

is the *Milnor number* of (X, z) and

$$\tau(X, z) := \dim_{\mathbb{C}} \mathbb{C}\{x_1, \ldots, x_n\}/\langle f, \frac{\partial f}{\partial x_1}, \ldots, \frac{\partial f}{\partial x_n}\rangle$$

is the *Tjurina number* of (X, z), which is the dimension of the base space of the semiuniversal deformation of (X, z).

For a reduced curve singularity (C, z) we call

$$\delta(C, z) := \dim_{\mathbb{C}}(\nu_* \mathcal{O}_{\overline{C}}/\mathcal{O}_C)_z$$

the *delta-invariant (δ-invariant)* of (C, z), where $\nu : \overline{C} \to C$ is the normalization of a representative C of (C, z). Let (C, z) be a reduced plane curve singularity defined by $f \in \mathbb{C}\{x, y\}$. The *kappa-invariant (κ-invariant)* of (C, z) is the intersection multiplicity of (C, z) with a generic polar, that is,

$$\kappa(C, z) := \dim_{\mathbb{C}} \mathbb{C}\{x, y\}/\langle f, \alpha\frac{\partial f}{\partial x} + \beta\frac{\partial f}{\partial y}\rangle, \tag{2.1}$$

with $(\alpha : \beta) \in \mathbb{P}^1$ generic. We also write $\mu(f)$, $\delta(f)$ and $\kappa(f)$. Recall for a plane curve singularity f the formulas (cf. [56] and [28, Propositions I. 3.35 and I. 3.38])

$$\mu(f) = 2\delta(f) - r(f) + 1,$$
$$\kappa(f) = \mu(f) + \mathrm{mt}(f) - 1,$$

where $r(f) = r(C, z)$ is the *number of branches* of (C, z) (irreducible factors of f) and $\mathrm{mt}(f) = \mathrm{mt}(C, z)$ the *multiplicity* of (C, z) (degree of lowest non-vanishing term of f).

We introduce further the *tau-es-invariant (τ^{es}-invariant)*

$$\tau^{es}(C, z) := \tau(C, z) - \dim_{\mathbb{C}} T^{1,es}(C, z) = \dim_{\mathbb{C}} \mathcal{O}_{\Sigma,z}/I^{es}(f),$$

with $I^{es}(f)$ the equisingularity ideal (cf. [31, Definition 1.1.63]) and $T^{1,es}(C, z)$ the tangent space to the equisingular stratum (= the μ-constant stratum) Δ^{μ} in the base of the semiuniversal deformation of (C, z). Since Δ^{μ} is smooth, $\tau^{es}(C, z)$ is equal to the codimension of the μ-constant stratum in the (τ-dimensional) base space of the semiuniversal deformation of (C, z), which coincides with the codimension of the μ-constant stratum in the (μ-dimensional) base space of the semiuniversal unfolding of f. We have also (cf. [24, Lemma 1.3])

$$\tau^{es}(C, z) = \mu(C, z) - \text{ modality}(f),$$

where *modality*(f) is the modality of the function f with respect to right equivalence. Note that $\tau^{es}(C, z)$ can be effectively computed in terms of the resolution invariants of (C, z)), an algorithm is implemented in SINGULAR [15]. For details we refer to [28, Remark to Corollary II.2.71] and to [31, Corollary 1.1.64].

Now we can explain more precisely the *T-smoothness property*. Let S be an analytic resp. topological singularity type of a plane curve singularity (C, z). The requirement that a curve of degree d has a singularity of type S imposes $\tau(S) := \tau(C, z)$ resp. $\tau^{es}(S) := \tau^{es}(C, z)$ conditions on the space of all curves of degree d (cf. [23] for anaytic types and [24] for toplogical types). Let S_1, ..., S_q be analytic types and S_{q+1}, ..., S_r topolgical types of singularities of the degree d-curve $C \subset \mathbb{P}^2$. Then we say that $V_d^{(irr)}(S_1, \ldots, S_r)$ has the *expected dimension* at C if its dimension at C is

$$\frac{d^2 + 3d}{2} - \sum_{i=1}^{q} \tau(S_i) - \sum_{i=q+1}^{r} \tau^{es}(S_i), \tag{2.2}$$

and $V_d^{(irr)}(S_1, \ldots, S_r)$ is *T-smooth at C* if it is smooth of expected dimension at C (in particular, the number (2.2) must be non-negative). We refer to [23, Corollary 6.3], [24, Theorem 3.6], and [31, Theorem 2.2.40] for this and for further properties of $V_d^{(irr)}(S_1, \ldots, S_r)$.

2.2 Singular Plane Curves: Restrictions

Various restrictions for the existence of plane curves of degree d with prescribed singularities $S_1, \ldots, S_r$ have been found. We recall the most important ones.

2.2.1 Genus Formula and Bézout's Theorem

First, one should mention the general classical bound

$$\sum_{i=1}^{r} \delta(S_i) \leq \frac{(d-1)(d-2)}{2}, \tag{2.3}$$

for the existence of an irreducible plane curve of degree d having r singularities of types $S_1, \ldots, S_r$, which results from the genus formula (2.9).

For a reduced (not necessarily irreducible) plane curve we get as necessary bound for the existence, i.e., for $V_d(S_1, \ldots S_r) \neq \emptyset$, the inequality

$$\sum_{i=1}^{r} \mu(S_i) \leq (d-1)^2. \tag{2.4}$$

This is a consequence of *Bézout's theorem* (see e.g. [31, Theorem II. 1.16]):

Two plane projective curves C, $D \subset \mathbb{P}^2$ of degrees c and d, respectively, which have no component in common, intersect at $c \cdot d$ points, counting intersection multiplicities. That is,

$$c \cdot d = \sum_{z \in C \cap D} \dim_{\mathbb{C}} \mathcal{O}_{\mathbb{P}^n, z}/\langle f, g\rangle, \tag{2.5}$$

with f resp. g being local equations of C resp. D at z.

To see (2.4) let C be given by a homogeneous polynomial $F \in \mathbb{C}[x_0, x_1, x_2]$ of degree d and let $F'_\alpha = \sum_{i=0}^{2} \alpha_i \partial F/\partial x_i$ and $F'_\beta = \sum_{i=0}^{2} \beta_i \partial F/\partial x_i$ with α_i, β_i generic, define two generic polars of C, both of degree $d-1$. The intersection points of $\{F'_\alpha = 0\}$ and $\{F'_\beta = 0\}$ include the singular points of C and the intersection multiplicities are just the corresponding Milnor numbers. Thus, we get (2.4).

For the proof of (2.3) let us recall first two genus formulae. The *arithmetic genus* of an arbitrary projective scheme X is defined as

$$p_a(X) := (-1)^{\dim X}(\chi(\mathcal{O}_X) - 1).$$

Here, for any coherent sheaf $\mathcal{F}$ on X, $\chi(\mathcal{F}) = \sum(-1)^i \dim_{\mathbb{C}} H^i(X, \mathcal{F})$ is the *(algebraic) Euler characteristic* of $\mathcal{F}$.

For a curve C we have $p_a(C) = 1 - \chi(\mathcal{O}_C) = 1 - \dim_{\mathbb{C}} H^0(C, \mathcal{O}_C) + \dim_{\mathbb{C}} H^1(C, \mathcal{O}_C)$. If C is reduced and connected, then we have $H^0(C, \mathcal{O}_C) = \mathbb{C}$ and hence we get for the arithmetic genus $p_a(C) = \dim_{\mathbb{C}} H^1(C, \mathcal{O}_C) \geq 0$. If C has s connected components C_1, ..., C_s, the additivity of the Euler characteristic implies $p_a(C) = 1 - s + \dim_{\mathbb{C}} H^1(C, \mathcal{O}_C) = 1 - s + \sum_{i=1}^{s} p_a(C_i)$, which may be negative for $s > 1$.

The *geometric genus* $g(C)$ of a reduced curve C is defined as the arithmetic genus of the normalization $\overline{C}$ of C, hence

$$g(C) := p_a(\overline{C}) = p_a(C) - \delta(C), \tag{2.6}$$

with $\delta(C) := \dim_{\mathbb{C}} H^0(\nu_* \mathcal{O}_{\overline{C}}/\mathcal{O}_C)$ the *total delta invariant of* C and $\nu : \overline{C} \to C$ the normalization map. (2.6) follows from applying χ to the exact sequence

$$0 \to \mathcal{O}_C \to \nu_* \mathcal{O}_{\overline{C}} \to \nu_* \mathcal{O}_{\overline{C}}/\mathcal{O}_C \to 0,$$

noting that $\chi(\nu_* \mathcal{O}_{\overline{C}}) = \chi(\mathcal{O}_{\overline{C}})$ and $H^i(\nu_* \mathcal{O}_{\overline{C}}/\mathcal{O}_C) = 0$ for $i > 0$. Moreover, $\delta(C) = \sum_{z \in \mathrm{Sing}(C)} \delta(C, z)$, with $\delta(C, z)$ the delta invariant of C at z. For a smooth curve C the arithmetic genus and the geometric genus coincide ($\delta(C) = 0$).

If C is irreducible, then $\overline{C}$ is connected and smooth and $g(C) = p_a(\overline{C}) = g(\overline{C}) \geq 0$. If C is a reduced curve with s irreducible components $C_1, \ldots, C_s$, we have

$$g(C) = 1 - s + \sum_{i=1}^{s} g(C_i) \tag{2.7}$$

and hence $g(C) + s - 1 \geq 0$. The general genus formulas (2.6) and (2.7) were first proved by Hironaka [40, Theorem 2] using the resolution of the singularities of C (he defines $g(C)$ as $\sum_{i=1}^{s} g(C_i)$).

If $C \subset \mathbb{P}^2$ is a *plane curve* of degree $d > 0$, then C is connected (by Bézout's theorem) and we have

$$p_a(C) = \frac{(d-1)(d-2)}{2}. \tag{2.8}$$

This follows from the exact sequence

$$0 \to \mathcal{O}_{\mathbb{P}^2}(-d) \to \mathcal{O}_{\mathbb{P}^2} \to \mathcal{O}_C \to 0,$$

giving $1 - p_a(C) = \chi(\mathcal{O}_C) = \chi(\mathcal{O}_{\mathbb{P}^2}) - \chi(\mathcal{O}_{\mathbb{P}^2}(-d)) = 1 - \chi(\mathcal{O}_{\mathbb{P}^2}(-d))$, and from $\chi(\mathcal{O}_{\mathbb{P}^2}(-d)) = \dim_{\mathbb{C}} H^2(\mathbb{P}^2, \mathcal{O}_{\mathbb{P}^2}(-d)) = \binom{d-1}{2}$ (see [38, Theorem III.5.1] for the cohomology of projective space). Below we compute the arithmetic genus via the topological Riemann-Hurwitz formula.

Now, if $C \subset \mathbb{P}^2$ is reduced and irreducible, then $\overline{C}$ is smooth and connected and the geometric genus $g(C) = g(\overline{C})$ is non-negative. The formulas (2.6) and (2.8) imply the *genus formula*

$$g(C) = \frac{(d-1)(d-2)}{2} - \delta(C). \tag{2.9}$$

Since $g(C)$ is non-negative for an irreducible curve of degree d we get the inequality (2.3).

Of course, Bézout's theorem leads to various further necessary conditions for the existence of the curve C such as, for instance by considering a line through 2 points or a conic through 5 points,

$$\max_{i \neq j} \left(\mathrm{mt}(S_i) + \mathrm{mt}(S_j) \right) \leq d, \quad \max_{\#(I)=5} \sum_{i \in I} \mathrm{mt}(S_i) \leq 2d.$$

Finally we mention the inequality

$$\sum_{i=1}^{q} \tau(S_i) + \sum_{i=q+1}^{r} \tau^{es}(S_i) \leq \frac{d^2+3d}{2} \tag{2.10}$$

for *regular existence*, that is, for the existence of a curve $C \subset \mathbb{P}^2$ of degree d with q analytic singularities $S_1, ..., S_q$ and $r-q$ topological singularities $S_{q+1}, ..., S_r$, such that $V_d(S_1, \ldots S_r)$ is T-smooth at C (cf. (2.2) and [23, Corollary 6.3 (ii)], [24, Corollary 3.9], [31, Theorem 2.2.40]).

2.2.2 Plücker Formulae

Besides the genus formula and Bézout's theorem, the Plücker formulae provide necessary bounds for the existence, which are often sharper. Let's deduce these formulae.

Let $C \subset \mathbb{P}^2$ be a reduced, irreducible curve of degree $d > 1$, given by a homogeneous polynomial $F \in \mathbb{C}[x_0, x_1, x_2]$. Denote by $C^* \subset (\mathbb{P}^2)^*$ its *dual curve*, that is, the Zariski closure of the quasi-projective curve

$$\left\{ (a_0 : a_1 : a_2) \in (\mathbb{P}^2)^* \;\middle|\; \begin{array}{l} \{a_0x_0 + a_1x_1 + a_2x_2 = 0\} \text{ is tangent} \\ \text{to } C \text{ at some smooth point } p \in C \end{array} \right\}$$

Here $(\mathbb{P}^2)^*$ is the *(dual) projective 2-space*, whose points $(a_0 : a_1 : a_2)$ are in 1-1 correspondence with the lines $\{a_0x_0 + a_1x_1 + a_2x_2 = 0\} \subset \mathbb{P}^2$.

We have a natural rational duality morphism $\boldsymbol{d} : C \dashrightarrow C^*$, mapping a (smooth) point z of C to its tangent at z: Let $z \in C$ and let P be an irreducible component of the germ (C, z). In local affine coordinates x, y such that $z = (0, 0)$ and the x-axis is tangent to P, this component admits a parametrization

$$\begin{cases} x = t^p \\ y = \lambda t^q + O(t^{q+1}) \end{cases} \quad 1 \leq p < q, \quad \lambda \neq 0, \quad t \in (\mathbb{C}, 0) .$$

and the tangent lines to the points of P are given by equations $y = b(t)x + c(t)$ with

$$b(t) = \frac{\lambda q}{p} t^{q-p} + O(t^{q-p+1}), \quad c(t) = \lambda t^q + O(t^{q+1}), \quad t \in (C, 0) . \tag{2.11}$$

It follows that generically the duality morphism $\boldsymbol{d}$ is 1–1, and hence birational.

Furthermore, C^* is an irreducible projective curve[1] of degree $d^* > 1$. The degree d^* of C^* is classically called the *class of* C.

Let $C = \{F = 0\}$ with $F(x_0, x_1, x_2)$ a homogeneous polynomial of degree d. At a smooth point $z \in C$, the coefficients of the tangent line are given by

$$a_0 = \frac{\partial F}{\partial x_0}, \quad a_1 = \frac{\partial F}{\partial x_1}, \quad a_2 = \frac{\partial F}{\partial x_2} .$$

Let $(\mathbb{C}, 0) \ni t \mapsto z(t) = (x_o(t), x_1(t), x_2(t))$ parametrize the germ (C, z). Since $F(z(t)) \equiv 0$, we have $\frac{\partial F(z(t))}{dt} = \frac{\partial F}{\partial x_0} \cdot \frac{dx_0}{dt} + \frac{\partial F}{\partial x_1} \cdot \frac{dx_1}{dt} + \frac{\partial F}{\partial x_2} \cdot \frac{dx_2}{dt} = 0$, that is, with $a_i(t) = a_i(z(t))$,

$$a_0(t) \cdot \frac{dx_0}{dt} + a_1(t) \cdot \frac{dx_1}{dt} + a_2(t) \cdot \frac{dx_2}{dt} = 0 . \tag{2.12}$$

Combining this with the Euler formula $d \cdot F = x_0 \partial F/\partial x_0 + x_1 \partial F/\partial x_1 + x_2 \partial F/\partial x_2$, which implies that $a_0(t)x_0(t) + a_1(t)x_1(t) + a_2(t)x_2(t) = 0$, we obtain

$$x_0(t) \cdot \frac{da_0}{dt} + x_1(t) \cdot \frac{da_1}{dt} + x_2(t) \cdot \frac{da_2}{dt} = 0 , \tag{2.13}$$

which is dual to (2.12). Thus, the dual to C^* is the original curve C.

We call a tangent line L to C a *singular tangent*, i f

(a) either L is tangent to C at a singular point,
(b) or L is tangent to C at more than one point,
(c) or L intersects C at a non-singular point with multiplicity > 2.

The set of singular tangents is finite, since the set $\mathrm{Sing}(C)$ is finite, and the conditions (b) and (c) determine L as a singular point of C^* (cf. Formula (2.11)). Hence, there exists a point $\boldsymbol{q} = (q_0 : q_1 : q_2) \in \mathbb{P}^2 \setminus C$ which does not lie on any singular tangent. Denote by $\Lambda_{\boldsymbol{q}}$ the pencil of lines through q. Recall that a line $L \in \Lambda_{\boldsymbol{q}}$ is tangent to C at the (non-singular) point $z \in C$ iff z lies also on the polar curve relative to $\boldsymbol{q}$, that is, iff

$$F(z) = 0 = q_0 \frac{\partial F}{\partial x_0}(z) + q_1 \frac{\partial F}{\partial x_1}(z) + q_2 \frac{\partial F}{\partial x_2}(z) .$$

We observe that d^* is the number of lines $L \in \Lambda_{\boldsymbol{q}}$ tangent to C at non-singular points, and that $\{q_0 \frac{\partial F}{\partial x_0} + q_1 \frac{\partial F}{\partial x_1} + q_2 \frac{\partial F}{\partial x_2} = 0\}$ is a generic polar of C. Applying Bézout's Theorem (2.5) to the non-singular intersection points of $\{F = 0\}$ with a generic

---

[1]An equation F^* for C^* can be obtained as follows: let

$$g(x_1, x_2) := a_0^d \cdot F\left(\frac{-(a_1x_1 + a_2x_2)}{a_0}, x_1, x_2, \right) \in \mathbb{C}(a_0, a_1, a_2)[x_1, x_2] ,$$

and compute the discriminant $D \in \mathbb{C}[a_0, a_1, a_2] \setminus \{0\}$ of $g(1, x_2)$. D is homogenous of degree $2d^2 - d$ and is the product of F^* with some number of linear factors. Hence, factorizing D and removing all linear factors, we get an equation for C^*. The computations can be carried out with the SINGULAR software [15].

polar (which gives d^* as total number) and the singular intersection points (which gives the kappa-invariant $\kappa(C,z)$ at each intersection point z), we obtain the *first Plücker formula*

$$d^* = d(d-1) - \sum_{z\in\mathrm{Sing}(C)} \kappa(C,z). \tag{2.14}$$

For κ see the Definition (2.1) and the subsequent formula from the Preliminaries section. In particular, if C has n nodes and k cusps as its only singularities one gets

$$d^* = d(d-1) - 2n - 3k. \tag{2.15}$$

Using again a point $\mathbf{q}$ as above, we derive now the Riemann-Hurwitz formula and give another proof of the genus formula (2.9). Let $\overline{C} \to C$ be the normalization map. Then the *topological Euler characteristic* of C satisfies (using Mayer-Vietoris[2]).

$$\begin{aligned}\chi_{\mathrm{top}}(C) &= \chi_{\mathrm{top}}\big(\overline{C}\big) - \sum_{z\in\mathrm{Sing}(C)} \big(r(C,z)-1\big)\\ &= 2-2g(C) - \sum_{z\in\mathrm{Sing}(C)} \big(r(C,z)-1\big),\end{aligned} \tag{2.16}$$

where $r(C,z)$ is the number of irreducible branches of the germ (C,z). Besides, considering the projection of C on some straight line $L_0 \not\ni \{\boldsymbol{q}\}$ from the point $\boldsymbol{q}$ leads to the following version of the topological *Riemann-Hurwitz formula*,

$$\chi_{\mathrm{top}}(C) = d\cdot\chi_{\mathrm{top}}(L) - d^* - \sum_{z\in\mathrm{Sing}(C)} \big(\mathrm{mt}(C,z)-1\big),$$

since a line $L \in \Lambda_{\boldsymbol{q}}$, which is tangent to C at a non-singular point, meets C at $d-1$ points, and a line $L \in \Lambda_{\boldsymbol{q}}$ through a point $z \in \mathrm{Sing}(C)$ meets C at $d - \mathrm{mt}(C,z)+1$ points. Combining the last equation with (2.14) and (2.16), we come to the genus formula (2.9),

$$g(C) = \frac{(d-1)(d-2)}{2} - \sum_{z\in\mathrm{Sing}(C)} \delta(C,z).$$

We also mention the *second Plücker formula*: for any reduced plane curve of degree $d \geq 3$ which does not contain lines as components, the following equality holds

$$\sum_{z\in\mathrm{Sing}(C)} h(C,z) = 3d(d-2) - \sum_{z\in C\setminus\mathrm{Sing}(C)} ((C\cdot T_zC)_z - 2), \tag{2.17}$$

[2]Consider a covering $C = (U\cap C)\cup(U'\cap C)$, U the union of non-intersecting open ε-balls and U' the complement of the union of closed ε'-balls, $\varepsilon' < \varepsilon$, around the singular points of C.

where $h(C, z)$ is the intersection multiplicity of the curve C with its Hessian determinant at the point z and $(C \cdot T_z C)_z$ stands for the intersection number of C with its tangent line $T_z C$ at z. According to [74], if $z \in \mathrm{Sing}(C)$, then

$$h(C, z) = 3\kappa(C, z) + \sum_{C'} ((C' \cdot T_z C')_z - 2\,\mathrm{mt}(C', z)), \tag{2.18}$$

where C' ranges over all local branches of C at z (i.e., irreducible components of the germ (C, z)).

The second Plücker equation for a reduced plane curve of degree $d \geq 3$ which does not contain lines as components and with n nodes and k cusps states

$$k^* = 3d(d-2) - 6n - 8k, \tag{2.19}$$

where k^* is the number of cusps of C^*. This follows from (2.17) and (2.18): indeed, when there are no flexes at the nodes, and at all smooth flexes we have a triple intersection with the tangent, then $h(A_1) = 3\kappa(A_1) + 0 = 6$, $h(A_2) = 3\kappa(A_2) - 1 = 8$.

2.2.3 Log-Miyaoka-Yau Inequality

Any smooth complex algebraic surface X of general type (i.e., of Kodaira dimension 2) satisfies the Bogomolov-Miyaoka-Yau inequality

$$c_1^2 \leq 3c_2,$$

where c_1, c_2 are the Chern classes of the complex tangent bundle $T(X)$, and the terms in the inequality are evaluated at the fundamental class [57, 103, 104] (see also [5, Theorem VII.4.1]). Sakai [69] noticed that $c_1^2 < 3c_2$ if the surface contains rational or elliptic curves and gave a strengthened inequality for this case (so-called log-Miyaoka-Yau inequality), which was improved further by Miyaoka [58]. In the form suggested by Hirzebruch [41, Theorem 3] it reads

$$c_1^2 - 3c_2 \geq \sum_{i=1}^{k} m(E_i) + \sum_{j=1}^{p} (-C_j^2), \tag{2.20}$$

where $E_1, ..., E_k \subset X$ are pairwise disjoint curves splitting into rational components, $C_1, ..., C_p$ are elliptic curves (disjoint to each other and to $E_1, ..., E_k$), and all the summands in the right-hand side are positive.

By applying the log-Miyaoka-Yau inequality (2.20), Sakai [69, Theorem A] obtained the necessary condition

$$\sum_{i=1}^{r} \mu(S_i) < \frac{2\nu}{2\nu+1} \cdot \left(d^2 - \frac{3}{2}d\right)$$

where ν denotes the maximum of the multiplicities $\mathrm{mt}(S_i)$, $i = 1, \ldots, r$. In [69] further bounds for the total Milnor number are given. In particular, if $S_1, \ldots, S_r$ are ADE-singularities then

$$\sum_{i=1}^{r} \mu(S_i) < \begin{cases} \frac{3}{4}d^2 - \frac{3}{2}d + 2 & \text{if } d \text{ is even}, \\ \frac{3}{4}d^2 - d + \frac{1}{4} & \text{if } d \text{ is odd}, \end{cases}$$

is necessary for the existence of a plane curve with r singularities of types $S_1, \ldots, S_r$.

Applying the inequality (2.20) to the desingularized double covering of the plane ramified along a curve with simple singularities, i.e., A_r, D_r, E_6, E_7, E_8, Hirzebruch and Ivinskis [41, 43] obtained the following bound for a reduced plane curve C of an even degree $d \geq 6$ having only simple singularities:

$$\sum_{z \in \mathrm{Sing}(C)} m(C, z) \leq \frac{d(5d-6)}{2}, \tag{2.21}$$

where the invariant $m(C, z)$ can be computed as follows:

$$m(A_r) = \frac{3r(r+2)}{r+1}, \quad m(D_r) = \frac{3(4r^2 - 4r - 9)}{4(r-2)}, \tag{2.22}$$

$$m(E_6) = \frac{167}{8}, \quad m(E_7) = \frac{383}{16}, \quad m(E_8) = \frac{1079}{40}.$$

Langer [51, Theorem 1] generalized the Bogomolov-Miyaoka-Yau inequality to orbifold Euler numbers and obtained an upper bound to the number of simple singularities of curves on surfaces. In particular (see [51, Theorem 9.4.2 and formula (11.1.1)]), for any reduced curves of degree $d \geq 10$ with n nodes and k cusps it yields the bound

$$(2-\alpha)n + \left(\frac{7}{2} - \frac{3}{2}\alpha - \frac{1}{24\alpha}\right)k \leq \left(1 - \frac{\alpha}{3}\right)d^2 - d \tag{2.23}$$

with an arbitrary $\frac{3}{10} \leq \alpha \leq \frac{5}{6}$, which is always better than Hirzebruch-Ivinskis' bound (2.21). Substituting $\alpha = \frac{\sqrt{73}-1}{24}$, one obtains the maximal coefficient of k in (2.23), and hence

$$\frac{6059 + 7\sqrt{73}}{10512} n + k \leq \frac{125 + \sqrt{73}}{432} d^2 - \frac{511 + 11\sqrt{73}}{1752} d. \tag{2.24}$$

2.2.4 Spectral Bound

Further necessary conditions can be obtained by applying the semicontinuity of the singularity spectrum (see [94]), which works in any dimension. The singularity spectrum of a hypersurface singularity $f : \mathbb{C}^{n+1} \to \mathbb{C}$ gathers the information about the eigenvalues of the monodromy operator T and about the Hodge filtration $\{F^p\}$ on its vanishing cohomology. The *(singularity) spectrum* is defined as an unordered $\mu(f)$-tuple of rational numbers $(a_1, ..., a_\mu)$ (counted with frequencies), where the frequency of the number a in the spectrum is equal to the dimension of the eigenspace of the semisimple part of T acting on F^p/F^{p+1}, $p = [n - a]$, with eigenvalue $exp(-2\pi ia)$. If $F : X \to S$ is a good representative of a deformation of f, let $\Sigma_{F^{-1}(s)}$ denote the union of all spectra of the singular points in the fiber $F^{-1}(s)$, where the frequency of a in $\Sigma_{F^{-1}(s)}$ is the sum of its frequencies in the spectra of all singular points of $F^{-1}(s)$.

The *semicontinuity of the spectrum* says that any half open interval $(t, t+1] \subset \mathbb{R}$ is a semicontinuity domain for F, that is, the sum $M_{F^{-1}(s)}$ of the frequencies of the elements of $(t, t+1]$ in $\Sigma_{F^{-1}(s)}$ is upper semicontinuous for $s \in S$ ([85, Theorem 2.4]). Before that Varchenko [94] had proved that for deformations F of low weight of a quasi-homogeneous f even every open interval $(t, t+1)$ is a semicontinuity domain.

The semicontinuity of the singularity spectrum can be used to compute effectively an upper bound for the number of isolated hypersurface singularities of a given type occurring on a hypersurface $V \subset \mathbb{P}^n$ of degree d. Observe the following:

- Any hypersurface of degree d with isolated singularities can be obtained as small deformation of $\{f_d = 0\}$, where $f_d(x_1, ..., x_n)$ is a nondegenerate d-form, which stays fixed in the deformation, while all variable terms have degree $< d$. To see this, choose coordinates in $\mathbb{P}^n$ so that the hyperplane $\{x_0 = 0\}$ avoids Sing(V) and meets V transversally. The affine hypersurface $V^a \subset \mathbb{A}^n = \{x_0 \neq 0\}$ has the same collection of singularities and is given by $\{f(x_1, ..., x_n) = 0\}$, where the highest form f_d of degree d is nondegenerate. Including V^a into the family of affine hypersurfaces $\{V_t^a\}_{|t|\le 1}$ of the same degree d given by[3]

$$t^d f(x_1/t, ..., x_n/t) = f_d(x_1, ..., x_n) + tf_{d-1}(x_1, ..., x_n) + ... = 0,$$

  we obtain the given collection of singularities in a deformation of the nondegenerate d-multiple singularity $\{f_d = 0\}$ at the origin. Since the space of nondegenerate d-forms is connected, by the semicontinuity of the spectrum it is sufficient to compute the spectrum of $x_1^d + \ldots + x_n^d$ to get a bound for V.
- If precisely M of the spectral numbers (counted with their frequencies) of the singularity defined by $f = x_1^d + \ldots + x_n^d$ are in the interval $(t, t+1]$, $t \in \mathbb{R}$, then the sum of the frequencies of the spectral numbers in the interval $(t, t+1]$ of the singularities (close to 0) of a small deformation F of f can be at most M,

[3] Such families are called *lower deformations*.

i.e., $M_{F^{-1}(s)} \leq M_{f^{-1}(0)} = M$ (cf. [85, Theorem 2.4]; if the deformation is of low weight, we can use even the open interval $(t, t+1)$ by [94]).

For instance, we can look for an upper bound for the number of cusps which may appear on a curve of degree 11 by using SINGULAR [15, 70]:

```
LIB ''gmssing.lib'';
ring r=0,(x,y),ds;
poly g=x^2-y^3;            // a cusp
list s1=spectrum(g);       // spectral numbers of a cusp (with mult's)
s1;
//-> [1]:
//->       _[1]=-1/6   _[2]=1/6
//-> [2]:
//->       1,1
```

That is, for a cusp we have $M_{g^{-1}(0)} = 2$ for each interval $(t, t+1)$ with $t < -1/6$, $t + 1 > 1/6$.

```
poly f = x^11+y^11;
list s2 = spectrum(f);   // spectral numbers of f (with mult's)
s2;
//-> [1]: (spectral numbers)
//->      _[1]=-9/11   _[2]=-8/11   _[3]=-7/11   _[4]=-6/11   _[5]=-5/11
//->      _[6]=-4/11   _[7]=-3/11   _[8]=-2/11   _[9]=-1/11   _[10]=0
//->      _[11]=1/11   _[12]=2/11   _[13]=3/11   _[14]=4/11   _[15]=5/11
//->      _[16]=6/11   _[17]=7/11   _[18]=8/11   _[19]=9/11
//-> [2]: (frequencies or multiplicities)
//->      1,2,3,4,5,6,7,8,9,10,9,8,7,6,5,4,3,2,1
```

Having computed the spectral numbers, we look for an appropriate interval $(t, t+1)$ to apply the semicontinuity theorem $M_{F^{-1}(s)} \leq M_{f^{-1}(0)}$, F a deformation of f. If $F^{-1}(s)$ contains k cusps, then $M_{g^{-1}(0)}k = 2k \leq M_{f^{-1}(0)}$. Choosing $t = -\frac{2}{11}$ we get $2k \leq 63$, i.e., at most 31 cusps can appear on a curve of degree 11.

The same result can be computed by using the SINGULAR procedure `spsemicont` to get the sharpest bound for the number of singularities obtainable in the above way:

```
spsemicont(s2,list(s1),1);
// -> [1]: 31
```

We recall that the spectrum is a topological invariant of the curve singularity, and, for example, according to [84] for the quasihomogeneous singularity $x^m + y^n = 0$ is the multiset (set with frequencies)

$$\left\{ \frac{i}{m} + \frac{j}{n} - 1 \ : \ 1 \leq i \leq m-1, \ 1 \leq j \leq n-1 \right\}. \tag{2.25}$$

A simple algorithm for computing the spectrum of an arbitrary isolated curve singularity was suggested in [44].

Varchenko [94] used the semicontinuity of the spectrum to give an upper bound for the number of nondegenerate singular points (i.e. of type A_1) on arbitrary hypersurfaces in $\mathbb{P}^n$. Let $N_n(d)$ be the maximal number of singular points of type A_1 which can exist on a hypersurface in $\mathbb{P}^n$ of degree d. He proves the following inequality (conjectured by Arnold),

$$N_n(d) \leq A_n(d) = a_n d^n + \text{ (lower degrees of } d\text{)}, \tag{2.26}$$

with $a_n \sim \sqrt{(6/\pi)n} = 1.3819...\sqrt{n}$, if $n \to \infty$.

2.3 Plane Curves with Nodes and Cusps

The simplest singularities, the node A_1 and the ordinary cusp A_2, typically occur in most of the questions related to singular curves. The case of curves with nodes and cusps is also the most studied case, both in classical and in modern algebraic geometry. It suggests beautiful results and challenging problems. Furthermore, the study of the particular case of curves with nodes and cusps led to the development of important techniques and the discovery of most interesting phenomena in the geometry of singular algebraic curves and their families. We shall demonstrate this for the problem of the existence of a plane curve of a given degree with a given collection of nodes and cusps, both in the complex and real case.

2.3.1 Plane Curves with Nodes

We start with the construction of complex plane curves with only nodes as singularities and with any prescribed number being allowed by the genus bound (2.3). The construction is due to Severi [73] and very simple. It uses, however, the T-smoothness of a family of nodal curves in an essential way, called classically the "completeness of the characteristic linear series" (see [59]). For a modern proof see [23] and [31, Sect. 4.5.2.1].

For real curves their existence with the number of nodes below or equal to the genus bound is also classically known and due to Brusotti [8], using the T-smoothness of the family of real nodal curves. But in the real case we have to distinguish between three kinds of nodes: hyperbolic, elliptic and non-real (coming in complex conjugate pairs). The fact that, subject to the genus bound, any prescribed distribution among the three different kinds can be realized was proved much later in [77], and with a different method by Pecker [65, 66, 68]. The construction is much more difficult than in the complex case. It uses a "patchworking construction" invented by Viro for non-singular real curves and extended to singular curves in [77, 79] (see also [31, Sects. 2.3 and 4.5.1]).

Complex Curves

Let C be a complex plane irreducible curve of degree d with n nodes. The genus bound (2.3) yields

$$n \le \frac{(d-1)(d-2)}{2}, \tag{2.27}$$

since $g(C) \ge 0$ and the δ-invariant of the node is 1. If we assume that C consists of s irreducible components, then we get

$$n \le \frac{(d-1)(d-2)}{2} + s - 1. \tag{2.28}$$

It goes back to Severi [73] that the bounds (2.27) and (2.28) are, in fact, necessary and sufficient for the existence of a plane curve of degree d with n nodes. More precisely,

Theorem 2.3.1 *The bound (2.27) is necessary and sufficient for the existence of a complex plane irreducible curve of degree d with n nodes as its only singularities.*

Furthermore, for any $s \ge 2$, any positive integers $d, d_1, ..., d_s$ satisfying $d = d_1 + ... + d_s$, and nonnegative integers $n_1, ..., n_s$, the inequalities

$$n_i \le \frac{(d_i-1)(d_i-2)}{2}, \quad i = 1, ..., s,$$

are necessary and sufficient for the existence of a complex plane reduced curve of degree d splitting into s irreducible components of degrees $d_1, ..., d_s$ and having

$$n = \sum_{i=1}^{s} n_i + \sum_{1 \le i < j \le s} d_i d_j$$

nodes as its only singularities, while the i-th component has precisely n_i nodes, $i = 1, ..., s$.

Proof Severi proved that, given a nodal plane curve C of degree d, the germ of the family of curves of degree d having a node in a neighborhood of an arbitrary singular point of C, is a smooth hypersurface germ in $|\mathcal{O}_{\mathbb{P}^2}(d)| \simeq \mathbb{P}^{d(d+3)/2}$, and, moreover, all these germs intersect transversally at C (for a modern treatment see [23, Corollary 6.3] and [31, Corollary 4.3.6]). This fact immediately yields that there exists a deformation of C in $\mathbb{P}^{d(d+3)/2}$ along which prescribed nodes are smoothed out, while the others persist (possibly changing their position). Thus, given n and d satisfying (2.27), we take the union of d straight lines in general position, which is a curve with $\frac{d(d-1)}{2}$ nodes (the maximum by (2.28)). Then choose some line and deform the curve by smoothing out all $d-1$ intersection points of this line with the other lines, obtaining an irreducible, rational curve with $\frac{(d-1)(d-2)}{2}$ nodes (see Fig. 2.1). Finally, we take another deformation by smoothing out $\frac{(d-1)(d-2)}{2} - n$ nodes and obtain an irreducible curve of degree d with n nodes as desired.

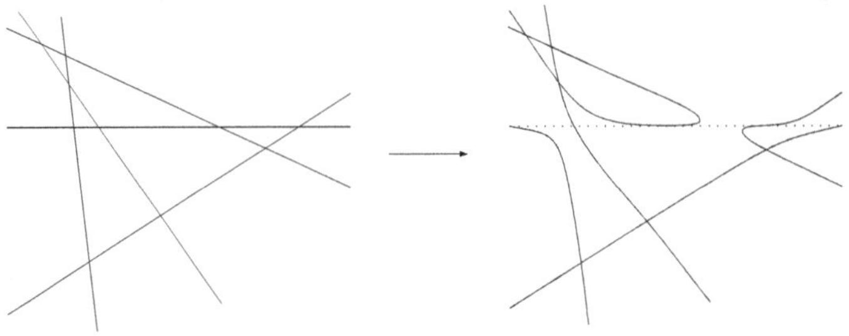

Fig. 2.1 Severi's construction of irreducible nodal curves

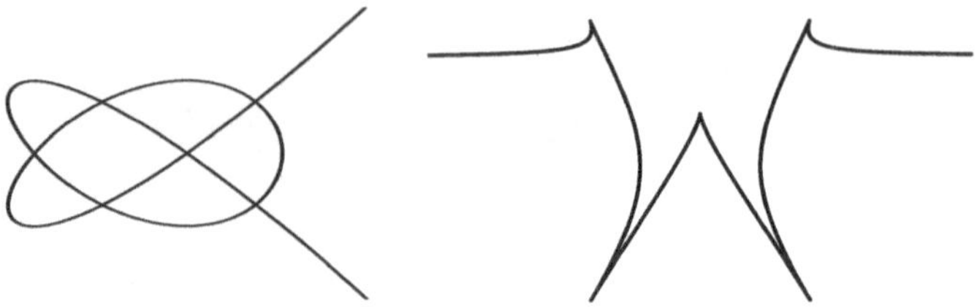

(a) (b)

Fig. 2.2 Real plane quintics **a** with 6 nodes: $x^5 - \frac{5}{4}x^3 + \frac{5}{16}x - \frac{1}{2}y^4 + \frac{1}{2}y^2 - \frac{1}{16} = 0$, **b** with 5 cusps: $\frac{129}{8}x^4y - \frac{85}{8}x^2y^3 + \frac{57}{32}y^5 - 20x^4 - \frac{21}{4}x^2y^2 + \frac{33}{8}y^4 - 12x^2y + \frac{73}{8}y^3 + 32x^2 = 0$

In the reducible case, we take irreducible curves of degrees $d_1, ..., d_s$ with $n_1, ..., n_s$ nodes respectively and place them in general position in the plane. □

Remark 2.3.2 We would like to underline the importance of the fact that each curve appearing in the Severi's construction was a member of a smooth equisingular family in $\mathbb{P}^{d(d+3)/2}$ of expected dimension (such families are called T-smooth), and the germ of the family was the transversal intersection of smooth equisingular families corresponding to individual singular points of the given curve. That is, given such a curve with a possibly maximal number of singularities, one immediately obtains the existence of curves with any smaller amount of singularities. We shall see later how efficient this property (which follows from T*-smoothness*) is for the construction of curves with arbitrary singularities.

Real Curves

Over the reals, a nodal singular point can be of one of the following three types:

- either *hyperbolic*, i.e., a real intersection point of two smooth real local branches, locally equivalent to $\{x^2 - y^2 = 0\}$,
- or *elliptic*, i.e., a real intersection point of two complex conjugate smooth local branches, locally equivalent to $\{x^2 + y^2 = 0\}$,

- or *imaginary*, i.e., a non-real node (which always comes in pairs of complex conjugate nodes).

Thus, a natural question is what amount of hyperbolic, elliptic and pairs of complex conjugate nodes a real plane curve can have. Note that Severi's construction is not of much help, since, for instance, a generic conjugation-invariant configuration of d lines in the plane can have at most $\left[\frac{d}{2}\right]$ elliptic nodes. The patchworking construction invented by O. Viro around 1980 for the study of the topology of smooth real algebraic varieties (see, for instance, the Appendix to [31]) was later applied to curves and hypersurfaces with singularities (see [31, Sect. 2.3]). It allowed one to completely answer the above question [77]. Another solution was later suggested by Pecker [65], who used explicit parameterizations of real rational curves, obtaining, for instance, the quintic shown in Fig. 2.2a. Here we demonstrate the patchworking construction, which also applies efficiently to curves with singularities of other types, while the methods of [65] are restricted to nodal curves only.

The version of the *patchworking construction* which we need was introduced in [77, 79] (see also [31, Sects. 2.3 and 4.5.1]). Let us be given a convex lattice polygon $\Delta \subset \mathbb{R}^2$ and a convex[4] subdivision of it into convex lattice polygons $\Delta_1, \ldots, \Delta_N$. Let $F_1, \ldots, F_N$ be bivariate complex or real polynomials with Newton polygons $\Delta_1, \ldots, \Delta_N$, respectively, such that

(i) the truncations of F_i and F_j on the common side σ of Δ_i and Δ_j coincide,
(ii) each polynomial F_i is peripherally nondegenerate, i.e., the truncation of F_i on any side of Δ_i defines a smooth curve in $(\mathbb{C}^*)^2$,
(iii) each polynomial F_i defines a curve with isolated singularities in $(\mathbb{C}^*)^2$.

Denote by $S(F_i)$ the multi-set of topological or analytic types of the singular points of the curve $F_i(x, y) = 0$ in $(\mathbb{C}^*)^2$.

Then we orient the adjacency graph of the polygons $\Delta_1, \ldots, \Delta_N$ without oriented cycles and verify the $\mathcal{S}$-transversality condition ($\mathcal{S}$ being the topological or analytic equivalence of singularities, see [31, Definition 2.3.3 and Definition 2.3.12]) for each patchworking pattern $(\Delta_i, \partial\Delta_{i,+}, F_i)$, where $\partial\Delta_{i,+}$ is the union of the sides of Δ_i corresponding to the incoming arcs of the adjacency graph and $\mathcal{S}$ stands for the chosen complex or real topological or analytic equivalence of singularities.

The following *patchworking theorem* for curves says that we can "glue" the polynomials F_i together to one polynomial F, which defines a curve with isolated singularities in $(\mathbb{C}^*)^2$ that inherits the singularity types from the F_i.

Theorem 2.3.3 *With the above notations and assumptions let $F_1, \ldots, F_N$ with properties* (i), (ii), (iii) *be given. Then there exists a polynomial $F(x, y)$ with Newton polygon Δ such that*

$$S(F) = \bigcup_{i=1}^{N} S(F_i).$$

[4] A convex subdivision is a subdivision into linearity domains of some convex piecewise linear function defined on the lattice triangle $T_d = \text{Conv}\{(0, 0), (d, 0), (0, d)\}$.

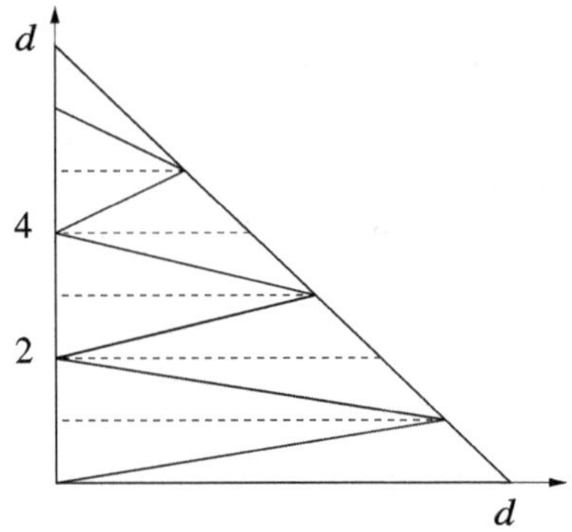

Fig. 2.3 Patchworking of a real plane curve with elliptic nodes

Moreover, the family of $\mathcal{S}$-equisingular curves defined by polynomials with Newton polygon Δ is T-smooth at $\{F = 0\}$.

We say that a family of curves is *$\mathcal{S}$-equisingular*, if the choosen types $\mathcal{S}$ of the singular points of the curves stay locally constant along some section.

One of the nicest sides of the patchworking construction is that it works equally well over the complex and the real fields. The first example illustrating this feature is as follows.

Theorem 2.3.4 *For any integer $d \geq 3$ and nonnegative integers a, b, c, the inequality*

$$a + b + 2c \leq \frac{(d-1)(d-2)}{2}, \tag{2.29}$$

is necessary and sufficient for the existence of a real plane irreducible curve of degree d having a hyperbolic nodes, b elliptic nodes and c pairs of complex conjugate nodes as its only singularities. Moreover, the constructed curves belong to a T-smooth family.

Proof The necessity follows from the genus bound (2.3). Thus, we focus on the construction. We only sketch the proof, which in full detail is presented in [77]. Namely, we shall prove the theorem in the case $a = c = 0$. As noticed in Remark 2.3.2, it is enough to construct a real rational curve with $\frac{(d-1)(d-2)}{2}$ elliptic nodes (since the curves belong to a T-smooth family). Consider the subdivision of the lattice triangle $T_d = \text{Conv}\{(0,0), (d,0), (0,d)\}$ into lattice triangles as shown in Fig. 2.3. Observe that the number of interior integral points in these triangles amounts to $\frac{(d-1)(d-2)}{2}$.

Each tile of the subdivision is a triangle of the form $T = \text{Conv}\{(0,0), (0,2), (m,1)\}$ (up to an automorphism of the lattice $\mathbb{Z}^2$). We claim that the real polynomial

$$F(x, y) = y^2 - 2yP_m(x + \lambda) + 1, \tag{2.30}$$

where $P_m(x) = \cos(m \arccos x)$ is the m-th Chebyshev polynomial and λ is a generic real number, has Newton polygon T and defines a real plane curve with $m - 1$ elliptic nodes in $(\mathbb{R}^*)^2$. Note that $m-1$ is the number of interior integral points in T. To prove the claim it is enough to rewrite the equation of the curve $F(x,y)=0$ in the form

$$y = P_m(x+\lambda) \pm \sqrt{P_m(x+\lambda)^2 - 1}$$

and recall that P_m has $m-1$ extrema: $\left[\frac{m}{2}\right]$ minima on the level -1 and $\left[\frac{m-1}{2}\right]$ maxima on the level 1 .

Now we associate with each triangle $T^{(i)}$ of the subdivision a polynomial $F^{i)}$ with Newton triangle $T^{(i)}$ which is obtained from a polynomial like (2.30) by the coordinate change matching an appropriate automorphism of the lattice $\mathbb{Z}^2$. A further transformation $F^{(i)}(x,y) \mapsto \alpha_i F^{(i)}(\beta_i x, \gamma_i y)$ with suitable positive $\alpha_i, \beta_i, \gamma_i \in \mathbb{R}$ equates the truncations of each pair of the neighboring polynomials $F^{(i)}$, $F^{(i+1)}$ on the common side of $T^{(i)}$ and $T^{(i+1)}$. To complete the proof, we apply Theorem 2.3.3, observing that, in the nodal case, every patchworking pattern $(T^{(i)}, \partial_+ T^{(i)}, F^{(i)})$ is $\mathcal{S}$-transversal ($\mathcal{S}$ being the topological or analytic equivalence of singularities), see [79, Theorem 4.2] or [31, Corollary 4.5.3]. □

2.3.2 *Plane Curves with Nodes and Cusps*

Questions concerning the number of nodes and cusps on a plane curve of a given degree are classical and highly nontrivial as compared to the purely nodal case, in particular over the reals. No complete answer is known in general, neither in the complex case, nor in the real one.

The general restrictions for their existence from Sect. 2.2, such as the genus formula, the Plücker formulas, and the bounds by Hirzebruch-Ivinskis [41, 43] and Langer [51] take a special simple form for curves with nodes and cusps. We compare the asymptotics of the bounds by Hirzebruch-Ivinskis and Langer if the degree goes to infinity.

In the second part of this section we report on the state of knowledge on curves with many cusps, complex as well as real. Special attention will be given to small degrees, with precise references to their construction.

The last part is devoted to the patchworking construction, which provides an asymptotically complete answer if we restrict to real and complex plane curves with nodes and cusp belonging to T-smooth equisingular families.

For the results of this section see also [31, Sects. 4.2.2.2, 4.5.2.1] (for restrictions) and [31, Sects. 4.5.2.2, 4.5.2.3] (for constructions).

In the whole section we consider complex or real curves which are irreducible over the complex numbers.

Restrictions for the existence

Let C be a curve with only nodes and cusps as singularities. For a node we have $\delta=\mu=1$, $\kappa=2$ and for a cusp $\delta=1$, $\mu=2$, $\kappa=3$. Hence the Formulas (2.4), resp. (2.9), give as a necessary condition for the existence of an irreducible, resp. not necessarily irreducible, curve with n nodes and k cusps the estimates

$$\begin{aligned} 2n + 2k &\le (d-1)(d-2), \text{ resp.} \\ n + 2k &\le (d-1)^2. \end{aligned}$$

By combining the Plücker Formulae (2.15) and (2.17) with Formula (2.18) and the fact that $d^*(d^*-1) \ge d$ (resulting from formula (2.14)), one gets the following necessary conditions for the existence of irreducible curves, that is, for the non-emptiness of $V_d^{irr}(nA_1 + kA_2)$, originally due to Lefschetz [52]:

$$\begin{aligned} 2n + 3k &\le d^2 - d - \sqrt{d}\,, \\ 6n + 8k &\le 3d^2 - 6d\,. \end{aligned} \tag{2.31}$$

Better bounds from Hirzebruch-Ivinskis, Langer, and spectral estimates are obtained as a consequence of deep results in algebraic geometry. The general Hirzebruch-Ivinskis inequality (2.21) reads for a curve of an even degree $d \ge 10$ with n nodes and k cusps as (cf. (2.22))

$$\frac{9}{8}n + 2k \le \frac{5}{8}d^2 - \frac{3}{4}d \quad \text{for all } d \text{ even}, \ d \ge 6 \tag{2.32}$$

Langer's inequality (2.24) applies to all degrees $d \ge 10$, and in this range it is always better then (2.32). Hirzebruch-Ivinskis (HI) compared to Langer (L) gives:

$$\begin{aligned} \text{(HI):} \quad & 0.5625n + k \le 0.3125d^2 - 0.375d, \\ \text{(L):} \quad & 0.5821n + k \le 0.3091d^2 - 0.3453d. \end{aligned}$$

In particular, Langer's inequality implies that the *maximal number of cusps* $k_{\max}(d)$ on a curve of degree d satisfies

$$\lim_{d\to\infty} \sup \frac{k_{\max}(d)}{d^2} \le \frac{125 + \sqrt{73}}{432} = 0.3091\ldots\,. \tag{2.33}$$

We also mention the spectral bound [94, Theorem, p. 164]

$$\frac{1}{2}n + k \le \frac{1}{2} \cdot \begin{pmatrix} \text{the number of integral points } (i, j) \text{ satisfying} \\ 0 < i, j < d, \quad \frac{d}{6} < i + j \le \frac{7}{6}d \end{pmatrix} \tag{2.34}$$

$$= \frac{23}{72}d^2 + O(d) \approx 0.3194d^2 + O(d),$$

which is asymptotically weaker than the Hirzebruch-Ivinskis' and Langer's bounds.

Curves with a large number of cusps

The problem of existence of plane curves with a large number of cusps attracted a special attention, due to the fact that the maximal possible number of cusps in general is not known yet. We shortly describe here several constructions.

Following Zariski [108, Formula (14), p. 220], consider curves $C_r^{(1)} \subset \mathbb{P}^2$ of degree $d = 6r$, $r \geq 1$, given by $F^2 + G^3 = 0$, where $F, G \in \mathbb{C}[x, y, x]$ are generic homogeneous polynomials of degree $3r$ and $2r$, respectively. The curves $F = 0$ and $G = 0$ then intersect transversally at $6r^2$ distinct points, and each of these intersection points is an ordinary cusp of the curve $C_r^{(1)}$. The total number of cusps is $6r^2 = \frac{d^2}{6}$, which is far from the upper bounds discussed above. However, choosing appropriate real F and G, one can obtain a real irreducible curve $C_r^{(1)}$ of degree $d = 6r$ with $\frac{d^2}{6}$ real cusps.

Ivinskis' construction [43] has provided a bigger number of cusps. Namely, he started with the sextic curve $F(x, y, z) = 0$ having 9 cusps in the torus $(\mathbb{C}^*)^2$ and considered the series of curves $C_r^{(2)} = \{F(x^r, y^r, z^r = 0\}$, $r \geq 1$, of degree $d = 6r$. Since the substitution of (x^r, y^r, z^r) for (x, y, z) defines an r^2-sheeted covering of the torus $(\mathbb{C}^*)^2$, we obtain that $C_r^{(2)}$ has $9r^2 = \frac{d^2}{4}$ cusps. This number of cusps is closer to the upper bounds, but the number of real cusps of $C_r^{(2)}$ is at most 12.

A new idea was suggested by Hirano [39]: she used a sequence of coverings defined by the substitution of (x^3, y^3, z^3) for (x, y, z), choosing each time the coordinate system with axes tangent to the current curve, and noticing that each tangent point like that lifts to three ordinary cusps. In particular, starting with the sextic curve having 9 cusps, she produced the sequence of curves $C_k^{(3)}$, $k \geq 1$, of degree $d = d_k = 2 \cdot 3^k$, having

$$s_k = \frac{9}{8}(9^k - 1)$$

cusps. The limit ratio of the number of cusps by the square of the degree equals here

$$\lim_{k\to\infty} \frac{s_k}{d_k^2} = \frac{9}{32} = 0.28125,$$

which is closer to Langer's bound (2.33). Moreover, this was the first example of an equisingular family having a negative expected dimension

$$\frac{d_k(d_k + 3)}{2} - 2s_k \sim -\frac{d_k^2}{16} \quad \text{as } k \to \infty,$$

implying that the cusps impose dependent conditions on the space of curves of degree d and that this ESF is not T-smooth.

The construction by Hirano was later refined by Kulikov [50] and then by Calabri et al. [9, Theorem 6], who found a sequence of curves $C_k^{(4)}$ of degree $d_k = 108 \cdot 9^k$, $k \geq 1$, having at least

$$s_k = \frac{69309}{20} 9^{2k} - 27 \cdot 9^k - \frac{9}{20}$$

cusps, which yields the (so far) best known limit ratio

$$\lim_{k\to\infty} \frac{s_k}{d_k^2} = \frac{2567}{8640} = 0.2971\ldots .$$

Notice also that the constructions by Hirano, Kulikov, and Calabri et al. give a rather small number of real cusps.

For small degrees d, the known upper bounds often coincide with the known maximal number $s_{\max}(d)$ of cusps of a plane curve of degree d or differ by 1. We present the results in the following table with an additional information on the known maximal number $s_{\max,\mathbb{R}}(d)$ of real cusps on a real curve of degree d.

Degree	3	4	5	6	7	8	9	10	11	12
Upper bound	1	3	5	9	10	15	21	26	31	40
$s_{\max}$	1	3	5	9	10	15	20	26	30	39
$s_{\max,\mathbb{R}}$	1	3	5	7	10	14	14	18	23	28

The upper bounds for $3 \le d \le 6$ follow from the second Plücker formula in the form (2.31). The upper bounds for $7 \le d \le 9$ were proved by Zariski [108, pp. 221, 222]: he showed that for the d-multiple cover of the plane ramified along a plane curve of degree d with only nodes and cusps, the irregularity vanishes [108, p. 213], and then he derived that the family of curves of degree $d - 3 - \left[\frac{d-1}{6}\right]$ passing through the cusps of the given curve was unobstructed; hence, the number of cusps does not exceed $\frac{(d-m)(d-3-m)}{2} + 1$, where $m = \left[\frac{d-1}{6}\right]$ (cf. [108, Formula (18) on page 221]). The bounds for $9 \le d \le 10$ follow also from the spectral estimate (2.34). The bound 31 for $d = 11$ follows from the semicontinuity of the spectrum for lower deformations of quasihomogeneous singularities [95, Theorem in page 1294] applied to the open interval $\left(\frac{2}{11}, \frac{13}{11}\right)$ (different from that in (2.34)). At last, the bound 40 for $d = 12$ follows from the Hirzebruch-Ivinskis estimate (2.32).

The cubic, quartic, quintic, sextic, and septic with the indicated number of cusps were classically known. For example, the quartic is dual to the nodal cubic, while the sextic is dual to a smooth cubic, and the quintic was known to Segre (an explicit construction can be found in [33], see also Fig. 2.2b). The maximal known cuspidal curves for $d = 10$ and 12 were constructed by Hirano [39]. The maximal known cuspidal curves for $d = 8, 9$, and 11 were constructed via patchworking respectively in [79, Theorem 4.3] and [9, Appendix], [31, Sect. 4.5.2.3]. The maximal cuspidal septic was constructed by Zariski [108, p. 222] (see also [45]). We comment on this result, which nicely combines duality of curves with the classical result on deformation of curves: the nodes and cusps of an irreducible plane curve of degree d can be independently deformed in a prescribed way or preserved as long as the number of cusps is less than $3d$ [71, 72]. Zariski starts with the sextic having 9 cusps, deforms it into the sextic with 7 cusps and one node, takes its dual which is a curve of degree 7 with 10 cusps and 3 nodes and, finally, smoothes out the nodes.

For $d \le 5$, there are real maximal cuspidal curves with only real cusps: such a quartic is dual to the cubic with an elliptic node, the cuspidal quintic constructed in [33] and shown in Fig. 2.2b is real and has only real cusps. The known maximal numbers of real cusps for degrees $6 \le d \le 8$ were reached in [42]. It is interesting that $s_{\max,\mathbb{R}}(6) = 7$ i s he actual maximum for real sextics as shown in [42]: the absence of a real sextic with 8 real cusps was derived from a delicate analysis of the moduli space of real K3 surfaces obtained as double covers of the plane ramified along a real plane sextic curve. The values of $s_{\max,\mathbb{R}}$ for $9 \le d \le 12$ are borrowed from Theorem 2.3.6.

Patchworking curves with nodes and cusps.

The patchworking construction gives an asymptotically proper answer to the existence problem for real and complex plane curves with nodes and cusps. If we restrict our attention to real and complex plane curves with nodes and cusp which belong to T-smooth equisingular families, then this construction, in view of (2.10), provides an asymptotically complete answer (see [79, Theorems 2.2, 3.3, and 4.1] and [31, Sect. 4.5.2.2]):

Theorem 2.3.5 *For any non-negative integers d, n, k such that*

$$n + 2k \le \frac{d^2 - 4d + 6}{2}\ , \tag{2.35}$$

there exists a (complex) plane irreducible curve with n nodes and k cusps as only singularities. Moreover, the result is asymptotically T-smooth optimal, i.e., up to linear terms in d no more nodes and cusps are possible on a curve belonging to a T-smooth ESF.

Theorem 2.3.6 *(1) For any $d \ge 3$ and any positive integer c such that*

$$c \le \frac{d^2 - 3d + 4}{4}\ , \tag{2.36}$$

there exists a real plane curve of degree d with c real cusps as its only singularities.

(2) There exists a linear polynomial $\psi(d)$ in the variable d such that, for any $d \ge 3$ and nonnegative integers $n_h, n_e, n_{im}, c_{re}, c_{im}$ with

$$n_h + n_e + 2n_{im} + 2c_{re} + 4c_{im} \le \frac{d^2}{2} + \psi(d), \tag{2.37}$$

there is a real plane curve of degree d having n_h hyperbolic nodes, n_e elliptic nodes, n_{im} pairs of complex conjugate nodes, c_{re} real cusps, and c_{im} pairs of complex conjugate cusps as its only singularities.

Moreover, these curves correspond to T-smooth germs of the respective equisingular families of curves with cusps in (1) resp. with nodes and cusps in (2) and the bounds in (2.36) and (2.37) are asymptotically optimal w.r.t. T-smooth ESF.

Proof We prove here the part (1) of Theorem 2.3.6, referring to the references above for the rest. Consider the subdivision of the lattice triangle $T_d = \text{Conv}\{(0,0), (d,0),$

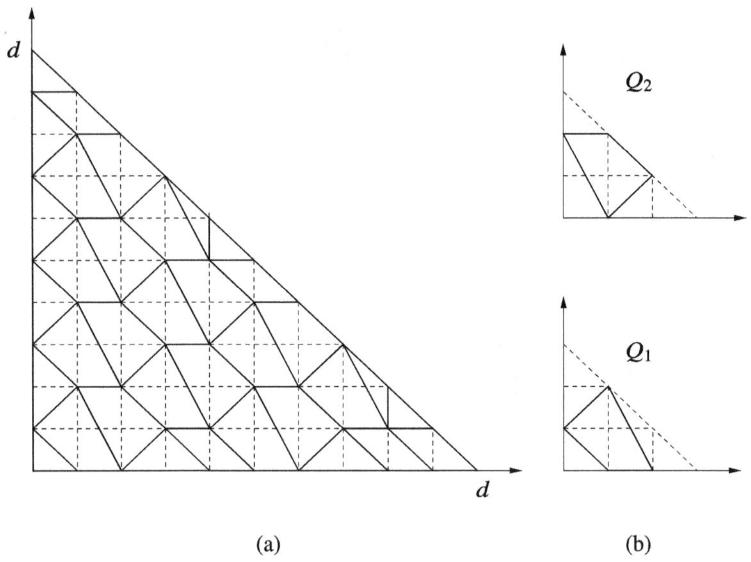

Fig. 2.4 Construction of real curves with real cusps

$(0, d)\}$ into the lattice quadrangles an triangles depicted in Fig. 2.4a. It is not difficult to show that the subdivision is convex.

It is easy to see that, for any given $\alpha, \beta, \gamma \in \mathbb{R}^*$, there exist real polynomials

$$F_1(x, y) = \alpha y + \beta x + \gamma x^2 + a_{11}xy + a_{12}xy^2,$$
$$F_2(x, y) = \alpha y^2 + \beta x + \gamma x^2 y + b_{11}xy + b_{12}xy^2,$$

with Newton quadrangles Q_1, Q_2 (see Fig. 2.4b) which define curves with a real cusp in $(\mathbb{R}^*)^2$. Both curves coincide with the real cuspidal cubic tangent to the coordinate axes in an appropriate way. Thus, we can associate compatible real polynomials with each tile of the subdivision so that the polynomials for the translates of Q_1 and Q_2 will define real curves with a real cusp in $(\mathbb{R}^*)^2$. Orient the adjacency graph of the subdivision so that, for each pattern $(\Delta_i, \partial_+\Delta_i, G_i)$, the part of the boundary $\partial_+\Delta_i$ will consist of the two lower sides of each translate of Q_1 and Q_2. The lattice length of the rest of the boundary is 2, which is greater than 1, the number of cusps in $(\mathbb{C}^*)^2$, which finally yields the transversality of each pattern (see [79, Theorem 4.1(1)] or [31, Proposition 4.5.2]). By Theorem 2.3.3 there exists a real curve of degree d with real cusps as its only singularities, where the number of cusps equals $\left[\frac{d^2-3d+4}{4}\right]$, the number of translates of Q_1 and Q_2 in the subdivision. Since the resulting curve belongs to a T-smooth equisingular family, it can be deformed with smoothing out prescribed cusps. □

We note that the inequalities (2.35), (2.36) and (2.37) differ, w.r.t. the number of cusps, from the restriction of the genus bound (2.3) by a factor $\frac{1}{2}$ at d^2 and a term linear in d, showing that the result is asymptotically proper.

2.4 Plane Curves with Arbitrary Singularities

In this section we discuss curves with special singularities (ordinary multiple points, simple singularities) as well as curves with arbitrary singularities up to topological or analytic equivalence. Moreover, the bounds based on the Bogomolov-Miyaoka-Yau inequality are mainly restricted to simple singularities and semi-quasihomogeneous singularities (we discuss this in more detail in Sect. 2.4.2). Thus, in general we are left only with the genus bound, the Plücker bounds, and the spectral bound.

2.4.1 Curves of Small Degrees

For degrees ≤ 6 the possible collections of singularities of irreducible complex plane curves are classified.

The fact that an irreducible real or complex cubic may have either a node A_1 or a cusp A_2 was known already to Newton [60–62] (for a modern treatment of Newton's study, see [49]).

Collections of singularities of complex irreducible quartic curves can easily be classified by manipulating equations or by using quadratic Cremona transformations, and all this has been known to algebraic geometers of the 19-th century. Moreover, it can be shown that each collection of singularities defines a smooth irreducible subvariety of expected dimension in the space $\mathbb{P}^{14}$ of plane projective quartics (see, for instance, [7, 98]). The classification of real singular quartic curves was completed in [34] (see also [48] for more details as well as for the classification of real singular affine quartics).[5]

Still, the classification of collections of singularities of irreducible plane quintic curves can be reached by elementary methods. The complete classification of singularities of plane quintics together with the statement that each collection of singularities defines a smooth irreducible equisingular family of expected dimension can be found in [16] and [17, Sect. 7.3] (se also [33, 99] for interesting particular examples of singular quintics).

The case of plane sextic curves is the first non-elementary one. The study of plane sextics heavily relies of a thorough investigation of K3 surfaces appearing as double covers of the plane ramified along the considered sextic curve. On the other hand, it reveals a highly interesting new phenomenon - the existence of the so-called *Zariski*

[5]Both papers contain much more material: they classify all possible dispositions of singular points on the real point set of the quartic curve.

pairs, i.e., pairs of curves having the same collection of singularities, but belonging to different components of the equisingular family. The first example presenting two sextic curves with 6 ordinary cusps that have non-homeomorphic complements in the projective plane was found by Zariski [108, p. 214]. The complete classification of collections of singularities (up to topological equivalence) can be found in [3] and [17, Sect. 7.2] (various particular cases have been investigated in [91–93]). The classiication of real singularities of real sextics is not completed yet (for the case of complex and real cusps, see Sect. 2.3.2 above).

2.4.2 Curves with Simple, Ordinary, and Semi-quasihomogeneous Singularities

In the present section we construct equisingular families of curves with many simple resp. ordinary resp. semi-quasihomogeneous singularities and compare their number with the known necessary bounds. Again, by means of the patchworking construction we are able to construct families such that the number of singularities is asymptotically optimal resp. proper. The results presented in Theorem 2.4.5, 2.4.6 and 2.4.7 are new.

Curves with simple singularities. The patchworking construction, essentially used in Sect. 2.3 for the construction of curves with real and complex nodes and cusps, works equally well in the case of arbitrary simple singularities $A_n, n \geq 1$, $D_n, n \geq 4$, E_n, $n = 6, 7, 8$. The results of [83, 100] on the existence of plane complex curves with simple singularities can be summarized in the following statement (cf. [31, Theorem 4.5.5])

Theorem 2.4.1 *(1) For any simple singularity S, there exists a linear polynomial $\varphi_S(d)$ such that the inequality*

$$k\mu(S) \leq \frac{d^2}{2} + \varphi_S(d) \tag{2.38}$$

is sufficient for the existence of an irreducible complex plane curve of degree d having k isolated singular points of type S as its only singularities and belonging to a T-smooth ESF.

(2) Furthermore, for any integer $m \geq 1$, there exists a linear polynomial $\psi_m^{simple}(d)$ such that the inequality

$$\sum_{i=1}^{r} \mu(S_i) \leq \frac{d^2}{2} + \psi_m^{simple}(d), \tag{2.39}$$

with arbitrary r and simple singularities $S_1, ..., S_r$ with Milnor numbers $\leq m$, is sufficient for the existence of an irreducible complex plane curve of degree d having

r isolated singular points of types S_1, ..., S_r, respectively, as its only singularities and belonging to a T-smooth ESF.

Notice that this existence statement is *asymptotically optimal* as long as we consider curves belonging to T-smooth ESF. Indeed, the codimension of a T-smooth ESF in the considered cases equals the left-hand side in (2.38) or (2.39) (see [23, Corollary 6.3(ii)] and [31, Theorem 2.2.40]), and hence satisfies

$$k\mu(S) \leq \frac{d(d+3)}{2}, \quad \text{resp.} \quad \sum_{i=1}^{r} \mu(S_i) \leq \frac{d(d+3)}{2}.$$

In each case, the difference between the bounds in the necessary and sufficient conditions is linear in d. For the proof we refer to [83, 100] (see also the proof of Theorem 2.4.5).

The existence of real curves with simple singularities was analyzed in [101] along the same lines, though the argument was incomplete: in particular, Lemma 3.1 in [101] is wrong as pointed out by E. Brugallé. So, in general the problem over the reals remains open.

While the patchworking construction basically resolves the existence problem for curves with simple singularities that impose independent conditions on the coefficients of the defining equation, there are examples of curves with extremely many simple singularities imposing thereby dependent conditions on the curve.

So, the construction invented by Hirano [39] (which we discussed in connection to plane curves with large number of cusps, Sect. 2.3.2) applies well also to more complicated cusps A_n (see [39, Theorem 2]). Namely, one starts with a smooth conic and, in each step, chooses axes tangent to the current curve and substitutes $(x^{n+1}, y^{n+1}, z^{n+1})$ for (x, y, z), which results in the following sequence of singular curves:

Theorem 2.4.2 *For any even $n \geq 4$, there exists a sequence of irreducible complex plane curves C_k, $k = 1, 2, ...,$ of degree $d_k = 2(n+1)^k$ having*

$$s_k = \frac{3(n+1)((n+1)^{2k} - 1)}{n(n+2)}$$

singular points of type A_n as their only singularities.

Remark 2.4.3 In fact, Hirano's construction works for odd $n \geq 3$ as well with the above formulas for the degree and for the number of A_n singularities, but the curves appear to be reducible.

Since a singularity A_n imposes in general $\tau(A_n) = \mu(A_n) = n$ conditions, we obtain that the total number of conditions compared with the dimension of the space of curves of the given degree reveals the following asymptotics:

$$\lim_{k\to\infty} \frac{ns_k}{d_k(d_k+3)/2} = \frac{3}{2} - \frac{3}{2(n+2)} > 1 \quad \text{for all } n \geq 2. \tag{2.40}$$

Thus, the A_n impose dependent conditions and the corresponding equisingular stratum is not T-smooth. For instance, in general, we cannot decide whether there exist curves of degree d_k with any number $m < s_k$ of singularities of type A_n.

Other series of extremal examples exhibit plane curves with one singularity A_m with $m = m(d)$ as large as possible for a given degree d. More precisely, each series consists of plane curves of degrees $d \to \infty$ with one $A_{m(d)}$ singularity: Gusein-Zade and Nekhorochev [36, Proposition 2] constructed a series of curves for which

$$\lim_{d\to\infty} \frac{m(d)}{d(d+3)/2} = \frac{15}{14} = 1.0714\ldots .$$

Later Cassou-Nogues and Luengo [13] obtained another sequence of curves with

$$\lim_{d\to\infty} \frac{m(d)}{d(d+3)/2} = 8 - 4\sqrt{3} = 1.0717\ldots .$$

The best known result is due to Orevkov [64, Sect. 4]:

Theorem 2.4.4 *There exists a sequence of plane curves of degrees $d_k \to \infty$ as $k \to \infty$ having a singular point of type A_{m_k} such that*

$$\lim_{k\to\infty} \frac{m_k}{d_k(d_k+3)/2} = \frac{7}{6} = 1.1666\ldots .$$

The ratios of the number of imposed conditions to the dimension of the space of curves of a given degree in all these examples appears to be > 1; hence, we again observe a non-T-smooth equisingular family. Also, in general, we cannot decide whether there exist curves of degree d_k with an A_r singularity for all $r < m_k$.

It is worth to compare the extremal examples by Hirano and Orevkov with the known restrictions. The genus bound (2.3) and Plücker formulas (2.14), (2.17) yield weaker bounds than the Hirzebruch-Ivinskis' and the spectral ones. Namely, the Hirzebruch-Ivinskis bound (2.21) combined with the first formula in (2.22) implies that the limit ratio of the total Milnor number to the dimension of the space of curves of the given degree does not exceed

$$\frac{5(n+1)}{3(n+2)} \quad \text{in the case of Hirano,}$$
$$\frac{5}{3} \quad \text{in the case of Orevkov.}$$

By (2.25) the spectra of the singularity A_n and of an ordinary d-fold singularity are

$$\left\{\frac{k}{n+1} - \frac{1}{2} \ :\ 1 \le k \le n\right\}, \quad \left\{\frac{i+j}{d} - 1 \ :\ 1 \le i, j \le d-1\right\},$$

respectively. Applying the semicontinuity of the spectrum on the interval $(\frac{n}{n+1} - \frac{3}{2}, \frac{n}{n+1} - \frac{1}{2}]$, one can easily derive an upper bound of the total Milnor number for the existence of a curve of fixed degree d with s singular points of type A_n:

$$sn \le \#\left\{(i,j) \in \mathbb{Z}^2 \mid 1 \le i, j < d,\ \frac{n}{n+1} - \frac{1}{2} < \frac{i+j}{d} \le \frac{n}{n+1} + \frac{1}{2}\right\}. \quad (2.41)$$

If we fix $n \ge 2$ and let $d \to \infty$, then we obtain

$$\limsup_{d\to\infty} \frac{sn}{d(d+3)/2} \le \frac{3}{2} - \frac{2}{(n+1)^2},$$

which is comparable with the limit ratio in examples of Hirano (2.40), showing that, for large n the spectral bound is almost sharp.

If we let $s = 1$ and $d \to \infty$ in (2.41), we will obtain

$$\limsup_{d\to\infty} \frac{n}{d(d+3)/2} = \frac{3}{2},$$

which differs from the asymptotical ratio attained in Orevkov's examples, Theorem 2.4.4, leaving open the question on the sharpness of the spectral bound in the case of one singularity A_n.

Curves with ordinary multiple singular points. By an *ordinary (multiple) singular point* we understand a singularity consisting of several smooth local branches intersecting each other transversally. It happens that the patchworking construction provides an asymptotically optimal existence condition for curves with ordinary multiple points as well, which, moreover, completely covers both the complex and the real case.

Theorem 2.4.5 *For a fixed positive integer m the following holds:*

(1) There exists a linear polynomial $\psi^{ordinary}_{m,fix}(d)$ such that, for an arbitrary sequence of integers $r_2, \ldots, r_m \ge 0$, the inequality

$$\sum_{i=2}^{m} \frac{i(i+1)}{2} r_i \le \frac{d^2}{2} + \psi^{ordinary}_{m,fix}(d) \quad (2.42)$$

is sufficient for the existence of an irreducible complex plane curve of degree d in a T-smooth equisingular family, having r_i ordinary singular points of multiplicity $i = 2, \ldots, m$ as its only singularities, all of them in general position.

(2) Furthermore, the same inequality (2.42) is sufficient for the existence of a real plane irreducible curve of degree d in a T-smooth equisingular family, having the given collection of ordinary multiple points in conjugation-invariant general position, when we prescribe the numbers of pairs of imaginary ordinary singular points for each multiplicity $2, \ldots, m$ and prescribe the number of real local branches for each real ordinary singular point.

Note that the necessary existence condition in the setting of Theorem 2.4.5 is

$$\sum_{i=2}^{m} \frac{i(i+1)}{2} r_i \le \frac{d(d+3)}{2},$$

as long as d is big enough, which follows from the Alexander-Hirschowitz theorem [4, Theorem 1.1] (see also [31, Theorem 3.4.22]). That is, the bound (2.42) is asymptotically optimal (even for non-T-smooth families).

Proof The first part of the theorem follows, in fact, from the Alexander-Hirschowitz theorem [4, Theorem 1.1] after some routine work ensuring that a generic bivariate polynomial of degree d, whose derivatives vanish up to appropriate order at the given points in general position, defines an irreducible curve with only ordinary singularities as prescribed. The second part, however, is not accessible within this framework, since the control over the real singularity types may require at least $\frac{d^2}{m}$ extra independent conditions, which is not bounded by a linear function of d. So, to prove the second statement (and thereby the first one), we apply a suitable version of the patchworking construction.

The main element of the construction consists of a collection of the following patchworking patterns. Fix any $2 \le i \le m$ and consider the lattice rectangle $R_i = \text{Conv}\{(0,0), (0,i), (i+1,0), (i+1,i)\}$. We claim that there exists a real irreducible polynomial $F_i(x,y)$ with Newton polygon R_i (see Fig. 2.5a) which defines a real plane curve having in $(\mathbb{C}^*)^2$ exactly two singular points:

- either two complex conjugate ordinary singularities of order i,
- or two real ordinary singularities with the prescribed number $j \le i/2$ of pairs of complex conjugate local branches.

Indeed, in the projective plane $\mathbb{P}^2$ with coordinates x, y, z consider the pencil of conics passing through the points $(1,0,0)$ and $(0,1,0)$ and through two more points $p_1, p_2 \in (\mathbb{C}^*)^2 = \mathbb{P}^2 \setminus \{xyz = 0\}$, either a pair of complex conjugate points, or a pair of real generic points. In the case of complex conjugate p_1, p_2, we pick i distinct smooth real conics $Q_1, ..., Q_i$ in our pencil, while in the case of real p_1, p_2, we pick i smooth conics $Q_1, ..., Q_i$ in our pencil so that $i = 2j$ of them are real and the others form j pairs of complex conjugate conics. Consider the projective curve $Q_1 \cdots Q_i \cdot L = 0$, where $L = x + \lambda z$ with a generic real number λ. This curve has ordinary singularities of order i at p_1, p_2, and $(0,1,0)$, an ordinary singularity of order $i+1$ at $(1,0,0)$, and i more nodes, which are intersection points of the line $L = 0$ with the conics. There exists a small real deformation of the considered curve that preserves the ordinary singularities at $p_1, p_2, (1,0,0)$, and $(0,1,0)$ and smoothes out all the extra nodes. This follows directly from [76, Theorem, p. 31] or, after the blowing up $\Sigma \to \mathbb{P}^2$ of the four ordinary singularities, from [23, Theorem 6.1(iii)], since each component C of the blown-up curve satisfies $-CK_\Sigma > 0$, and the nodes do not contribute to the right-hand side of the required inequalities (see also [31, Theorem 4.4.1(b) (formula (4.4.1.3)), Proposition 4.4.3(b) (formula (4.4.1.10)), and

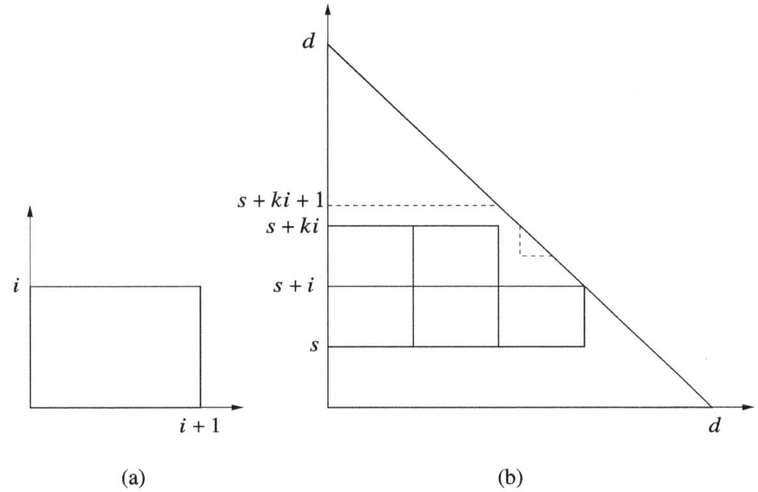

Fig. 2.5 Patchworking of real curves with ordinary singularities

Remark 4.4.212(ii)]). So, we obtain the desired polynomial F_i by substituting $z = 1$ into the equation of the above deformed curve.

Then we subdivide the lattice triangle $T_d = \mathrm{Conv}\{(0,0),(d,0),(0,d)\}$ into convex lattice polygons, in which the desired number of patches corresponding to any fixed real ordinary singularity type should be arranged as shown in Fig. 2.5b. The polynomials for each rectangle are obtained from just one suitable polynomial F_i constructed above, which should be multiplied by an appropriate monomial and undergo the coordinate change $x \mapsto x^{-1}$ and/or $y \mapsto y^{-1}$ in order to make any two neighboring polynomials agree along the common side of their Newton rectangles. Between the unions of rectangles corresponding to different types of ordinary singularities, we leave the space of vertical size at most $m - 1$ (shown by dashed lines in Fig. 2.5b) which should be subdivided into lattice triangles with the associated polynomials defining smooth real curves in $(\mathbb{C}^*)^2$ so that finally one obtains a convex subdivision of T_d. To apply the patchworking theorem [79, Theorem 3.6] we have to verify the topological transversality conditions. With an appropriate orientation of the adjacency graph of the tiles of the subdivision, we get $\partial_+ R_i$ in each rectangle R_i to be the union of the bottom and the left sides. The sufficient transversality condition stated in [79, Theorem 4.1(1), the first formula] (see also [31, Proposition 4.5.2 and Corollary 1.2.22]) reads as

$$\text{the contribution of the two } i - \text{multiple points } = 2i$$

$$< 2i + 1 = \text{the total length of the upper and the right sides of } R_i.$$

Finally, we note that the resulting curve (obtained by the patchworking construction) admits by T-smoothness a deformation smoothing out any prescribed singularities and keeping the remaining ones. Thus, one can realize any values in the right-hand side of (2.42) with $\phi^{ordinary}_{m,fix}(d) \sim m^2 d$. □

Curves with semi-quasihomogeneous singularities. Consider curve singularities topologically equivalent to $x^m + y^n = 0$, $2 \le m \le n$. A slightly modified Zariski construction mentioned in Sect. 2.3.2 provides a series of curves with a large number of semi-quasihomogeneous singularities. Combining Zariski's with the patchworking construction we obtain even a series of curves belonging to a T-smooth family.

Theorem 2.4.6 *Let* $2 \le m \le n$.

(1) If $\gcd(m,n) = 1$, *then for any* $r \ge 1$, *the curve* C *of degree* $d = rmn$ *given by* $F^n + G^m = 0$, *where* $F, G \in \mathbb{C}[x,y,z]$ *are generic homogeneous polynomials of degree* rm *and* rn, *respectively, is irreducible and has* $r^2 mn = \frac{d^2}{mn}$ *singular points of type* $x^m + y^n = 0$ *as its only singularities.*

(2) If $\gcd(m,n) > 1$, *then for any* $r \ge 1$, *the curve* C *of degree* $d = rmn + 1$ *given by* $L_1 F^n + L_2 G^m = 0$, *where* $F, G \in \mathbb{C}[x,y,z]$ *are generic homogeneous polynomials of degree* rm *and* rn, *respectively and* L_1, L_2 *are generic linear polynomials, is irreducible and has* $r^2 mn = \frac{(d-1)^2}{mn}$ *singular points of type* $x^m + y^n = 0$ *as its only singularities.*

Moreover, in both cases we can assume that the constructed curves are real and all their singular points are real.

Proof By construction, the curves $F = 0$ and $G = 0$ intersect transversally at $r^2 mn$ distinct points, and the curve C has the topological singularity $x^m + y^n = 0$ at each of these intersection points. □

Note that the number of singularities obtained on these "Zariski curves" is close to the genus bound (2.3): for instance, under the conditions of Theorem 2.4.6, the number of the considered singular points does not exceed

$$\frac{(d-1)(d-2)}{(m-1)(n-1) + \gcd(m,n) - 1} < \frac{(d-1)^2}{(m-1)(n-1)},$$

which is comparable with the actual numbers $\frac{d^2}{mn}$ and $\frac{(d-1)^2}{mn}$ of singularities. On the other hand, we cannot guarantee that the conditions imposed on the curve by singular points are independent; hence, we cannot ensure that there exist curves of the given degree with any intermediate amount of singular points of the given type.

One can modify Zariski's construction further and obtain a curve with different semi-quasihomogeneous singularities, and, moreover, obtain a curve belonging to a T-smooth equisingular stratum.

Theorem 2.4.7 *(1) Given integers*

$$d_0, d'_0 \ge 0,\ d_1, ..., d_k, d'_1, ..., d'_l \ge 1,\ m_1, ..., m_k, n_1, ..., n_l \ge 2$$

such that

$$d = \sum_{i=1}^{k} m_i d_i + d_0 = \sum_{j=1}^{l} n_j d'_j + d'_0.$$

Then the plane curve of degree d

$$H \prod_{i=1}^{k} F_i^{m_i} + H' \prod_{j=1}^{l} G_J^{n_j} = 0, \tag{2.43}$$

where $F_1, ..., F_k, G_1, ..., G_l, H, H'$ *are generic homogeneous polynomials of degree* $d_1, ..., d_k, d'_1, ..., d'_l, d_0, d'_0$, *respectively, has* $d_i d'_j$ *singular points of topological type* $x^{m_i} + y^{n_j} = 0$ *for all* $i = 1, ..., k$, $j = 1, ..., l$. *Further on, we can achieve that the constructed curve is real and all its singular points are real.*

(2) In addition, if either

(i)

$$\begin{cases} \left[d_0 \neq d'_0 \ \ \textit{or} \ \ \gcd(m_1, ..., m_k, n_1, ..., n_l) = 1\right] \\ \textit{and} \ \ \sum_{i=1}^{k} \sum_{j=1}^{l} d_i d'_j (m_i + n_j - \gcd(m_i, n_j) - \varepsilon(m_i, n_j)) < \widetilde{d}(d' + 3), \end{cases}$$

where $d' = \min\{d_0, d'_0\}$, $\widetilde{d} = d - d'$,

$$\varepsilon(a, b) = \begin{cases} 1, & a \equiv 1 \mod b, \textit{ or } b \equiv 1 \mod a, \\ 0, & \textit{otherwise}, \end{cases}$$

or

(ii)

$$\sum_{i=1}^{k} \sum_{j=1}^{l} d_i d'_j (m_i + n_j - \gcd(m_i, n_j) - \varepsilon(m_i, n_j)) < \widetilde{d}(d' + 2), \tag{2.44}$$

then the curve constructed above belongs to a T-smooth equisingular family, and it admits a deformation moving all its singular points to a general position.

Proof We have to explain only part (2). Consider the case (2ii). We can assume that $d_0 = d'_0 = d' > 0$. We shall prove that the curve given by

$$\widehat{H} \left(L \prod_{i=1}^{k} F_i^{m_i} + L' \prod_{j=1}^{l} G_J^{n_j} \right) = 0, \tag{2.45}$$

where $F_1, ..., F_k, G_1, ..., G_l, L, L', \widehat{H}$ are generic homogeneous polynomials of degree $d_1, ..., d_k, d'_1, ..., d'_l, 1, 1, d' - 1$, respectively, belongs to a T-smooth equisingular family with respect to the singular points located at the set $\bigcup_{i,j}\{F_i =$

$0\} \cap \{G_j = 0\}$, and all these singular points can be moved to a general position (while the nodes located on H may disappear). Since these properties are open, they will hold for the original curve (2.43) as well. On the other hand, by [76, Theorem in page 31] the required properties will follow from [76, Inequality (4)] which in our situation takes the form of (2.44) (cf. the definition of the invariant b in [76, Definition 2]). In the same manner, one can settle the case (2i). □

Remark 2.4.8 (1) Under the hypotheses of part (2) of Theorem 2.4.7, one can deform the curve (2.43) smoothing out prescribed singularities and keeping the other ones.

(2) The hypotheses of Theorem 2.4.7(2) can be relaxed to the following one:

$$d' \geq \widetilde{d} \cdot \max_{i,j} \left\{ \frac{1}{m_i} + \frac{1}{n_j} \right\} - 2\,.$$

2.4.3 *Curves with Arbitrary Singularities*

Note that none of the constructions discussed in Sects. 2.3 and 2.4.2 can be applied directly, if we ask about the existence of plane curves of a given degree with a prescribed collection of topological or analytic singularities, which are not further specified. For instance, the patchworking construction requires to find a patchworking pattern for and prescribed singularity, i.e., a bivariate polynomial that defines a curve with a given singularity and whose Newton polygon can be used as a tile in the subdivision of the triangle $T_d = \mathrm{Conv}\{(0,0), (d,0), (0,d)\}$. To get a reasonable existence result, one needs such tiles of a possibly minimal area, and it is not clear at all how to find these minimal tiles for arbitrary topological or analytic singularities.

However, there is another approach that combines some features of the patchworking construction with suitable H^1-vanishing criteria for the ideal sheaves of zero-dimensional subschemes of the plane placed in general position. The first attempt like that, undertaken in [26], has led to the following existence criterion: for any positive integer d and topological singularity types $S_1, \ldots, S_r$, the inequality

$$\sum_{i=1}^{r} \mu(S_i) \leq \frac{d^2}{392} \tag{2.46}$$

is sufficient for the existence of an irreducible plane curve of degree d having r singular points of types $S_1, \ldots, S_r$, respectively, as its only singularities.

This criterion already possessed two important properties: it was *universal*, i.e., uniformly applicable to arbitrary topological singularities, and *asymptotically proper* i.e. comparable with the necessary condition (2.4). On the other hand, the analytic singularity types were left aside, and the coefficient of d^2 in (2.46) was too small. An improved method was suggested in [81] (for a further improvement and a detailed exposition see [31, Sect. 4.5.5]). The main result was (cf. [81, Theorem 3 and Remark 5] and [31, Corollary 4.5.15]):

Theorem 2.4.9 *For any positive integer d and an arbitrary sequence of complex, resp. real, analytic singularity types* $S_1, \ldots, S_r$, *the inequality*

$$\sum_{i=1}^{r} \mu(S_i) \le \frac{1}{9}(d^2 - 2d + 3) \tag{2.47}$$

is sufficient for the existence of a plane irreducible complex, resp. real, curve of degree d having r singular points of types $S_1, \ldots, S_r$, *respectively, as its only singularities. Moreover, the positions of the singular points together with the tangent directions for unibranch singularities can be chosen generically.*

We point out that condition (2.47) has a much larger coefficient of d^2 in the right-hand side than (2.46) and covers both arbitrary topological and analytic singularity types, being universal with respect to the choice of singularities.

We also remark that condition (2.47) is a weaker form of the following stronger sufficient existence conditions (see [31, Theorem 4.5.14]):

$$6n + 10k + \frac{49}{6}t + \frac{625}{48} \sum_{S_i \ne A_1, A_2} \delta(S_i) \le d^2 - 2d + 3,$$

for topological singularity types $S_1, \ldots, S_r$, and

$$6n + 10k + \sum_{S_i \ne A_1, A_2} \frac{7\mu(S_i) + 2\delta(S_i))^2}{6\mu(S_i) + 3\delta(S_i)} \le d^2 - 2d + 3,$$

for analytic singularity types $S_1, \ldots, S_r$. In these inequalities, n is the number of nodes A_1, k is the number of cusps A_2, and t is the number of singularities A_{2m}, $m \ge 2$, occurring in the list $S_1, \ldots, S_r$. Notice that the coefficients in front of n and k in the above two formulas are the best possible, since a node at a prescribed point imposes 3 conditions, while a cusp at a fixed point with a fixed tangent direction imposes 5 conditions.

The case of just one singularity is of special interest. The corresponding result sounds as follows ([81, Theorem 2 and Remark 5] and [31, Theorem 4.5.19]):

Theorem 2.4.10 *For an arbitrary analytic singularity type S, there exists a plane curve of degree*

$$d \le 3\sqrt{\mu(S)} - 1 \tag{2.48}$$

having a singular point of type S as its only singularity.

Remark 2.4.11 A necessary condition for d as in Theorem 2.4.10 comes, for instance, from (2.4): $d \ge \sqrt{\mu(S)} + 1$. Thus, the sufficient condition (2.48) is of the same order $\sqrt{\mu(S)}$ as the necessary one.

We comment on the main ideas behind Theorems 2.4.9 and 2.4.10:

The first important ingredient is to reduce the existence problem to an H^1-vanishing condition for the ideal sheaf of a suitable zero-dimensional subscheme of the plane. Namely, to each reduced plane curve germ (C, p) we associate two zero-dimensional schemes $Z^s_{st}(C, p)$ and $\widetilde{Z}^a_{st}(C, p)$ in $\mathbb{P}^2$ supported at p that are defined as follows (cf. [31, Sects. 1.1.4 and 4.5.5.1]):

- Take the complete resolution tree $\mathcal{T}^\infty(C, p)$, choose the subtree $\mathcal{T}^*(C, p)$ containing all infinitely near points which are not the nodes of the union of the strict transform of (C, p) with the exceptional locus, and then define $Z^s_{st}(C, z)$ by the ideal $I^s_{st} \subset \mathcal{O}_{\mathbb{P}^2,p}$ generated by the elements $\varphi \in \mathcal{O}_{\mathbb{P}^2,p}$ having the multiplicity $\mathrm{mt}(C, p) + 1$ at p, and the multiplicity of the strict transform of (C, p) at each infinitely near point $q \in \mathcal{T}^*(C, p) \setminus \{p\}$.
- Let (C, p) be given by $f(x, y) = 0$ with $f \in \mathcal{O}_{\mathbb{P}^2,p}$ square-free, x, y affine coordinates in a neighborhood of p such that $p = (0, 0)$. Define $\widetilde{Z}^a_{st}(C, z)$ by the ideal $\mathfrak{m}_p \widetilde{I}^a \subset \mathcal{O}_{\mathbb{P}^2,p}$, where

$$\widetilde{I}^a = \{g \in \mathcal{O}_{\mathbb{P}^2,p} \ : \ g, g_x, g_y \in \langle f, f_x, f_y \rangle\}.$$

The importance of these schemes comes from the following claim (cf. [31, Proposition 4.5.12]):

Lemma 2.4.12 *Let Z denote $Z^s_{st}(C, p)$, resp. $\widetilde{Z}^a_{st}(C, p)$. If a positive integer d satisfies*

$$H^1(\mathbb{P}^2, \mathcal{J}_{Z/\mathbb{P}^2}(d)) = 0,$$

where $\mathcal{J}_{Z/\mathbb{P}^2}$ is the ideal sheaf of the subscheme $Z \subset \mathbb{P}^2$, then there exists a curve $C' \subset \mathbb{P}^2$ of degree d such that the germ (C', p) is topologically, resp. analytic equivalent to (C, p). Furthermore, the corresponding equisingular family $V^{irr}_d(S)$ (S being the topological, resp. analytic type of (C, z)) is T-smooth at C'.

The length of $Z^s_{st}(C, z)$ and $\widetilde{Z}^a_{st}(C, z)$ can be estimated from above by a linear function in $\delta(C, p)$ and $\mu(C, p)$ (see [31, Corollary 1.1.4 and Lemma 1.1.78]).

It is not difficult to see that, for a randomly chosen germ (C, p), the minimal d in Lemma 2.4.12 can be of order $\deg Z$, i.e., of order $\mu(C, p)$, but $\sqrt{\mu(C, p)}$ is required in Theorems 2.4.9 and 2.4.10. So, the second important idea is to replace the germ (C, p), or, more precisely, the corresponding zero-dimensional scheme Z by a generic element in $\mathrm{Iso}(Z)$, the orbit of Z by the action of the group $\mathrm{Aut}(\mathcal{O}_{\mathbb{P}^2,p})$. The principal bound is as follows (cf. [81, Propositions 8 and 10, Remark 3] and [31, Proposition 3.6.1 and Corollary 3.6.4]). Given an irreducible zero-dimensional scheme $Z \subset \mathbb{P}^2$ supported at $z \in \mathbb{P}^2$, denote by $M_2(Z)$ the intersection multiplicity of two generic elements of the ideal $I(Z) \subset \mathcal{O}_{\mathbb{P}^2,z}$. If Z consists of irreducible components $Z_1, ..., Z_k$, we set $M_2(Z) = M_2(Z_1) + ... + M_2(Z_k)$.

Lemma 2.4.13 *For an arbitrary zero-dimensional Z of degree* $\deg Z > 2$, *there exists $Z' \in \mathrm{Iso}(Z)$ and*

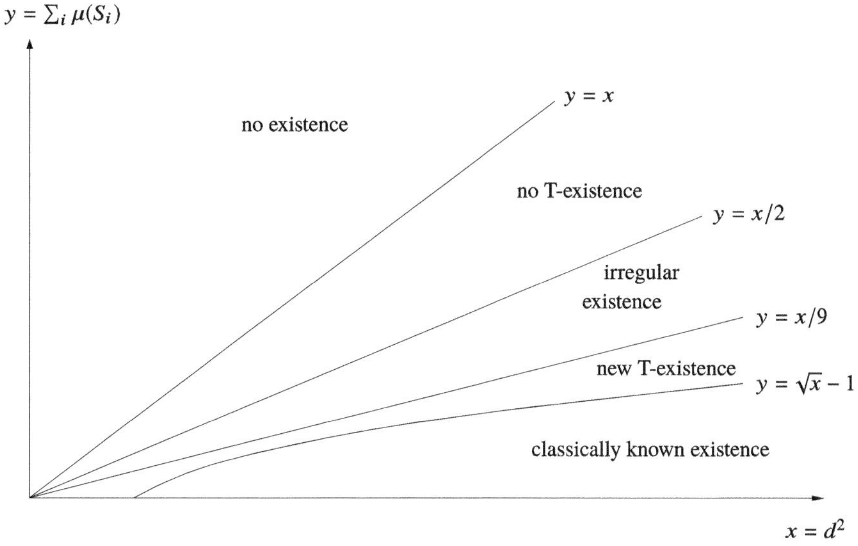

Fig. 2.6 Necessary and sufficient conditions for existence: The statement holds asymptotically for the region between the corresponding boundary curves

$$d \leq \frac{\deg Z}{\sqrt{\frac{4}{3}M_2(Z)}} + \sqrt{\frac{4}{3}M_2(Z)} - 2$$

such that

$$H^1(\mathbb{P}^2, \mathcal{J}_{Z'/\mathbb{P}^2}(d)) = 0\,.$$

In the case of a reducible scheme $Z = Z_1 \cup \ldots \cup Z_r$*, we move its supporting points to a general position and choose generic element* $Z_i' \in \mathrm{Iso}(Z_i)$ *for each component* Z_i*.*

A proper combination of these two ideas (Lemmas 2.4.12 and 2.4.13) leads to Theorems 2.4.9 and 2.4.10.

The following diagram illustrates the present knowledge about the existence of plane curves of degree d with arbitrary (analytical or topological) singularities $S_1, \ldots, S_r$ (Fig. 2.6).

2.5 Related and Open Problems

2.5.1 Existence Versus T-Smoothness and Irreducibility

The existence problem for singular algebraic curves is tightly related to the geometry of the corresponding equisingular family, especially, to the T-smoothness property, which was crucial in the patchworking construction and also used in other constructions. In this section, we construct singular plane algebraic curves of two sorts: (i) those which demonstrate the sharpness o f the known T-smoothness criteria and (ii) those which yield examples of reducible equisingular families.

Recall the *universal sufficient conditions for the T-smoothness* (see [75, Theorem 1], [24, Corollary 3.9(d)], [28, Theorems 1 and 2] and [31, Theorems 4.3.8 and 4.3.9]):

Theorem 2.5.1 *Let C be an irreducible plane curve of degree d with singular points* $p_1, ..., p_r$ *of topological or analytic types* $S_1, ..., S_r$ *respectively. Then the equisingular family* $V_d^{irr}(S_1, ..., S_r)$ *is* T*-smooth at C if either*

$$\sum_{i=1}^{r} \tau'(S_i) < 4d - 4, \tag{2.49}$$

where $\tau' = \tau^{es}$ *if* S_i *is a topological type and* $\tau' = \tau$ *if* S_i *is an analytic type,* $i = 1, ..., r$*, or*

$$\sum_{i=1}^{r} \gamma'(C, p_i) \le (d+3)^2\ , \tag{2.50}$$

where $\gamma' = \gamma^{es}$ *for* S_i *is a topological type and* $\gamma' = \gamma^{ea}$ *for* S_i *an analytic type,* $i = 1, ..., r$.

The symbols τ^{es}, τ, γ^{es}, γ^{ea} are topological or analytic singularity invariants. For precise definitions we refer to [31, Sect. 1.2.3.1 and Definition 1.1.63], and for the detailed study of their properties to [31, Corollary 1.1.64, Proposition 1.2.26]. Here we provide only the following information used below (see also Preliminaries):

- τ is the Tjurina number, i.e., the dimension of the Tjurina algebra

$$\tau(C, p) = \tau(f) = \dim_{\mathbb{C}} \mathcal{O}_{\mathbb{P}^2,z}/\langle f, f_x, f_y\rangle,$$

  where a square-free element $f \in \mathcal{O}_{\mathbb{P}^2,z}$ defines the curve germ (C, p), and x, y are local coordinates;
- $\tau^{es}(C, p)$ is the codimension of the μ-const stratum in the versal deformation base of the curve germ (C, p); it always satisfies

$$\tau^{es}(S) \le \tau(S) \le \mu(S),$$

with equalities for a simple singularity type S.

- $\gamma^{es}(C, p) \leq \gamma^{ea}(C, p)$ for any curve germ with equality if the singularity is simple, furthermore

$$\gamma^{ea}(S) \begin{cases} = (\mu(S)+1)^2, & \text{if } S = A_n,\ n \geq 1, \\ \leq (\mu(S)+1)^2, & \text{otherwise.} \end{cases}$$

We also recall important particular cases (see [31, Corollaries 4.3.6, 4.3.11, formula (1.2.3.1)] and Theorem 2.5.1):

Theorem 2.5.2 *(1) An irreducible plane curve of degree d with n nodes and k cusps as its only singularities belongs to a T-smooth equisingular family if either*

$$k < 3d\ , \tag{2.51}$$

(any $n \geq 0$, i.e., nodes do not count) or

$$4n + 9k \leq (d+3)^3\ . \tag{2.52}$$

(2) An irreducible plane curve of degree d with r singular points of simple singularity types $S_1, \ldots, S_r$ belongs to a T-smooth equisingular family if

$$\sum_{S_i \in A} (\mu(S_i)+1)^2 + \sum_{S_i \in D,E} \max\left\{(\mu(S_i)-1)^2, \frac{1}{2}(\mu(S_i)+2)^2\right\} \leq (d+3)^3. \tag{2.53}$$

Remark 2.5.3 Observe that, first, the right-hand sides in the *4d-criterion* (2.49) and the *3d-criterion* (2.51) for T-smoothness are only linear in d, while the sufficient condition (2.47) for existence is quadratic in d in the right-hand side. Second, the T-smoothness restrictions (2.51) and (2.52) are far away from the existence conditions in Theorem 2.3.6, and, third, the invariants assigned to singular points in the left-hand side of (2.50) and (2.53) are in general of order $\mu^2(S)$, while in (2.47) we have just $\mu(S)$. That is, for general analytic or topological types there exists a wide range of nonempty equisingular families, which do not fall to the limits of Theorems 2.5.1 and 2.5.2. On the other hand, for curves with simple (Theorem 2.4.1) , ordinary (Theorem 2.4.5), and semi-quasihomogeneous singularites (Theorem 2.4.7), we have T-smooth equisingular families with asymptotically proper bounds for existence.

Thus, natural questions arise for arbitrary singularities:

(1) *Does the difference between the left-hand sides of the sufficient T-smoothness criteria in Theorems* 2.5.1 *and* 2.5.2 *and the left-hand side of the sufficient existence criterion* (2.47) *reflect the lack of T-smoothness outside the limits pointed in Theorems* 2.5.1 *and* 2.5.2*?*

(2) *Or, can the singularity invariants in the left-hand side of the sufficient T-smoothness criteria be essentially diminished?*

We exhibit a series of singular plane algebraic curves and their families demonstrating that the linear bounds (2.49) and (2.51) are sharp, the coefficients in the left-hand side of (2.52) are sharp, and the singularity invariants in the other T-smoothness conditions can, in principle, be improved only by a constant factor, while their order with respect to the Milnor number persists. We mainly use the constructions discussed in Sects. 2.3 and 2.4. For all details and more examples see [31, Sects. 4.2.3, 4.3.3] and references therein.

The following example, which is due to du Plessis and Wall [18] (elaborated further in [21]), shows that the bound (2.49) is sharp.

Theorem 2.5.4 *For any $d \ge 5$ the irreducible curve $C \subset \mathbb{P}^2$ given by $x_1^d + x_2^5 x_0^{d-5} + x_2^d = 0$ has the unique singular point $z = (1, 0, 0)$ with Tjurina number $4d - 4$, and the equisingular family $V_d^{irr}(S)$, where S is the analytic type of the germ (C, z), is not T-smooth. Furthermore, the family $V_d^{irr}(S)$ is nonreduced for $d \le 6$, consists of two intersecting components for $d = 7$, and is reduced, irreducible with a singular locus containing C for $d \ge 8$.*

The examples found in [28, Proposition 4.5] (see also [31, Theorem 4.3.23]) show that the classical Severi-Segre-Zariski bound (2.51) is sharp and that the coefficients 4 and 9 in (2.52) are sharp.

Theorem 2.5.5 *Let $p \ge 6$, $q \ge 9$. Then the variety $V_d^{irr}(n \cdot A_1, k \cdot A_2)$ has a non-T-smooth component if*

(a) $d = 3p$, $n = 0$, $k = p^2 + 3p$, or
(b) $d = 2q$, $n = q^2 - 9q$, $k = 6q$.

In the series (a), $9k = d^2 + 9d$, and hence the coefficient 9 in (2.52) is sharp. In the series (b), $4n = d^2 - 18d$ and also $k = 3d$, which yields the sharpness of (2.51) and of the coefficient 4 in (2.52). A curve $C \in V_d^{irr}(n \cdot A_1, k \cdot A_2)$ at which the T-smoothness fails can be constructed by Zariski's method as in Theorem 2.4.6: namely, we set C to be given by

$$AP^2R^2 + BQ^2 = 0,$$

where A, B, P, Q, R are generic polynomials of degrees a, b, p, q, r, respectively, such that

$$a = d - 3p - 2r \ge 0, \quad b = d - 2q \ge 0.$$

The series (a) corresponds to $r = 0, d = 3p, q = p + 3, p \ge 6$, while the series (b) corresponds to $d = 2q$, $p = 6$, $r = q - 9$. For the failure of the T-smoothness see the proof of [31, Theorem 4.3.23].

The next series of examples were found in [25, Theorems 5 and 6] (see also [31, Theorems 4.3.24 and 4.3.25]). They show that the coefficients of the A_n and D_n singularities in (2.53) may, in principle, be reduced, but only by a constant factor $\ge \frac{1}{4}$.

Theorem 2.5.6 *Let $l \geq 2$, $0 \leq s \leq l-2$, $q \geq \frac{3}{l-s-1}$ be integers.*

(1) Let $k > 2l - s$ be an integer. Then there exists an irreducible plane curve C of degree $d = q(k+s)$ having precisely q^2 singular points, all of type A_{kl+s-1}, such that the family $V_d^{irr}(q^2 \cdot A_{kl+s-1})$ is not T-smooth at C. Moreover,

- *if $k > 4l - s$, then C belongs to a component of $V_d^{irr}(q^2 \cdot A_{kl+s-1})$ of expected dimension which is singular at C;*
- *if $k \geq \max\{l^2 + 2l, 4l + 4 - s\}$, then the germ of $V_d^{irr}(q^2 \cdot A_{kl+s-1})$ at C is a singular, normal complete intersection.*

(2) Let $k > 2l - s + 1$ be integer. Then there exists an irreducible plane curve C of degree $d = q(k+s)$ having precisely q^2 singular points, all of type D_{kl+s+1}, such that the family $V_d^{irr}(q^2 \cdot D_{kl+s+1})$ is not T-smooth at C. Moreover,

- *if $k > 4l - s$, then C belongs to a component of $V_d^{irr}(q^2 \cdot D_{kl+s+1})$ of expected dimension which is singular at C;*
- *if $k \geq \max\{l^2 + 2l + 2, 4l + 4 - s\}$, then $V_d^{irr}(q^2 \cdot D_{kl+s+1})$ is a singular, normal locally complete intersection at C.*

In particular, for $l = 2$, $s = 0$, we have in part (1)

$$d^2 = q^2k^2 \quad \text{and} \quad q^2(\mu(A_{2k-1}) + 1)^2 = 4q^2k^2,$$

and in part (2)

$$d^2 = q^2k^2 \quad \text{and} \quad q^2(\mu(D_{2k+1}) - 1)^2 = 4q^2k^2.$$

In part (1) the construction is as follows. Take the affine curve

$$(y + y^l - x^l)^2(1 + \lambda_1 x^{k+s-2l} + \lambda_2 x^s y^k + y^{k+s}) = 0,$$

where $\lambda_1, \lambda_2 \in \mathbb{C}$ are generic. It is easy to check that it is irreducible with the unique singularity A_{kl+s-1} at the origin. Then we take its projective closure, choose a generic projective coordinate system (x_0, x_1, x_2) and apply the transformation $(x_0, x_1, x_2) = (y_0^q, y_1^q, y_2^q)$ (cf. Ivinskis' and Hirano's constructions [39, 43]). In part (2), we start with the affine curve

$$x(y + y^l - x_l)^2(1 + \lambda_1 x^{k+s_2l-1}) + \lambda_2 x^s y^k + y^{k+s} = 0,$$

which is irreducible and has the unique singularity D_{kl+s+1} at the origin. Then similarly take the projective closure and apply the transformation $(x_0, x_1, x_2) = (y_0^q, y_1^q, y_2^q)$ in generic projective coordinates x_0, x_1, x_2. For the lack of the T-smoothness, we refer to [31, Sect. 4.3.3.2].

Remark 2.5.7 The first ever (finitely many) examples of reduced equisingular families of expected dimension which are not smooth are due to Luengo [54, 55]. In particular, one of his examples comes with a curve of degree 9 with a unique singularity A_{35}. We recover this curve in Theorem 2.5.6(1) for $l = 4$, $s = 0$, $q = 1$, $k = 9$.

Another application of the construction methods discussed in Sects. 2.3 and 2.4 is to find interesting examples of reducible equisingular families. There are several ways to verify that an equisingular family is reducible:

- An explicit computation of the equisingular family. This is available only for very specific situations for relatively small degrees, see, for example, [21, Theorem 1.1(ii)], where the family $V_7^{irr}(S)$ (S being the analytic type of the singularity $x^5 + y^7 = 0$) was shown to be reducible.
- Exhibiting two (or more) components of an equisingular family, whose generic members differ from the algebraic-geometric point of view. In the classical example by Zariski [108, Sects. VIII.3 and VIII.5], the family $V_6^{irr}(6 \cdot A_2)$ contains (at least) two components: in one of them the generic curve has 6 cusps on a conic, while on the other one this is not the case.
- Exhibiting two (or more) components of the equisingular family, whose generic members are embedded into the plane in a topologically different way. The aforementioned Zariski example is of such kind [106], since the complements to these generic curves in the plane are not homeomorphic (they have different Alexander polynomials and different fundamental groups).
- Exhibiting two (or more) components of the equisingular family that have different dimensions, or such that one is reduced (for instance, T-smooth) and the other is not.

We present here examples of the last kind. In fact, equisingular families with components of dimension higher than the expected one were known for a while: Segre [72] (see also [89]) showed that the dimension of the component of $V_{6m}^{irr}(6m^2 \cdot A_2)$ containing the curves $F^3 + G^2 = 0$, $\deg F = 2m$, $\deg G = 3m$, exceeds the expected dimension by at least

$$\frac{(m-1)(m-2)}{2} > 0 \quad \text{as long as} \quad m \geq 3,$$

Wahl [96] showed that the family $V_{104}^{irr}(3636 \cdot A_1, 9000 \cdot A_2)$ contains a nonreduced component. The problem is to show that there exists another (say, T-smooth) component of the considered equisingular family.

According to [108, Sects. VIII.3 and VIII.5], [78, Theorem 2.1], [28, Proposition 5.4], [29, Proposition 1.1] (see also [31, Examples 4.2.0.9 and 4.2.0.10, Propositions 4.6.10 and 4.6.11]), we have the following statement.

Theorem 2.5.8 *(1) Each of the families $V_6^{irr}(6 \cdot A_2)$ and $V_{12}(24 \cdot A_2)$ has (at least) two distinct components of the expected dimension.*

(2) Let p, d be integers satisfying

$$p \geq 3, \quad 6p \leq d \leq 12p - \frac{3}{2} - \sqrt{35p^2 - 15p + \frac{1}{4}}\,.$$

Then the family $V_d^{irr}(6p^2 \cdot A_2)$ has components of different dimensions. Moreover, if $d > 6p$, then $\pi_1(\mathbb{P}^2 \setminus C) \simeq \mathbb{Z}/d\mathbb{Z}$ for all curves $C \in V_d^{irr}(6p^2 \cdot A_2)$.

(3) Let $m \geq 9$ Then there exists $k_0 = k_0(m)$ such that for any $k \geq k_0$ and any integer s atisfying

$$\frac{k-1}{2} \leq s \leq k\left(1 - \sqrt{\frac{2}{m}}\right) - \frac{3}{2}\,,$$

the equisingular family $V_{km+s}^{irr}(k^2 \cdot S(m))$ of irreducible plane curves of degree $d = km + s$ with k^2 ordinary m-fold points (topological type $S(m)$) has components of different dimensions. Moreover, $\pi_1(\mathbb{P}^2 \setminus C) \simeq \mathbb{Z}/d\mathbb{Z}$ for all curves $C \in V_d^{irr}(k^2 \cdot S(m))$.

In part (1) the former family is the classical Zariski's example discussed above. The latter family $V_{12}^{irr}(24 \cdot A_2)$ contains a component of expected dimension 42 formed by the curves given by

$$F^3 + G^2 = 0, \quad \deg F = 4, \quad \deg G = 6,$$

whose 24 cusps lie on a plane quartic curve. However, according to Theorem 2.3.6(1) there exists a T-smooth component of the family $V_{12}^{irr}(28 \cdot A_2)$. Smoothing out any four cusps of a curve $C \in V_{12}^{irr}(28 \cdot A_2)$, one obtains curves in T-smooth components of $V_{12}^{irr}(24 \cdot A_2)$, and it is not difficult to verify that there is a 24-tuple of cusps of $C \in V_{12}^{irr}(28 \cdot A_2)$ which does not lie on any plane quartic curve.

In parts (2), resp. (3), one obtains components of the equisingular families of dimension above the expected one formed by the Zariski type curves (cf. Theorem 2.4.7) given by $C_{2p}^2 C'_{d-6p} + C_{3p}^2 C''_{d-6p} = 0$, where

$$\deg C_{2p} = 2p, \ \deg C_{3p} = 3p, \ \deg C'_{d-6p} = \deg C''_{d-6p} = d - 6p,$$

resp. $\sum_{i=0}^m R_s^{(i)} F_k^i G_k^{m-k} = 0$, where

$$\deg F_k = \deg G_k = k, \ \deg R_s^{(i)} = s, \ i = 0, \ldots, m.$$

However, in both the cases there exists a T-smooth component (of expected dimension): in part (2) by Theorem 2.3.6(i), in part (3) this can be derived by means of the Alexander-Hirschowitz theorem [4, Theorem 1.1] (see also [31, Theorem 3.4.22]). The fact that the fundamental group of the complement to the considered curves is always abelian follows from Nori's theorem [63, Proposition 3.27]. Namely, we use the following very particular case of Nori's result:

Theorem 2.5.9 *Let $C \subset \mathbb{P}^2$ be a reduced, irreducible curve of degree d, and let the blowing-up $\beta : X \to \mathbb{P}^2$ resolve all the non-nodal singular points of C so that* *the union of the strict transform D of C with the reduced exceptional divisor E is nodal. If $D^2 > 2r(C)$, where $r(C)$ is the number of the nodes of C, then $\pi_1(\mathbb{P}^2 \setminus C)$ is abelian (and equal to $\mathbb{Z}/d\mathbb{Z}$).*

So, given a curve $C \in V_d^{irr}(6p^2 \cdot A_2)$ with $d > 6p^2$ as in Theorem 2.5.8(2), we blow up each cusp of C three times obtaining a smooth strict transform D on the blown up plane X such that the union of $D \cup E$ is nodal; hence Theorem 2.5.9 applies and yields that $\pi_1(\mathbb{P}^2 \setminus C)$ is abelian due to $D^2 = d^2 - 6 \cdot 6p^2 > 0$. Similarly, given a curve $C \in V_d^{irr}(k^2 \cdot S(M))$ as in Theorem 2.5.8(3), we blow up each multiple point of C obtaining a smooth strict transform D on the blown up plane X, while the union $D \cup E$ is nodal. Again Theorem 2.5.9 applies and yields that $\pi_1(\mathbb{P}^2 \setminus C)$ is abelian due to $D^2 = d^2 - k^2m^2 = (km + s)^2 - k^2m^2 > 0$.

2.5.2 Curves on Other Algebraic Surfaces

For other algebraic surfaces than $\mathbb{P}^2$, we consider only the case of nodal curves, which is the most important one, since it is directly related to the vanishing/nonvanishing of Gromov-Witten invariants.

In the following cases we know complete answers in which the equisingular family is T-smooth [11, 12, 14, 27, 88] (see also [31, Sect. 4.5.6.3]).

Theorem 2.5.10 *(1) Let Σ be a toric surface associated with the planar nondegenerate lattice polygon Δ, $\mathcal{L}(\Delta)$ the corresponding tautological line bundle. Then the inequality*

$$0 \le n \le \#(\mathrm{Int}(\Delta) \cap \mathbb{Z}^2)$$

is necessary and sufficient for the existence of an irreducible curve with n nodes (as its only singularities) in the linear system $|\mathcal{L}(\Delta)|$.

(2) Let $\Sigma = \mathbb{P}^2_k$, $1 \le k \le 9$, be the plane blown up at k distinct generic points, D an effective divisor class of type $D = dL - d_1E_1 - ... - d_kE_k$, where L is the lift of a general line on $\mathbb{P}^2$, $E_1, ..., E_k$ are exceptional divisors, $d \ge d_1 \ge ... \ge d_k > 0$. Suppose that $-DK_\Sigma > 0$. An irreducible curve $C \in |D|$ with n nodes as its only singularities exists if and only if either

$$k = 1, \quad 0 \le n \le p_a(D) = \frac{D^2 + DK_\Sigma}{2} + 1,$$

or

$$k = 2, \quad 0 \le n \le p_a(D), \quad \begin{cases} \text{either } d \ge d_1 + d_2, \\ \text{or } d = d_1 = d_2 = 1, \end{cases}$$

or

$$k \ge 3, \quad D^2 > 0, \quad 0 \le n \le p_a(D).$$

(3) For any $g \geq 3$, given a general smooth K3 surface Σ of the principal series in $\mathbb{P}^g$, and $m > 0$ and n satisfying

$$0 \leq n \leq \dim |\mathcal{O}_\Sigma(m)|,$$

there exists an irreducible curve in the linear system $|\mathcal{O}_\Sigma(m)|$ with n nodes as its only singularities.

(4) Let $\Sigma \subset \mathbb{P}^3$ be a generic smooth surface of degree $d \geq 5$. Then, for all

$$m \geq d, \quad 0 \leq n \leq \dim |\mathcal{O}_\Sigma(m)|$$

there exists an irreducible curve in the linear system $|\mathcal{O}_\Sigma(m)|$ having n nodes as its only singularities.

Remark 2.5.11 Part (1) is actually well-known, one can find details in [31, Theorem 4.5.32].

Part (2) admits an extension to the generic surfaces $\mathbb{P}^2_k$, $k > 10$, with extra restrictions to the divisor D and the number of nodes n (see [27, Theorem 5] or [31, Tearem 4.5.30 and Corollary 4.5.31]).

Part (4) is proved by the method resembling the patchworking construction. Namely, the proof goes by induction with the case $d = 4$ (settled in part (3)) as the base. The induction step consists in a pair of deformations:

- the union of a generic surface Σ_{d-1} of degree $d - 1$ with a generic tangent plane π to it deforms in a family into a generic smooth surface Σ_d of degree d;
- an inscribed deformation of a curve in the central fiber that consists of an irreducible curve in the linear system $|\mathcal{O}_{\Sigma_{d-1}}(m)|$ on Σ_{d-1} having $\dim |\mathcal{O}_{\Sigma_{d-1}}(m)|$ nodes, and of a nodal curve in the plane π of degree m having $\frac{1}{2}(m - d + 2)(m - d + 3)$ nodes.

Under certain transversality conditions, the above central curve can be deformed into an irreducible curve $C \in |\mathcal{O}_{\Sigma_d}(m)|$ having

$$\dim |\mathcal{O}_{\Sigma_{d-1}}(m)| + \frac{(m - d + 2)(m - d + 3)}{2} = \dim |\mathcal{O}_{\Sigma_d}(m)|$$

nodes.

It should be noted that Chiantini and Ciliberto [14, Sect. 1] exhibit examples of superabundant nodal curves on surfaces in $\mathbb{P}^3$: in particular, for $d \geq 20$ and $m = 3$ there are curves in $|\mathcal{O}_\Sigma(3)|$ with $n > \dim |\mathcal{O}_\Sigma(3)|$ nodes, and for $d \geq 8$ and $m \gg 0$ there exists a component of the equisingular family of curves with $n < \dim |\mathcal{O}_\Sigma(m)|$ nodes that has a dimension greater than the expected one.

One can obtain some sufficient existence conditions for curves with arbitrary singularities on smooth projective surfaces. For a topological or analytic singularity type S, denote by $e(S)$ the minimal degree of a reduced plane curve C having a singular point of type S as its only singularity, belonging to a T-smooth equisingular family, which intersects transversally with the space of curves passing through the intersection of C with a generic fixed line. Then the following holds (see [31, Proposition 4.5.26]).

Lemma 2.5.12 *Let Σ be a smooth projective algebraic surface, D an effective divisor on Σ, L a very ample divisor on Σ. Given topological or analytic singularity types $S_1, \ldots, S_r$ and a zero-dimensional scheme $Z \subset \Sigma$ defined in some distinct r points $z_1, \ldots, z_r \in \Sigma$ by the powers $\mathfrak{m}_{z_i}^{e(S_i)} \subset \mathcal{O}_{\Sigma,z_i}$ of the maximal ideals so that*

$$H^1(\Sigma, \mathcal{J}_{Z/\Sigma}(D-L)) = 0 \quad \textit{and} \quad \max_{1 \le i \le r} e(S_i) < L(D - L - K_\Sigma) - 1,$$

then there exists an irreducible curve $C \in |\mathcal{O}_\Sigma(D)|$ with r singular points of types $S_1, \ldots, S_r$, respectively, as its only singularities.

The proof is based on a version of the patchworking construction as it appears in [80, 82] (see also [31, Sect. 2.3.5]). Some numerical conditions, based on h^1-vanishing criteria [102], can be found in [31, Sect. 4.5.6.2].

2.5.3 Other Related Problems

Rational cuspidal curves. A rational cuspidal curve is a complex rational plane curve homeomorphic to a sphere, equivalently, a rational plane curve having only irreducible singularities (called (generalized) cusps). They attracted much attention due to their interesting properties and tight links to the Jacobian conjecture, affine algebraic geometry, and birational geometry (see [1, 46, 105]). The subject definitely deserves a separate full size survey. We only mention one result directly related to the existence problem for singular plane curves [47, Theorem 1.1]:

Theorem 2.5.13 *A rational cuspidal curve has at most* 4 *singular points.*

There is a series of classification results for rational cuspidal curves (see references in [47]).

Curves in the higher-dimensional projective spaces. Each reduced projective curve can be embedded into $\mathbb{P}^n$ with $n \ge 3$. The question on the number of nodes of an irreducible curve in $\mathbb{P}^n$, $n \ge 3$, of degree d and genus g was studied in [86, 87] over the complex field and in [65] over the real field. For $n \ge 3$, the genus of an irreducible nondegenerate (i.e., not contained in a hyperplane) curve of degree $d \ge n$ in $\mathbb{P}^n$ is bounded from above by

$$C(d,n) = \frac{1}{2} m((m-1)(n-1) + 2e), \quad \text{where } d - 1 = m(n-1) + e, \ 0 \le e < n - 1,$$

(see [10] or [32, p. 57]).

Theorem 2.5.14 *For any $d \ge n \ge 3$ and any $\delta \le C(d,n)$, there exists a real irreducible nondegenerate curve of degree d in $\mathbb{P}^n$ with δ real nodes as its only singularities.*

For the proof, Pecker [65] constructs a suitable real plane rational curve with $C(d, n)$ real nodes in the affine plane, then maps it by

$$\psi(x, y) = (x, x^2, ..., x^{n-k}, y, yx, yx^2, ..., yx^{k-1}), \quad k = n - \left[\frac{d-1}{m+1}\right],$$

to $\mathbb{P}^n$ with the image of degree d. It is not difficult to see that prescribed nodes of the obtained curve can be smoothed out (cf. also [87]).

Deformations of plane curves singularities. A local version of the problems discussed in this survey is the following local adjacency problem:

Given a reduced plane curve singular germ (C, p), what collections of singularities can appear in its (versal) deformation?

The question on the existence of a global plane curve of a given degree d with prescribed singularities can be considered as the above local deformation question for an ordinary d-fold singular point.

We show only two specific examples, both over the real field and both concerning the nodal deformations of arbitrary real plane curve singular points.

The first result is due to Pecker [67]. Recall that the maximal number of nodes appearing in a deformation of a plane curve singularity (C, p) equals $\delta(C, p)$, which in case of an irreducible (i.e., unibranch) germ (C, p) can be written as $\frac{1}{2}\mu(C, p)$.

Theorem 2.5.15 *Given an irreducible real plane curve singularity (C, p) and any nonnegative $\alpha \le \frac{1}{2}\mu(C, p)$. Then there exists a real deformation of (C, p), whose general member has α real elliptic nodes as its only singularities.*

Due to the openness of versality (see, for instance, [30, Theorem I.1.15]), given a deformation of a singularity (C, p) with a singular general member, there exists a deformation of (C, p) in which the singularities of that general member can independently be deformed in a prescribed way. That is, to prove the theorem it is enough to find a deformation realizing $\alpha = \frac{1}{2}\mu(C, p)$ elliptic nodes. For the latter deformation, Pecker explicitly constructs a deformation of the parametrization of (C, p).

An in a sense opposite question is to find a deformation with the maximal possible number of hyperbolic nodes. Such deformations are called *morsifications*, and they carry out an important information on the topology of the singularity (C, p) [2, 35] (see also [19] for the relation of morsifications to mutations of quivers). A'Campo and Gusein-Zade [2, 35] proved the following claim.

Theorem 2.5.16 *Every totally real plane curve singularity (i.e., a real plane curve singularity (C, p) all of whose local branches are real) possesses a morsification, and each morsification exhibits $\delta(C, p)$ hyperbolic nodes.*

A'Campo and Gusein-Zade gave different proofs, using sequences of blow-ups and contractions on one side, and explicit formulae involving Tchebycheff polynomials on the other side.

The quintic shown in Fig. 2.2a represents, in fact, a morsification of the singularity $y^4 - 2x^5 = 0$.

The question on the existence of morsifications for real singularities (C, p) having complex conjugate local branches turns to be much harder. A partial answer to this question was suggested in [53].

2.5.4 Some Questions and Conjectures

In principle, every time we presented a partial answer to a specific or general existence problem, we encourage the reader to improve or even complete the answer. However, several questions deserve a more detailed comment.

Cuspidal plane curves. In Sect. 2.3.2 we discussed one of the most challenging questions: what is the maximal number $k_{\max}(d)$ of ordinary cusps of a plane curve of degree d? Langer [51] conjectures that the coefficient of d^2 in the right-hand side of (2.33) is sharp. More precisely,

Conjecture 2.5.17

$$\limsup_{d\to\infty} \frac{k_{\max}(d)}{d^2} = \frac{125+\sqrt{73}}{432}.$$

Concerning the maximal number $k_{\max,\mathbb{R}}(d)$ of real cusps on a real plane curve of degree d, the best existence result is Theorem 2.3.6. We conjecture that the coefficient of d^2 in (2.36) is sharp, i.e.,

Conjecture 2.5.18

$$\limsup_{d\to\infty} \frac{k_{\max,\mathbb{R}}(d)}{d^2} = \frac{1}{4}.$$

The following version of the problem was pointed by Vik. Kulikov. Choose an almost complex structure on the plane tamed by the standard symplectic structure.

Question 1
What is the maximal number of cusps of a pseudo-holomorphic plane curve of degree d? Does there exists a cuspidal plane pseudoholomorphic curve of degree d with the number of cusps breaking consequences of the Bogomolov-Miyaoka-Yau inequality (e.g., the Hirzebruch-Ivinskis bound (2.32))?

The latter question reflects the fact that there is no analogue of the Bogomolov-Miyaoka-Yau inequality for symplectic fourfolds.

Reducible equisingular families of plane curves. In contrast to the sufficient T-smoothness conditions of equisingular families of plane curves, which were shown to be sharp (or close to sharp) in several important cases (see Sect. 2.5.1), the known examples of reducible equisingular families like in Theorem 2.5.8 are very far from the available general sufficient irreducibility conditions, which consist of three inequalities (see [31, Theorem 4.6.4])

$$\max_{1\le i\le r}\nu'(S_i)\le\frac{2}{5}d-1,$$

$$\sum_{i=1}^{r}(\nu'(S_i)+2)^2<\frac{9}{10}d^2,$$

$$\frac{25}{2}\cdot\#(\text{nodes})+18\cdot\#(\text{cusps})+\sum_{S_i\ne A_1,A_2}(\tau'(S_i)+2)^2<d^2, \tag{2.54}$$

where the singularity invariants ν' and τ' are of the order of the Tjurina number τ, and hence the coefficients assigned to the singularities in (2.54) are of order τ^2. For instance, an ordinary m-fold singular point (considered up to topological equivalence) enters the left-hand side of (2.54) with the coefficient $\frac{1}{4}m^2(m+1)^2$ (see [31, Corollary 4.6.7]), while in the series of reducible equisingular families of curves with ordinary singularities from Theorem 2.5.8(3), the ratio of d^2 to the number of ordinary m-fold singularities does not exceed $(m+1)^2$. This leaves completely open the following question

Question 2
How sharp are the sufficient irreducibility conditions (2.54)*?*

Another specific feature of the examples in Theorem 2.5.8, namely, the fact that the curves in different components of the equisingular family have the same fundamental group of the complement (i.e., form a so-called *anti-Zariski pair*) raises the following interesting question.

Question 3
Can the curves in an anti-Zariski pair be transferred to each other by a homeomorphism of the plane onto itself?

Sharpness of restrictions to curves with arbitrary singularities. We have discussed above the sharpness of the known restrictions, notably, of Langer's bound in the case of curves with ordinary cusps. On the other hand, in Sect. 2.4.2 we have seen that, for A_n singularities, with large Milnor number n, Hirano's examples (Theorem 2.4.2 and Remark 2.4.3) have almost the same asymptotics as the spectral bound does. Beyond the range of simple or ordinary multiple singularities, the spectral bound and the genus and Plücker formulas are the only universal bounds applicable to arbitrary singularities, and the spectral bound is much stronger than the genus and Plücker bounds. So, it is natural to ask.

Question 4
For which singularity types (say, semiquasihomogeneous, irreducible, etc.) with large Milnor numbers is the spectral bound (asymptotically) sharp, or almost sharp?

As said above, so far this is known to be true only for A_n singularities.

Gromov-Witten invariants of rational surfaces. Let $\mathbb{P}^2_r$ be the plane blown up at $r>0$ generic points. For $r\le 9$, we know a complete answer about the existence of nodal curves of arbitrary genus in an arbitrary linear system on $\mathbb{P}^2_r$ (see Theorem 2.5.10(2)).

If $r > 9$, one can find in the literature only partial answers, see [27, Theorem 5 and Corollary 3.1.7] or [31, Theorem 4.5.30 and Corollary 4.5.31]. For a divisor class $D \in \text{Pic}(\mathbb{P}^2_r)$, the expected dimension of the moduli space $\mathcal{M}_{0,g}(\mathbb{P}^2_r, D)$ of stable maps of (unmarked) curves of genus g to $\mathbb{P}^2_r$ representing the class D equals (cf. [20])

$$-DK_{\mathbb{P}^2_r} + g - 1 .$$

The following question arises

Question 5
Suppose that $r > 9$ and $D \in \text{Pic}(\mathbb{P}^2_r)$ satisfies the conditions $-DK_{\mathbb{P}^2_r} > 0$ and $D^2 \geq -1$. Does there exist a nodal rational curve $C \in |D|$?

The restriction $D^2 \geq -1$ comes from the fact that $\mathbb{P}^2_r$ does not contain $(-k)$-curves with $k > 1$. The above question is directly related to the non-vanishing of genus zero Gromov-Witten invariants of $\mathbb{P}^2_r$: it is shown in [22, Theorem 4.1 and Section 5.2] that these Gromov-Witten invariants do count rational curves in $|D|$ if either $-DK_{\mathbb{P}^2_r} > 1$, or $d \leq 10$, or some d_i equals 1 or 2, where $D = dL - d_1E_1 - ... - d_rE_r$ (L being the lift of a generic line in $\mathbb{P}^2$, $E_1, ..., E_r$ the exceptional divisors of the blowing up). We note also that, in view of the condition $-DK_{\mathbb{P}^2_r} > 0$, an affirmative answer to Question 5 yields the existence of a nodal curve $C' \in |D|$ with any nonnegative number of nodes fewer than for the rational curve C; hence, the nonvanishing of the corresponding Gromov-Witten invariants of positive genus. Furthermore, Question 5 can be extended in the following way.

Question 5'
Suppose that $r > 9$, $g \geq 0$, and $D \in \text{Pic}(\mathbb{P}^2_r)$ satisfies the conditions $-DK_{\mathbb{P}^2_r} + g > 0$ and $D^2 > 0$. Does there exist a curve $C \in |D|$ of genus g? What is the enumerative meaning of the corresponding genus g Gromov-Witten invariants of $\mathbb{P}^2_r$?

Acknowledgements We are very grateful to the unknown referee for the careful reading of our paper and numerous remarks and corrections. The second author was supported by the Israel Science Foundation grant no. 501/18 and by the Bauer-Neuman Chair in Real and Complex Geometry.

References

1. Abhyankar, S. S., Moh, T. T.: Embeddings of the line in the plane. J. Reine Angew. Math. **276**, 148–166 (1975)
2. A'Campo, H.: Le groupe de monodromie des singularités isolées des courbes planes, I. Math. Ann. **213**, 1–32 (1975)
3. Akyol, A., Degtyarev, A.: Geography of irreducible plane sextics. Proc. Lond. Math. Soc. (3) **111**, no. 6, 1307–1337 (2015)
4. Alexander, J., Hirschowitz, A.: An asymptotic vanishing theorem for generic unions of multiple points. Invent. Math. **140**, no. 2, 303–325 (2000)

5. Barth, V. P., Hulek, K., Peters, C. A. M., and Van de Ven, A.: Compact complex surfaces. Springer, Berlin (2004)
6. Brieskorn, E., Knörrer, H.: Plane algebraic curves. Birkhäuser-Verlag, Switzerland (1986)
7. Bruce, J. W., Giblin, P. J.: A stratification of the space of plane quartic curves. Proc. London Math. Soc. **42**, 270–298 (1981)
8. Brusotti, L.: Sulla "piccola variazione" di una curva piana algebrica reali. Rend. Rom. Ac. Lincei (5) **30**, 375–379 (1921)
9. Calabri, A., Paccagnan, D., Stagnaro, E.: Plane algebraic curves with many cusps with an appendix by Eugenii Shustin. Annali di Matematica Pura ed Applicata (4) **193**, no. 3, 909–921 (2014)
10. Castelnuovo, G.: Ricerche di geometria sulle curve algebriche. Atti Reale Accademia delle Scienze di Torino **24**, 346–373 (1889)
11. Chen, Xi: Rational curves on K3 surfaces. J. Alg. Geom. **8**, 245–278 (1999)
12. Chen, Xi: Nodal Curves on K3 Surfaces. New York J. Math. **25**, 168–173 (2019)
13. Cassou-Nogues, P., Luengo, I.: On A_k singularities on plane curves of fixed degree. Preprint (2000)
14. Chiantini, L., Ciliberto, C.: On the Severi varieties of surfaces in $\mathbb{P}^3$. J. Alg. Geom. **8**, no. 1, 67–83 (1999)
15. Decker, W., Greuel, G.-M., Pfister, G., Schönemann, H.: Singular 4-1-1 – A computeralgebra system for polynomial computations (2018). https://www.singular.uni-kl.de
16. Degtyarev, A.: Isotopic classification of complex plane projective curves of degree 5. Algebra i Analiz **1**, no. 4, 78–101 (1989)
17. Degtyarev, A.: Topology of algebraic curves. An approach via dessins d'enfants. Walter de Gruyter, Berlin/Boston (2012)
18. Du Plessis, A. A., Wall, C. T. C.: Singular hypersurfaces, versality and Gorenstein algebras. J. Algebraic Geom. **9**, no. 2, 309–322 (2000)
19. Fomin, S., Pylyavskyy, P., Shustin, E., Thurston, D.: Morsifications and mutations. Preprint at arXiv:1711.10598
20. Fulton, W., Pandharipande, R. Notes on stable maps and quantum cohomology. Algebraic geometry—Santa Cruz 1995, Proc. Sympos. Pure Math., 62, Part 2, Amer. Math. Soc., Providence, RI, pp. 45–96 (1997)
21. Gourevich, A. Geometry of obstructed families of curves. J. Pure Appl. Algebra **210**, no. 3, 721–734 (2007)
22. Göttsche, L., Pandharipande, R.: The quantum cohomology of blow-ups of $\mathbb{P}^2$ and enumerative geometry. J. Differential Geom. **48**, no. 1, 61–90 (1998)
23. Greuel, G.-M., Karras, U.: Families of varieties with prescribed singularities. Compos. Math. **69**, no. 1, 83–110 (1989)
24. Greuel, G.-M., Lossen, C.: Equianalytic and equisingular families of curves on surfaces. Manuscr. Math. **91**, 323–342 (1996)
25. Greuel, G.-M., Lossen, C., Shustin, E.: New asymptotics in the geometry of equisingularfamilies of curves. Int. Math. Res. Not. **13**, 595–611 (1997)
26. Greuel G.-M., Lossen, C., Shustin, E.: Plane curves of minimal degree with prescribed singularities. Invent. Math. **133**, no. 3, 539–580 (1998)
27. Greuel, G.-M., Lossen, C., Shustin, E.: Families of nodal curves on the blown-up projective plane. Trans. Amer. Math. Soc. **350**, no. 1, 251–274 (1998)
28. Greuel, G.-M., Lossen, C., Shustin, E.: Castelnuovo function, zero-dimensional schemes and singular plane curves. J. Algebr. Geom. **9**, no. 4, 663–710 (2000)
29. Greuel, G.-M., Lossen, C., Shustin, E.: The Variety of Plane Curves with Ordinary Singularities is not irreducible. Intern. Math. Res. Notices **11**, 542–550 (2001)
30. Greuel, G.-M., Lossen, C., Shustin, E.: Introduction to singularities and deformations. Springer, Berlin (2007)
31. Greuel, G.-M., Lossen, C., Shustin, E.: Singular algebraic curves. Springer, Switzerland (2018)
32. Griffiths, P., Harris, J.: Principles of algebraic geometry. John Wiley, NY (1978)

33. Gudkov, D. A.: On the curve of 5th order with 5 cusps. Funct. Anal. Appl. **16**, 201–202 (1982)
34. Gudkov, D. A., Tai, M. L., Utkin, G. A.: Complete classification of irreducible curves of the 4th order. Mat. Sbornik **69**, no. 2, 222–256 (1966)
35. Gusein-Zade, S. M.: Dynkin diagrams for singularities of functions of two variables. Func. Anal. Appl. **8**, 295–300 (1974)
36. Gusein-Zade, S. M., Nekhoroshev, N. N.: Singularities of type A_k on simple curves of chosen degree. Funct. Anal. Appl. **34**, no. 3, 214–215 (2000)
37. Harris, J.: On the Severi problem. Invent. Math. **84**, 445–461 (1985)
38. Hartshorne, R.: Algebraic geometry. Springer, NY (1977)
39. Hirano, A.: Constructions of plane curves with cusps. Saitama Math. J. **10**, 21–24 (1992)
40. Hironaka, H.: On the arithmetic genera and the effective genera of algebraic curves. Mem. Coll. Sci., Univ. of Kyoto **30**, 177–195 (1957)
41. Hirzebruch, F.: Singularities of algebraic surfaces and characteristic numbers. Contemp. Math. **58**, 141–155 (1986)
42. Itenberg, I., Shustin, E.: Real algebraic curves with real cusps. Amer. Math. Soc. Transl. (2) **173**, 97–109 (1996)
43. Ivinskis, K.: Normale Flächen und die Miyaoka-Kobayashi Ungleichung. Diplomarbeit, Bonn (1985)
44. Kerner, D.: Recombination formulas for the spectrum of curve singularities and some applications. Preprint at arXiv:1403.5789
45. Koelman, R. J.: Over de cusp. University of Leiden, Diplomarbeit (1986)
46. Koras, M., Palka, K.: The Coolidge-Nagata conjecture. Duke Math. J. **166**, no. 16, 3085–3145 (2017)
47. Koras, M., Palka, K.: Complex planar curves homeomorphic to a line have at most four singular points. Preprint at arXiv:1905.11376
48. Korchagin, A. B., Weinberg, D. A.: The isotopy classification of affine quartic curves. Rocky Mountain J. Math. **32**, no. 1, 255–347 (2002)
49. Korchagin, A. B., Weinberg, D. A.: Quadric, cubic, and quartic cones. Rocky Mountain J. Math. **35**, no. 5, 1627–1656 (2005)
50. Kulikov, Vik. S.: The generalized Chisini conjecture. Proc. Steklov Inst. Math. **241**, no. 2, 110–119 (2003)
51. Langer, A.: Logarithmic orbifold Euler numbers of surfaces with applications. Proc. Lond. Math. Soc. (3) **86**, 358–396 (2003)
52. Lefschetz, S.: On the existence of loci with given singularities. Trans. Amer. Math. Soc. **14**, 23–41 (1913)
53. Leviant, P., Shustin, E.: Morsifications of real plane curve singularities. J. Singularities **18**, 307–328 (2018)
54. Luengo, I.: The μ-constant stratum is not smooth. Invent. Math. **90**, 139–152 (1987)
55. Luengo, I.: On the existence of complete families of projective plane curves, which are obstructed. J. London Math. Soc. **2-36**, no. 1, 33–43 (1987)
56. Milnor, J.: Singular points of complex hypersurfaces. Princeton Univ. Press, Princeton (1968)
57. Miyaoka, Y.: On the Chern numbers of surfaces of general type. Invent. Math. **42**, 225–293 (1977)
58. Miyaoka, Y.: The maximal number of quotient singularities on surfaces with given numerical invariants. Math. Ann. **268**, 159–171 (1984)
59. Mumford, D.: Intuition and Rigor and Enriques's Quest. Notices of the AMS **58**, No. 2 (2011)
60. Newton, I.: Analysis of the properties of cubic curves and their classification by species. The mathematical papers of Isaac Newton (D.T. Whiteside, ed.), Cambridge Univ. Press, vol. 2, pp. 389–34 (1968)
61. Newton, I.: Enumeratio linearum tertij ordinis. The mathematical papers of Isaac Newton (D.T. Whiteside, ed.), Cambridge Univ. Press, vol. 7, pp. 565–645 (1976)
62. Newton, I.: The final "Geometriæ libri duo". The mathematical papers of Isaac Newton (D.T. Whiteside, ed.), Cambridge Univ. Press, vol. 7, pp. 402–469 (1976)

63. Nori, M.: Zariski conjecture and related problems. Ann. Sci. Ec. Norm. Sup. **4** (16), 305–344 (1983)
64. Orevkov, S. Yu.: Some examples of real algebraic and real pseudoholomorphic curves. *Perspectives in Analysis, Geometry, and Topology*. Progress in Mathematics, vol. 296, pp. 355–387, Birkhäuser, Basel (2012)
65. Pecker, D.: Simple constructions of algebraic curves with nodes. Compos. Math. **87**, 1–4 (1993)
66. Pecker, D.: Note sur la réalité des points doubles des courbes gauches. C. R. Acad. Sci.Paris, Sér. I **324**, 807–812 (1997)
67. Pecker, D.: Sur le thèoréme local de Harnack. C. R. Acad. Sci. Paris, Sér. I **326**, no. 5, 573–576 (1998)
68. Pecker, D.: Sur la réalité des points doubles des courbes gauches. Ann. Inst. Fourier (Grenoble) **49**, 1439–1452 (1999)
69. Sakai, F.: Singularities of Plane Curves. Geometry of Complex Projective Varieties (Seminars and Conferences 9), Mediterranian Press, Rende, pp. 257–273 (1993)
70. Schulze, M.: A singular 3-1-6 library for computing invariants related to the Gauss-Maninsystem of an isolated hypersurface singularity (2012) (gmssing.lib.)
71. Segre, B.: Dei sistemi lineari tangenti ad un qualunque sistema di forme. Atti Acad. naz. Lincei Rendiconti serie 5, **33**, 182–185 (1924)
72. Segre, B.: Esistenza e dimensione di sistemi continui di curve piane algebriche con dati caraterri. Atti Acad. naz. Lincei Rendiconti serie 6, **10**, 31–38 (1929)
73. Severi, F.: Vorlesungen über Algebraische Geometrie (Anhang F). Leipzig, Teubner (1921)
74. Shustin, E. I.: On invariants of algebraic curve singular points. Math. Notes of Acad. Sci. USSR **34**, 962–963 (1983)
75. Shustin, E.: Versal deformations in the space of plane curves of fixed degree. Func. Anal. Appl. **21**, 82–84 (1987)
76. Shustin, E.: On manifolds of singular algebraic curves. Selecta Math. Sov. **10**, no.1, 27–37 (1991)
77. Shustin, E.: Real plane algebraic curves with prescribed singularities. Topology **32**, no. 4, 845–856 (1993)
78. Shustin, E.: Smoothness and irreducibility of varieties of algebraic curves with nodes and cusps. Bull. Soc. Math. France **122**, 235–253 (1994)
79. Shustin, E.: Gluing of singular and critical points. Topology **37**, no. 1, 195–217 (1998)
80. Shustin, E.: Lower deformations of isolated hypersurface singularities. St. Petersburg Math. J. **11**, no. 5, 883–908 (2000)
81. Shustin, E.: Analytic order of singular and critical points. Trans. Amer. Math. Soc. **356**, 953–985 (2004)
82. Shustin, E., Tyomkin, I.: Patchworking singular algebraic curves, I. Israel J. Math. **151**, 125–144 (2006)
83. Shustin, E., Westenberger, E.: Projective hypersurfaces with many singularities of pre-scribed types. J. Lond. Math. Soc. (2) **70**, no. 3, 609–624 (2004)
84. Steenbrink, J. H. M.: Intersection form for quasihomogeneous singularities. Compos. Math. **34**, no. 2, 211–223 (1977)
85. Steenbrink, J. H. M.: Semicontinuity of the singularity spectrum. Invent. Math. **79**, 557–566 (1985)
86. Tannenbaum, A.: On the geometric genera of projective curves. Math. Ann. **240**, no. 3, 213–221 (1979)
87. Tannenbaum, A.: Families of algebraic curves with nodes. Compos. Math. **41**, 107–126 (1980)
88. Tannenbaum, A.: Families of curves with nodes on K3 surfaces. Math. Ann. **260**, 239–253 (1982)
89. Tannenbaum, A.: On the classical characteristic linear series of plane curves with nodes and cuspidal points: two examples of Beniamino Segre. Compositio Math. **51**, 169–183 (1984)
90. Teissier, B.: The hunting of invariants in the geometry of discriminants. Real and Complex Singularities (P.Holm, ed.), Oslo. Sijthoff-Noordhoff Publ., Alphen aan den Rijn, pp. 565–677 (1977)

91. Urabe, T.: On quartic surfaces and sextic curves with certain singularities. Proc. Japan. Acad. Ser. A. **59**, 434–437 (1983)
92. Urabe, T.: On quartic surfaces and sextic curves with singularities of type $\widetilde{E}_8$, $T_{2,3,7}$, $\widetilde{E}_{12}$. Publ. Inst. Math. Sci. Kyoto Univ. **80**, 1185–1245 (1984)
93. Urabe, T.: Singularities in a certain class of quartic surfaces and sextic curves and Dynkin graphs. Proc. Vancouver Conf. Algebraic Geom., July 2-12, 1984. Providence, pp. 477–497 (1986)
94. Varchenko, A. N.: Asymptotic of integrals and Hodge structures. Modern Problems of Math., Vol. 22 (Itogi nauki i tekhniki VINITI), Moscow, pp. 130–166 (1983)
95. Varchenko, A. N.: On semicontinuity of the spectrum and an upper estimate for the number of singular points of a projective hypersurface. Sov. Math. Dokl. **27**, 735–739 (1983); translation from Dokl. Akad. Nauk SSSR **270**, no. 6, 1294–1297 (1983)
96. Wahl, J.: Deformations of plane curves with nodes and cusps. Amer. J. Math. **96**, 529–577 (1974)
97. Wahl, J.: Equisingular deformations of plane algebroid curves. Trans. Amer. Math. Soc. **193**, 143–170 (1974)
98. Wall, C. T. C.: Geometry of quartic curves. Math. Proc. Camb. Phil. Soc. **117**, 415–423 (1995)
99. Wall, C. T. C.: Highly singular quintic curves. Math. Proc. Camb. Phil. Soc. **119**, 257–277 (1996)
100. Westenberger, E.: Existence of hypersurfaces with prescribed singularities. Commun. Algebra **31**, no. 1, 335–356 (2003)
101. Westenberger, E.: Real hypersurfaces with many simple singularities. Rev. Mat. Complut. **18**, no. 2, 455–464 (2005)
102. Xu, G.: Ample line bundles on smooth surfaces. J. Reine Angew. Math. **469**, 199–209 (1995)
103. Yau, S.-T.: Calabi's conjecture and some new results in algebraic geometry. Proc. Nat. Acad. Sci. USA **74**, 1798–1799 (1977)
104. Yau, S.-T.: On the Ricci curvature of a complete Kaehler manifold and the complex Monge-Ampère equation. Commun. Puer Appl. Math. **31**, 339–411 (1978)
105. Zaidenberg, M. G., Lin, V. Ya.: An irreducible simply connected algebraic curve in $\mathbb{C}^2$ is equivalent to a quasihomogeneous curve. Dokl. Akad. Nauk SSSR **271**, no. 5, 1048–1052 (1983)
106. Zariski, O.: On the Problem of Existence of Algebraic Functions of Two Variables Possessing a Given Branch Curve. Amer. J. Math. **51**, no. 2, 305–328 (1929)
107. Zariski, O.: Studies in equisingularity I-III. Amer. J. Math. **87**, 507–536, (1965); Amer. J. Math. **87**, 972–1006 (1965); Amer. J. Math. **90**, 961–1023 (1968)
108. Zariski, O.: Algebraic surfaces, 2nd edn. Springer, Heidelberg (1971)

Chapter 3
Limit of Tangents on Complex Surfaces

Tráng Dũng Lê and Jawad Snoussi

Abstract In these notes we give an introduction on the limits of tangents to a complex analytic surface. We first describe the case of hypersurfaces, using integral dependence on ideals and equisingularity controlled by Milnor number, and then we discuss the case of general surfaces of $\mathbb{C}^N$ using Whitney equisingularity and equivalent criteria.

Contents

T. D. Lê (✉)
University of Aix-Marseille, Marseille, France
e-mail: ledt@ictp.it

J. Snoussi
Universidad Nacional Autónoma de México, Instituto de Matemáticas, Unidad Cuernavaca, Avenida Universidad S/n, Lomas de Chamilpa, 62210,Cuernvaca, México
e-mail: jsnoussi@im.unam.mx

Unidad Mixta Internacional CNRS, Laboratorio Solomon Lefschetz, Cuernavaca, Mexico

J. L. Cisneros-Molina et al. (eds.), *Handbook of Geometry and Topology of Singularities II*, https://doi.org/10.1007/978-3-030-78024-1_3

3.1 Introduction

After H. Whitney introduced a regular condition for complex analytic stratified sets (see [54] Sect. 19) a natural question was to understand the set of limits of tangents at singular points. In these notes we give a survey in the case of complex analytic surfaces.

Let X be a reduced complex analytic space. Let x be a point of X. If the point x is a non-singular point of X, there is only a tangent space at x and if (x_n) is a sequence of points of X tending to x, the limit of tangent spaces $T_{x_n}(X)$ is the tangent space $T_x(X)$. However when x is a singular point of X, there might be several limits of tangent spaces $\lim_{x_n \to x} T_{x_n}(X)$.

Let us define.

Definition 3.1.1 A space T is a limit of tangent spaces at a point x on X if there exists a sequence of non-singular points (x_n) of X which tends to x such that the tangent spaces $T_{x_n}(X)$ tend to T.

Suppose X is reduced, equidimensional of dimension d at the point x and call X^0 the subset of non-singular points of X. When the germ (X, x) is not equidimensional, one needs to work with each irreducible component.

Let $(X, x) \subset (\mathbb{C}^N, x)$ be a local embedding in an Euclidean space. Let $\mathcal{U}$ be an open neighborhood of x in $\mathbb{C}^N$ and $U = \mathcal{U} \cap X$ a neighborhood of x in X.

We have a map:

$$\gamma_0 : U \cap X^0 \to \mathbf{G}(d, N)$$

into the Grassmanian $\mathbf{G}(d, N)$ of d-linear spaces of $\mathbb{C}^N$ which maps a point y of $U \cap X^0$ into the linear subspace $T_y(X)$ of $\mathbb{C}^N$.

By a theorem of Nobile ([35, Theorem 1]) and a Theorem of Remmert-Stein ([41, Satz 13 p. 299]) the closure of the graph of γ_0 in $\mathcal{U} \times \mathbf{G}(d, N)$ is an analytic subspace $\tilde{X}_U$. The projection of $\mathcal{U} \times \mathbf{G}(d, N)$ onto $\mathcal{U}$ induces a holomorphic map:

$$\nu : \tilde{X}_U \to X \cap \mathcal{U}$$

which is a proper modification of $X \cap \mathcal{U}$ of dimension d. The map ν is called *the Nash modification* of X at x.

Then the fiber of ν over x is the set of limits of tangent spaces of X at x. Since ν is a complex analytic modification, this fiber is analytic, but by the theorem of Chow (see [5, Theorem V p. 910]) it is algebraic because it is a closed analytic subspace of the projective variety $\{x\} \times \mathbf{G}(d, N)$.

If X is not equidimensional at x, the set of limits of tangents of X at x is a finite union of projective varieties living in different Grassmannian spaces. However, along this work, we shall focus on reduced pure dimensional spaces.

Proposition 3.1.2 *The set of limits of tangents at a point x of a reduced analytic space X is a projective algebraic set.*

We shall often make the abuse of language that instead of talking of the linear space $T_y(X)$ we consider the affine space $T_y(X)$ at the point y. Therefore sometimes a limit of tangents at x will be an affine space at the point x.

The set of limits of tangents to X at x, is strongly related to another set: *the tangent cone* of X at x.

Let $\mathcal{I}$ be the ideal of $\mathcal{O}_{\mathbb{C}^N,x}$ defining the germ (X, x) in $(\mathbb{C}^N, x)$. Let $\mathrm{in}_x(f)$ denote the initial form of a function $f \in \mathcal{O}_{\mathbb{C}^N,x}$ at x, i.e., the sum of terms of lowest degree in the Taylor expansion of f at x. We call the initial ideal of $\mathcal{I}$ at x, the ideal of $\mathcal{O}_{\mathbb{C}^N,x}$ generated by all the initial forms $\mathrm{in}_x(f)$ of functions in $\mathcal{I}$. It is a homogeneous ideal in $\mathbb{C}[X_1, \ldots, X_N]$ that we denote by $\mathrm{In}_x(\mathcal{I})$. It defines a cone in $\mathbb{C}^N$ with vertex in x that we call the tangent cone of X at x and denote it by $C_{X,x}$.

When X is equidimensional at x the tangent cone $C_{X,x}$ is an algebraic space of the same dimension as X.

One can check that the generatrices of the tangent cone $C_{X,x}$ are the lines in $\mathbb{C}^N$ containing x obtained as limits of secants to X at x (see e.g [54, Sect. 8 p. 510]). That is, a line $x \in \ell \subset \mathbb{C}^N$ is a generatrix of the reduced subspace $|C_{X,x}|$ subjacent to $C_{X,x}$ if and only if there exists a sequence of points (x_n) in X converging to x such that the sequence of lines (xx_n) converges to ℓ.

The tangent cone determines completely the set of limits of tangent lines to a curve. In fact, in curves, thanks to l'Hospital's rule and to the existence of local parametrization, we know that the limits of tangent lines and of secant lines coincide (see Lemma 6.2.1 of [27]). Since algebraic cones of dimension 1 are finite unions of lines, the limits of tangent lines to a curve at a given point are precisely the lines of the tangent cone at that point.

All along this work, we shall call a tangent line to a curve, either a tangent line at a non-singular point, or a limit of tangent lines at a singular point.

Another situation where the limits of tangents are quite easy to understand and describe is when the considered space is a two-dimensional reduced cone.

In fact, a cone C of dimension 2 is a cone over a projective curve $\mathrm{Proj}(C)$. So for a non-singular point x of C, the tangent plane to C at x is tangent to C along the generatrix ℓ containing x. It is the plane obtained as the cone over the tangent line to $\mathrm{Proj}(C)$ at the point $\mathrm{Proj}(\ell)$.

When the point $x \in C$ is singular, the whole generatrix ℓ is singular and so is the point $\mathrm{Proj}(\ell) \in \mathrm{Proj}(C)$. The homogeneity of the equations of C allows to prove that the limits of tangent planes to C at any point of ℓ different from the vertex, are exactly the cones over the tangent lines to $\mathrm{Proj}(C)$ at $\mathrm{Proj}(\ell)$.

Since tangent lines to a curve at singular points are clearly characterized, there is no ambiguity in talking about tangent planes to a cone of dimension two at a singular point. In particular, the set of limits of tangent planes to a cone at its vertex is the set of all the cones over all tangent lines to the associated projective curve at each of its points.

In higher dimensions, the relation with the tangent cone is not so direct, but still there is a relation.

We shall begin by proving a theorem due to Hironaka saying that for equidimensional complex analytic sets, tangent spaces to the reduced tangent cone, are among the set of limits of tangent spaces to the complex set at the given point.

This theorem is a consequence of another theorem due to Hironaka showing the existence of Whitney regular stratifications with the Thom condition.

Then, we give a theorem of B. Teissier which gives a necessary and sufficient condition for a hyperplane H to be a limit of tangent hyperplanes to a complex analytic hypersurface at an isolated singular point. To establish this result we need to link limits of tangent hyperplanes to the theory of integral dependence on ideals.

As a corollary of the theorem of Teissier we give a characterization of the limits of tangents of a complex analytic hypersurface of dimension 2 at an isolated singular point.

This characterization is given in terms of the Milnor number of the hyperplane sections of the surface. In case of hypersurfaces of $\mathbb{C}^3$, the hyperplane sections are plane curves, and the Milnor number is a strong tool for measuring the equisingularity in families of curves. This allows to link limits of tangent planes to the equisingularity of the surface obtained by blowing-up the origin, along the exceptional fiber.

At this stage, an important object related to the limits of tangents is introduced: *the exceptional tangents*.

The characterization of exceptional tangents allows to complete the description of the set of limits of tangents to hypersurfaces of $\mathbb{C}^3$ with isolated singularities, since every hyperplane of $\mathbb{C}^3$ containing an exceptional tangent is a limit of tangent planes.

Then we define polar curves and prove that exceptional tangents of a hypersurface in $\mathbb{C}^3$ are precisely the common tangents to the general elements of the family of polar curves.

The following step is to generalize the study of limits to reduced equidimensional surfaces in $\mathbb{C}^N$ without conditions on the codimension or on the singular locus.

Here it is useful to define tangent and limits of tangent hyperplanes. These are hyperplanes containing a tangent or a limit of tangent planes.

Using again Hironaka's result on the tangents to the tangent cone, we explain how the limits of tangent hyperplanes are the hyperplanes that are either tangent to the tangent cone, or that contain a special generatrix of the tangent cone, that we define, in this context, to be the exceptional tangents.

In order to explain that this definition of exceptional tangents generalizes conveniently the definition given in hypersurface case, we prove that in the general case, the exceptional tangents are still the common tangents to the general polar curves.

Afterwards, we establish the relation with equsingularity. The first step in that direction is to relate the limits of tangents, to tangent lines to discriminants of generic projections to $\mathbb{C}^2$. This part of the work is based on the paper [25] by D. T. Lê, dealing with hypersufaces of $\mathbb{C}^3$ that we generalize to surfaces in $\mathbb{C}^N$.

Using Whitney regularity, Zariski's discriminant criterion and strong simultaneous resolution, we establish the relation between limits of tangents and exceptional tangents with equisingularity. This part of the text relates and generalizes the paper [44] where J. Snoussi works on normal surfaces.

Finally we describe surfaces which do not have exceptional tangents.

Along the work we give different examples to illustrate the results presented.

In the text, as it is usually done, we shall denote by $|Z|$ the reduced complex analytic space underlying a complex analytic space Z.

3.2 An Application of a Theorem of Hironaka

3.2.1 The Thom Stratification

We do as in [30, Sect. (1.4) for the complex analytic case].

Let $f : X \to Y$ be a holomorphic morphism between reduced complex analytic spaces.

For general results and properties on stratifications and Whitney stratifications we refer to [52]. We say that f is a *stratified* morphism if there are a Whitney stratification $\mathcal{S} = (S_i)$ of X and a Whitney stratification $\mathcal{T} = (T_j)$ of Y such that, for any stratum S_i, there is an index $j(i)$ for which f induces a surjective submersion of S_i onto $T_{j(i)}$.

Definition 3.2.1 1. Let f be a holomorphic morphism stratified by the stratifications $\mathcal{S}$ of X and $\mathcal{T}$ of Y. We say that the pair (S_α, S_β) satisfies the *Thom condition at* $x \in X$ if $x \in S_\alpha$, $S_\alpha \subset \overline{S_\beta}$ and for a fixed embedding of (X, x) in $(\mathbb{C}^N, 0)$ for any sequence of points $(x_n \in S_\beta)$ tending to x for which the sequence $(T_{x_n}(f^{-1}(f(x_n)) \cap S_\beta)$ converges to T then $T \supset T_x(f^{-1}(f(x)) \cap S_\alpha)$.
2. We say that the stratified map f satisfies Thom A_f condition if for any point $z \in X$, and S_α the stratum of X which contains z, for any stratum S_β such that $S_\alpha \subset \overline{S_\beta}$, the pair (S_α, S_β) satisfies the Thom condition at $z \in X$.

A remarkable result of Hironaka [20, Corollary 1 of Theorem 2 of Sect. 5 Thom A_f-condition and flattening] (or [2, Théorème 4.2.1]) is that for any germ of holomorphic function $(X, x) \to (\mathbb{C}, 0)$ there is a representative $f : X \to U \subset \mathbb{C}$ such that:

Theorem 3.2.2 *There are Whitney stratifications of X and U for which f is stratified and satisfies the Thom condition.*

As a consequence we have the following corollary also attributed to Hironaka.

Corollary 3.2.3 *Suppose that X is equidimensional at x. For a generic generatrix ℓ of $|C_{X,x}|$ and for any sequence of non-singular points (x_n) in X for which the sequence of lines (xx_n) tends to ℓ and the sequence of tangent planes $T_{x_n}(X)$ tends to a plane T, the plane T is the tangent plane to $|C_{X,x}|$ along ℓ.*

In particular the tangent spaces to the reduced tangent cone $|C_{X,x}|$ belong to the set of limits of tangents of X at x.

By generic generatrix of $|C_{X,x}|$ we mean a line which corresponds to a point of a Zariski dense set of non-singular points of $\mathrm{Proj}(|C_{X,x}|)$.

3.2.2 Deformation on the Tangent Cone

In order to prove the Corollary 3.2.3, we need to define a 1-parameter family which specializes on the tangent cone (see [10], [12, 2.4] or [47, 1.7.1 Theorem 2]).

Let X be a reduced complex analytic space and x be a point of X. The germs of complex analytic functions on (X, x) define an analytic local ring $O_{X,x}$. The maximal ideal $\mathfrak{m}_{X,x}$ of $O_{X,x}$ defines the graded algebra:

$$Gr_{\mathfrak{m}}(O_{X,x}) := \oplus_{n\in\mathbb{N}} \frac{\mathfrak{m}^n_{X,x}}{\mathfrak{m}^{n+1}_{X,x}}$$

This graded algebra defines a subscheme $\mathrm{Proj}(C_{X,x})$ of $\mathrm{Proj}(\mathbb{C}^N)$, where:

$$N = \dim_{\mathbb{C}}(\mathfrak{m}/\mathfrak{m}^2)$$

is the *Zariski dimension* of the local ring $O_{X,x}$. The graded algebra $Gr_{\mathfrak{m}}(O_{X,x})$ also defines a cone $C_{X,x}$ in $\mathbb{C}^N$ called the tangent cone of X at x.

Notice that the tangent cone of X at x is not reduced in general.

The tangent cone $C_{X,x}$ is defined by a homogeneous ideal of $\mathbb{C}[X_1, \dots, X_N]$ that can be obtained as follows:

Let J be the ideal of $O_{\mathbb{C}^N,x}$ which defines the germ (X, x) in $(\mathbb{C}^N, x)$. Since $O_{X,x}$ is noetherian (see e.g. [21, Corollaire 2 p. 18–07]), the graded algebra $Gr_{\mathfrak{m}}(O_{X,x})$ is also noetherian and we can find generators $a_1, \dots a_r$ of J such that their initial forms i.e., the non-constant homogeneous polynomials of lowest degree of their Taylor expansions at x (see e.g. [56, p. 249]) $\mathrm{in}_x(a_0), \dots, \mathrm{in}_x(a_r)$, generate the ideal $\mathrm{In}_x J$ defining $C_{X,x}$.

We may assume that the germ (X, x) is embedded in $(\mathbb{C}^N, 0)$. Consider $f_1, \dots, f_r$ to be functions defined on a neighbourhood U of 0 in $\mathbb{C}^N$ which induce the elements $a_1, \dots, a_r$ in the local ring $O_{\mathbb{C}^N,x}$. The ideal of $O_{\mathbb{C}^N,x}$ generated by the initial forms $\mathrm{in}_x f_1, \dots, \mathrm{in}_x f_r$ of $f_1, \dots, f_r$ defines a cone isomorphic to $C_{X,x}$.

As in the proof of Théorème (1.5) of [19], we can define the 1-parameter deformation given by:

$$F_1(z,t) = \mathrm{in}_x f_1(z) + t f_{1,n_1+1} + \dots$$
$$\dots$$
$$F_r(z,t) = \mathrm{in}_x f_r(z) + t f_{r,n_r+1} + \dots$$

where $f_{i,j}$ is the sum of the terms of f_i of degree j and n_i is the multiplicity of f_i at the point x.

The complex analytic set W defined by $F_1 = \dots = F_r = 0$ is an analytic subset of $U \times \mathbb{C}$. The projection on the second factor induces a holomorphic map:

$$\varphi : W \to \mathbb{C}.$$

Since:

$$F_i(z,t) = \frac{1}{t^{n_i}} f_i(tz)$$

The fiber over $t_0 \neq 0$ of φ has a germ at $(0, t_0)$ isomorphic to (X, x) and the fiber of φ above 0 has a germ at $(0, 0)$ which is isomorphic to the germ of $C_{X,x}$ at its vertex.

This is why one can call the morphism φ a deformation of (X, x) on its tangent cone.

For the convenience of our proof, we shall assume that the open set U is invariant by homothety with center x.

3.2.3 *Proof of Corollary 3.2.3*

Remember that we assume that the space X is equidimensional at x. Let d be the dimension of X at x. We also suppose that the point x is the origin 0 of $\mathbb{C}^N$.

Let (x_n, t_n) be a sequence of non-singular points of the fibers $W_{t_n} := \varphi^{-1}(t_n)$ of φ tending to $(z, 0) \in \varphi^{-1}(0)$.

Since for each $t \neq 0$, the fiber W_t has a germ at $(0, t)$ isomorphic to $(X, 0)$ by the homothety $z \mapsto tz$, the sequence (x_n, t_n) of W corresponds to the sequence of non-singular points $(x'_n = t_n x_n)$ of X which tends to 0.

In fact the sequence of tangent spaces $(T_{(x_n,t_n)}(W_{t_n}))$ is equal to the sequence of tangent spaces $(T_{t_n x_n}(X))$ of X. By choosing a subsequence of (x_n, t_n) we may suppose that the sequence $(T_{(x_n,t_n)}(W_{t_n}))$ has a limit T_1 and the sequence $(T_{t_n x_n}(X))$ has a limit T_2. Since the two sequences $(T_{(x_n,t_n)}(W_{t_n}))$ and $(T_{t_n x_n}(X))$ coincide in the Grassmannian space $\mathbf{G}(d, N)$, w e have $T_1 = T_2$.

By the result of Hironaka stated in Theorem 3.2.2 there are Whitney stratifications of W and of $\mathbb{C}$ such that they satisfy the Thom condition. We can choose the stratification of $\mathbb{C}$ to be $\{0\}$, $\mathbb{C} - \{0\}$ since the germs of fibers of φ at the points $(0, t)$, for $t \neq 0$, are isomorphic.

Let S_α be a stratum of the Whitney stratification of $|W_0| \cap U$ of dimension dim W_0. Assume that z is a point of this stratum.

The Thom condition implies that the limit of $T_{(x_n,t_n)}(W_{t_n})$ when (x_n, t_n) tends to $(z, 0)$ is the tangent space to the stratum S_α at z. In fact, since we have supposed that X is equidimensional at the point x, the points of S_α are non-singular points of $|C_{X,0}| \cap U$. Since this observation is true for a finite number of strata of $|W_0| \cap U$ of dimension dim $|C_{X,0}|$, and since the closures of these strata are the irreducible components of $|C_{X,0}| \cap U$, the limit of tangent spaces $T_1 (= T_2)$ is then tangent to the tangent cone along the generatrix limit of secants $\lim_n 0(t_n x_n)$.

In this way all the tangent spaces to the strata of $|W_0| \cap U$ of dimension dim $|W_0|$ are limit of tangents of X at 0. This proves Corollary 3.2.3.

Remark 3.2.4 In the case X is a reduced complex hypersurface in $\mathbb{C}^{n+1}$ the set of limits of tangent hyperplanes to the hypersurface is an algebraic subset of the space $\check{\mathbb{P}}^n$ of projective hyperplanes. In that case Corollary 3.2.3 says that the dual variety of $\mathrm{Proj}(|C_{X,0}|)$ is in the set of limits of tangent spaces to X at 0.

3.3 The Theorem of Teissier

3.3.1 Statement

Theorem 3.3.1 *Let (X, x) be a germ of complex analytic hypersurface at the point x with an isolated singularity at x. A hyperplane H through x is not a limit of tangent hyperplanes to X at x if and only if the intersection $X \cap H$ is a hypersurface with an isolated singularity at x with the minimum Milnor number at x among all the Milnor numbers at x of the intersections with hyperplanes through x.*

Proof The proof uses the notion of integral dependence over an ideal (see [56, p. 347], or the book [38], here we use the paper [22] or [48] where the results are adapted to the case of local analytic algebras that we need in these notes).

We first prove:

Lemma 3.3.2 *Let (X, x) be a germ of complex analytic hypersurface with an isolated singularity at the point x and defined by a germ of holomorphic function f at x in $(\mathbb{C}^{n+1}, x)$. Let H be a hyperplane through x which is not a limit of tangent hyperplanes of X at x. Then, the multiplicity $e(I)$ of the ideal I induced by the Jacobian ideal $J(f) = (\partial f/\partial z_0, \ldots \partial f/\partial z_n)$ on the analytic local ring $\mathcal{O}_{X,x}$ is equal to $\mu(X, x) + \mu(X \cap H, x)$:*

$$e(I) = \mu(X, x) + \mu(X \cap H, x).$$

Proof Remember that $\mu(X, x)$ is the Milnor number of a hypersurface with isolated singularity. It was defined by J. Milnor in [34, Chap. 8 Theorem 7.2 p. 60]. The Milnor number o fX at x is given by:

$$\mu(X, x) = \dim_{\mathbb{C}} \frac{O_{\mathbb{C}^{n+1},x}}{Jac(f)}$$

where $Jac(f)$ is the ideal of the ring $O_{\mathbb{C}^{n+1},x}(\simeq \mathbb{C}\{z_0, \dots, z_n\})$ generated by the partial derivatives of f.

In particular the Lemma says that, if H is not a limit of tangents of X at x, the singularity of $X \cap H$ is isolated at x.

Instead of considering germs, we shall consider representatives of these germs.

Assume x is the origin 0 of $\mathbb{C}^{n+1}$, and choose an open neighborhood $\mathcal{U}$ of the origin in $\mathbb{C}^{n+1}$. Call $U := \mathcal{U} \cap X$. We can choose coordinates in $\mathbb{C}^{n+1}$ such that the hyperplane H is given by $z_0 = 0$.

Since the affine hyperplane $\{z_0 = 0\}$ is not a limit of tangent hyperplanes, for any sequence (x_n) of non-singular points of X tending to 0, the sequence of tangent hyperplanes $(T_{x_n}(X))$ does not tend to the hyperplane $z_0 = 0$. In fact, since the space $\check{X}$ of the Nash modification ν is complex analytic, it is equivalent to say that, for any germ of complex analytic curve $p : (\mathbb{D}, 0) \to (X, 0)$ whose image of $\mathbb{D} - \{0\}$ is in the non-singular part X^0 of X, the hyperplane $\{z_0 = 0\}$ is not the limit $\lim_{t\to 0} T_{p(t)}(X)$.

Since $X = \{f = 0\}$ is a hypersurface, the gradient of f at a non-singular point z of X is orthogonal to the tangent hyperplane $T_z(X)$. In [34, Sect. 4 p.33] one finds that the gradient of a complex analytic function f at a point z is given by the conjugates of the partial derivatives of f at z:

$$\textbf{grad}\ f(z) = (\overline{\partial f/\partial z_0}(z), \dots \overline{\partial f/\partial z_n}(z)).$$

Therefore, the hyperplane $\{z_0 = 0)\}$ is not a limit of tangent hyperplanes to X at x if and only if:

$$\lim_{t\to 0}(\text{line}(\overline{\partial f/\partial z_0}(p(t)), \dots, \overline{\partial f/\partial z_n}(p(t)))) \text{ is not line}(1, 0, \dots, 0) \tag{3.1}$$

for any germ of complex analytic curve $p : (\mathbb{D}, 0) \to (X, x)$ such that $p(t)$ is a non-singular point for $t \neq 0$.

Since $\partial f/\partial z_i(p(t))$ is an analytic function of one variable of the open disk $\mathbb{D}$ at 0, this means that one of terms $\partial f/\partial z_1(p(t)), \dots, \partial f \partial z_n(p(t))$ has an entire series at 0 with a lower degree less or equal to the lower degree of the power series $\partial f/\partial z_0(p(t))$.

This is where we use the notion of integral dependence over an ideal.

Since we need this notion of integral dependence in the complex analytic case, we consider a local analytic ring O, i.e., the quotient of a ring of convergent series by an ideal, and an ideal I in O.

Recall that an element α of the ring O is integral over the ideal I if there is a relation:

$$\alpha^n + \sum_1^n a_i \alpha^{n-i} = 0$$

where a_i belongs to I^i, for i, $1 \leq i \leq n$.

This algebraic notion gives a bridge between geometry and analysis.

Precisely consider a germ of analytic space (X, x) such that $\mathcal{O}_{X,x} \simeq \mathcal{O}$. In [22, Théorème 2.1 equivalence of (i) and (iii)], one can find the proof of: $\alpha \in \mathcal{O}_{X,x}$ is integral over the ideal $I \subset \mathcal{O}_{X,x}$ if and only if, for any germ of morphisms $p : (\mathbb{D}, 0) \to (X, x)$, i.e., any germ of complex analytic path of X at x, we have $p^*\alpha \in p^*I\mathcal{O}_{\mathbb{D},0}$, where $p^*\alpha$ is the function $\alpha \circ p$ and $p^*I\mathcal{O}_{\mathbb{D},0}$ is the pull back in $\mathcal{O}_{\mathbb{D},0}$ of the ideal I of $\mathcal{O}_{X,x}$.

Therefore the element $\partial f/\partial z_0$ in the local analytic algebra:

$$\mathcal{O}_{X,0} = \frac{\mathbb{C}\{z_0, \ldots, z_n\}}{(f)}$$

is integral over the ideal I_0 generated by $(\partial f/\partial z_1, \ldots, \partial f/\partial z_n)$ in $\mathcal{O}_{X,0}$, because the preceding Condition (3.1) above means that for any analytic path p on X at 0, we have:

$$p^*(\partial f/\partial z_0) \in p^* I_0 \mathcal{O}_{\mathbb{D},0}.$$

In particular, since the singularity of X at 0 is isolated, the ideals I and I_0 are primary for the maximal ideal $\mathfrak{m}$ of $\mathcal{O}_{X,0}$ and they have the same multiplicity (see [22, Proposition 1.18]).

In fact, a Theorem of Rees (see [37]) says that in a local analytic ring the primary ideals $I \subset J$ for the maximal ideal have the same multiplicity if and only if the integral closures $\overline{I}$ and $\overline{J}$ are equal. It yields that the hyperplane $z_0 = 0$ is not a limit of tangents of X at the point 0 if and only if the multiplicity $e(I_0)$ of the ideal I_0 of the local ring $\mathcal{O}_{X,0}$ equals the multiplicity $e(I)$ of the ideal I of the local ring $\mathcal{O}_{X,0}$.

The $\mathfrak{m}$-primary ideal I_0 of $\mathcal{O}_{X,0}$ is generated by n generators in the Cohen-Macaulay local analytic ring $\mathcal{O}_{X,0}$ of dimension n (in fact here $\mathcal{O}_{X,0}$ is a hypersurface local ring, in particular it is Cohen-Macaulay because it is the local ring of a complete intersection), so the multiplicity $e(I_0)$ of I_0 is equal to:

$$\dim_{\mathbb{C}} \frac{\mathcal{O}_{X,0}}{I_0}.$$

Since:

$$\dim_{\mathbb{C}} \frac{\mathcal{O}_{X,0}}{I_0} = \dim_{\mathbb{C}} \frac{\mathbb{C}\{z_0, \ldots, z_n\}}{(f, \partial f/\partial z_1, \ldots, \partial f/\partial z_n)},$$

the multiplicity $e(I_0)$ equals the intersection number of f and $(\partial f/\partial z_1, \ldots, \partial f/\partial z_n)$ at the point 0. In particular it implies that the complex analytic set:

$$\{\partial f/\partial z_1 = \ldots = \partial f/\partial z_n = 0\}$$

is a curve at the point 0.

Since the number $e(I_0) = e(I)$ is finite, the curve

$$\Gamma := \{\partial f/\partial z_1 = \ldots = \partial f/\partial z_n = 0\}$$

intersects the hypersurface $\{f = 0\}$ locally at 0. Consider:

$$\Phi := (z_0, f) : (\mathbb{C}^{n+1}, 0) \to (\mathbb{C}^2, 0).$$

Notice that Γ is the critical locus of Φ.

The restriction of Φ to $(\Gamma, 0)$ is finite because locally at 0 we have:

$$\Phi^{-1}(\mathbb{C} \times \{0\}) \cap \Gamma = \{0\}$$

with a good choice of the representatives of the germs at 0. Since a finite analytic map is proper, a theorem of Remmert ([40] Satz 23) gives that the image of $|\Gamma|$ by Φ is a complex analytic curve $|\Delta|$ at 0.

As the multiplicity $e(I)$ equals the intersection number of $\{f = 0\}$ and the curve $\Gamma = \{\partial f/\partial z_1 = \ldots = \partial f/\partial z_n = 0\}$ at 0, since the direct image $\Phi_*\Gamma$ of Γ by the map $\Phi := (z_0, f) : (\mathbb{C}^{n+1}, 0) \to (\mathbb{C}^2, 0)$ is the discriminant Δ of Φ (see the definition in [49, 2.6 p. 588]), the intersection number $(\{f = 0\}.\Gamma)_0$ is given by the projection formula (see [6]):

$$(\{f = 0\}.\Gamma)_0 = (\Phi^{-1}(\{\mathbb{C} \times \{0\}).\Gamma)_0 = (\mathbb{C} \times \{0\}.\Phi_*\Gamma)_0 = (\mathbb{C} \times \{0\}.\Delta)_0.$$

Beware that Γ or $\Phi_*\Gamma$ might not be reduced. Each branch of $|\Delta|$ might be counted with a multiplicity.

We have:

Lemma 3.3.3 *The intersection number* $(\mathbb{C} \times \{0\}.\Delta)_0$ *equals* $\mu(X, 0) + \mu(X \cap H)$.

By the invariance of the discriminant by pull-back (see [49, Remarque 3.3 Chap. III p. 341–342]) the intersection number $(\mathbb{C} \times \{0\}.\Delta)_0$ equals the discriminant of the map $(X, 0) \to (\mathbb{C}, 0)$ induced by the linear form z_0, since in this case the discriminant is a point whose multiplicity is precisely $(\mathbb{C} \times \{0\}.\Delta)_0$.

Proof of Lemma 3.3.3 Let U be a sufficiently small neighborhood of the origin 0 in $\mathbb{C}^{n+1}$ and a sufficiently small neighborhood V of $(0, 0)$ in $\mathbb{C}^2$ such that the mapping Φ defined above induces a map $\Phi_{U,V} : U \to V$.

We observe first that the multiplicity of a point v of $\Delta \cap V$ is the sum of the Milnor numbers of critical points of Φ over v in U .

Then, choose U and V small enough. Now consider line $\mathbb{C} \times \{t\}$ such that t is chosen small enough such that:

1. the number of intersection points of $V \cap (\mathbb{C} \times \{t\})$ with $|\Delta|$ is the intersection number $(\mathbb{C} \times \{0\}.|\Delta|)_0$;
2. by choosing U conveniently, $U \cap \Phi^{-1}(\mathbb{C} \times \{t\})$ is the Milnor fiber of f at 0 and $U \cap \Phi^{-1}((0, t))$ the Milnor fiber of $(X \cap H, 0)$.

Then we do a proof similar to the one of [26, Theorem 4.5 particularly of Lemma 4.9] (or [27, Proof of Lemma 6.6.9 Chap. 6 of this handbook vol. I]). In [48, Proposition 1.2 Chap II] Teissier gives a n ảgebraic proof of the Lemma 3.3.3.

Lemma 3.3.3 implies that $(\Phi^{-1}(\mathbb{C} \times \{0\}).\Gamma)_0 = \mu(X, 0) + \mu(X \cap H)$ which yields:

$$(\{f = 0\}.\Gamma)_0 = \dim_{\mathbb{C}} \frac{\mathbb{C}\{z_0, \ldots, z_n\}}{(f, \partial f \partial z_1, \ldots, \partial f \partial z_n)} = \mu(X, 0) + \mu(X \cap H) = e(I).$$

This proves Lemma 3.3.2.

Using Lemma 1.4 of [48] and the upper semi-continuity of Milnor number in the family of hyperplane sections through 0, one obtains the minimality of Milnor number $\mu(X \cap H)$ as stated in Theorem 3.3.1. This proves Theorem 3.3.1.

Our arguments above lead to:

Corollary 3.3.4 *Consider a hypersurface defined by the germ of complex analytic function $f \in \mathcal{O}_{\mathbb{C}^{n+1},0}$ having at the origin 0 an isolated singularity. Let H be a hyperplane of $\mathbb{C}^{n+1}$ through the origin 0. Choose the coordinates of $\mathbb{C}^{n+1}$ such that $H = \{z_0 = 0\}$. The hyperplane H is not a limit of tangents of X at 0 if and only if the ideal I_0 generated by $\partial f/\partial z_1, \ldots, \partial f/\partial z_n$ in $\mathcal{O}_{X,0}$ has the same multiplicity than the ideal generated by $\partial f/\partial z_0, \partial f/\partial z_1, \ldots, \partial f/\partial z_n$ in $\mathcal{O}_{X,0}$.*

Remark 3.3.5 In [50, Théorème 2, Appendice p. 291] B. Teissier proves that if for a hyperplane $\mu(X \cap H, 0)$ is the minimum among all the Milnor numbers of intersections by hyperplanes at 0, then there is a flag $L_1 \subset \ldots \subset L_{n-1}$ of subspaces of H such that $(\mu(X, 0), \mu(X \cap H, 0), \mu(X \cap L_{n-1}), \ldots, \mu(X \cap L_1))$ is the $\mu^*(X, 0)$ sequence defined by Teissier in [48, Définition 1.5 Chap. 1].

For instance, L_1 has dimension one and $\mu(X \cap L_1) = m(X, 0) - 1$, where $m(X, 0)$ is the multiplicity of x at the point 0, since the intersection number $(L_1.X)_0 = m(X, 0)$. Therefore L_1 is not contained in the tangent cone $C_{X,0}$ of X at the point 0.

3.4 Hypersurfaces of Dimension 2

3.4.1 Consequences of Teissier's Theorem

In the case we consider complex surfaces in $\mathbb{C}^3$, i.e., complex hypersurfaces of dimension 2, with isolated singularities, the Theorem of Teissier allows us to describe the set of limits of tangent planes at an isolated singular point because the section of a surface by a plane is a curve.

Let S be a complex analytic surface in $\mathbb{C}^3$ with an isolated singularity at 0 defined by $f : \mathbb{C}^3 \to \mathbb{C}$. Let P be a plane through 0. The result of Teissier tells us:

Corollary 3.4.1 *The plane P is not a limit of tangent planes of S at 0 if and only if the Milnor number $\mu(S \cap P, 0)$ is minimum among all the Milnor numbers at 0 of intersections of S by planes.*

As a consequence we have:

Proposition 3.4.2 *If P_1 and P_2 are two planes which are not limit of tangents of the surface S at 0, we can embed $S \cap P_1$ and $S \cap P_2$ in a family of plane curves satisfying μ-constant.*

Proof Consider Ω the open set of the space of planes through 0 such that for any plane P in Ω, P is not a limit of tangents ta S at 0.

Over Ω we have the tautological bundle $\mathbb{T}(\Omega)$ where the fiber over a point τ of Ω is the plane which corresponds to the point τ. This bundle is a sub-fibration of the trivial bundle $\mathbb{C}^3 \times \Omega$:

$$\begin{array}{ccc} \mathbb{T}(\Omega) & \hookrightarrow & \mathbb{C}^3 \times \Omega \\ & {\scriptstyle\theta}\searrow \quad \swarrow & \\ & \Omega & \end{array} \tag{3.2}$$

Let $F : (\mathbb{C}^3, 0) \times \Omega \to \mathbb{C}$ be the trivial extension of $f : (\mathbb{C}^3, 0) \to \mathbb{C}$. The intersection of a representative of $\{F = 0\}$ with $\mathbb{T}(\Omega)$ endowed with the projection onto Ω defines a flat deformation on Ω.

Since the planes P_1 and P_2 belong to Ω and Ω is connected, we can prove that $S \cap P_1$ and $S \cap P_2$ belong to a μ-constant family of plane curves.

As a consequence of the main result of [28] we obtain:

Lemma 3.4.3 *If P_1 and P_2 are two planes which are not limit of tangents of the surface S at 0, the plane curves $S \cap P_1$ and $S \cap P_2$ are topologically equisingular.*

Now the problem is to find a way to compute the Milnor number of a generic plane section of the surface S.

In what follows, we shall give results to obtain the Milnor number of a hyperplane section of the surface singularity and give results to compare two general hyperplane sections.

The idea is to blow-up the singular point and observe that there is a way to obtain conditions on the plane section to have the generic Milnor number. For generalities on blowing-ups we refer to [46, Sect. 3].

When one blows up a point x in $\mathbb{C}^N$, one obtains a complex analytic manifold Z of dimension N. The blowing-up of the point is the map $e : Z \to \mathbb{C}^N$.

Namely we have a map h from $\mathbb{C}^N - \{x\}$ into $\mathbb{P}^{N-1}$ which sends a point $y \in \mathbb{C}^N - \{x\}$ to the affine line which connects x and y. The space of affine lines of $\mathbb{C}^N$ through x is isomorphic to $\mathbb{P}^{N-1}$.

The graph of h is a subset of $\mathbb{C}^N \times \mathbb{P}^{N-1}$ which is not closed. The closure of this graph is precisely the manifold Z.

The projection e of Z onto $\mathbb{C}^N$ is the *blowing-up* of the point x in $\mathbb{C}^N$. By abuse of language it is often said that Z is the blowing-up of x in $\mathbb{C}^N$.

The fiber of e over x is isomorphic to $\mathbb{P}^{N-1}$ and called the *exceptional divisor* of the blowing-up e.

Similarly one can define the blowing-up e_U of a point x in an open subset U of $\mathbb{C}^N$. In fact e_U is the morphism induced by e from $e^{-1}(U)$ into U.

Let X be a reduced equidimensional closed complex analytic subset of an open subset U of $\mathbb{C}^N$. Let $e_U : Z_U \to U$ be the blowing-up of the point $x \in X$ in U. The *strict transform* X' of X by the blowing-up e_U is the closure of $e_U^{-1}(X - \{x\})$ in Z_U. The blowing-up e_X of x in X is the map induced by e_U from the strict transform X' of X to X. One can construct the blowing-up e_X by considering the map h_X from $X - \{x\}$ into $\mathbb{P}^{N-1}$ and by taking the closure of the graph of h_X in $U \times \mathbb{P}^{N-1}$.

One notices that the set underlying the exceptional divisor of the blowing-up e_X is equal to the space $\mathrm{Proj}(|C_{X,x}|)$ of lines through x in the tangent cone of X at x.

One can prove that the inverse image by e_X of the maximal ideal $\mathfrak{m}$ of $\mathcal{O}_{X,x}$ is a sheaf of $\mathcal{O}_{X'}$-modules on X' which is locally of rank one.

More generally the blowing-up of an ideal I in X, where X is a small representative of (X, x) where I is generated by the holomorphic functions $f_1, \ldots, f_k$ defined on X is the following: Let $V(I)$ be the complex analytic subspace of X defined by $f_1 = \ldots = f_k = 0$. On $X - V(I)$ we have a map into $\mathbb{P}^{k-1}$ defined by $z \mapsto (f_1(z), \ldots, f_k(z))$. The closure of the graph of this map in $X \times \mathbb{P}^{k-1}$ is an analytic space X_I. The projection of X_I onto X is the blowing-up e_I of X with center the ideal I. One can prove that the inverse image of I by e_I is a locally invertible $\mathcal{O}_{X_I}$-module.

We have the following universal property: For any map $f : Z \to X$ such that the inverse image f^*I is a locally invertible $\mathcal{O}_Z$-module, the morphism f factorizes uniquely through e_I, i.e., there is a unique morphism $f_I : Z \to X_I$ such that $f = e_I \circ f_I$:

$$\begin{array}{ccc} Z & \xrightarrow{\ f_I\ } & X_I \\ & {\scriptstyle f}\searrow \quad \swarrow{\scriptstyle e_I} & \\ & X & \end{array} \tag{3.3}$$

If one considers a complex analytic surface S closed in an open subset U of $\mathbb{C}^3$, the blowing-up of the point $0 \in S$ in S is a complex analytic map $e : S' \to S$ and the reduced exceptional divisor is the projective curve $\mathrm{Proj}(|C_{S,0}|)$.

For instance, for the blowing-up $e_C : C' \to C$ of a cone C at 0 of dimension 2 the fiber $e_C^{-1}(0)$ is the projective curve $\mathrm{Proj}(C)$.

Let P be a complex plane of $\mathbb{C}^3$ containing 0 and U an open subset of $\mathbb{C}^3$ also containing 0. The strict transform of $P \cap U$ by the blowing-up $e_0 : U' \to U$ of 0 in U is a surface P' whose intersection with the exceptional divisor $e_0^{-1}(0)$ of e_0 is a projective line in $e_0^{-1}(0)$.

When the intersection $S \cap P$ is a reduced curve, the intersection $S' \cap P'$ is the blowing-up at 0 of $S \cap P$. Therefore the points of $P' \cap |e^{-1}(0)|$ correspond to the distinct tangents of $S \cap P$ at the singular point 0.

There is a relation between the Milnor number $\mu(S \cap P, 0)$ of the plane curve at 0 and the Milnor numbers of $S' \cap P'$ at the points $t_1, \ldots, t_k$ of $P' \cap |e^{-1}(0)|$.

Namely we have:

Lemma 3.4.4 *Let $(C, 0)$ be the germ at 0 of a reduced plane curve having k distinct tangents at 0. Let $C' \to C$ be the blowing-up at 0 of a sufficiently small representative C of $(C, 0)$. We have:*

$$\mu(C, 0) = m(C, 0)(m(C, 0) - 1) + \sum_{i=1}^{k} \mu(C', t_i) - k + 1$$

where $t_1, \ldots, t_k$ are the points of C' above 0 and $m(C, 0)$ is the multiplicity of the plane curve C at 0.

Proof Let $\delta(C, 0)$ be the following number:

$$\delta(C, 0) := \dim_{\mathbb{C}} \frac{\bar{O}_{C,0}}{O_{C,0}}$$

where $\bar{O}_{C,0}$ is the normalization of the local ring $O_{C,0}$. It is known that $\delta(C, 0)$ is a finite number.

In [34, Theorem 10.5, p. 85] Milnor proves that:

$$\mu(C, 0) + r - 1 = 2\delta(C, 0),$$

where r is the number of branches of C at 0.

It is known that (see e.g [53] Remark p.151):

$$\delta(C, 0) = \frac{m(C, 0)(m(C, 0) - 1)}{2} + \sum_{i=1}^{k} \delta(C', t_i).$$

Therefore using Milnor formula given above we have:

$$\mu(C,0) = 2\delta(C,0) - r + 1 = m(C,0)(m(C,0)-1) + \sum_{i=1}^{k} 2\delta(C',t_i) - r + 1$$
$$= m(C,0)(m(C,0)-1) + \sum_{i=1}^{k} (\mu(C',t_i) + r_i - 1) - r + 1$$
$$= m(C,0)(m(C,0)-1) + \sum_{i=1}^{k} \mu(C',t_i) + r - k - r + 1$$

where r_i is the number of branches of C' at the point t_i and $r = \sum_{i=1}^{k} r_i$. This yields:

$$\mu(C,0) = m(C,0)(m(C,0)-1) + \sum_{i=1}^{k} \mu(C',t_i) - k + 1$$

as we claim in Lemma 3.4.4.

If we apply this lemma to the plane curve $S \cap P$, we have:

$$\mu(S \cap P, 0) = m(S \cap P, 0)(m(S \cap P) - 1) + \sum_{i=1}^{k} \mu(S' \cap P', t_i) - k + 1. \quad (3.4)$$

Since two plane sections with minimal Milnor number are topologically equisingular, we can apply the following results:

Lemma 3.4.5 *If two germs of plane curve $(C_1, 0)$ and $(C_2, 0)$ are topologically equisingular, i.e., there are good neighbourhoods U_1 and U_2 of 0 in $\mathbb{C}^2$ relatively to C_1 and C_2 and a homeomorphism $h : U_1 \to U_2$ such that $h(C_1 \cap U_1) = C_2 \cap U_2$, then:*

1. *they have the same number of branches, the homeomorphism h sends a branch of C_1 at 0 on a branch of C_2 at 0;*
2. *the intersection number of the branch Γ_i and the branch Γ_j $(i \neq j)$ at 0 is equal to the intersection number of $h(\Gamma_i)$ and $h(\Gamma_j)$ at 0.*

Proof The notion of good neighborhood due to Prill ([36] or see [17, 9.3.1]) implies that one can choose small open balls $\mathring{\mathbb{B}}_1$ and $\mathring{\mathbb{B}}_2$ centered at 0 as good neighbourhoods in $\mathbb{C}^2$ relatively to C_1 and C_2.

The analytic components of C_1 and C_2 at 0 are given by the connected components of $\mathring{\mathbb{B}}_1 \cap C_1 - \{0\}$ and $\mathring{\mathbb{B}}_2 \cap C_2 - \{0\}$. This gives 1 .

By choosing the balls $\mathbb{B}'_1$ and $\mathbb{B}'_2$ centered at 0, closed and smaller than $\mathring{\mathbb{B}}_1$ and $\mathring{\mathbb{B}}_2$ the property of being good neighbourhoods implies that $\mathbb{B}'_1 - (\Gamma_i \cup \Gamma_j)$ and $\mathbb{B}'_2 - (h(\Gamma_1) \cup h(\Gamma_2))$ have the same homotopy type. It implies that on the spheres $\mathbb{S}'_1 = \partial\mathbb{B}'_1$ and $\mathbb{S}'_2 = \partial\mathbb{B}'_2$ the complements of the links defined by $\Gamma_1 \cup \Gamma_2$ and $h(\Gamma_1) \cup h(\Gamma_2)$ have isomorphic fundamental groups, so the corresponding links have the same Alexander polynomials. A result of Burau [3] shows that this implies that the intersection number of Γ_i and Γ_j at 0 equals the intersection number of $h(\Gamma_1)$ and $h(\Gamma_2)$ at 0.

Another reasoning consists to use a result of Reeve [39]: the intersection number of a branch Γ_i and a branch Γ_j of C_1 at 0 is equal to the linking number of the knot $\Gamma_i \cap \mathbb{S}$ with the knot $\Gamma_j \cap \mathbb{S}$, where $\mathbb{S}$ is a sufficiently small sphere centered at 0.

This proves 2.

We shall need another result:

Lemma 3.4.6 *Let* $(C_1, 0)$ *and* $(C_2, 0)$ *be two germs of complex analytic plane curves which are topologically equisingular. Then*

1. *the germs* $(C_1, 0)$ *and* $(C_2, 0)$ *have the same multiplicity;*
2. *their strict transforms by a blowing-up of* 0 *in* $\mathbb{C}^2$ *have the same number* k *of points over* 0 *and there is an order of these points such that the germs* $(C_1', t_{1,i})$ *and* $(C_2', t_{2,i})$ *of the strict transform at these points are topologically equisingular, for* $1 \leq i \leq k$.

Proof If $(C_1, 0)$ and $(C_2, 0)$ are topologically equisingular by a local homeomorphism h of $\mathbb{C}^2$ an analytic component Γ_i of C_1 at 0 is topologically equisingular to the component $h(\Gamma_i)$ of C_2 at the point 0. By using Corollary 1 of 3.1.1 and Proposition 3.2.1 of [47] one proves that two topologically equisingular branches can be embedded in a one parameter μ constant deformation. With the result of [23] (or [28]) the multiplicity of Γ_i and $h(\Gamma_i)$ at 0 must be the same

The multiplicity of $(C_1, 0)$ is the sum of the multiplicities at 0 of the components Γ_i and the multiplicity of $(C_2, 0)$ is the sum of the multiplicities at 0 of $h(\Gamma_i)$, which are the components of $(C_2, 0)$, so the two germs must have the same multiplicity. This proves 1 .

Since $(C_1, 0)$ and $(C_2, 0)$ are equisingular, their Lipschitz saturation (see [14, Definition 1.4.18]) are the same ([14, Theorem 1.4.23]), therefore, since they are the generic projections of their Lipschitz saturation curve embedded in some $\mathbb{C}^N$, by considering the family of projection onto $\mathbb{C}^2$ representatives of the germs of plane curve $(C_1, 0)$ and $(C_2, 0)$ can be embedded in an equisingular family which satisfies the hypotheses of [55, Theorem 7 p. 529].

By definition of the equivalence of C_1 and C_2 ([55, Definitions 2, 3, 4 and Theorem 1]), C_1 and C_2 have the same number of distinct tangents and their strict transforms at the points of the blowing-up of 0 which correspond to their tangents are equisingular. This proves 2 .

3.4.2 Limit of Tangents of Surfaces of $\mathbb{C}^3$ with Isolated Singularity

By using the Formula 3.4 a plane P is not limit of tangent planes to S at 0 if and only if the line $\mathrm{Proj}(P)$ intersects $\mathrm{Proj}(|C_{S,0}|)$ at points where the blowing-up S' of

S at 0 is equisingular along $\mathrm{Proj}(|C_{S,0}|)$ and since the plane P is not contained in the tangent cone $C_{S,0}$, this implies that the multiplicity of $S \cap P$ at 0 is equal to the multiplicity of S at 0.

This gives:

Theorem 3.4.7 *Let $(S, 0)$ be the germ of surface in $(\mathbb{C}^3, 0)$ with an isolated singularity. The limit of tangents of S at 0 are the planes tangent to the tangent cone $|C_{S,0}|$ and a finite number of pencils of planes around lines of $|C_{S,0}|$ that we call the exceptional tangents of S at 0.*

Proof Corollary 3.2.3 already tells us the the tangent planes to the reduced tangent cone $|C_{S,0}|$ are limit of tangent planes of S at 0.

Projective lines through a singular point of $\mathrm{Proj}(|C_{S,0}|)$ correspond to limits of tangent planes P of S at 0 because these lines intersect the curve $\mathrm{Proj}(|C_{S,0}|)$ in less points than a general projective line. This shows that the number of distinct tangents of $(S \cap P)$ is not the generic one.

Therefore singular points of $\mathrm{Proj}(|C_{S,0}|)$ corresponds to exceptional tangents as defined in the Theorem.

We are left with non-singular points of the curve $\mathrm{Proj}(|C_{S,0}|)$ and projective lines ℓ transverse to the curve $\mathrm{Proj}(|C_{S,0}|)$.

We first notice:

Lemma 3.4.8 *Let $(X, 0)$ be a germ of complex hypersurface embedded in $(\mathbb{C}^{n+1}, 0)$. Suppose that the singular locus $(\Sigma, 0)$ of $(X, 0)$ has dimension 1 and is non-singular. Assume that $(X, 0)$ defines a μ-constant family of complex hypersurfaces along $(\Sigma, 0)$. Then, for any function f transverse to Σ at 0 the family of germs of complex hypersurfaces $(X \cap \{f = f(y)\}, y)$ is μ-constant with the same Milnor number at y in an open neighbourhood U_0 of 0 in Σ.*

Proof Let $x_1, \ldots, x_n, t$ be local complex analytic coordinates of $\mathbb{C}^{n+1}$ such that the equation of $(X, 0)$ is $F(x_1, \ldots, x_n, t) = 0$ and $x_1 = \ldots = x_n = 0$ defines $(\Sigma, 0)$. The family $F_t = F(x_1, \ldots, x_n, t) = 0$ is the μ-constant family along Σ at 0.

Since f is transverse to Σ at 0, we can choose in an open neighbourhood $\mathcal{U}$ of 0 in $\mathbb{C}^{n+1}$ the local coordinates $x_1, \ldots, x_n, f$.

Because $\{F(x_1, \ldots, x_n, t) = 0\}$ defines in an open neighbourhood of 0 in $\mathbb{C}^{n+1}$ a μ-constant family of hypersurfaces, by [29] (see also [48, Remarque 3.10]) the hyperplane $\{t = 0\}$ is not a limit of tangent hyperplanes to X at 0 and in the ring $\mathcal{O}_{\mathbb{C}^{n+1},0}$ the fonction $\partial F/\partial t$ is integral over the ideal $(\partial F/\partial x_1, \ldots, \partial F/\partial x_n)$ (compare with the beginning of the proof of Lemma 3.3.2).

Therefore the set $\{\partial F/\partial z_1 = \ldots = \partial F/\partial z_n = 0\}$ coincides with $\{\partial F/\partial z_1 = \ldots = \partial F/\partial z_n = \partial F/\partial t = 0\}$, i.e., with the critical locus of F in an open neighbourhood of 0 in $\mathbb{C}^{n+1}$.

Similarly $\partial F/\partial f$ is integral over the ideal $(\partial F/\partial x_1, \ldots, \partial F/\partial x_n)$ and the set $\{\partial F/\partial z_1 = \ldots = \partial F/\partial z_n = 0\}$ coincides with $\{\partial F/\partial z_1 = \ldots = \partial F/\partial z_n = \partial F/\partial f = 0\}$.

It implies that the relative polar curve of $\{F = 0\}$ relatively to $t = 0$ at 0, i.e., the critical locus of $\Phi_0 = (t, F) : (\mathbb{C}^{n+1}, 0) \to (\mathbb{C} \times \mathbb{C}, 0)$ outside $\{F = 0\}$ (see [27, Theorem 6.6.3]), is empty. Similarly the relative polar curve of $\{F = 0\}$ relatively to $f = 0$ at 0 is empty.

From [24] the Milnor numbers of $X \cap \{t = 0\}$ and $X \cap \{f = 0\}$ at 0 are both equal to the $(n-1)$-Betti number of the Milnor fiber of F at 0. By hypothesis the Milnor number of $X \cap \{t = 0\}$ is μ.

Since $(X \cap \{t = u\}, (0, \ldots, 0, u))$ is the germ of a μ-constant family of hypersurfaces, the Milnor number of $(X \cap \{f = f(u)\}, (0, \ldots, 0, u))$ is also μ for u small enough, since by continuity f is also transverse to Σ at $(0, \ldots, 0, u)$ for u small enough. This gives a proof of Lemma 3.4.8.

Let us come back to the proof of Theorem 3.4.7.

Because of Lemma 3.4.8, for any projective line D which intersects Proj$(|C_{S,0}|)$ at a point x where the blowing-up surface S' is equisingular along Proj$(|C_{S,0}|)$ at x, the Milnor number of $S' \cap P'$ at x, where P' is the strict transform of the plane P defined by the projective line D, is the same among all the germ of curves $S' \cap S_0$ at x. where S_0 is any non-singular surface transverse at x with Proj$(|C_{S,0}|)$.

By Lemma 3.4.3, Lemma 3.4.6 and with Lemma 3.4.4 one can prove that any projective line which intersects transversally Proj$(|C_{S,0}|)$ at points which are smooth points of Proj$(|C_{S,0}|)$ and where S' is equisingular along Proj$(|C_{S,0}|)$ corresponds to a plane P of $\mathbb{C}^3$ through 0 such that the plane curve $S \cap P$ has the minimal Milnor number at 0 among all planes through 0.

In particular we obtain:

Corollary 3.4.9 *Let $(S, 0)$ be a germ of complex hypersurface of dimension 2 with isolated singularity at 0. The exceptional tangents of S at 0 correspond to the singular points of Proj$(|C_{S,0}|)$ and the non-singular points $y_1, \ldots, y_k$ of Proj$(|C_{S,0}|)$ where the blowing-up surface S' does not have a locally constant Milnor number along Proj$(|C_{S,0}|)$.*

3.5 Polar Varieties of a Hypersurface of Dimension 2

We shall provide another way to obtain exceptional tangents of a hypersurface of dimension 2 at an isolated singular point.

3.5.1 Polar Varieties

We first give a general definition of *Polar varieties* of a reduced equidimensional complex analytic space. For properties of Polar varieties we advise the reader to consult [31].

Let X be an equidimensional complex analytic subset of dimension d of an open subset U of $\mathbb{C}^N$. Let $p : X \to \mathbb{C}^k$ be the restriction to X of a projection $\mathbb{C}^N \to \mathbb{C}^k$, where $2 \leq k \leq d+1$.

Definition 3.5.1 Let x be a point of X. There is a non empty Zariski open subset Ω of the space of projections of $\mathbb{C}^N$ onto $\mathbb{C}^k$, for $2 \leq k \leq \dim X + 1$ such that, for $p \in \Omega$, the closure $P_{k-1}(p)$ of the critical locus of the restriction of p to the non-singular part $X - \Sigma$ of X is either empty or an analytic subspace of dimension $k-1$ whose equisingularity class of $(P_{k-1}(p), x)$ only depends on the germ (X, x). We call the germ $(P_{k-1}(p), x)$ the germ of polar variety defined at x by the general projection p.

Here we consider the equisingularity class as the one defined by equiresolution or more simply by Whitney conditions of the family parametrized by Ω.

Notice that here we have given the definition of polar varieties of the remark (2.2.3) of [31].

Remark 3.5.2 1. In the case of a swallow tail, *i.e* the surface discriminant of the general equation of degree 4, the limit of tangents at the origin is only a plane. If one considers a projection p onto that plane, the polar curve $P_1(p)$ is empty.
2. The polar variety $P_d(p)$ is always X.
3. In the case of a surface one has only to consider the polar curve $P_1(p)$ and the whole surface $P_2(p)$. When the surface is a hypersurface S of $\mathbb{C}^3$, we shall choose the Zariski open dense set Ω of projections $\mathbb{C}^3 \to \mathbb{C}^2$ defining the polar curve in such a way that the kernel l of a projection in Ω satifies the condition $l \not\subset |C_{S,0}|$

There is a relation between the tangent cone of the polar curve at a point 0 of a surface S and the exceptional tangents of S at 0. Now, we are going to specify this relation.

3.5.2 Exceptional Tangents of a Hypersurface of Dimension 2

In the proof of Theorem 3.4.7 we have obtained that the exceptional tangents of a germ of hypersurface (X, x) of dimension 2 with isolated singularity correspond to the singular points of $\mathrm{Proj}(|C_{X,x}|)$ and the non-singular points of $\mathrm{Proj}(|C_{X,x}|)$ at which the blowing-up X' of X at x is not equisingular along $\mathrm{Proj}(|C_{X,x}|)$.

There is another way to find these exceptional tangents at an isolated singular point of X. Namely:

Lemma 3.5.3 *Let S be a complex analytic hypersurface of dimension 2 with isolated singularities. Let 0 be a point of S. The lines which are in the tangent cone of the Polar curve $P_1(p)$ at 0, for almost all $p \in \Omega$, are exceptional tangents of S at 0.*

Proof Consider $p : \mathbb{C}^3 \to \mathbb{C}^2$ a sufficiently general projection. Let D be a line of the tangent cone $C_{P_1(p),0}$ of the polar curve $P_1(p)$ a t 0.

Let $z_n \in P_1(p)$ be a sequence of points $\neq 0$ tending to 0 such that the lines $0z_n$ converge to D. Since $P_1(p)$ is the closure in S of the critical locus of the restriction of p to $S - \Sigma$ and since we may assume that z_n is a non-singular point of $P_1(p)$ and of S, the tangent plane of S at z_n is a plane containing $T_{z_n}(P_1(p))$ and the kernel of p. Since the limit T of $T_{z_n}(S)$ contains D (see Lemma 6.2.1 of [27]), the plane T contains D and the kernel of p and is a limit of tangent planes to S at 0.

If the line D is in all the tangent cones $C_{P_1(p),0}$ of the polar curves $P_1(p)$ for an infinity of p, the line D must be an exceptional tangent as stated in the Lemma.

In fact we have:

Theorem 3.5.4 *Let S be complex analytic hypersurface of dimension 2 with an isolated singularity at the point 0. A line D is an exceptional tangent of S at 0 if and only if it is in the tangent cone of almost all polar curves of S at 0.*

Proof One of the implications of this proposition is given by Lemma 3.5.3. Let us show the converse, i.e., any exceptional tangent is in the tangent cone of all polar curves of S at 0.

We shall first explain a relation between limits of tangent planes and tangent lines to discriminants of finite projections to $\mathbb{C}^2$.

Let l be a line through 0 in $\mathbb{C}^3$. Consider a projection $\Pi_l : \mathbb{C}^3 \to \mathbb{C}^2$ having l as its kernel. For a generic choice of the line l the restriction π_l of the projection Π_l to S is finite, defined by two functions f and g in $\mathcal{O}_{S,0}$. The ideal (f, g), generated by f and g in $\mathcal{O}_{S,0}$ is primary for the maximal ideal of $\mathcal{O}_{S,0}$.

The blowing-up e_1 of the ideal (f, g) of S is given by the closure S_1 in $S \times \mathbb{P}^1$ of the graph of the map $S - \{0\} \to \mathbb{P}^1$ defined by:

$$q \mapsto (f(q) : g(q)).$$

Then, the blowing-up $e_1 : S_1 \to S$ of the ideal (f, g) of $\mathcal{O}_{S,0}$ is induced by the projection of the graph into S.

Let e_Z be the blowing-up of $\mathbb{C}^2$ at the point 0:

$$e_Z : Z \to \mathbb{C}^2.$$

By the universal property of the blowing-up e_Z there is a map p_1 from S_1 to Z such that $e_Z \circ p_1 = \pi_l \circ e_1$.

We have the following pull-back diagram of π_l by e_Z:

$$\begin{array}{ccc} S_1 & \xrightarrow{e_1} & S \\ {\scriptstyle p_1}\downarrow & & \downarrow{\scriptstyle \pi_l} \\ Z & \xrightarrow{e_Z} & \mathbb{C}^2 \end{array} \tag{3.5}$$

Let L be a line of $\mathbb{C}^2$ going through 0. The inverse image of L by the projection Π_l is a plane Q containing the line l. One can view S_1 as the family of curves $S \cap Q$ as L is in the family of lines of $\mathbb{C}^2$ through 0.

Let $t = (0, (u : v)) \in e_Z^{-1}(0)$ ($\simeq \mathbb{P}^1$). A line $L_t = \{uy - vx = 0\}$ of $\mathbb{C}^2$ going through 0 defines a strict transform L'_t by the blowing-up e_Z that intersects the exceptional divisor at the point $(0, t)$. The inverse image $\pi_l^{-1}(L_t)$ is the curve $S \cap Q_t$ where $Q_t = \Pi_l^{-1}(L_t)$.

One can see in our context that each section $S \cap Q_t$ is isomorphic to its strict transform in S_1 by e_1.

Let $P_1(\pi_l)$ be the polar curve at 0 defined by the projection π_l on S. Notice that the polar curve $P_1(\pi_l)$ is the critical space of π_l. The discriminant Δ_l is the image of $P_1(\pi_l)$ by π_l. Since discriminant and critical spaces are kept under base change in a pull-back diagram (see [49, Remarque 3.3 Chap. III p. 341–342]), their inverse images $e_Z^{-1}(\Delta_l)$ and $e_1^{-1}(P_1(\pi_l))$ by e_Z and e_1 are the discriminant and critical space of the map p_1.

Remark 3.5.5 The surface $S_1 \subset S \times \mathbb{P}^1$ is endowed with the restriction λ_1 of the second projection:

$$\lambda_1 : S_1 \to \mathbb{P}^1.$$

For any point $t = (u : v) \in \mathbb{P}^1$ the fiber $\lambda_1^{-1}(t)$ is the strict transform of the section $S \cap Q_t$ by e_1.

Then the surface S_1 is a flat family of plane curves, each of them is isomorphic to the hyperplane section $S \cap Q_t$ by the plane Q_t containing the line l. The family is parametrized by $\mathbb{P}^1$.

The Theorem 7 p. 529 of [55] tells that the discriminant $e_Z^{-1}(\Delta_l)$ has components other than the exceptional divisor $e_Z^{-1}(0)$ which intersect the exceptional divisor $e_Z^{-1}(0)$ at points $(0, t)$ over which the curve $\lambda_1^{-1}(t)$ does not have the generic equisingularity type, so by [28] it does not have the minimum Milnor number.

It yields the following Lemma:

Lemma 3.5.6 *The strict transform of the polar curve $P_1(\pi_l)$ by the blow-up e_1, intersects the exceptional divisor $e_1^{-1}(0)$ at the points $(0, t) \in S \times \mathbb{P}^1$ where the fiber $\lambda_1^{-1}(t)$ has a Milnor number at 0 strictly larger than the minimum Milnor number of the hyperplane sections of S at 0.*

In particular, a plane Q_t containing the line l, is a limit of tangent planes to S at 0 if and only if its image $\Pi_l(Q_t)$ is a tangent line to the discriminant $|\Delta_l|$ at 0.

Proof of Lemma 3.5.6. A point $(0, t_0)$ of the exceptional fiber $e_1^{-1}(0)$ is a point where S_1 is not equisingular along $|e_1^{-1}(0)|$ if and only if the image $p_1(0, t_0)$ is an intersection point of at least two components of the discriminant $e_Z^{-1}(\Delta_l)$, since $e_Z^{-1}(0)$ is already one component of $e_Z^{-1}(\Delta_l)$. Equivalently, $(0, t)$ is an intersection point of the strict transform of $P_1(\pi_l)$ by e_1 with the exceptional divisor.

These intersection points are the points $(0, t_0)$ for which the fiber $\lambda_1^{-1}(t_0)$ does not have the generic equisingularity type. Equivalently, the Milnor number of $\lambda_1^{-1}(t_0)$ is not minimum among Milnor numbers of general fibers.

Since, for every $t \in \mathbb{P}^1$, each fiber $\lambda_1^{-1}(t)$ is isomorphic to the hyperplane section $S \cap Q_t$, the considered Milnor numbers are Milnor numbers of hyperplane sections $S \cap Q_t$.

By the Theorem of Teissier 3.3.1 any such point $(0, t_0)$ of $e_1^{-1}(0)$ is the intersection of $e_1^{-1}(0)$ with the strict transform by e_1 of the section $S \cap H$ by a limit H of tangent planes o fS at 0.

The images of such points $(0, t_0)$ by the morphism p_1 are intersections of the strict transform of Δ_l by e_Z with the exceptional divisor of $e_Z^{-1}(0)$.

The commutativity of the Diagram 3.5 implies that a plane Q_{t_0} is a limit of tangent planes to S at 0 if and only if its image $\Pi_l(Q_{t_0})$ is a tangent line to $|\Delta_l|$ at 0.

Let us now come back to the proof of Theorem 3.5.4. We are finishing with the following Lemma:

Lemma 3.5.7 *Let D be an exceptional tangent of S at 0. For any general line $0 \in l \subset \mathbb{C}^3$ not contained in $|C_{S,0}|$, and such that a linear projection $\Pi_l : \mathbb{C}^3 \to \mathbb{C}^2$ with kernel l is in the open dense set Ω, call $P_1(\pi_l)$ the polar curve defined by the restriction π_l of Π_l to S.*

There is a branch Γ_l of $P_1(\pi_l)$ at 0, such that for any sequence of points $(x_n) \subset \Gamma_l$ tending to 0 and for which the limit $\lim_{n\to\infty} T_{x_n}(S)$ is a plane T we have:

1. *the plane T is the plane spanned by l and D;*
2. *the reduced tangent cone of Γ_l is D.*

Proof Let D be an exceptional tangent of S at 0. Let $l \subset \mathbb{C}^3$ be a line through 0 which is not in the tangent cone $|C_{S,0}|$. Call H_l the two-dimensional plane containing both l and D. By definition of exceptional tangent the plane H_l is a limit of tangent planes to S at 0.

First we are going to prove that H_l is the limit of tangent planes to S along a branch of the polar curve associated to the line l.

For a general choice of l, a linear projection $\Pi_l : \mathbb{C}^3 \to \mathbb{C}^2$ with kernel l is contained in the open dense set Ω defining the general polar curve. Call π_l its restriction to a representative S of $(S, 0)$. It is a finite map. We shall denote Δ_l its discriminant space and $P_1(\pi_l)$ the corresponding polar curve. In our setting we have that $|\pi_l(P_1(\pi_l))| = |\Delta_l|$.

Since the plane H_l is a limit of tangent planes to S at 0, by Lemma 3.5.6 its image $\Pi_l(H_l)$ is a tangent line D_0 to Δ_l at 0. But, since H_l is spanned by $l(= ker(\Pi_l))$ and D, we have:

$$\Pi_l(H_l) = \Pi_l(D) = D_0.$$

On the other hand, D_0 is tangent to a branch of Δ_l which is the image of a branch Γ_l of the polar curve $P_1(\pi_l)$.

Consider a sequence of points $(x_n) \subset \Gamma_l$ converging to 0 and $\neq 0$. By definition of polar curves, at each point x_n, we have:

$$l \subset T_{x_n}(S).$$

Assume that (x_n) is a subsequence such that, the limit of planes $T_{x_n}(S)$ is a plane T and the sequence of lines $(T_{\pi_l(x_n)}(|\Delta_l|))$ converges. Then:

$$\Pi_l(T) = \lim_{n\to\infty} \Pi_l(T_{x_n}(S)) = \lim_{n\to\infty} T_{\pi_l(x_n)}(|\Delta_l|) = D_0.$$

We have then $\Pi_l(H_l) = \Pi_l(T) = D_0$ which implies, as desired, that:

$$H_l = T = \lim_{n\to\infty} T_{x_n}(S).$$

Let us prove now that the line D is tangent to the branch Γ_l at 0.

Consider a sequence of points $(x_n) \subset \Gamma_l$ tending to 0, $\neq 0$, and such that:

- the sequence of secants $(0x_n)$ tends to the line $l' = |C_{\Gamma_l,0}|$
- and the sequence of planes $(T_{x_n}(S))$ converges to the plane H_l.

A combination of Corollary 3.2.3 and Theorem 3.4.7 can be stated as follows:

Let (z_n) be a sequence of non-singular points of a representative S of $(S, 0)$ converging to 0, such that the sequence of lines $(0z_n)$ converges to a line ℓ and the sequence of tangent planes $T_{z_n}(S)$ converges to a plane T. Then, either T is tangent to $|C_{S,0}|$ along ℓ, or ℓ is an exceptional tangent and $\ell \subset T$.

For a general choice of the line l, the plane H_l is not tangent to $|C_{S,0}|$ and D is the unique exceptional tangent contained in H_l.

Therefore, by the observation above, the limit l' is an exceptional tangent of S at 0 contained in H_l. Since D is the unique exceptional tangent contained in H_l, we have $l' = D$. As limits of secants and limits of tangents coinside for curves, it implies that D is the tangent to the branch Γ_l at 0. Which ends the proof of Lemma 3.5.7.

The line D is then tangent to almost all the polar curves of S at 0, which conludes the proof of Theorem 3.5.4.

3.6 Surfaces in $\mathbb{C}^N$

Up to now we have mainly discussed the case of hypersurface singularities, and in particular hypersurfaces of $\mathbb{C}^3$ with isolated singularities. In that context the main tools have been the relation between integral dependence on ideals, limits of tangents, equisingularity and Milnor number of hyperplane sections.

When we deal with general surfaces in $\mathbb{C}^N$ some changes have to be made. Tangent planes are no more hyperplanes, and for that reason it is convenient to work with limits of tangent hyperplanes. It allows to have clearer formulations and more precise results. Lê and Teissier showed [32] that it is an appropriate manner to deal with the problem of limits of tangents in a general context, using the conormal space.

In this work, we choose not to generalize the use of limits of tangent hyperplanes, avoiding to work with the conormal space, conormal morphism and Lagrangian methods; this will be done in a different work dedicated to general case. We keep using the Nash modification, but we shall use hyperplane sections in order to work with equisingularity.

The equisingularity criterion we shall be working with is Whitney regularity and its equivalent concepts in our context: Zariski's criterion and strong simultaneous resolution. Unlike the case of hypersurfaces, the family of curves are no more plane curves, making the equisingularity criteria weaker. For instance the number of tangents need not be constant in a Whitney equisingular family of space curves. Example 3.6.27 illustrates this fact.

Another important difference comes from the possibility for the surface to be non Cohen-Macaulay at a point. When this happens, the surfaces obtained by blowing-up certain ideals, that we use in order to study the equisingularity of the hyperplane sections, need not be Cohen-Macaulay at some points; so the special fibers through those points are not reduced. For this reason we need to work with equisingularity in families of generically reduced curves, following what has been done in [9, 16, 42].

A one-dimentional singular locus is another difference. In order to deal with the singular locus we need to work with the vertical apparent contour instead of the critical locus, following the way it has been used in [25]. This approach prevents us from using equisingularity criteria at a point of the singular locus.

The presentation of this section is based on the work of D.T. Lê in [25] where he works with non-isolated hypersurface singularities of $\mathbb{C}^3$, on the work of J. Snoussi in [44] where he studies limits of tangents on normal surfaces of $\mathbb{C}^N$ and the paper [32] where D.T. Lê and B. Teissier study limits of tangents to any analytic space. The main results stated and proved here are already available in these works. However, some generalizations to non Cohen-Macaulay surfaces, and non-isolated singularities are original results even if the proofs are straight forward extensions of the existing techniques.

From now on, a surface is an equidimensional and reduced complex analytic space of dimension 2, embedded in an open subset of $\mathbb{C}^N$. It may have non-isolated singularities.

3.6.1 Description of the Limits

We shall start by describing the set of limits of tangent spaces to a surface at a singular point. It is a generalization of the situation of isolated singularities of hypersurfaces in $\mathbb{C}^3$. We shall introduce exceptional tangents in a different manner than in the hypersurface case. Afterwards we shall show that this definition is a generalization of the one in hypersurfaces, by using the Whitney regularity as equisingularity criterion in Sect. 3.6.4.

Let $(S, 0)$ be a germ of surface embedded in $(\mathbb{C}^N, 0)$. Recall that the Nash modification $\nu : \tilde{S} \to S$ on a representative S of the germ $(S, 0)$ allows to view the limits of tangent spaces at a given point $x \in S$ as the fiber $\nu^{-1}(x)$. We want now to describe the fiber $\nu^{-1}(0)$. We know by Corollary 3.2.3, that the tangent planes to the tangent cone at 0 are in this set of limits. In order to complete the description of this set, consider the following commutative diagram, that we will call secant/tangent diagram:

$$\begin{array}{ccccccc} S\times\mathbb{P}^{N-1}\times\mathbf{G}(2,N) & \hookleftarrow & X & \xrightarrow{e'_0} & \tilde{S} & \hookrightarrow & S\times\mathbf{G}(2,N)\\ & & \downarrow{\scriptstyle\nu'} & \searrow{\scriptstyle\xi} & \downarrow{\scriptstyle\nu} & & \\ S\times\mathbb{P}^{N-1} & \hookleftarrow & S' & \xrightarrow{e_0} & S & \hookrightarrow & \mathbb{C}^N \end{array} \tag{3.6}$$

where $\mathbf{G}(2, N)$ is the Grassmannian of 2-dimensional vector spaces in $\mathbb{C}^N$, e_0 and e'_0 are respectively the blow-up of the origin in S and the blow-up of the pull-back of the maximal ideal $\mathfrak{m}_{S,0}$ of the local ring $\mathcal{O}_{S,0}$ in $\tilde{S}$. The morphism ξ is $e_0 \circ \nu' = \nu \circ e'_0$. The morphism ν' is given by the universal property of the blowing-up.

Now, in our situation all the spaces S, S', $\tilde{S}$ and X are surfaces, and the respective exceptional fibers, $e_0^{-1}(0)$ and $\xi^{-1}(0)$ are curves. Note that when the singular locus Σ of S has dimension one, the inverse image $\nu^{-1}(\Sigma)$ is also one dimensional, however the fiber $\nu^{-1}(0)$ might be finite. We consider the decomposition into irreducible components of the fibers over 0:

$$|e_0^{-1}(0)| = \bigcup_i V_i, \; |\nu^{-1}(0)| = \bigcup_i W_i, \text{ and } |\xi^{-1}(0)| = \bigcup_i D_i,$$

since we are interested in the limits of tangents at the point 0 we do not consider all the inverse image $|\nu^{-1}(\Sigma)|$ but only $|\nu^{-1}(0)|$.

Recall that the components V_i are precisely the irreducible components of the projective curve $\mathrm{Proj}(|C_{S,0}|)$ and the W_i are the irreducible components of the set of limits of tangent spaces to S at 0.

A point in the fiber $|\xi^{-1}(0)|$ is of the form $(0, \eta, \tau)$, where η is a point of $\mathbb{P}^{N-1}$ corresponding to a line ℓ, generatrix of the tangent cone $|C_{S,0}|$, and τ is a point in $\mathbf{G}(2, N)$ corresponding to a limit T of tangent spaces to S at 0, both of them taken with respect to the same limit, i. e. there exists a sequence of non-singular points $(x_n) \subset S$ such that $\ell = \lim_n (0x_n)$ and $T = \lim_n T_{x_n}(S)$. A result due to Whitney [54, Theorem 22.1], can be restated as $\ell \subset T$.

Consider an irreducible component D_i of $|\xi^{-1}(0)|$. Since $\dim D_i = 1$, its image by ν' is either a component V_i of dimension 1 or a point $\eta_i \in \mathrm{Proj}(|C_{S,0}|)$.

Definition 3.6.1 Let $\eta_i \in \mathrm{Proj}(|C_{S,0}|)$ be a point such that $\eta_i = \nu'(D_i)$ for some one-dimensional irreducible component D_i of $|\xi^{-1}(0)|$. The line ℓ_i corresponding to η_i is a generatrix of $|C_{S,0}|$, that we call an *exceptional tangent* of S at 0.

The following theorem shows that the exceptional tangents play a similar role in the description of the set of limits of tangents as in the case of hypersurfaces of dimension 2:

Theorem 3.6.2 *Let S be a representative of a germ of surface singularity $(S, 0) \subset (\mathbb{C}^N, 0)$. The set of limits of tangent planes to S at 0 is made of the tangent planes* *to the tangent cone $|C_{S,0}|$ along its generatrices, and a finite number of curves $\mathcal{L}_i \subset \mathbf{G}(2, N)$ such that for any plane T in $\mathcal{L}_i$, we have $\ell_i \subset T$, where the ℓ_i's are the exceptional tangents of S at 0.*

Proof Consider an irreducible component D_i of $|\xi^{-1}(0)|$ that maps by ν' onto a one-dimensional component V_i of $\mathrm{Proj}(|C_{S,0}|)$. By Corollary 3.2.3, the inverse image of an open dense subset of V_i by ν' is an open dense subset of D_i, and therefore the image $e'_0(D_i)$ corresponds to the set of limits of tangent spaces to S at 0 that are tangent to the irreducible component of $|C_{S,0}|$ corresponding to V_i along one of its generatrices.

When this component is a plane, $e'_0(D_i)$ is a single point which represents the plane itself.

When $\nu'(D_i) = \{\eta_i\}$ is a point, then, by Definition 3.6.1, it corresponds to an exceptional tangent ℓ_i of S at 0. The component D_i will be of the form $\{(0, \eta_i)\} \times \mathcal{L}_i$, where $\mathcal{L}_i$ is an irreducible curve in $\mathbf{G}(2, N)$. Any point $\tau \in \mathcal{L}_i$ corresponds to a plane T containing the line ℓ_i.

The image $e'_0(D_i)$ will be a 1-dimensional component $W_i = \{0\} \times \mathcal{L}_i$ of $|\nu^{-1}(0)|$. This completes the description of the limits of tangent spaces to S at 0.

Remark 3.6.3 In the case of surfaces that are not hypersurfaces, it is convenient to consider tangent hyperplanes and limits of tangent hyperplanes.

A hyperplane H is tangent to S at a non-singular point if it contains the tangent plane at that point. It is a limit of tangent hyperplanes if it contains a limit of tangent planes, or equivalently, it is a limit of tangent hyperplanes to S at non-singular points.

In this presentation, for equidimensional complex analytic spaces we can obtain a description of the set of limits of tangent hyperplanes, as it is done in [32, Sect. 2]. In this case of surface singularity the set of limit hyperplanes is the union of the dual variety of the projective curve defined by the tangent cone and a finite union of hyperplanes which are the duals of the points corresponding to the exceptional tangents.

We can now describe the limits of tangent spaces to the surface at a generic point of the singular locus.

Corollary 3.6.4 *Let us suppose that the dimension at 0 of the singular locus Σ of S is one. There is an open neighborhood U of 0 in S such that at any point x of $\Sigma \cap (U - \{0\})$ the limits of tangent spaces are a finite number of two-dimensional planes, that coincide with the planes of the tangent cone $|C_{S,x}|$ which all intersect along the tangent line to Σ at x.*

Proof We are in the case where the singular locus Σ of S has dimension one. Then inverse image $\nu^{-1}(\Sigma)$ has dimension one.

First, there is an open neighborhood U_1 of 0 in S, such that x is a non-singular point of $U_1 \cap \Sigma - \{0\}$. Then there is an open neighborhood U_2 of 0 in S such that the fiber $\nu^{-1}(x)$ of a point x in $U_2 \cap \Sigma - \{0\}$ is a finite set. By Theorem 3.6.2, we know how is this fiber. Since it is of dimension 0, there is no exceptional tangent and the components of the reduced tangent cone $|C_{S,x}|$ consist in a finite number of planes.

Finally there exists an open neighborhood U_3 of 0 in S such that a representative S of $(S, 0)$ satisfies Whitney conditions along $U_3 \cap \Sigma$ at x and x is a non-singular point of that component of Σ. Therefore, the tangent space to Σ at x is contained in any limit of tangent spaces to S at x.

If we choose the open neighborhood U as $U_1 \cap U_2 \cap U_3$, then for any point $x \in U \cap \Sigma - \{0\}$, all the planes of the reduced tangent cone $|C_{S,x}|$ intersect along the line $T_x(\Sigma)$, and these planes are the limits of tangent planes to S a x.

Example 3.6.5 Consider the surface S defined by the parametrization:

$$\varphi(u, v) = (u, v^3, v^5, uv^2).$$

Using the parametrization one can compute limits of secants and limits of tangents by choosing sequences of points (u_n, v_n) in $\mathbb{C}^2$ converging to $(0, 0)$ with different relative speed of convergence.

If $\mathbb{C}^4$ has local coordinates (x, y, z, w), then the singular locus Σ of S is the x-axis. One can find that its reduced tangent cone at the origin is defined by $z = w = 0$.

Consider a smooth point $p = (u, v^3, v^5, uv^2) \in S$, where $v \neq 0$. The direction of tangent space $T_p(S)$ is spanned by the vectors $(1, 0, 0, v^2)$ and $(0, 3v^2, 5v^4, 2uv)$. In order to compute limits of tangent spaces to S at 0, one chooses sequences of points $\alpha_n = (u_n, v_n)$ tending to 0 and computes the limits of $T_{\varphi(\alpha_n)}(S)$. The choice of sequences with different speed of convergence for (u_n) and (v_n) allows to describe all the possible limits.

In this case we find that the limits are all the planes in $\mathbb{C}^4$ defined by the equations $z = 0$ and $ay - bw = 0$; where $(a : b) \in \mathbb{P}^1$. Note that the plane of the reduced tangent cone is one of these limits. All the others are planes containing the x-axis, which corresponds to an exceptional tangent.

Consider now a point $q = (u_0, 0, 0, 0)$ with $u_0 \neq 0$; it is a point of Σ different from 0. At the point q the reduced tangent cone is a plane with direction $y = z = 0$. One can see that the unique limit of tangent planes to S at q is precisely the plane with direction $y = z = 0$.

This example has been taken from [9, 4.6].

Let us summarize the results that we have obtained. There is an open neighborhood U of 0 in S such that the limits of tangent planes at any point x of the singular locus Σ in $U - \{0\}$ are the planes of the reduced tangent cone. At the origin, a hyperplane is a limit of tangent hyperplanes if and only if it is tangent to the reduced tangent cone or it contains an exceptional tangent of the surface at that origin.

3.6.2 Polar Curves

We have seen that in the case of 2-dimensional hypersurfaces with isolated singularities, the exceptional tangents are the fixed tangents of the generic polar curves associated to the singularity. We are going to see that this correspondence holds for general surfaces.

Recall from Sect. 3.5.1, that a polar curve associated to a surface $S \subset \mathbb{C}^N$ at the origin is defined as follows: first consider a general $(N-2)$-plane L of $\mathbb{C}^N$ containing the origin and a linear projection $\Pi_L : \mathbb{C}^N \to \mathbb{C}^2$ whose kernel is L. When $L \cap |C_{S,0}| = \{0\}$ the restriction π_L of Π_L to a sufficiently small representative S of the germ of complex surface $(S, 0)$, is a finite map. The polar curve $P_1(\pi_L)$ is the closure in S of the critical locus of the restriction of π_L to the non-singular locus of S. For a general L this space is either always empty or of dimension one and, if the representative S of the germ $(S, 0)$ is small enough, $P_1(\pi_L) - \{0\}$ is non-singular.

Let us describe it more carefully. Following the way it was done in [31, Sect. 2], a non-singular point $x \in S$ is in the polar curve $P_1(\pi_L)$ if and only if $\dim(T_x(S) \cap L) \geq 1$. Consider now the subspace $C_1(L)$ of the Grassmannian $\mathbf{G}(2, N)$ made of planes whose intersection with L has dimension at least one:

$$C_1(L) = \{T \in \mathbf{G}(2, N) | \dim(T \cap L) \geq 1\}.$$

One can prove that $C_1(L)$ is of codimension 1 in $\mathbf{G}(2, N)$. It is one of the Schubert varieties in the Grassmannian.

Recall that the surface $\tilde{S}$ obtained by Nash modification can be endowed with the Gauss map $\gamma : \tilde{S} \to \mathbf{G}(2, N)$, restriction of the second projection of $S \times \mathbf{G}(2, N)$. We can now view the polar curve $P_1(\pi_L)$ as the closure in S of the image:

$$\nu(\gamma^{-1}(C_1(L)) \cap (\tilde{S} - \nu^{-1}(\Sigma))),$$

where Σ is the singular locus of S for a sufficiently small representative of the germ $(S, 0)$.

This description allows us to generalize the correspondence between exceptional tangents and fixed tangents of the polar curves.

Theorem 3.6.6 *Let $(S, 0) \subset (\mathbb{C}^N, 0)$ be a germ of surface. When the general polar curve of S a t0 is not empty, a line in the reduced tangent cone $|C_{S,0}|$ is an exceptional tangent of S at 0 if and only if it is tangent to all polar curves $P_1(\pi_L)$ for L in an open dense set of the Grassmannian $\mathbf{G}(N-2, N)$.*

When the general polar curve is empty, the surface has no exceptional tangent at 0 and its tangent cone is a finite union of planes.

Proof We shall first prove that, when the general polar curve $P_1(\pi_L)$ is not empty then its strict transform $\tilde{P}_1(\pi_L)$ by ν intersects every 1-dimensional irreducible component of $\nu^{-1}(0)$ in at least one point.

In [51, IV. 1.3. Proposition 2 and Corollary 1.3.2], Teissier proves that for a generic $(N-2)$-plane L, the space $\gamma^{-1}(C_1(L))$ is either empty or 1-dimensional, and that $\gamma^{-1}(C_1(L)) \cap (\tilde{S} - \nu^{-1}(\Sigma))$ is dense in $\gamma^{-1}(C_1(L))$. Therefore the strict transform $\tilde{P}_1(\pi_L)$ of the polar curve by ν is $|\gamma^{-1}(C_1(L))|$.

On the other hand we can prove that the Schubert variety $C_1(L)$ intersects every one-dimensional subvariety of $\mathbf{G}(2, N)$ (see Proposition 3.7.1 in the appendix). As a consequence, $C_1(L)$ will intersect the image by the Gauss map γ of any 1-dimensional irreducible component of $\nu^{-1}(0)$.

When the general polar curve is not empty, its strict transform by ν in $\tilde{S}$ is $|\gamma^{-1}(C_1(L))|$, and it will then intersect every one-dimensional component of $\nu^{-1}(0)$.

When the general polar curve is empty, $\gamma^{-1}(C_1(L))$ is empty and hence $|\nu^{-1}(0)|$ is zero-dimensional. In this case, the tangent cone to S at 0 is a finite union of planes, that are precisely the limits of tangent planes to S at 0.

In particular when the general polar curve is empty, the surface has no exceptional tangent.

Assume now that the general polar curve is not empty.

$$\begin{array}{ccc} X & \xrightarrow{e_0'} & \tilde{S} \\ {\scriptstyle \nu'}\big\downarrow & \searrow^{\xi} & \big\downarrow{\scriptstyle \nu} \\ S' & \xrightarrow{e_0} & S \end{array} \tag{3.7}$$

Let ℓ be an exceptional tangent of S at 0. By Definition 3.6.1, there exists a curve $\mathcal{L} \subset \mathbf{G}(2, N)$ such that $\{(0, l)\} \times \mathcal{L}$ is a component of $|\xi^{-1}(0)|$, where l is the point of $\mathbb{P}^{N-1}$ which represents the line ℓ. This component will then intersect almost all the strict transforms of the polar curves by ξ. But this component is contracted by ν' to the point l (see the Diagram 3.7).

It yields that almost all the polar curves have their strict transform by the blow-up e_0 of the origin in S containing the point l. The line ℓ is then a common tangent to almost all the polar curves.

Conversely, we shall prove that a common tangent to almost all polar curves is an exceptional tangent.

Suppose ℓ is such a common tangent. Then there exists a dense set Ω in the Grassmannian $\mathbf{G}(N-2, N)$, such that for every $L \in \Omega$, l is a point of the strict transform of $P_1(\pi_L)$ in S' by the blow-up e_0.

Suppose l does not correspond to an exceptional tangent. By Definition 3.6.1, since the morphism ν' is surjective, the inverse image $(\nu')^{-1}(l)$ is a finite set $\{\theta_1, \dots \theta_r\}$. Let us define the subset $Z_i \subset \mathbf{G}(N-2, N)$ as follows: an $(N-2)$-plane L is in Z_i if and only if the strict transform of the polar curve $P_1(\pi_L)$ in X by ξ contains the point θ_i. In this way we have $\Omega \subset Z_1 \cup \dots \cup Z_r$.

Since Ω is dense in $\mathbf{G}(N-2, N)$, necessarily at least one of the Z_i's is dense. Hence, there exists a dense subset $Z_{i_0} \subset \mathbf{G}(N-2, N)$, such that for any $L \in Z_{i_0}$ the strict transform of the polar curve $P_1(\pi_L)$ in X by ξ contains $\theta_{i_0} := (0, l, \tau_0) \in |\xi^{-1}(0)|$.

The image $(0, \tau_0) = e'_0(\theta_{i_0})$ will be a point in $\nu^{-1}(0)$ that belongs to the strict transform in $\tilde{S}$ by ν of almost all the polar curves.

Then for every $L \in Z_{i_0}$, there exists a sequence $(x_{n,L}) \subset P_1(\pi_L)$ of non-singular points of S, such that $\lim_{n\to\infty} T_{x_{n,L}}(S) = T_0$, where T_0 is the two-plane represented by $\tau_0 \in \mathbf{G}(2, N)$.

By definition of polar curves, every tangent space $T_{x_{n,L}}(S)$ contains a line $\ell_n \subset L$. The limit contains then a line $\ell_L \subset L$. For every $(N-2)$-plane $L \in Z_{i_0}$, the intersection with the plane T_0, has dimension at least 1 which is impossible, because for a fixed two-plane in $\mathbb{C}^N$ the intersection with a general $(N-2)$-plane is a point. This concludes the proof of Theorem 3.6.6.

Remark 3.6.7 *(a)* The space $C_1(L) \subset \mathbf{G}(2, N)$ used for describing the polar curve, is a particular case of Schubert varieties that have been used in [31, Section 2], to define the polar varieties associated to an equidimensional analytic space. See also [51, Chap. IV].

(b) The property that the polar curves have no fixed point by the Nash modification, is actually a property satisfied by Nash modification in any dimension (see [15, I.2] or [45, III.1.2]).

We have seen that if the general polar curve is empty the surface has no exceptional tangent. The converse is false. A surface may have no exceptional tangent and still have a non-empty generic polar curve. In fact any cone over a (reduced) non-singular projective curve o f degree at least 2 has no exceptional tangent and has a non-empty general polar curve. The limits of tangent planes at the vertex of the cone are the cones over the tangent lines to the projective curve. A generic projection to $\mathbb{C}^2$ of such a cone always has a critical locus.

Example 3.6.8 Consider the surface S of Example 3.6.5. One can consider linear projections given by functions $(f_a := x + a_1y + a_2z + a_3w,\ f_b := b_1y + b_2z + b_3w)$, and compute the polar curve for general values a_i's and b_i's. For simplicity we shall do it for chosen values known to be generic.

Consider the projections P and $Q : \mathbb{C}^4 \to \mathbb{C}^2$ defined respectively by:

$$P(x, y, z, w) = (x, y + w) \text{ and } Q(x, y, z, w) = (x + y, y - z + w).$$

Call p and q their respective restrictions to S. One can compute the critical loci of the compositions $p \circ \varphi$ and $q \circ \varphi$, where φ is the parametrization map defined in Example 3.6.5. Then one takes the images of this critical loci in S and removes from it the singular locus. The remaining spaces are the respective polar curves.

The polar curve associated to p is defined by the equations

$$y = -\left(\frac{2}{3}\right)^3 x^3,\ z = -(\frac{2}{3})^5 x^5,\ w = -\frac{3}{2}z.$$

Its tangent line is the x-axis.

For the projection Q, the associated polar curve to q is defined by the equations:

$$(-x+4y)^3 = \left(\frac{3}{2}\right)^3 y, (-x+4y)^5 = \left(\frac{3}{2}\right)^5 z, w = -\frac{3}{2}y + 4z.$$

Its tangent line is the x-axis.

The x-axis is the unique fixed tangent of both polar curves.

We have seen in Example 3.6.5, that this line is the unique exceptional tangent of S at 0. In this case it coincides with the tangent line to the singular locus.

3.6.3 Relation with Discriminants of Projections to $\mathbb{C}^2$

We want now to relate limits of tangents to the geometry of discriminants of generic projections. In [25, Sect. 2], D.T. Lê studied this relation in the case of hypersurfaces of $\mathbb{C}^3$ and established the link between limits of tangent planes and tangents to discriminants. The proof in the hypersurface case extends straight forward to general surfaces, and this is what we show in the first part of this subsection.

Consider a projection $\pi_L : S \to \mathbb{C}^2$ induced by a linear projection $\Pi_L : \mathbb{C}^N \to \mathbb{C}^2$ whose kernel L is an $(N-2)$-linear space. In what follows we shall say that π_L is a *good projection* at 0 , or L is a *good* $(N-2)$*-plane* at 0 , whenever $L \cap |C_{S,0}| = \{0\}$, or equivalently, when the degree of the projection π_L at 0 equals the multiplicity of the surface S at 0.

Definition 3.6.9 The *vertical apparent contour*, of the projection π_L is the closure of the set of points x in a sufficiently small representative S of $(S, 0)$, for which the intersection of L with the tangent cone of S at x has dimension at least 1. It will be denoted by $\mathcal{V}_L$.

The image of the vertical apparent contour by π_L will be called the *vertical part of the discriminant*, and will be denoted by $\mathcal{V}_{\Delta_L}$.

In other words, a non-singular point is in the vertical apparent contour if and only if it is critical, and a singular point x is in the vertical apparent contour if and only if either $\dim|C_{S,x}| \cap L \geq 1$ or x is a limit of non-singular points of S in the vertical apparent contour.

Notice that there exists an open dense set $\Omega \subset \mathbf{G}(N-2, N)$ such that, for $L \in \Omega$ the vertical apparent contour $\mathcal{V}_L$ coincides with the polar curve $P_1(\pi_L)$ associated to a projection with kernel L.

Let L be an $(N-2)$-plane of $\mathbb{C}^N$ such that $L \cap |C_{S,0}| = \{0\}$. Call $\Pi_L : \mathbb{C}^N \to \mathbb{C}^2$ and $\pi_L : S \to \mathbb{C}^2$ the induced maps.

Theorem 3.6.10 *Let H be a hyperplane of $\mathbb{C}^N$ containing L. Then H is a limit of tangent hyperplanes to S at 0 if and only if the image $\Pi_L(H)$ is a limit of tangent lines to the vertical part of the discriminant $\mathcal{V}_{\Delta_L}$.*

Proof Consider a linear projection Π_L inducing a finite map $\pi_L : S \to \mathbb{C}^2$, and let $\mathcal{V}_{\Delta_L}$ be the vertical part of its discriminant. Let D be a limit of directions of tangent lines D_n to $\mathcal{V}_{\Delta_L}$ at points $y_n \neq 0$ converging to 0.

Call $x_n \in S$ a point in the vertical apparent contour $\mathcal{V}_L$ such that $\pi_L(x_n) = y_n$.

When x_n is a non-singular point in S, the hyperplane $\Pi_L^{-1}(D_n)$ contains the tangent space $T_{x_n}(S)$.

When x_n is a singular point of S, we can assume it to be $\neq 0$ in a sufficiently small neighborhood U of 0. Recall from Corollary 3.6.4, that the tangent cone $|C_{S,x_n}|$ is a finite union of planes, so $\Pi_L^{-1}(D_n)$ contains a component of the tangent cone $|C_{S,x_n}|$.

In both cases, $\Pi_L^{-1}(D_n)$ contains either a tangent plane or a limit of tangent planes to S at x_n. So, when x_n tends to 0, this sequence of hyperplanes tends to a limit of tangent hyperplanes that coincides with $\Pi_L^{-1}(D)$.

Conversely, Let T be a limit of tangents to S at 0 contained in a hyperplane $H \supset L$. Since $L \cap |C_{S,0}| = \{0\}$, the two-plane T is not an irreducible component of the tangent cone $|C_{S,0}|$, and L intersects T along a line ℓ which is not in $|C_{S,0}|$.

Consider the restriction of the Gauss map γ to the germ $(\tilde{S}, (0, \tau))$. We shall call it $\gamma_T : (\tilde{S}, (0, \tau)) \to (\mathbf{G}(2, N), \tau)$, where $\tau \in \mathbf{G}(2, N)$ is the point representing the plane T.

If the fiber $\gamma_T^{-1}(\tau)$ has dimension two, then T will be tangent to S at every point of an irreducible component of S. In that case T is an irreducible component of S at 0 and therefore a component of $|C_{S,0}|$, which is excluded by the hypothesis.

If $\gamma_T^{-1}(\tau)$ has dimension 1, call $x = (y, \tau)$ a point of $\gamma_T^{-1}(\tau)$ with $y \in S$. When y is non-singular, T is tangent to S at y. When y is singular, close to 0 but $\neq 0$, then by Corollary 3.6.4, the plane T is a component of $|C_{S,y}|$. So in any case, the point y is in the vertical apparent contour of π_L. Therefore the image $\nu(\gamma_T^{-1}(\tau))$ is contained in the vertical apparent contour of π_L, and at any point $y \neq 0$ of that image, T is a component of $|C_{S,y}|$. The image $\Pi_L(T)$ is then tangent to $\mathcal{V}_{\Delta_L}$ at $\pi_L(y)$, for every $0 \neq y \in \nu(\gamma_T^{-1}(\tau))$; taking the limit we obtain that $\Pi_L(H) = \Pi_L(T)$ is a limit of tangents to $\mathcal{V}_{\Delta_L}$.

Now, suppose that γ_T is finite. Recall that $L \cap T = \ell$, where ℓ is a line which is not in the tangent cone $|C_{S,0}|$.

Call, as before (Sect. 3.6.2), $C_1(L) \subset \mathbf{G}(2, N)$ the set of two-dimensional linear subspaces of $\mathbb{C}^N$ containing a line of L. It is a Schubert variety of codimension 1 in $\mathbf{G}(2, N)$ that contains τ.

We have seen in the proof of Theorem 3.6.6, that for a general $(N-2)$-plane L, the inverse image $\gamma^{-1}(C_1(L))$ is either empty or equal to the strict transform by ν of the polar curve $P_1(\pi_L)$. Furthermore $\gamma^{-1}(C_1(L))$ is empty if and only if $|\nu^{-1}(0)|$ is a finite set. In our situation, T is a limit of tangent planes, that is not a plane of the reduced tangent cone. This implies that $|\nu^{-1}(0)|$ is of dimension one, and hence the general polar curve is not empty.

Then, the inverse image $\gamma_T^{-1}(C_1(L))$ contains at $(0, \tau)$ the strict transform $\mathcal{B}'_L$ by ν of a branch $\mathcal{B}_L$ of the polar curve $P_1(\pi_L)$.

Consider a sequence of points $(x_n) \subset \mathcal{B}_L$, converging to 0 and $\neq 0$. Since the strict transform $\mathcal{B}'_L$ intersects $|\nu^{-1}(0)|$ at $(0, \tau)$, the sequence of tangent planes $(T_{x_n}(S))$ converges to T.

On the other hand, since each $x_n \neq 0$ is on the polar curve $P_1(\pi_L)$, we have that $\dim(T_{x_n}(S) \cap L) = 1$, and then the image $l_n := \Pi_L(T_{x_n}(S))$ is a tangent line to $\mathcal{V}_{\Delta_L}$.

Passing to the limit we obtain:

$$\Pi_L(H) = \Pi_L(T) = \lim_{n\to\infty} \Pi_L(T_{x_n}(S)) = \lim_{n\to\infty} l_n$$

which is a limit of tangent lines to $\mathcal{V}_{\Delta_L}$. This ends the proof of the Theorem.

Notice that, if $\pi_L(H \cap S)$ is tangent to $\mathcal{V}_{\Delta_L}$ for some general $L \subset H$ of codimension 1, then $\pi_{L'}(H \cap S)$ will be tangent to $\mathcal{V}_{\Delta_{L'}}$ for any other general $(N-2)$-plane $L' \subset H$.

Example 3.6.11 Consider again the surface S of Example 3.6.5, and two hyperplanes H_1 and H_2 defined respectively by $y + w = 0$ and $x = 0$.

None of these hyperplanes contains the reduced tangent cone $|C_{S,0}|$.

Consider the 2-plane L defined by $x = y + w = 0$. We have $L \cap |C_{S,0}| = \{0\}$, $L \subset H_1$ and $L \subset H_2$. The kernel of projection P of Example 3.6.8 is precisely L.

We shall describe the vertical apparent contour and the vertical part of the discriminant of P. Recall that if $x \neq 0$ is a point of the singular locus of S, sufficiently close to 0, then the reduced tangent cone $|C_{S,x}|$ is defined by $y = z = 0$, and it is the only limit of tangent planes to S at x. This implies that the point x is not in the vertical apparent contour, i.e., the singular locus of S is not contained in the vertical apparent contour. Therefore, in this case, the vertical apparent contour of P coincides with the polar curve $P_1(P)$.

The equations of the polar curve are given in Example 3.6.8. The vertical apparent contour is defined by $t = (*)s^3$, where (s, t) is a local system of coordinates in $\mathbb{C}^2$ and $(*)$ is a unit in $\mathcal{O}_{\mathbb{C}^2,0}$. Its tangent line at 0 is the s-axis.

The image of H_1 by P is the s-axis, meanwhile the image of H_2 is the t-axis.

The first hyperplane contains the exceptional tangent and the second one does not. H_1 is a limit of tangent hyperplanes and H_2 is not.

Theorem 3.6.10 relates tangent hyperplanes to tangents of the vertical part of discriminants of projections into $\mathbb{C}^2$. This will allow us to establish a relation between limits of tangent hyperplanes and equisingularity via Zariski's discriminant criterion.

We want now to characterize the limits of tangent hyperplanes to a surface at a point, in terms of equisingularity of the family of their sections on S. In order to do that, we need to visualize some related surface as the family of hyperplane sections of S. This can be achieved by blowing-up ideals generated by 2 generic linear forms of the ambient space, as it has been done in the hypersurface case in the proof of Theorem 3.5.4. More precisely:

Consider a good $(N-2)$-plane $L \subset \mathbb{C}^N$ such that $L \cap |C_{S,0}| = \{0\}$. It is defined by two linear equations F_L and G_L. Call f_L and $g_L \in \mathcal{O}_{S,0}$ their respective restrictions to the surface $(S, 0)$, and $\pi_L = (f_L, g_L) : S \to \mathbb{C}^2$ the finite map induced on a representative of $(S, 0)$.

Call $e_L : S_L \to S$ the blow-up of the ideal $(f_L, g_L) \subset \mathcal{O}_{S,0}$.

At this stage it is relevant to mention the difference when $(S, 0)$ is Cohen-Macaulay and when it is not.

When $(S, 0)$ is Cohen-Macaulay, the functions f_L and g_L form a regular sequence, and the blown-up surface S_L is defined in $S \times \mathbb{P}^1$ by the equation $f_L v - g_L u = 0$, where $(u : v)$ is a homogeneous system of coordinates in $\mathbb{P}^1$, see for Example [7, 17.14]. This surface is Cohen-Macaulay at each of its points.

When the surface $(S, 0)$ is not Cohen-Macaulay, consider the surface S^* defined in $S \times \mathbb{P}^1$ by the equation $f_L v - g_L u = 0$. Every section of this surface S^* by subspace of $S \times \mathbb{P}^1$ defined by $(u : v) = (\alpha : \beta)$ is isomorphic to the curve defined on S by the equation $\beta f_L - \alpha g_L = 0$ which has an embedded component, so that the surface S^* has an embedded component supported on $\{0\} \times \mathbb{P}^1$. Since the surface obtained by the blow-up of the ideal (f_L, g_L) is the closure of the graph of the map $(f_L : g_L)$, it is then the reduced surface associated to the one defined in $S \times \mathbb{P}^1$ by $f_L v - g_L u = 0$. This blown-up surface may have isolated points at which it is not Cohen-Macaulay.

Consider now a hyperplane $H \supset L$. Therefore the hyperplane H is defined by a linear equation of the form $bF_L - aG_L = 0$ where $(a : b) \in \mathbb{P}^1$. The hyperplane section $H \cap S$ is defined in S by $bf_L - ag_L = 0$; we shall denote it by $h(a : b)$. When the surface $(S, 0)$ is Cohen-Macaulay, the hyperplane sections never have embedded components. Conversely, when the surface $(S, 0)$ is not Cohen-Macaulay, all its hyperplane sections have an embedded component at the origin. In both cases, some of the hyperplane sections may have non-reduced one-dimensional components. We shall treat this situation later on.

Consider the map $\lambda_L : S_L \to \mathbb{P}^1$ induced by the second projection of $S \times \mathbb{P}^1$.

$$\begin{array}{ccc} S_L & \xrightarrow{e_L} & S \\ \Big\downarrow{\scriptstyle \lambda_L} & & \\ \mathbb{P}^1 & & \end{array} \tag{3.8}$$

For any point $(a : b) \in \mathbb{P}^1$ the fiber $\lambda_L^{-1}(a : b)$ coincides with the strict transform of $h(a : b)$ by e_L. We can visualize the surface S_L as the family of hyperplane sections of S given by all the hyperplanes of $\mathbb{C}^N$ containing L.

When the surface S_L is not Cohen-Macaulay at a point $(0, (a : b))$, then the strict transform of the hyperplane section $h(a : b)$ by e_L will have an embedded component at that point. Still, the hyperplane defining $h(a : b)$ may not be a limit of tangent hyperplanes. See Example 3.6.27.

Remark 3.6.12 When the surface $(S, 0)$ is Cohen-Macaulay, a fiber $\lambda_L^{-1}(a : b)$ is the curve in $S \times \mathbb{P}^1$ defined by $(u : v) = (a : b)$. This curve is the subspace of $S \times \{(a : b)\}$ defined by $bf_L - ag_L = 0$.

So every fiber $\lambda_L^{-1}(a : b)$ is isomorphic to the respective hyperplane section $h(a : b)$ defined above. In this case we can view the surface S_L as the unfolding of the family of the hyperplane sections on S given by all the hyperplanes containing L.

Following the way it has been done in Sect. 3.5.2, let us consider the blow-up of the origin in $\mathbb{C}^2$. We obtain a map $e_Z : Z \to \mathbb{C}^2$, and we notice that the map e_L is precisely the blow-up of the pull-back by π_L of the maximal ideal of $O_{\mathbb{C}^2,0}$. The universal property of the blowing-up gives rise to a unique morphism $\phi_L : S_L \to Z$ which commutes the following diagram:

$$\begin{array}{ccc} S_L & \xrightarrow{e_L} & S \\ {\scriptstyle \phi_L}\downarrow & & \downarrow{\scriptstyle \pi_L} \\ Z & \xrightarrow{e_Z} & \mathbb{C}^2 \end{array} \tag{3.9}$$

It is the same Diagram as (3.5), with different notations in order to adapt it to this new situation.

Lemma 3.6.13 *Let H be a hyperplane of $\mathbb{C}^N$ containing the good $(N-2)$-plane L as above. The hyperplane H is a limit of tangent hyperplanes to S at 0 if and only if the image by ϕ_L of the strict transform of $H \cap S$ by e_L, intersects the strict transform by e_Z of the vertical part of the discriminant $\mathcal{V}_{\Delta_L}$ of π_L.*

Proof The lemma is a direct consequence of Theorem 3.6.10. In fact, since H contains a good $(N-2)$-plane L, it is a limit of tangent hyperplanes if and only if $D := \pi_L(H \cap S)$ is tangent to $\mathcal{V}_{\Delta_L}$ at 0. Equivalently, the strict transform D' of D by e_Z intersects the strict transform $\mathcal{V}'_{\Delta_L}$ of $\mathcal{V}_{\Delta_L}$ by e_Z. Since the Diagram (3.9) commutes, the image by ϕ_L of the strict transform of $H \cap S$ by e_L coincides with D'.

We shall now relate the fact of being a non tangent hyperplane to the equisingularity of the family of curves $\lambda_L : S_L \to \mathbb{P}^1$ defined before Remark 3.6.12. The criterion of equisingularity that we shall use is Whitney regularity .

Consider a one-parameter flat family of curves $\psi : X \to T$; X being a sufficiently small representative of a germ of surface $(X, 0)$ and T an open neighborhood of 0 in $\mathbb{C}$. Assume the singular locus Y of X is smooth, of dimension one and is such that for any $t \in T$ the fiber $\psi^{-1}(t)$ intersects Y in a single point. It is usual to call $\psi^{-1}(0)$ the special fiber and $\psi^{-1}(t)$ for $t \in T$ sufficiently close to the origin, a generic fiber.

In general, one considers families of reduced curves, i.e., each fiber $\psi^{-1}(t)$ is a reduced curve. However, we want to emphasize that in our context we allow the surface X to be non Cohen-Macaulay at the special point $0 \in Y$, which means that the fiber $\psi^{-1}(0)$ may be non-reduced at 0, having an embedded component. But still we require it to be reduced elsewhere. This is what we call a family of generically reduced curves in [9], see also [16].

Definition 3.6.14 Let $\psi : X \to T$ be a one-parameter flat family of curves, where X is a sufficiently small representative of a germ of surface $(X, 0)$. Let Y be the singular locus of X, assumed to be one-dimensional.

We say that the family of curves is Whitney regular at 0, or that the surface X is Whitney regular along Y at 0 if:

- Y is non-singular,
- for every $t \in T$ the curve $\psi^{-1}(t)$ is generically reduced and has a single intersection point with Y, and
- the surface X satisfies Whitney's condition along Y.

Notice that when the fiber $\psi^{-1}(0)$ has a non-reduced one-dimensional component, then the family of curves is not Whitney regular at 0.

In order to relate Whitney regularity to limits of tangent hyperplanes, we shall need to first relate Whitney equisingularity to the geometry of discriminants and then use Lemma 3.6.13. The relation between Whitney regularity and discriminants is given by the so called Zariski's discriminant criterion. In case of families of curves, Zariski's criterion has a simple statement in terms of constancy of the multiplicity of the discriminant of a generic projection to $\mathbb{C}^2$. We refer to [1] for precise definitions and properties in the case of families of reduced curves.

Since in our case the hyperplane sections may have embedded components we are going to need to use the equivalence between Whitney regularity and Zariski's criterion in that context. This equivalence was proved in [9, Theorem 3.1], and can be stated as follows:

Proposition 3.6.15 *A surface X as above is Whitney regular along Y at 0, if and only if for any finite map $\pi : X \to \mathbb{C}^2$ induced by a good linear projection, the reduced space associated to the discriminant of π is a smooth curve.*

The reader must be aware that the equivalence between Zariski criterion and Whitney conditions does not hold in higher dimensions.

We shall then use the relation we have established between limits of tangents and discriminant of good projections to relate the limits of tangents to Whitney equisingularity.

Let L be a good $(N-2)$ plane and $\lambda_L : S_L \to \mathbb{P}^1$ as above.

Proposition 3.6.16 *Let $H \supset L$ be a hyperplane defined by a linear equation $bF_L - aG_L = 0$. Call $\eta = (0, (a:b)) \in |e_L^{-1}(0)|$ the intersection point of the strict transform of $H \cap S$ by e_L with the exceptional divisor of e_L. Assume η is not a point of the strict transform of the singular locus Σ of S by e_L. Then the hyperplane H is not a limit of tangent hyperplanes to S at 0 if and only if the surface S_L is Whitney regular along $|e_L^{-1}(0)|$ at η.*

Proof Let H be as in the hypothesis. By Lemma 3.6.13, H is a limit of tangent hyperplanes if and only if $\phi_L(\eta)$ is an intersection point of the strict transform of $\mathcal{V}_{\Delta_L}$ by e_Z with the exceptional divisor of e_Z (see Diagram 3.9).

Since η is not in the strict transform by e_L of Σ, the reduced discriminant of ϕ_L at η coincides with its vertical part, and therefore it is precisely $|e_Z^{-1}(\mathcal{V}_{\Delta_L})|$. By Proposition 3.6.15, H is a limit of tangent hyperplanes to S at 0 if and only if S_L is not Whitney regular along $e_L^{-1}(0)$ at η.

In the case of surfaces with isolated singularities, the vertical part of the discriminant and the discriminant of π_L coincide, so Proposition 3.6.16 can be stated as follows:

Corollary 3.6.17 *If $(S, 0)$ is an isolated surface singularity, let L be a good $(N-2)$-plane and H a hyperplane containing L. Call η the intersection point of the strict transform by e_L of $H \cap S$ with the exceptional divisor of e_L. Then H is a limit of tangent hyperplanes to S at 0 if and only if the surface S_L is not Whitney regular along $|e_L^{-1}(0)|$ at η.*

Corollary 3.6.18 *Let H be a hyperplane of $\mathbb{C}^N$ which does not contain any component of the tangent cone $|C_{S,0}|$ and is not tangent to the singular locus at* 0. *Then H is a limit of tangent hyperplanes to S at* 0 *if and only if, there exists a good linear space $L \subset H$ of dimension $N-2$, i.e., for which $L \cap |C_{S,0}| = \{0\}$, such that S_L is not Whitney regular along $|e_L^{-1}(0)|$ at the intersection point of the strict transform of $H \cap S$ by e_L with the exceptional divisor of e_L.*

Proof Call D_i (respectively R_j) the tangent lines to Σ (respectively to $H \cap S$) at 0 with $1 \leq i \leq n$ and $1 \leq j \leq r$. Let $P_{i,j}$ be the two-planes generated by D_i and R_j. Since the hyperplane H is not tangent to Σ, the linear spaces $P_{i,j}$ are in fact two-dimensional. Each of these planes intersect H along the line R_j. For almost all $(N-2)$-planes $L \subset H$, we have $L \cap P_{i,j} = \{0\}$. This emplies that the images $\Pi_L(D_i)$ and $\Pi_L(R_j)$ are distinct and therefore the strict transform of $H \cap S$ and of Σ by e_L in S_L do not intersect. By Proposition 3.6.16, we obtain the required equivalence

Example 3.6.19 Consider the surface $S \subset \mathbb{C}^4$ defined by the parametrization

$$n : (u, v) \mapsto (u, uv, v^2, v^3)$$

which is the normalization of S. The surface S has an isolated singularity at 0, it is not Cohen-Macaulay at 0.

The limits of secants to S at 0 are limits of lines generated by vectors $(u_n, u_n v_n, v_n^2, v_n^3)$ when (u_n) and (v_n) are non-zero sequences in $\mathbb{C}$ converging to 0.

As in the previous examples, one varies the relative speed of convergence of such sequences and obtain that the reduced tangent cone is the (x, z)-space defined by the equations $y = w = 0$; where (x, y, z, w) is a system of coordinates in $\mathbb{C}^4$.

In a similar way, the tangent space at a point $\neq 0$ is spanned by the vectors $(1, v_n, 0, 0)$ and $(0, u_n, 2v_n, 3v_n^2)$, for u_n and v_n in $\mathbb{C}$ non simultaneously zero.

Making these sequences converge to zero, and taking different relative speed of convergence, one obtains that the set of limits of tangent spaces is the set of planes given by the equations $w = 0$ and $ay + bz = 0$ for $(a : b) \in \mathbb{P}^1$.

The x-axis is the only exceptional tangent of S at 0.

Consider now two hyperplanes H_z and H_x defined respectively by $z = 0$ and $x = 0$. None of them contains the plane $|C_{S,0}|$. The hyperplane H_z is a limit of tangent hyperplanes and H_x is not.

Consider the 2-plane L defined by $x = z = 0$; it is contained in both H_x and H_z and $L \cap |C_{S,0}| = \{0\}$. We shall describe the surface S_L obtained by the blow-up of the ideal (x, z) in S.

The surface S is defined by the equations $y^2 - x^2 z = 0$, $yz - xw = 0$, $xz^2 - yw = 0$ and $z^3 - w^2 = 0$. Since S is not Cohen-Macaulay at 0, the blown-up surface $S_L \subset S \times \mathbb{P}^1$ is defined as the reduced surface subjacent to the surface S_1 defined by the equation $xt = zs$ in $S \times \mathbb{P}^1$, where $(s : t)$ is a system of homogeneous coordinates in $\mathbb{P}^1$.

One can use a software such as Singular in order to obtain primary decompositions, or compute by hands.

In the chart $t \neq 0$, the surface S_1 is defined by the ideal

$$< y - w\left(\frac{s}{t}\right), w^2 - z^3 > \cap < y, z, w > .$$

It has an embedded component along the $\frac{s}{t}$-axis. Then, the surface $S_L(= |S_1|)$ is defined, in this chart, by the ideal

$$< y - w\left(\frac{s}{t}\right), w^2 - z^3 > .$$

It is isomorphic to the product of the plane curve $w^2 - z^3 = 0$ by the $\frac{s}{t}$-axis. It is then Whitney regular along the $\frac{s}{t}$-axis.

In the chart $s \neq 0$, The surface S_1 is defined by the ideal:

$$< y^2 - x^3\left(\frac{t}{s}\right), w - y\left(\frac{t}{s}\right) > \cap < x, y, w > .$$

which implies that the surface $S_L = |S_1|$ is defined in this chart by the ideal:

$$< y^2 - x^3\left(\frac{t}{s}\right), w - y\left(\frac{t}{s}\right) > .$$

In this chart, the surface S_L is isomorphic to the hypersurface of $\mathbb{C}^3$ defined by the ideal

$$< y^2 - x^3\left(\frac{t}{s}\right) > .$$

The normalization of this hypersurface of $\mathbb{C}^3$ is a singular surface defined in $\mathbb{C}^3$ by the equation $r^2 = x\left(\frac{t}{s}\right)$. So The surface S_L is not Whitney regular along the exceptional fiber at the point $((0, 0, 0, 0), (1 : 0))$. This is precisely the intersection point of the strict transform of $H_z \cap S$ by e_L with the exceptional fiber in S_L.

The intersection point of the strict transform of $H_x \cap S$ by e_L with the exceptional fiber is $((0, 0, 0, 0), (0 : 1))$; we have just seen that S_L is Whitney regular along $|e_L^{-1}(0)|$ at that point.

This example was used by Chavez [4, Example 4.2.9].

Remark 3.6.20 When a hyperplane section $H \cap S$ does not have isolated singularities at 0, then at least one of its branches at 0 is non-reduced. Let us call C such a branch.

When the curve C is a component of the singular locus Σ of S, the hyperplane H may not be a limit of tangent hyperplanes.

The example of the surface S defined by $y^2 - x^3 = 0$ in $\mathbb{C}^3$ illustrates this situation. In fact, S is a product of a cusp by a line. Its unique limit of tangents is the plane $y = 0$. If one takes the plane H_x defined by $x = 0$, the the section $H_x \cap S$ is not reduced, and it is not a limit of tangent hyperplanes.

When the curve C is not contained in Σ, then H is not transverse to S along C. The hyperplane H is tangent to S along C. The hyperplane H is then a limit of tangent hyperplanes.

If we want to apply Theorem 3.6.10 to this context, we can see that when H contains a component C of Σ, the section $H \cap S$ may not be tangent to the vertical apparent contour o f ageneral projection to $\mathbb{C}^2$.

Meanwhile, if $C \not\subset \Sigma$, then H is not transverse to S along C, which implies that H contains the tangent plane to S at any point $0 \neq x \in C$ close to 0. For any $(N-2)$-plane $L \subset H$, the intersection $L \cap T_x(S)$ has dimension at least one. The curve C is then in the vertical apparent contour of the projection π_L.

In Proposition 3.6.16 and Corollaries 3.6.17 and 3.6.18, the situation where the hyperplane contains a component of the singular locus is omitted. When the non-reduced component of the hyperplane section is not in the singular locus, Whitney regularity, as Defined in 3.6.14, is not satisfied.

We have thus established a link between limits of tangent hyperplanes and Whitney regularity along the exceptional fiber, of surfaces of type S_L, obtained by blowing-up ideals generated by two generic linear forms. We shall use this characterization in the following subsection to determine the exceptional tangents viewed as points in the exceptional fiber of the blow-up of the origin.

3.6.4 Exceptional Tangents and Equisingularity

In the case of surfaces in $\mathbb{C}^3$ the exceptional tangents at a given point were defined in terms of equisingularity of the surface obtained by blowing-up the point, along the exceptional divisor of the blow-up. In the general case of surfaces, the relation is not straightforward. That is why we have chosen to define the exceptional tangents using the secant/tangent Diagram (3.6). But still, we want to explain the relation between the exceptional tangents and the equisingularity, which allows, in the case of surfaces with isolated singularities, to characterize them completely.

Recall that a line $0 \in \ell \subset |C_{S,0}|$ is an exceptional tangent of S at 0 if and only if every hyperplane containing it is a limit of tangent hyperplanes to S at 0 (see Remark 3.6.3). Such a line corresponds to a point in the exceptional fiber $e_0^{-1}(0)$ of the blow-up of the origin of S, $e_0 : S' \to S$.

Remark 3.6.21 Recall from page xxx that for an $(N-2)$-plane L we have defined the blow-up $e_L : S_L \to S$. Now we can construct a map $\alpha_L : S' \to S_L$ as follows: Let L be an $(N-2)$-plane such that $L \cap |C_{S,0}| = \{0\}$. We have that $S' \subset S \times \mathbb{P}^{N-1}$ and $S_L \subset S \times \mathbb{P}^1$. Consider the map

$$A_L : S \times (\mathbb{P}^{N-1} - \text{Proj}(L)) \to S \times \mathbb{P}^1$$

defined as $Id_S \times A'_L$, where A'_L is the projection from $\mathbb{P}^{N-1} - \text{Proj}(L)$ onto $\mathbb{P}^1$ with center $\text{Proj}(L)$.

Since $\mathrm{Proj}(L) \cap \mathrm{Proj}(|C_{S,0}|) = \emptyset$, the projection A_L induces a finite map:

$$\alpha_L : S' \to S_L.$$

Proposition 3.6.22 *Let $l \in Proj(C_{S,0})$ be a point corresponding to an exceptional tangent ℓ of a surface S at 0. Then for every $(N-2)$-plane L such that $L \cap |C_{S,0}| = \{0\}$, the surface S_L is not Whitney regular along $e_L^{-1}(0)$ at $\alpha_L(l)$.*

Proof Let ℓ be an exceptional tangent and let L be an $(N-2)$-plane such that $L \cap |C_{S,0}| = \{0\}$. Then, by Remark 3.6.3, the hyperplane H containing L and ℓ is a limit of tangent hyperplanes to S at 0.

When ℓ is not tangent to the singular locus of S, then by Proposition 3.6.16, $\alpha_L(l)$ is a point in S_L in which the surface S_L is not Whitney regular along the exceptional divisor of e_L.

When ℓ is tangent to the singular locus Σ of S, then the image $\alpha_L(l)$ is a point of the strict transform of Σ by e_L. In this case, the discriminant of $\phi_L : S_L \to Z$ at $\phi_L(\alpha_L(l))$ (see Diagram 3.9) has at least two components: the exceptional fiber and part of the strict transform of $\pi_L(\Sigma)$ by e_Z. By Proposition 3.6.15, the surface S_L is not Whitney regular along $e_L^{-1}(0)$ at $\alpha_L(l)$.

For the converse of the previous proposition, we need to take into acount the singular locus. As we have seen in the previous sub-section, since we characterize limits of tangents with respect to the vertical part of the discriminant, and not the whole discriminant, we only can characterize exceptional tangents by equisingularity at points that are not in the strict transform of the singular locus by the blowing-up. More precisely:

Proposition 3.6.23 *Let $\eta \in Proj(C_{S,0})$ be a point that does not belong to the strict transform of the singular locus of S by e_0. If for every good $(N-2)$-plane L, i.e., such that $L \cap |C_{S,0}| = \{0\}$, the surface S_L is not Whitney regular along $e_L^{-1}(0)$ at $\alpha_L(\eta)$, then η corresponds to an exceptional tangent of S at 0.*

Proof Let $\eta \in \mathrm{Proj}(C_{S,0})$ be a point which is not in the strict transform of Σ by e_0. Let H be a hyperplane containing the line corresponding to η and a good $(N-2)$-plane in H.

The strict transform of $H \cap S$ by e_L intersects the strict transform of the singular locus Σ by e_L if and only if H contains a tangent line, or a limit of tangent lines, to Σ at 0. In fact, if the strict transform of $H \cap S$ by e_L intersects the strict transform of Σ by e_L, then, by commutativity of Diagram (3.9), the tangent lines of Σ and of $H \cap S$ have a common image D by Π_L, and therefore $H = \Pi_L^{-1}(D)$ contains a tangent line to Σ at 0.

For a generic choice of L, the hyperplane H does not contain any tangent to Σ, then $\alpha_L(\eta)$ is not in the strict transform of Σ by e_L.

By hypothesis, the surface S_L is not Whitney regular along $e_L^{-1}(0)$ at $\alpha_L(\eta)$. By Proposition 3.6.16, H is a limit of tangent hyperplanes to S at 0. This implies that η is an exceptional tangent.

Thanks to this equisingular description of the exceptional tangents we can characterize them directly on the blown-up surface S' in relation with the normalization morphism of S' . It is a partial characterization when the original singularity is not isolated, since Proposition 3.6.23 does not allow to determine whether a tangent line to the singular locus is an exceptional tangent.

The following result generalizes Theorem 5.8 of [44] to the case of non-necessarily normal surfaces. The proof is very similar to the one in the case of normal surfaces.

Let $e_0 : S' \to S$ be the blow-up of the origin in a representative S of a germ of surface $(S, 0)$, and call $n' : \overline{S'} \to S'$ its normalization.

Definition 3.6.24 We shall call *special point* in S', any point $\eta \in e_0^{-1}(0)$ satisfying one or more of the following properties:

1. η is image by n' of a singular point of $\overline{S'}$
2. η is image by n' of a singular point of the reduced exceptional fiber $|(e_0 \circ n')^{-1}(0)|$
3. η is a critical value of the restriction of the normalization to the exceptional fiber.

Remark 3.6.25 Here and elsewhere in the text, by a critical point of a finite map $f : X \to Y$ we mean a point of X at which the map does not induce a local analytic isomorphism into its image. In particular, when the target Y is non-singular, any singular point of X is critical. When $x \in X$ is non-singular and the image $f(x)$ of x is a singular point of $f(X)$, then x is critical.

Theorem 3.6.26 *Let $\eta \in Proj(C_{S,0}) = e_0^{-1}(0)$ be a point which does not belong to the strict transform of the singular locus by e_0. Then η corresponds to an exceptional tangent to S at 0 if and only if it is a special point in S'.*

In order to prove this equivalence, we shall need to use another concept of equisingularity, equivalent in our context to Whitney regularity, namely strong simultaneous resolution.

A flat morphism $f : X \to T$ is a one parameter family of generically reduced curves where T is a smooth curve, X is a reduced surface, the generic fiber is reduced and the special fibers are generically reduced, i.e., with isolated singularities and possibly embedded components. We consider situations where such a morphism has a section $\sigma : T \to X$ such that for any $t \in T$ the fiber $f^{-1}(t) - \{\sigma(t)\}$ is a non-singular curve.

Call $\nu : \overline{X} \to X$ the normalization of X. We say that the family of curves admits a normalization in family if $(f \circ \nu)^{-1}(t)$ are non-singular for all t in T.

The family has a weak simultaneous resolution if it has a normalization in family and the reduced inverse image $|\nu^{-1}(\sigma(T))|$ is isomorphic to $|\nu^{-1}(x_0)| \times \sigma(T)$. The family has a strong simultaneous resolution when it has a normalization in family and the inverse image $\nu^{-1}(\sigma(T))$ is isomorphic to the product $\nu^{-1}(x_0) \times \sigma(T)$.

In the context of families of generically reduced curves, G.-M. Greuel proved in [16, theorem 9.3], that weak simultaneous resolution is equivalent to topological triviality, and in [42, Theorems 4.7 and 5.2], O.N. da Silva and J. Snoussi prove that strong simultaneous resolution is equivalent to Whitney regularity and also equivalent to weak simultaneous resolution together with the constancy of the multiplicity of the fibers along $\sigma(T)$.

For the case of families of reduced curves see [1], where a general panorama is presented.

Proof of Theorem 3.6.26 We shall first prove that if a point of $\mathrm{Proj}(C_{S,0}) = e_0^{-1}(0)$ is not special then it does not correspond to an exceptional tangent of S at 0.

Let η be such a point. One can choose a good $(N-2)$-plane L such that the hyperplane H, generated by L and the line l corresponding to η, is not tangent to the tangent cone $C_{S,0}$, does not contain any line corresponding to a special point, does not contain any tangent line to the singular locus of S and the hyperplane section $H \cap S$ has isolated singularities (generically reduced).

We shall prove that H is not a limit of tangent hyperplanes. Let $\alpha_L : S' \to S_L$ be the morphism defined in Remark 3.6.21 (see the Diagram 3.10 below).

$$\begin{array}{ccccc}
W & \hookrightarrow & \overline{S'} & \xrightarrow{\;n'\;} & S' \\
\downarrow & & {\scriptstyle n_L}\downarrow & \overset{\alpha_L}{\swarrow} & \downarrow{\scriptstyle e_0} \\
U & \hookrightarrow & S_L & \xrightarrow{\;e_L\;} & S \\
{\scriptstyle \lambda_U}\downarrow & & {\scriptstyle \lambda_L}\downarrow & & \\
T & \hookrightarrow & \mathbb{P}^1 & &
\end{array} \tag{3.10}$$

Call $\tau = \alpha_L(\eta)$. Since η is not in the strict transform by e_0 of the singular locus of S, there are neighborhoods U of τ in S_L and T of $t_0 := \lambda_L(\tau)$ in $\mathbb{P}^1$, such that λ_L induces the flat morphism $\lambda_U : U \to T$. The morphism λ_U admits a section $\sigma : T \to U$ such that $\lambda_U^{-1}(t) - \{\sigma(t)\}$ is non-singular for all $t \in T$. Recall that $\lambda_L^{-1}(t_0)$ is the strict transform of $H \cap S$ by e_L.

Since L is a good $(N-2)$-plane, the degree at 0 of the projection π_L equals the multiplicity of the surface S at 0. If we call I_L the ideal of $\mathcal{O}_{S,0}$ whose blow-up is e_L, then by [6], the degree of π_L at 0 equals the multiplicity of I_L. Therefore the ideals I_L and $\mathfrak{m}$ have the same multiplicity, where $\mathfrak{m}$ is the maximal ideal of $\mathcal{O}_{S,0}$. We also have $I_L \subset \mathfrak{m}$ and it is $\mathfrak{m}$-primary. By a Theorem of Rees [38, 3.2] these ideals have the same integral closure, and then by [22, Theorem 2.1], these two ideals have the same normalized blowing-up.

Therefore, both surfaces S' and S_L have the same normalization. Call $n_L : \overline{S'} \to S_L$ the normalization map. We have $n_L = \alpha_L \circ n'$.

Call $W := n_L^{-1}(U) \subset \overline{S'}$. Let D be the intersection $W \cap (e_0 \circ n')^{-1}(0)$. If the open set U is chosen small enough, and since η is not a special point, then all the points in W are smooth points of $\overline{S'}$ and of $|D|$.

The image $|n'(D)|$ is a smooth curve in a neighborhood of η in S', otherwise the restriction of n' to D would be critical according to the definition we gave in Remark 3.6.25.

Let us first prove normalization in family.

Let $\theta \in D \subset W$ be a point such that $n'(\theta) = \eta$. By hypothesis, the projective hyperplane $\mathrm{Proj}(H)$ is not tangent to $\mathrm{Proj}(|C_{S,0}|)$. Since η is not a special point, then by Definition 3.6.24.3., θ is not critical for the restriction of n' to $|D|$. It implies that the strict transform of $H \cap S$ by $e_0 \circ n'$ intersects transversally $|D|$ at θ. In other words, the curve $(\lambda_L \circ n_L)^{-1}(t_0)$ is transversal to $|D|$ in W, hence it is non-singular. This proves that $\lambda_L : U \to T$ has a normalization in family.

In order to prove that λ_L has a weak simultaneous resolution at $\tau \in S_L$, we shall prove that the restriction of the normalization n_L to $|D|$ does not ramify over τ.

Since $|D|$ and $|n_L(D)|$ are non-singular, we shall prove the constancy of the number o f pre-images by n_L of points in $\sigma(T)$.

Let $x \in \sigma(T)$, recall that $n_L^{-1}(x) = n'^{-1}(\alpha_L^{-1}(x))$. We shall first understand the fibers $\alpha_L^{-1}(x)$.

For every $x \in \sigma(T)$ there exists a unique hyperplane H_x containing the $(N-2)$-plane L such that the strict transform of $H_x \cap S$ by e_L intersects $e_L^{-1}(0)$ at x. Note that under this notation the hyperplane H is H_τ. By commutativity of the Diagram (3.10), we have $\alpha_L^{-1}(x) = \mathrm{Proj}(H_x) \cap \mathrm{Proj}(|C_{S,0}|)$.

For $x \in \sigma(T)$, $x \neq \tau$, and for U and T sufficiently small, the intersection of $\mathrm{Proj}(H_x)$ and $\mathrm{Proj}(|C_{S,0}|)$ is transverse in $\mathbb{P}^{N-1}$ and consists in exactly as many points as the degree of the reduced curve $\mathrm{Proj}(|C_{S,0}|)$.

Now, consider an intersection point $\eta_i \in \mathrm{Proj}(H) \cap \mathrm{Proj}(|C_{S,0}|)$. The point η is one of the η_i's. The $(N-2)$-plane L can be chosen so that all the points η_i, except maybe η, are non-singular points of $\mathrm{Proj}(|C_{S,0}|)$ at which the intersection with $\mathrm{Proj}(H)$ is transverse. We need to be more specific about the possibility for η to be a singular point of $\mathrm{Proj}(|C_{S,0}|)$.

Since η is not a special point, the inverse images of η by the normalization n' are non-singular points of $|D|$ and the images of the germs of $|D|$ at these points by n' are also non-singular. Therefore the only possibility for η to be a singular point of $\mathrm{Proj}(|C_{S,0}|)$, is when it is the intersection point of two, or more, non-singular branches of the germ $(\mathrm{Proj}(|C_{S,0}|), \eta)$. Each one of these branches is contained in a different analytically irreducible component of the germ (S', η). It may happen that a branch is contained in several analytic components of (S', η).

In this way, the normalization n' separates all the analytic components of (S', η) and the branches of $(\mathrm{Proj}(|C_{S,0}|), \eta)$, and there is no critical point of the restriction of n' to $|D|$ over η.

Then if η is a singular point of $\mathrm{Proj}(|C_{S,0}|)$, and x is a point in $\sigma(T)$ sufficiently close to τ, consider a small neighborhood V of η. The number of points in the intersection $V \cap \mathrm{Proj}(H_x) \cap \mathrm{Proj}(|C_{S,0}|) = \{x_1, \ldots, x_r\}$, is equal to the number of branches of $\mathrm{Proj}(|C_{S,0}|)$, say $\beta_1, \ldots, \beta_r$, intersecting at η, with $x_i \in \beta_i$.

Each branch β_i is contained in some irreducible components of S' at η, say $S'_1, \ldots, S'_{r_i}$, with $r_i \geq 1$.

We have then:

$$\#(n'^{-1}(\eta)) = \sum_{i=1}^{r} \#(n'^{-1}(x_i)) = \sum_{i=1}^{r} r_i.$$

On the other hand, if $\eta_j \neq \eta$, then, when T is small enough, for x in $\sigma(T) - \{\tau\}$, the projective hyperplane $\mathrm{Proj}(H_x)$ intersects $\mathrm{Proj}(|C_{S,0}|)$ only in one point y_j in a small neighbourhood of η_j. The point y_j belongs to s_j irreducible components of (S', η_j). It yields:

$$\#(n'^{-1}(\eta_j)) = \#(n'^{-1}(y_j)) = s_j.$$

Therefore:

$$\#(n_L^{-1}(\tau)) = \sum_{i,\eta_i \neq \eta} \#(n'^{-1}(\eta_i)) + \#(n'^{-1}(\eta))$$

$$= \sum_{j,\eta_j \neq \eta} s_j + \sum_{i=1}^{r} r_i.$$

If $x \in \sigma(T)$ with $x \neq \tau$ close enough to τ, then we have

$$\#(n_L^{-1}(x)) = \sum_{j,\eta_j \neq \eta} \#(n'^{-1}(y_j)) + \sum_{i=1}^{r} \#(n'^{-1}(x_i))$$

$$= \sum_{j,\eta_j \neq \eta} s_j + \sum_{i=1}^{r} r_i = \#(n_L^{-1}(\tau)).$$

Which implies that $\lambda_L : S_L \to \mathbb{P}^1$ admits a weak simultaneous resolution at η.

In order to prove strong simultaneous resolution, we only need to prove that the hyperplane section $H \cap S$ has the same multiplicity as a generic hyperplane section.

The choice of the good $(N-2)$-plane $L \subset H$ in the beginning of this proof was made such that the hyperplane section $H \cap S$ is with isolated singularity. It is proved in ([42, Lemma 4.8]) that the multiplicity of a curve with isolated singularities does not depend on the possible embedded component.

Therefore the multiplicity at 0 of the hyperplane section $H \cap S$ equals the degree at 0 of the prjection π_L, which is equal to the multiplicity of S at 0, since the projection Π_L is good.

The multiplicity at 0 of the hyperplane section $H \cap S$ is the the same as the one of a generic hyperplane section $H' \cap S$.

The strict transform $(H \cap S)'$ of $H \cap S$ by e_L has the same multiplicity at the intersection point with $e_L^{-1}(0)$ as the section $H \cap S$ at 0, which is equal to the degree of e_L at the intersection point. And the same is true for a generic hyperplane H'.

It implies that strict transforms by e_L, $(H \cap S)'$ and $(H' \cap S)'$ have the same multiplicity at their respective intersection point with $e_L^{-1}(0)$.

The family $\lambda_L : U \to T$ has then constant multiplicity along $\sigma(T)$, and therefore it is a Whitney regular family.

So H is not a limit of tangent hyperplanes to S at 0, and l is not an exceptional tangent.

Conversely, let η be a special point, and consider a good $(N-2)$-plane L. Define $\alpha_L, n_L, \tau := \alpha_L(\eta)$ and $\lambda_L : U \to T$ as before. If $\eta = n'(z)$ with z singular point for $\overline{S'}$, then $\tau = n_L(z)$ which makes it impossible for λ_L to have a normalization in family. It implies that the surface S_L is not Whitney regular along $|e_L^{-1}(0)|$ at η. By Proposition 3.6.23 the line l is an exceptional tangent.

If $\eta = n'(z)$ with z singular for $|(e_0 \circ n')^{-1}(0)|$, then z is a ramification point of the restriction of n_L to the exceptional fiber. So $\lambda_L : U \to T$ cannot have à weak simultaneous resolution. Again, by Proposition 3.6.23, the line l is an exceptional tangent.

Now let $\eta = n'(z)$ with z critical point of the restriction of n' to the exceptional fiber $|(e_0 \circ n')^{-1}(0)|$. Consider a sufficiently small neighborhood W of z in $\overline{S'}$ and call $D = W \cap |(e_0 \circ n')^{-1}(0)|$. Call W' the image of W in S' and $D' = |n'(D)|$. Notice that W' is a neighborhood of η in an analytically irreducible component of S' at η, so that D' may not be $\mathrm{Proj}(|C_{S,0}|) \cap W'$, but only one, or some, of the branches of the exceptional fiber. Suppose that λ_L has a weak simultaneous resolution at τ then $|n'^{-1}(D')| = |n_L^{-1}(\sigma(T)|$ is isomorphic to $T \times |n_L^{-1}(\tau)|$ which contradicts the hypothesis that z is critical.

In the theorem above, an intersection point of two components of the exceptional fiber in S' which is not a special point does not correspond to an exceptional tangent. The following example illustrates such a situation

Example 3.6.27 Let $(S, 0) \subset (\mathbb{C}^4, 0)$ be a germ of surface with two irreducible components defined as follows: $(S_1, 0)$ is defined by the equations $Z - X^2 = 0$ and $T - Y^2 = 0$ and $(S_2, 0)$ defined by $Y + T^2 = 0$ and $Z + X^2 = 0$, with $(S, 0) = (S_1, 0) \cup (S_2, 0)$.

The surface $(S, 0)$ is reduced, equidimensional and its irreducible components are smooth and intersect exactly at 0.

The tangent cone $|C_{S,0}|$ is the union of two planes P_1 defined by $Z = T = 0$ and P_2 defined $Z = Y = 0$. These two planes intersect along the line $Y = Z = T = 0$.

The limits of tangent planes to S at 0 are the planes P_1 and P_2. The surface does not have any exceptional tangent.

When one blows-up the surface S at the origin, the exceptional fiber in S' is $\mathrm{Proj}(C_{S,0})$ which is the union of two projective lines intersecting in $(1 : 0 : 0 : 0)$; this point does not correspond to an exceptional tangent.

Remark 3.6.28 The situation of the example above does not occur in hypersurfaces of $\mathbb{C}^3$. In fact, the blow-up of a point in a hypersurface gives rise to a space that is locally a hypersurface and therefore Cohen-Macaulay at every point. Meanwhile, the Example 3.7.1 is a typical example of a surface that is not Cohen-Macaulay at the origin and its blowing-up at the origin is again a non Cohen-Macaulay surface at some point.

It would be interesting to know if in the case of normal surface singularities, an intersection line of two components of the tangent cone is always an exceptional tangent, as was asked in [44, Remark 5.10].

A different way to pose that question, is to ask whether the number of tangents to general hyperplane sections is constant for normal surfaces. It is known that this number of tangents is not necessarily constant in Whitney regular families of curves, see for example [13]. However it is constant in Lipschitz equisingular families of curves. Then the question is whether the family of general hyperplane sections of a germ of normal surface, have a generic projection to $\mathbb{C}^2$ with the same topological type. for more information on Lipschitz equisingularity and equisaturation we refer to [14].

At an intersection point η of the strict transform of the singular locus with the exceptional divisor of e_0 in S', equisingularity criteria do not specify whether η corresponds to an exceptional tangent. Using Theorems 3.6.10 and 3.6.6 a tangent line to the singular locus is an exceptional tangent if and only if it is tangent to the vertical apparent contour of almost all projections induced by linear projections to $\mathbb{C}^2$. The example of the swallow tail, illustrates this situation:

Example 3.6.29 The swallow tail is the surface defined in $\mathbb{C}^3$ by the equation

$$256x^3 - 27y^4 - 128x^2z^2 + 144xy^2z + 16xz^4 - 4y^2z^3 = 0.$$

It has a singular locus of dimension one. The Nash modification ν of this surface is the blow-up of its Jacobian ideal. When one computes this blow-up, one sees that the fiber $\nu^{-1}(0)$ is a single point. The only limit of tangent planes is the plane of the tangent cone defined by $x = 0$.

So the intersection of the strict transform of the singular locus by the blow-up of 0 with the exceptional fiber does not correspond to an exceptional tangent.

3.6.5 Surfaces Without Exceptional Tangents

We have seen that exceptional tangents of a surface $(S, 0)$ contribute significantly in the composition of the sets of limits of tangents and they are an obstruction to the equisingularity of the surface obtained by the blowing-up of 0. In particular when the set of limits of tangents is finite, the surface has no exceptional tangent.

In [33], D.T. Lê and B. Teissier showed that under some conditions, a surface in $\mathbb{C}^3$ with no exceptional tangent can be deformed in a Whitney equisingular manner to its tangent cone. So a surface with no exceptional tangent is very close to be a cone.

We want here to describe very shortly the situation of surfaces without exceptional tangents.

We have seen in Sect. 3.2.2 that one can deform a surface into its tangent cone through the specialization to the tangent cone. More precisely, for a surface $(S, 0) \subset (\mathbb{C}^N, 0)$ there exists a map $\varphi : W \subset \mathbb{C}^N \times \mathbb{C} \to \mathbb{C}$ such that the fiber $\varphi^{-1}(0)$ has a germ at $(0, 0)$ isomorphic to the germ of the tangent cone $(C_{S,0}, 0)$ and $\varphi^{-1}(t)$ has a germ a $(0, t)$ isomorphic to $(S, 0)$ for every $t \neq 0$ close to the origin. Moreover, if we call $T := \{0\} \times \mathbb{C}$ then we have $T \subset W$.

Theorem 3.6.30 *Let $(S, 0)$ be a germ of surface for which the tangent cone $C_{S,0}$ is a reduced space. Let $\varphi : W \to \mathbb{C}$ be the specialization to the tangent cone as above. Call W^0 the non-singular locus of W.*

If the surface $(S, 0)$ has no exceptional tangent, then W^0 satisfies Whitney conditions along T.

This result was first proved in [33, Theorem 2.1.1] in the case of hypersurfaces of $\mathbb{C}^3$ and then generalized by Giles Flores [11, Theorem 8.11]. The reader can find there detailed proofs that we do not reproduce here.

As consequence of Theorem 3.6.30, when the surface has a reduced tangent cone and no exceptional tangent, then the germ of the surface is homeomorphic to its tangent cone. The surface is topologically a complex cone.

One would like to skip the hypothesis of having a reduced tangent cone to deduce some properties on surfaces with no exceptional tangents. This has been done for surfaces with isolated singularities.

When the surface $(S, 0)$ is normal, the exceptional tangents are characterized in Sect. 3.6.26 and therefore its non-existence implies the following:

Proposition 3.6.31 *If $(S, 0)$ is a normal surface without exceptional tangent, then the normalized blow-up of 0 and the normalized Nash modification are isomorphic and they are a resolution of the singularity. Furthermore their reduced exceptional fiber is smooth.*

This result was proved in [43, Theorem 5.4]. The idea of the proof is that when the surface has no exceptional tangent, then there are no base points of the polar curves by the blow-up of 0. It is also proved that in this situation there are no base points of the hyperplane sections by the Nash modification. As a consequence each of the considered normalized modifications factors through the other one.

In the case of hypersurfaces of $\mathbb{C}^3$ with isolated singularity, a sharper result was previously obtained in [33, Theorem 2.2.1]:

Proposition 3.6.32 *Let $(S, 0)$ be an isolated singularity of hypersurface in $\mathbb{C}^3$. Then the following are equivalent:*

(ii) *$(S, 0)$ has no exceptional tangent.*
(ii) *The specialization to the tangent cone has a strong simultaneous resolution.*
(iii) *The blow-up of the origin is a resolution of the singularity and the reduced fiber $|e_0^{-1}(0)|$ is non-singular.*
(iv) *The tangent cone $C_{S,0}$ is reduced and has an isolated singularity.*
(v) *The blown-up surface S' is equisingular along the curve $|e_0^{-1}(0)|$.*

In this context note that the fact of having a reduced tangent cone is a consequence of not having any exceptional tangent. In particular an isolated singularity of hypersurface of $\mathbb{C}^3$ with no exceptional tangent is homeomorphic to the cone over a non-singular (reduced) curve.

We wonder if such a homeomorphism can be obtained for any surface with isolated singularity. Note that in Example 3.6.27, the surface is not homeomorphic to its tangent cone, but still is homeomorphic to a complex cone over two lines in $\mathbb{C}^3$ that do not intersect.

Acknowledgements The authors would like to thank the referee for his careful reading of the manuscript and for his comments and suggestions. We also thank J. Elizondo and C. Alcántara for fruitful discussions on the geometry of Grassmannians. The second author acknowledges support from CONACyT (Mexico) grant 282937.

3.7 Appendix: Intersections in Grassmannians

In this appendix, we state and prove a result on intersections with Schubert varieties in Grassmannians. This was needed in the proof of Theorem 3.6.6.

Recall that we have defined the first Schubert variety in $\mathbf{G}(2, N)$ as follows:

Let L be an $(N-2)$-linear subspace of $\mathbb{C}^N$. We call $C_1(L)$ the subset of $\mathbf{G}(2, N)$ such that $T \in C_1(L)$ if and only if $\dim(T \cap L) \geq 1$. It is a subvariety of codimension one in $\mathbf{G}(2, N)$.

Proposition 3.7.1 *Let $C \subset \mathbf{G}(2, N)$ be an irreducible projective curve. Then the intersection $C_1(L) \cap C$ is not empty.*

Proof Instead of working with the Grassmannian of linear subspaces of $\mathbb{C}^N$ of dimension 2, we shall work with the equivalent setting of projective linear subspaces of dimension 1 in $\mathbb{P}^{N-1}$, that we denote by $\mathbf{G}(1, \mathbb{P}^{N-1})$.

We have a natural isomorphism of $\mathbf{G}(2, N)$ onto $\mathbf{G}(1, \mathbb{P}^{N-1})$.

Call $\tilde{C} \subset \mathbf{G}(1, \mathbb{P}^{N-1})$, the space made of projective lines of $\mathbb{P}^{N-1}$ representing the planes of $\mathbb{C}^N$ in the curve C. The space $\tilde{C}$ is an irreducible projective curve in $\mathbf{G}(1, \mathbb{P}^{N-1})$.

We shall denote by $l(\tilde{\Lambda}) \subset \mathbb{P}^{N-1}$ the projective line represented by the point $\tilde{\Lambda} \in \mathbf{G}(1, \mathbb{P}^{N-1})$.

Following the idea developed in [8, 4.2.3], define:

$$X = \bigcup_{\tilde{\Lambda} \in \tilde{C}} l(\tilde{\Lambda}) \subset \mathbb{P}^{N-1}.$$

We want to prove that X is an irreducible projective subset of dimension 2 in $\mathbb{P}^{N-1}$.

Consider the subspace:

$$\tilde{\mathcal{I}} := \{(\tilde{\Lambda}, p) \in \mathbf{G}(1, \mathbb{P}^{N-1}) \times \mathbb{P}^{N-1} | \, p \in l(\tilde{\Lambda})\}.$$

The space $\tilde{\mathcal{I}}$ is an algebraic projective subset of $\mathbf{G}(1, \mathbb{P}^{N-1}) \times \mathbb{P}^{N-1}$, see [8, Sect. 3.2.3].

The first and second projections on $\tilde{\mathcal{I}}$ induce respectively the morphisms:

$$\alpha : \tilde{I} \to \mathbf{G}(1, \mathbb{P}^{N-1}) \text{ and } \beta : \tilde{I} \to \mathbb{P}^{N-1}.$$

We have then:

$$X = \beta(\alpha^{-1}(\tilde{C}));$$

which gives X a structure of projective subset of $\mathbb{P}^{N-1}$.

Call α_0 the restricted morphism $\alpha_0 := \alpha_{|\alpha^{-1}(\tilde{C})} : \alpha^{-1}(\tilde{C}) \to \tilde{C}$.

The fibers of α_0 are $\{\tilde{\Lambda}\} \times l(\tilde{\Lambda})$, for $\tilde{\Lambda} \in \tilde{C}$. These are irreducible projective algebraic subsets of dimension 1 in $\tilde{I}$. Since $\tilde{C}$ is also irreducible, then by [18, Theorem 11.14], the inverse image $\alpha^{-1}(\tilde{C})$ is an irreducible projective set. Furthermore, since $\tilde{C}$ has dimension one and the fibers of α_0 are of dimension 1, then by [18, Theorem 11.12], the set $\alpha^{-1}(\tilde{C})$ has dimension 2. We have seen that $X = \beta(\alpha^{-1}(\tilde{C}))$ which implies that X is an irreducible projective subset of $\mathbb{P}^{N-1}$.

Consider now the restricted map $\beta_0 := \beta_{|\alpha^{-1}(\tilde{C})} : \alpha^{-1}(\tilde{C}) \to X$. Let p be a point of X. We have:

$$\beta_0^{-1}(p) = \beta^{-1}(p) \cap \alpha^{-1}(\tilde{C}) = \{(\tilde{\Lambda}, p) \in \tilde{C} \times \{p\} \mid p \in l(\tilde{\Lambda})\},$$

which shows that we can consider the fibers of β_0 as algebraic subsets of $\tilde{C}$. Since the space $\tilde{C}$ has dimension 1, the fibers of β_0 have dimension at most one. As a first consequence, again by [18, Theorem 11.12], X has dimension at least one.

If a fiber $\beta_0^{-1}(p)$ has dimension 1, then it corresponds to the curve $\tilde{C}$ itself.

There is at most one point p in X such that the fiber $\beta_0^{-1}(p)$ has dimension one. In fact, for a point $p \in X \subset \mathbb{P}^{N-1}$ we shall denote by p^* the set of lines of $\mathbb{P}^{N-1}$ that contain the point p. In this setting the fibers of β_0 can be written as:

$$\beta_0^{-1}(p) = (p^* \cap \tilde{C}) \times \{p\}.$$

For two distinct points p and q in X, if the respective fibers $\beta_0^{-1}(p)$ and $\beta_0^{-1}(q)$ have dimension 1, they correspond to the same irreducible curve $\tilde{C}$, then this curve would be the intersection $p^* \cap q^* \cap \tilde{C}$ which is either empty or the point in $\tilde{C}$ corresponding to the line (pq) joining p and q. It is never a one dimensional space.

Since the projective algebraic set X is irreducible of dimension at least 1, then by the preceding observation, a generic fiber of $\beta_0 : \alpha^{-1}(\tilde{C}) \to X$ has dimension 0. Again by [18, Theorem 11.12], The space X is a projective algebraic subset of $\mathbb{P}^{N-1}$ of dimension $\dim(\alpha^{-1}(\tilde{C})) = 2$.

In order to prove the proposition above, we need to consider a Schubert variety associated to an $(N-2)$-space in $\mathbb{C}^N$. Consider now an $(N-2)$-plane $L \subset \mathbb{C}^N$. The corresponding projective space $\text{Proj}(L) \subset \mathbb{P}^{N-1}$ has dimension $N-3$.

The projective subsets X and $\text{Proj}(L)$ have complementary dimensions in $\mathbb{P}^{N-1}$, therefore:

$$X \cap \text{Proj}(L) \neq \emptyset.$$

Call $\tilde{p}$ a point in $X \cap \mathrm{Proj}(L)$, and recall that $X = \bigcup_{\tilde{\Lambda} \in \tilde{C}} l(\tilde{\Lambda}) \subset \mathbb{P}^{N-1}$.

Then there exists $\tilde{\Lambda} \in \tilde{C}$ such that $\tilde{p} \in l(\tilde{\Lambda})$. Therefore:

$$\tilde{p} \in l(\tilde{\Lambda}) \cap \mathrm{Proj}(L).$$

Coming back to the affine space $\mathbb{C}^N$ and to the Grassmannian of two-planes in $\mathbb{C}^N$, the statement above can be formulated as follows:

The projective line $l(\tilde{\Lambda})$ corresponds to a two-dimensional plane $T(\Lambda) \subset \mathbb{C}^N$ represented by a point $\Lambda \in C \subset \mathbf{G}(2, N)$. The intersection of the plane $T(\Lambda)$ with the $(N-2)$-plane L contains the line $l(p) \subset \mathbb{C}^N$ represented by $\tilde{p} \in \mathbb{P}^{N-1}$. In other words, The plane $T(\Lambda)$ corresponds to a point in the Schubert variety $C_1(L)$, equivalently:

$$\Lambda \in C \cap C_1(L),$$

proving that the intersection is not empty.

For a non irreducible projective curve in $\mathbf{G}(2, N)$, every irreducible component intersects the Schubert variety $C_1(L)$ for any $(N-2)$-plane L of $\mathbb{C}^N$.

References

1. J. Briançon, A. Galligo, M. Granger, *Déformations équisingulières des germes de courbes gauches réduites*, Mém. Soc. Math. France (N.S.) No. **1** (1980/81), 69 pp.
2. J. Briançon, P. Maisonobe, M. Merle, *Localisation de systèmes différentiels, stratifications de Whitney et condition de Thom*, Invent. math. **117** (1994) 531–550.
3. W. Burau, *Kennzeichnung den Schlauchverkettungen,* Abh. Hamburg Sem. **10** (1934), 285–297.
4. E. Chávez Martínez, *Variedades tóricas sin la condición de normalidad*, Master thesis, Universidad Nacional Autónoma de México, 2016. Available at https://repositorio.unam.mx/.
5. Wei-Liang, Chow, *On compact complex analytic varieties,* Amer. J. Math. **71** (1949), 893–914.
6. Draper, R.N. *Intersection theory in analytic geometry,* Math. Ann. **180**, 175–204 (1969).
7. D. Eisenbud, *Commutative algebra, with a view toward algebraic geometry*, Graduate Texts in Mathematics **150**, Springer-Verlag, 1995.
8. D. Eisenbud, J. Harris, *3264 and all that - a second course in algebraic geometry.* Cambridge University Press, Cambridge, 2016.
9. J. Fernández de Bobadilla, J. Snoussi and M. Spivakovsky, *Equisingularity in one parameter families of generically reduced curves*. Int. Math. Res. Not. **5**, 1589–1609 (2017).
10. M. Gerstenhaber, *On the deformations of rings and algebras*, Ann. of Math. **79** (1964) 59–103.
11. A. Giles Flores, *Specialization to the tangent cone and Whitney equisingularity,* Bull. Soc. Math. France **141** (2013), no. 2, 299–342.
12. A. Giles Flores and B. Teissier, *Local polar varieties in the geometric study of singularities*, Annales de la Faculté des Sciences de Toulouse, Vol. 27, no. **4**, (2018), 679–775.
13. A. Giles Flores, O. N. Silva, J. Snoussi, *On tangency in equisingular families of curves and surfaces,* Q. J. Math. **71** (2020), no. 2, 485–505.
14. A. Giles Flores, Otoniel N. Silva, B. Teissier, *The bi-Lipschitz geometry of complex curves: an algebraic approach,* to be published in Springer LNM series.
15. G. González-Sprinberg, *Résolution de Nash des points doubles rationnels,* Annales de l'Institut Fourier **32** (1982), 111–78.

16. G.-M.Greuel, *Equisingular and Equinormalizable deformations of Isolated Non Normal Singularities*. Methods Appl. Anal. **24**, 215–276 (2017).
17. H.A, Hamm - Lê D.T. *The Lefschetz theorem on hyperplane sections,* Vol. 1 of this handbook, to be published, Springer.
18. J. Harris, *Algebraic geometry, a first course*, Graduate Texts in Mathematics **133**, Springer-Verlag, 1992.
19. J. P. G. Henry - Lê Dũng Tráng, *Limites d'espaces tangents,* in Fonctions de plusieurs variables complexes, II (Sém. François Norguet, 1974-1975), pp. 251 - 265. Lecture Notes in Math., Vol. **482**, Springer, Berlin (1975).
20. H. Hironaka, *Stratification and flatness,* In "Real and complex singularities" (Proc. Ninth Nordic Summer School/NAVF Sympos. Math., Oslo, 1976), p. 199–265, Sijthoff and Noordhoff (1977).
21. C. Houzel, *Géométrie analytique locale, I,* in Séminaire H. Cartan 1960-1961, Familles d'espaces complexes, Exposé 18, Secrétariat Mathématique, IHP, 11 rue Pierre et Marie Curie, Paris (1962).
22. M. Lejeune - B. Teissier, *Cloture des idéaux et équisingularité,* avec 7 Compléments, Annales de la Faculté des Sciences de Toulouse, Vol XVII, No.4 (2008), 781–859.
23. Lê Dũng Tráng, *Sur un critère d'équisingularité*, C. R. Acad. Sci. Paris Sér. A-B 272 (1971), A138–A140.
24. Lê Dũng Tráng, *Calcul du nombre de cycles évanouissants d'une hypersurface complexe,* Ann. Inst. Fourier **23** (1973), 261–270.
25. Lê Dũng Tráng, *Limites d'espaces tangents sur les surfaces,* Nova Acta Leopoldina NF 52 **240**(1981), 119–137.
26. Lê Dũng Tráng - J. J. Nuño Ballesteros, *A remark on the topology of complex polynomial functions*, Rev. R. Acad. Cienc. Exactas Fis. Nat. Ser. A Mat. RACSAM **113** (2019), 3977–3994.
27. Lê Dũng Tráng - J. J. Nuño Ballesteros - J. Seade, *The topology of the Milnor fibration,* to be published in Handbook of Geometry and Topology of Singularities.
28. Lê Dũng Tráng - C.P. Ramanujam, *The invariance of Milnor's number implies the invariance of the topological type,* Amer. J. Math. 98 (1976), no. 1, 67–78.
29. Lê Dũng Tráng - K. Saito, *La constance du nombre de Milnor donne des bonnes stratifications*, C. R. Acad. Sci. Paris Sér. A - B **277** (1973), A793–A795.
30. Lê Dũng Tráng - B. Teissier, *Cycles évanescents et conditions de Whitney II.* In Proc. Symp. Pure Math, 40 (Part 2), 1983, 65–103.
31. Lê Dũng Tráng - B. Teissier, *Variétés Polaires Locales et Classes de Chern des Variétés Singulières,* Ann. of Math. **114** (1981), 457–491.
32. Lê Dũng Tráng - B. Teissier, *Limites d'espaces tangents en géométrie analytique,* Com. Mat. Helv. **63** (1988), 540–578.
33. Lê Dũng Tráng and B. Teissier, *Sur la géométrie des surfaces complexes. I. Tangentes exceptionnelles,* Amer. J. Math. **101** (1979), no. 2, 420–452.
34. J. Milnor, *Singular points of complex hypersurfaces*, Annals of Mathematics Studies **68**, Princeton University Press, Princeton, N.J. 1968.
35. A. Nobile, *Some properties of the Nash blowing-up*, Pacific Jr of Maths **60** (1975) 297–305.
36. D. Prill, *Local classification of quotients of complex manifolds by discontinuous groups,* Duke Math. J. **34** (1967) 375–386.
37. D. Rees, *a-transforms of local rings and a theorem on multiplicities of ideals,* Proceedings Camb. Philos., **57**, 1, 8–17.
38. D. Rees, *Asymptotic properties of ideals,* London Math. Soc. Lecture Series, 113, 1988.
39. J. Reeve, *A summary of results in the topological classification of plane algebroid curves,* Rend. Sem. Mat. Torino **14** (1955), 159–187.
40. R. Remmert, *Holomorphe und meromorphe Abbildungen komplexer Räume,* Math. Ann. **133** (1957), 328–370.
41. R. Remmert, K. Stein, *Über die wesentlichen Singularitäten analytischer Mengen,* Mathematische Annalen, **126** (1953) 263–306.

42. O.N. Silva, J. Snoussi, *Whitney equisingularity in one parameter families of generically reduced curves*, manuscripta math. **163** (2020), no. 3-4, 463–479.
43. J. Snoussi, *The Nash modification and hyperplane sections on surfaces,* Bull. Braz. Math. Soc. (N.S.) **36** (2005), no. 3, 309–317.
44. J. Snoussi. Limites d'espaces tangents à une surface normale. *Comment. Math. Helv.*, 76(1):61–88, 2001.
45. M. Spivakovsky, *Sandwiched singularities and desingularization of surfaces by normalized Nash transformations,* Ann. of Math. **131** (1990), 411–491.
46. M. Spivakovsky, *Resolution of singularities: an introduction*, Handbook of geometry and topology of singularities, ed. J.L. Cisneros-Molina, D.T. Lê and J. Seade.
47. B. Teissier, *Appendix to : O. Zariski, Le problème des modules pour les branches planes*, Course given at the Centre de Mathématiques de l'Ecole Polytechnique, Paris, October–November 1973. Second edition. Hermann, Paris, 1986. x+212 pp. ISBN: 2-7056-6036-4. English translation by Ben Lichtin, University Lecture Series, 39 (2006).
48. B. Teissier, *Cycles évanescents, sections planes et conditions de Whitney,* Astérisque 7–8 (1973), 285–362.
49. B. Teissier, *The hunting of invariants in the geometry of discriminants,* Real and complex singularities (Proc. Ninth Nordic Summer School NAVF Sympos. Math., Oslo, 1976), pp. 565–678. Sijthoff and Noordhoff, Alphen aan den Rijn, 1977.
50. B. Teissier, *Variétés polaires. I. Invariants polaires des singularités d'hypersurfaces,* Invent. Math. **40** (1977), no. 3, 267–292.
51. B. Teissier, *Variétés polaires 2: Multiplicités polaires, sections planes, et conditions de Whitney.* Actes de la conférence de géométrie algébrique à La Rábida 1981, Springer Lecture Notes **961** (1982), p. 314–491.
52. D. Trotman, *Stratification theory,* Handbook of geometry and topology of singularities, ed. J.L. Cisneros-Molina, D.T. Lê and J. Seade.
53. C.T.C. Wall, *Singular points of plane curves*, London Math. Soc., Student texts **63** (2004).
54. H. Whitney, *Tangents to an Analytic Variety*, Annals of Math. **81**, No. 3, (1965).
55. O. Zariski, *Studies in Equisingularity I. Equivalent plane algebroid curves,* Amer. J. Math. **87** (1965) 507–536.
56. O. Zariski, P. Samuel, *Commutative algebra*, Volume II, Graduate Texts in Mathematics **29**, Springer-Verlag (1975).

Chapter 4
Algebro-Geometric Equisingularity of Zariski

Adam Parusiński

Abstract This is a survey on Zariski equisingularity. We recall its definition, main properties, and a variety of applications in Algebraic Geometry and Singularity Theory. In the first part of this survey, we consider Zariski equisingular families of complex analytic or algebraic hypersurfaces. We also discuss how to construct Zariski equisingular deformations. In the second part, we present Zariski equisingularity of hypersurfaces along a nonsingular subvariety and its relation to other equisingularity conditions. We also discuss the canonical stratification of such hypersurfaces given by the dimensionality type.

Contents

Partially supported by ANR project LISA (ANR-17-CE40-0023-03).

A. Parusiński (✉)
Université Côte d'Azur, CNRS, LJAD, UMR 7351, 06108 Nice, France
e-mail: adam.parusinski@univ-cotedazur.fr

J. L. Cisneros-Molina et al. (eds.), *Handbook of Geometry and Topology of Singularities II*, https://doi.org/10.1007/978-3-030-78024-1_4

4.1 Introduction

A singularity is the germ of a complex or real analytic space (V, p) that is not regular at p. Equisingularity means equivalent or similar singularity and it is always necessary to make precise which equivalence of singularities we have in mind. Thus two singularities (X, x) and (Y, y) are analytically equivalent if there is an analytic isomorphism germ $\phi : (X, x) \to (Y, y)$. If ϕ is only a homeomorphism then we say that (X, x) and (Y, y) are topologically equivalent. If (X, x) and (Y, y) are both subspaces of the affine space $(\mathbb{K}^n, 0)$, $\mathbb{K} = \mathbb{R}$ or $\mathbb{C}$, then we may require ϕ to be the restriction of an isomorphism (resp. homeomorphism) of the ambient spaces $\Phi : (\mathbb{K}^n, 0) \to (\mathbb{K}^n, 0)$. If such Φ exists we say then that (X, x) and (Y, y) are ambient analytically (resp. topologically) equivalent.

Let V be a real or complex analytic space. Then there exists a stratification $\mathcal{S}$ of V, that is a decomposition of V into analytic manifolds that, moreover, are usually required to satisfy some additional properties. For the notion of stratification and a historical account of stratification theory, we refer the reader to the paper of D. Trotman [70] in the first volume of this handbook and the references therein. It is known that there always exists a stratification of V that is topologically equisingular along each stratum, that is if p_1 and p_2 belong to the same stratum then (V, p_1) and (V, p_1) are topologically equivalent. If V is a subspace of $\mathbb{K}^n$ then one may, moreover, require this stratification to be ambient topologically equisingular. This can be achieved by constructing a Whitney stratification of V. Another and entirely independent way of constructing such a stratification is Zariski equisingularity, which is the subject of this survey.

Recall that, in general, there is no stratification that is analytically equisingular along each stratum, as a classical example of Whitney [78, Example 13.1] shows: $V = \{(x, y, z) \in \mathbb{K}^3; xy(y + x)(y - zx) = 0\}$ admits a continuous family of analytically (or even C^1-diffeomorphically) non-equivalent singularities along the z-axis, due to the phenomenon of continuous moduli (the cross-ratio in this example).

In 1971 in "Some open questions in the theory of singularities" [85], O. Zariski proposed a general theory of equisingularity for complex algebraic and analytic hypersurfaces. Zariski's approach was based on a new version of equisingularity that Zariski called algebro-geometric equisingularity, since it was defined by purely algebraic means but it reflected many geometric properties. For instance, as Varchenko shows in [72–74] answering a question posed by Zariski, Zariski equisingularity, which we now call the algebro-geometric equisingularity of Zariski, implies ambient topological triviality. Zariski equisingularity under an additional genericity of projection assumption implies Whitney's conditions as shown by Speder [64].

This notion of equisingularity extended Zariski's earlier work on the singularities of plane curves, their equivalence and their families, see [82–84]. For the general case of hypersurface singularities over an algebraically closed field of characteristic zero Zariski presented his program in [87, 88]. The paper [88] "Foundations of a general theory of equisingularity on r-dimensional algebroid and algebraic varieties, of embedding dimension $r + 1$", published in 1979, contains a complete foundation of this theory, stated for algebroid varieties over an algebraically closed field of characteristic zero. (Recall that algebroid varieties are the varieties defined by ideals of the rings of formal power series, see [32, Chap. 4] and [88, Sect. 2]) Since then Zariski equisingularity has been widely applied in the theory of singularities. We present in this survey an account, certainly incomplete, of this development.

Intuitively, Zariski's notion can be characterized by two properties:

1. If (V, P_1) and (V, P_2) are equisingular, then P_1 is a regular point of V if and only if P_2 is a regular point of V.
2. If $W \subset \operatorname{Sing} V$ is non-singular then V is equisingular along W at $P \in W$ if and only if for all sufficiently general projections $\pi : \mathbb{K}^{r+1} \to \mathbb{K}^r$ the discriminant locus of $\pi|_V$ is equisingular along $\pi(W)$ at $\pi(P)$.

Formally, one may talk about two notions of Zariski equisingularity, of a hypersurface along its nonsingular subvariety and of a family of hypersurface singularities parameterized by a finite number of parameters. This is already present, implicitly, in [85], where the former one is motivated by the latter one. We shall follow this path in this survey as well. In Sect. 4.2 we describe the equisingular families of complex plane curve singularities. This description is based on Puiseux with parameter theorem, Theorem 4.2.1. Then we introduce Zariski equisingularity of families, Sects. 4.2 and 4.3. As we have mentioned, Zariski equisingular families are topologically trivial. Therefore Zariski equisingularity implies the generic topological equisingularity of real or complex, algebraic varieties or analytic spaces (not necessarily hypersurfaces). We present this principle in Sect. 4.3.5. As a consequence Zariski equisingularity provides an algorithmic construction of a topologically equisingular stratification. More applications to Algebraic Geometry are presented in Sects. 4.3.6 and 4.3.7.

In Sect. 4.4 we show how to construct equisingular deformations of a given singularity. This construction appears in many applications, in particular it is used to show that a (real or complex) analytic singularity is homeomorphic to an algebraic one, see subsection 4.4.4, and, moreover we may assume that the latter one is defined over the field of algebraic numbers $\overline{\mathbb{Q}}$, see Sect. 4.4.2. At the end of Sect. 4.4 we discuss how Zariski equisingularity can be used to trivialize families of analytic function and map germs, see Sects. 4.4.5 and 4.4.6.

In Sect. 4.5 we present the original notion of Zariski equisingularity of a hypersurface V along a nonsingular subvariety W, and a related notion of the dimensionality type. Zariski equisingularity along a hypersurface is defined by taking successive co-rank 1 projections and their discriminants, and a similar construction is used to define Zariski equisingularity in families. The main, and to some extent still open problem, is to decide what projection to take to verify whether such an equisingularity holds. As follows from Zariski work, in the case of families of plane curves singularities, the equisingularity given by a single projection implies equisingularity for all transverse projections, for this notion see Sect. 4.5.1. Therefore, originally, Zariski considered transverse projections as sufficient for such verification, see [85]. In [37] Luengo gave an example of a family of surface singularities in $\mathbb{C}^3$ that is Zariski equisingular for one transverse projection but not for a generic or generic linear projection. Therefore, in [88], Zariski proposed to build this theory on the notion of "generic" projection. The definition of such generic projection given in [88] is therefore crucial. It involves adding all the coefficients of a generic formal change of coordinates as indeterminates to the ground field. As Zariski also showed in [88] a generic (in a more standard meanining) polynomial projection gives the same theory, that is to say the same notion of generic Zariski equisingularity along a nonsingular subvariety. But it is not known how to verify which polynomial projections are generic in this sense or even whether there is a bound on the minimal degree of such polynomial generic projections. This makes algorithmic computations of the dimensionality type and related notions of generic Zariski equisingularity and Zariski's canonical stratification impossible at the moment. The algebraic case was studied in more detail by Hironaka [26], where the algebraic semicontinuity of the minimal degree of such polynomial projection is shown. The question whether a generic linear projection is always sufficient is still open for dimensionality type ≥ 2, though the case of the dimensionality type 2 is fairly well understood thanks to [9].

General set-up

In this survey we present Zariski's theory in the complex analytic set-up, which seems to be the most common and of the biggest interest for singularity theory. There are two obvious extensions that one has to keep in mind. The first one, as the original definition of Zariski, is the theory of algebroid varieties over an arbitrary algebraically closed field of characteristic zero, when one works with the varieties defined by the ideals in the ring of formal power series. The second one is the real analytic set-up. Many results on Zariski equisingularity, such as topological triviality for example, are valid in both complex analytic and real analytic set-ups. The real analytic set up sometimes requires more careful statements, for instance, by replacing analytic sets by the equations or ideals defining them. In general, for Zariski equisingularity, the assumption on the ground field to be algebraically closed seems not to be essential, unlike the assumption to be of characteristic zero, which is necessary.

In Sect. 4.2, which can be considered as a motivation for the general definition, we discuss the equisingularity of complex plane curves. Sections 4.3 and 4.4 are presented for complex and real analytic or algebraic spaces. For the definitions, theorems and proofs of these two sections there is no essential difference between the real and the complex case. The second part of Sect. 4.5, the dimensionality type, is presented in the algebroid set-up, like Zariski's original definition. Every statement of this section holds in complex analytic case. We also believe that it can be carried over to the real analytic set-up, but this has yet to be done.

Notation and terminology.

We denote by $\mathbb{K}$ either $\mathbb{R}$ or $\mathbb{C}$. Thus, by $\mathbb{K}$-analytic we mean either real analytic or holomorphic (complex analytic). Sometimes we abbreviate it saying that a space or a map is analytic if the ground field, $\mathbb{C}$ or $\mathbb{R}$, is clear from the context or if the result holds in both cases.

By an analytic space, we mean one in the sense of [45]. As we work mostly in the local analytic case, it suffices to consider only analytic set germs. For an analytic space X by $\operatorname{Sing} X$ we denote the set of singular points of X, i.e. the support of the singular subspace of X. By $\operatorname{Reg} X$ we denote its complement $X \setminus \operatorname{Sing} X$, the set of regular points of X. For an analytic function germ F we denote by $V(F)$ its zero set and by F_{red} its reduced (i.e. square free) form. By a real analytic arc, we mean a real analytic map $\gamma : I \to X$, where $I = (-1, 1)$ and X is a real or a complex analytic space.

For a polynomial monic in z, $F(x, z) = z^d + \sum_{i=1}^{d} a_i(x) z^{d-i}$, with coefficients analytic functions in x, we denote by $D_F(x)$ its discriminant, and by $\Delta_F(x)$ its discriminant locus, the zero set of $D_F(x)$. The discriminant of F, and more generally the generalized discriminants of F, are recalled in Appendix, Sect. 4.6.

We say that $f \in \mathbb{K}\{x\}$ is *a unit* if $f(0) \neq 0$. We often use Weierstrass Preparation Theorem. Recall briefly its statement, see for instance [45, Theorem 2, p. 12], [34, Chap. 3, Sect. 2] for more details. Let $F(x, z) \in \mathbb{K}\{x, z\}$ be regular in the variable z, that is $F(0, z) = z^d \operatorname{unit}(z)$. Then there are $a_i(x) \in \mathbb{K}\{x\}$ such that $a_i(0) = 0$ and

$$F(x, z) = \operatorname{unit}(x, z)\,(z^d + \sum_{i=1}^{d} a_i(x) z^{d-i}).$$

We call the monic polynomial $z^d + \sum_{i=1}^{d} a_i(x) z^{d-i}$, *the Weierstrass polynomial associated to* F. An analogous statement holds for formal power series, i.e. for $F(x, z) \in \mathbb{K}[[x, z]]$.

4.2 Equisingular Families of Plane Curve Singularities

We recall the notion of equisingular families of complex plane curve singularities. There are several equivalent definitions that are proposed by Zariski in [82, 83]. We use the one based on the discriminant of a local projection. Firstly, this is the definition that Zariski generalizes to the higher-dimensional case. Secondly, by Puiseux with parameter theorem, it gives an equiparameterization of such singularities by fractional power series.

Let

$$F(t, x, y) = y^d + \sum_{i=1}^{d} a_i(t, x) y^{d-i} \tag{4.1}$$

be a monic polynomial in $y \in \mathbb{C}$ with complex analytic coefficients $a_i(t, x)$, defined on $U_{\varepsilon,r} = U_\varepsilon \times U_r$, where $U_\varepsilon = \{t \in \mathbb{C}^l;\ \|t\| < \varepsilon\}$, $U_r = \{x \in \mathbb{C};\ |x| < r\}$. Here $t = (t_1, \ldots, t_l)$ is considered as a parameter. One also often assumes that F is reduced (has no multiple factors) so that its discriminant D_F is not identically equal to zero. For arbitrary F we either consider $D_{F_{red}}$ or, equivalently, the first not identically equal to zero generalized discriminant of F, see Appendix, Sect. 4.6.

Theorem 4.2.1 (Puiseux with parameter) *Suppose that the discriminant of F_{red} is of the form $D_{F_{red}}(t, x) = x^M$ unit(t, x) where unit(t, x) is a complex analytic function defined and nowhere vanishing on $U_{\varepsilon,r}$. Then there is a positive integer N and complex analytic functions $\tilde{\xi}_i(t, u)$ defined on $U_\varepsilon \times U_{r^{1/N}}$ such that*

$$F(t, u^N, y) = \prod_{i=1}^{d} (y - \tilde{\xi}_i(t, u)).$$

Let θ be an Nth root of unity. Then for each i there is j such that $\tilde{\xi}_i(t, \theta u) = \tilde{\xi}_j(t, u)$.

If F is irreducible then one can take $N = d$. In general, $N = d!$ always works, but it is not minimal.

If $M = 0$ then, by the Implicit Function Theorem (IFT), the roots of F, which we denote by $\xi_1(t, x)$, ..., $\xi_d(t, x)$, are $\mathbb{C}$-analytic functions of (t, x). Moreover two such ξ_i and ξ_j either coincide or are distinct everywhere. In general, for arbitrary M, Theorem 4.2.1 implies that the projection of the zero set $V = V(F)$ of F onto U_ε, given by $(x, y, t) \to t$ is topologically trivial. To see it one may use the following corollary.

Corollary 4.2.2 *For x_0 fixed, the roots of F, $\xi_1(t, x_0), \ldots, \xi_d(t, x_0)$, can be chosen complex analytic in t. Moreover, if $\xi_i(0, x_0) = \xi_j(0, x_0)$ then $\xi_i(t, x_0) \equiv \xi_j(t, x_0)$. Thus the multiplicity of each $\xi_i(t, x_0)$ as a root of F is independent of t.*

Proof It suffices to show it for F reduced. Then for $x_0 \neq 0$ it follows from the IFT. Let us show it for $x_0 = 0$. The family $\xi_1(t, 0), \ldots, \xi_d(t, 0)$ coincides (as unordered sets) with $\tilde{\xi}_1(t, 0), \ldots, \tilde{\xi}_d(t, 0)$. If $\tilde{\xi}_i(0, 0) = \tilde{\xi}_j(0, 0)$ then $\tilde{\xi}_i(t, u) - \tilde{\xi}_j(t, u)$ is either identically zero or divides u^{NM} and hence equals a power of u times a unit. □

Using Corollary 4.2.2 we may trivialize topologically V with respect the parameter t by

$$\Phi(t, x, \xi_i(0, x)) = (t, x, \xi_i(t, x)) \quad i = 1, \ldots, d.$$

The map Φ is, by Corollary 4.2.2, complex analytic in t and one can show, moreover, that it is a local homeomorphism. (It follows, for instance, from much more general [52, Theorem1.2].)

The parameterized Puiseux Theorem, Theorem 4.2.1, can be proven in the same way as the classical Puiseux Theorem by considering the finite covering of $V(F)$ over $U_\varepsilon \times U_r^*$, where $U_r^* = U_r \setminus \{0\}$. Then, for a positive integer N, the pullback of this covering by $(t, u) \to (t, u^N) = (t, x)$ is trivial, its sheets define the roots $\tilde{\xi}_i(t, u)$ that extend analytically to $U_\varepsilon \times \{0\}$ by Riemann's Removable Singularity Theorem. For details we refer for instance to [57]. Note also that this theorem is a special case of the Jung-Abhyankar Theorem, see [1, 27], that in complex analytic case can be proven exactly along the same lines, see [55, Proposition 2.1].

Theorem 4.2.3 (Jung-Abhyankar) *Let k be an algebraically closed field of characteristic zero and let $f \in k[[x_1, \ldots, x_{r+1}]]$ be of the form*

$$f(x_1, \ldots, x_{r+1}) = x_{r+1}^d + \sum_{i=1}^{d} a_i(x_1, \ldots, x_r) x_{r+1}^{d-i}.$$

Suppose the discriminant D_f of f equals a monomial $\prod_{i=1}^k x_i^{n_i}$ times a unit. Then the roots of f are fractional power series in $x_1, \ldots, x_r$, More precisely, there is a positive integer N, such that the roots of $f(u_1^N, \ldots, u_k^N, x_{k+1}, \ldots, x_{r+1})$ belong to $k[[u_1, \ldots, u_k, x_{k+1}, \ldots, x_r]]$.

4.2.1 Equisingular Families of Plane Curve Singularities. Definition

Let us fix a local projection $pr : \mathbb{C}^l \times \mathbb{C}^2 \to \mathbb{C}^l$ and suppose that F is a complex analytic function defined in a neighborhood of the origin in $\mathbb{C}^l \times \mathbb{C}^2$ and vanishing identically on $T = \mathbb{C}^l \times \{0\}$. We assume that F is reduced and consider its zero set $V = V(F) = F^{-1}(0)$ as a family of plane curve singularities

$$t \rightsquigarrow (V_t, 0) = (V \cap pr^{-1}(t), 0)$$

parameterized by $t \in (\mathbb{C}^l, 0)$. We say that a local system of coordinates $t_1, \ldots, t_l, x, y$ is *pr-compatible* if $pr(t, x, y) = t$, where $t = (t_1, \ldots, t_l)$, and $T = \{x = y = 0\}$. Suppose that in such a system of coordinates F is regular in variable y, i.e. $F(0, 0, y) \not\equiv 0$. Then, by the Weierstrass Preparation Theorem, we may assume that, up to a multiplication by an analytic unit, F is of the form (4.1) with all $a_i(0, 0) = 0$. Note also that because $T \subset V(F)$, we have $F(t, 0, 0) \equiv 0$.

Definition 4.2.4 We say that $V = V(F)$ is *an equisingular family of plane curve singularities* if there are a pr-compatible system of coordinates t, x, y, such that F is regular in variable y, and a non-negative integer M, such that the discriminant $D_F(t, x)$ of F is of the form

$$D_F(t, x) = x^M \operatorname{unit}(t, x). \tag{4.2}$$

Equisingular families of plane curve singularities were studied in the algebroid set-up (i.e. defined by F being a formal power series) over an algebraically closed field of characteristic zero by Zariski [82–84] mainly by means of (equi)resolution. All the results of [82–84], properly stated, are valid for the complex analytic case. In particular, Zariski has shown that in such families the special fiber $(V_0, 0)$ and the generic fiber $(V_{t_{gen}}, 0)$ are equivalent plane curve singularities, see [82, Sect. 6], see also [82, Sect. 3] for several equivalent definitions of equivalent plane curve singularities. In the complex analytic set-up, two complex plane curve singularities are equivalent if and, only if they are ambient topologically equivalent. By [82, Theorem 7], Zariski equisingular families of plane curve singularities are equimultiple, that is to say $\operatorname{mult}_{(t,0,0)} V$ is independent of t. If this multiplicity equals $d = \deg_y F$, then we say that the associated projection $\pi(x, y, t) = (x, t)$ is *transverse*. Geometrically it means that the kernel of π is not included in the tangent cone $C_0(V)$. Because the equimultiple families are normally pseudo-flat (continuity of the tangent cone), it is enough to check the transversality for the special fiber V_0 and if it holds for the special fiber then it holds also for the generic one. Zariski shows in Theorem 7 of [82] also the following result.

Theorem 4.2.5 (*[82]*, Theorem 7) *If a family of plane curve singularities is equisingular (for a not necessarily transverse projection) then it is equisingular for all transverse projections.*

Note that if $V = V(F)$ is equisingular then the singular locus $\operatorname{Sing} V$ of V is $T = \mathbb{C}^l \times \{0\}$. I n β3, Sect. 8] Zariski shows that a family $V(F)$ of plane curve singularities is equisingular if and only if $\operatorname{Sing} V = T$ and $V \setminus T, T$ is a Whitney stratification of V. In the complex analytic case, this gives another proof of the fact that the equisingular families of plane curve singularities are topologically trivial and the following holds.

Corollary 4.2.6 *The Puiseux pairs of the roots $\xi_i(t, x)$ and the contact exponents between different branches of V_t are independent of t.*

In the complex analytic set-up, in [65, p. 623] B. Teissier gives 12 characterizations of equisingular families of plane curve singularities, including (equi)resolutions, constancy of the Milnor number, Whitney's conditions, and topological triviality.

4.2.2 *Equisingular Families of Plane Curve Singularities and Puiseux with Parameter*

The Puiseux with parameter theorem, Theorem 4.2.1, gives the following criterion of equisingularity of families of plane curve singularities.

Theorem 4.2.7 *Let F be reduced and of the form* (4.1) *in a pr-compatible system of coordinates t, x, y. We also assume $a_i(0,0) = 0$ for all i. Then $V(F)$ is an equisingular family of plane curve singularities for this system of coordinates if and only if there are $\tilde{\xi}_i \in \mathbb{C}\{t,u\}, i = 1, \dots, d$, and strictly positive integers N, $k_{ij}, i < j$, such that*

$$F(t, u^N, y) = \prod_{i=1}^{d} (y - \tilde{\xi}_i(t,u)) \tag{4.3}$$

and $\tilde{\xi}_i - \tilde{\xi}_j = u^{k_{ij}}$ unit(t,u) *or ξ_i and ξ_j coincide everywhere (the latter possibility may occur only if F is not reduced).*

The above observation implies, in particular, Corollary 4.2.2. We also note that it implies that all a_i of (4.1) satisfy $a_i(t,0) \equiv 0$. Indeed, by the assumption $T \subset V(F)$ there is a root $\tilde{\xi}_j$ of F such that $\tilde{\xi}_j(t,0) \equiv 0$. Since $\tilde{\xi}_i(0,0) = 0$ for all i the last claim of the above theorem implies our assertion.

4.3 Zariski Equisingularity in Families

Zariski equisingularity of families of singular varieties was introduced by Zariski in [85] in the context of equisingularity of a hypersurface along a smooth subvariety, that we discuss in Sect. 4.5. This is a direct generalization of Definition 4.2.4 but instead of a single co-rank one projection one considers a system of such successive projections. It can be formulated over any field, in particular, in the analytic case over $\mathbb{K} = \mathbb{R}$ or $\mathbb{C}$. For $x = (x_1, \dots, x_n) \in \mathbb{K}^n$ we denote $x^i = (x_1, \dots, x_i) \in \mathbb{K}^i$.

Definition 4.3.1 By a *local system of pseudopolynomials in $x = (x_1, ..., x_n) \in \mathbb{K}^n$ at* $(0,0) \in \mathbb{K}^l \times \mathbb{K}^n$, *with a parameter $t \in U \subset \mathbb{K}^l$*, we mean a family of $\mathbb{K}$-analytic functions

$$F_i(t, x^i) = x_i^{d_i} + \sum_{j=1}^{d_i} a_{i-1,j}(t, x^{i-1}) x_i^{d_i - j}, \quad i = 0, \dots, n, \tag{4.4}$$

defined on $U \times U_i$, where U_i is a neighborhood of the origin in $\mathbb{K}^i$, with the coefficients $a_{i,j}$ vanishing identically on $T = U \times \{0\}$. This includes $d_i = 0$, in which case we mean $F_i \equiv 1$.

Definition 4.3.2 Let $V = F^{-1}(0)$ be an analytic hypersurface in a neighborhood of the origin in $\mathbb{K}^l \times \mathbb{K}^n$. We say that V is *Zariski equisingular with respect to the parameter t (and the system of coordinates $x_1, \ldots, x_n$)* if there are $k \geq 0$ and a system of pseudopolynomials $F_i(t, x^i)$ such that

1. F_n is the Weierstrass polynomial associated to F.
2. for every i, $k \leq i \leq n-1$, the discriminant of $(F_{i+1})_{red}$ (or, equivalently, the first not identically equal to zero generalized discriminant of F_{i+1}, see Appendix, Sect. 4.6) divides F_i.
3. $F_k \equiv 1$ (and then we put $F_i \equiv 1$ for all $0 \leq i < k$).

Remark 4.3.3 In the above definition, we suppose that the system of local coordinates $x_1, \ldots, x_n$ is fixed. Of course one may say that V is *Zariski equisingular with respect to the parameter t*, if such a system exists. This raises a variety of interesting questions, for instance, how to check whether such a system exists. We will discuss it in Sect. 4.5 in a slightly different set-up, Zariski singularity along a nonsingular subspace.

Remark 4.3.4 In the original definition of Zariski equisingularity [85] and also in [72, 73], the condition 2. was stated in an apparently more restrictive way:

2'. for every i, $k \leq i \leq n-1$, F_i is the Weierstrass polynomial associated to the discriminant of $(F_{i+1})_{red}$.

The definition given here comes from [52] and is often more convenient to work with than the original one. Probably, both definitions are equivalent.

4.3.1 Topological Equisingularity and Topological Triviality

In [85] Zariski asked the following question.
Does algebro-geometric equisingularity (i.e. Zariski equisingularity), in complex analytic case, imply topological equisingularity or even differential equisingularity?

By the latter one, Zariski meant Whitney's conditions (a) and (b). The answer to this part of Zariski's question depends on how generic the system of coordinates giving Zariski equisingularity is, or equivalently how generic the projections defining the successive discriminants are, see Sect. 4.5.2 below. In 1972 Varchenko [73] gave the affirmative answer to the first part of the question, see also [72, 74] for the statement of results.

Theorem 4.3.5 *Suppose that V is Zariski equisingular with respect to the parameter t. Then there are neighborhoods U of the origin in $\mathbb{K}^l$, Ω_0 of the origin in $\mathbb{K}^n$, and Ω of the origin in $\mathbb{K}^{l+n}$, and a homeomorphism*

$$\Phi : U \times \Omega_0 \to \Omega, \tag{4.5}$$

such that

(i) $\Phi(t, 0) = (t, 0)$, $\Phi(0, x_1, \dots, x_n) = (0, x_1, \dots, x_n)$;
(ii) Φ *has a triangular form*

$$\Phi(t, x_1, \dots, x_n) = (t, \Psi_1(t, x_1), \dots, \Psi_{n-1}(t, x_1, \dots, x_{n-1}), \Psi_n(t, x_1, \dots, x_n)); \tag{4.6}$$

(iii) $\Phi(U \times (V \cap \Omega_0)) = V \cap \Omega$.

We note that Varchenko's result gives local topological triviality, a property stronger than the topological equisingularity. Here by topological equisingularity we mean the constancy of local topological types of $V_t := V \cap (\{t\} \times \mathbb{K}^n)$ at the origin, i.e. the existence of homeomorphism germs $h_t : (V_0, 0) \to (V_t, 0)$, possibly given by ambient homeomorphisms $H_t : (\mathbb{K}^n, 0) \to (\mathbb{K}^n, 0)$. The (ambient) topological triviality, that is the existence of Φ of (4.5), implies that such $H_t(x) = \Phi(t, x)$ depends continuously on t.

The details of the proof of Theorem 4.3.5 are published in [73]. Strictly speaking the proof in [73] is in the global polynomial case but it can be adapted easily to the local analytic case. The homeomorphism Φ is constructed in the complex case $\mathbb{K} = \mathbb{C}$. The real case follows from the complex one under a standard argument using the invariance by complex conjugation. The functions Ψ_i are constructed inductively so that every

$$\Phi_i(t, x_1, \dots, x_i) = (t, \Psi_1(t, x_1), \dots, \Psi_i(t, x_1, \dots, x_i)) \tag{4.7}$$

induces topological triviality of $F_i^{-1}(0)$. Given Φ_i, then Φ_{i+1} is constructed in two steps.
Step 1. One lifts Φ_i to the zero set of $F_{i+1}^{-1}(0)$. Such a continuous lift exists and is unique thanks to the following lemma, cf. the multiplicity preservation lemmas of Sect. 2 of [73] or Lemma on p. 429 of [72]. This lemma and the standard argument of the continuity of roots show that such a lift is continuous.

Lemma 4.3.6 *Let*

$$F(t, x) = x_n^d + \sum_{j=1}^{d} a_{i-1,j}(x^{n-1}) x_n^{d-j}, \tag{4.8}$$

be a pseudopolynomial defined in a neighborhood of $p = (p', p_n) \in \mathbb{C}^n$. Let $H_t : (\mathbb{C}^{n-1}, p') \to (\mathbb{C}^{n-1}, p'_t)$, $t \in [0, 1]$, be a continuous family of local homeomorphisms preserving the discriminant locus of F, that is $H_t(\Delta_F, p') = (\Delta_F, p'_t)$. Then the number of distinct roots of F over p'_t, as well as their multiplicities, are independent of t.

Step 2. As soon as Φ_{i+1} is defined on the zero set of $F_{i+1}^{-1}(0)$ it suffices to extend it to the ambient space. This is obtained in Section 1 of [73] by the covering isotopy lemma, see also Fundamental Lemma of [74]. The construction of such extension is based on a triangulation of the base space, so that the finite branched covering $F_{i+1}^{-1}(0) \to \mathbb{C}^{l+i}$ is trivial over each open simplex, and a simplicial extension argument.

4.3.2 Arc-Wise Analytic Triviality

Zariski equisingularity implies much stronger triviality property than just the topological one. The following result was shown in [52, Theorem 3.1].

Theorem 4.3.7 *Suppose that V is Zariski equisingular with respect to the parameter t. Then there are neighborhoods U of the origin in $\mathbb{K}^l$, Ω_0 of the origin in $\mathbb{K}^n$, and Ω of the origin in $\mathbb{K}^{l+n}$, and a homeomorphism*

$$\Phi : U \times \Omega_0 \to \Omega, \tag{4.9}$$

such that

(i) $\Phi(t, 0) = (t, 0)$, $\Phi(0, x_1, \dots, x_n) = (0, x_1, \dots, x_n)$*;*
(ii) Φ *has a triangular form* (4.6) *;*
(iii) *there is $C > 0$ such that for all $(t, x) \in U \times \Omega_0$*

$$C^{-1}|F_n(\Phi(0, x))| \le |F_n(\Phi(t, x))| \le C|F_n(\Phi(0, x))|;$$

(iv) *For $(t, x_1, \dots, x_{i-1})$ fixed, $\Psi_i(t, x_1, \dots, x_{i-1}, \cdot) : \mathbb{K} \to \mathbb{K}$ is bi-Lipschitz and the Lipschitz constants of Ψ_i and Ψ_i^{-1} can be chosen independent of $(t, x_1, \dots, x_{i-1})$;*
(v) Φ *is an arc-wise analytic trivialization of the projection $\Omega \to U$.*

(Note that (iii) of Theorem 4.3.7 implies (iii) Theorem 4.3.5.)

Let us recall after [52] the notion of arc-wise analytic trivialization. First, we need to recall, after [31], the notion of arc-analytic map. Let Y, Z be real analytic spaces. We say that a map $g(z) : Z \to Y$ is *arc-analytic* if for every real analytic arc $z(s) : I \to Z$, $g(z(s))$ is analytic in s. Suppose now that T, Y, Z are $\mathbb{K}$-analytic spaces, T nonsingular. We say that a map $f(t, z) : T \times Z \to Y$ is *arc-wise analytic in t* if it is $\mathbb{K}$-analytic in t and arc-analytic in z, that is for every real analytic arc $z(s) : I \to Z$, the map $f(t, z(s))$ is analytic in both t and s. Note that in the complex analytic case it means that $f(t, z(s))$ can be written as a convergent power series $\sum_{\alpha=(\alpha_1,\dots,\alpha_l)} \sum_k a_{\alpha,k} t^\alpha s^k$ in t complex and s real.

Remark 4.3.8 In the complex analytic case, it is in general impossible to have the complex analytic dependence of Φ on x, even only on the complex arcs. This rigidity property already appears for moduli spaces of elliptic curves.

Suppose now that, moreover, $\pi : Y \to T$ is $\mathbb{K}$-analytic. We say

$$\Phi(t, z) : T \times Z \to Y$$

is an *arc-wise analytic trivialization of* π, see [52, Definition 1.2], if it satisfies the following properties

1. Φ is a subanalytic homeomorphism (semi-algebraic in the algebraic case),
2. Φ is arc-wise analytic in t (in particular it is $\mathbb{K}$-analytic with respect to t),
3. $\pi \circ \Phi(t, z) = t$ for every $(t, z) \in T \times Z$,
4. the inverse of Φ is arc-analytic,
5. there exist $\mathbb{K}$-analytic stratifications $\{Z_i\}$ of Z and $\{Y_i\}$ of Y, such that for each i, $Y_i = \Phi(T \times Z_i)$ and $\Phi_{|T\times Z_i} : T \times Z_i \to Y_i$ is a real analytic diffeomorphism.

The proof of Theorem 4.3.7 follows Varchenko's strategy [73], which we recalled briefly in Sect. 4.3.1. It is technically simpler since in Step 2 of the proof, the extension of the trivialization to the ambient space, is based on Whitney Interpolation Formula, see [78], [52, Appendix I]. The homeomorphism Φ_{i+1} is given by a precise algebraic formula (formula (3.5) of [52]) in terms of the roots of the pseudopolynomial F_{i+1} and Φ_i. (This algebraic formula is a real rational map, it involves, in particular, the square of the distance to the roots of F_{i+1}. There is no such a complex rational formula and no hope, of course, to make Φ_{i+1} complex arc-analytic, because by Hartog's Theorem complex arc-analytic means just complex analytic).

The fact that thus obtained trivialization Φ_{i+1} is arc-wise analytic is proven by induction on i. The inductive step is obtained by a reduction to the Puiseux with parameter theorem, Theorem 4.2.1. Let $x^i(s)$ be a real analytic arc. By the inductive assumption $\Phi_i(t, x^i(s))$ is analytic in t, s. Therefore $P(t, s, x_{i+1}) = F_{i+1}(\Phi_i(t, x^i(s), x_{i+1}))$ is a pseudopolynomial with respect ot x_{i+1} depending analytically on s and t. The main point of the proof is to show that $P(t, s, x_{i+1})$ defines a Zariski equisingular family of plane curve singularities parameterized by t. It follows in essence by the stability of the discriminant by a base change, though technically it is more involved, P is not necessarily reduced even if so is F_{i+1}, see the proof of [52, Theorem 3.1] for more details.

Remark 4.3.9 Arc-wise analytic triviality is, in part, motivated by the relation of Zariski equisingularity and equiresolution of singularities and the theory of blow-analytic equivalence, see the last paragraphs of Sect. 4.5.8.

4.3.3 Whitney Fibering Conjecture

In [52], Theorem 4.3.7 is used to show Whitney fibering conjecture.

Whitney stated this conjecture in the context of the regularity conditions (a) and (b) introduced in [79]. These conditions on stratification imply the topological triviality along each stratum. This trivialization is obtained by the flow of "controlled" vector fields as follows from proofs of Thom-Mather Isotopy Lemmas. By stating the Fibering Conjecture, Whitney wanted a stronger version of triviality, namely that the stratified set locally fibers into submanifolds isomorphic to strata.

Conjecture 4.3.10 (Whitney fibering conjecture, [78] Sect. 9, p. 230) Any analytic subvariety $V \subset U$ (U open in $\mathbb{C}^n$) has a stratification such that each point $p_0 \in V$ has a neighborhood U_0 with a semi-analytic fibration.

By a semi-analytic fibration Whitney meant a local trivialization as in (4.9) that depends complex analytically on the parameter t. Whitney does not specify the dependence on x, besides that he requires it to be continuous and that the existence of such fibration should imply Whitney's regularity conditions (a) and (b) (Whitney's semi-analytic fibration should not be confused with the notion of semi-analytic set introduced about the same time by Łojasiewicz in [35]). Partial results on Whitney fibering conjecture were obtained in [22], and in the smooth case in [43].

Whitney fibering conjecture was proven in [52] in the local complex and real analytic cases and in global algebraic cases by means of Zariski equisingularity and arc-wise analytic triviality. More precisely, by Theorem 4.3.7, every such set has a stratification that locally admits arc-wise analytic trivializations (see the previous subsection) along each stratum. Existence of such trivialization guarantees Whitney's regularity condition (a) but not necessarily condition (b) (see Briançon-Speder example, Example 4.5.3 below). Here we touch for the first time in this survey an interesting and important feature, some properties of Zariski equisingular families depend on the genericity of the system of coordinates $x_1, \ldots, x_n$. In order to guarantee Whitney's condition (b) we consider transverse Zariski equisingularity, see [52, Definition 4.1]. We call Zariski equisingularity *transverse (or transversal)* if at each inductive stage the kernel of the projection $(t, x^i) \to (t, x^{i-1})$ is not included in the tangent cone to $F_i = 0$ at the origin. If we have a family that is transverse Zariski equisingular then, by Theorem 4.3 of [52], the arc-wise analytic trivialization constructed in [52] satisfies additionally the property, called regularity,

$$C^{-1}\|x\| \le \|\Phi(t,x) - \Phi(t,0)\| \le C\|x\|,$$

for a constant C independent of t and x. Geometrically it means that the trivialization Φ preserves the magnitude of the distance to $U \times \{0\}$. It is proven in [52, Proposition 7.4] that this regularity implies Whitney's condtion (b), and even Verdier's condition (w), along $U \times \{0\}$.

We discuss the relation of Zariski equisingularity, the plain one or with extra conditions such as transversality and genericity, and Whitney's conditions in Sect. 4.5.2.

4.3.4 Algebraic Case

In the papers [72, 74] Varchenko considers the families of analytic singularities while the paper [73] deals with the families of affine or projective algebraic varieties. Similarly, the families of algebraic varieties were considered in Sects. 5 and 9 of [52]. The version presented below is stated in [56]. It follows from the proof of the main theorem, Theorem 3.3, of [52], see also Theorems 3.1 and 4.1 of [73] in the complex case, Theorems 6.1 and 6.3 of [73] in the real case, and Proposition 5.2 and Theorem 9.2 of [52] where the global algebraic case is treated.

Theorem 4.3.11 *Let $\mathcal{V}$ be an open connected neighborhood of* $\mathbf{t}$ *in* $\mathbb{K}^r$ *and let* $\mathcal{O}_{\mathcal{V}}$ *denote the ring of* $\mathbb{K}$*-analytic functions on* $\mathcal{V}$*. Let* $t = (t_1, \ldots, t_r)$ *denote the variables in* $\mathcal{V}$ *and let* $x = (x_1, \ldots, x_n)$ *be a set of variables in* $\mathbb{K}^n$*. Suppose that for* $i = k_0, \ldots, n$*, there are given*

$$F_i(t, x^i) = x_i^{d_i} + \sum_{j=1}^{d_i} a_{i-1,j}(t, x^{i-1}) x_i^{d_i - j} \in \mathcal{O}_{\mathcal{V}}[x^i], \tag{4.10}$$

with $d_i > 0$*, such that*

(i) *for every* $i > k_0$*, the first non identically equal to zero generalized discriminant of* $F_i(t, x^{i-1}, x_i)$ *divides* $F_{i-1}(t, x^{i-1})$*.*
(ii) *the first non identically equal to zero generalized discriminant of* F_{k_0} *is independent of* x *and does not vanish on* $\mathcal{V}$*.*

Then, for every $\mathbf{q} \in \mathcal{V}$ *there is a homeomorphism*

$$h_{\mathbf{q}} : \{\mathbf{t}\} \times \mathbb{K}^n \to \{\mathbf{q}\} \times \mathbb{K}^n$$

such that $h_{\mathbf{q}}(V_{\mathbf{t}}) = V_{\mathbf{q}}$*, where for* $\mathbf{q} \in \mathcal{V}$ *we denote* $V_{\mathbf{q}} = \{(\mathbf{q}, \mathbf{x}) \in \mathcal{V} \times \mathbb{K}^n \mid F_n(\mathbf{q}, \mathbf{x}) = 0\}$*.*

Moreover, if $F_n = G_1 \cdots G_s$ *then for every* $j = 1, \ldots, s$

$$h_{\mathbf{q}}\left(G_j^{-1}(0) \cap (\{\mathbf{t}\} \times \mathbb{K}^n)\right) = G_j^{-1}(0) \cap (\{\mathbf{q}\} \times \mathbb{K}^n).$$

Remark 4.3.12 By construction of [52, 72, 73, 75], the homeomorphisms $h_{\mathbf{q}}$ can be obtained by a local topological trivialization. That is there are a neighborhood $\mathcal{W}$ of $\mathbf{t}$ in $\mathbb{K}^r$ and a homeomorphism

$$\Phi : \mathcal{W} \times \mathbb{K}^n \to \mathcal{W} \times \mathbb{K}^n,$$

so that $\Phi(\mathbf{q}, \mathbf{t}, x)) = h_{\mathbf{q}}(\mathbf{t}, x)$. This Φ is triangular of the form (4.6). If we write $\Phi(\mathbf{q}, x) = (\mathbf{q}, \Psi_{\mathbf{q}}(x))$, i.e. as a family of homeomorphisms $h_{\mathbf{q}} = \Psi_{\mathbf{q}} : \mathbb{K}^n \to \mathbb{K}^n$, then, as follows from [52], we may require that:

1. The homeomorphism Φ is subanalytic. In the algebraic case, i.e. if we replace in the assumptions $\mathcal{O}_{\mathcal{V}}$ by the ring of regular or $\mathbb{K}$-valued Nash functions on $\mathcal{V}$, Φ can be chosen semialgebraic.
2. Φ is arc-wise analytic. In particular, each $h_{\mathbf{q}}$ and its inverse $h_{\mathbf{q}}^{-1}$ are arc-analytic.

Remark 4.3.13 If F_i are homogeneous in x, then the functions $\Psi_{\mathbf{q}}$ satisfy, by construction,

$$\forall \lambda \in \mathbb{K}^*, \forall x \in \mathbb{K}^n \quad \Psi_{\mathbf{q}}(\lambda x) = \lambda \Psi_{\mathbf{q}}(x).$$

Hence if we define $\mathbb{P}(V_{\mathbf{q}}) = \{(\mathbf{q}, \mathbf{x}) \in \mathcal{V} \times \mathbb{P}^n_{\mathbb{K}} \mid F_n(\mathbf{q}, \mathbf{x}) = 0\}$, the homeomorphism $h_{\mathbf{q}}$ induces a homeomorphism between $\mathbb{P}(V_{\mathbf{t}})$ and $\mathbb{P}(V_{\mathbf{q}})$.

4.3.5 Principle of Generic Topological Equisingularity

Varchenko applies Theorem 4.3.5 to establish in [73, Sects. 5 and 6] generic topological equisingularity for families of real or complex, affine or projective, algebraic sets. The principle of generic topological equisingularity says that in an algebraic family X_t of algebraic sets, parameterized by $t \in T$, where T is not necessarily nonsingular, irreducible algebraic variety, there is a proper algebraic subset Y of T such that the fibers X_t have constant topological type for t from each connected component of $T \setminus Y$. In the complex algebraic case $T \setminus Y$ is connected by the irreducibility of T, in the real algebraic case it has finitely many connected components. For analytic spaces or sets, a similar principle holds locally. In both analytic and algebraic cases, the results give actually local topological triviality of the family X_t over $T \setminus Y$. We give examples of possible precise statements below. Let us first make some remarks.

Generic topological equisingularity can be proven, in general, either by Zariski equisingularity or by stratification theory using Whitney stratification and Thom-Mather Isotopy Lemmas, see [70, Theorem 4.2.17]. Whitney stratification approach is independent of the choice of coordinates and simple to define. But the trivializations obtained by this method are not explicit since they are flows of "controlled" vector fields. Even if such vector fields can be chosen subanalytic or semialgebraic, not much can be said about the regularity of their flows. Zariski's equisingularity method is more explicit and in a way constructive. It uses the actual equations and coordinate systems. This can be considered either as a drawback or as an advantage. The trivializations can be chosen subanalytic (semialgebraic in the algebraic case), as shown in [52]. Actually, the trivializations are given there by explicit formulas in terms of the coefficients of the polynomials and their roots.

In the real case, the triangulation provides another method for proving generic topological triviality. The classical triangulation procedures are based on a similar construction as Zariski equisingularity, i.e. successive co-rank 1 projections and their discriminants. For instance, a beautiful result on semialgebraic triviality was shown by Hardt using this approach in [23]. For a fairly complete account on this approach, the reader can consult [7] and the references therein.

It is fairly straightforward to apply Theorem 4.3.5 to obtain generic topological equisingularity for families of hypersurfaces. In the case of varieties and spaces of arbitrary codimension the argument goes as follows. If $F = G_1 \cdots G_k$, then, under the assumptions of Lemma 4.3.6, for x^{n-1} fixed, the number of roots of each $G_j(\Phi_{n-1}(t, x^{n-1}), x_n) = 0$ is independent of t, see Lemma 2.2 of [73] or Proposition 3.6 of [52]. In particular Φ trivializes not only $V(F) = F^{-1}(0)$ but also each of $V(G_j) = G_j^{-1}(0)$. Thus [73] implies the following.

Theorem 4.3.14 *If $F = G_1 \cdots G_k$ then for each $j = 1, \ldots, k$, the homeomorphisms Φ of Theorem 4.3.5 satisfies $\Phi(T \times (V(G_j) \cap \Omega_0)) = V(G_j) \cap \Omega$, where $V_t(G_j) = (G_j^{-1}(0) \cap (\{t\} \times \mathbb{K}^n)$. In particular Φ trivializes $\{G_1 = \cdots = G_k = 0\}$.*

Now let us give two possible exact statements for this principle taken from [52]. Note that they give not only generic topological equisingularity but much stronger generic arc-wise analytic triviality.

Theorem 4.3.15 *([52, Theorem 9.3],* cf. *[73, Theorems 5.2 and 6.4]) Let T be an algebraic variety (over $\mathbb{K}$) and let $\mathcal{X} = \{X_k\}$ be a finite family of algebraic subsets $T \times \mathbb{P}^{n-1}_{\mathbb{K}}$. Then there exists an algebraic stratification $\mathcal{S}$ of T such that for every stratum S and for every $t_0 \in S$ there is a neighborhood U of t_0 in S and a semialgebraic arc-wise analytic trivialization of π, preserving each set of the family $\mathcal{X}$,*

$$\Phi : U \times \mathbb{P}^{n-1}_{\mathbb{K}} \to \pi^{-1}(U), \tag{4.11}$$

$\Phi(t, x) = (t, \Psi(t, x))$, $\Phi(t_0, x) = (t_0, x)$, where $\pi : T \times \mathbb{P}^{n-1}_{\mathbb{K}} \to T$ denotes the projection.

Theorem 4.3.16 *([52, Theorem 6.2]) Let T be a $\mathbb{K}$-analytic space, $U \subset \mathbb{K}^n$ an open neighborhood of the origin, $\pi : T \times U \to T$ the standard projection, and let $\mathcal{X} = \{X_k\}$ be a finite family of $\mathbb{K}$-analytic subsets of $T \times U$. Let $t_0 \in T$. Then there exist an open neighborhood T' of t_0 in T and a proper $\mathbb{K}$-analytic subset $Z \subset T'$, containing* Sing T'*, such that for every $t \in T' \setminus Z$, $\mathcal{X}$ is regularly arc-wise analytically equisingular along $T \times \{0\}$ at t.*

Moreover, there is an analytic stratification of an open neighborhood of t_0 in T such that for every stratum S and every $t \in S$, $\mathcal{X}$ is regularly arc-wise analytic equisingular along $S \times \{0\}$ at t.

In the above theorem by saying that $\mathcal{X}$ is *arc-wise analytically equisingular along $T \times \{0\}$ at $t \in$* Reg T we mean that there are neighborhoods B of t in Reg T and Ω of $(t, 0)$ in $T \times \mathbb{K}^n$, and an arc-wise analytic trivialization $\Phi : B \times \Omega_t \to \Omega$, where $\Omega_t = \Omega \cap \pi^{-1}(t)$, such that $\Phi(B \times \{0\}) = B \times \{0\}$ and for every k, $\Phi(T \times X_{k,t}) = X_k$, where $X_{k,t} = X_k \cap \pi^{-1}(t)$. We say that $\mathcal{X}$ is *regularly arc-wise analytically equisingular along $T \times \{0\}$ at $t \in T$* if, moreover, Φ preserves, up to a constant, the distance to $T \times \{0\}$, as we explained at the end of Sect. 4.3.3. The latter property is related to Whitney's conditions, see Sect. 7 of [52].

4.3.6 Zariski's Theorem on the Fundamental Group

Varchenko in [73] applies topological triviality of Zariski equisingular projective algebraic varieties to prove Zariski's theorem on the fundamental group of the complement. This theorem says that the fundamental group of the complement $\mathbb{P}^n_{\mathbb{C}} \setminus V_{n-1}$

of a complex projective hypersurface $V_{n-1}, n > 2$, coincides with the corresponding group obtained from a general hyperplane section.

This theorem was announced by Zariski in [81], but the proof published in it is not considered as complete. Another complete proof of this theorem, different from the one of Varchenko, is given in [20, 21].

4.3.7 General Position Theorem

In [40] Zariski equisingular families of affine or projective algebraic varieties are used, together with Whitney interpolation, to prove stratified general position and transversality theorems for semialgebraic subsets of algebraic stratifications.

In classical algebraic topology, general position of chains was used by Lefschetz to define the intersection pairing on the homology of a manifold. This approach is based on a possibility of moving a "subvariety" Z of a C^∞ manifold M, by a family of diffeomorphisms, so that the image Z becomes transverse to a given another "subvariety" W of M. This principle was made precise by Trotman [71] and, independently, by Goresky [18]. They proved that by a diffeomorphism one can put in a stratified general position two Whitney stratified closed subsets Z and W of M.

The main theorem of [40] is expressed in terms of a submersive family of diffeomorphisms introduced in [17, I.1.3.5]. Let T and M be C^∞ manifolds and let $\Psi : T \times M \to M$ be a C^∞ map. Consider $\Psi_t : M \to M$, $\Psi_t(x) = \Psi(t, x)$, and $\Psi^x : T \to M$, $\Psi^x(t) = \Psi(t, x)$. We say Ψ is *a family of diffeomorphisms* if for all $t \in T$ the map Ψ_t is a diffeomorphism. The family Ψ is called *submersive* if, for each $(t, x) \in T \times M$, the differential $D\Psi^x$ is surjective. By Theorem [17, I.1.3.6], if $\Psi : T \times M \to M$ is submersive and both Z and W are Whitney stratified closed subsets of M then the set of $t \in T$ such that $\Psi_t(Z)$ is transverse to W is dense in T and open provided Z is compact. A good example of a submersive family is a transitive action $\Psi : G \times M \to M$ of a Lie group. Note that in this case Theorem [17, I.1.3.6] gives characteristic 0 part of Kleiman's transversality of a general translate theorem [28]. For a stratified set $X = \bigsqcup S_i$ we say that $\Psi : T \times X \to X$ is *a stratified submersive family of diffeomorphisms* if for each stratum S_j, we have $\Psi(T \times S_j) \subset S_j$, and the map $\Psi : T \times S_j \to S_j$ is a submersive family of diffeomorphisms.

In algebraic geometry the intersection of cycles can be defined via a moving lemma that allows to move the cycle of nonsingular varieties, see [15, Sect. 11.4]. But there is no moving lemma nor algebraic general position theorem for singular varieties. In the original construction of Intersection Cohomology [16] in order to define the intersection pairing on singular complex algebraic varieties equipped with a Whitney stratification Goresky and MacPherson used a piecewise linear general position theorem of McCrory [38]. The main theorem of [40] shows the existence of such stratified submersive family in the arc-wise analytic category of [52], see also Sect. 4.3.2.

Theorem 4.3.17 *([40, Theorem 1.1]) Let $\mathcal{V} = \{V_i\}$ be a finite family of algebraic subsets of projective space $\mathbb{P}^n_{\mathbb{K}}$. There exists an algebraic stratification $\mathcal{S} = \{S_j\}$ of $\mathbb{P}^n_{\mathbb{K}}$ compatible with each V_i and a semialgebraic stratified submersive family of diffeomorphisms $\Psi : U \times \mathbb{P}^n_{\mathbb{K}} \to \mathbb{P}^n_{\mathbb{K}}$, where U is an open neighborhood of the origin in $\mathbb{K}^{n+1}$, such that $\Psi(0, x) = x$ for all $x \in \mathbb{P}^n_{\mathbb{K}}$. Moreover, the map $\Phi : U \times \mathbb{P}^n_{\mathbb{K}} \to U \times \mathbb{P}^n_{\mathbb{K}}$, $\Phi(t, x) = (t, \Psi(t, x))$, is an arc-wise analytic trivialization of the projection $U \times \mathbb{P}^n_{\mathbb{K}} \to U$.*

A similar result holds for affine varieties; see [40, Corollary 3.2].

The proof of Theorem 4.3.17 is rather tricky. It uses the formulas used in [52] in Step 2 of the construction topological trivialization of Zariski equisingular families, see Sect. 4.3.1. These formulas are based on Whitney interpolation and can be perturbed by introducing complex parameters, these are $t \in U$ of the theorem. The whole construction is applied to a trivial family, that is to the product $U \times \mathbb{P}^n_{\mathbb{K}}$, thus producing a non-trivial arc-wise analytic trivialization of a trivial family.

Theorem 4.3.17 implies the general position in terms of the expected dimension of the intersection and the general transversality. The general position in terms of dimension is exactly what is needed to define the intersection pairing for the intersection homology, cf. [16]. The general position in terms of dimension can be expressed as follows, the dimension means the real dimension since we consider semialgebraic sets.

Corollary 4.3.18 *([40, Proposition 1.3]) Let $\Psi : U \times \mathbb{P}^n_{\mathbb{K}} \to \mathbb{P}^n$ be a stratified family as in Theorem 4.3.17, and let $\mathcal{S}$ be the associated algebraic stratification of $\mathbb{P}^n_{\mathbb{K}}$. Let Z and W be semialgebraic subsets of $\mathbb{P}^n_{\mathbb{K}}$. There is an open dense semialgebraic subset U' of U such that, for all $t \in U'$ and all strata $S \in \mathcal{S}$,*

$$\dim(Z \cap \Psi_t^{-1}(W) \cap S) \le \dim(Z \cap S) + \dim(W \cap S) - \dim S.$$

If $\mathcal{S}$ is a stratification of a semialgebraic set X, and $\mathcal{T}$ is a stratification of a semialgebraic subset Y of X, then $(Y, \mathcal{T})$ is a *substratified object* of $(X, \mathcal{S})$ if each stratum of $\mathcal{T}$ is contained in a stratum of $\mathcal{S}$. Two substratified objects $(Z, \mathcal{A})$ and $(W, \mathcal{B})$ of $(X, \mathcal{S})$ are *transverse* in $(X, \mathcal{S})$ if, for every pair of strata $A \in \mathcal{A}$ and $B \in \mathcal{B}$ such that A and B are contained in the same stratum $S \in \mathcal{S}$, the manifolds A and B are transverse in S.

Corollary 4.3.19 *([40, Proposition 1.5]) Let $\Psi : U \times \mathbb{P}^n \to \mathbb{P}^n$ be a stratified family as in Theorem 4.3.17, and let $\mathcal{S}$ be the associated algebraic stratification of $\mathbb{P}^n$. Let Z and W be semialgebraic subsets of $\mathbb{P}^n$, with semialgebraic stratifications $\mathcal{A}$ of Z and $\mathcal{B}$ of W such that $(Z, \mathcal{A})$ and $(W, \mathcal{B})$ are substratified objects of $(\mathbb{P}^n, \mathcal{S})$. There is an open dense semialgebraic subset U' of U such that, for all $t \in U'$, $(Z, \mathcal{A})$ is transverse to $\Psi_t^{-1}(W, \mathcal{B})$ in $(\mathbb{P}^n, \mathcal{S})$.*

In a recent paper [39] Corollary 4.3.19 is used to define an intersection pairing for *real intersection homology*, an analog of intersection homology for real algebraic varieties.

4.4 Construction of Equisingular Deformations

Let f be either a polynomial or the germ of an analytic function, and let $V = V(f)$ denote the zero set of f. We explain below how to construct Zariski equisingular deformations of V (or more precisely of its equation f). The idea comes from [41], where the local complex analytic case was considered. We begin with the global polynomial case as considered in [56] since it is conceptually simpler and does not require Artin approximation. Note also that this method can be applied as well to construct equisingular deformations of sets given by a system of several equations as was explained at the end of Sect. 4.3.5.

4.4.1 *Global Polynomial Case*

Given an algebraic subset V of $\mathbb{K}^n$ and let the polynomials $g_1,\dots, g_s \in \mathbb{K}[x]$ generate the ideal defining V. Let

$$g_i = \sum_{\alpha \in \mathbb{N}^n} g_{i,\alpha} x^{\alpha}.$$

In general, a deformation of the $g_{i,\alpha}$, even arbitrarily small, destroys the topological structure of V due to the presence of singularities (and "singularities at infinity" in the global case). In this method we construct a finite number of constraints satisfied by the coefficients $g_{i,\alpha}$, these are the Eqs. (4.15), (4.16), (4.18), (4.19) and the in Eqs. (4.13) and (4.17) below, that satisfy the following property. Any deformation $t \mapsto g_{i,\alpha}(t)$ with $g_{i,\alpha}(0) = g_{i,\alpha}$, that satisfies the same constraints (4.13) and (4.15)–(4.19) is, by construction, Zariski equisingular. In particular any such deformation is topologically trivial. Moreover, the entries of (4.13) and (4.15)–(4.19) are rational functions in $g_{i,\alpha}$ with rational coefficients, that is they belong to $\mathbb{Q}(u_{i,\alpha})$, for some new indeterminates $u_{i,\alpha}$.

Let us fix a finite set of coefficients $g_{i,\alpha} \in \mathbb{K}$ that contains all nonzero of them. In what follows we will perturb only these coefficients and keep all the other equal to zero.

After a linear change with rational coefficients of coordinates x we can assume that

$$g_i = c_i x_n^{p_i} + \sum_{j=1}^{p_i} b_{n-1,r,j}(x^{n-1}) x_n^{p_i - j} = \sum_{\beta \in \mathbb{N}^n} a_{n,i,\beta} x^{\beta}, \quad \forall i = 1, \dots, s, \tag{4.12}$$

with

$$c_i \neq 0, \qquad i = 1, \dots, s. \tag{4.13}$$

By multiplying each g_i by $1/c_i$ we can assume that $c_i = 1$ for every i. Denote by $f = f_n$ the product of the g_i and by a_n the vector of coefficients $a_{n,i,\beta}$. The entries of a_n are rational functions in the original $g_{i,\alpha}$ (i.e. before the linear change of coordinates x) with rational coefficients, say

$$a_n = A_n(g_{i,\alpha}), \tag{4.14}$$

where $A_n = (A_{n,i,\beta})_{i,\beta} \in \mathbb{Q}(u_{i,\alpha})^{N_n}$ for some integer $N_n > 0$. Let the integer l_n be defined by

$$D_{n,l_n}(a_n) \not\equiv 0 \text{ and } D_{n,l}(a_n) \equiv 0, \quad \forall l < l_n,$$

where $D_{n,l}$ denotes the l-th generalized discriminant of f_n, see Appendix. After a new linear change of coordinates x^{n-1} with rational coefficients, $D_{n,l_n}(a_n) = e_{n-1} f_{n-1}$ with $e_{n-1} \neq 0$ and

$$f_{n-1} = \sum_{\beta \in \mathbb{N}^n} a_{n-1,\beta} x^\beta = x_{n-1}^{d_{n-1}} + \sum_{j=1}^{d_{n-1}} b_{n-2,j}(x^{n-2}) x_{n-1}^{d_{n-1}-j}$$

for some constants $a_{n-1,\beta}$ and polynomials $b_{n-2,j}$. We repeat this construction and define recursively a sequence of polynomials $f_j(x^j)$, monic in x_j, such that

$$D_{j+1,l_{j+1}}(a_{j+1}) = e_j \left(x_j^{d_j} + \sum_{k=1}^{d_j} b_{j-1,k}(x^{j-1}) x_j^{d_j-k} \right) = e_j \left(\sum_{\beta \in \mathbb{N}^n} a_{j,\beta} x^\beta \right) = e_j f_j \tag{4.15}$$

is the first non identically equal to zero generalized discriminant of f_{j+1} and a_j denotes the vector of coordinates $a_{j,\beta}$. This way we get a system of equations

$$D_{j+1,l}(a_{j+1}) \equiv 0 \;\; \forall l < l_{j+1}, \tag{4.16}$$

and inequations

$$e_j \neq 0, \tag{4.17}$$

for $j = n, n-1, \ldots, k_0$, until we get

$$f_{k_0} = 1 \text{ for some } k_0 \geq 0. \tag{4.18}$$

By (4.13), (4.14) and (4.15) the entries of the c_i, a_k and e_j are rational functions in the $g_{i,\alpha}$ with rational coefficients, let us say

$$c_i = C_i(g_{i,\alpha}), \;\; a_k = A_k(g_{i,\alpha}), \;\; e_j = E_j(g_{i,\alpha}), \tag{4.19}$$

for some $C_i \in \mathbb{Q}(u_{i,\alpha})$, $A_k \in \mathbb{Q}(u_{i,\alpha})^{N_k}$ and $E_j \in \mathbb{Q}(u_{i,\alpha})$. Thus (4.13) and (4.15)–(4.19) are equations and inequations, with rational coefficients, on the original coefficients $g_{i,\alpha}$.

Let $\mathcal{V}$ be an open connected neighborhood of a point $\mathbf{t} \in \mathbb{K}^l$ and let $\mathcal{O}_{\mathcal{V}}$ denote the ring of $\mathbb{K}$-analytic functions on $\mathcal{V}$. Suppose that $g_{i,\alpha}(t) \in \mathcal{O}_{\mathcal{V}}$, where $t \in \mathcal{V}$, satisfy $g_{i,\alpha} = g_{i,\alpha}(\mathbf{t})$. For $t \in \mathcal{V}$ and $i = 1, \dots, s$, we define

$$\tilde{g}_i(t, x) := \sum_{\alpha \in \mathbb{N}^n} g_{i,\alpha}(t) x^\alpha.$$

We claim that if the $g_{i,\alpha}(t)$ satisfy the identities and the inequations (4.13) and (4.15)–(4.19), then the family $t \to \{\tilde{g}_1(t, x) = \cdots = \tilde{g}_s(t, x) = 0\}$ is topologically trivial for t in a small neighborhood of $\mathbf{t}$ in $\mathcal{V}$. For this we construct a system $F_j(t, x^j)$ satisfying the assumptions of Theorem 4.3.11. We set

$$F_n(t, x) = \prod_{i=1}^{s} G_i(t, x), \text{ where } G_i(t, x) = \sum_{\beta \in \mathbb{N}^n} A_{n,i,\beta}(g_{i,\alpha}(t)) x^\beta, i = 1, \dots, s,$$

and $A_n = (A_{n,i,\beta})_{i,\beta} \in \mathbb{Q}(u_{i,\alpha})^{N_n}$ given in (4.14). Similarly for $j = k_0, \dots, n-1$ we set

$$F_j(t, x^j) = \sum_{\beta \in \mathbb{N}^j} A_{j,\beta}(g_{i,\alpha}(t)) x^\beta. \tag{4.20}$$

Note that $G_i(\mathbf{t}, x)$ concide with g_i after the linear change of coordinates made during the construction. It is clear from the above construction that the family $(F_j(t, x^j))$ satisfies the assumptions of Theorem 4.3.11. Let us summarize it in the following.

Theorem 4.4.1 *Suppose that $g_{i,\alpha}(t)$ satisfy the identities and the inequations* (4.13) *and* (4.15)–(4.19). *Then $F_n(t, x)$ defines a Zariski equisingular family with respect to the parameter t.*

4.4.2 Application: Algebraic Sets are Homeomorphic to Algebraic Sets Defined Over Algebraic Number Fields

The following result was proven in [56].

Theorem 4.4.2 *Let $V \subset \mathbb{K}^n$ (resp. $V \subset \mathbb{P}^n_{\mathbb{K}}$) be an affine (resp. projective) algebraic set, where $\mathbb{K} = \mathbb{R}$ or $\mathbb{C}$. Then there exist an affine (resp. projective) algebraic set $W \subset \mathbb{K}^n$ (resp. $W \subset \mathbb{P}^n_{\mathbb{K}}$) and a homeomorphism $h : \mathbb{K}^n \longrightarrow \mathbb{K}^n$ (resp. $h : \mathbb{P}^n_{\mathbb{K}} \longrightarrow \mathbb{P}^n_{\mathbb{K}}$) such that:*

(i) the homeomorphism h maps V onto W,
(ii) W is defined by polynomial equations with coefficients in $\overline{\mathbb{Q}} \cap \mathbb{K}$,
(iii) the variety W is obtained from V by a Zariski equisingular deformation. In particular the homeomorphism h can be chosen semialgebraic and arc-analytic.

Suppose, as in the previous section, that the ideal defining V is generated by the polynomials $g_1,\ldots, g_s \in \mathbb{K}[x]$. In order to prove Theorem 4.4.2 one constructs in [56] a deformation $t \mapsto g_{i,\alpha}(t)$ of the coefficients $g_{i,\alpha} \in \mathbb{K}$ of the g_i that preserves all polynomial relations over $\mathbb{Q}$ satisfied by these coefficients. Therefore this deformation preserves the identities (4.15), (4.16), (4.18) and (4.19). If it is sufficiently small the inequations (4.13), (4.17) are also preserved and, by Theorem 4.4.1, the deformation is equisingular in the sense of Zariski.

This construction is particularly simple if the field extension k of $\mathbb{Q}$ generated by the coefficients $g_{i,\alpha}$ is a purely transcendental extension of $\mathbb{Q}$. For the general case we refer the reader to [56]. Thus assume that $k = \mathbb{Q}(\mathbf{t}_1, \ldots, \mathbf{t}_r)$, where the $\mathbf{t}_i \in \mathbb{K}$ are algebraically independent over $\mathbb{Q}$. Then there are rational functions $g_{i,\alpha}(t) \in \mathbb{Q}(t)$, $t = (t_1, \ldots, t_r)$, such that $g_{i,\alpha} = g_{i,\alpha}(\mathbf{t})$. Let $\mathcal{V}$ be a neighborhood of $\mathbf{t} = (\mathbf{t}_1, \ldots, \mathbf{t}_r)$ that does not contain the poles of the $g_{i,\alpha}(t)$. Since $\mathbf{t}_i \in \mathbb{K}$ are algebraically independent any polynomial relation with coefficients in $\mathbb{Q}$, satisfied by $g_{i,\alpha} = g_{i,\alpha}(\mathbf{t})$, is also satisfied by $g_{i,\alpha}(t)$. In particular, $g_{i,\alpha}(t)$ satisfy the identities (4.15), (4.16), (4.18) and (4.19) as we wanted. Choose $\mathbf{q} \in (\overline{\mathbb{Q}})^r \cap \mathcal{V}$ sufficiently close to $\mathbf{t}$. Then all $g_{i,\alpha}(\mathbf{q}) \in \overline{\mathbb{Q}}$. Therefore the family (F_j), defined by (4.20), satisfies the hypothesis of Theorem 4.3.11 and the hypersurfaces $X_0 := \{F_n(\mathbf{q}, x) = 0\}$ and $X_1 := \{F_n(\mathbf{t}, x) = 0\}$ are homeomorphic. Moreover, thus constructed homeomorphism maps every component of X_0 defined by $G_i(\mathbf{q}, x) = 0$ onto the component of X_1 defined by $G_i(\mathbf{t}, x) = 0$, as in Theorem 4.3.14. This proves that the algebraic variety $V = \{g_1 = \cdots = g_s = 0\}$ is homeomorphic to the algebraic variety $\{G_1(\mathbf{q}, x) = \cdots = G_s(\mathbf{q}, x) = 0\}$ defined by polynomial equations over $\overline{\mathbb{Q}}$.

A result analogous to Theorem 4.4.2 in the local case, for singularities of analytic spaces or analytic functions was proven by G. Rond in [61].

Remark 4.4.3 Note that, by the above proof, in the special case when k is a purely transcendental extension of $\mathbb{Q}$, we may replace, in the statement of Theorem 4.4.2, $\overline{\mathbb{Q}}$ by $\mathbb{Q}$ if $\mathbb{K} = \mathbb{R}$, resp. $\mathbb{Q}[i]$ if $\mathbb{K} = \mathbb{C}$. In general, this is an open problem, whether every algebraic variety is homeomorphic to a variety defined over $\mathbb{Q}$, resp. $\mathbb{Q}[i]$. In [67] B. Teissier gave an example of a complex analytic surface singularity defined over $\mathbb{Q}(\sqrt{5})$, which is not Whitney equisingular to any singularity defined over $\mathbb{Q}$.

? Open problem 1.

Is every complex algebraic variety homeomorphic to a variety defined over $\mathbb{Q}[i]$? Is every real algebraic variety homeomorphic to a variety defined over $\mathbb{Q}$?

? Open problem 2.

Is every complex analytic set germ homeomorphic to a set germ defined over $\mathbb{Q}[i]$? Is every real analytic set germ homeomorphic to a set germ defined over $\mathbb{Q}$?

4.4.3 Analytic Case

Suppose now that V is the germ at the origin of an analytic subset of $\mathbb{K}^n$ and let $g_1,\ldots, g_s \in \mathbb{K}\{x\}$ generate the ideal defining V. We describe below, following [41], the construction of Zariski equisingular deformations of V. The main idea is similar to that of Sect. 4.4.1, that is to use the discriminants of successive linear projections to construct a system of "constrains", that is equations and inequations satisfied by the g_i. These are the equations and inequations (4.22), (4.23) defined below. Then any deformation of the g_i that satisfies the same constraints is Zariski equisingular. The main difference comes from the fact that now we are not going to use the coefficients of the g_i, since there are infinitely many of them. Instead we treat the equation of (4.22), (4.23), as a system of equations on the functions $u_i(x^i)$, $a_{i,j}(x^i)$, that is the coefficients of these successive discriminants.

Let us consider a finite set of distinguished polynomials $g_1, \ldots, g_s \in \mathbb{K}\{x\}$:

$$g_i(x) = x_n^{r_i} + \sum_{j=1}^{r_i} a_{n-1,i,j}(x^{n-1})x_n^{r_i-j},$$

i.e. we suppose $a_{n-1,i,j}(0) = 0$ for all i, j. Arrange $a_{n-1,i,j}$ in a row vector $a_{n-1} \in \mathbb{K}\{x^{n-1}\}^{p_n}$, where $p_n := \sum_i r_i$. Let f_n be the product of the g_i's. The generalized discriminants $D_{n,i}$ of f_n are polynomials in the entries of a_{n-1}. Let l_n be a positive integer such that

$$D_{n,l}(a_{n-1}) \equiv 0 \qquad l < l_n, \tag{4.21}$$

and $D_{n,l_n}(a_{n-1}) \not\equiv 0$. Then, after a linear change of coordinates x^{n-1}, by the Weierstrass Preparation Theorem, we may write

$$D_{n,l_n}(a_{n-1}) = u_{n-1}(x^{n-1})\Big(x_{n-1}^{p_{n-1}} + \sum_{j=1}^{p_{n-1}} a_{n-2,j}(x^{n-2})x_{n-1}^{p_{n-1}-j}\Big),$$

where $u_{n-1}(0) \neq 0$ and for all j, $a_{n-2,j}(0) = 0$. We denote

$$f_{n-1} = x_{n-1}^{p_{n-1}} + \sum_{j=1}^{p_{n-1}} a_{n-2,j}(x^{n-2})x_{n-1}^{p_{n-1}-j}$$

and the vector of its coefficients $a_{n-2,j}$ by $a_{n-2} \in \mathbb{K}\{x^{n-2}\}^{p_{n-1}}$. Let l_{n-1} be the positive integer such that the first $l_{n-1} - 1$ generalized discriminants $D_{n-1,l}$ of f_{n-1} are identically zero and $D_{n-1,l_{n-1}}$ is not. Then again we define $f_{n-2}(x^{n-2})$ as the Weierstrass polynomial associated to $D_{n-1,l_{n-1}}$.

We continue this construction and define a sequence of pseudopolynomials $f_i(x^i)$, $i = 1, \dots, n-1$, such that $f_i = x_i^{p_i} + \sum_{j=1}^{p_i} a_{i-1,j}(x^{i-1})x_i^{p_i-j}$ is the Weierstrass polynomial associated to the first non-identically zero generalized discriminant $D_{i+1,l_{i+1}}(a_i)$ of f_{i+1}, where we denote in general $a_i = (a_{i,1}, \dots, a_{i,p_{i+1}})$,

$$D_{i+1,l_{i+1}}(a_i) = u_i(x^i)\Big(x_i^{p_i} + \sum_{j=1}^{p_i} a_{i-1,j}(x^{i-1})x_i^{p_i-j}\Big), \quad i = 0, ..., n-1. \tag{4.22}$$

Thus, for $i = 0, ..., n-1$, the vector of functions a_i satisfies

$$D_{i+1,l}(a_i) \equiv 0 \text{ for } l < l_{i+1}, \quad D_{i+1,l_{i+1}}(a_i) \not\equiv 0. \tag{4.23}$$

This means in particular that

$$D_{1,k}(a_0) \equiv 0 \quad \text{for } l < l_1 \text{ and } D_{1,l_1}(a_0) \equiv u_0,$$

where u_0 is a non-zero constant.

The following theorem follows from the construction of the family $u_i(t, x^i)$, $a_{i,j}(t, x^i)$.

Theorem 4.4.4 *Suppose that we extend all function $u_i(x^i)$, $a_{i,j}(x^i)$ to analytic families $u_i(t, x^i), a_{i,j}(t, x^i) \in \mathbb{K}\{t, x\}$, $u_i(0, x^i) = u_i(x^i)$, $a_{i,j}(0, x^i) = a_{i,j}(x^i)$, where $t \in \mathbb{K}^l$ is considered as a parameter. If the identities and the inequations of* (4.22), (4.23) *are still satisfied by these extensions $u_i(t, x^i), a_{i,j}(t, x^i)$ then the family $f_n(t, x) = 0$ is Zariski equisingular.*

4.4.4 Application: Analytic Set Germs are Homeomorphic to Algebraic Ones

The problem of approximation of analytic objects (sets or mappings) by algebraic ones has a long history, see e.g. [8] and the bibliography therein. In particular, several results were obtained in the case of isolated singularities. The local topological algebraicity of analytic set germs, in the general set-up, was first established in [41] by Mostowski. Given an analytic set germ $(V, 0) \subset (\mathbb{K}^n, 0)$, Mostowski shows the existence of a local homeomorphism $\tilde{h} : (\mathbb{K}^{2n+1}, 0) \to (\mathbb{K}^{2n+1}, 0)$ such that, after the embedding $(V, 0) \subset (\mathbb{K}^n, 0) \subset (\mathbb{K}^{2n+1}, 0)$, the image $\tilde{h}(V)$ is algebraic. It is easy to see that Mostowski's proof together with Theorem 2 of [8] gives the following result.

Theorem 4.4.5 *Let* $\mathbb{K} = \mathbb{R}$ *or* $\mathbb{C}$. *Let* $(V, 0) \subset (\mathbb{K}^n, 0)$ *be an analytic germ. Then there is a homeomorphism* $h : (\mathbb{K}^n, 0) \to (\mathbb{K}^n, 0)$ *such that* $h(V)$ *is the germ of an algebraic subset of* $\mathbb{K}^n$.

We remark that in [41] Mostowski states his results only for $\mathbb{K} = \mathbb{R}$ but his proof also works for $\mathbb{K} = \mathbb{C}$.

The proof of Theorem 4.4.5 is, in principle, similar to the one of Theorem 4.4.2, but is techniquely much more demanding. The main idea is to use Theorem 4.4.4 and deform analytic solutions of (4.22) and (4.23) to algebraic ones. Here by algebraic solutions we mean given by the ones defined by algebraic power series (an algebraic power series is a power series algebraic over $\mathbb{K}[x_1, ..., x_n]$—for example the power series $u(x)$ such that $u(0) = 1$ and $u(x)^2 = 1 + x$). Recall that the Artin approximation theorem states that convergent power series solutions of algebraic equations can be approximated by algebraic power series solutions. Clearly, we need a stronger result, not only an approximation but also a parameterized deformation from the old, convergent solutions to the new, algebraic ones. This is provided by Płoski's version of Artin approximation, see [59]. Finally, in order to apply Theorem 4.4.4 we need the nested Artin approximation, i.e. solutions $u_i(t, x^i), a_{i,j}(t, x^i) \in \mathbb{K}\{t, x^i\}$, of (4.22) and (4.23), that depend only on $x_1, \ldots, x_i$ and not on x_k for $k > i$. Nested Artin Approximation Theorem follows from the Néron Desingularization, proven by Popescu [60], and was not available at the time Mostowski's paper [41] was written. Instead, Mostowski proposes a recursive construction of the system of Eqs. (4.22) and (4.23) giving Zariski equisingularity conditions by local linear changes of coordinates and, at the same time, step by step, provides the deformation-approximation by algebraic power series solutions following the recipe given in [59].

One may shorten considerably Mostowski's construction using a stronger result, the nested variant of Płoski's version Artin Approximation. This is done in [6], where such Nested Artin-Płoski-Popescu Approximation Theorem is proven. This theorem was used in [6] to deform $u_i(x^i), a_{i,j}(x^i)$ to algebraic power series solutions of (4.22) and (4.23). Furthermore, a result of Bochnak-Kucharz [8], based on Artin-Mazur Theorem of [2], allows one to approximate the zeros of algebraic power series (or equivalently germs of Nash functions) by the zeros of polynomial functions.

A stronger version of Theorem 4.4.5 was given in [5] where it was shown that such a homeomorphism h can be found with any prescribed order of tangency at the origin.

? Open problem 3.

What is the best level of regularity of homeomorphisms for which the statement of Theorem 4.4.5 holds? It is known, for instance, that Theorem 4.4.5 is no longer true if one replaces "homeomorphism" by "diffeomorphism", for examples see [8] and the last section of [6]. It is not known whether Theorem 4.4.5 holds true if one requires the homeomorphism h to be bi-Lipschitz.

4.4.5 Equisingularity of Function Germs

Zariski Equisingularity can also be used to construct topologically trivial deformations of analytic map germs, see [73]. Let us consider first the case of functions as studied in [6], that is the mappings with values in $\mathbb{K}$. Given a family $g_t(y) = g(t, y_1, \ldots, y_{n-1})$ of such germs parameterized by $t \in (T, t_0)$ We consider the associated family of set germs defined by the graph of g, the zero set of $F(t, x_1, \ldots, x_n) := x_1 - g(t, x_2, \ldots, x_n)$, and construct a topological trivialization h_t of $V = V(F)$ that does not move the variable x_1

$$h_t(x_1, \ldots x_n) = (x_1, \hat{h}_t(x_1, x_2, \ldots, x_n)) \tag{4.24}$$

so that $V \ni (t_0, x)$ if and only if $(t, h_t(x)) \in V$. Set $\sigma_t(y) := \hat{h}_t(g(y), y)$. Then

$$g_t \circ \sigma_t = g_{t_0},$$

that is g_t and g_{t_0} are right (i.e. by a homeomorphism of the source) topologically equivalent. Moreover, since σ_t depends continuously on t the family g_t is topologically trivial.

We now follow the main ideas of [6] in order to explain the construction of topological trivialization of a family V_t of analytic subspaces of $(\mathbb{K}^n, 0)$ that preserves the variable x_1. For this we adapt the definition of Zariski equisingular families, Definition 4.3.2, by changing it slightly, and also by changing accordingly the construction of equisingular deformations. The point is that, when we make linear changes of coordinates in order to replace a function by its Weierstrass polynomial, now we are no longer allowed to change the variable x_1 and mix it with the other variables. So if one of the successive discriminants is divisible by x_1 we cannot proceed the way we have done it before. Therefore we replace the assumptions (2) and (3) of Definition 4.3.2 by

(2a) There are $q_i \in \mathbb{N}$ such that the discriminant of $(F_i)_{red}$ divides $x_1^{q_i} F_{i-1}(t, x^{i-1})$.
(3a) $F_1 \equiv 1$.

Then the construction of the homeomorphisms that we presented in Sect. 4.3.1 gives the following version of Theorem 4.3.5, that is a simplified statement of [6, Theorem 5.1].

Theorem 4.4.6 *Suppose that V is Zariski equisingular with respect to the parameter t in the sense of Definition 4.3.2 with the conditions (2) and (3) replaced by conditions (2a) and (3a). Then we may require that the homeomorphisms Φ of* (4.9) *satisfies additionally $\Psi_1(t, x_1) = x_1$.*

Proof Idea of proof.

Because $F_1 \equiv 1$, by (2a), the discriminant locus of F_2 is either empty or given by $x_1 = 0$. Therefore we may take $\Psi_1(t, x_1) = x_1$. Then we show by induction on i that each Φ_i can be lifted so that the lift Φ_{i+1} preserves the zero set of F_{i+1} and the values of x_1. The former condition follows by inductive assumption and the fact that Φ_i preserves the discriminant locus of F_{i+1}. The latter condition is satisfied trivially since Φ_{i+1} is a lift of Φ_i. □

As a corollary we obtain the following result.

Theorem 4.4.7 *([6, Theorem 1.2]) Let $\mathbb{K} = \mathbb{R}$ or $\mathbb{C}$. Let $g : (\mathbb{K}^n, 0) \to (\mathbb{K}, 0)$ be an analytic function germ. Then there is a homeomorphism $\sigma : (\mathbb{K}^n, 0) \to (\mathbb{K}^n, 0)$ such that $g \circ \sigma$ is the germ of a polynomial.*

For the proof of Theorem 4.4.7 one follows the proof of Theorem 4.4.5 that gives such homeomorphism to a Nash function and not directly to a polynomial, since we cannot get a better result just using the Artin approximation. Recall that a function is *Nash* if it is analytic and satisfies an algebraic equation. Thus $f : (\mathbb{K}^n, 0) \to \mathbb{K}$ is the germ of a Nash function if and only if its Taylor series is an algebraic power series. For more details on real and complex Nash functions and sets see [7, 8]. The final step of the proof of Theorem 4.4.7, a homeomorphism of a Nash germ to a polynomial germ follows from [8], that is in essence from the Artin-Mazur Theorem, and a Thom stratification argument, see Sect. 5.5 of [6] for details.

There is a common generalization of Theorems 4.4.7 and 4.4.5.

Theorem 4.4.8 *([6, Theorem 1.3]) Let $(V_i, 0) \subset (\mathbb{K}^n, 0)$ be a finite family of analytic set germs and let $g : (\mathbb{K}^n, 0) \to (\mathbb{K}, 0)$ be an analytic function germ. Then there is a homeomorphism $\sigma : (\mathbb{K}^n, 0) \to (\mathbb{K}^n, 0)$ such that $g \circ \sigma$ is the germ of a polynomial, and for each i, $\sigma^{-1}(V_i)$ is the germ of an algebraic subset of $\mathbb{K}^n$.*

Theorem 4.4.8 cannot be extended to many functions or to maps with values in $\mathbb{K}^p$ for $p > 1$, see [6, Example 6.3].

Corollary 4.4.9 *([6, Corollary 1.4]) Let $g : (V, p) \to (\mathbb{K}, 0)$ be an analytic function germ defined on the germ (V, p) of an analytic space. Then there exists an algebraic affine variety V_1, a point $p_1 \in V_1$, the germ of a polynomial function $g_1 : (V_1, p_1) \to (\mathbb{K}, 0)$ and a homeomorphism $\sigma : (V_1, p_1) \to (V, p)$ such that $g_1 = g \circ \sigma$.*

We do not know whether the above results hold true with "homeomorphism" replaced by "bi-Lipschitz homeomorphism".

? Open problem 4.

Is an analytic function germ bi-Lipschitz homeomorphic to a Nash or a polynomial germ?

Unlike the analogous open problem for analytic set germs, it is more likely that the answer to this one is negative. The reason is the following. By the existence of Lipschitz stratification, cf. [42, 50], the bi-Lipschitz equivalence of analytic set germs does not have continuous moduli (the principle of generic bi-Lipschitz triviality, analogous to the one described in Sect. 4.3.5, holds true). On the other hand, the bi-Lipschitz right equivalence of analytic function germs admits continuous moduli, see [24, 25].

Using Theorem 4.4.6 one may show that the principle of generic topological equisingularity of analytic function germs holds true. (An alternative proof follows again by stratification theory, more precisely by Thom stratification and Thom-Mather Isotopy Lemma.) Let T be a $\mathbb{K}$-analytic space and let $g_t(y) = g(t, y) : (T, t_0) \times (\mathbb{K}^n, 0) \to (\mathbb{K}, 0)$ be a $\mathbb{K}$-analytic family of $\mathbb{K}$-analytic function germs. We say that the family $g_t(y)$ is *topologically trivial at* t_0 (for topological right equivalence) if there are an open neighborhood T' of t_0 in T and neighborhoods Ω_0 of the origin in $\mathbb{K}^n$, and Ω of $(t_0, 0)$ in $T \times \mathbb{K}^n$, and a homeomorphism

$$\Phi : U \times \Omega_0 \to \Omega,$$

such that $g(\Phi(t, y)) = g(0, y)$. Then the following statement holds.

Corollary 4.4.10 (Principle of generic topological equisingularity of analytic function germs, *[52, Theorem 8.5]*) *Let T be a $\mathbb{K}$-analytic space and let $g_t(y) : (T, t_0) \times (\mathbb{K}^n, 0) \to (\mathbb{K}, 0)$ be a $\mathbb{K}$-analytic family of $\mathbb{K}$-analytic function germs. Let $t_0 \in T$. Then there exist an analytic stratification of an open neighborhood of t_0 in T such that for every stratum S and every $t_0' \in S$, the family $g_t(y), t \in S$ is topologically trivial at t_0'.*

4.4.6 Local Topological Classification of Smooth Mappings

The principle of generic topological equisingularity does not hold for the germs of mappings. That is it is known by an example of Thom [68], see also [44], that the topological classification of real or complex, analytic or polynomial map germs admits continuous moduli. This means that there are, polynomial in t, families of polynomial map germs $f_t : (\mathbb{K}^n, 0) \to (\mathbb{K}^p, 0)$ that have different topological types for different t, provided $n \geq 3$, $p \geq 2$, see [44]. Recall that we say that two germs $f_i : (\mathbb{K}^n, 0) \to (\mathbb{K}^p, 0)$, $i = 1, 2$, have *the same topological type* if there exist homeomorphisms germs $h : (\mathbb{K}^n, 0) \to (\mathbb{K}^n, 0)$ and $g : (\mathbb{K}^p, 0) \to (\mathbb{K}^p, 0)$ such that $f_1 \circ h = g \circ f_2$, that is, in other words, they are right-left topologically equivalent.

A smooth map germ $f : (\mathbb{R}^n, 0) \to (\mathbb{R}^p, 0)$ is *topologically r-determined* if every smooth map germ with the same r-jet as f is topologically equivalent to f. In [69] Thom proposed a stabilization theorem:

For any positive integer r, there i s a closed semialgebraic subset Σ_r of the r-jet space $J_r(n, p)$ such that

(i) codim $\Sigma_r \to \infty$ *as* $r \to \infty$, *and*
(ii) if the r*-jet of a map-germ* f *belongs to* $J_r(n, p) \setminus \Sigma_r$, *then* f *is topologically* r*-determined.*

In other words "most" smooth mappings, that is up to a set of infinite codimension in the jet space, look algebraic and are finitely determined. Thom gave a sketch of proof in [69]. The first complete proof was given by Varchenko in [75, 76] using very different ideas that the ones of Thom, namely Zariski equisingularity. Actually, Varchenko proved a much stronger result.

Theorem 4.4.11 *([72, Theorem 2]) There exists a partition of the space of* r*-jets* $J_r(n, p)$ *in disjoint semialgebraic sets* $V_0, V_1, \dots$ *having the following properties.*

1. *Maps whose jets live in the same* V_i, $i > 0$, *are (right-left) topologically equivalent.*
2. *Any germ whose* r*-jet is in* V_i *for* $i > 0$ *is simplicial for suitable triangulations of* $\mathbb{R}^n$ *and* $\mathbb{R}^p$.
3. *The codimension of* V_0 *in* $J_r(n, p)$ *tends to infinity as* r *tends to infinity.*

The stabilization theorem of Thom was also shown by du Plessis in [12]. The proof given there follows the original Thom's ideas, stratification theory, transversality, isotopy lemmas and Mather's ideas about versal unfoldings. Another application of Zariski equisingularity method to finite determinacy was given in [13], where the function case ($p = 1$) was considered. Note that topologically finitely determined function germs $f : (\mathbb{K}^n, 0) \to (\mathbb{K}, 0)$ have isolated singularities (or are regular). In [13] Bobadilla gives a meaningful version of Theorem 4.4.11 for non-isolated singularities by considering functions belonging to a fixed ideal I instead of the whole space of analytic germs at the origin. We refer the reader to [13] for details.

Remark 4.4.12 If the target space of f_t is of dimension bigger than one then the method, we applied in the previous subsection to trivialize families of function germs may not work. In general we cannot trivialize the family of graphs starting from the variables in the target, as we did by taking $F(t, x_1, \dots, x_n) := x_1 - g(t, x_2, \dots, x_n)$, if this graph is not included in the zero set of a Weierstrass polynomial in a variable in the source. This is related to the presence of fibers of dimension bigger than the expected dimension (dimension of the source minus dimension of the target), not only for f but for every function (discriminant) obtained during the construction process.

Even if this phenomenon of "blowing-up" of the special fiber is not present, that is we can apply Zariski method without mixing the variables of the source and of the target, we cannot, in general, construct topological trivialization that is the identity on the target. That means that, if $p \geq 2$, we get the right-left equivalence instead of the right one as in Corollary 4.4.10.

4.5 Equisingularity Along a Nonsingular Subspace. Zariski's Dimensionality Type

In [85, Definition 3], see also [72], Zariski introduced the notion of algebro-geometric equisingularity, now called Zariski equisingularity, of an algebroid hypersurface $V \subset \mathbb{K}^{r+1}$ along a nonsingular subspace of Sing V. This notion can be easily adapted to the complex and real analytic set-ups.

Let $V = f^{-1}(0)$ be an analytic hypersurface defined in a neighborhood of a point $P \in \mathbb{K}^{r+1}$. As before we assume that f is reduced. Let W be a nonsingular analytic subspace of Sing V containing P. Let $x_1, x_2, \ldots, x_{r+1}$ be a local coordinate system at P. Consider a set of r elements $z_1, z_2, \ldots, z_r$ of the local ring of V at P:

$$z_i = z_i(x_1, x_2, \ldots, x_{r+1}) = z_{i,1} + z_{1,2} + \cdots, \quad i = 1, 2, \ldots, r,$$

where the z_i are convergent power series in the x's, and $z_{i,\alpha}$ is homogeneous of degree α. We say that the r elements z_i form *a set of parameters* if the following two conditions are satisfied :

(a) $x = 0$ *is an isolated solution of the* $r + 1$ *equations* $z_1(x) = z_2(x) = \cdots = z_r(x) = f(x) = 0$.

(b) The r *linear forms* $z_{i,1}$ *are linearly independent.*

If condition (b) is satisfied, then the r linear equations $z_{i,1}(x) = 0, i = 1, 2, \ldots, r$ define a line l_z through P and the parameters z_i define a co-rank one projection π_z of a neighborhood of P in $\mathbb{K}^{r+1}$ onto a neighborhood of $\overline{P} = \pi(P)$ in $\mathbb{K}^r$. This projection π_z is called *permissible* if the fiber $\pi^{-1}(\pi(P))$, that is a nonsingular curve, is transverse to W (here "transverse" means that the the tangent line to the fiber is not tangent to W). If this curve is transverse to V at P, that is l_z does not belong to the tangent cone $C_P(V)$ to V at P, then the projection is called *transversal* (or transverse) and the $z_i(x), i = 1, \ldots, r$, are *transversal parameters*.

Let π_z be a permissible projection and let $\pi_{z,V}$ denote the restriction of π to V. Thus we may suppose that locally f is a suitable reduced Weierstrass polynomial whose discriminant D_f is an analytic function in $(z_1, z_2, \ldots, z_r)$. Denote by Δ_z its zero set, that is the discriminant locus of $\pi_{z,V}$.

The projection $\pi_z(W)$ is a nonsingular variety $\overline{W}$, of the same dimension as W. Since we have assumed that $W \subset$ Sing V we have $\overline{W} \subset \Delta_z$. I f dm $W = \dim \Delta_z = r - 1$ then we say that V is *Zariski equisingular at* P *along* W if $\overline{P}$ is a non-singular point of Δ_z. In the general case, Zariski's definition is inductive and goes as follows.

Definition 4.5.1 We say that V is *Zariski equisingular at* P *along* W if there exists a permissible projection π_z such that Δ_z is Zariski equisingular along $\overline{W}$ at $\overline{P}$ (or if $\overline{P}$ is a nonsingular point of Δ_z).

4.5.1 *Equimultiplicity. Transversality of Projection*

As Zariski states on page 489 of [85] the algebro-geometric equisingularity, i.e. Zariski equisingularity as defined in Definition 4.5.1, implies equimultplicity.

Proposition 4.5.2 *If V is Zariski equisingular at P along W then the multiplicity of V is constant along W.*

Zariski proves it when $\dim W = \dim V - 1 = r - 1$, see [82, Theorem 7], and in the general algebroid case in [86]. For a proof in the complex, and also real analytic case, see [52, Proposition 3.6].

Similarly to Definition 4.5.1 one may define *transverse Zariski equisingularity along a nonsingular subspace* as the one given by transverse projections. By Proposition 4.5.2, because the equimultiple families are normally pseudo-flat (continuity of the tangent cone), the transversality of π_z at P implies the transversality at all points of W in a neighborhood of P.

One can also define *generic or generic linear Zariski equisingularity along a nonsingular subspace*. For generic linear it means that we require at each stage the projection to be chosen from a Zariski open non-empty set of linear projection. Note that a priori this notion depends on the choice of coordinates and it is not clear whether it is preserved by nonlinear changes of coordinates. We discuss the notion of generic projection in Sects. 4.5.4 and 4.5.5.

4.5.2 *Relation to Other Equisingularity Conditions. Examples*

As we mentioned before Varchenko showed in [73], see also [72, 74], that in the complex or real analytic case Zariski equisingularity implies topological triviality.

In [85, Question E], Zariski asked as well whether Zariski equisingularity implies Whitney's conditions. This has been disproved by Briançon end Speder in [11] for the equisingularity as defined in Definition 4.5.1. In [64] Speder shows that if V is Zariski equisingular along a nonsingular variety W for sufficiently generic projections, then the pair $(\mathrm{Reg}(V), W)$ satisfies Whitney's conditions. For instance, generic linear projections, that is from a Zariski open non-empty subset of such projections, are generic in the sense of Speder. This result was improved in [52], where it was shown that transverse Zariski equisingularity, both in real and complex analytic cases, implies Whitney's conditions, see Theorems 4.3 and 7.1 of [52] for precise statements.

There are several classical examples describing the relation between Zariski equisingularity and Whitney's conditions. The general set up for these examples is the following. Consider a complex algebraic hypersurface $X \subset \mathbb{C}^4$ defined by a polynomial $F(t, x, y, z) = 0$ such that $\mathrm{Sing}\, X = T$, where T is the t-axis. Let $\pi : \mathbb{C}^4 \to T$ be the standard projection. In all these examples $X_t = \pi^{-1}(t)$, $t \in T$, is a family of isolated singularities, topologically trivial along T. These examples relate the following conditions:

1. X is Zariski equisingular along T, Definition 4.5.1.
2. X is Zariski equisingular along T for a transverse projection.
3. X is Zariski equisingular along T for a generic system of coordinates. Here we consider "generic" in the sense of [9]. It is equivalent, see loc. cit. to be generic linear, or generic in the sense of Zariski [88], that we recall in Sect. 4.5.5 below.
4. The pair $(X \setminus T, T)$ satisfies Whitney's conditions (a) and (b).

Clearly (3)⇒(2)⇒(1). Speder showed (3)⇒(4) in [64] and (2)⇒(4) for families of complex analytic hypersurfaces with isolated singularities in $\mathbb{C}^3$ in his thesis [63] (not published). Theorem 7.1 of [52] gives (2)⇒(4) in the general case. As the examples below show, all the other implications are false.

Example 4.5.3 ([11])

$$F(x, y, z, t) = z^5 + ty^6z + y^7x + x^{15} \tag{4.25}$$

This example satisfies (1) for the projections $(x, y, z) \to (y, z) \to x$ but (4) fails. As follows from Theorem 7.1 of [52], (2) fails as well.

Example 4.5.4 ([10])

$$F(x, y, z, t) = z^3 + tx^4z + y^6 + x^6 \tag{4.26}$$

In this example (4) is satisfied and (3) fails. This example satisfies (1) for the projections $(x, y, z) \to (x, z) \to x$.

Example 4.5.5 ([37])

$$F(x, y, z, t) = z^{16} + tyz^3x^7 + y^6z^4 + y^{10} + x^{10} \tag{4.27}$$

In this example (2) is satisfied and (3) fails.

Example 4.5.6 ([47])

$$F(x, y, z, t) = x^9 + y^{12} + z^{15} + tx^3y^4z^5 \tag{4.28}$$

In this example (4) is satisfied and (1) fails. This also shows that (4) does not imply (2).

4.5.3 Lipschitz Equisingularity

In 1985 Mostowski [42] introduced the notion of Lipschitz stratification and showed the existence of such stratification for germs of complex analytic subsets of $\mathbb{C}^n$. For complex algebraic varieties, such stratification exists globally. The existence of Lipschitz stratification for real analytic spaces and algebraic varieties was shown [48, 50]. Lipschitz stratification satisfies the extension property of stratified Lipschitz vector fields from lower-dimensional to higher-dimensional strata, and therefore implies local bi-Lipschitz triviality along each stratum (and hence Lipschitz equisingularity as well). Mostowski's construction is similar to the one of Zariski, but involves considering many co-rank one generic projections at each stage of construction. For more on Lipschitz stratification, we refer the interested reader to [19, 42, 49].

By Lipschitz saturation, see [58], an equisingular family of complex analytic plane curves is bi-Lipschitz trivial, i.e. trivial by a local ambient bi-Lipschitz homeomorphism. In general, there is a conjectural relation between Lipschitz and Zariski equisingularity, at least in the complex analytic set up.

? Open problem 5.

Are generically Zariski equisingular families of complex hypersurfaces bi-Lipschitz equisingular? Does Zariski equisingularity provide a "natural" way of construction of Lipschitz stratification in the sense of Mostowski?

For families of complex surface singularities, that is along a nonsingular subspace of codimension 2 the following results have been announced in [46, 53]. In [46] it was shown that generically Zariski equisingular families of normal complex surface singularities are bi-Lipschitz trivial. In [53] was shown that a natural stratification given by successive generic (or generic linear) projections of a complex hypersurface satisfies Mostowski's Conditions in codimension 2. In particular, the latter result implies that generic Zariski equisingulariy of families (not necessarily isolated) of complex surface hypersurface singularities is Lipschitz equisingular.

4.5.4 Zariski Dimensionality Type. Motivation

When $\dim W = \dim V - 1$, V is Zariski equisingular at P along W if and only if V is isomorphic to the total space of an equisingular family of plane curve singularities along W, see [83, Theorem 4.4]. Then, moreover, Zariski equisingularity can be realized by any transversal projection π_z.

Guided by this example, Zariski conjectured in [85, Question I], that Zariski equisingularity for a single permissible projection implies the equisingularity for generic projection (or for almost all projections that we recall later in Sect. 4.5.6). An affirmative answer to this question would imply, in particular, that if there exists an equisingular projection then there exist a transversal equisingular projection. Both turned out not to be true. In [37] Luengo gave an example of a family of surface singularities in $\mathbb{C}^3$ that is Zariski equisingular for one projection, that is even transversal, but is not equisingular for the generic projection, see Example 4.5.5. Briançon end

Speder gave in [11] an example that is equisingular for one projection but there is no transversal projection that gives Zariski equisingularity, see Example 4.5.3.

Therefore Zariski in [88] proposes a different strategy. Instead of arbitrary permissible projections, or even transversal projections, Zariski uses generic projections to define the equisingularity relation. (We recall what "generic" means for Zariski in the next subsection.) Having fixed such an equisingularity relation, Zariski introduces the notion of dimensionality type. For this the equisingularity relation should first satisfy the following property.

The set of points of equivalent singularities form a locally nonsingular subspace of V of codimension that depends only on this equisingularity class.

Thus, for a point $P \in V$ the set of points equivalent to (V, P) is nonsingular and its codimension in V characterizes how complicated the singularity is. This codimension is then called *the dimensionality type of* (V, P). The points of dimensionality type 0 are the nonsingular points of V. The simplest singular points of V, of dimensionality type 1, are those at which V is isomorphic to the total space of an equisingular family of plane curves. The closure of the set of points of fixed equisingularity type may contain points of different equisingularity type but only of the higher dimensionality type and only on finitely many such equisingular strata.

The very definition of what is meant by the word "generic" is the main point of Zariski's definition. Let us make a quick comment on an apparently obvious choice. Similarly to Definition 4.5.1 one may define *generic linear Zariski equisingularity* as the one given by linear projections belonging to a Zariski open non-empty subset of linear projections. Except the case of the dimensionality type 1, it is not clear whether such notion of generic linear Zariski equisingularity is preserved by non-linear local changes of coordinates, nor whether it implies the generic Zariski equisingularity.

4.5.5 Zariski Dimensionality Type

Formally Zariski's original definition of the dimensionality type requires the field extension by infinitely many indeterminates. Therefore, in this subsection exceptionally, we work over an arbitrarily algebraically closed field of characterisitc zero that will be denoted by k, and instead of the category of complex analytic spaces we consider the category of algebraic or algebroid varieties. Recall that the algebroid varieties are the varieties defined by ideals of the rings of formal power series, see [32, Chap. 4] and [88, Sect. 2]. We note, however, that in [88, Proposition 5.3] Zariski shows that his definition involving such a field extension can be replaced by a condition that is based on the notion of almost all projections that does not require a field extension. We recall the approach via almost all projections in the next subsection.

Let k be an algebraically closed field of characteristic zero. Consider an algebroid hypersurface $V = f^{-1}(0) \subset (k^{r+1}, P)$ at $P \in k^{r+1}$ defined in a local system of coordinates by a formal power series $f \in k[[x_1, \ldots, x_{r+1}]]$ (that we assume reduced). Zariski's definition of the dimensionality type is based on the following notion of generic projection. *The generic projection*, in the sense of [88], is the map $\pi_u(x) = (\pi_{u.1}(x), \dots, \pi_{u,r}(x))$, with

$$\pi_{u,i}(x) = \sum_{d\geq 1} \sum_{\nu_1+\cdots\nu_{r+1}=d} u^{(i)}_{\nu_1,\cdots,\nu_{r+1}} x^{\nu}. \tag{4.29}$$

This map is defined over k^*, any field extension of k that contains all coefficients $u^{(i)}_{\nu_1,\cdots,\nu_{r+1}}$ as interdeterminates, thus formally $\pi_u : ((k^*)^{r+1}, P) \to ((k^*)^r, P_0^*)$, where $P_0^* = \pi_u(P)$. Denote by $\Delta_u^* \subset (k^{*r}, P_0^*)$ the discriminant locus of $(\pi_u)_{|V^*}$, where $V^* = f^{-1}(0) \subset (k^{*(r+1)}, P)$.

Let W be a nonsingular algebroïd subspace of Sing V and let $\overline{W^*} = \pi_u(W)$. If dim $W =$ dim $V - 1$ then we say that V is generically Zariski equisingular at P along W if $\bar{P}^*$ is a non-singular point of Δ_u^*. In general, the definition is similar to Definition 4.5.1.

Definition 4.5.7 We say that V is *generically Zariski equisingular at P along W* if Δ_u^* is generically Zariski equisingular along $\overline{W^*}$ at $\bar{P}^*$ (or if $\bar{P}^*$ is a non-singular point Δ_u^*).

The definition of dimensionality type of [88] is again recursive. It is defined for any point Q of V, not only the closed point P.

Definition 4.5.8 Any simple (i.e. non-singular) point Q of V is of dimensionality type 0. Let Q be a singular point V and let $Q_0^* = \pi_u(Q)$. Then *the dimensionality type of V at Q*, denoted by d. t.(V, Q), is equal to

$$\text{d. t.}(V, Q) = 1 + \text{d. t.}(\Delta_u^*, Q_0^*).$$

The notions of generic Zariski equisingularity and of dimensionality type are independent of the choice of this field extension k^*, see [88, 89].

As follows from [88] the set of points where the dimensionality type is constant, say equal to σ, is either empty or a nonsingular locally closed subvariety of V of codimension σ. The dimensionality type defines a stratification $V = \sqcup_\alpha S_\alpha$ of V that satisfies the frontier condition, i.e. if $S_\alpha \cap S_\beta \neq \emptyset$ then $S_\alpha \subset \overline{S_\beta}$, and V is generically Zariski equisingular along W at P if and only if W is contained in the stratum containing P.

A singularity is of dimensionality type 1 if and only if it is isomorphic to the total space of an equisingular family of plane curve singularities, see [83] Theorem 4.4. Moreover, if this is the case, then V is such an equisingular family for any transverse system of coordinates.

4.5.6 Almost all Projections

In [88, Proposition 5.3] Zariski shows that, in Definitions 4.5.7 and 4.5.8, the generic projection π_u can be replaced by a condition that involves almost all projections $\pi_{\bar{u}} : k^{r+1} \to k^r$ (so it does not require a field extension).

One says that a property holds *for almost all projections* if there exists a finite set of polynomials $\mathcal{G} = \{G_\mu\}$ in the indeterminates $u^{(i)}_{\nu_1,\cdots,\nu_{r+1}}$ and coefficients in k such that this property holds for all projections $\pi_{\bar{u}}$ for $\bar{u}$ satisfying $\forall_\mu G_\mu(\bar{u}) \neq 0$. Here the bar denotes the specialization $u \to \bar{u}$, i.e. we replace all indeterminates $u^{(i)}_{\nu_1,\cdots,\nu_{r+1}}$ by elements of k, $\bar{u}^{(i)}_{\nu_1,\cdots,\nu_{r+1}} \in k$. Thus, for almost all projections $\pi_{\bar{u}}$ the dimensionality type d. t.(V, P) equals

$$\text{d. t.}(V, P) = 1 + \text{d. t.}(\Delta_{\bar{u}}, \pi_{\bar{u}}(P)), \tag{4.30}$$

where $\Delta_{\bar{u}}$ denotes the discriminant locus of $\pi_{\bar{u}|V}$. Since the finite set of polynomials $\mathcal{G}$ involves nontrivially only finitely many indeterminates $u^{(i)}_{\nu_1,\cdots,\nu_{r+1}}$, we may specialize the remaining ones to 0, and then the projection $\pi_{\bar{u}}$ becomes polynomial. This means that, as soon as we know the set of polynomials $\mathcal{G}$, we may compute the dimensionality type of V at P just by computing d. t.$(\Delta_{\bar{u}}, \pi_{\bar{u}}(P))$, for only one polynomial projection $\pi_{\bar{u}}$, satisfying $\forall_\mu G_\mu(\bar{u}) \neq 0$. Similarly, in order to check whether V is generic Zariski equisingular at P along S, it suffices to check it for $\Delta^*_{\bar{u}}$ along $\pi_{\bar{u}}(S)$ at $\pi_{\bar{u}}(P)$.

4.5.7 Canonical Stratification of Hypersurfaces

The dimensionality type defines a canonical stratification of a given algebroid or algebraic hypersurface over an algebraically closed field of characteristic zero. Unfortunately, in general, no specific information on the polynomials of $\mathcal{G}$ is given in [88]. Zariski's construction is purely transcendental, and there is no explicit bound on the degree of such polynomial projections. This makes, for instance, an algorithmic computation of Zariski's canonical stratification impossible. The algebraic case was studied in more detail by Hironaka [26], where the semicontinuity of such a degree in Zariski topology is shown. This implies in particular that the dimensionality type induces, in the algebraic case, a stratification by locally closed algebraic subvarieties.

For complex analytic singularities, we can define the dimensionality type using generic polynomial or generic analytic projections. It follows from Speder [64, Theorems 1–4], that in the complex analytic case thus defined canonical stratification satisfies Whitney's conditions (a) and (b). It also follows from Theorems 3.7 and 7.3 of [52], where a stronger regularity condition, called (arc-w) was proven.

It is an open question whether generic linear projections are generic in the sense of Zariski.

? Open problem 6.

Are generic linear projections sufficient to define generic Zariski equisingularity and the dimensionality type? More precisely, is the formula (4.30) valid for a Zariski open non-empty set of linear projections $\pi_{\bar{u}}$?

Even if the answers to the above questions were positive it would not give an algorithm to compute the dimensionality type and the canonical stratification automatically, but the positive answer to this question would probably help to consider other related open problems that we summarize below.[1]

? Open problem 7.

Characterize geometrically or algebraically generic polynomial projections in the sense of Zariski?

In the case of strata of dimensionality type 2 a partial answer to both questions of problem 6 was obtained in [9]. In this paper Briançon and Henry characterized generically Zariski equisingular families of isolated surface singularities in the 3-space in terms of local analytic invariants: Teissier's numbers (multiplicity, Milnor number, and Milnor number of a generic plane section), the number of double points and the number of cusps of the apparent contours of the generic projection of the generic fibre of a mini-versal deformation. All these numbers are local analytic invariants, and therefore linearly generic change of coordinates is generic in the sense of Zariski. This shows in particular that, if Sing V is of codimension 2 in V at P, then the generic linear projections are sufficient to verify whether $\mathrm{d.t.}(V, p) = 2$.

Note that the answer to problem 7 should be quite subtle even in the case of the dimensionality type 2. Let us remind that in [37] Luengo gave an example of a family of surface singularities in $\mathbb{C}^3$ that is Zariski equisingular for one transverse projection but not for the generic ones.

? Open problem 8.

Is the canonical Zariski's stratification of a complex analytic hypersurface Lipschitz equisingular?

This problem is a version of problem 5. As we have mentioned in Sect. 4.5.3, the positive answer for the strata of codimension 2 was announced [53] and, if moreover $\mathrm{codim}(\mathrm{Sing}\, V) = 2$, in [46].

[1] Added in proofs. The answer to Problem 6 is positive in the case of dimensionality type 2 see [54].

4.5.8 Zariski Equisingularity and Equiresolution of Singularities

In the case of dimensionality type 1, i.e. equivalently, for families of plane curves singularities, Zariski equisingularity can be expressed in terms of blowings-up (monoidal transformations) and equiresolution. More precisely, firstly, the following property of stability by blowings-up holds.

Theorem 4.5.9 *([83, Theorem 7.4]) Assume that the singular locus W of V is of codimension 1 and let P be a regular point of W. Let $\pi : V' \to V$ be the blowing-up of W, and let P_{gen} denote a general point of W. Then V is of dimensionality type 1 along W at P if and only if the following two conditions hold*

(1) *$\pi^{-1}(P)$ is a finite set of the same cardinality as $\pi^{-1}(P_{gen})$,*
(2) *each $P' \in \pi^{-1}(P)$ is either a nonsingular point of V' or a point of dimensionality type 1.*

In the complex analytic set up the conditions (1) and (2) mean that over a neighborhood of P, $W' := \pi^{-1}(W) \to W$ is a finite analytic covering and V' is Zariski generically equisingular along each connected component of W' (this includes the case that V' is nonsingular along this connected component).

Secondly, the repeating process of such blowings-up leads not only to a resolution $\tilde{\pi} : \tilde{V} \to V$ of V but also to an equiresolution in the following sense. Fix a local projection of $pr : \mathbb{K}^{r+1} \to W$, such that $pr^{-1}(P)$ is nonsingular and transverse to W. The fibers of this projection restricted to W are plane curve singularities V_t parameterized by $t \in W$. Then the restrictions of $\tilde{\pi}$, $\tilde{V}_t := \tilde{\pi}^{-1}(V_t) \to V_t$ are the resolutions of V_t, see e.g. [83, Corollary 7.5] and the paragraph after it, and the induced projections of $\tilde{V}$ and of the exceptional divisor E of $\tilde{\pi}$ onto W are submersions. Note that $\tilde{V}$ coincides with a normalization of V, and hence it is also an equinormalization of the family V_t. this can be deduced as well from Puiseux with parameter Theorem 4.2.1).

? Open problem 9.

Do the two properties, stability by blowing-up and equiresolution, hold for arbitrary codimension strata of Zariski's stratification?

The first part of this question was stated by Zariski in [87] : "Now, one test that any definition of equisingularity must meet is the test of its stable behavior along W under blowing-up of W". It also appears in questions F, G and H of [85]. An example of Luengo [36] shows that the generic Zariski equisingularity does not satisfy the stability under blowings-up property. That is $V = \{z^7 + y^7 + ty^5x^3 + x^{10} = 0\} \subset \mathbb{C}^4$ is generically Zariski equisingular along $W = \text{Sing}\, V = \{x = y = z = 0\}$ but the blow-up $\tilde{V}$ of V along W is not generically Zariski equisingular along $\tilde{W} = \text{Sing}\, \tilde{V}$ ($\tilde{W}$ is a nonsingular curve in this example).

Moreover, reciprocally, a blowing-up may make non-equisingular families equisingular as shows another example from [36]. In this example $V = \{z^4 + y^6 + tz^2y^3 + x^8 = 0\} \subset \mathbb{C}^4$ is not generically Zariski equisingular along $W = \operatorname{Sing} V = \{x = y = z = 0\}$ (the origin is of dimensionality 3), but the strata of Zariski canonical stratification of the blow-up of W, $\tilde{V} \to V$, project submersively onto W (no point of the highest possible dimensionality 3 in $\tilde{V}$).

In order to answer the second part of this question one has to make precise what the equiresolution, also often called simultaneous resolution, means. To start with there are embedded resolutions (modifications of the ambient space containing V) and the non-embedded ones. The concept of equiresolution was largely studied and clarified in [66] within the context of non-embedded resolutions of complex analytic surface singularities, and in [33] in the context of embedded equiresolution of complex analytic or algebraic varieties.

The relation between generic Zariski equisingularity and equiresolution depends which notion of the equiresolution is adapted. In the first cited above example of Luengo [36], V understood as a family V_t does not have strong simultaneous resolution in the sense of Teissier [66], if we require, moreover, that this resolution is given by a sequence blowings-up of non-singular equimultiple centers following Hironaka's algorithm, see [36] for details. In [33] Lipman proposes a strategy to prove a weaker version of equiresolution for such families. This proof in the algebraic case was completed by Villamayor [77]. Villamayor's equiresolution is a modification of the ambient nonsingular space containing V more general than the ones obtained as compositions of sequences of nonsingular centers blowings-up. Moreover, it is not required that the induced resolution of V is an isomorphism over $V \setminus \operatorname{Sing} V$.

There is another open problem related to the equiresolutions of families of singularities. Namely, it is not clear whether, in general, equiresolution can be used to construct topological trivializations. Let us explain it on a simplified example. Suppose that V is a hypersurface of a nonsingular (real or complex) analytic manifold M, $\tilde{\pi} : \tilde{M} \to M$ is a modification, the composition of blowings-up of smooth centers for instance, and that

(i) there is a local (at $P \in W$) analytic projection $pr : M \to W$, such that $pr^{-1}(P)$ is nonsingular and transverse to W and whose fibers restricted to V define a family of reduced hypersurfaces V_t, $t \in W$.
(ii) $\tilde{\pi}^{-1}(V)$ is a divisor with normal crossings that is the union of the strict transform $\tilde{V}$ of V, assumed non-singular, and the exceptional divisor E.
(iii) The strata of the canonical stratification of $\tilde{\pi}^{-1}(V)$ (as a divisor with normal crossings) project by $\tilde{pr} := pr \circ \tilde{\pi}$ submersively onto W.

Then, by a version of Ehresmann fibration theorem, there is a trivialization of $\tilde{pr}$ that preserves the strata of $\tilde{\pi}^{-1}(V)$ and hence also $\tilde{V}$. Moreover, by [30], this trivialization can be made real analytic. If this trivialization is a lift by $\tilde{\pi}$ of a trivialization of pr then we call the latter *a blow-analytic trivialization*.

$$\begin{array}{ccc} \tilde{M} & \xrightarrow{\tilde{\pi}} & M \\ & \tilde{pr} \searrow \quad \swarrow pr & \\ & W & \end{array}$$

In [30] Kuo developed the theory of blow-analytic equivalence of real analytic function germs. Kuo shows that for families of isolated hypersurface (or function) singularities such blow-analytic trivializations exist under the following additional assumptions: $W = \text{Sing}$ and $\tilde{\pi}$ is an isomorphism over the complement of W. In this case he constructs a nice (real analytic) trivialization of the resolution space that projects to a topological trivialization of pr. In particular, it shows that the principle of generic blow-analytic equisingularity of real analytic function germs holds, see [30, Theorem 1]. But in the general, non-isolated singularity case it is not even clear whether there is a topological trivialization that lifts to the resolution space. The existence of a blow analytic trivialization of family of non-isolated singularities remains the main open problem of Kuo's Theory, see [14, 29].

Blow-analytic trivializations are, in particular, arc-analytic, and, actually, at least in the algebraic (i.e. Nash) case, blow-analytic and arc-analytic maps coincide, see [4, 51]. Thus there is a hope that Theorem 4.3.7, proven using Zariski equisingularity, can help in developing blow-analytic theory of non-isolated singularities.

Acknowledgements I would like to thank Jean-Baptiste Campesato, Clint McCrory, Laurentiu Paunescu, and Guillaume Rond for several remarks and suggestions concerning the earlier versions of this survey. I would like to thank as well the anonymous referee for many precise and helpful suggestions.

4.6 Appendix. Generalized Discriminants

We recall the notion of generalized discriminants, see Appendix IV of [3, 62, 80], or Appendix B of [52]. Let k be an arbitrary field of characteristic zero and let

$$F(Z) = Z^d + \sum_{i=1}^{d} a_i Z^{d-i} = \prod_{i=1}^{d} (Z - \xi_i) \in k[Z], \tag{4.31}$$

be a polynomial with coefficients $a_i \in k$ and the roots $\xi_i \in \overline{k}$. Recall that the discriminant of F is a polynomial in the coefficients a_i that can be defined either in terms of the roots

$$D_F = \prod_{1 \le j_1 < j_2 \le d} (\xi_{j_2} - \xi_{j_1})^2,$$

or as the determinant of the Jacobi-Hermite matrix

$$D_F = \begin{vmatrix} s_0 & s_1 & \cdots & s_{d-1} \\ s_1 & s_2 & \cdots & s_d \\ \cdots & \cdots & \cdots & \cdots \\ s_{d-1} & s_l & \cdots & s_{2d-2} \end{vmatrix},$$

where $s_i = \sum_{j=1}^{d} \xi_j^i$, for $i \in \mathbb{N}$, are Newton's symmetric functions. Thus $D_F = 0$ if and only if F has a multiple root in $\overline{k}$.

The generalized discriminants, or subdiscriminants, D_{d+1-l} of F, $l = 1, \ldots, d$, can be defined as the principal minors of the Jacobi-Hermite matrix

$$D_{d+1-l} := \begin{vmatrix} s_0 & s_1 & \cdots & s_{l-1} \\ s_1 & s_2 & \cdots & s_l \\ \cdots & \cdots & \cdots & \cdots \\ s_{l-1} & s_l & \cdots & s_{2l-2} \end{vmatrix},$$

and we put $D_d = 1$ by convention. Thus D_{d+1-l} are polynomials in the coefficients a_i. The generalized discriminants can be defined equivalently in terms of the roots

$$D_{d+1-l} = \sum_{r_1 < \cdots < r_l} \prod_{j_1 < j_2;\ j_1, j_2 \in \{r_1, \ldots, r_j\}} (\xi_{j_2} - \xi_{j_1})^2.$$

In particular $D_1 = D_F$ and F admits exactly l distinct roots in $\overline{k}$ if and only if $D_1 = \cdots = D_{d-l} = 0$ and $D_{d-l+1} \neq 0$.

If F is not reduced, that means in this case that F has multiple roots, the generalized discriminants can replace the (classical) discriminant of F_{red}. Here F_{red} equals $\prod (Z - \xi_i)$, where the product is taken over all distinct roots of F. The following lemma is easy.

Lemma 4.6.1 *Suppose F has exactly $l > 1$ distinct roots in $\overline{k}$ of multiplicities $\mathbf{m} = (m_1, \ldots, m_l)$. Then there is a positive constant $C = C_{l,\mathbf{m}}$, depending only on $\mathbf{m} = (m_1, \ldots, m_l)$, such that the generalized discriminant D_{d-l+1} of F and the standard discriminant $D_{F_{red}}$ of F_{red} are related by*

$$D_{d-l+1} = C D_{F_{red}}.$$

We conclude with the following obvious consequence of the IFT.

Lemma 4.6.2 *Let $F \in \mathbb{C}\{x_1, \ldots, x_n\}[Z]$ be a monic polynomial in Z such that the discriminant $D_{F_{red}}$ does not vanish at the origin. Then, there is a neighborhood U of $0 \in \mathbb{C}^n$ such that the complex roots F are analytic on U, distinct, and of constant multiplicities (as the roots of F).*

References

1. S. Abhyankar, *On the ramification of algebraic functions*, Amer. J. Math., 77 (1955), pp. 575–592.
2. M. Artin and B. Mazur, *On periodic points*, Ann. of Math. (2), 81 (1965), pp. 82–99.
3. S. Basu, R. Pollack, and M.-F. Roy, *Algorithms in real algebraic geometry*, vol. 10 of Algorithms and Computation in Mathematics, Springer-Verlag, Berlin, second ed., 2006.
4. E. Bierstone and P. D. Milman, *Arc-analytic functions*, Invent. Math., 101 (1990), pp. 411–424.
5. M. Bilski, K. Kurdyka, A. Parusiński, and G. Rond, *Higher order approximation of analytic sets by topologically equivalent algebraic sets*, Math. Z., 288 (2018), pp. 1361–1375.
6. M. Bilski, A. Parusiński, and G. Rond, *Local topological algebraicity of analytic function germs*, J. Algebraic Geom., 26 (2017), pp. 177–197.
7. J. Bochnak, M. Coste, and M.-F. Roy, *Real algebraic geometry*, vol. 36 of Ergebnisse der Mathematik und ihrer Grenzgebiete (3) [Results in Mathematics and Related Areas (3)], Springer-Verlag, Berlin, 1998. Translated from the 1987 French original, Revised by the authors.
8. J. Bochnak and W. Kucharz, *Local algebraicity of analytic sets*, J. Reine Angew. Math., 352 (1984), pp. 1–14.
9. J. Briançon and J. P. G. Henry, *Équisingularité générique des familles de surfaces à singularité isolée*, Bull. Soc. Math. France, 108 (1980), pp. 259–281.
10. J. Briançon and J.-P. Speder, *Familles équisingulières de surfaces à singularité isolée*, C. R. Acad. Sci. Paris Sér. A-B, 280 (1975), pp. Aii, A1013–A1016.
11. J. Briançon and J.-P. Speder, *La trivialité topologique n'implique pas les conditions de Whitney*, C. R. Acad. Sci. Paris Sér. A-B, 280 (1975), pp. Aiii, A365–A367.
12. A. du Plessis, *On the genericity of topologically finitely-determined map-germs*, Topology, 21 (1982), pp. 131–156.
13. J. Fernández de Bobadilla, *Topological finite-determinacy of functions with non-isolated singularities*, Comment. Math. Helv., 79 (2004), pp. 659–688.
14. T. Fukui, S. Koike, and T.-C. Kuo, *Blow-analytic equisingularities, properties, problems and progress*, in Real analytic and algebraic singularities (Nagoya/Sapporo/Hachioji, 1996), vol. 381 of Pitman Res. Notes Math. Ser., Longman, Harlow, 1998, pp. 8–29.
15. W. Fulton, *Intersection theory*, vol. 2 of Ergebnisse der Mathematik und ihrer Grenzgebiete (3) [Results in Mathematics and Related Areas (3)], Springer-Verlag, Berlin, 1984.
16. M. Goresky and R. MacPherson, *Intersection homology theory*, Topology, 19 (1980), pp. 135–162.
17. M. Goresky and R. MacPherson, *Stratified Morse theory*, vol. 14 of Ergebnisse der Mathematik und ihrer Grenzgebiete (3) [Results in Mathematics and Related Areas (3)], Springer-Verlag, Berlin, 1988.
18. R. M. Goresky, *Whitney stratified chains and cochains*, Trans. Amer. Math. Soc., 267 (1981), pp. 175–196.
19. I. Halupczok and Y. Yin, *Lipschitz stratifications in power-bounded o-minimal fields*, J. Eur. Math. Soc. (JEMS), 20 (2018), pp. 2717–2767.
20. H. Hamm and L. D. ung Tráng, *Un théorème du type de Lefschetz*, C. R. Acad. Sci. Paris Sér. A-B, 272 (1971), pp. A946–A949.
21. H. A. Hamm and L. D. ung Tráng, *Un théorème de Zariski du type de Lefschetz*, Ann. Sci. École Norm. Sup. (4), 6 (1973), pp. 317–355.
22. R. Hardt and D. Sullivan, *Variation of the Green function on Riemann surfaces and Whitney's holomorphic stratification conjecture*, Inst. Hautes Études Sci. Publ. Math., (1988), pp. 115–137 (1989).
23. R. M. Hardt, *Semi-algebraic local-triviality in semi-algebraic mappings*, Amer. J. Math., 102 (1980), pp. 291–302.
24. J.-P. Henry and A. Parusiński, *Existence of moduli for bi-Lipschitz equivalence of analytic functions*, Compositio Math., 136 (2003), pp. 217–235.

25. J.-P. Henry and A. Parusiński, *Invariants of bi-Lipschitz equivalence of real analytic functions*, in Geometric singularity theory, vol. 65 of Banach Center Publ., Polish Acad. Sci. Inst. Math., Warsaw, 2004, pp. 67–75.
26. H. Hironaka, *On Zariski dimensionality type*, Amer. J. Math., 101 (1979), pp. 384–419.
27. H. W. Jung, *Darstellung der funktionen eines algebraischen körpers zweier unabhängigen veränderlichen x, y in der umgebung einer stelle x = a, y = b.*, Journal für die reine und angewandte Mathematik, 133 (1908), pp. 289–314.
28. S. L. Kleiman, *The transversality of a general translate*, Compositio Math., 28 (1974), pp. 287–297.
29. S. Koike, *Finiteness theorems on blow-Nash triviality for real algebraic singularities*, in Geometric singularity theory, vol. 65 of Banach Center Publ., Polish Acad. Sci. Inst. Math., Warsaw, 2004, pp. 135–149.
30. T. C. Kuo, *On classification of real singularities*, Invent. Math., 82 (1985), pp. 257–262.
31. K. Kurdyka, *Ensembles semi-algébriques symétriques par arcs*, Math. Ann., 282 (1988), pp. 445–462.
32. S. Lefschetz, *Algebraic geometry*, Princeton University Press, Princeton, N. J., 1953.
33. J. Lipman, *Equisingularity and simultaneous resolution of singularities*, in Resolution of singularities (Obergurgl, 1997), vol. 181 of Progr. Math., Birkhäuser, Basel, 2000, pp. 485–505.
34. S. Łojasiewicz, *Introduction to complex analytic geometry*, Birkhäuser Verlag, Basel, 1991. Translated from the Polish by Maciej Klimek.
35. S. Łojasiewicz, *Ensembles semi-analytiques*, Preprint IHES (1965), see also https://perso.univ-rennes1.fr/michel.coste/.
36. I. Luengo, *A counterexample to a conjecture of Zariski*, Math. Ann., 267 (1984), pp. 487–494.
37. I. Luengo, *An example concerning a question of Zariski*, Bull. Soc. Math. France, 113 (1985), pp. 379–386.
38. C. McCrory, *Stratified general position*, in Algebraic and geometric topology (Proc. Sympos., Univ. California, Santa Barbara, Calif., 1977), vol. 664 of Lecture Notes in Math., Springer, Berlin, 1978, pp. 142–146.
39. C. McCrory and A. Parusiński, *Real intersection homology*, Topology Appl., 234 (2018), pp. 108–129.
40. C. McCrory, A. Parusiński, and L. Păunescu, *Algebraic stratified general position and transversality*, J. Algebraic Geom., 28 (2019), pp. 139–152.
41. T. Mostowski, *Topological equivalence between analytic and algebraic sets*, Bull. Polish Acad. Sci. Math., 32 (1984), pp. 393–400.
42. T. Mostowski, *Lipschitz equisingularity*, Dissertationes Math. (Rozprawy Mat.), 243 (1985), p. 46.
43. C. Murolo, A. du Plessis, and D. Trotman, *On the smooth whitney fibering conjecture.*, 12 2018.
44. I. Nakai, *On topological types of polynomial mappings*, Topology, 23 (1984), pp. 45–66.
45. R. Narasimhan, *Introduction to the theory of analytic spaces*, Lecture Notes in Mathematics, No. 25, Springer-Verlag, Berlin-New York, 1966.
46. W. D. Neumann and A. Pichon, *Lipschitz geometry of complex surfaces: analytic invariants and equisingularity*, arXiv:1211.4897, 2012.
47. A. Parusiński, *Algebraic equisingularity and Whitney's conditions*, Bull. Polish Acad. Sci. Math., 33 (1985), pp. 247–253.
48. A. Parusiński, *Lipschitz properties of semi-analytic sets*, Ann. Inst. Fourier (Grenoble), 38 (1988), pp. 189–213.
49. A. Parusiński, *Lipschitz stratification*, in Global analysis in modern mathematics (Orono, ME, 1991; Waltham, MA, 1992), Publish or Perish, Houston, TX, 1993, pp. 73–89.
50. A. Parusiński, *Lipschitz stratification of subanalytic sets*, Ann. Sci. École Norm. Sup. (4), 27 (1994), pp. 661–696.
51. A. Parusiński, *Subanalytic functions*, Trans. Amer. Math. Soc., 344 (1994), pp. 583–595.
52. A. Parusiński and L. Păunescu, *Arc-wise analytic stratification, Whitney fibering conjecture and Zariski equisingularity*, Adv. Math., 309 (2017), pp. 254–305.

53. A. Parusiński and L. Păunescu, *Lipschitz stratification of complex hypersurfaces in codimension 2*, arXiv:1909.00296v3, 2019. J Eur Math Soc (JEMS), to appear
54. A. Parusinski and L. Paunescu, *Zariski dimensionality type of singularities. Case of dimensionality type 2*. arXiv:2104.07156, 2021.
55. A. Parusiński and G. Rond, *The Abhyankar-Jung theorem*, J. Algebra, 365 (2012), pp. 29–41.
56. A. Parusiński and G. Rond, *Algebraic varieties are homeomorphic to varieties defined over number fields*, Comment. Math. Helv., in press.
57. W. Pawłucki, *Le théorème de Puiseux pour une application sous-analytique*, Bull. Polish Acad. Sci. Math., 32 (1984), pp. 555–560.
58. F. Pham and B. Teissier, *Fractions lipschitziennes d'une algebre analytique complexe et saturation de zariski*, Prépublications Ecole Polytechnique No. M17.0669 (1969). https://hal.archives-ouvertes.fr/hal-00384928.
59. A. Płoski, *Note on a theorem of M. Artin*, Bull. Acad. Polon. Sci. Sér. Sci. Math. Astronom. Phys., 22 (1974), pp. 1107–1109.
60. D. Popescu, *General Néron desingularization and approximation*, Nagoya Math. J., 104 (1986), pp. 85–115.
61. G. Rond, *Local topological algebraicity with algebraic coefficients of analytic sets or functions*, Algebra Number Theory, 12 (2018), pp. 1215–1231.
62. M.-F. Roy, *Subdiscriminant of symmetric matrices are sums of squares*, in Mathematics, Algorithms, Proofs, T. Coquand, H. Lombardi, and M.-F. Roy, eds., no. 05021 in Dagstuhl Seminar Proceedings, Dagstuhl, Germany, 2006, Internationales Begegnungs- und Forschungszentrum für Informatik (IBFI), Schloss Dagstuhl, Germany.
63. J. P. Speder, *Équisingularité et conditions de Whitney*, Thèse Université de Nice, (1971).
64. J. P. Speder, *Équisingularité et conditions de Whitney*, Amer. J. Math., 97 (1975), pp. 571–588.
65. B. Teissier, *The hunting of invariants in the geometry of discriminants*, in Real and complex singularities (Proc. Ninth Nordic Summer School/NAVF Sympos. Math., Oslo, 1976), 1977, pp. 565–678.
66. B. Teissier, *Résolution simultanée, ii*, in Séminaire sur les singularités des surfaces, vol. 777 of Lecture Notes in Math., Springer, Berlin, 1980, pp. 86–146.
67. B. Teissier, *Un exemple de classe d'équisingularité irrationnelle*, C. R. Acad. Sci. Paris Sér. I Math., 311 (1990), pp. 111–113.
68. R. Thom, *La stabilité topologique des applications polynomiales*, Enseign. Math. (2), 8 (1962), pp. 24–33.
69. R. Thom, *Local topological properties of differentiable mappings*, in Differential Analysis, Bombay Colloq, Oxford Univ. Press, London, 1964, pp. 191–202.
70. D. Trotman, *Stratification theory*, in Handbook of Geometry and Topology of Singularities, vol. I, J. L. C. Molina, J. Seade, and L. D. Tráng, eds., in press, pp. 231–260.
71. D. J. A. Trotman, *Stability of transversality to a stratification implies Whitney (a)-regularity*, Invent. Math., 50 (1978/79), pp. 273–277.
72. A. N. Varchenko, *Algebro-geometrical equisingularity and local topological classification of smooth mappings*, in Proceedings of the International Congress of Mathematicians (Vancouver, B.C., 1974), Vol. 1, 1975, pp. 427–431.
73. A. N. Varčenko, *Theorems on the topological equisingularity of families of algebraic varieties and families of polynomial mappings*, Izv. Akad. Nauk SSSR Ser. Mat., 36 (1972), pp. 957–1019.
74. A. N. Varčenko, *The connection between the topological and the algebraic-geometric equisingularity in the sense of Zariski*, Funkcional. Anal. i Priložen., 7 (1973), pp. 1–5.
75. A. N. Varčenko, *Local topological properties of analytic mappings*, Izv. Akad. Nauk SSSR Ser. Mat., 37 (1973), pp. 883–916.
76. A. N. Varčenko, *Local topological properties of differentiable mappings*, Izv. Akad. Nauk SSSR Ser. Mat., 38 (1974), pp. 1037–1090.
77. O. Villamayor U., *On equiresolution and a question of Zariski*, Acta Math., 185 (2000), pp. 123–159.

78. H. Whitney, *Local properties of analytic varieties*, in Differential and Combinatorial Topology (A Symposium in Honor of Marston Morse), Princeton Univ. Press, Princeton, N. J., 1965, pp. 205–244.
79. H. Whitney, *Tangents to an analytic variety*, Ann. of Math. (2), 81 (1965), pp. 496–549.
80. H. Whitney, *Complex analytic varieties*, Addison-Wesley Publishing Co., Reading, Mass.-London-Don Mills, Ont., 1972.
81. O. Zariski, *A theorem on the Poincaré group of an algebraic hypersurface*, Ann. of Math. (2), 38 (1937), pp. 131–141.
82. O. Zariski, *Studies in equisingularity. I. Equivalent singularities of plane algebroid curves*, Amer. J. Math., 87 (1965), pp. 507–536.
83. O. Zariski, *Studies in equisingularity. II. Equisingularity in codimension* 1 *(and characteristic zero)*, Amer. J. Math., 87 (1965), pp. 972–1006.
84. O. Zariski, *Studies in equisingularity. III. Saturation of local rings and equisingularity*, Amer. J. Math., 90 (1968), pp. 961–1023.
85. O. Zariski, *Some open questions in the theory of singularities*, Bull. Amer. Math. Soc., 77 (1971), pp. 481–491.
86. O. Zariski, *On equimultiple subvarieties of algebroid hypersurfaces*, Proc. Nat. Acad. Sci. U.S.A., 72 (1975), pp. 1425–1426.
87. O. Zariski, *The elusive concept of equisingularity and related questions*, in Algebraic geometry (J. J. Sylvester Sympos., Johns Hopkins Univ., Baltimore, Md., 1976), 1977, pp. 9–22.
88. O. Zariski, *Foundations of a general theory of equisingularity on r-dimensional algebroid and algebraic varieties, of embedding dimension $r+1$*, Amer. J. Math., 101 (1979), pp. 453–514.
89. O. Zariski, *Addendum to my paper: "Foundations of a general theory of equisingularity on r-dimensional algebroid and algebraic varieties, of embedding dimension $r+1$" [Amer. J. Math.* **101** *(1979), no. 2, 453–514; MR 81m:14005]*, Amer. J. Math., 102 (1980), pp. 649–651.

Chapter 5
Intersection Homology

Jean-Paul Brasselet

Abstract The famous duality theorems for compact oriented manifolds: Poincaré duality between cohomology and homology, and Poincaré-Lefschetz duality, intersection between cycles, are no longer true for a singular variety. A huge and fantastic step forward was taken by Mark Goresky and Robert MacPherson by the simple but brilliant idea of rediscovering duality by restricting oneself to chains only meeting the singular part of a stratified singular variety in controlled dimensions. Intersection homology was born. In this survey, we recall the first geometric definition as well as the theoretical sheaf definition allowing to describe the main properties of the intersection homology. Fruitful and unexpected developpments have been obtained in the context of singular varieties. For instance de Rham's theorem and Lefschetz's fixed point theorem find their place in the theory of intersection homology. The same is true for Morse's theory (see the Mark Goresky's survey in this Handbook, Chap. 5, Vol. I). In the last section, we provide some applications of intersection homology, for example concerning toric varieties or asymptotic sets. It must be said that the main application and source, itself, of innumerable applications is the fascinating and fruitful topic of perverse sheaves, which unfortunately it is not possible to develop in such a survey.

Contents

J.-P. Brasselet (✉)
I2M, CNRS and Université d'Aix-Marseille, Marseille, France
e-mail: jean-paul.brasselet@univ-amu.fr

J. L. Cisneros-Molina et al. (eds.), *Handbook of Geometry and Topology of Singularities II*, https://doi.org/10.1007/978-3-030-78024-1_5

5.1 Introduction

In the case of manifolds, global homological invariants like Betti numbers enjoy remarkable duality properties as stated by Poincaré (1893) and Lefschetz (1926). For smooth manifolds, the de Rham theorem (1931) and Morse theory (1934) show that it is possible to compute such topological invariants using differential forms and smooth functions. Unfortunately, all these beautiful results fail to hold for singular varieties. In an attempt to generalize the powerful theory of characteristic numbers to the singular case, Mark Goresky and Robert MacPherson noticed about 1973 that the failure of Poincaré duality is caused by the lack of transversal intersection of cycles on the singular locus. As a remedy, they introduced chains with well controlled intersection behaviour on the singular locus. These "intersection chains" form a complex that yields a new (co-) homology theory, called "intersection homology". As a key point, the new theory yields an intersection product with suitable duality properties.

That new approach turned out to be extremely fruitful, far beyond the original purpose at hand, and stimulated a whole wealth of unexpected developments. Already at an early stage in the development of intersection homology, the original geometric construction of intersection chains has been recast into the formalism of sheaf theory and (hyper-) cohomology. The powerful machinery thus made available has been indispensable for the development of the theory; yet it bears the risk to hide the beautiful geometry that lies at the bottom.

The article is divided into five main sections: In the first Sect. 5.2 the classical results in the manifold case are presented and examples show their failure for singular varieties.

Section 5.3 is devoted to the main tools in the frameworks of sheaf theory and derived categories. Definitions are provided and notations are fixed.

Section 5.4 is devoted to the definition of intersection homology, both in the PL and in the topological situations. The local calculus eventually leads to "sheafify" the original geometric approach, thus obtaining the intersection sheaf complexes. In this context, the Deligne sheaf complex is of fundamental importance.

Section 5.5 shows how several important concepts and results carry over from the usual (co-) homology of manifolds to intersection homology of singular varieties: first basic properties Sect. 5.5.1, functoriality Sect. 5.5.2, the Lefschetz fixed point theorem Sect. 5.5.3, Morse theory Sect. 5.5.4, de Rham theorem Sect. 5.5.5, and cohomology operations like Steenrod squares, cobordism and Wu classes Sect. 5.5.6.

Section 5.6 is a supplement and thus of a different nature: Here are collected various applications and generalizations that deserve mention, but where an appropriate introduction would by far exceed the scope of the present survey. Therefore brief sketches and suitable references are presented.

There is a vast literature consisting of research articles, conference papers, surveys, books, course notes etc. dealing with intersection homology and its implications and generalizations, some including historical comments. The first mention is for the surveys by MacPherson [127, 128], Goresky [89], Kleiman [117], the conferences in the Bourbaki Seminar by Brylinski [45] and Springer [171], and surveys by Friedman [79], Klinger [118]. This short list is far from being exhaustive.

Among the books dedicated to intersection homology and perverse sheaves, let mention those by Borel et al. [18], Kirwan [116], Goresky-MacPherson [96], Schürmann [167], Maxim [130] Dimca [66].

More specialized surveys are for instance: on de Rham theorem [24], on Morse theory [126], on combinatorial toric intersection homology [70], on perverse sheaves [125, 161], etc. This list presents only a small selection.

This concise overview of such an extensive theory is of mainly introductory character and remains thus necessarily incomplete; yet the author hopes that the reader will deepen the interest in this fascinating subject.

The author is very grateful to Gottfried Barthel, Karl-Hienz Fieseler, Ludger Kaup, Alice Libardi, Thủy Nguyễn Thị Bích, Eliris Rizziolli, Marcelo Saia, and to the anonymous referee for many helpful corrections and suggestions on an earlier version of this paper.

Important note: The concept of intersection homology was first defined by Goresky and MacPherson in the framework of PL-spaces and PL-stratifications ([92] see also [18, Chaps. I–III]). In this framework, definitions are intuitive and geometric, it is possible to make explicit figures and proofs are nice, ingenious, delicate but often technical. In their paper [94] Goresky and MacPherson consider the more general framework of topological spaces and topological stratifications. The use of tools such as sheaf theory and derived categories, together with the notion of Deligne sheaf, makes proofs easier and opens the door to deeper results and to more applications (see also [18, Chaps. IV and V]).

Both viewpoints are important and have their own advantages and disadvantages and it would be a mistake to hide one of them. The first one provides the motivation and the meaning of the concept, the second one provides simpler proofs, as well as a huge amount of extensions and applications. The introduction of Habegger's chapter in the Borel book [18, Chap. II] explicits these roles.

5.2 Classical Results—Poincaré and Poincaré-Lefchetz

In order to understand the introduction of intersection homology it is useful to recall some elementary properties for manifolds.

An *n-manifold* (or n-topological manifold) is a (non-empty, Hausdorff) topological space X such that each point admits a neighborhood homeomorphic with an open subset of the Euclidean space $\mathbb{R}^n$.

Let X be an n-dimensional compact, connected, oriented and without boundary smooth manifold. The Poincaré duality says that the p and $(n-p)$ Betti numbers of X agree. It first stated without proof in [152] then Poincaré gave a proof of the theorem using topological intersection theory in his 1895 paper Analysis Situs [153]. Heegaard [104], provided a counter-example to Poincaré's formula and, finally, Poincaré provided a new proof performed in terms of dual cell decompositions [154, 155]. For historical details, see for example [157]. In order to simplify, in this introduction, homology and cohomology groups are with $\mathbb{Z}$ coefficients.

The Poincaré result is presented in two ways: using intersection of cycles, i.e. showing the existence of an intersection morphism

$$H_{n-p}(X) \otimes H_{n-q}(X) \xrightarrow{\bullet} H_{n-(p+q)}(X)$$

(see Sect. 5.2.5) and showing the existence of a Poincaré duality isomorphism

$$PD : H^p(X) \xrightarrow{\cong} H_{n-p}(X),$$

using dual cells (see Sect. 5.2.6). The two definitions are linked by the commutative diagram, where $\cup$ denotes the usual cup-product.

$$\begin{array}{ccc} H^p(X) \otimes H^q(X) & \xrightarrow{\ \cup\ } & H^{p+q}(X) \\ \Big\downarrow{\scriptstyle PD \otimes PD} & & \Big\downarrow{\scriptstyle PD} \\ H_{n-p}(X) \otimes H_{n-q}(X) & \xrightarrow{\ \bullet\ } & H_{n-(p+q)}(X). \end{array} \tag{5.1}$$

The first part of this section consists of some useful definitions and notations. Then the classical dualities (Poincaré and Poincaré-Lefschetz) for manifolds are recalled and counter-examples in the situation of singular varieties are provided.

5.2.1 *PL*-Structures

A piecewise-linear structure, PL-structure on a topological space X is a class of locally finite (simplicial) triangulations such that any subdivision of one of them belongs to the class and two of them admit a common subdivision.

When endowed with a PL-structure, the space is said a PL-space. Not all topological space admit a PL-structure and when such structure exists it is, in general, not unique.

A triangulation of a PL-space X is a triangulation of the corresponding class. That is a simplicial complex K whose geometric realization $|K|$ is homeomorphic to X. The space is said triangulated and one writes $X = |K|$.

The advantage of having a whole class of triangulations is that any open subset $U \subset X$ inherits a PL-structure. This property is convenient for the construction of sheaves (see in particular Example 5.3.2).

A manifold equipped with a structure of PL-space is called PL-manifold. In a triangulation K of a PL-manifold, every $(n-1)$-simplex is a face of exactly two n-simplices. This property is one of the conditions for a PL-space to be a PL-pseudomanifold.

5.2.2 *Pseudomanifolds*

Definition 5.2.1 The (non-empty, paracompact, Hausdorff) topological space X is an *n-pseudomanifold* if there is a closed subspace $\Sigma \subset X$ such that:

1. $X \setminus \Sigma$ is an n-dimensional manifold dense in X.
2. $\dim \Sigma \leq n - 2$.

The subspace Σ of the pseudomanifold X contains the subset of singular points of X i.e. the points which do not admit a neighborhood homeomorphic to a ball and whose boundary is homeomorphic to a sphere.

A PL-pseudomanifold X of dimension n is an n-dimensional PL-space X containing a closed PL-subspace Σ of codimension at least 2 such that $X - \Sigma$ is an n-dimensional PL-manifold dense in X.

Equivalently, given a triangulation $X = |K|$, then $|K|$ is the union of the n-simplices and each $n-1$-simplex is face of exactly two n-simplices.

A connectivity condition of the set $X - \Sigma$ is sometimes added. The connected PL-pseudomanifold X is oriented if there exists a compatible orientation of all n-simplices. In the connected and oriented situation of a PL-pseudomanifold, the conditions ensure existence of a fundamental class $[X]$. Namely, given a triangulation $X = |K|$ of an n-dimensional connected and oriented PL-pseudomanifold, the sum of all (oriented) n-simplices is a cycle whose class is the fundamental class. The original article by Goresky and MacPherson suppose the PL-pseudomanifold to be compact and oriented. These hypothesis are dropped in the further articles.

The pinched torus (Fig. 5.1 and Example 5.2.6) and the suspension of the torus (Example 5.5.10) are examples of connected and oriented PL-pseudomanifolds.

5.2.3 *Stratifications*

Dealing with singular spaces, the notion of stratification is one of the most important tool. The main reference for the definitions and results is the Trotman's survey in this Handbook, vol I [179, Chap. 4] (see also [128]).

A (topological) stratification $\mathcal{S}$ of the n-dimensional pseudomanifold X is the data of a filtration

$$(\mathcal{S}) \qquad X = X_n \supset X_{n-1} = X_{n-2} \supset \cdots \supset X_1 \supset X_0 \supset X_{-1} = \emptyset \tag{5.2}$$

by closed subspaces such that

- every stratum $S_i = X_i - X_{i-1}$ is either empty or a finite union of i-dimensional smooth submanifolds of X,
- each point x in S_i admits a *distinguished neighborhood* $U_x \subset X$ together with a homeomorphism

$$\phi_x : U_x \to \mathbb{B}^i \times \mathring{c}(L) \tag{5.3}$$

  (local triviality property) where:

  - $\mathbb{B}^i$ is an open ball in $\mathbb{R}^i$,
  - the "link" L (called the link of the stratum S_i) is a compact $(n-i-1)$-dimensional pseudomanifold independent (up to homeomorphism) of the point x in the stratum S_i and filtered by:

$$L = L_{n-i-1} \supset L_{n-i-3} \supset \cdots \supset L_0 \supset L_{-1} = \emptyset,$$

- $\mathring{c}(L)$ is the open cone over L defined by $\mathring{c}(L) = L \times [0,1[/(x,0) \sim (x',0)$, and filtred by

$$(\mathring{c}(L))_0 = \{0\} \quad \text{and} \quad (\mathring{c}(L))_k = \mathring{c}(L_{k-1}) \quad \text{if } k > 0.$$

By definition, one has $\mathring{c}(\emptyset) = \{point\}$.
Moreover, the homeomorphism ϕ_x preserves the stratifications of U_x and $\mathbb{B}^i \times \mathring{c}(L)$ respectively, that is there are restriction homeomorphisms

$$\phi_x|_{X_j} : U_x \cap X_j \to \mathbb{B}^i \times \mathring{c}(L_{j-i-1}), \quad \text{for } j \geq i.$$

In particular stratifications which satisfy the Whitney conditions (see [89], [179]) satisfy the topological local triviality property (A' Campo [18, Chap. IV]):

A PL-stratification $\mathcal{S}$ of the n-dimensional PL-pseudomanifold X is a stratification such that all involved subspaces are PL-subspaces and the local triviality property holds in the PL-category.

5.2.4 Borel-Moore Homology

In the following, G will denote an R-module, for R a PID. For example, G can be $\mathbb{Z}$, $\mathbb{Q}$, $\mathbb{R}$ or $\mathbb{C}$. In this (sub)section, X is a connected, oriented, not necessarily compact n-dimensional PL-manifold or PL-pseudomanifold.

Given a triangulation $X = |K|$, the complex of possibly infinite simplicial chains of K with coefficients in G is denoted by $C_*(K; G)$. A chain ξ in $C_i(K; G)$ is written $\xi = \sum \xi_\sigma \sigma$ where σ are oriented i-simplices in K and ξ_σ are elements of G. It has a canonical image in $C_i(K'; G)$ for any subdivision K' of K. Two chains in $C_i(K_1; G)$ and $C_i(K_2; G)$ are identified if their image in a common subdivision coincide. The group $C_i(X; G)$ of PL-geometric chains with closed supports of X is the direct limit under refinement of the groups $C_i(K; G)$ over all triangulations of X.

The support of $\xi \in C_i(K; G)$ is the union of the closed simplices such that $\xi_\sigma \neq 0$ and is denoted by $|\xi|$, i t does not depend on subdivision, thus the support of $\xi \in C_i(X; G)$ is well defined.

Using the usual boundary operator, the complex of chains $C_*(X; G)$ is well defined and its homology, denoted by $H_*(X; G)$ is called homology with closed supports of X, or Borel-Moore homology of X [17] (see also "Homologie de deuxième espèce" in Cartan [48, Exposé 5, Sect. 6]).

The subcomplex of chains with compact supports is denoted by $C^c_*(X; G)$ and its homology $H^c_*(X; G)$ is the homology with compact support. If X is compact, the two homology groups coincide.

5.2.5 Poincaré Duality Homomorphism

The idea, for defining Poincaré duality "à la Poincaré", is to associate to a given triangulation K of an n-dimensional manifold X a decomposition of X into "cells", in such a way that there is a one-to-one correspondence between p-dimensional simplices and $(n-p)$-dimensional dual cells (called "Polyèdre réciproque" in [154, Sect. VII]).

Let X be a triangulated, compact and oriented n-dimensional PL-pseudomanifold $X = |K|$ such that the triangulation K itself is the first barycentric subdivision of a triangulation of X.

A p-elementary cochain is an (oriented) p-simplex σ, denoted by σ^* when considered as a p-cochain. A p-cochain with coefficients in G, element of $C^p(K;G)$, is a formal sum $\sum g_i\sigma_i^*$ where $g_i \in G$ and the σ_i are (oriented) p-simplices in K.

The coboundary $\delta\sigma^*$ of the p-elementary cochain σ^* is defined to be the $(p+1)$-cochain

$$\delta\sigma^* = \sum [\sigma : \tau]\tau^*$$

where the sum involves all $(p+1)$-simplices τ such that σ is a face of τ (denoted $\sigma < \tau$). The incidence number $[\sigma : \tau]$ is $+1$ if the orientation of σ is the one as boundary of τ and -1 otherwise. This defines the homomorphism

$$\delta^p : C^p(K;G) \to C^{p+1}(K;G)$$

by linearity.

Considering a first barycentric subdivision K' of K, the barycenter of every simplex σ in K is denoted by $\hat{\sigma}$. The simplices in K' whose first vertex is $\hat{\sigma}$ are all simplices on the form $(\hat{\sigma}, \hat{\sigma}_{i_1}, \ldots, \hat{\sigma}_{i_q})$ with $\sigma < \sigma_{i_1} < \cdots < \sigma_{i_q}$. The union of these simplices, is called the the dual block of σ and is denoted by $D(\sigma)$. One has

$$D(\sigma) = \{\tau \in K' \, : \, \tau \cap \sigma = \{\hat{\sigma}\}\}. \tag{5.4}$$

The dual block $D(\sigma)$ has dimension $(n-p)$, i t i s adowed with an orientation such that the orientation of $D(\sigma)$ followed by the orientation of σ is the orientation of X (see [23, 177]).

The Poincaré homomorphism (at the level of chains and cochains) is the map

$$PD : C^p(K;G) \to C_{n-p}(K';G)$$

defined by $PD(\sigma^*) = D(\sigma)$ and extended by linearity. One has

$$PD(\delta\sigma^*) = \partial PD(\sigma^*).$$

The correspondence "simplex" → "dual block" sends K-cochains to K'-chains. By this correspondence, cocycles are sent to cycles and coboundaries to boundaries.

The Poincaré homomorphism is then well defined:

$$PD : H^p(K;G) \longrightarrow H_{n-p}(K'G).$$

As it is well known, the homology and cohomology groups of X do not depend on the given triangulation. The Poincaré duality morphism is realised by the cap-product by the fundamental class $[X]$:

$$PD : H^p(X;G) \overset{\bullet\cap[X]}{\longrightarrow} H_{n-p}(X;G)$$

The main Poincaré's result is:

Theorem 5.2.2 • *In a compact oriented manifold, the dual blocks are cells, i.e. the dual block of a p-simplex σ is homeomophic to an $(n-p)$-ball and its boundary is homeomorphic to an $(n-p-1)$-sphere.*

• *In a compact oriented manifold, the Poincaré morphism is an isomorphism.*

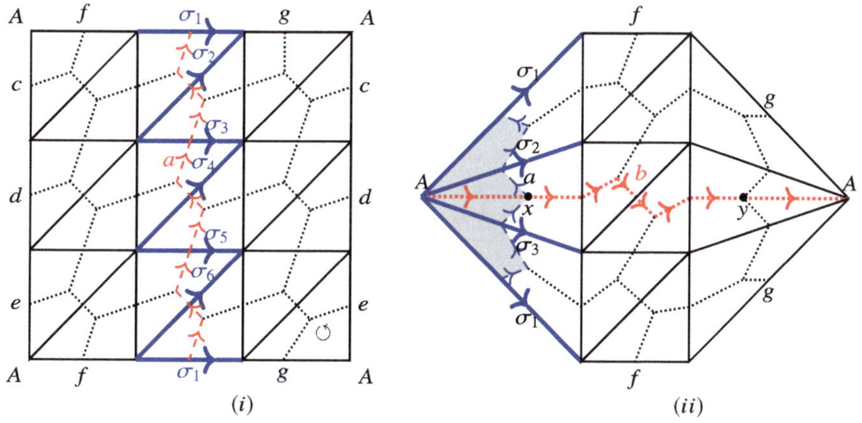

Fig. 5.1 Cycles and cocycles on the torus (i) and the pinched torus (ii)

Example 5.2.3 Examples of computations for a manifold and a pseudomanifold.

Figure 5.1i and ii are planar representations of the torus and the pinched torus, with suitable identification of the simplices of the boundary and with given compatible orientation of all 2-simplices.

In the torus, the cochain $\alpha = \sigma_1^* + \sigma_2^* + \ldots + \sigma_6^*$ is a cocycle, not a coboundary. The dual chain $a = PD(\alpha)$ is a cycle, not a boundary. In the same way, by symmetry with respect to the first diagonal, one has a "horizontal" cocycle β and dual cycle $b = PD(\beta)$. In the torus, the Poincaré homomorphism is an isomorphism (here $G = \mathbb{Z}$):

$$H^1(T;\mathbb{Z}) \to H_1(T;\mathbb{Z}), \qquad \mathbb{Z}\alpha \oplus \mathbb{Z}\beta \to \mathbb{Z}a \oplus \mathbb{Z}b.$$

On the pinched torus, which is a singular variety, the Poincaré homomorphism is no longer an isomorphism.

On the one hand, the cocycle $\alpha = \sigma_1^* + \sigma_2^* + \sigma_3^*$ is not a coboundary. Indeed, the coboundary of the vertex A consists of all 1-dimensional simplices of which A is a vertex and α is only half of it. The cohomology class of α is not zero. The dual of the cocycle α is the cycle a. This cycle is a boundary (the boundary of the gray part), its homology class is zero: the Poincaré morphism maps the non-zero class of α on the zero class of a. It is not injective.

On the other hand, the (red) cycle b going from A to A is not a boundary, its homology class is not zero and generates the 1-dimensional homology. But it is easy to see that b is not the dual of a cochain. The Poincaré morphism is not surjective.

The Poincaré morphism of the pinched torus is neither injective, nor surjective, although it is a morphism from $\mathbb{Z}$ to $\mathbb{Z}$:

$$\mathbb{Z}\alpha \cong H^1(X;\mathbb{Z}) \overset{\bullet\cap[X]}{\longrightarrow} H_1(X;\mathbb{Z}) \cong \mathbb{Z}\, b.$$

5.2.6 *Poincaré—Lefschetz Homomorphism*

In his 1895 paper [153, Sect. 9] (corrected in [154, 155]), Poincaré gave a definition of intersection of two oriented and complementary dimensional cycles in a compact oriented manifold. Lefschetz, in 1936 [122] defined the intersection of an i-chain a and a j-chain b in a compact oriented n-manifold M whenever $|a| \cap |b|$ contains simplices of dimension at most $i + j - n$, and gave a formula for the multiplicity in $a \cap b$ of an $i + j - n$-simplex $\sigma \subset |a| \cap |b|$ which is local, in the sense that it depends only on the behavior of a and b near an interior point of σ.

Let X be a smooth PL-manifold, two cycles a and b are said *dimensionally transverse* if either they do not meet or their dimensions satisfy the formula:

$$\text{codim}\ (|a| \cap |b|) = \text{codim}\ |a| + \text{codim}\ |b|.$$

Theorem 5.2.4 (See [120–122] and [176, Sect. 5] for a summary) *In a compact oriented smooth PL-manifold, the intersection of two dimensionally transverse cycles with appropriately defined orientations and multiplicities is a cycle.*

In a compact oriented PL-manifold X, if two dimensionally transverse cycles a and b have complementary dimensions, then the intersection $a \cap b$ is a finite number of points $\{x_i\}$. The cycles being oriented, in each of the points x_i Lefschetz defines the local intersection index $I(a, b; x_i)$, depending on orientations and multiplicities [122, 153, 154]. For elementary cycles (i.e. with multiplicities $+1$), the index $I(a, b; x_i)$ is $+1$ if the orientation of a followed by the orientation of b is the orientation of X and -1 otherwise, then extend by linearity. The intersection index

$$I(a,b) = \sum_{x_i \in a \cap b} I(a,b;x_i).$$

defines an intersection product

$$\begin{array}{ccc} C_{n-p}(X;\mathbb{Z}) \times C_p(X;\mathbb{Z}) & \longrightarrow & \mathbb{Z} \\ a \quad , \quad b & \leadsto & I(a,b) \end{array} \tag{5.5}$$

which associates to each pair of oriented, dimensionally transverse and complementary dimensional cycles (a,b) the intersection index $I(a,b)$.

The intersection index $I(a,b)$ does not depend on the representative of the homology classes of the cycles a and b.

Theorem 5.2.5 (Poincaré-Lefschetz duality [120, 122, 153, 154]) *In a compact oriented smooth* PL*-manifold, the intersection product (5.5) induces a bilinear map*

$$H_{n-p}(X;\mathbb{Z}) \times H_p(X;\mathbb{Z}) \longrightarrow H_0(X;\mathbb{Z}) \xrightarrow{\epsilon} \mathbb{Z}$$

which is non-degenerate when tensored by the rational numbers. Here ϵ *is the evaluation map* $\epsilon : \sum n_i\{x_i\} \mapsto \sum n_i$.

Let X be a singular variety, then the Poincaré-Lefschetz duality is no longer true. The pinched torus is a classical example:

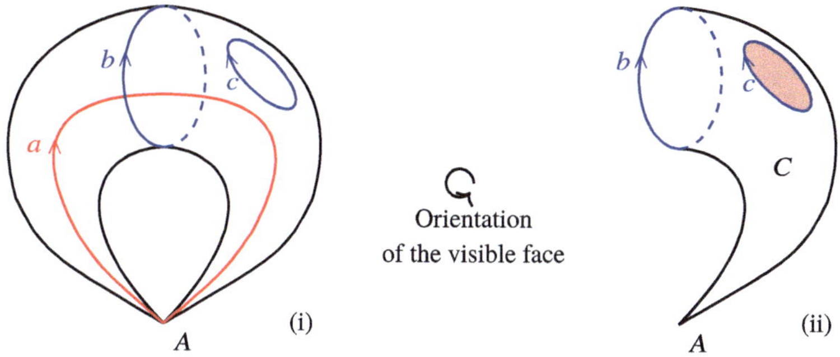

Fig. 5.2 Transverse cycles in the pinched torus

Example 5.2.6 (The pinched torus) Consider the pinched torus.

The area C in Fig. 5.2ii is a chain whose boundary is $c - b$. The cycles b and c are homologous. However, the intersection indices are $I(a,b) = +1$ and $I(a,c) = 0$. There is no intersection product at the level of homology classes.

5.3 The Useful Tools: Sheaves—Derived Category

In the previous section, the spaces considered were mainly compact and non singular. The notions of duality and intersection can be generalized to singular spaces using Borel-Moore homology on the one hand and sheaf theory on the other hand.

General useful references for these sections are Godement and Bredon (see [86, Chapitre II] and [36]). The interested reader will find a history of sheaf theory in the Christian Houzel article [106], in particular the passage from closed supports (Jean Leray) to open supports (Henri Cartan).

5.3.1 Sheaves

Let X be a topological PL-pseudomanifold. Let R be a PID, that may be sometimes $\mathbb{Z}$ or even a field such as $\mathbb{R}$, $\mathbb{Q}$ or $\mathbb{C}$. A sheaf on X will be a sheaf $\mathcal{A}$ of R-modules. The category of sheaves on X is denoted by $Sh(X)$. The constant sheaf is denoted by $\mathbf{R}_X$.

The set of sections of the sheaf $\mathcal{A}$ over an open subset U of X is denoted by $\Gamma(U, \mathcal{A})$. Given a family of supports Φ, the subset of elements $s \in \Gamma(X, \mathcal{A})$ for which support of s belongs to Φ is denoted by $\Gamma_\Phi(X, \mathcal{A})$. The families considered are mainly the family of closed supports, the family of compact supports c and for a subspace $A \subset X$ the family (A) of supports whose elements are closed subsets contained in A.

The stalk at a point $x \in X$ of the sheaf $\mathcal{A}$ is denoted by $\mathcal{A}_x$. The restriction of $\mathcal{A}$ to a subspace $Y \subset X$ is denoted by $\mathcal{A}|_Y$ or simply $\mathcal{A}_Y$.

A sheaf $\mathcal{L}$ on the topological space X is called *locally constant* if there is an open covering $\{U_i\}$ of X and a family of R-modules $\{L_i\}$ such that $\mathcal{L}|_{U_i}$ is the constant sheaf on U_i represented by the R-module $\{L_i\}$. Equivalently, every point $x \in X$ has a neighborhood U such that the restriction maps

$$\mathcal{L}_x \leftarrow \Gamma(U; \mathcal{L}) \rightarrow \mathcal{L}_y$$

are isomorphisms for all $y \in U$.

5.3.2 System of Local Coefficients

The notion of system of local coefficients comes from Steenrod [174]. In fact in his introduction, Steenrod wrote that he generalizes an idea originating from Whitney (1940), who, in turn credits the idea to de Rham (1932). Also Steenrod claims that the notion is equivalent to the Reidemeister Überdeckung (1935) [159]. Steenrod provides applications of the notion, in particular full duality and intersection theory in a non-orientable manifold [174, Sect. 14]. Later, Steenrod applied that notion in 1951 in his book "The Topology of Fibre Bundles" when defining characteristic classes (Stiefel-Whitney and Chern classes) by obstruction theory (see [175, Sect. 30–31]).

The interest of local systems is well demonstrated by the example given by MacPherson [128, p. 19] of a local system which makes intersection homology interesting even when the space is nonsingular.

A *local coefficient system* (or local system) of R-modules on a topological space X is a locally constant sheaf $\mathcal{L}$ of R-modules.

If X is connected, then it is possible to use a single R-module L instead of a family L_i. If X is not connected, a local coefficient system is determined by the data of a base point x_i in C_i and a representation $\rho : \pi_1(X, x_i) \to Aut(L_i)$ for each connected component C_i of X.

Example 5.3.1 An exemple of local system is given by the orientation sheaf $\mathcal{O}_X$ on a (not necessarily orientable) n-dimensional manifold. That is the sheaf associated to the presheaf

$$U \mapsto H_n(X, X \setminus U; R).$$

If $\partial X = \emptyset$, then $\mathcal{O}_X$ is a locally constant sheaf with stalks isomorphic to R. It is constant if X is orientable.

5.3.3 Complexes of Sheaves

A bounded complex of sheaves $\mathcal{A}^\bullet$ is a sequence

$$\dots \longrightarrow \mathcal{A}^{p-1} \xrightarrow{d^{p-1}} \mathcal{A}^p \xrightarrow{d^p} \mathcal{A}^{p+1} \longrightarrow \dots \qquad p \in \mathbb{Z}$$

such that $d^p \circ d^{p-1} = 0$ for all p and $\mathcal{A}^p = 0$ for $|p|$ sufficiently large. If necessary to specify the complex, the differential will be denoted by $d^p_{\mathcal{A}^\bullet}$.

A sheaf $\mathcal{A}$ can be regarded as a complex of sheaves $\mathcal{A}^\bullet$ with $\mathcal{A}^0 = \mathcal{A}$, $\mathcal{A}^p = 0$ for $p \neq 0$, and $d^p = 0$ for all p. In this case, the complex $\mathcal{A}^\bullet$ is said to be concentrated in degree 0.

Given a complex of sheaves, the shifted complex $\mathcal{A}[n]^\bullet$ is defined by $\mathcal{A}[n]^p = \mathcal{A}^{n+p}$ and $d_{\mathcal{A}[n]^\bullet} = (-1)^n d_{\mathcal{A}^\bullet}$.

The sheaf of sections associated with a complex of sheaves $\mathcal{A}^\bullet$ assigns to every open subset $U \subset X$ the chain complex

$$\dots \to \Gamma(U; \mathcal{A}^{p-1}) \to \Gamma(U; \mathcal{A}^p) \to \Gamma(U; \mathcal{A}^{p+1}) \to \dots.$$

The p-th cohomology sheaf $\mathcal{H}^p(\mathcal{A}^\bullet)$ associated with $\mathcal{A}^\bullet$ is the sheafification ([86, Chapitre II, Sect. 1.2]) of the presheaf whose group of sections over U is the p-th homology group of $\mathcal{A}^\bullet$. The stalk at a point $x \in X$ of the sheaf $\mathcal{H}^p(\mathcal{A}^\bullet)$ is $\mathcal{H}^p(\mathcal{A}^\bullet)_x \cong \mathcal{H}^p(\mathcal{A}^\bullet_x)$.

Borel and Moore [17] define *cohomologically locally constant* (denoted CLC) complex of sheaves if the associated cohomology sheaves are locally constant. A complex of sheaves $\mathcal{A}^\bullet$ is said *(cohomologically) constructible* with respect of a filtration (5.2) of X if all $\mathcal{A}^\bullet|_{(X_i - X_{i-1})}$ are CLC and their stalk cohomology is finitely generated.

The complex $\mathcal{A}^\bullet$ is said *PL-(cohomologically) constructible* if it is bounded and (cohomologically) constructible with respect of a filtration of X by closed PL-subsets. Finally, the complex $\mathcal{A}^\bullet$ is said *topologically (cohomologically) constructible* if it is bounded and (cohomologically) constructible with respect to a topological filtration of X.

Henceforth, we will adopt the modern shorthand of replacing the words "(cohomologically) constructible" simply with "constructible". As in [94], all complexes of sheaves considered will be topologically constructible.

Example 5.3.2 The *sheaf complex of PL-chains with closed suports.*

Let X be a connected, oriented, not necessarily compact n-dimensional PL-manifold or PL-pseudomanifold. In Sect. 5.2.4 the Borel-Moore homology chains $C_i(X; G)$ with coefficients in an abelian group have been defined. The same definition applies with coefficients in a local system $\mathcal{L}$, denoted by $C_i(X; \mathcal{L})$.

In a first step the presheaf $U \mapsto C_i(U; \mathcal{L})$ for U open in X, is defined as follows. Let $V \subset U$ be two open subsets in X, the natural restriction maps

$$\rho_{VU} : C_i(U; \mathcal{L}) \to C_i(V; \mathcal{L}) \tag{5.6}$$

are defined in the following way: (see also [94, Sect. 2.1]) For a chain $\xi \in C_i(U; \mathcal{L})$, there is a locally finite triangulation K_U of U such that ξ can be written $\sum_{\sigma \in K_U} \xi_\sigma \sigma$ with $\xi_\sigma \in \mathcal{L}_\sigma$ and $\mathcal{L}_\sigma$ is the (constant) value of $\mathcal{L}$ on σ. Any triangulation of V admits a subdivision K_V such that every simplex υ in K_V is contained in a simplex $\sigma(\upsilon)$ of a subdivision of K_U and such that $\dim \upsilon = \dim \sigma(\upsilon)$. Considering orientations of all simplices of the triangulations, the chain $\rho_{VU}(\xi) \in C_i(V; \mathcal{L})$ is defined by

$$\rho_{VU}(\xi) = \sum_{\upsilon \in K_V} (-1)^{(\upsilon : \sigma(\upsilon))} \xi_{\sigma(\upsilon)} \upsilon$$

where the sign is $+1$ i fυ and $\sigma(\upsilon)$ have the same orientation and -1 otherwise.

The boundary $\partial_i : C_i(U; \mathcal{L}) \to C_{i-1}(U; \mathcal{L})$ is defined in the following way: A chain $\xi \in C_i(U; \mathcal{L})$ is written $\sum_{\sigma \in K_U} \xi_\sigma \sigma$ for a locally finite triangulation K_U of U and $\xi_\sigma \in \mathcal{L}_\sigma$. Let τ be a face of σ and $\rho^\sigma_\tau : \mathcal{L}_\sigma \to \mathcal{L}_\tau$ the natural morphism, then

$$\partial_i(\xi) = \sum_\sigma \sum_{\tau < \sigma} [\tau : \sigma] \rho^\sigma_\tau(\xi_\sigma).\tau.$$

where the incidence number $[\tau : \sigma]$ is $+1$ if the orientation of τ is the one as boundary of σ and -1 otherwise.

Definition 5.3.3 ([94, Sect. 2.1], *see Remark* 5.4.10 *for the notation*) The Borel-Moore complex of sheaves of PL-chains $\mathcal{C}^\bullet$ on X with coefficients in $\mathcal{L}$ is defined by

$$\Gamma(U; \mathcal{C}^{-i}) = C_i(U; \mathcal{L})$$

with the above boundary.

5.3.4 *Injective Resolutions*

The injective resolutions are particularly important, by the fact that for any abelian category with enough injective objects, each R-module admits an injective resolution.

Definition 5.3.4 ([94, Sect. 1.5], [18, II, Sect. 5]) A map of complexes of sheaves $\varphi^\bullet : \mathcal{A}^\bullet \to \mathcal{B}^\bullet$ which commutes with the differentials

$$\varphi^{i+1} \circ d^i_{\mathcal{A}^\bullet} = d^i_{\mathcal{B}^\bullet} \circ \varphi^i$$

is called a *quasi-isomorphism* if it induces isomorphisms $\mathcal{H}^i(\varphi^\bullet) : \mathcal{H}^i(\mathcal{A}^\bullet) \to \mathcal{H}^i(\mathcal{B}^\bullet)$ of the cohomology sheaves of the complexes.

Definition 5.3.5 ([49, IV, Sect. 3]) Two morphisms of complexes $\varphi^\bullet : \mathcal{A}^\bullet \to \mathcal{B}^\bullet$ and $\psi^\bullet : \mathcal{A}^\bullet \to \mathcal{B}^\bullet$ are *homotopic* if there exists a collection $\{h^i : \mathcal{A}^i \to \mathcal{B}^{i-1}\}$, $i \in \mathbb{Z}$ of sheaf maps, called a homotopy, so that:

$$d^{i-1}_{\mathcal{B}^\bullet} \circ h^i + h^{i+1} \circ d^i_{\mathcal{A}^\bullet} = \phi^i - \psi^i$$

for all $i \in \mathbb{Z}$.

Definition 5.3.6 Let $K(X)$ denote the category whose objects $\mathcal{A}^\bullet$ are topologically constructible bounded complexes of sheaves on X and whose morphisms $\varphi^\bullet : \mathcal{A}^\bullet \to \mathcal{B}^\bullet$ are homotopy classes of sheaf maps which commute with the differentials (Definition 5.3.5).

Definition 5.3.7 ([86, II, Sect. 7.1], [49, V, Sect. 1], [17, Sect. 1], [36, II, Sect. 9]) A sheaf $\mathcal{I}$ is *injective* if, for any sheaf monomorphism $\mathcal{F} \to \mathcal{G}$ and any sheaf map $\mathcal{F} \to \mathcal{I}$ there exists an extension $\mathcal{G} \to \mathcal{I}$.

Coefficients are important in the definition: the sheaf of integers $\mathbb{Z}$ on a point is fine (see Definition 5.3.10) but not injective, because $\mathbb{Z}$ is not injective over itself.

If $\varphi^\bullet : \mathcal{A}^\bullet \to \mathcal{I}^\bullet$ is a quasi-isomorphism of complex of sheaves on X and if each $\mathcal{I}^\bullet$ is injective, then $\mathcal{I}^\bullet$ is called an *injective resolution* of $\mathcal{A}^\bullet$.

Main properties of injective resolutions are (see [17, Sect. 1], [36, II, Sect. 9], [86, II Sect. 7.1], [89, Sect. 4.2], and [94, Sect. 1.5]):

Proposition 5.3.8 *1. If $\varphi^\bullet : \mathcal{A}^\bullet \to \mathcal{I}^\bullet$ is a quasi-isomorphism of complex of sheaves on X and $\mathcal{I}^\bullet$ is an injective resolution of $\mathcal{A}^\bullet$, then there exists a morphism $\psi^\bullet : \mathcal{I}^\bullet \to \mathcal{A}^\bullet$ that is a homotopy inverse to $\varphi^\bullet$. Therefore $\varphi^\bullet$ is invertible in the category $K(X)$.*

2. Injective resolutions exist for any complex of sheaves of R-modules and are uniquely determined up to chain homotopy.

3. Every bounded complex of sheaves admits a canonical bounded injective resolution (see given references).

5.3.5 Hypercohomology

The p-th *hypercohomology group* $\mathbb{H}^p(X; \mathcal{A}^\bullet)$ of a complex of sheaves $\mathcal{A}^\bullet$ is the p-th cohomology group of the cochain complex

$$\ldots \to \Gamma(X; \mathcal{I}^{p-1}) \to \Gamma(X; \mathcal{I}^p) \to \Gamma(X; \mathcal{I}^{p+1}) \to \ldots,$$

where $\mathcal{I}^\bullet$ is the canonical injective resolution of $\mathcal{A}^\bullet$ ([49, XVII, Sect. 2], [94, 1.6]).

Considering sections with supports in a family of supports Φ, one defines *hypercohomology* $\mathbb{H}^p_\Phi(X; \mathcal{A}^\bullet)$ *with support in the family* Φ as

$$\mathbb{H}^p_\Phi(X; \mathcal{A}^\bullet) = H^p(\Gamma_\Phi(X; \mathcal{I}^\bullet)).$$

A quasi-isomorphism induces an isomorphism on hypercohomology. In particular, the hypercohomology groups are naturally isomorphic to the cohomology group of the single complex which is associated to the double complex $C^p(X; \mathcal{A}^q)$ (see [86, II Sect. 4.6]).

Definition 5.3.9 ([86, II, Sect. 3.5 and 3.6]) A sheaf $\mathcal{A}$ is called *soft* ("faisceau mou" in french) if any section over any closed subset of X can be extended to a global section, i.e. the restriction maps

$$\Gamma(U, \mathcal{A}) \to \Gamma(B, \mathcal{A})$$

are surjective for all open $U \subset X$ and closed subset $B \subset U$.

Definition 5.3.10 ([86, II, Sect. 3.7]) A sheaf $\mathcal{A}$ over a paracompact Hausdorff space X is called *fine* ("faisceau fin" in french) if for every locally finite open cover $\{U_i\}$ of X there are endomorphisms φ_i of $\mathcal{A}$ such that:

- for every i, φ_i is zero outside a closed subset contained in U_i,
- one has $\sum_i \varphi_i = id$.

Here, locally finite means that every point $x \in X$ admits an open neighborhood which meets a finite number of elements U_i.

Every fine sheaf is soft, but the converse is not true [86, II, Sect. 3.7].

Proposition 5.3.11 ([18, II, Sect. 5], [86]) *Let X be a paracompact topological space, if the sheaf complex $\mathcal{A}^\bullet$ consists of injective, fine or soft sheaves, then*

$$\mathbb{H}^i(X; \mathcal{A}^\bullet) = H^i(\Gamma(X; \mathcal{A}^\bullet)). \tag{5.7}$$

Examples of Hypercohomology

Let X be a n-dimensional PL-space, the following examples provide particular cases of hypercohomology groups which will be useful for considering the properties of intersection homology. Coefficients are either a R-module G or a sheaf of local coefficients $\mathcal{L}$.

Example 5.3.12 (a) Hypercohomology of the sheaf complex of PL-chains with closed supports.
The complex of sheaves of PL-chains $\mathcal{C}^\bullet$ is a complex of fine sheaves on X (see [18, Sect. 5, Note]). Hence the complex $\mathcal{C}^\bullet$ satisfies Proposition 5.3.11. One has:

$$\mathbb{H}^{-i}(X; \mathcal{C}^\bullet) \cong H^{-i}(\Gamma(X; \mathcal{C}^\bullet)) = H_i(X).$$

For every family of supports Φ one has:

$$\mathbb{H}_\Phi^{-i}(X; \mathcal{C}^\bullet) = H_i^\Phi(X). \tag{5.8}$$

(b) Hypercohomology of the constant sheaf.
Consider the *constant sheaf* $\mathbf{R}_X$ on X, viewed as a complex concentrated in degree 0, then,

$$\mathbb{H}^i(X; \mathbf{R}_X) = H^i(X)$$

is cohomology of X with closed supports. For every family of supports Φ one has:

$$\mathbb{H}_\Phi^i(X; \mathbf{R}_X) = H_\Phi^i(X). \tag{5.9}$$

5.3.6 *The (Constructible) Derived Category*

The derived category was defined by Verdier [183, 184]. An object in the derived category is a complex of sheaves. In this category, new morphisms are added so that every quasi-isomorphism has an inverse and, consequently, every quasi-isomorphism becomes an isomorphism in the derived category (Property 5.3.13). Verdier found he was able to prove his duality theorems only for complexes of sheaves $\mathcal{A}^\bullet$ whose cohomology sheaves are constructible. Since then, it has become common to focus on the *constructible* derived category, in which each object is a complex of sheaves with constructible cohomology.

The reader is assumed to be familiar with the notions of categories and functors [100, Chapitre 1]. Text books which provide useful notions are, for instance, Kashiwara-Shapira [112]) or Gelfand-Manin [85]. Classsical references for this section are [103, 183], [184, Chap. 8]. The useful tools are well presented in [94, Sects. 1.8–1.15] and the reader will find there all necessary tools and material.

For the convenience of the reader, as far as possible, conventions and notations of main references [92, 94] and [18] are used. However in the case of possible doubt the notations of [94] are privileged (see Remark 5.4.10).

The Derived Category

Let $\mathcal{A}$ and $\mathcal{B}$ be sheaves on X, let $Hom(\mathcal{A}, \mathcal{B})$ denote the abelian group of all sheaf maps $\mathcal{A} \to \mathcal{B}$. Let $\mathcal{H}om(\mathcal{A}, \mathcal{B})$ be the sheaf whose sections over an open set U are the sections $\Gamma(U; \mathcal{H}om(\mathcal{A}, \mathcal{B})) = Hom(\mathcal{A}|_U, \mathcal{B}|_U)$. If $\mathcal{A}^\bullet$ and $\mathcal{B}^\bullet$ are complexes of sheaves, $\mathcal{H}om(\mathcal{A}^\bullet, \mathcal{B}^\bullet)$ is the single complex of sheaves which is obtained from the double complex $\mathcal{H}om^{p,q}(\mathcal{A}^\bullet, \mathcal{B}^\bullet) = \mathcal{H}om(\mathcal{A}^p, \mathcal{B}^q)$.

The *derived category* $D^b(X)$ was introduced by J.L. Verdier by localization of $K(X)$. The objects in $D^b(X)$ are still topologically constructible bounded complexes of sheaves on X but morphisms $\mathcal{A}^\bullet \to \mathcal{B}^\bullet$ are defined as equivalence class of diagrams of chain maps:

$$\mathcal{A}^\bullet \xleftarrow{\text{q.i.}} \mathcal{E}^\bullet \longrightarrow \mathcal{B}^\bullet$$

where “q.i.” means a quasi-isomorphism. Two such diagrams

$$\mathcal{A}^\bullet \xleftarrow{\text{q.i.}} \mathcal{E}_1^\bullet \longrightarrow \mathcal{B}^\bullet, \qquad \mathcal{A}^\bullet \xleftarrow{\text{q.i.}} \mathcal{E}_2^\bullet \longrightarrow \mathcal{B}^\bullet$$

are equivalent is there is a commutative diagram in $K(X)$ (meaning a diagram that commutes up to homotopy).

$$\begin{array}{ccccc} & & \mathcal{E}_1^\bullet & & \\ & \overset{\text{q.i.}}{\swarrow} & \uparrow & \nwarrow & \\ \mathcal{A}^\bullet & \xleftarrow{\text{q.i.}} & \mathcal{E}_3^\bullet & \longrightarrow & \mathcal{B}^\bullet . \\ & \overset{\text{q.i.}}{\nwarrow} & \downarrow & \nearrow & \\ & & \mathcal{E}_2^\bullet & & \end{array}$$

Property 5.3.13 The derived category converts quasi-isomorphisms to isomorphisms: If $\varphi^\bullet : \mathcal{A}^\bullet \to \mathcal{B}^\bullet$ is a quasi-isomorphism (that is, a morphism of bounded complexes of sheaves whose induced map on cohomology is an isomorphism) then it has an inverse in the derived category $D^b(X)$.

This because $\varphi^\bullet$ may be composed with an injective resolution $\psi^\bullet : \mathcal{B}^\bullet \to \mathcal{I}^\bullet$. Then Proposition 5.3.8 (1) implies that $\psi^\bullet \circ \varphi^\bullet$ has a homotopy inverse $\theta^\bullet : \mathcal{I}^\bullet \to \mathcal{A}^\bullet$ which is therefore also a quasi-isomorphism, so $\theta^\bullet \circ \psi^\bullet$ is an inverse to $\varphi^\bullet$ in the derived category.

5.3.7 Derived Functors

An exact functor $F : Sh(X) \to Sh(Y)$ gives rise to a functor $D^b(X) \to D^b(Y)$ on derived categories. In this case, the homotopy category functor $F : K(X) \to K(Y)$ transforms quasi-isomorphisms into quasi-isomorphisms. However important functors such as $\mathcal{H}om(\mathcal{A}, \bullet)$, $\mathcal{A} \otimes \bullet$, $\Gamma(X, \bullet)$, direct image f_*, are not exact. The way to extend such functors in D^b is the Verdier's notion of derived functor. That will be very useful to express properties of intersection homology, in particular using formulae (5.12) and (5.13).

A covariant additive functor T from complexes of sheaves to an abelian category gives rise to its *right derived functor* RT defined on $D^b(X)$ by defining

$$RT(\mathcal{A}^\bullet) = T(\mathcal{I}^\bullet).$$

where $\mathcal{I}^\bullet$ is the canonical injective resolution of $\mathcal{A}^\bullet$ (see references in (5.3.5) in particular [49, Chap. V] and [94, Sect. 1.5]).

Classical Derived Functors

(a) The functor $\mathcal{H}om(\mathcal{A}, \bullet) \to K(X)$ has a (right) derived functor $R\mathcal{H}om^\bullet$ [49, Ch. VI]. Let $\mathcal{A}^\bullet$ and $\mathcal{B}^\bullet$ be bounded complexes of sheaves on X. To define $R\mathcal{H}om^\bullet(\mathcal{A}^\bullet, \mathcal{B}^\bullet)$, consider $\mathcal{H}om^\bullet(\mathcal{A}^\bullet, \mathcal{B}^\bullet)$ as a functor of $\mathcal{B}^\bullet$, and take its right derived functor. The functor

$$\mathcal{A}^\bullet \to R\mathcal{H}om^\bullet(\mathcal{A}^\bullet, \mathcal{B}^\bullet)$$

is a functor from $D^b(X)$ into itself [18, V.5.17].

(b) The (left) derived tensor product functor $\mathcal{A}^\bullet \overset{L}{\otimes} \bullet : D^b(X) \to D^b(X)$ is defined in a similar way to the right derived functors by the formula

$$\mathcal{A}^\bullet \overset{L}{\otimes} \mathcal{B}^\bullet = \mathcal{A}^\bullet \otimes \mathcal{J}^\bullet$$

where $\mathcal{J}^\bullet \to \mathcal{B}^\bullet$ is a resolution of $\mathcal{B}^\bullet$ whose stalks are flat R-modules (see [49, Ch. VI], [168]). If R is a field then $\mathcal{A}^\bullet \overset{L}{\otimes} \mathcal{B}^\bullet = \mathcal{A}^\bullet \otimes \mathcal{B}^\bullet$.

If $\mathcal{A}^\bullet$ and $\mathcal{B}^\bullet \in D^b(X)$ are constructible with respect to a given stratification of X then so are $R\mathcal{H}om^\bullet(\mathcal{A}^\bullet, \mathcal{B}^\bullet)$ and $\mathcal{A}^\bullet \overset{L}{\otimes} \mathcal{B}^\bullet$ ([94, 1.9]).

(c) By definition, the i-th hypercohomology group $\mathbb{H}^i(X; \mathcal{A}^\bullet)$ of $\mathcal{A}^\bullet \in D^b(X)$ is the i-th derived functor of the global section functor $\Gamma(X, \bullet)$.

Functors associated to a map

Consider now a continuous map $f : X \to Y$ between locally compact topological spaces. Complete definitions and properties of the following functors are presented in the Grivel chapter in [18, Chapter VI].

(d) The functor direct image $f_* : Sh(X) \to Sh(Y)$.

If $\mathcal{A}$ is a sheaf on X, the presheaf defined by

$$\Gamma(V, f_*\mathcal{A}) = \Gamma(f^{-1}(V); \mathcal{A}) \qquad \text{for all } V \text{ open in } Y$$

is a sheaf on Y denoted by $f_*\mathcal{A}$. If $j : X \hookrightarrow Y$ is a closed immersion and $\mathcal{A}$ a sheaf on X, then $j_*\mathcal{A} = \mathcal{A}^Y$ is the extension of the sheaf $\mathcal{A}$ by zero. One defines the derived functors $Rf_* : D^b(X) \to D^b(Y)$ as in the definition.

(e) The functor $f_! : Sh(X) \to Sh(Y)$ direct image with proper supports.

If $V \subset Y$ is open, the family of subsets $C \subset X$ which are closed in $f^{-1}(V)$ and such that the map $f|_C : C \to V$ is proper is a family of supports in $f^{-1}(V)$ denoted by Φ_V. If $\mathcal{A}$ is a sheaf on X, the presheaf defined by

$$\Gamma(V; f_!\mathcal{A}) = \Gamma_{\Phi_V}(f^{-1}(V); \mathcal{A}) \qquad \text{for all } V \text{ open in } Y$$

is a sheaf on Y [18, VI, 2.2].

If Y is a point, then $f_!\mathcal{A} = \Gamma_c(X; \mathcal{A})$, where c denotes the family of compact subsets in X. If $j : X \hookrightarrow Y$ is an open (or closed) immersion, then $j_!\mathcal{A} = \mathcal{A}^Y$ and the functor $j_!$ is exact, one has $Rj_!\mathcal{A}^\bullet = j_!\mathcal{A}^\bullet$. Finally, the functor $f_!$ is exact in the subcategory of injective sheaves on X.

The right derived functor of $f_!$ denoted $Rf_! : D^b(X) \to D^b(Y)$ has stalks

$$\mathcal{H}^\bullet(Rf_!\mathcal{A}^\bullet)_y \cong \mathbb{H}^\bullet_c(f^{-1}(y); \mathcal{A}^\bullet) \qquad \forall y \in Y.$$

If $f : X \to Y$ is stratified with respect of stratifications of X and Y ([179, Sect. 4.4]), then $Rf_*\mathcal{A}^\bullet$ and $Rf_!\mathcal{A}^\bullet$ are constructible with respect to the stratification of Y. This is a consequence of the topological locally trivial nature of a stratification, see (5.3).

(f) The functor pull-back $f^* : Sh(Y) \to Sh(X)$.

The functor $f_* : Sh(X) \to Sh(Y)$ admits a left adjoint, denoted by $f^* : Sh(Y) \to Sh(X)$. There is an isomorphism

$$\mathcal{H}om_{Sh(X)}(f^*\mathcal{B}, \mathcal{A}) \cong \mathcal{H}om_{Sh(Y)}(\mathcal{B}, f_*\mathcal{A}). \tag{5.10}$$

for $\mathcal{A} \in Sh(X)$ and $\mathcal{B} \in Sh(Y)$. The functor f^* is exact and $Rf^*\mathcal{B} = f^*\mathcal{B}$ for all $\mathcal{B} \in Sh(Y)$.

For every point $x \in X$ and $\mathcal{B} \in Sh(Y)$, there is an isomorphism at the level of stalks

$$(f^*\mathcal{B})_x = \mathcal{B}_{f(x)} \qquad \forall x \in X.$$

For an inclusion $j : X \hookrightarrow Y$, then $j^*\mathcal{B} = \mathcal{B}|_X$ is the restriction of the sheaf $\mathcal{B}$ to X.

Denote by $p : X \to \{\mathrm{pt}\}$ the map to a point. The constant sheaf $\mathbf{R}_X$ is equal to

$$\mathbf{R}_X = p^*\mathbf{R}_{\mathrm{pt}}. \tag{5.11}$$

(g) Unlike the adjonction (5.10) between the functors f_* and f^*, in general there is no functor $f^! : Sh(Y) \to Sh(X)$ with a sheaf isomorphism $\mathcal{H}om(f_!\mathcal{A}, \mathcal{B}) \cong f_*\mathcal{H}om(\mathcal{A}, f^!\mathcal{B})$.

The functor $f^! : D^b(Y) \longrightarrow D^b(X)$ is defined at the level of derived categories (see [94, 1.12] and [18, V, 5.12]). If $\mathcal{I}^\bullet$ is a complex of injective sheaves on Y, then $f^!(\mathcal{I}^\bullet)$ is defined to be the sheaf associated to the presheaf whose sections over an open set $U \subset X$ are $\Gamma(U; f^!\mathcal{I}^\bullet) = Hom^\bullet(f_!\mathcal{K}^\bullet_U, \mathcal{I}^\bullet)$ where $\mathcal{K}^\bullet_U$ is the canonical injective resolution of the constant sheaf $\mathbf{R}_U$.

Example 5.3.14 • For an open immersion $j : X \hookrightarrow Y$, one has $j^! = j^*$.
• For a closed immersion $j : X \hookrightarrow Y$, one has

$$j^!(\mathcal{G}^\bullet)(U) = \Gamma_{(X)}(V; \mathcal{G}^\bullet)$$

where V is an open subset in Y such that $U = V \cap X$ Here (X) denotes the family of supports whose elements are closed subsets contained in X.

The (local) Verdier duality theorem ([94, 1.12]) is a canonical isomorphism in $D^b(Y)$,

$$Rf_*R\mathcal{H}om^\bullet(\mathcal{A}^\bullet, f^!\mathcal{B}^\bullet) \cong R\mathcal{H}om^\bullet(Rf_!\mathcal{A}^\bullet, \mathcal{B}^\bullet)$$

for any $\mathcal{A}^\bullet \in D^b(X)$ and $\mathcal{B}^\bullet \in D^b(Y)$.

5.3.8 Dualizing Complex ([94, 1.12], [18, V, 7.1])

Borel and Moore first defined the dual $\mathcal{D}(\mathcal{A}^\bullet)$ of a complex of sheaves $\mathcal{A}^\bullet$ [17], and they showed that for any open set $U \subset X$ the hypercohomology groups $\mathbb{H}^i_c(U; \mathcal{A}^\bullet)$ and $\mathbb{H}^i(U; \mathcal{D}(\mathcal{A}^\bullet))$ are dual. Here, $\mathbb{H}^i_c$ denotes the hypercohomology with compact supports, i.e., $R^i\Gamma_c$. This property characterizes $\mathcal{D}(\mathcal{A}^\bullet)$ up to quasi-isomorphism. It implies, for example, that if X is compact and R is a field, then

$$\mathbb{H}^i(X;\mathcal{A}^\bullet) = Hom(\mathbb{H}^{-i}(X;\mathcal{D}(\mathcal{A}^\bullet)), R).$$

On a PL-manifold, Borel and Moore considered the dual of the constant sheaf, and showed that this is the Borel-Moore sheaf of chains $\mathcal{C}^\bullet$ (cf. Definition 5.3.3).

Later, Verdier [183, 184] defined a complex of sheaves $\mathbf{D}^\bullet_X$, the *dualizing complex* such that

$$\mathcal{D}(\mathcal{A}^\bullet) \cong R\mathcal{H}om^\bullet(\mathcal{A}^\bullet, \mathbf{D}^\bullet_X)$$

for any bounded complex $\mathcal{A}^\bullet$. Verdier identified $\mathbf{D}^\bullet_X = \mathcal{D}(\mathbf{R}_X)$ and showed that, in D^b, the sheaf $\mathcal{C}^\bullet$ is isomorphic to the dualizing sheaf. Therefore, the Borel-Moore dual of $\mathcal{A}^\bullet$ may be identified with $\mathcal{H}om(\mathcal{A}^\bullet, \mathcal{C}^\bullet)$. While defining the dualizing sheaf, Verdier provided the good language to express the duality and showed an isomorphism in the derived category between $\mathcal{A}^\bullet$ and the double dual of $\mathcal{A}^\bullet$, i.e. if $\mathcal{A}^\bullet$ is a bounded topologically constructible complex of sheaves on X, then there is a natural isomorphism in $D^b(X)$

$$\mathcal{A}^\bullet \cong \mathcal{D}(\mathcal{D}(\mathcal{A}^\bullet)).$$

If $\mathcal{B}^\bullet \cong \mathcal{D}(\mathcal{A}^\bullet)$, then the corresponding pairing

$$\mathcal{B}^\bullet \overset{L}{\otimes} \mathcal{A}^\bullet \to \mathbf{D}^\bullet_X$$

is called a Verdier dual pairing.

The associated cohomology sheaves of $\mathbf{D}^\bullet_X$ are nonzero in negative degree only, with stalks $\mathcal{H}^{-i}(\mathbf{D}^\bullet_X)_x = H_i(XX - \{x\}; R)$. If X is an n-dimensional PL-manifold, the shifted complex $\mathbf{D}^\bullet_X[-n]$ is naturally isomorphic to (in fact, an injective resolution of) the orientation sheaf of X (see Example 5.3.1 and [94, Sect. 1.12], also [18, V, Sect. 7.3] but taking care of notations cf. Remark 5.4.10).

The hypercohomology groups $\mathbb{H}^{-i}(X;\mathbf{D}^\bullet_X)$ equal the ordinary homology groups with closed support $H_i(X;R)$ and $\mathbb{H}^{-i}_\Phi(X;\mathbf{D}^\bullet_X) = H_i^\Phi(X;R)$ for any family of supports Φ on X [18, V,7.1-2-3].

Consider the projection $p : X \to \{\mathrm{pt}\}$ and the sheaf $\mathbf{R}_{\mathrm{pt}}$. The dualizing sheaf satisfies [18, V,7.18]:

$$\mathbf{D}^\bullet_X \cong p^!\mathbf{R}_{\mathrm{pt}}.$$

Hence for every map $f : X \to Y$ one has a canonical isomorphism $\mathbf{D}^\bullet_X \cong f^!\mathbf{D}^\bullet_Y$ (compare with (5.11)).

Let $f : X \to Y$ be a continuous map between topological manifolds, $\mathcal{A}^\bullet \in D^b(X)$ and $\mathcal{B}^\bullet \in D^b(Y)$. The functors satisfy the following duality formulae

$$f^!\mathcal{B}^\bullet \cong \mathcal{D}_X(f^*\mathcal{D}_Y(\mathcal{B}^\bullet))$$
$$Rf_!\mathcal{A}^\bullet \cong \mathcal{D}_Y(Rf_*\mathcal{D}_X(\mathcal{A}^\bullet)).$$

If $\mathcal{A}^\bullet$ is a topologically constructible complex of sheaves on X, $f_x\colon \{x\} \to X$ is the inclusion of a point, and U_x is a distinguished neighborhood of x (see 5.3), then [94, p.91] (see also [18, V, Sect. 4.5])

$$H^j(f_x^* \mathcal{A}^\bullet) \cong \mathbb{H}^j(U_x; \mathcal{A}^\bullet) \tag{5.12}$$

$$H^j(f_x^! \mathcal{A}^\bullet) \cong \mathbb{H}_c^j(U_x; \mathcal{A}^\bullet). \tag{5.13}$$

These groups are respectively called the stalk homology and the costalk homology of $\mathcal{A}^\bullet$ at x.

The following geometric interpretations are taken from [94, Sect. 4, p. 106] and will be useful to interpreting the Theorem 5.4.9.

If a class $\xi \in \mathbb{H}^j(X; \mathcal{A}^\bullet)$ does not vanish under the homomorphism

$$\mathbb{H}^j(X; \mathcal{A}^\bullet) \to H^j(f_x^* \mathcal{A}^\bullet)$$

then any cycle representative of ξ must contain the point x. Thus, $H^j(f_x^* \mathcal{A}^\bullet)$ represents local classes which "cannot be pulled away from the point x". The set

$\{x \in X | H^j(f_x^* \mathcal{A}^\bullet) \neq 0\}$ is called the *local j-support of the complex* $\mathcal{A}^\bullet$.

Similarly, a class $\eta \in \mathbb{H}^j(X; \mathcal{A}^\bullet)$ is in the image of the homomorphism

$$H^j(f_x^! \mathcal{A}^\bullet) \to \mathbb{H}^j(X; \mathcal{A}^\bullet).$$

if some cycle representative of η is completely contained in a neighborhood of x. Thus $H^i(f_x^! \mathcal{A}^\bullet)$ represents local classes which are "supported near x". The set

$\{x \in X | H^j(f_x^! \mathcal{A}^\bullet) \neq 0\}$ is called the *local j-cosupport of the complex* $\mathcal{A}^\bullet$.

5.4 Intersection Homology—Geometric and Sheaf Definitions

In order to recover duality properties for singular varieties, the idea of intersection homology, due to Mark Goresky and Robert MacPherson, is to restrict the consideration to cycles which meet the singular part of the variety with a "controlled" dimension. That makes sense if the variety is endowed with a suitable stratification. The considered singular varieties are pseudomanifolds.

As observed by Goresky and MacPherson [92], in a PL-pseudomanifold of dimension n, if two cycles of respective dimensions i and j are in general position, then their intersection can be given canonically the structure of an $i + j - n$

chain. However, their intersection is (in general) no longer a cycle and Theorems 5.2.4 and 5.2.5 do not hold. That is the motivation for the following definitions.

The first definition has been given by Goresky and MacPherson in the framework of stratified compact oriented PL-pseudomanifolds (see [91, 92]) and also [18, Chaps. I–IV]). The compactness is not required here and the considered chains are the PL-geometric chains (Sect. 5.2.4). The second definition, using sheaves and, in particular the Deligne sheaf complex, has been given by the same authors in [94] (see also [18, Chaps. V–IX]).

5.4.1 *The Definition for PL-Stratified Pseudomanifolds ([91], 53)*

Let X be a PL-stratified pseudomanifold. If a chain ξ meets transversaly an element $X_{n-\alpha}$ of the PL-filtration, then one has

$$\dim(|\xi| \cap X_{n-\alpha}) = i - \alpha.$$

The allowed chains and cycles will be those which meet each element $X_{n-\alpha}$ of the singular part with a controlled and fixed transversality defect p_α. This defect is called the *perversity* (in French: *Perversité*, in German: *Toleranz*).

A *perversity*, also called GM-perversity for Goresky-MacPherson perversity, is an integer value function

$$\bar{p} : [0, \dim X] \cap \mathbb{Z} \to \mathbb{N}, \qquad p_\alpha := \bar{p}(\alpha)$$

such that $p_0 = p_1 = p_2 = 0$ and

$$p_\alpha \leq p_{\alpha+1} \leq p_\alpha + 1 \quad \text{for} \quad \alpha \geq 2. \tag{5.14}$$

This condition is the one given originally by Goresky and MacPherson in order to ensure the main properties of the theory. More general perversities have been considered by various authors (see [128] and Sect. 5.6.4) providing other aspects for the theory (Fig. 5.3).

Example 5.4.1 Examples of perversities are

- the zero perversity $\bar{0} = (0, 0, \dots, 0)$,
- the maximal (or top) perversity $\bar{t} = (0, 0, 0, 1, 2, \dots, n-2)$,
- for n even, $n \geq 4$, the upper middle $\bar{n} = (0, 0, 0, 1, 1, 2, 2, \dots, \frac{n}{2} - 1, \frac{n}{2} - 1)$ and the lower middle perversities $\bar{m} = (0, 0, 0, 0, 1, 1, \dots, \frac{n}{2} - 2, \frac{n}{2} - 1)$.

Let $\bar{p} = (p_0, p_1, p_2, \dots, p_n)$ be a perversity, the complementary perversity $\bar{q} = (q_0, q_1, q_2, \dots, q_n)$ is defined by $p_\alpha + q_\alpha = t_\alpha$ for all $\alpha \geq 2$.

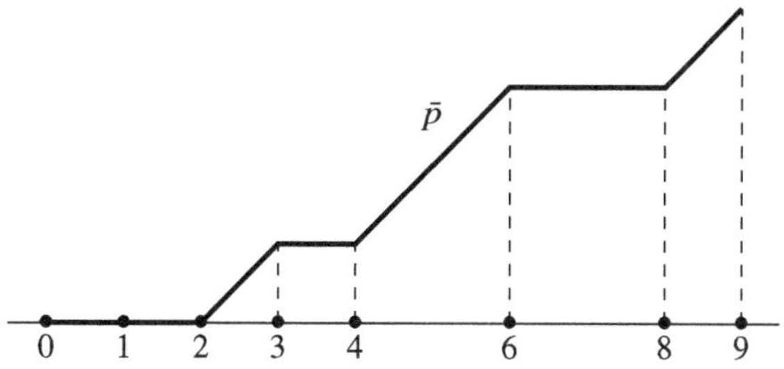

Fig. 5.3 The perversity $\bar{p} = (0, 0, 0, 1, 1, 2, 3, 3, 3, 4)$

Given a stratification (5.2) of a n-dimensional pseudomanifold X, Goresky and MacPherson [92, Sect. 1.3] call *$(\bar{p}, i)$-allowable* an i-chain $\xi \in C_i(X; G)$ such that

$$\dim(|\xi| \cap X_{n-\alpha}) \leq i - \alpha + p_\alpha \qquad \forall \alpha \geq 0$$

The condition means that the perversity is the maximum admissible defect of transversality. The boundary of a $\bar{p}$-allowable chain is not necessarily $\bar{p}$-allowable (easy examples). In order to define a complex of chains, one has to set:

Definition 5.4.2 The intersection chains $IC_i^{\bar{p}}(X; G)$ is the subset of $C_i(X; G)$ consisting of chains ξ such that ξ and $\partial\xi$ are $\bar{p}$-allowable, that is

$$IC_i^{\bar{p}}(X; G) = \left\{ \xi \in C_i(X; G) \left| \begin{array}{l} \dim(|\xi| \cap X_{n-\alpha}) \leq i - \alpha + p_\alpha \\ \dim(|\partial\xi| \cap X_{n-\alpha}) \leq (i-1) - \alpha + p_\alpha \end{array} \right. \forall \alpha \geq 2 \right\}$$

Using the usual boundary of chains, the obtained chain complex is denoted by $(IC_*^{\bar{p}}(X; G), \partial_*)$.

Definition 5.4.3 The *intersection homology* groups $IH_*^{\bar{p}}(X; G)$ are the homology groups of the complex $(IC_*^{\bar{p}}(X; G), \partial_*)$.

Using, in the definition, the subcomplex $C_*^c(X; G)$ of chains with compact supports (see Sect. 5.2.4) provides the intersection homology groups with compact supports, denoted by $IH_*^{\bar{p},c}(X; G)$. Notice that on the one hand, the intersection homology defined in [92] agrees with the intersection homology with compact supports as defined in [94]. On the other hand, the intersection homology defined in [94] agrees with the Borel-Moore intersection homology (with closed supports) of [92].

5.4.2 Definition with Local Systems

To make the construction of homology with coefficients in a local system, work in intersection homology, one only needs the local system $\mathcal{L}$ to be defined on the dense open part $X - \Sigma$ of X.

Let $\mathcal{L}$ be a local coefficient system of R-modules on $X - \Sigma$. Given an open subset $U \subset X$ and a locally finite triangulation K of U, since $\mathcal{L}$ may not be defined on all of U, it is impossible to define a group $C_i^K(U, \mathcal{L})$ of i-chains ξ with coefficients in $\mathcal{L}$. Nevertheless, [94, 2.2] or [89, 9.4] observes that, for any perversity $\bar{p}$, and for any $(\bar{p}, i)$-allowable chain ξ, if σ is any i-simplex with nonzero coefficient in ξ, both the interior of σ and the interiors of all the $i-1$ dimensional faces of σ lie entirely in $X - \Sigma$ by the allowability conditions. That justifies the definition:

$$IC_i^{\bar{p},K}(U;\mathcal{L}) = \left\{ \xi \in C_i^K(U;\mathcal{L}) \left| \begin{array}{c} \dim(|\xi| \cap U \cap X_{n-\alpha}) \leq i - \alpha + p_\alpha \\ \\ \dim(|\partial\xi| \cap U \cap X_{n-\alpha}) \leq (i-1) - \alpha + p_\alpha \end{array} \right. \quad \forall \alpha \geq 0 \right\}$$

The map $IC_i^{\bar{p},K}(U;\mathcal{L}) \to IC_{i-1}^{\bar{p},K}(U;\mathcal{L})$ is well defined and the intersection homology groups $IH_*^{\bar{p}}(X;\mathcal{L})$ are defined as in Sect. 5.2.4 (with $U = X$).

Many examples of computation of intersection homology groups can be found for instance in the Goresky-MacPherson's chapter [18, Chap. III] and in [27, 63, 71, 128]. See also Example 5.5.10. Here are two elementary examples, where the coefficient sheaf $\mathcal{L}$ is the constant sheaf $\mathbb{Z}_X$.

Example 5.4.4 (The pinched torus) (see Fig. 5.1) The singular set is a point: the pinched point $\{0\}$. The considered stratification is given by the filtration

$$X \supset \Sigma = \{0\} \supset \emptyset$$

The only possible perversity is the perversity $\bar{0}$. The 1-dimensional intersection homology of the pinched torus is zero, while its 1-dimensional homology does not vanish.

$$IH_0^{\bar{0}}(X) = \mathbb{Z}[pt] \qquad IH_1^{\bar{0}}(X) = 0 \qquad IH_2^{\bar{0}}(X) = \mathbb{Z}[X].$$

(compare with Example 5.2.3).

Example 5.4.5 (The double cone) Though it is similar to the previous example and it is not a connected pseudomanifold, the example of the double cone is instructive. One may compare with the example of the suspension of two circles [128].

The double cone X is obtained by pinching the cylinder $\mathbb{S}^1 \times \mathbb{R}$ at level $\{0\}$ into a point $\{a\}$. The line ℓ (see Fig. 5.4) goes through the singular point $\{a\}$ and C_1 and C_2 are the two 2-dimensional components of the double cone. Poincaré duality fails for the double cone X. The only possible perversity is the perversity $\bar{0}$. Two points x_1 and x_2 contained in different connected components of $X \backslash \{a\}$ are not homologous, in intersection homology, as any 1-chain linking these two points contain the vertex $\{a\}$ and is not permitted. Poincaré duality is recovered with intersection homology (see 5.23) (Table 5.1).

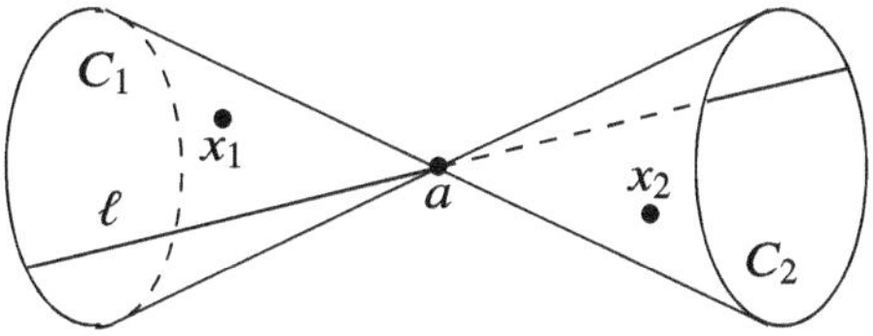

Fig. 5.4 The double cone

Table 5.1 Homology and intersection homology of the (double) cone

i	$H_i(X)$	$H_i^c(X)$	$IH_i(X)$	$IH_i^c(X)$
0	0	$\mathbb{Z}\{pt\}$	0	$\mathbb{Z}\{x_1\} \oplus \mathbb{Z}\{x_2\}$
1	$\mathbb{Z}[\ell]$	0	0	0
2	$\mathbb{Z}[C_1] \oplus \mathbb{Z}[C_2]$	0	$\mathbb{Z}[C_1] \oplus \mathbb{Z}[C_2]$	0

5.4.3 Witt Spaces

For many applications, the class of spaces with even dimension strata is too restrictive. The largest class of Witt spaces still enjoys Poincaré duality of the middle intersection homology, but allows for some strata of odd dimension (see for instance the following references (5.26), (5.5.3), (5.5.3), (5.5.46)).

Definition 5.4.6 A stratified pseudomanifold X is a R-*Witt space* ([94, Sect. 5.6.1] [169]) if, for each stratum of odd codimension $\alpha = 2k + 1$, then $IH_k^{\bar{m}}(L_\alpha; R) = 0$, where L_α is the link of the stratum (5.3) For such a space, the intersection homology groups of the two middle perversities coincide (see Example 5.4.1):

$$IH_*^{\bar{m}}(X; R) \cong IH_*^{\bar{n}}(X; R).$$

5.4.4 The Intersection Homology Sheaf Complex

The intersection homology sheaf complex is defined in the context of PL-pseudomanifolds. In the following section, the Deligne complex will be defined in the more general context of topological pseudomanifolds. When both defined, the intersection homology sheaf complex and the Deligne complex are quasi-isomorphic and their hypercohomology computes intersection homology. That is made precise in the following sections.

Definition 5.4.7 Let $\mathcal{IC}_{\bar{p}}^{-i}$ be the subsheaf of $\mathcal{C}^{-i}$ (see Example 5.3.2) whose sections over an open subset $U \subset X$ consist of all locally finite PL-chains $\xi \in \Gamma(U; \mathcal{C}^{-i})$ such that $|\xi|$ is $(\bar{p}, i)$-allowable and $|\partial\xi|$ is $(\bar{p}, i-1)$-allowable with respect to the filtration of U

$$U \supset U \cap X_{n-2} \supset \cdots \supset U \cap X_1 \supset U \cap X_0.$$

That is

$$\Gamma(U;\mathcal{I}C_{\bar{p}}^{-i}) = \left\{ \xi \in \Gamma(U;\mathcal{C}^{-i}) \;\middle|\; \begin{array}{c} \dim(|\xi| \cap U \cap X_{n-\alpha}) \leq i-\alpha+p_\alpha \\ \dim(|\partial\xi| \cap U \cap X_{n-\alpha}) \leq (i-1)-\alpha+p_\alpha \end{array} \quad \forall \alpha \geq 0 \right\}$$

The sheaf $\mathcal{I}C_{\bar{p}}^{-i}$ is well defined and $\Gamma(U;\mathcal{I}C_{\bar{p}}^{-i}) = IC_i^{\bar{p}}(U)$. If $V \subset U$ are two open subsets in X, then for every perversity $\bar{p}$, there are natural restriction maps $\rho_{VU} : IC_i^{\bar{p}}(U) \to IC_i^{\bar{p}}(V)$ as in (5.6) (see [94, Sect. 2.1]).

Using the restriction of the usual boundary, one obtains a complex of sheaves $\mathcal{I}C_{\bar{p}}^{\bullet}$ on X. This complex is soft [18, II, Sect. 5] so that the complex satisfies Proposition 5.3.11. The hypercohomology groups $\mathbb{H}^{-i}(X;\mathcal{I}C_{\bar{p}}^{\bullet})$ are canonically isomorphic to the intersection homology groups $IH_i^{\bar{p}}(X;R)$ defined in [92, Sect. 1.3] for $R=\mathbb{Z}$. Also, one has

$$\mathbb{H}_c^{-i}(X;\mathcal{I}C_{\bar{p}}^{\bullet}) = IH_i^{\bar{p},c}(X;R),$$

intersection homology with compact supports, and more generally

$$\mathbb{H}_\Phi^{-i}(X;\mathcal{I}C_{\bar{p}}^{\bullet}) = IH_i^{\bar{p},\Phi}(X;R) \tag{5.15}$$

for any family of supports Φ on X (see [18, II, 5]).

The associated cohomology sheaves $\mathcal{H}^{-i}(\mathcal{I}C_{\bar{p}}^{\bullet})$ are called the local intersection homology sheaf. The stalk at $x \in X$ of this sheaf is $IH_i^{\bar{p}}(X, X-\{x\}; R)$.

Definition with Local Systems

Considering local systems provide many useful examples as well as powerful applications.

Let $\mathcal{L}$ be a local coefficient system of R-modules on $X-\Sigma$. Given an open subset $U \subset X$ and a locally finite triangulation K of U, the group $IC_i^{\bar{p},K}(U;\mathcal{L})$ is well defined (see Sect. 5.4.2) as well as maps $IC_i^{\bar{p},K}(U;\mathcal{L}) \to IC_{i-1}^{\bar{p},K}(U;\mathcal{L})$.

Definition 5.4.8 ([94, Sect. 2.2], [89, 9.4]) Let X be a PL-stratified PL-pseudomanifold and $\bar{p}$ a perversity, the sheaf complex $\mathcal{I}C_{\bar{p}}^{\bullet}(\mathcal{L})$ of intersection chains with local coefficients in $\mathcal{L}$ is defined by

$$\Gamma(U,\mathcal{I}C_{\bar{p}}^{-i}(\mathcal{L})) = \lim_{K} IC_i^{\bar{p},K}(U;\mathcal{L})$$

where the limit is taken over locally finite compatible triangulations of U. The intersection homology groups of X with coefficients in $\mathcal{L}$, denoted $IH_i^{\bar{p}}(X;\mathcal{L})$, are the hypercohomology groups $\mathbb{H}^{-i}(X;\mathcal{I}C_{\bar{p}}^{\bullet}(\mathcal{L}))$.

Combining formula (5.15) with (5.12) and (5.13), one obtains the following theorem:

Theorem 5.4.9 *Let $f_x : \{x\} \to X$ be the inclusion of a point x in X, U_x a distinguished neighborhood of x and let $\mathcal{IC}_X^{\bar{p},\bullet}(\mathcal{L})$ be the sheaf complex of intersection chains, one has:*

$$H^j(f_x^* \mathcal{IC}_X^{\bar{p},\bullet}(\mathcal{L})) \cong \mathbb{IH}^j(U_x; \mathcal{IC}_X^{\bar{p},\bullet}(\mathcal{L})) = IH_{-j}^{\bar{p}}(U_x; \mathcal{L}) \tag{5.16}$$

$$H^j(f_x^! \mathcal{IC}_X^{\bar{p},\bullet}(\mathcal{L})) \cong \mathbb{IH}_c^j(U_x; \mathcal{IC}_X^{\bar{p},\bullet}(\mathcal{L})) = IH_{-j}^{\bar{p},c}(U_x; \mathcal{L}). \tag{5.17}$$

Remark 5.4.10 An important remark is that the index for the dimension of the homology and intersection homology groups differs according to the authors. That can be considered as unfortunate but shows the diversity of the theory and diversity of applications.

In [94, Sect. 2.3] Goresky and MacPherson explicit four different indices in the literature and the reader has to take care of the convention used in the concerning article.

(a) Homology subscripts, as in [92] or [63]: a subscript k indicates chains of dimension k.
(b) Homology superscripts, as in [94] and this survey: a superscript $-k$ indicates chains of dimension k.
(c) Cohomology superscripts, as in [18, 19, 84]: a superscript j indicates chains of dimension $n - j$.
(d) The Beilinson-Bernstein-Deligne-Gabber scheme [10]: a superscript j indicates chains of codimension $\frac{n}{2} + j$.

For an n–dimensional compact oriented pseudomanifold these schemes compare as follows: $H_k(X)$ in scheme (a) is isomorphic to $H^{-k}(X)$ in scheme (b), $H^{n-k}(X)$ in scheme (c), and $H^{\frac{n}{2}-k}(X)$ in scheme (d).

5.4.5 The Deligne Construction

In a conversation at the IHES, in the fall of 1976, R. MacPherson explained about intersection homology to P. Deligne. P. Deligne had been thinking about variation of Hodge structures on a smooth algebraic curve where truncation arises naturally. When R. MacPherson explained the intersection homology of a cone, it looked like this truncation so P. Deligne conjectured that perhaps intersection homology might be explained by repeated truncation. His conjecture was proven by Goresky and MacPherson [94], who pointed out that this construction could be used to prove the topological invariance of intersection homology, and to give a definition that works in characteristic p. In the meantime, on 20 April 1979, P. Deligne had written to D. Kazhdan and G. Lusztig about the theory, describing his interpretation using truncation, and conjecturing that intersection homology might be pure [94, Sect. 3] or [18, V, Sect. 2.2].

The Deligne idea is to start with the constant sheaf (or a local system of coefficients) on the non-singular part and extending stratum by stratum by alternate operations of "pushing" and "truncating". While requiring technical tools, the idea of Deligne construction is relatively simple. One starts with (all) chains on the regular stratum, then pushing the complex on the "next" stratum, and then "cutting" (truncating) according to the perversity in order to retain only allowed chains. One continues the process by induction on decreasing dimension of the strata.

Therefore, the Deligne construction uses two tools: the "pushing" attaching property and the "truncating" operation.

The Attaching Map

Let Y be a closed subspace of X and i the inclusion of $U = X - Y$ into X. For a sheaf $\mathcal{A}^\bullet$, the composition of the natural morphisms

$$\mathcal{A}^\bullet \to i_* i^* \mathcal{A}^\bullet \to R i_* i^* \mathcal{A}^\bullet$$

is the *attaching map*.

Consider a stratification (5.2)

$$X = X_n \supset X_{n-1} = X_{n-2} \supset \cdots \supset X_1 \supset X_0 \supset X_{-1} = \emptyset$$

and, denoting by $U_k = X - X_{n-k}$ the complementary open subsets, consider the filtration

$$U_1 = U_2 \subset U_3 \subset \cdots \subset U_{n+1} = X.$$

One has $U_2 = X - \Sigma$. Denote by $i_k : U_k \hookrightarrow U_{k+1}$ the inclusion. Then

$$\mathcal{A}^\bullet_k = \mathcal{A}^\bullet|_{U_k} = i_k^* \mathcal{A}^\bullet|_{U_{k+1}}.$$

The following result is one of the main ingredients of the sheaf axiomatic construction of intersection homology (see Sect. 5.4.7).

Theorem 5.4.11 ([94, Proposition 2.5], [18, II, Theorem 6.1]) *The natural homomorphism*

$$\mathcal{IC}^\bullet|_{U_{k+1}} \to R i_{k*} \mathcal{IC}^\bullet|_{U_k} = R i_{k*} i_k^* \mathcal{IC}^\bullet|_{U_{k+1}}$$

induces an isomorphism

$$\mathcal{H}^j(\mathcal{IC}_{n-\bullet})_x \to \mathcal{H}^j(R i_{k*} \mathcal{IC}_{n-\bullet})_x \qquad \text{for } x \in U_{k+1} - U_k$$

for all $j \leq p(k) - n$.

The Deligne Truncation Functor

If $k \in \mathbb{Z}$, the *truncation* of a complex of sheaves $\mathcal{A}^\bullet$ on X is a new complex ([94, 1.14], [18, V,1.10]):

$$(\tau_{\leq k}\mathcal{A}^\bullet)^i = \begin{cases} \mathcal{A}^i & \text{if } i < k \\ \ker d^i & \text{if } i = k \\ 0 & \text{if } i > k. \end{cases}$$

The functor $\tau_{\leq k}$ determine a truncation functor on the derived category $D^b(X)$. See [94, 1.14] for more detailed properties.

The Deligne Sheaf [94, Sect. 3.1], [18, V, Sect. 2.2]

In this section X is a topological pseudomanifold and $\mathcal{L}$ denotes a system of local coefficients on the regular part $X - \Sigma$.

Let $\bar{p}$ a fixed perversity, the Deligne complex of sheaves (or Deligne sheaf) $\mathbb{P}_k^\bullet(\mathcal{L}) \in D^b(U_k)$ is defined inductively by

$$\begin{aligned} \mathbb{P}_2^\bullet(\mathcal{L}) &= \mathcal{L}[n] \\ \mathbb{P}_{k+1}^\bullet(\mathcal{L}) &= \tau_{\leq p(k)-n} Ri_{k*}\mathbb{P}_k^\bullet(\mathcal{L}) \qquad \text{for } k \geq 2. \end{aligned}$$

The resulting complex $\mathbb{P}^\bullet(\mathcal{L}) = \mathbb{P}_{n+1}^\bullet(\mathcal{L})$ is called the Deligne intersection homology chain complex with coeffcients in $\mathcal{L}$.

Starting with a regular Noetherian ring R of finite Krull dimension and the constant sheaf $\mathbf{R}$ on $X - \Sigma$ instead of $\mathcal{L}$, i.e. starting with

$$\mathbb{P}_2^\bullet = \mathbf{D}_{U_2}^\bullet \cong \mathbf{R}_{U_2}[n]$$

the complex $\mathbb{P}^\bullet = \mathbb{P}_{n+1}^\bullet$ is written

$$\mathbb{P}^\bullet = \tau_{\leq p(n)-n} Ri_{n*} \cdots \tau_{\leq p(3)-n} Ri_{3*}\tau_{\leq p(2)-n} Ri_{2*}\mathbf{R}_{X-\Sigma}[n].$$

5.4.6 Local Calculus and Consequences

The local calculus, and precisely the computation in formulae (5.18) and (5.19) below, are the starting points for the characterization of intersection homology.

Let $x \in X$ be a point in the stratum $S_{n-\alpha}$ with codimension α in X. Let U be a neighborhood of x homeomorphic to $\mathbb{B}^{n-\alpha} \times \mathring{c}(L_x)$, where $\dim L_x = \alpha - 1$ (see 5.3). The following result is proved in [18, II, Sect. 3–4] in the context of PL-

pseudomanifolds and in [94, Sect. 2.4], [93, 1.7] in the context of topological pseudomanifolds.

Proposition 5.4.12 *Let X be a locally compact stratified pseudomanifold and $\bar{p}$ any perversity. Let x be a point in a stratum with codimension α in X and let U be a neighborhood of x homeomorphic to $\mathbb{B}^{n-\alpha} \times \mathring{c}(L_x)$, then one has:*

$$IH_i^{\bar{p}}(U) \cong IH_{i-(n-\alpha)}^{\bar{p}}(\mathring{c}(L_x)) \cong \begin{cases} 0 & i < n - p_\alpha \\ IH_{i-(n-\alpha)-1}^{\bar{p}}(L_x) & i \geq n - p_\alpha . \end{cases} \tag{5.18}$$

$$IH_i^{\bar{p},c}(U) \cong IH_i^{\bar{p},c}(\mathring{c}(L_x)) \cong \begin{cases} IH_i^{\bar{p}}(L_x) & i < \alpha - p_\alpha - 1 \\ 0 & i \geq \alpha - p_\alpha - 1 \end{cases} \tag{5.19}$$

The link L_x is compact and its homology groups, with compact and closed supports coincide.

Here is an useful and important notation for the sequel:

Denoting by $\bar{p}$ a perversity and $\bar{q}$ the complementary perversity, one recalls that $p_k + q_k = k - 2$ for all $k \geq 2$. If $j \in \mathbb{N}$, one defines the inverse perversity function

$$p^{-1}(j) = \min\{k \mid p_k \geq j\}$$

and $p^{-1}(j) = \infty$ if $j > p_n$ (Fig. 5.5).

Using the "inverse perversity function"; the properties (5.18) and (5.19) are written

$$\dim\{x \in X \mid IH_i^{\bar{p}}(U_x) \neq 0\} \leq n - p^{-1}(n-i) \qquad \text{for } i \leq n-1,$$
$$\dim\{x \in X \mid IH_i^{c,\bar{p}}(U_x) \neq 0\} \leq n - q^{-1}(i) \qquad \text{for } i \geq 1,$$

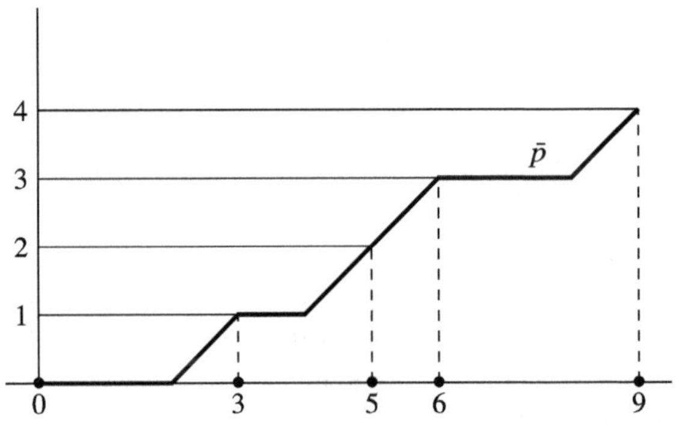

Fig. 5.5 An example of the function p^{-1} (for the perversity of Fig. 4) One has $p^{-1}(1) = 3$, $p^{-1}(2) = 5$, $p^{-1}(3) = 6$, $p^{-1}(4) = 9$, $p^{-1}(5) = +\infty$

and in terms of hypercohomology (see (5.15) and with $j = -i$)

$$\dim\{x \in X \mid \mathbb{IH}^j(U_x; \mathcal{IC}_X^{\bar{p},\bullet}) \neq 0\} \leq n - p^{-1}(j+n) \qquad \text{for } j \geq 1-n. \quad (5.20)$$

$$\dim\{x \in X \mid \mathbb{IH}_c^j(U_x; \mathcal{IC}_X^{\bar{p},\bullet}) \neq 0\} \leq n - q^{-1}(-j). \qquad \text{for } j \leq -1. \quad (5.21)$$

As observed by Goresky and MacPherson [93, 1.8, Theorem] the results of this section are valid for intersection homology with coefficients in a local system.

5.4.7 Characterizations of the Intersection Complex

In the introduction of their Sect. 3 [94], Goresky and MacPherson provide the motivation for the characterizations of the intersection complex, in particular topological invariance of intersection homology. The theorem [94, Theorem 3.5] shows that if X has a PL structure and is stratified by a PL stratification and if $\bar{p}$ denotes a fixed perversity then the Deligne complex $\mathbb{P}^\bullet(\mathcal{L})$ and the complex of PL intersection chains $\mathcal{IC}_X^{\bar{p},\bullet}(\mathcal{L})$ are canonically isomorphic in $D^b(X)$ whenever they are both defined. That justifies the use of the notation $\mathcal{IC}_X^{\bar{p},\bullet}(\mathcal{L})$ to denote this isomorphism class of objects, for any topological pseudomanifold.

In this section, X is a topological pseudomanifold. The first characterization of the intersection complex, as a system of axioms called $[AX_1]_{\bar{p}}$, is given in [94, Sect. 3.3] and [18, V, Sect. 4]. In [18, V, Sect. 4.20], Borel discusses some points concerning the "differences" between [18, 94], in particular the usefulness of the hypothesis "topologically constructible".

If $\mathcal{S}$ denotes a filtration (5.2) of the space X, let $U_k = X - X_{n-k}$ denote the complementary increasing filtration by open sets. There are inclusions

$$i_k : U_k \hookrightarrow U_{k+1} \quad \text{and} \quad j_k : S_{n-k} = (U_{k+1} - U_k) \hookrightarrow U_{k+1}.$$

Definition 5.4.13 ([94, Sect. 3.3]) Let $\bar{p}$ be a perversity and $\mathcal{L}$ is a local system defined on the regular part of X. A complex of sheaves $\mathcal{A}^\bullet$ on X satisfies axioms $[AX_1]_{\bar{p}}(\mathcal{L})$ if it satisfies:

(1a) $\mathcal{A}^\bullet$ is constructible with respect to the given stratification and $\mathcal{A}^\bullet|_{U_2}$ is quasi-isomorphic to $\mathcal{L}[n]$.
(1b) $\mathcal{H}^i(\mathcal{A}^\bullet) = 0$ for $i < -n$.
(1c) $\mathcal{H}^i(\mathcal{A}^\bullet|_{U_{k+1}}) = 0$ for $i > p(k) - n$.
(1d) The attaching maps (see Theorem 5.4.11) induce isomorphisms

$$\mathcal{H}^i(j_k^* \mathcal{A}^\bullet|_{U_{k+1}}) \to \mathcal{H}^i(j_k^* Ri_{k*} i_k^* \mathcal{A}^\bullet|_{U_{k+1}})$$

for all $k \geq 2$ and $i \leq p(k) - n$.

Theorem 5.4.14 ([94, Sect. 3.5] [18, V, Theorem 2.5]), *The sheaf $\mathbb{P}^\bullet_{\bar{p}}(\mathcal{L})$ satisfies properties $[AX_1]_{\bar{p}}(\mathcal{L})$. Any complex of sheaves $\mathcal{A}^\bullet$ satisfying $[AX_1]_{\bar{p}}(\mathcal{L})$ is quasi-isomorphic to $\mathbb{P}^\bullet_{\bar{p}}(\mathcal{L})$.*

Theorem 5.4.15 ([94, Sect. 3.5, Corollary]) *Let $\bar{p}$ be a perversity and $\mathcal{A}^\bullet$ be a constructible complex of fine (or soft) sheaves on X satisfying axioms $[AX_1]_{\bar{p}}(R)$, then the cohomology groups of the complex*

$$\ldots \to \Gamma(X; \mathcal{A}^{j-1}) \to \Gamma(X; \mathcal{A}^j) \to \Gamma(X; \mathcal{A}^{j+1}) \to \ldots$$

i.e., the hypercohomology groups $\mathbb{H}^j_c(X; \mathcal{A}^\bullet)$, are naturally isomorphic to the intersection homology groups $IH^{\bar{p}}_{n-j}(X; R)$.

In fact, Goresky and MacPherson prove the following main result, of which follow the main properties of intersection homology (Sect. 5.5.1).

Theorem 5.4.16 ([94, Sect. 3.5]) *The functor $\mathbb{P}^\bullet_{\bar{p}}$ which assigns to any locally trivial sheaf $\mathbb{F}$ on $X^0 = X - \Sigma$, the complex*

$$\mathbb{P}^\bullet_{\bar{p}}(\mathbb{F}) = \tau_{\leq p(n)-n} Ri_{n*} \cdots \tau_{\leq p(3)-n} Ri_{3*} \tau_{\leq p(2)-n} Ri_{2*} \mathbb{F}[n].$$

defines an equivalence of categories between

(a) the category of locally constant sheaves on $X^0 = X - \Sigma$ and
(b) the full subcategory of $D^b(X)$ whose objects are all complexes of sheaves which satisfy the axioms $[AX_1]_{\bar{p}}$.

Example 5.4.17 The orientation sheaf $\mathcal{O}$ on X^0 is quasi-isomorphic to the dualizing sheaf $\mathbf{D}^\bullet_{X^0}[-n]$. Then $\mathbb{P}^\bullet_{\bar{p}}(\mathcal{O})$ is the intersection homology sheaf and its cohomology is:

$$H^{-i}(\mathbb{P}^\bullet(\mathcal{O})) = IH^{\bar{p}}_i(X; \mathbb{Z})$$

for any $r \geq 0$.

Example 5.4.18 Let $\mathbf{R}_{X^0}$ be the constant sheaf on X^0, placed in degree 0. Then

$$H^j(\mathbb{P}^\bullet_{\bar{p}}(\mathbf{R}_{X^0})) = IH^j_{\bar{p}}(X; R)$$

is the intersection cohomology.

Theorem 5.4.19 ([94, Sect. 3.6] [18, II, Theorem 6.1]) *Let X be a PL-pseudomanifold with a fixed PL-stratification then the sheaf of PL-intersection chains $\mathcal{IC}^{\bar{p},\bullet}_X$ satisfies the axioms $[AX_1]_{\bar{p}}(R)$ with respect to the given stratification. It is naturally quasi-isomorphic to $\mathbb{P}^\bullet_{\bar{p}}(\mathbf{R})$.*

The second characterization of the intersection complex of sheaves goes as follows, as a consequence of the local calculus (see [18, V, Sect. 2.12] and formulae (5.20), (5.21), (5.16), (5.17)).

Definition 5.4.20 ([94, Sect. 4.1], [127, Sect. 9] and [18, V Sect. 4.13].) Let $\mathcal{L}$ be a local system on an open dense submanifold U of codimension at least 2 in X and let $f_x\colon \{x\} \to X$ be the inclusion of a point x in X. One says that the sheaf complex $\mathcal{A}^\bullet$ satisfies the axioms $[AX_2]_{\bar{p}}(\mathcal{L})$ for the perversity $\bar{p}$ if one has:

(2a) $\mathcal{A}^\bullet$ is a topologically constructible complex and $\mathcal{A}^\bullet|_U = \mathcal{L}[n]$ for some open dense submanifold U of codimension at least 2 in X and over which the local system $\mathcal{L}$ is defined.
(2b) $\mathcal{H}^j(\mathcal{A}^\bullet) = 0$ if $j < -n$
(2c) $\dim\{x \in X | H^j(f_x^* \mathcal{A}^\bullet) \neq 0\} \leq n - p^{-1}(j+n)$ for every $j \geq -n+1$.
(2d) $\dim\{x \in X | H^j(f_x^! \mathcal{A}^\bullet) \neq 0\} \leq n - q^{-1}(-j)$ for every $j \leq -1$.

where is $\bar{q}$ the complementary perversity of $\bar{p}$

The uniqueness theorem, proved in Goresky and MacPherson [94, 4.1] (see also [18, V, 4.17]) states that up to canonical isomorphism, there exists an unique complex in $D^b(X)$ which satisfies axioms $[AX_2]_{\bar{p}}(\mathcal{L})$. It is given by the sheaf $\mathcal{IC}_X^{\bar{p},\bullet}(\mathcal{L})$, constructed as before with any stratification of X. As a corollary, the intersection homology groups $IH_*^{\bar{p}}(X)$ are topological invariant and exist independently of the choice of the stratification of X. One has:

Theorem 5.4.21 ([94, Sect. 4.1] [18, V, 4.17]) *Let $\mathcal{A}^\bullet$ be a fine (or soft) sheaf complex on X satisfying Axioms $[AX_2]_{\bar{p}}$ for a perversity $\bar{p}$ and Φ a family of supports on X, then the cohomology groups of the complex*

$$\ldots \to \Gamma_\Phi(X; \mathcal{A}^{j-1}) \to \Gamma_\Phi(X; \mathcal{A}^j) \to \Gamma_\Phi(X; \mathcal{A}^{j+1}) \to \ldots$$

i.e., the hypercohomology groups $\mathbb{H}_\Phi^j(X; \mathcal{A})$, are isomorphic to the intersection homology groups $IH_{-j}^{\bar{p},\Phi}(X; \mathcal{L})$.

In the common setting, equivalence of the systems of axioms $[AX_1]_{\bar{p}}(\mathcal{L})$ and $[AX_2]_{\bar{p}}(\mathcal{L})$ is proved in [94, 4.3], [18, V Sect. 4.10].

5.5 Main Properties of Intersection Homology

The first properties have been proved by Goresky and MacPherson [92] in the framework of PL-pseudomanifolds (see also [18, Chaps. I–IV]). They have been proved in the topological setting, using sheaves and in particular the Deligne sheaf complex, by the same authors in [94] (see also [18, Chaps. V–IX]).

5.5.1 First Properties

In general, results and proofs of this section can be found in various books or surveys concerning intersection homology. However, references will be given to the original papers of Goresky and MacPherson [92] for the PL situation and [94] for the case with sheaves and systems of local coefficients. References to the Borel book [18] will be provided as well, always taking care of the difference of notations.

The intersection homology groups are not homotopy invariant. The intersection homology groups of a cone do not (all) vanish (see (5.18)), while the cone is homotopic to a point, whose non-zero homology groups vanish. However, one has.

Topological Invariance

In [92, Corollary, p.148] Goresky and MacPherson show that the PL intersection homology groups are PL-invariants, i.e. independent of the PL-stratification (see also [18, V, 4.19]). In [94, 4.1] (see also [18, V, 4.18]), Goresky and MacPherson show independence of the topological stratification as a consequence of the Deligne construction performed for the canonical $\bar{p}$-filtration they defined ([94, 4.2]) and the system of axioms $[AX_2]_{\bar{p}}(\mathcal{L})$. The canonical $\bar{p}$-filtration is a homological stratification, the coarsest one for which the intersection homology sheaf is cohomologically constructible.

Theorem 5.5.1 ([94, Sect. 4, Introduction and Sect. 4.1, Corollary 1]) *Let X be a locally compact pseudomanifold and $\bar{p}$ a perversity. Let $\mathcal{L}$ be a local system on the regular part $X^0 = X - \Sigma$.*

- *The intersection homology groups $IH_*^{\bar{p}}(X; \mathcal{L})$ and $IH_*^{\bar{p},c}(X; \mathcal{L})$ are topological invariants and exist independently of the choice of a stratification of X,*
- *For any homeomorphism $f : X \to Y$, the complexes $\mathcal{IC}_X^{\bar{p},\bullet}$ and $f^*\mathcal{IC}_Y^{\bar{p},\bullet}$ are isomorphic in the derived category.*

In [113] King proves topological invariance without sheaves in the case of GM perversities. He also provides a generalization of the intersection homology groups using singular theory and general perversities ("loose perversities"). King claims that the PL intersection homology theory of [92] agrees with his singular theory for any loose perversity and PL stratified set (see the discussion [113, p. 158]). "One can define intersection homology for topological pseudomanifolds, independently of PL structures". A modification of the King's method is provided by Friedman [79, 5.6.2].

In [162] Rourke and Sanderson use homology stratifications to present a simplified version of the Goresky-MacPherson proof valid for PL-spaces.

Products in Intersection Homology

Definition 5.5.2 ([94, Sect. 5.0]) An R-orientation for X is a chosen quasi-isomorphism

$$\mathbf{R}_{X-\Sigma}[n] \to \mathbf{D}^{\bullet}_{X-\Sigma}.$$

If char(R) $\neq 2$ then an R-orientation of X is equivalent to an orientation of $X - \Sigma$ in the usual topological sense.

Suppose X is a (not necessarily orientable) n-dimensional pseudomanifold. For any local system $\mathcal{L}$ on the regular part $X^0 = X - \Sigma$ and any perversity $\bar{p}$ the Deligne's sheaf $\mathbb{P}^\bullet_{\bar{p}}(\mathcal{L})$ is defined on X. In [94, Sect. 5.2] and [18, V, Sect. 9, C] it is shown that any pairing of local systems $\mathcal{L}_1 \otimes \mathcal{L}_2 \to \mathcal{L}_3$ induces a pairing

$$\mathbb{P}^\bullet_{\bar{p}}(\mathcal{L}_1) \otimes \mathbb{P}^\bullet_{\bar{q}}(\mathcal{L}_2) \to \mathbb{P}^\bullet_{\bar{r}}(\mathcal{L}_3) \qquad \text{for } \bar{r} \geq \bar{p} + \bar{q}. \tag{5.22}$$

by induction using the construction of Deligne "attaching–truncating" from the multiplication on the regular subset X^0.

The generalized Poincaré duality, Poincaré-Lefschetz theorem as well as intersection pairing, and cup and cap-products follow from particular cases of formula 5.22, mainly in the case of Examples 5.4.17 and 5.4.18 (see also [18, V, Sect. 9.15] taking care of difference of indices—see Remark 5.4.10).

For instance, cup products $IH^a \otimes IH^b \to IH^{a+b}$ and cap products $IH^a \otimes IH_b \to IH_{b-a}$ in intersection cohomology follow from the canonical pairings

$$\mathbf{R}_X \overset{L}{\otimes} \mathbf{R}_X \to \mathbf{R}_X \qquad \text{and} \qquad \mathbf{R}_X \overset{L}{\otimes} \mathbf{D}^\bullet_X \to \mathbf{D}^\bullet_X.$$

This constructions works over any commutative ring R of finite cohomological dimension.

Intersection Pairing

One of the most important properties of intersection homology is the generalization of the Poincaré-Lefschetz duality, i.e. the intersection pairing (Sect. 5.2.6).

The following Proposition ([92, Sect. 2]) has been first stated by Goresky and MacPherson in the PL setting, using a McCrory Lemma [131, 132], itself using the Zeeman technique to move cycles into general position (see [92, Sect. 2.2]).

Proposition 5.5.3 ([92, Sect. 2.3]) *Let X a compact oriented PL-pseudomanifold and let $\bar{p}$, $\bar{q}$ and $\bar{r}$ perversities such that $\bar{p} + \bar{q} \leq \bar{r}$, one has canonical bilinear pairings*

$$IH_i^{\bar{p}}(X;\mathbb{Z}) \times IH_j^{\bar{q}}(X;\mathbb{Z}) \to IH_{i+j-n}^{\bar{r}}(X;\mathbb{Z}),$$

These pairings are compatible with the cup and cap products ([92, Sect. 7, Appendix]).

Note that, in the non compact situation, the preceding construction gives rise to the pairings

$$IH_i^{\bar{p}}(X) \times IH_j^{\bar{q},c}(X) \to IH_{i+j-n}^{\bar{r},c}(X).$$

Goresky and MacPherson generalized the results in the topological setting [94, Sects. 5.2 and 5.3], using the intersection sheaf complex (see also [18, V, 9.15] taking care of difference of indices—see Remark 5.4.10).

Starting with local coefficient systems $\mathcal{L}_1, \mathcal{L}_2, \mathcal{L}_3$ on $X - \Sigma$, a product $\mathcal{L}_1 \otimes \mathcal{L}_2 \to \mathcal{L}_3$ gives rise to intersection pairings (cf 5.22) [94, 5.2]

$$\mathcal{IC}^\bullet_{\bar{p}}(\mathcal{L}_1) \overset{L}{\otimes} \mathcal{IC}^\bullet_{\bar{q}}(\mathcal{L}_2) \to \mathcal{IC}^\bullet_{\bar{r}}(\mathcal{L}_3)[n].$$

and the Theorem:

Theorem 5.5.4 ([94, 5.2], [18, I, 4.2; V, 9.14]) *Let X be a topological pseudomanifold. If $\bar{p} + \bar{q} \leq \bar{r}$ there are canonical intersection pairings*

$$IH_i^{\bar{p}}(X; \mathcal{L}_1) \times IH_j^{\bar{q}}(X; \mathcal{L}_2) \to IH_{i+j-n}^{\bar{r}}(X; \mathcal{L}_3).$$

These pairings are compatible with the cup and cap products.

Goresky and MacPherson remark that it is not necessary to have an orientation in the preceding construction [94, Sect. 5.2].

Verdier Duality—The Generalized Poincaré-Lefschetz

In their original article, in a delicate and very geometric proof, using so-called "basic sets" $Q_i^{\bar{p}}$, Goresky and MacPherson prove the generalized Poincaré duality:

Theorem 5.5.5 ([92, 3.3, Theorem]) *Let X be a compact, oriented pseudomanifold and let $\bar{p}$ and $\bar{q}$ be two complementary perversities, then the pairing*

$$IH_i^{\bar{p}}(X; \mathbb{Z}) \times IH_{n-i}^{\bar{q}}(X; \mathbb{Z}) \to IH_0^{\bar{t}}(X; \mathbb{Z}) \xrightarrow{\varepsilon} \mathbb{Z}$$

followed by the evaluation map ε (which counts points with their multiplicity order) is non-degenerate, when tensorised by the rationals $\mathbb{Q}$.

Note that, in the non compact situation, the preceding construction gives rise to the pairing (see Example 5.4.5).

$$IH_i^{\bar{p}}(X; \mathbb{Z}) \times IH_{n-i}^{\bar{q},c}(X; \mathbb{Z}) \to IH_0^{\bar{t},c}(X; \mathbb{Z}) \xrightarrow{\varepsilon} \mathbb{Z} \tag{5.23}$$

In a more general way, let k be a field, then the pairing

$$IH_i^{\bar{p}}(X; k) \times IH_{n-i}^{\bar{q}}(X; k) \to IH_0^{\bar{t}}(X; k) \xrightarrow{\varepsilon} k$$

is non-degenerate and induces isomorphisms

$$IH_i^{\bar{p}}(X; k) \cong \mathrm{Hom}(IH_{n-i}^{\bar{q}}(X; k), k)$$

Note that, in the non-compact case, one has isomorphisms:

$$IH_i^{\bar{p}}(X;k) \cong \mathrm{Hom}(IH_{n-i}^{\bar{q},c}(X;k),k)$$

In [94] these results have been generalized by Goresky and MacPherson, assuming that the coefficient ring R is a field k and that X is k-orientable (with a choice of k-orientation). The following results follow from the property of duality between $\mathcal{IC}^\bullet_{\bar{p}}$ and $\mathcal{IC}^\bullet_{\bar{q}}$ for complementary pervesities $\bar{p}+\bar{q}=\bar{t}$. In particular if X has even codimension strata, then $\mathcal{IC}^\bullet_{\bar{m}}$ is self dual (for example, if X is a complex analytic variety).

Definition 5.5.6 ([94, 5.3]) Let $n=\dim(X)$, a pairing $\mathcal{A}^\bullet \overset{L}{\otimes} \mathcal{B}^\bullet \to \mathbf{D}^\bullet_X[n]$ of objects in $D^b(X)$ is called a Verdier dual pairing if it induces an isomorphism in $D^b(X)$

$$\mathcal{A}^\bullet \longrightarrow R\mathcal{H}om^\bullet(\mathcal{B}^\bullet, \mathbf{D}^\bullet_X)[n].$$

Theorem 5.5.7 ([94, 5.3, Theorem]) *Suppose $\bar{p}$ and $\bar{q}$ are complementary perversities, then the intersection pairing followed by the map to homology*

$$\mathcal{IC}^\bullet_{\bar{p}} \overset{L}{\otimes} \mathcal{IC}^\bullet_{\bar{q}} \to \mathcal{IC}^\bullet_{\bar{t}}[n] \to \mathbf{D}^\bullet_X[n]$$

is a Verdier dual pairing.

Corollary 5.5.8 ([94, 5.3, Corollary]) *Let X be a compact, oriented stratified pseudomanifold and let $\bar{p}$ and $\bar{q}$ be two complementary perversities, then the pairing*

$$IH_i^{\bar{p}}(X;k) \times IH_{n-i}^{\bar{q}}(X;k) \to IH_0^{\bar{t}}(X;k) \xrightarrow{\varepsilon} k$$

followed by the evaluation map ε (which counts points with their multiplicity order) induces isomorphisms

$$IH_i^{\bar{p}}(X;k) \cong \mathrm{Hom}(IH_{n-i}^{\bar{q}}(X;k),k)$$

Dropping the assumption that X is oriented, let $\mathcal{O}$ be the orientation local system of k-modules on $X-\Sigma$. A pairing $\mathcal{L}_1 \otimes \mathcal{L}_2 \to \mathcal{O}$ of local systems on $X-\Sigma$ is called perfect if the induced mapping $\mathcal{L}_1 \to Hom(\mathcal{L}_2,\mathcal{O})$ is an isomorphism.

Theorem 5.5.9 ([94, 5.3, last Theorem]) *Suppose $\bar{p}$ and $\bar{q}$ are complementary perversities and the pairing $\mathcal{L}_1 \otimes \mathcal{L}_2 \to \mathcal{O}$ is perfect. Then the intersection pairing followed by the map to homology*

$$\mathcal{IC}^\bullet_{\bar{p}}(\mathcal{L}_1) \overset{L}{\otimes} \mathcal{IC}^\bullet_{\bar{q}}(\mathcal{L}_2) \to \mathcal{IC}^\bullet_{\bar{t}}(\mathcal{O})[n] \to \mathbf{D}^\bullet_X[n]$$

is a Verdier dual pairing.

Note that, using Alexander-Whitney chains, Friedman and McClure [80] re-prove these results in the special case of field coefficients.

Example 5.5.10 (Suspension of the torus) That is the original Goresky-MacPherson example (see [92]) for which Poincaré duality fails for usual homology. The suspension of the torus (Fig. 5.6) is the join of the torus with two points A and B. It is a 3-dimensional singular variety with two singular points A and B: the link of A (or B) is a torus, not a sphere. See the alternative very nice picture of the suspension of the torus in [92].

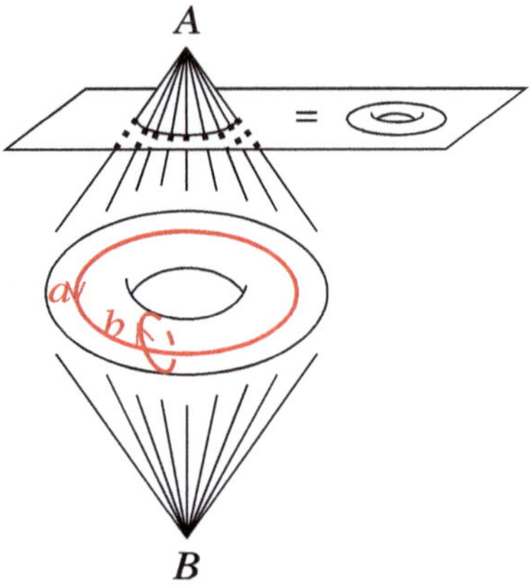

Fig. 5.6 Suspension of the circle $\mathbb{S}^1$ and of the torus

The natural stratification of the suspension of the torus is

$$X \supset X_0 = \{A, B\} \supset \emptyset.$$

There are two possible perversities:

$$\bar{p} = \bar{0} = (0, 0, 0, 0) \qquad \text{and} \qquad \bar{p} = \bar{t} = (0, 0, 0, 1)$$

An i-dimensional chain c containing one (or two) of the singular points A and B is allowable if

$$0 = \dim(|c| \cap X_0) \leq i - 3 + p_3,$$

that means, if $\bar{p} = \bar{0}$, then $i \geq 3$ and if $\bar{p} = \bar{t}$, then $i \geq 2$.

The intersection homology groups $IH_i^{\bar{p}}(X; \mathbb{Q})$ corresponding to the two perversities are easily computed and are resumed in the Table 5.2. The cycle a is boundary of the cycle $c(a)$, cone with vertex a. The suspension of a is a 2-dimensional cycle denoted by $\Sigma(a)$.

Table 5.2 Intersection homology of the suspension of the torus

Perversities	$\bar{p} = \bar{0}$	$\bar{p} = \bar{t}$
i	$IH_i^{\bar{0}}(X)$	$IH_i^{\bar{t}}(X)$
0	$\mathbb{Q}_{\{x\}}$	$\mathbb{Q}_{\{x\}}$
1	$\mathbb{Q}_a \oplus \mathbb{Q}_b$	0
2	0	$\mathbb{Q}_{\Sigma(a)} \oplus \mathbb{Q}_{\Sigma(b)}$
3	$\mathbb{Q}_{[X]}$	$\mathbb{Q}_{[X]}$

The intersection matrix of the intersection product

$$IH_1^{\bar{0}}(X;\mathbb{Q}) \times IH_1^{\bar{t}}(X:\mathbb{Q}) \to \mathbb{Q}$$

$$\begin{array}{cc} & \begin{array}{cc}\Sigma(a) & \Sigma(b)\end{array} \\ \begin{array}{c} a \\ b \end{array} & \begin{pmatrix} 0 & \mp 1 \\ \pm 1 & 0 \end{pmatrix} \end{array}$$

is non-degenerate.

Factorization of Poincaré Homomorphism. See [92, Sect. 1.4], [94, Sect. 5.1], [18, I, Sect. 4.1; I, Sect. 3.2]

Poincaré Duality—Return to the smooth case

The Poincaré duality can be proved by using sheaf complexes: R is still a regular Noetherian ring with finite Krull dimension, which can be $\mathbb{Z}$, $\mathbb{Q}$ or $\mathbb{R}$. Assuming that X is an n-dimensional oriented manifold, the quasi-isomorphism of complexes of sheaves $R_X[n] \cong \mathcal{C}_X^\bullet$ induces isomorphisms of hypercohomology groups:

$$\mathbb{H}_\Phi^{-i}(X;\mathbf{R}_X[n]) \cong \mathbb{H}_\Phi^{-i}(X;\mathcal{C}_X^\bullet)$$

i.e., (5.8), (5.9)

$$H_\Phi^{n-i}(X) \cong H_i^\Phi(X).$$

In particular, one has (see Sect. 5.2.2):

$$H^{n-i}(X) \cong H_i(X) \quad \text{and} \quad H_c^{n-i}(X) \cong H_i^c(X).$$

Poincaré Duality—Singular case

An orientation on X is an isomorphism (5.5.2)

$$\mathbf{R}_{X-\Sigma}[n] \to \mathbf{D}^\bullet_{X-\Sigma}.$$

The (unique in D^b) canonical lift of the orientation

$$\mathbf{R}_X[n] \to Ri_*\mathbf{R}_{X-\Sigma}[n] \overset{\cong}{\to} Ri_*\mathbf{D}^\bullet_{X-\Sigma}$$

induces the cap-product with the orientation (fundamental) class

$$PD_X : \mathbf{R}_X[n] \to \mathbf{D}^\bullet_X.$$

Proposition 5.5.11 ([94, Sect. 5.1]) *Assume X oriented, and let $i : X - \Sigma \to X$ denote the inclusion. For any perversity $\bar{p}$ there is a unique morphism in D^b*

$$\mathbf{R}_X[n] \to \mathcal{IC}^\bullet_{\bar{p}} \to \mathbf{D}^\bullet_X$$

such that the induced morphism $i^*\mathbf{R}_X[n] \to i^*\mathcal{IC}^\bullet$ is the evident one and $i^*\mathcal{IC}^\bullet \to i^*\mathbf{D}^\bullet_X$ is given by the orientation. These morphisms factor the cap-product with the fundamental class $[X]$, i.e. $PD_X : \mathbf{R}_X[n] \to \mathbf{D}^\bullet_X$.

For any perversity $\bar{p}$, denote the previous morphisms by

$$\alpha_X : \mathbf{R}_X[n] \to \mathcal{IC}^\bullet_{\bar{p}} \qquad \text{and} \qquad \omega_X : \mathcal{IC}^\bullet_{\bar{p}} \to \mathbf{D}^\bullet_X.$$

By taking hypercohomology, one obtain the classical comparison morphisms

$$H^\bullet(X) \xrightarrow{\alpha_X} IH^{\bar{p}}_{n-\bullet}(X) \quad \text{and} \quad IH^{\bar{p}}_{n-\bullet}(X) \xrightarrow{\omega_X} H_{n-\bullet}(X).$$

The composition $\omega_X \circ \alpha_X : \mathbf{R}_X[n] \to \mathbf{D}^\bullet_X$

$$\begin{array}{ccc} \mathbf{R}_X[n] & \xrightarrow{\quad PD_X \quad} & \mathbf{D}^\bullet_X \\ & {\scriptstyle \alpha_X}\searrow \qquad \nearrow{\scriptstyle \omega_X} & \\ & \mathcal{IC}^\bullet_{\bar{p}} & \end{array}$$

induces at the global level, i.e., taking hypercohomology, the "classical" Poincaré duality homomorphism

$$H^\bullet(X) \to H_{n-\bullet}(X)$$

that is factorized by intersection homology

$$\begin{array}{ccccc} H^{n-i}(X) & & \xrightarrow{\quad \bullet\cap[X] \quad} & & H_i(X) \\ {\scriptstyle \alpha^{\bar{0}}_X}\downarrow & \searrow^{\alpha^{\bar{p}}_X} & & \nearrow^{\omega^{\bar{t}}_X} & \uparrow{\scriptstyle \omega^{\bar{t}}_X} \\ IH^{\bar{0}}_i(X) & \longrightarrow & IH^{\bar{p}}_i(X) & \longrightarrow & IH^{\bar{t}}_i(X). \end{array} \tag{5.24}$$

Poincaré Duality—Singular case—geometry
In this (sub)-section, X is an oriented compact PL-pseudovariety.

First, remark that if $\bar{p}$ and $\bar{q}$ are two perversities such that $\bar{p} \leq \bar{q}$, that is $p_\alpha \leq q_\alpha$ for all α, then one has a natural morphism

$$IC_*^{\bar{p}}(X) \hookrightarrow IC_*^{\bar{q}}(X) \tag{5.25}$$

for every support family and it induces a morphism $IH_*^{\bar{p}}(X) \to IH_*^{\bar{q}}(X)$.

In particular, one has a morphism $IH_*^{\bar{0}}(X) \to IH_*^{\bar{p}}(X)$ and a morphism $IH_*^{\bar{p}}(X) \to IH_*^{\bar{t}}(X)$ for every perversity $\bar{p}$.

The morphism $\alpha_X^{\bar{0}} : H^{n-i}(X) \to IH_i^{\bar{0}}(X)$ can be described in the following way: Assuming that X is embedded in a smooth m-dimensional PL oriented manifold M, the stratification of X can be extended to a stratification of M by taking $M \setminus X$ as the regular stratum. Let K be a locally finite triangulation of M compatible with the stratification. For each $p = n - i$-simplex $\sigma \in K$ contained in X, the dual cell of σ in M, denoted by $D\sigma$ has dimension $m - p$ (see 5.4) and is transverse to all strata. The Poincaré homomorphism

$$C_{(K)}^{n-i}(X) \to C_i^{(K')}(X)$$

associates to the elementary $(n - i)$-cochain σ^* which corresponds to the simplex σ in K, the i-chain $\xi = D\sigma \cap X$ of (K'), which is $\bar{0}$-allowed. Therefore, for each perversity $\bar{p}$, one has the factorisation (5.24).

Relative Homology, See [93, 1.3], [18, I, 2.2.2]

Let X be a stratified pseudomanifold, and U an open subset in X, then U inherits a structure of stratified pseudomanifold induced by the one of X. For every perversity $\bar{p}$, the complex of intersection chains of U with compact supports $IC_*^{\bar{p}}(U)$, is a sub-complex of $IC_*^{\bar{p}}(X)$.

Defining $IC_*^{\bar{p}}(X, U) = IC_*^{\bar{p}}(X)/IC_*^{\bar{p}}(U)$, one obtains a relative complex and one has a long exact sequence:

$$\cdots \to IH_i^{\bar{p}}(U) \to IH_i^{\bar{p}}(X) \to IH_i^{\bar{p}}(X, U) \to IH_{i-1}^{\bar{p}}(U) \to \cdots$$

The property is also valid with local systems (see [93, 1.8]).

Excision, See [93, 1.5]

Lemma 5.5.12 *Let X be a locally compact stratified pseudomanifold and U and V two open subsets in X, then the inclusion $(U, U \cap V) \hookrightarrow (U \cup V, V)$ induces an isomorphism of intersection homology groups with compact supports*

$$IH_i^{\bar{p}}(U, U \cap V) \cong IH_i^{\bar{p}}(U \cup V, V).$$

Proposition 5.5.13 *Let X be a locally compact stratified pseudomanifold, U an open subset in X and A a closed subset in U. Let $\bar{p}$ be any perversity, then the inclusion $(X - A, U - A) \hookrightarrow (X, U)$ induces an isomorphism of intersection homology groups with compact supports*

$$IH_i^{\bar{p}}(X, U) \cong IH_i^{\bar{p}}(X - A, U - A).$$

The property is also valid with local systems (see [93, 1.8]).

Künneth Formulae, See [94, 6.3]

Künneth formulae in homology If X and Y are topological spaces, and R a PID (principal ideal domain) the Künneth formula is written as short exact sequence (where all homology groups have R coeffivients)

$$0 \to \bigoplus_{a+b=i} H_a(X) \otimes H_b(Y) \to H_i(X \times Y) \to \bigoplus_{a+b=i-1} \mathrm{Tor}^R(H_a(X), H_b(Y)) \to 0,$$

that (not canonically) splits.

If R is a field k, then the Künneth formula is written

$$\bigoplus_{a+b=i} H_a(X; k) \otimes H_b(Y; k) \cong H_i(X \times Y; k).$$

Künneth formula in intersection homology In general, the Künneth formula is no longer true for intersection homology (see counterexamples in [61, Sect. 5]). However, there are some situations for which the the Künneth formula is true.

(1) Cheeger [59] observes that the Künneth formula holds for the middle intersection cohomology, and for Witt spaces X, Y (see Sect. 5.4.6) with $k = \mathbb{R}$.

$$IH_i^{\bar{m}}(X \times Y; \mathbb{R}) \cong \bigoplus_{a+b=i} IH_a^{\bar{m}}(X; \mathbb{R}) \otimes IH_b^{\bar{m}}(Y; \mathbb{R}). \tag{5.26}$$

The formula is extended in [94, 6.3] in the context of "middle homology sheaves". A middle homology sheaf is a complex of sheaves $\mathcal{S}^\bullet$ such that for some local coefficient system $\mathcal{F}$ on $X^0 = X \backslash \Sigma$,

$$\mathcal{S}^\bullet = \mathcal{IC}^\bullet_{\bar{m}}(\mathcal{F}) = \mathcal{IC}^\bullet_{\bar{n}}(\mathcal{F}).$$

Let $\pi_1 : X \times Y \to X$ and $\pi_2 : X \times Y \to Y$ be the projections. Let $\mathcal{F}_1$ and $\mathcal{F}_2$ be local coefficients systems on the regular parts of X and Y respectively, satisfying the previous formula, then

$$IH_i^{\bar{m}}(X \times Y; \pi_1^* \mathcal{F}_1 \otimes \pi_2^* \mathcal{F}_2) \cong \bigoplus_{a+b=i} IH_a^{\bar{m}}(X; \mathcal{F}_1) \otimes IH_b^{\bar{m}}(Y; \mathcal{F}_2).$$

(2) When one of the element of the product is a smooth manifold, the formula is verified (see [93, Sect. 1.6], [61]).

Proposition 5.5.14 *Let X be a locally compact stratified pseudomanifold and M a manifold. Let $\bar{p}$ be a perversity, one has a split exact sequence:*

$$0 \to (IH_*^{\bar{p}}(X) \otimes H_*(M))_i \to IH_i^{\bar{p}}(X \times M) \to (IH_*^{\bar{p}}(X) * H_*(M))_{i-1} \to 0$$

(3) Fix a coefficient ring R which is a principal ideal domain and suppose X and Y are compact pseudomanifolds. Cohen, Goresky and Ji show [61] more general results showing, for instance that if the perversity $\bar{p}$ satisfies

$$p(a) + p(b) \leq p(a+b) \leq p(a) + p(b) + 1$$

for all a and b, then there is a split short exact sequence for intersection cohomology with the perversity $\bar{p}$ and coefficients in R:

$$0 \to \bigoplus_{a+b=i} IH^a(X) \otimes IH^b(Y) \to IH^i(X \times Y) \to \bigoplus_{a+b=i-1} \mathrm{Tor}^R(IH^a(X), IH^b(Y)) \to 0.$$

The condition on the perversity $\bar{p}$ means that the graph of the perversity function does not deviate far from some straight line through the origin (see also [28, Corollary 9.3]).

(4) G. Friedman, in [74], considers biperversities $(\bar{p}, \bar{q})$ and obtains a Künneth theorem relating $IH_*^{\bar{p},\bar{q}}(X \times Y)$ and $IH_*^{\bar{p}}(X)$ and $IH_*^{\bar{q}}(Y)$ for all choices of $\bar{p}$ and $\bar{q}$. and this recovers the result of Cohen,Goresky and Ji.

(5) Let k be a field, Friedman and McClure [80] define a perversity $\bar{Q}(\bar{p}, \bar{q})$ on the product $X \times Y$, whose value depends on regularness or not of the elements of the filtration of the product. They obtain an isomorphism

$$IH_*^{\bar{p}}(X; k) \otimes IH_*^{\bar{q}}(Y; k) \to IH_*^{\bar{Q}}(X \times Y; k).$$

Normalization and Intersection Homology, [92, Sect. 4], [18, I, 1.6, I, 3.2, V,2.8 and 2.12]

The oriented n-dimensional pseudomanifold X is *normal* if

$$H_n(X, X - \{x\}; \mathbb{Z}) = \mathbb{Z} \quad \text{for all} \quad x \in X.$$

Equivalently, each point x admits a fundamental system of neighborhoods U whose regular part $U \setminus \Sigma$ is connected.

Proposition 5.5.15 ([92, Sect. 4.3], [94, Sect. 5.6] [18, I, Sect. 4.1]) *Let X be a normal pseudomanifold, the morphisms α_X and ω_X (see Sect. 5.5.1) induce isomorphisms:*

$$\mathbf{R}_X \cong \mathcal{IC}^{\bullet}_{\bar{0}} \quad \textit{and} \quad \mathcal{IC}^{\bullet}_{\bar{t}} \cong \mathbf{D}^{\bullet}_X$$

respectively for the zero perversity $\bar{0}$ and the total one $\bar{t}$:

$$H^{n-i}(X) \cong IH_i^{\bar{0}}(X), \qquad IH_i^{\bar{t}}(X) \cong H_i(X).$$

The vertical arrows in diagram (5.24) are isomorphisms.

Proposition 5.5.16 ([92, Sect. 4.2], [18, I, Sect. 3.2]) *Let $\tilde{X}$ be the normalization of a pseudomanifold X, then one has:*

$$IH_i^{\bar{p}}(\tilde{X}) = IH_i^{\bar{p}}(X).$$

Homology Manifolds

Goresky and MacPherson conjecture in [92, Sect. 6.6] that if X is a normal pseudovariety such that $IH_*^{\bar{p}}(X) \to IH_*^{\bar{q}}(X)$ are isomorphisms for all $\bar{p} \leq \bar{q}$, then X is a $\mathbb{Z}$-homology manifold, i.e. there is an integer n such that for each point $\{x\}$ in X, the local homology group satisfies

$$H_i(X, X \setminus \{x\}; \mathbb{Z}) = \begin{cases} 0 & \text{if } i \neq n \\ \mathbb{Z} & \text{if } i = n. \end{cases} \tag{5.27}$$

The conjecture is false, by a counter-example of King [114] who shows that it is true if one considers more general perversities, so-called "loose perversities". In [32], Brasselet and Saralegi show that the conjecture is true with a supplementary hypothesis, namely if there are tubular neighborhoods of the strata without homological monodromy. On the other hand, Fieseler and Kaup define in [73, 115] invariants linked to properties of the fibers of the Deligne sheaf. Using these invariants Brasselet, Fieseler and Kaup provide computable criteria for X being a homology manifold [25].

Case of Isolated Singularities

Proposition 5.5.17 ([18, Sect. 5.1]) *Let X be an n-dimensional pseudomanifold with an isolated singularity at $\{x\}$. The integer p_n is the only one pertinent element of the perversity, one has $0 \leq p_n \leq n-2$ and:*

$$IH_i^{\bar{p},c}(X) = \begin{cases} H_i^c(X \setminus \{x\}) & i < n - p_n - 1 \\ \mathrm{Im}(H_i^c(X \setminus \{x\}) \to H_i^c(X)) & i = n - p_n - 1 \\ H_i^c(X) & i > n - p_n - 1, \end{cases} \tag{5.28}$$

$$IH_i^{\bar{p}}(X) = \begin{cases} H_i^\phi(X \setminus \{x\}) & i < n - p_n - 1 \\ \mathrm{Im}(H_i^\phi(X \setminus \{x\}) \to H_i(X)) & i = n - p_n - 1 \\ H_i(X) & i > n - p_n - 1 \end{cases} \tag{5.29}$$

where ϕ denotes the family of closed subsets in X which are contained in $X \setminus \{x\}$.
If n is even and $\bar{p}$ is the middle perversity, one has $n - p_n - 1 = n/2$.

The Proposition is valid with local coefficieny systems.

Example of Thom Spaces, [18, I, Sect. 5.3], [26]

Let B a compact $2n$-dimensional manifold and $\pi : E \to B$ a real oriented vector bundle with even rank r on B. The Thom space $\mathfrak{T}$ associated to E is the Alexandroff compactification of E by adjonction of a point at infinity. It is also the quotient $T(E)/S(E)$ where $T(E)$ and $S(E)$ are the fibre bundles associated to E whose fibers are respectively closed balls and spheres in the fibers of E. The Thom space is a pseudomanifold with an isolated singular point and its dimension is $2s = 2n + r$.

Let $[\mathfrak{T}]$ be the fundamental class of $\mathfrak{T}$ and $e \in H^r(B)$ the Euler class of the bundle E. For every i, different from 0 and $2s$, one has a commutative diagram ([27], see also [18, I, Sect. 5.3]):

$$\begin{array}{ccc} H^{2s-i}(\mathfrak{T}) & \xrightarrow{\cdot \cap [\mathfrak{T}]} & H_i(\mathfrak{T}) \\ \downarrow \cong & & \downarrow \cong \\ H_i(B) & \xrightarrow{\cdot \cap e} & H_{i-r}(B) \end{array}$$

and for the middle (lower) perversity:

$$IH_i(\mathfrak{T}) = \begin{cases} H_i(\mathfrak{T}) & i < s \\ \mathrm{Im}(H_i(B) \xrightarrow{\cdot \cap e} H_{i-r}(B)) & i = s \\ H_{i-r}(B) & i > s. \end{cases} \tag{5.30}$$

In [92, Sect. 6.3] Goresky and MacPherson illustrate the behavior of torsion in the intersection homology by the example of Thom space for which the universal coefficient theorem fails and the generalized Poincaré duality theorem is not true over $\mathbb{Z}$.

Exemples of computations of Thom spaces associated to the Segre and Veronese embeddings are provided by Brasselet and Gonzalez-Sprinberg in [27].

5.5.2 Functoriality

In general, for a map $f : X \to Y$, there is no functoriality, i.e. no maps If^* and If_* such that the diagrams below (5.32) and (5.33) commute. The functoriality problem has been proposed by Goresky and MacPherson in [18, IX, C, Problem 4]): "Find the most general category of spaces and maps (perhaps with additional data) on which intersection homology is functorial."

Goresky and MacPherson earlier proved functoriality for Normally Nonsingular Maps [94, Sect. 5.4]:

(a) A *normally nonsingular map* ([82, Sect. 4.1]) $f : X \to Y$ between oriented topological spaces, is a map such that there is a diagram

$$\begin{array}{ccc} N & \xrightarrow{\;i\;} & Y \times \mathbb{R}^n \\ {\scriptstyle s}\big\uparrow\big\downarrow{\scriptstyle \pi} & & \big\downarrow {\scriptstyle p} \\ X & \xrightarrow[\;f\;]{} & Y \end{array} \tag{5.31}$$

in which $\pi : N \to X$ is a rank d vector bundle with zero-section s, the map i is an open embedding, p is the first projection and $f = p \circ i \circ \pi$. The integer $d - n$ is the relative codimension of f. As said in [82], "Geometrically, that says that the singularities of X at any point x are no better or worse than the singularities of Y at $f(x)$." Topological pseudomanifotds and normally nonsingular maps form a category (see [82]).

Theorem 5.5.18 ([94, 5.4.3]) *Let $f : X \to Y$ be a proper normally nonsingular map of relative dimension ν. Then there are homomorphisms*

$$If_* : IH_k^{\bar{p}}(X) \to IH_k^{\bar{p}}(Y) \qquad \textit{and} \qquad If^* : IH_k^{\bar{p}}(Y) \to IH_{k-\nu}^{\bar{p}}(X).$$

$IH_k^{\bar{p}}$ is both a covariant functor (via If_*) and a contravariant functor (via If^*) on the category of topological pseudomanifotds and normally nonsingular maps.

In their discussion in [18, IX, C], Goresky and MacPherson give and discuss several classes of maps $f : X \to Y$ for which there are natural homomorphisms between $IH_*^{\bar{m}}(X)$ and $IH_*^{\bar{m}}(Y)$ (where $\bar{m}$ is the middle (lower) perversity). In particular, they give the following examples:

(b) The placid maps. A continuous map $f : X \to Y$ between stratified spaces is said placid if it is stratum preserving (i.e. the image of every stratum of X is contained in a single stratum of Y) and for each stratum S in Y, the inequality holds:

$$\operatorname{codim}_X f^{-1}(S) \geq \operatorname{codim}_Y(S).$$

Proposition 5.5.19 *[98, Proposition 4.1] Assume that $f : X \to Y$ is a placid map. Then pushforward of chains and pullback of generic chains induce homomorphisms on intersection homology:*

$$If_* : IH_k^{\bar{m}}(X) \to IH_k^{\bar{m}}(Y) \quad \text{and} \quad If^* : IH_{n-k}^{\bar{m}}(Y) \to IH_{m-k}^{\bar{m}}(X).$$

where $m = \dim(X)$ and $n = \dim(Y)$.

(c) Small maps. [94, Sect. 6.2]
A proper surjective algebraic map $f : X \to Y$ between irreducible complex n-dimensional algebraic varieties is small if X is nonsingular and for all $r > 0$,

$$\mathrm{cod}_{\mathbb{C}}\{y \in Y | \ \dim_{\mathbb{C}} f^{-1}(y) \geq r\} > 2r.$$

If Y is one or two dimensional then a small map $f : X \to Y$ must be a finite map. If Y is a threefold then the fibres of a small map f must be zero dimensional except possibly over a set of isolated points in Y where the fibres may be at most curves.

(d) A more general result has been proved by Barthel, Brasselet, Fieseler, Gabber and Kaup.

Theorem 5.5.20 ([5, Théorème 2.3]) *Let $f : X \to Y$ be a map between algebraic complex varieties of respective pure (real) dimensions m and n, and consider $R = \mathbb{Q}$. Then*
(1) There are contravariant homomorphisms (with closed supports)

$$If^* : IH_{n-\bullet}(Y) \to IH_{m-\bullet}(X)$$

and covariant homomorphisms with compact supports

$$If_* : IH_\bullet^c(X) \to IH_\bullet^c(Y)$$

such that the following diagrams commute:

$$\begin{array}{ccccccc} IH_{n-\bullet}(Y) & \xrightarrow{If^*} & IH_{m-\bullet}(X) & & IH_\bullet^c(X) & \xrightarrow{If_*} & IH_\bullet^c(Y) \\ \uparrow \alpha_Y & & \uparrow \alpha_X & & \downarrow \omega_X & & \downarrow \omega_Y \\ H^\bullet(Y) & \xrightarrow{f^*} & H^\bullet(X) & & H_\bullet^c(X) & \xrightarrow{f_*} & H_\bullet^c(Y). \end{array} \tag{5.32}$$

(2) Assume that the map $f : X \to Y$ is proper, then there are contravariant homomorphisms with compact supports

$$If^* : IH_{n-\bullet}^c(Y) \to IH_{m-\bullet}^c(X)$$

and covariant homomorphisms with closed supports

$$If_* : IH_\bullet(X) \to IH_\bullet(Y)$$

such that the following diagrams commute:

$$\begin{array}{ccccccc} IH^c_{n-\bullet}(Y) & \xrightarrow{If^*} & IH^c_{m-\bullet}(X) & & IH_\bullet(X) & \xrightarrow{If_*} & IH_\bullet(Y) \\ \uparrow{\scriptstyle \alpha_Y} & & \uparrow{\scriptstyle \alpha_X} & & \downarrow{\scriptstyle \omega_X} & & \downarrow{\scriptstyle \omega_Y} \\ H^\bullet_c(Y) & \xrightarrow{f^*} & H^\bullet_c(X) & & .H_\bullet(X) & \xrightarrow{f_*} & H_\bullet(Y). \end{array} \tag{5.33}$$

The results can be summarized by the following commutative diagram (the reader is invited to write the similar diagram with compact supports):

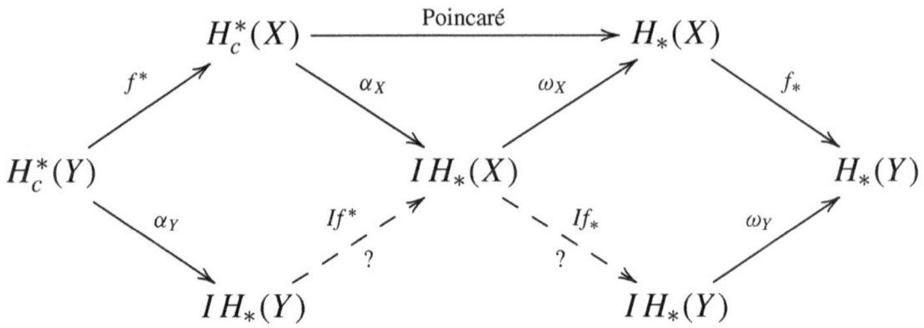

Note that the notations used in [5] are μ^f for If^* and ν_f for If_*. In general, the associated maps If^* and If_* in intersection homology are not uniquely determined. They are uniquely determined by f in the following particular cases:

- if Y is smooth. In that case, α_Y and $If^* \circ \alpha_Y$ are isomorphisms and $If^*$is determined by α_X,
- if f is an equidimensional dominant map or, more generally, a placid map (see [5, (3.3)]),
- if f is the embedding of a closed submanifold X with codimension 1 in Y such that Y is locally analytically irreducible along X (see [5, (3.6)]),
- if f is a homologically small map in the sense of [94, Sect. 6.2].

Based on the previous results, Weber [185] assumes that a map of analytic varieties is an inclusion of codimension one. He shows that the existence of an associated morphism in intersection homology follows from Saito's decomposition theorem. For varieties with conical singularities he shows that the existence of intersection homology morphism is equivalent to the validity of the Hard Lefschetz Theorem for links.

Lifting of Algebraic Cycles

The notion of intersection homology $IH^{(C)}_\bullet(Y)$ of Y with supports in a closed subvariety C of Y, i.e. relative intersection homology $IH_\bullet(Y, Y \setminus C)$ (see Sect. 5.3.1) is useful for this section.

Theorem 5.5.21 ([5, Théorème 2.4]). *Let C be a closed subvariety of Y, with pure dimension n, then the homology class $[C]$, with rational coefficients, is in the image of the morphism*

$$\omega_Y : IH_n^{(C)}(Y) \longrightarrow H_n^{(C)}(Y) .$$

The classes corresponding to algebraic cycles in an algebraic variety can be lifted in intersection homology, however the lifting is not unique.

Coming to the original question asked by Goresky and MacPherson (see [18, Goresky-MacPherson, Chap. IX, Sect. H, Problem 10]) the homology Chern-Schwartz-MacPherson classes of an algebraic variety can be lifted to intersection homology, for the middle perversity and with rational coefficients [5, Corollaire 2.6]. On the one hand, Goresky constructed an example for which there is no lifting when using $\mathbb{Z}$ coefficients, on the other hand Verdier constructed an example for which the lifting is not unique even with rational coefficients (see [26, 27] for these examples). Also, using the previous results obtained for the middle perversity (and higher ones) it is not possible to multiply more than two homology classes. This gives an obstruction to the definition of general characteristic numbers for singular complex algebraic varieties.

The Classification Theorem

The morphisms If^* and If_* in intersection homology are not uniquely determined by the morphism f. The following result provides a measure of the ambiguity. It gives also a geometric meaning of the motivation and completes the principal result:

Theorem 5.5.22 ([5, Théorème 2.7]) *There is a one-to-one correspondence between the morphisms If^*, resp. If_*, such that the diagrams (5.32) and (5.33) commute and classes $\gamma \in IH_n^{(\Gamma_f)}(X \times Y)$ which are liftings of the homology class $[\Gamma_f] \in H_n^{(\Gamma_f)}(X \times Y)$ of the graph of f.*

5.5.3 *Lefschetz Fixed Points and Coincidence Theorems*

Lefschetz Fixed Points Theorem

The smooth case

Let M be an n-dimensional oriented smooth manifold, and $f : M \to M$. One of the possible definitions of the Lefschetz number $L(f)$ (known as Lefschetz fixed point formula [122]) is:

$$L(f) = \sum_{k=0}^{n} (-1)^k \text{Trace}(f_k : H_k(M; \mathbb{Q}) \to H_k(M; \mathbb{Q})). \tag{5.34}$$

Let $G(f) \subset M \times M$ be the graph of f. In general, $G(f)$ is not transverse to the diagonal Δ_M in $M \times M$. However, one can find a map $f' : M \to M$ homotopic to f such that the graph $G(f')$ is transverse to Δ_M. The (oriented) cycles $G(f')$ and Δ_M are transverse and complementary dimensional in $M \times M$. Moreover, there are finitely many intersection points $b_j \in G(f') \cap \Delta_M$. In such a point, the intersection number $I(G(f'), \Delta_M; b_j)$ is well defined (see Sect. 5.2.6) and one has

$$L(f) = \sum_{b_j} I(G(f'), \Delta_M; b_j). \tag{5.35}$$

That number does not depend on the map f' homotopic with f and such that $G(f')$ is transverse to Δ_M.

The main properties of the Lefschetz number are the following: If $L(f) \neq 0$, then f admits fixed points. If $f = \mathrm{id}_M$ then $L(f) = \chi(M)$. If f and g are two homotopic maps from M to M, then $L(f) = L(g)$.

The singular case

Goresky and MacPherson proved in [98] the Lefschetz fixed point theorem in the context of placid (Sect. 5.5.2 b) self maps of Witt spaces (see Sect. 5.4.6) and by using intersection homology with middle lower perversity.

The intersection homology Lefschetz number of a placid self-map $f : X \to X$ is defined by the formula [98, Sect. 4 Definition]:

$$IL(f) = \sum_{i=0}^{\dim X} (-1)^i \mathrm{Trace}(f_i : IH_i^{\bar{m}}(X; \mathbb{Z}) \to IH_i^{\bar{m}}(X; \mathbb{Z})). \tag{5.36}$$

In [98, Proposition 4.2], Goresky and MacPherson show that if $f : X \to Y$ is a placid map between two compact oriented $\mathbb{Q}$-Witt spaces, with $n = \dim X$, then the graph of f determines a canonical homology class $[G(f)] \in IH_n^{\bar{m}}(X \times Y; \mathbb{Q})$.

For a placid self map of a $\mathbb{Q}$-Witt space, both the graph of f and the diagonal carry fundamental classes in intersection homology of $X \times X$ and one has:

Theorem 5.5.23 ([98], Theorem I) *Let $f : X \to X$ be a placid self map of an n-dimensional $\mathbb{Q}$-Witt space. Let $[G(f)]$ and $[\Delta]$ be the homology classes of the graph of f and of the diagonal in $IH_n^{\bar{m}}(X \times X; \mathbb{Q})$. Then the Lefschetz number $IL(f)$ is given by*

$$IL(f) = [G(f)] \bullet [\Delta]$$

where $\bullet$ denotes the intersection product of cycles in intersection homology.

The formula 5.35 has been extended in the singular situation by Goresky and MacPherson (see [97],[98, Sects. 7–12]) in terms of local Lefschetz numbers of a placid map $f : X \to X$ at isolated fixed points.

Theorem 5.5.24 ([98, Theorem II]) *The intersection Lefschetz number is the sum of the local contributions taken over all the fixed points.*

Another way to define local Lefschetz numbers is developed by Bisi et al. [14] using Čech-de Rham theory. The coincidence of this later notion with Goresky and MacPherson ones is shown in Brasselet-Suwa [33].

The Coincidence Theorem

The smooth case

In [122] Lefschetz defined the coincidence number of two maps $f : M \to N$ and $g : M \to N$ where M and N are compact oriented smooth n-dimensional manifolds without boundaries. The coincidence set $C(f, g)$ is defined to be

$$C(f, g) = \{x \in M \mid f(x) = g(x)\}.$$

The Lefschetz coincidence number is defined as

$$L(f, g) = \sum_{k=0}^{n} (-1)^k \mathrm{Trace}(PD_M \circ g^{n-k} \circ PD_N^{-1} \circ f_k) \tag{5.37}$$

$$\begin{array}{ccc} H_k(M; \mathbb{Q}) & \xrightarrow{\ f_k\ } & H_k(N; \mathbb{Q}) \\ {\scriptstyle PD_M}\uparrow{\scriptstyle \cong} & & {\scriptstyle PD_N}\uparrow{\scriptstyle \cong} \\ H^{n-k}(M; \mathbb{Q}) & \xleftarrow{\ g^{n-k}\ } & H^{n-k}(N; \mathbb{Q}) \end{array}$$

where vertical arrows are Poincaré duality isomorphisms. If $L(f, g)$ is not zero, then there is at least one coincidence point: $C(f, g)$ is not empty.

The singular case

In the case of singular varieties, Goresky and MacPherson defined the notion of placid correspondences C between n-dimensional Witt spaces X and Y as beeing an n-dimensional compact oriented pseudomanifold $C \subset X \times Y$ such that each of the projections $\pi_X : C \to X$ and $\pi_Y : C \to Y$ is placid. According to the Proposition 5.5.19, one has homomorphisms on intersection homology:

$$(\pi_Y)_*(\pi_X)^* : IH_i^{\bar{m}}(X) \to IH_i^{\bar{m}}(Y) \quad \text{and} \quad (\pi_X)_*(\pi_Y)^* : IH_i^{\bar{m}}(Y) \to IH_i^{\bar{m}}(X).$$

If $C_1$and C_2 are two correspondences between the Witt spaces X and Y, the Lefschetz number $IL(C_1, C_2)$ is defined to be the alternating sum of traces of the induced map

$$(\pi_X^2)_*(\pi_Y^2)^*\pi_Y^1)_*(\pi_X^1)^* : IH_i^{\bar{m}}(X) \to IH_i^{\bar{m}}(X).$$

Each correspondence defines a canonical intersection homology class

$$[C_i] \in IH_n^{\bar{m}}(X \times Y; \mathbb{Q})$$

and the Lefschetz number $IL(C_1, C_2)$ is equal to the intersection product $[C_1] \bullet [C_2]$ ([98, Theorem I']). Moreover, Goresky and MacPherson show ([98, Theorem II']) that it is equal to the sum of the local linking numbers suitably defined (see [98, Sect. 8]).

In the particular case of coincidences, given $f, g : X \to Y$ placid maps between n-dimensional oriented compact $\mathbb{Q}$-Witt spaces, the Lefschetz coincidence number is defined by [35]

$$IL(f, g) = \sum_i (-1)^i \mathrm{Trace}(g^i f_i),$$

where $f_i : IH_i^{\bar{m}}(X) \to IH_i^{\bar{m}}(Y)$ and $g^i : IH_i^{\bar{m}}(Y) \to IH_i^{\bar{m}}(X)$ are defined for the lower middle perversity $\bar{m}$ (Proposition 5.5.19).

Theorem 5.5.25 *[98, Sect. 14], [35] The Lefschetz coincidence number of (f, g) is determined by the intersection of the canonical homology classes of the graphs, $[G(f)]$ and $[G(g)]$.*

$$IL(f, g) = (-1)^n [G(f)] \bullet [G(g)].$$

If $IL(f, g) \neq 0$ then there is (at least one point) $x \in X$ such that $f(x) = g(x)$.

Examples of coincidence of maps are provided in [35] (J.-P. Brasselet, A.K.M. Libardi, T.F.M. Monis, E.C. Rizziolli and M.J. Saia) with local and global explicit computations.

5.5.4 *Morse Theory*

A complete history of Morse theory can be found, for instance in the Introduction of the Goresky-MacPherson's book [96], Sect. 1.7. A complete survey is given by Mark Goresky in the Chap. 5 of this Handbook (Vol. 1), see [90].

The Smooth Case

The main results of classical Morse theory for ordinary homology and for a compact smooth variety M can be summarized as follows [136]:

A critical point of a smooth function $f : M \to \mathbb{R}$ on a manifold M is a point where the differential of f vanishes, its image by f is a critical value. A non-degenerate critical point of f is a point for which the Hessian matrix of second partial derivatives of f is non-singular.

A smooth function $f : M \to \mathbb{R}$ on a manifold M is a Morse function if it has only non-degenerate critical points. According to a result by René Thom [178], the Morse functions form an open, dense subset of all smooth functions $f : M \to \mathbb{R}$ (for the $\mathcal{C}^2$- Whitney topology).

Considering a smooth function $f : M \to \mathbb{R}$ and a point $c \in \mathbb{R}$, let $M_{<c}$ denote the inverse image by f of the open interval $]-\infty, c[$.

The Morse Lemma says:

- For small enough ε, if the interval $]v-\varepsilon, v+\varepsilon[$ does not contain any critical value, then $M_{<v+\varepsilon}$ is homeomorphic to $M_{<v-\varepsilon}$.
- If p is a non-degenerate critical point of $f : M \to \mathbb{R}$, then there exists a chart $(x_1, x_2, \dots, x_n)$ in a neighborhood U_p of p such that $x_i(p) = 0$ for all i and $f(x) = f(p) - x_1^2 - x_2^2 - \cdots - x_k^2 + x_{k+1}^2 + \cdots + x_n^2$ in U_p. The integer k is the Morse index of f at p. For small enough ε, one has, with $v = f(p)$,

$$H_i(M_{<v+\varepsilon}, M_{<v-\varepsilon}) = \begin{cases} 0 & \text{for } i \neq k \\ \mathbb{Z} & \text{for } i = k \end{cases} \tag{5.38}$$

The Singular Case

In the case of a singular variety, there is no longer a Morse index for ordinary homology. Goresky and MacPherson [93, Sect. 4.5 (3)] provide a nice counter-example.

In fact, the concept of Morse function in the case of isolated singular varieties has been introduced by F. Lazzeri [119]. Some conditions for beeing a Morse function on a stratified space have been stated by Benedetti [12] and Pignoni [151] (see [96, Introduction, Sect. 1.4]).

Goresky and MacPherson assume that X is a purely n-dimensional complex analytic variety, endowed with a Whitney stratification (with complex analytic strata), and embedded in a complex analytic manifold M. In [96, Introduction, Sect. 1.4 What is a Morse function ?] A C^∞ function $f : M \to \mathbb{R}$ is called a Morse function for X provided

- For each stratum S of X, the function $f|_S$ has only nondegenerate critical points. The critical points of f are the critical points of $f|_S$ and the critical values of f are the values of f at these points.
- At each critical point $p \in X$, the differential $df(p)(\tau) \neq 0$ whenever τ is a limit of tangent planes from some larger stratum containing S in its closure.
- All critical values are distinct.

If p is a critical point in the stratum S, then the Morse index k of f at p is defined to be $c + \lambda$ where c is the complex codimension of S in X and λ is the classical Morse index of $f|_S$.

In order to recover Morse theory in the context of intersection homology, Goresky and MacPherson define the following ingredients [95, 96], see also [90].

The first one is the complex link of a stratum S. Choose a manifold N meeting S transversaly at p and a generic projection $\pi : N \cap X \to \mathbb{C}$ sending p to 0. For

$0 < \varepsilon \ll \delta \ll 1$ denote by $B(p, \delta)$ the ball of radius δ centered at p and $B_\delta = X \cap B(p, \delta)$, $\partial B_\delta = X \cap \partial B(p, \delta)$. The complex link $\mathcal{L}_\mathbb{C}$ of S is a pseudomanifold with:

$$\mathcal{L}_\mathbb{C} = \pi^{-1}(t) \cap B_\delta, \qquad \partial\mathcal{L}_\mathbb{C} = \pi^{-1}(t) \cap \partial B_\delta.$$

where $0 < |t| < \varepsilon$.

Denote by μ he monodromy transformation obtained by carrying a chain Z in $\mathcal{L}_\mathbb{C}$ with in $\partial\mathcal{L}_\mathbb{C}$, over a small loop around 0 in $\mathbb{C}$. The second ingredient is the Morse group A_p, image of the variation map (see [93, Sect. 3.7] and [96, Part II, Sect. 6.3]):

$$(1 - \mu) : IH_{c-1}(\mathcal{L}_\mathbb{C}, \partial\mathcal{L}_\mathbb{C}; \mathbb{Z}) \longrightarrow IH_{c-1}(\mathcal{L}_\mathbb{C}; \mathbb{Z})$$

where $(1 - \mu)$ vanishes on $IH_{c-1}(\partial\mathcal{L}_\mathbb{C})$.

Using intersection homology, Goresky and MacPherson recover Morse theory for a compact Whitney stratified singular complex analytic variety X analytically embedded in a smooth variety M, as follows:

Theorem 5.5.26 ([96]) *For an open dense set of Morse functions $f : M \to \mathbb{R}$ (in the sense of Lazzeri and Pignoni), all values $v \in \mathbb{R}$ have exactly one of the following properties (and only finitely many values have property 2):*

(1) For small enough ε, then $X_{<v+\varepsilon}$ is homeomorphic to $X_{<v-\varepsilon}$ in a stratum preserving way,
(2) There is a Morse index k of the critical point p with critical value v such that for small enough ε,

$$IH_i(X_{<v+\varepsilon}, X_{<v-\varepsilon}; \mathbb{Z}) = \begin{cases} 0 & \text{for } i \neq k \\ A_p & \text{for } i = k \end{cases} \tag{5.39}$$

In [93, 95, 96] Goresky and MacPherson provide various applications of stratified Morse theory :

- The Lefschetz hyperplane theorem holds for the intersection homology of a (singular) projective algebraic variety [93, Sect. 5.4].
- The intersection homology of a complex n-dimensional Stein space vanishes in dimensions $> n$ [93, Sect. 5.3].
- (3) The sheaf of intersection chains on a general fibre specializes (over a curve) to a perverse object on the special fibre [93, Sect. 6.1].

As the authors write on [93, Sect. 0.2], “other methods have been used to obtain some of these results ..., however the method of Morse theory has several advantages: it can be used to study homology with $\mathbb{Z}$ coefficients (as well as $\mathbb{Q}$ coefficients) and it applies to analytic (as well as algebraic) varieties.”

5.5.5 De Rham Theorems

By relating differential geometry to topology, de Rham's theorem (1931) opened the door to "countless" new results, applications, conjectures, and many alternative proofs.

The passage from the smooth case to the singular case is due to Cheeger, Goresky and MacPherson. From "geometric" results, they mainly developed the theory within the framework of sheaves (see [88]). Several important conjectures have resulted in various fields.

Although implicit in the previous works, the explicit and geometric translation in terms of order of poles corresponding to the perversity was given in [28] (see Sect. 5.5.5).

This section is divided into four parts: de Rham's theorem in the smooth case, de Rham's theorem in the singular case, conjectures and applications, geometric translation. In this section, all intersection homology groups are written with the middle perversity the notation of which is omitted.

The Smooth Case

The de Rham Theorem (de Rham thesis [160]) provides a very useful relationship between the topology and the differentiable structure of a PL-manifold. The de Rham complex is the complex of smooth differential forms on a manifold M with exterior derivative as the differential:

$$0 \to \Omega^0(M) \xrightarrow{d} \Omega^1(M) \xrightarrow{d} \Omega^2(M) \xrightarrow{d} \ldots$$

The de Rham Theorem says that the cohomology $H^j_{dR}(M)$ of the de Rham complex is isomorphic to the PL-cohomology $H^j(M; \mathbb{R})$. There are many proofs in the literature. The Whitney's book "Geometric integration theory" provides a nice geometric proof of the Theorem [186, Chap. IV, Theorem 29A].

Let M be a Riemannian (compact) oriented manifold endowed with a metric g. The metric induces an inner product on fibers $T^*_x(M)$ of the cotangent bundle and then an $\mathcal{L}^2$-metric on $\Omega^j(M) = \Gamma(\Lambda^j(T^*(M)))$. Let $\delta : \Omega^j(M) \to \Omega^{j-1}(M)$ be the formal adjoint of d relatively to the inner product and $* : \Omega^j(M) \to \Omega^{n-j}(M)$ the Hodge star operator [20]. The Hodge Theorem, first proved by Hodge (1933–1936) with final proof by Hermann Weyl and Kunihiko Kodaira, says that every de Rham cohomology class is represented by a unique harmonic form, i.e. a differential form of which the Laplacian Δ is zero:

$$\Delta(\omega) = (d\delta + \delta d)(\omega) = 0.$$

A compact complex projective manifold is a Kähler manifold. The cohomology groups admit a decomposition (pure Hodge decomposition)

$$H^r(M;\mathbb{C}) = \bigoplus_{p+q=r} H^{p,q}(M;\mathbb{C}),$$

as direct sum of complex vector spaces and $\overline{H^{p,q}(M;\mathbb{C})} = H^{q,p}(M;\mathbb{C})$. The (p,q) components of a harmonic form are again harmonic.

The complex of sheaves is exact and is a soft resolution of the constant sheaf $\mathbb{R}_M$. It follows that the sheaf cohomology is the singular cohomology with $\mathbb{R}$ coefficients.

The Sheaf of $\mathcal{L}^2$ Differential Forms

In order to extend the theory to singular varieties, one considers a pseudomanifold X whose regular part $X^0 = X \setminus \Sigma$ is a smooth (possibly incomplete) Riemannian manifold with a metric g ([60], Sect. 3).

One define a presheaf on X by assigning by assigning to each open set $U \subset X$ the subset $\Gamma(U, \Omega^i_{(2)})$ in $\Omega^i(U \cap X^0)$ of differential forms ω such that for any point x in U, there is a neighbourhood V of x in U such that

$$\int_{V\cap X^0} \omega \wedge *\omega < \infty \qquad \text{and} \qquad \int_{V\cap X^0} d\omega \wedge *d\omega < \infty$$

This presheaf is filtered by differential form degree and the exterior derivative makes it into a complex of presheaves. The associated sheaf complex, obtained by "sheafification" is the sheaf of $\mathcal{L}^2$ differential forms denoted by $\Omega^*_{(2)}$. It is a complex of fine sheaves whose cohomology is denoted by $H^*_{(2)}(X)$.

The definition makes sense in the case of a local system $\mathcal{L}$ on X^0 provided that $\mathcal{L}$ has a smoothly varying positive definite inner product on each fiber. The restriction of the sheaf to X^0 is the sheaf of all smooth differential forms (with arbitrary growth) on X^0 (and coefficients in $\mathcal{L}$).

The Cheeger-Goresky-MacPherson's Conjecture

The study of $\mathcal{L}^2$ cohomology on the non-singular part of a variety with conical singularities was initiated by Cheeger in the context of the study of analytic torsion. In 1976 Sullivan observes similarity between IH and $\mathcal{L}^2$, namely similarity between local results in Proposition 5.4.12 and forthcoming Lemma 5.5.30 The same year, Deligne proposes to consider variation of Hodge structure on intersection homology [60, p. 308] (see Sect. 5.5.5).

These observations led Cheeger, Goresky and MacPherson's to the famous CGM conjecture which concerns complex n-dimensional projective varieties and is in fact made up of three conjectures [60, Sect. 4].

The conjecture are written for the middle perversity (here the two middle perversities agree) which will be omitted.

Conjecture 5.5.27 states that the intersection homology groups $IH_*(X)$ satisfy the following 5 conditions of the "*Kähler package*". That is :

1. Poincaré duality (see Sect. 5.5.8). The intersection pairing

$$IH_i(X) \times IH_{2n-i}(X) \to \mathbb{C} \tag{5.40}$$

is non singular for all i.

2. Pure Hodge decomposition. There is a natural direct sum Hodge decomposition

$$IH_r(X) \cong \oplus_{p+q=r} IH_{p,q}(X)$$

such that

$$IH_{p,q}(X) \cong \overline{IH_{q,p}(X)}.$$

The decomposition is compatible with maps If_* and If_* when they exist, for example if $f : Y \to X$ is normally nonsingular with relative dimension m then

$$If_* : IH_{p,q}(Y) \to IH_{p,q}(X) \qquad \text{and} \qquad If^* : IH_{p,q}(X) \to IH_{p-m,q-m}(Y).$$

The map from cohomology $H^i(X) \to IH_{2n-i}(X)$ is a morphism of Hodge structures.

3. Hard Lefschetz. Let H be a hyperplane in the ambient projective space, which is transverse to a Whitney stratification of X. Let $N \in H^2(X)$ denote the cohomology class represented by $H \cap X$ and let $L : IH_i(X) \to IH_{i-2}(X)$ denote multiplication by this class. then the map

$$L^k : IH_{n+k}(X) \to IH_{n-k}(X)$$

is an isomorphism for each k.

Let define $P_{n+k}(X) = \ker(L^{k+1})$, then the Lefschetz decomposition

$$IH_m(X) = \oplus_k L^k(P_{m+2k}(X))$$

is compatible with the Hodge decomposition.

4. Lefschetz Hyperplane Theorem, Let H be a hyperplane in the ambient projective space, which is transverse to a Whitney stratification of X. The homomorphism induced by inclusion

$$IH_k(X \cap H; \mathbb{Z}) \to IH_k(X; \mathbb{Z})$$

is an isomorphism for $k < n - 1$ and a surjection for $k = n - 1$. (for instance see [72, 102]).

5. Hodge Signature Theorem. If $\sigma(X)$ denotes the signature of the intersection pairing (5.40) on $IH_n(X)$, then

$$\sigma(X) = \sum_{p+q \equiv 0 (\mathrm{mod} 2)} (-1)^p \dim IH_{(p,q)}(X).$$

As written in [60], Conjecture 5.5.27 follows from the stronger following Conjectures 5.5.28 and 5.5.29.

Conjecture 5.5.28 The $\mathcal{L}^2$ cohomology group $H^k_{(2)}(X)$ is finite dimensional and is isomorphic to the subspace $\mathcal{H}^k$ of $\Omega^k \cap \mathcal{L}^2$ which consists of the square summable differential k-forms which are closed and co-closed $d\omega = \delta\omega = 0$. Furthermore, the operator "integration" preserves this subspace $\mathcal{H}^k$.

Conjecture 5.5.29 For almost any chain $\xi \in C_k(X)$ and almost any differential form $\theta \in \mathcal{H}^k$, the integral $\int_\xi \theta$ is finite and $\int_{\partial\eta} \theta = \int_\eta d\theta$ whenever both sides are defined. The induced homomorphism

$$H^j_{(2)}(X) \xrightarrow{\int} \mathrm{Hom}(IH^{\bar{m}}_j(X); \mathbb{C})$$

is an isomorphism.

Cheeger, Goresky and MacPherson conjectured that each class contains an unique harmonic (closed and co-closed) representative and that splitting the harmonic forms into their (p, q)–pieces yields a (pure) Hodge decomposition, compatible with Deligne's mixed Hodge structure on the ordinary cohomology groups of X. They noted that the Hodge decomposition would exist if the metric on U were complete, and they suggested that another approach to constructing a Hodge decomposition of $IH^*(X)$ is to construct a complete (Kähler) metric. Moreover, they gave a lot of evidence for the validity of the conjectures. This fundamental work of Cheeger, Goresky, and MacPherson has lead to a great deal of work by many people.

Poincaré Lemma for $\mathcal{L}^2$-cohomology
Let L be an $n-1$-dimensional Riemannian (compact) manifold endowed with a metric g_L. For $h > 0$, the metric cone on L, denoted by $c^h(L)$, is the completion of the incomplete Riemannian manifold $L \times [0, \infty[$ endowed with the metric $g = dr \otimes dr + r^{2h} g_L$.

As before, $\Omega^\bullet_{(2)}(c^h L)$ denotes the subset of differential forms $\omega \in \Omega^\bullet(c^h(L) \setminus \{0\})$ such that

$$\int_{c^h(L)\setminus\{0\}} \omega \wedge *\omega < \infty \qquad \text{and} \qquad \int_{c^h(L)\setminus\{0\}} d\omega \wedge *d\omega < \infty$$

where $d : \Omega^i \to \Omega^{i+1}$ is induced by the external derivative and the operator $*$ is the Hodge operator [20, 58, 59].

The $\mathcal{L}^2$-cohomology groups of the cone $c(L)$, denoted by $H^j_{(2)}(c^h(L))$ are cohomology groups of the complex $\Omega^\bullet_{(2)}(c^h(L))$.

Lemma 5.5.30 ([59, Lemma 3.4]) *The $\mathcal{L}^2$-cohomology groups of the cone $c(L)$ satisfy*

$$H^j_{(2)}(c^h(L)) = \begin{cases} H^j_{\mathrm{dR}}(L) & \textit{if } j < \dfrac{n-1}{2} + \dfrac{1}{2h} \\ \\ 0 & \textit{if } j \geq \dfrac{n-1}{2} + \dfrac{1}{2h}. \end{cases}$$

Varieties with isolated conical singularities

Let X be a singular variety whose singularities are isolated points $\{a_i\}$ admitting each one a neighborhood U_i in X, which is isometric to the (open) metric cone $\mathring{c}(L_i)$ whose basis is a smooth manifold L_i. I n he particular case n is even, $h = 1$ and $\bar{p}$ is the middle perversity $\bar{m}$, then $\bar{p}(n) = \frac{n}{2} - 1$, and one has:

$$n - 1 - p_n = \frac{n-1}{2} + \frac{1}{2h} = \frac{n}{2}$$

Cheeger, Goresky and MacPherson study the two following cases:

(1) [59, Theorem 6.1] and [60, Sect. 3.4]. Let X be a pseudomanifold embedded as PL-subvariety in $\mathbb{R}^N$ and let Σ the singular subset in X. There is, on $X^0 = X \setminus \Sigma$, a metric $\tilde{g}$ which endows the manifold $X \setminus \Sigma$ of a structure of flat Riemannian manifold, i.e. every point x in the $n-1$-skeleton admits a neighborhood U_x isometric to an open subset in $\mathbb{R}^N$.

(2) [59, Sect. 3.5], [60, Sect. 3.5]. X is a compact analytic variety embedded in a Kählerian manifold. Then $X \setminus \Sigma$ is endowed with the metric induced by restriction of the Kählerian metric. One assume that Σ is locally analytically conical, that means the following:

A variety X is locally analytically conical if each point $p \in X$ has a neighborhood U and an analytic embedding $\rho : U \to \mathbb{C}^N$ such that $\rho(U)$ is a cone at $\rho(p)$ (see [60, Sect. 3.5 Definition and Examples]).

Theorem 5.5.31 ([60]) *In the two previous cases, the integration map induces an isomorphism:*

$$H^j_{(2)}(X) \xrightarrow{\int} \mathrm{Hom}(IH^{\bar{p}}_j(X); \mathbf{R}) \tag{5.41}$$

The idea is to prove that the direct image of the presheaf on U formed of the appropriate $\mathcal{L}^2$-forms of degree i has a "fine" associated sheaf and that, as i varies, those associated sheaves form a (de Rham) complex that satisfies the axioms that characterize $\mathcal{IC}^*(X)$; the cohomology groups of the complex are equal to its hypercohomology groups because the sheaves are fine.

Other proofs of CGM conjectures

The conjectures of Cheeger, Goresky and MacPherson were also treated with some success in the case that X^0 is the smooth part of a complex projective variety X with isolated singularities.

Let X be a normal singular algebraic surface (over $\mathbb{C}$) embedded in the projective space $\mathbb{P}^N(\mathbb{C})$ and let Σ be its singularity set, which consists of isolated singular points. Restricting the Fubini-Study metric of $\mathbb{P}^N(\mathbb{C})$ to $X^0 = X \setminus \Sigma$, provides an incomplete Riemannian manifold (X^0, g). Wu-Chung Hsiang and Vishwambhar Pati proved in [107] that the $\mathcal{L}^2$-cohomology $H^i_{(2)}(X^0)$ is naturally isomorphic to the dual of the middle intersection homology $IH_i^{\overline{m}}(X)$. However their proof has a certain gap corrected by Nagase [139] (see also [137, 138]). The "non- normal" case can be proved in the same way by making its normalization, as asserted in [107].

Saper [163, 164] who was inspired by the case of the Zucker conjecture (Sect. 5.5.5), constructed a complete Kähler metric on X^0 whose $\mathcal{L}^2$-cohomology groups are dual to the intersection homology groups of X.

Finally, Ohsawa [147, 148] proved the conjecture in dimension $\dim X \leq 2$: If X in $\mathbb{P}^n(\mathbb{C})$ is a projective variety of dimension $\dim X \leq 2$, then the $\mathcal{L}^2$ de Rham cohomology groups of the regular part X^0, with respect to the Fubini-Study metric are canonically isomorphic to the intersection cohomology groups of X.

The Deligne Conjecture: Variation of Hodge Structures

Over a compact Kähler manifold X, Deligne (unpublished manuscript, see [188]) has constructed canonical Hodge structures on the cohomology groups $H^p(X, \mathcal{L})$, of weight $p + k$. When the basis X is non-compact, Deligne's arguments still put Hodge structures on the $\mathcal{L}^2$ cohomology groups of the completion $\overline{X}$ provided they are finite dimensional.

Let X be a nonsingular algebraic variety and D is a divisor with normal crossings in X, which may be interpreted as giving a stratification of X whose largest stratum is $X \setminus D$. The considered local system is underlying a polarizable variation of Hodge structure.

A variation of Hodge structure, considered as local system $\mathcal{L}$ on $X \setminus D$, has an $\mathcal{IC}$ extension to all of X. Deligne conjectures that $\mathcal{IC}(X; \mathcal{L})$ is isomorphic to the sheaf of $\mathcal{L}^2$ differential forms on X, where the Riemannian metric on $X \setminus D$, is the *complete* metric that is hyperbolic near each codimension 1 divisor.

In the case of one dimensional base, Zucker [188] has obtained a natural identification

$$H^*_{(2)}(\overline{X}, \mathcal{L}) \cong H^*(\overline{X}, i_*\mathcal{L})$$

here $i_*\mathcal{L}$ is the direct image of $\mathcal{L}$ on $\overline{X}$. The $\mathcal{L}^2$ cohomology groups are then finite dimensional and come equipped with Hodge structures.

Cattani et al. [51], and independently, Kashiwara and Kawai [111] proved the Deligne conjecture, for higher dimensions:

Theorem 5.5.32 *The complex of sheaves $\mathcal{L}^2$ differential forms on X satisfies the axioms of middle intersection cohomology sheaf with values in the local system $\mathcal{L}$. In particular*

$$H^*_{(2)}(\overline{X}, \mathcal{L}) \cong IH^*(\overline{X}, \mathcal{L})$$

Corollary 5.5.33 *The intersection cohomology groups $IH^*(\overline{X}, \mathcal{L})$ carry canonical (pure) Hodge structures of weight $p + k$.*

The Zucker Conjecture

Zucker was aware of the work of Cheeger, Goresky, and MacPherson that appears in [59, 60] when he made the following conjecture, which first appeared in a 1980 preprint ([189]):

Conjecture 5.5.34 Let X be the Satake, Baily-Borel compactification of the quotient space U of a Hermitian symmetric domain modulo a proper action of an arithmetic group Γ. Let U be provided with the natural complete metric, then the sheaf of $\mathcal{L}^2$-differential forms on X with coefficients in a metrized local system $\mathcal{L}$ on U is isomorphic (in the derived category) to the sheaf $\mathcal{IC}(X; \mathcal{L})$ (see Introduction in [123]).

Zucker was led to this conjecture by some examples that he worked out [188, Sect. 6] of his general results [189, (3.20) and (5.6)] about the $\mathcal{L}^2$-cohomology groups of an arithmetic quotient of a symmetric space. In the examples, the compactification is obtained by adjoining a finite number of isolated singular points, and Zucker was struck by the values of the local $\mathcal{L}^2$-cohomology groups at these points: they are equal to the singular cohomology groups of the link in the bottom half dimensions and to 0 in the middle and in the top half dimensions (compare with Lemma 5.5.30).

Borel [15], Borel and Casselman [16] proved the Zucker conjecture in the particular case of a group of $\mathbb{Q}$-rank one or two (see also [50]).

The conjecture has been fully proved by Looijenga [123], Saper and Stern [165]. Looijenga uses Mumford's (1975) desingularization of X and the decomposition theorem. Saper and Stern use a more direct method, which they feel will also yield a generalization of a conjecture due to Borel (see [165]).

One reason for the great interest in Zucker's conjecture is that it makes it possible to extend the "Langlands program" to cover the important non compact case, as Zucker indicates in [190].

Other Related Results

There is a lot of results related to the previous ones. The intersted reader may consult Nagase [139], Saper [163], Pardon and Stern [150], etc.

The Geometric Viewpoint

In this section, all the homology and cohomology groups will be with real coefficients.

Shadow forms

The shadow forms have been defined by Brasselet, Goresky and MacPherson [28].

The idea is to associate a differential form $\omega(\xi)$ to simplices of a barycentric subdivision K' of a given triangulation K, so that there is a clear relationship between the defect of transversality of the simplices relatively to the simplices σ of K and the order of the pole of the corresponding differential form on σ.

Various equivalent definitions of the shadow forms are provided in [28]. One of them goes as follows: Let $\Delta = \Delta^n$ be the standard n-simplex.

$$\Delta = \{(x_1, x_2, \dots, x_{n+1}) \in \mathbb{R}^{n+1} | \ \ 0 \le x_i \le 1, \sum x_I = 1\}.$$

The shadow forms are defined for k-simplices ξ of the barycentric subdivision Δ' which do not lie in the boundary of Δ. Such a barycentric subdivision can be defined for each point p in the interior of Δ, requiring that for each pair $F' < F$ of faces of Δ, the barycenters of F, F' and of the face opposite to F' in F are collinear. The corresponding barycentric subdivision of Δ will be denoted $\Delta'(p)$. Every k-simplex ξ admits a geometrical realization $\xi(p)$ in this subdivision.

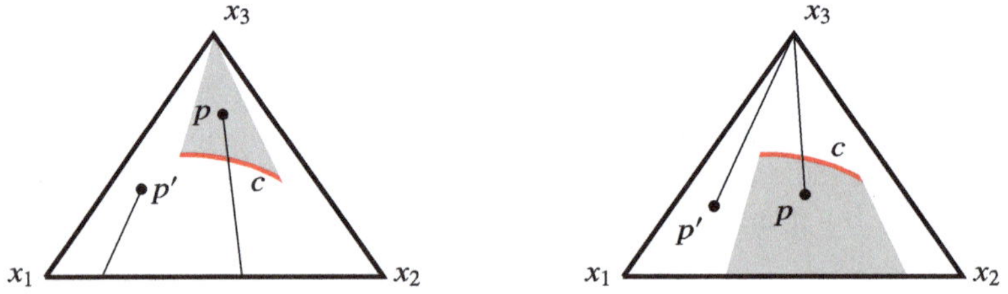

Vertices of ξ are barycenters of $(x_1 x_2)$ and of Δ Vertices of ξ are x_3 and the barycenter of Δ

Fig. 5.7 The shadow is the dotted area. The point p is in the shadow $S_\xi(c)$ but p' is not

Let c be a singular chain in the interior of Δ, the shadow $S_\xi(c)$ cast by an $(n-k)$-chain c with respect to ξ is the set of all points p such that $\xi(p)$ intersects c (Fig. 5.7).

Definition 5.5.35 The shadow form $\omega(\xi)$ is the unique differential form such that the value of its integral over any $(n-k)$-chain c is the volume of the shadow $S_\xi(c)$:

$$\int_c \omega(\xi) = \text{ volume } (S_\xi(c)).$$

An explicit equivalent definition goes as follows: denote by D_ξ the incidence variety

$$D_\xi = \{(p, x) \in int(\Delta) \times int(\Delta) : x \in \xi(p)\}.$$

Let i be the inclusion $i : D_\xi \hookrightarrow \Delta \times \Delta$ and let π_1 and π_2 be the projections on the first and second factors of $\Delta \times \Delta$.

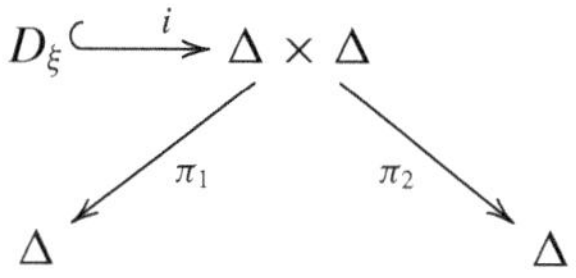

If $(x_1, \ldots, x_{n+1})$ are the barycentric coordinates of Δ^n, the Whitney form $W(\Delta^n)$ is the volume form of Δ^n,

$$W(\Delta^n) = W(x_1, \ldots, x_{n+1}) = n! \sum_{i=1}^{n+1} (-1)^{i+1} x_i dx_1 \wedge \cdots \wedge \widehat{dx_i} \wedge \cdots \wedge dx_{n+1}.$$

Proposition 5.5.36 *The shadow form $\omega(\xi)$ is the $(n-k)$-differential form defined by*

$$\omega(\xi) = \int_{\pi_2} i^* \pi_1^* (W(\Delta^n))$$

where $\int_{\pi_2}$ denotes integration along the fibres of π_2 (see [20]).

The differential form $\omega(\xi)$ is C^∞ on $int(\Delta)$. Indeed D_ξ is a smooth manifold and the fibres $\pi_2^{-1}(x) \cap D_\xi$ are relatively compact.

Generalizing the definition to polyhedra provides:

Theorem 5.5.37 ([28, Corollary 9.3]) *Let X be a polyhedron in the Euclidean space $\mathbf{R}^n$. Fix q, $1 \le q \le \infty$, and denote by $\bar{p}(q)$ the highest perversity whose graph is situated strictly below the line from origin and with slope $1/q$. Then the intersection homology of X, for the perversity $\bar{p}(q)$, is isomorphic to $\mathcal{L}^q$-cohomology of X:*

$$IH_k^{\bar{p}(q)}(X) \cong H_{(q)}^{n-k}(X).$$

Conjecture 5.5.38 (Brasselet et al. [28]) Let X be a stratified space with a Riemannian metric and conical singularities. Let Σ be the singular set, $q \ge 2$, and $\mathcal{L}^q$-cohomology of $X \setminus \Sigma$ is finite dimensional, then it is isomorphic to intersection cohomology of X.

The conjecture has been proved by Youssin [187] who also extends the result to spaces with horn-singularities.

Belkacem Bendifallah [11] provided an explicit formula for the coefficients of shadow forms as integrals of Dirichlet type, obtaining an alternative proof of Theorem 5.5.37. He gave a duality formula and a product formula for shadow forms and constructed the correct underlying algebraic structure.

The Brasselet-Legrand approach.

J.-P. Brasselet and A. Legrand consider the situation of an n-dimensional pseudovariety X endowed with a Thom-Mather stratification, and whose strata are smooth manifolds.

The idea is to prove a de Rham type theorem by considering a complex of differential forms whose coefficients are C^∞ functions on the regular part of X which may have poles on the singular strata but whose behavior in a neighborhood of the strata is controlled. The control is performed through two parameters, associated with each stratum $S_{n-\alpha}$. The first control β_α corresponds to an admissible maximum order of poles of the functions on the stratum, the second c_α is related to the local conical metric in the neighborhood of the stratum. An admissible differential form can have a pole on a stratum, but the the order of the poles should not be too large for the Poincaré lemma to be verified to some degree. Also, the quotients $[\beta_\alpha/c_\alpha]$ should satisfy the same inequalities than the GM-perversities (see formula 5.14).

The obtained complex $\Omega^\bullet_{\beta,c}$ is a complex of soft sheaves satisfying axioms $[AX1]_{\bar{p}}$ with

$$p_\alpha = \alpha - 2 - \left[\frac{\beta_\alpha}{c_\alpha}\right]$$

whose hypercohomology is intersection homology for the complementary perversity.

On the one hand the complex $\Omega^\bullet_{\beta,c}$ is a generalization of the complex of shadow forms.

On the other hand, it allows to define a suitable algebra in order to generalize the Hochschild-Kostant-Rosenberg theorem to the case of singular varieties, more precisely to manifolds with boundary and to varieties with isolated singularities. The classical result of Hochschild et al. [105] asserts that the Hochschild homology of a finitely generated, smooth complex algebra A equals the space of Kähler differentials over A. In 1982, Connes [62] extended this result in a topological setting and in [31] the authors generalize the Connes's idea to the case of singular varieties with isolated singularities.

The relation between the defects of transversality (perversity) of a cycle and the order of the poles of the associated differential form are explicit in the context of shadow forms and the context of the complex $\Omega^\bullet_{\beta,c}$. The smallest is the dimension of the stratum, the greater the admissible order of the poles of the differential forms. The physicist Alain Connes (private conversation) says that "there is a higher concentration of energy in the smaller singular strata".

The Goresky-MacPherson's complex $\Omega^\bullet_{\bar{q}}$.
The Goresky-MacPherson complex has been described by Brylinski [46] (for an interpretation in terms of sheaf defined on the resolution of the stratified space see [1, Sect. 6.5]).

Let $\pi : M \to B$ be a smooth fibration of smooth manifolds. A filtration (Cartan's filtration) of the de Rham complex $\Omega^\bullet_M$ is defined as follows :

Definition 5.5.39 For $k \geq 0$, $F_k\Omega^\bullet_M$ is the sub-complex of $\Omega^\bullet_M$ consisting of the differential forms ω such that ω and $d\omega$ satisfy: if $\xi_1, \xi_2, \dots, \xi_{k+1}$ are $k+1$ vector fields on M, tangent to the fibres of π, then $i(\xi_1) \circ \cdots \circ i(\xi_{k+1})(\omega) = 0$.

Let X be a pseudomanifold with $\mathcal{C}^\infty$-structure, equipped with a Thom-Mather stratification (see [179, 4.2.17]). If $S_i \subset \bar{S_j}$, $(\pi_i)|_{T_i \cap S_j}$ is $\mathcal{C}^\infty$ and $\pi_i \circ \pi_j = \pi_i$ on $T_i \cap T_j$. One denotes by X^0 the smooth stratum of X.

Definition 5.5.40 Let $\bar{q}$ be a perversity, denote by $\Omega^\bullet_{\bar{q}}$ the sub-complex of $\Omega^\bullet_{X^0}$ consisting of the differential forms ω such that every point of $S_{n-\alpha}$ admits a neighborhood $V \subset T_{n-\alpha}$ on which the restriction of ω is in $F_{q(\alpha)}\Omega^\bullet_{V\cap X^0}$, relative to the projection $V \cap X^0 \to S_{n-\alpha}$ induced by $\pi_{n-\alpha}$.

This means that, near $S_{n-\alpha}$, ω satisfies $i(\xi_1) \circ \cdots \circ i(\xi_{q(\alpha)+1})\omega = 0$ if the ξ_i are vector fields defined on X^0 and are tangent to the fibers of $\pi_{n-\alpha}$.

Proposition 5.5.41 *[46, Proposition 1.2.6] The complex of sheaves $\Omega^\bullet_{\bar{q}}$ satisfies $(AX_1)_{\bar{q}}$.*

As a corollary, the hypercohomology groups of the complex of sheaves $\Omega^\bullet_{\bar{q}}$ are isomorphic to $\mathrm{Hom}(IH^{\bar{p}}_j(X); \mathbb{R})$, where $\bar{p}$ and $\bar{q}$ are complementary perversities. See also the survey by Pollini [156].

$\mathcal{L}^\infty$-cohomology.
Let X be a subanalytic compact pseudomanifold. In [182] Valette shows a de Rham theorem for $\mathcal{L}^\infty$-cohomology forms on the nonsingular part of X. The obtained cohomology is isomorphic to the intersection cohomology of X for the top perversity. There is a Lefschetz duality theorem relating the $\mathcal{L}^\infty$-cohomology to the so-called Dirichlet $\mathcal{L}^1$-cohomology. As a corollary, the Dirichlet $\mathcal{L}^1$-cohomology is isomorphic to intersection cohomology in the zero perversity.

Morse functions.
Let X be a space with isolated conical singularities. In [124] U. Ludwig establishes, using anti-radial Morse functions on X, a combinatorial complex which computes the intersection homology of X. The complex constructed is generated by the smooth critical points of the Morse function and representatives of the de Rham cohomology (in low degree) of the link manifolds of the singularities of X. It can be seen as an analogue of the famous Thom-Smale complex for smooth Morse functions and singular homology on a compact manifold.

5.5.6 *Steenrod Squares, Cobordism and Wu Classes*

In this section, the coefficients are the mod 2 integers $\mathbb{Z}_2$.

Steenrod Squares and Wu Classes

Goresky and Pardon [99] define four classes of singular spaces for which they define various characteristic numbers and for which these characteristic numbers determine the cobordism groups. In the four cases, they construct characteristic numbers by lifting Wu classes to intersection homology. Then they can multiply them.

In the singular case, the mod 2 *Steenrod square operations* have been defined in intersection cohomology by Goresky in [87] (see also [99, Sect. 4]), as operations

$$Sq^i : IH^j_{\bar{c}}(X) \to IH^{i+j}_{2\bar{c}}(X)$$

for perversities $\bar{c}$ such that $2\bar{c} \leq \bar{t}$. Via Poincaré duality one has similar operations in intersection homology (with compact supports).

Definition 5.5.42 ([99, Sect. 5.1]) Let X be an n-dimensional pseudomanifold. Assume $\bar{c}$ is a perversity such that $2\bar{c} \leq \bar{t}$. Let $\bar{b} = \bar{t} - \bar{c}$ be the complementary perversity. For any i with $0 \leq i \leq [n/2]$ the Steenrod square operation

$$Sq^i : IH_i^{\bar{c}}(X) \to IH_0^{2\bar{c}}(X) \to \mathbb{Z}_2$$

is given by multiplication with the intersection cohomology i^{th}-Wu class of X:

$$v^i(X) = v^i_{\bar{b}}(X) \in IH^i_{\bar{b}}(X).$$

One defines $v^i(X) = 0$, for $i > [n/2]$.

If X is a $\mathbb{Z}_2$-Witt space (see Sect. 5.4.6), then the middle intersection homology group is self-dual, i.e., satisfies the Poincaré duality over $\mathbb{Z}_2$. Also the natural homomorphism

$$IH^i_{\bar{m}}(X) \to IH^i_{\bar{n}}(X)$$

is an isomorphism.

Definition 5.5.43 ([99, Sect. 8.1]) A stratified pseudomanifold X is locally orientable if, for each stratum, the link is an orientable pseudomanifold. A stratified pseudomanifold X is a locally orientable Witt space if it is both locally orientable and a $\mathbb{Z}_2$-Witt space.

In the situation of a locally orientable Witt space, the Wu classes which are defined to be middle intersection homology classes, can be multiplied to construct characteristic Wu numbers

$$\varepsilon(v_i(X) \cdot v_j(X)) = \langle v^{n-i}(X) \cup v^{n-j}(X), [X] \rangle \in \mathbb{Z}_2$$

where $i + j = n$. The map $\varepsilon : H_0(X, \mathbb{Z}_2) \to \mathbb{Z}_2$ denotes the augmentation and the following diagram commutes:

$$\begin{array}{ccccc} IH_i^{\bar{m}}(X) \times IH_j^{\bar{m}}(X) & \xrightarrow{\cdot} & IH_0^{\bar{t}}(X) & \xrightarrow{\varepsilon} & \mathbb{Z}_2 \\ \uparrow{\scriptstyle \cong \times \cong} & & \uparrow{\scriptstyle \cong} & & \\ IH_{\bar{m}}^{n-i}(X) \times IH_{\bar{m}}^{n-j}(X) & \xrightarrow{\cup} & IH_{\bar{0}}^{n}(X). & & \end{array}$$

Theorem 5.5.44 ([99, Theorem 10.5]) *A locally orientable Witt space X of dimension n is a boundary of a locally orientable Witt space Y if and only if each of the characteristic Wu numbers*

$$v^{ij}(X) = \varepsilon(v^i(X)v^j(X)v^1(X)^{n-i-j}) \in \mathbb{Z}_2$$

vanish, where $\varepsilon : H_0(X; \mathbb{Z}_2) \to \mathbb{Z}_2$ denotes the augmentation.

Here, the class v^1 is a cohomology class and $v^i v^j$ is a (intersection) homology class, so the product is a well defined cobordism invariant.

In [99] M. Goresky and W. Pardon provide further important results concerning cobordism of singular spaces (see also [56]).

Cobordism of Maps in the Singular Case

Generalizing the results of R. Stong in the smooth case, J.-P. Brasselet, A. Libardi, E. Rizziolli and M. Saia define the cobordism of maps in the following way:

Definition 5.5.45 ([34]) Let $f : X \to Y$ be a map between pseudomanifolds of dimensions m and n respectively. The triple (f, X, Y) is null-cobordant if there exist:

1. pseudomanifolds V and W with dimensions $m + 1$ and $n + 1$, respectively, and $\partial V = X$ and $\partial W = Y$.
2. a map $F : V \to W$ such that the following diagram commutes.

$$\begin{array}{ccc} U_X & \xrightarrow{F_{|U_X}} & U_Y \\ {\scriptstyle \cong}\downarrow{\scriptstyle \phi} & & {\scriptstyle \psi}\downarrow{\scriptstyle \cong} \\ \partial V \times [0,1) & \xrightarrow{f \times Id} & \partial W \times [0,1), \end{array}$$

   where U_X and U_Y are collared neighborhoods of X and Y in V and W respectively, and ϕ and ψ are PL-diffeomorphisms such that $\phi(x) = (x, 0)$, $x \in \partial V$ and $\psi(y) = (y, 0)$, $y \in \partial W$.
3. $F_{|\partial V} = f : \partial V \to \partial W$.

Let $f : X \longrightarrow Y$ be a map, with X a compact locally orientable Witt space of pure dimension m and Y a closed n-dimensional smooth manifold. Then the map $f_! : IH_i^{\bar{p}}(X) \to IH_i^{\bar{p}}(Y)$ is defined in such a way that the following diagram commutes

$$\begin{array}{ccc} H_i(X) & \xrightarrow{f_*} & H_i(Y) \\ \uparrow{\scriptstyle \omega_X} & & \uparrow{\scriptstyle \omega_Y}\ \simeq \\ IH_i^{\bar{p}}(X) & \xrightarrow{f_!} & IH_i^{\bar{p}}(Y) \end{array}$$

i.e. $f_! = (\omega_Y)^{-1} \circ f_* \circ \omega_X$, where the map ω_Y is an isomorphism since Y is smooth.

Denote by $\widetilde{f_!}$ the composition map $\widetilde{f_!} = \alpha_Y^{-1} \circ f_!$, i.e. the composition map

$$IH_i^{\bar{p}}(X) \xrightarrow{\omega_X} H_i(X) \xrightarrow{f_*} H_i(Y) \xrightarrow{P_Y^{-1}} H^{n-i}(Y)$$

where the last arrow denotes the inverse Poincaré isomorphism.

Theorem 5.5.46 ([34]) *Let X be a compact locally orientable Witt space of pure dimension m and Y a closed n-dimensional smooth manifold. Given a map $f : X \longrightarrow Y$, if the triple (f, X, Y) is null-cobordant, with $(f, X, Y) = \partial(F, V, W)$ and W is a smooth manifold, then for any partition ℓ and r numbers $u_1, \dots, u_r$ satisfying $u_i \leq [m/2]$ for all i and $(\ell_1 + \ell_2 + \dots \ell_s) + u_1 + \dots + u_r + r(m - n) = n$, the Stiefel-Whitney–Wu numbers*

$$\langle w_\ell(Y).\widetilde{f_!}(v_{m-u_1}(X)).\cdots.\widetilde{f_!}(v_{m-u_r}(X)), [Y]\rangle$$

are zero.

Let $f : X \to Y$ be a proper and normally nonsingular map of pseudomanifolds, there is an unique Gysin map

$$If_i : IH_i^{\bar{m}}(X) \to IH_i^{\bar{m}}(Y)$$

such that the following diagram commutes (Theorem 5.5.18, see [94, Sect. 5.4.3]).

$$\begin{array}{ccc} H_i(X) & \xrightarrow{f_*} & H_i(Y) \\ \uparrow{\scriptstyle \omega_X} & & \uparrow{\scriptstyle \omega_Y} \\ IH_i^{\bar{p}}(X) & \xrightarrow{If_i} & IH_i^{\bar{p}}(Y). \end{array} \tag{5.42}$$

The same result holds for placid maps as well (Proposition 5.5.19, see [94] and [5, Proposition 3.2]).

Theorem 5.5.47 ([34]) *Let $f : X \longrightarrow Y$ be a normally nonsingular (or placid) map, with X and Y compact locally orientable Witt spaces of pure dimension m and n respectively. If (f, X, Y) is null-cobordant, then for any u with $0 \leq u \leq n$, the following Wu numbers vanish:*

$$\langle v_{n-u}(Y).If_i(v_u(X)), [Y]\rangle = 0.$$

5.6 Supplement: More Applications and Developments

5.6.1 *Toric Varieties*

Max Brückner (for the octatope) [43, 44], Max Dehn (in 1905, for dimensions 4 and 5) [65] and Duncan Sommerville [170] (in 1927, in all dimensions) proved certain relations involving numbers of faces for simplicial polytopes.

Let P be an n-dimensional simplicial polytope. For $i = 0, \dots, d-1$, let f_i denote the number of i-dimensional faces of p. The sequence

$$(f_0, f_1, \dots, f_{d-1})$$

is called the f-vector of the polytope P. Additionally, set $f_{-1} = f_d = 1$. Then for any $k = 0, \dots, d-2$ the following Dehn-Sommerville equation holds:

$$\sum_{j=k}^{d-1} (-1)^j \binom{j+1}{k+1} f_j = (-1)^{d-1} f_k.$$

When $k = -1$, it expresses the fact that Euler characteristic of an $(d-1)$-dimensional simplicial sphere is equal to $1 + (-1)^{d-1}$.

For $k = 0, 1, \dots, d+1$, let

$$h_k = \sum_{i=0}^{k} (-1)^{k-i} \binom{n-i}{k-i} f_{i-1}.$$

The $(d+2)$-uple

$$h(P) = (h_0, h_1, \dots, h_{d+1})$$

is called the h-vector of P.

The generalized lower bound conjecture (McMullen-Walkup) [135] is the following:

Conjecture 5.6.1 Let P be a simplicial n-dimensional polytope. Then

1. $1 = h_0 \le h_1 \le \cdots \le h_{[\frac{d}{2}]}$.
2. for an integer $1 \le r \le \frac{d}{2}$, the following are equivalent:
   a. $h_{r-1} = h_r$.
   b. there is a triangulation K of P all of whose faces of dimension at most $d - r$ are faces of P.

In McMullen [134] conjectured that the Dehn-Sommerville relations together with the generalized lower bound conjecture provide sufficient conditions for the existence of a simplicial polytope with a given h-vector.

If k and i are positive integers, then k can be written uniquely in the form

$$k = \binom{n_i}{i} + \binom{n_{i-1}}{i-1} + \cdots + \binom{n_j}{j},$$

where $n_i > n_{i-1} > \cdots > n_j \geq j \geq 1$. Define

$$k^{<i>} = \binom{n_i + 1}{i + 1} + \binom{n_{i-1} + 1}{i - 1} + \cdots + \binom{n_j + 1}{j + 1}.$$

Also define $0^{<i>} = 0$. A vector $k_0, k_1, \ldots, k_d$ of integers is an M-vector if $k_0 = 1$ and $0 \leq k_{i-1} \leq k_i^{<i>}$ for $1 \leq i \leq d - 1$. McMullen conjectured that a sequence $(h_0, \ldots, h_d)$ of integers is the h-vector of a simplicial convex d-polytope if and only if $h_0 = 1$, $h_i = h_{d-i}$ for $0 \leq i \leq d$ and the following sequence is an M-vector:

$$(h_0, h_1 - h_0, h_2 - h_1, \ldots, h_{[\frac{d}{2}]} - h_{[\frac{d}{2}]-1}).$$

The "if" part was proven by Billera and Lee [13].

The "only if" part was proven by Stanley [173] in a very surprising paper, as a consequence of the inequalities of Betti numbers provided by the hard Lefschetz theorem, and considering the cohomology of an associated toric variety, which is non-singular. By this paper deep results from algebraic geometry are related to the study of combinatorics.

A simplicial polytope is always rational so there exists an associated toric variety. In the non-simplicial (but still rational) case the associated toric variety is singular. In 1981 R. MacPherson showed how to compute the (rational) intersection cohomology of the (possibly singular) toric variety associated to any rational convex polytope and spoke about it in many conferences. This calculation was popularized by J. Bernstein and A. Khovanski. Proofs were published by: Fieseler [68] and by Denef and Loeser [64]

In [172] Stanley used this calculation together with the hard Lefschetz theorem for IH to prove the generalized lower bound conjectures for rational convex polytopes and conjectured that the same result holds in the non-rational case as well. The calculations are simplified if one considers the torus-equivariant intersection cohomology instead.

In the case of a non-rational polytope, no toric variety exists. This led to the possibility of proving the same result for non-rational polytopes by constructing the torus-equivariant intersection cohomology, in a purely combinatorial manner, together with a proof that it satisfies the hard Lefschetz theorem. The theory was successfully developed by Barthel et al. [8, 9], Bressler [37], Bressler and Lunts [38, 39], Karu [110] (see also Fieseler [69] and Braden [21, 22]). This completed the proof of Stanley's conjectures for non-rational polytopes.

5.6.2 The Asymptotic Set

Let $F : X = \mathbb{C}^n \to Y = \mathbb{C}^n$ be a polynomial mapping. In the study of geometrical or topological properties of polynomial mappings, the set of points at which those maps fail to be proper plays an important role. The asymptotic set

$$S_F = \{a \in Y \quad s.t. \quad \exists \{\xi_k\} \subset X, |\xi_k| \to \infty, F(\xi_k) \to a\}$$

is the smallest set S_F such that the map

$$F : X \setminus F^{-1}(S_F) \to Y \setminus S_F$$

is proper. In a topological approach of the Jacobian conjecture, it reduces to show that the asymptotic set of a complex polynomial mapping with non zero constant Jacobian is empty. It is then natural to study the topology of the asymptotic set.

Define by $Sing(F)$ the singular locus of F (the zero set of its Jacobian determinant) and denote by $K_0(F)$ the set of critical values of F, i.e. the set $F(Sing(F))$. Define the Riemannian manifold M_F as $\mathbb{C}^n \setminus Sing(F)$ with the pull back of Euclidean Riemannian metric on $\mathbb{R}^{2n} = \mathbb{C}^n$. This metric is non degenerate outside the singular locus of F.

Proposition 5.6.2 ([180, Proposition 2.3]) *Let $F : \mathbb{C}^n \to \mathbb{C}^n$ be a polynomial map. There exists a real semi-algebraic pseudomanifold $N_F \subset \mathbb{R}^\nu$, for some $\nu \geq 2n$, such that*

$$Sing(N_F) \subset (S_F \cup K_0(F)) \times \{0_{\mathbb{R}^p}\}.$$

with $p = \nu - 2n$, and there exists a semi-algebraic bi-Lipschitz map:

$$h_F : M_F \to Reg(N_F),$$

where N_F is equipped with the metric induced by $\mathbb{R}^\nu$.

Case $F : \mathbb{C}^2 \to \mathbb{C}^2$.

The first result comes from Anna and Guillaume Valette. In [180], they associate the singular pseudomanifolds N_F to polynomial mappings $F : \mathbb{C}^2 \to \mathbb{C}^2$. They prove that the map F with non-vanishing Jacobian is not proper if and only if the intersection homology of N_F is nontrivial in dimension 2 and for any (or some) perversity. The intersection homology of N_F describes the geometry of the singularities at infinity of the mapping F. This provides a new and original approach to the Jacobian conjecture.

Case $F : \mathbb{C}^n \to \mathbb{C}^n$.

Thủy Nguyễn Thị Bích, with Anna and Guillaume Valette [144] consider the leading forms $\widehat{F}_i$ of the components of a polynomial mapping

$$F = (F_1, \dots, F_n) : \mathbb{C}^n \to \mathbb{C}^n.$$

They obtain:

Theorem 5.6.3 ([144]) *Let $F : \mathbb{C}^n \to \mathbb{C}^n$ be a polynomial mapping with nowhere vanishing Jacobian. If* $\operatorname{rank}(D_{\mathbb{C}} \widehat{F}_i)_{i=1,\dots,n} > n - 2$ *then F is not proper if and only if $IH_2^{\bar{p}}(N_F) \neq 0$ for any (or some) perversity $\bar{p}$.*

In [140] Thủy Nguyễn T.B. shows that for a class of non-proper generic dominant polynomial mappings, the results in [144, 180] hold also without hypothesis of non emptyness of the set $K_0(F)$. In her thesis, [141], she provides explicit stratifications of the asymptotic set S_F and of the critical set $K_0(F)$ of polynomial map $F : \mathbb{C}^n \to \mathbb{C}^n$ by a new method, that she called the method of "façons". That method appears to be a very powerful and a promising method not only for the computation of intersection homology. A large number o f examples is provided.

In [142], Thủy Nguyễn T.B. describes explicitely such a variety N_F associated to the Pinchuk's map and calculate its intersection homology. The result describes the geometry of singularities at infinity of the Pinchuk's map. She also shows that the real version of the A. and G. Valette's results in [180] does not hold.

Case $F : \mathbb{C}^n \to \mathbb{C}^{n-1}$

Given a polynomial mapping $G : \mathbb{C}^n \to \mathbb{C}^{n-1}$, with $n \geq 2$, in [143], Thủy Nguyễn T.B. and M.A. Soares Ruas construct singular varieties V_G, similarly to the previous N_F. They prove that if the intersection homology with total perversity (with compact or closed supports) in dimension two of (any of the corresponding) V_G is trivial then G is a fibration.

5.6.3 *Factorization of Poincaré Morphism for Toric Varieties*

In this section, all homology and cohomology groups are with $\mathbb{Z}$ coefficients. References for this section are [6–8] (see also [83]).

The Cartier and Weil divisors play an important role for normal varieties. Given a Cartier divisor of a complex n-dimensional variety X one can associate its Chern class in $H^2(X; \mathbb{Z})$. Given a Weil divisor , one can associate its class in $H_{2n-2}(X; \mathbb{Z})$. In the two cases, the class of a principal divisor is zero. Denoting by $\mathrm{Div}_{\mathrm{C}}(\mathrm{X})$ and $\mathrm{Div}_{\mathrm{W}}(\mathrm{X})$ the abelian groups of classes of algebraic divisors of Cartier and Weil modulo the principal divisors, there are homomorphisms

$$c^1 : \mathrm{Div_C(X)} \to \mathrm{H^2(X)} \qquad \text{and} \qquad \kappa : \mathrm{Div_W(X)} \to \mathrm{H_{2n-2}(X)}.$$

For $k = 2$ the Poincaré morphism,

$$P_2 : H^2(X) \longrightarrow H_{2n-2}(X),$$

is cap-product by the fundamental class $[X]$ of X. Let D be a Cartier divisor, then

$$P_2(c^1(D)) = \kappa(D).$$

In the smooth case, the Poincaré morphism is an isomorphism and the two notions of divisors coincide.

For a normal toric variety X, the divisors classes admit invariant representatives under action of the torus $\mathbb{T}$ (see [81, 3.4]); There are isomorphisms

$$\mathrm{Div_C(X)} \cong \mathrm{Div_C^{\mathbb{T}}(X)} \qquad \text{and} \qquad \mathrm{Div_W(X)} \cong \mathrm{Div_W^{\mathbb{T}}(X)}$$

where $\mathrm{Div_C^{\mathbb{T}}(X)}$ and $\mathrm{Div_W^{\mathbb{T}}(X)}$ denote the groups of invariant divisors classes.

A non-degenerated toric varieties is a toric variety which is not isomorphic to the product of a toric variety of dimension $d < n$ and an $n-d$-dimensional torus. Its fundamental group is finite.

For every perversity $\bar{p}$, let denote by $i(p)$ the highest integer $i \leq n$ such that $p(2i) \leq 1$ and V_p the invariant open subset of X union of orbits with dimension at least $n - i(p)$. Then the group $IH^{\bar{p}}_{2n-2}(X)$ is isomorphic to the group

$$\mathrm{Div}^{\mathbb{T}}_{\bar{p}}(X) = \{[D] \in \mathrm{Div_W^{\mathbb{T}}(X)} : \mathrm{D}|_{\mathrm{V_p}} \in \mathrm{Div_C^{\mathbb{T}}(V_p)}\}$$

and one has.

Theorem 5.6.4 ([6, Satz 2]) *Let X be a non degenerated toric variety, then:*

$$H^1(X) \cong H^{cld}_{2n-1}(X) = 0$$

and one has a commutative diagram

$$\begin{array}{ccccc}
\mathrm{Div_C^{\mathbb{T}}(X)} & \hookrightarrow & \mathrm{Div}^{\mathbb{T}}_{\bar{p}}(X) & \hookrightarrow & \mathrm{Div_W^{\mathbb{T}}(X)} \\
{\scriptstyle c^1}\downarrow \cong & & \downarrow \cong & & {\scriptstyle \kappa}\downarrow \cong \\
H^2(X) & \xrightarrow{\alpha_X} & IH^{\bar{p}}_{2n-2}(X) & \xrightarrow{\omega_X} & H_{2n-2}(X).
\end{array}$$

where the composition of maps in the lowest line is the Poincaré homomorphism.

In the case of degenerate toric variety, one has a more general result taking into account the torus factor [6, 8].

5.6.4 General Perversities

Several authors are breaking away from the conditions on perversities as defined by Goresky and MacPherson in their original articles; see the (non exhaustive list of) papers of Cappell and Shaneson [47], Chataur et al. [55], Friedman [75–77, 79], Habegger and Saper [101], King [113], Saralegi-Aranguren [166].

These provides some interesting generalizations and results. that are sketched out at certain points in this survey. Friedman's article [79] itself provides a very good survey on the subject. Quoting Friedman, his article is an expository survey of the different notions of perversity in intersection homology and how different perversities require different definitions of intersection homology theory.

"With more general perversities than GM-perversities, one usually loses topological invariance of intersection homology (though this should be seen not as a loss but as an opportunity to study stratification data), but duality results remain, at least if one chooses the right generalizations of intersection homology. Complicating this choice is the fact that there are a variety of approaches to intersection homology."

With previous notation of strata, perversities such that $p_\alpha \leq codim_X(S_{n-\alpha}) - 2$ have been studied in detail by Friedman (see [78]) who proved in particular Poincaré duality for general perversities, Lefschetz duality for pseudomanifolds with boundary and Mayer-Vietoris sequence.

The Lefschetz duality for pseudomanifolds with boundary is also aim of the paper [181] by G. Valette. On a pseudomanifold X with boundary, two perversities are considered, the one for X and the other for the boundary ∂X. If the difference between the chosen perversities is constant, then Lefschetz duality holds on X. Here, allowable chains of the boundary ∂X are allowable on X.

5.6.5 Equivariant Intersection Cohomology

Equivariant intersection cohomology has been mainly studied by J.L. Brylinski, M. Brion and R. Joshua, by T. Oda and, in the circle case, by J.I.T. Prieto, G. Padilla and M. Saralegi-Aranguren.

Brylinski [46] provides an explicit complex in order to compute intersection homology in the equivariant setting. T. Oda considers the situation of toric action [145, 146]. Brion [40], Brion and Joshua [41], Joshua [108] provide a relationship between the vanishing of the odd dimensional intersection cohomology sheaves and of the odd dimensional global intersection cohomology groups. The authors provide a geometric proof of the vanishing of odd dimensional local and global intersection cohomology for Schubert varieties and complex spherical varieties. For a survey on these works, see [109]. In their paper [42] the authors extend their methods to algorithmically compute the intersection cohomology Betti numbers of reductive varieties.

In the papers [149, 158] G. Padilla, J.I.T. Prieto and M. Saralegi-Aranguren study circle actions on pseudomanifolds by using intersection cohomology and equivariant intersection cohomology. The orbit space and the Euler class of the action determine the equivariant intersection cohomology of the pseudomanifold as well as its localization.

5.6.6 Intersection Spaces

In [67], Timo Essig assigns cell complexes to certain topological pseudomanifolds depending on a perversity function in the sense of intersection homology. The main property of the intersection spaces is Poincaré duality over complementary perversities for the reduced singular (co)homology groups with rational coefficients. In the paper [3] of M. Banagl, using differential forms, the resulting generalized cohomology theory for pseudomanifolds was extended to 2-strata pseudomanifolds with a geometrically flat link bundle.

The resulting homology theory HI is well-known not to be isomorphic to intersection homology (see Banagl and Hunsicker [4]) but mirror symmetry "tends to" interchange IH and HI ([3]). A new duality theory for pseudomanifolds is obtained, which addresses certain needs in string theory related to the existence of massless D-branes in the course of conifold transitions and their faithful representation as cohomology classes (see Banagl [2]).

5.6.7 Blown-Up Intersection Homology

In [29, 30] Brasselet et al. use a notion of "déplissage" apparented to blow-up in order to define integration of differential forms on simplices and to prove a de Rham theorem for stratified varieties.

A similar method has been used by D. Chataur, M. Saralegi and D. Tanré to define the so called "blown-up intersection homology". The initial aim [52] is to extend Sullivan's minimal models theory to the framework of pseudomanifolds. The authors prove also a conjecture of M. Goresky and W. Pardon on Steenrod squares in intersection homology [99]. The relation with rational homotopy has been extended in [53]. The authors work in a context of simplicial sets in the sense of Rourke and Sanderson [162]. This provides a definition of formality in the intersection setting.

In [54] the authors prove the topological invariance of the blown-up intersection cohomology with compact supports in the case of a paracompact pseudomanifold with no codimension one strata.

Based upon simplicial blow-up, Chataur and Tanré construct in [57] Eilenberg-MacLane spaces for the intersection cohomology groups of a stratified space, answering a problem asked by M. Goresky and R. MacPherson ([18, Chap. IX, Problem 11]).

5.6.8 Real Intersection Homology

Whether there is a good analog of intersection homology for real algebraic varieties was stated as a problem by Goresky and MacPherson in [18, Chap. IX, Problem 7)]. They observed that if such a theory exists then it cannot be purely topological; indeed the groups constructed by McCrory and Parusiński in [133] are not homeomorphism invariants. These authors consider a class of algebraic stratifications that have a natural general position property for semialgebraic subsets. They define the real intersection homology groups $IHS_k(X)$ and show that they are independent of the stratification. If X is nonsingular and pure dimensional then $IHS_k(X) = H_k(X; \mathbb{Z}_2)$, classical homology with $\mathbb{Z}_2$ coefficients. An intersection pairing is defined.

5.6.9 Perverse Sheaves and Applications

Perverse sheaves and applications deserve a survey for the subject itself. Various authors wrote surveys concerning perverse sheaves and applications, they are clear and informative. In the MacPherson papers [127, 128], MacPherson and Vilonen [129], Massey survey [125] and Klinger survey [118], many references and results are given concerning in particular three main applications of perverse sheaves: Decomposition theorem, Weak and Hard Lefschetz theorems.

It is fair to mention in these papers other important applications such as Kazdhan-Lusztig conjecture, D-modules and Riemann-Hilbert correpondence, characteristic p and Weil conjecture, etc.

The story is far from over. Today there are many books and papers: an extensive literature on perverse sheaves in various fields of mathematics, showing the interest and diversity of the subject. The interested reader will find in the perverse sheaves a subject of fascinating discovery and exploratory innovation.

References

1. P. Albin, *On the Hodge theory of stratified spaces.* arXiv:1603.04106 [math.DG].
2. M. Banagl, *Intersection Spaces, Spatial Homology Truncation, and String Theory*, Lecture Notes, Springer-Verlag Berlin and Heidelberg GmbH and Co. KG, Germany (2010).
3. M. Banagl, *Foliated stratified spaces and a De Rham complex describing intersection space cohomology*, J. Diff. Geom. 104 (2016) 1, 1–58.
4. M. Banagl and E. Hunsicker, *Hodge Theory For Intersection Space Cohomology,* Geometry and Topology 23 (2019) 2165–2225.
5. G. Barthel, J.-P. Brasselet, K.-H. Fieseler, O.Gabber et L. Kaup, *Relèvement des cycles algébriques et homomorphismes associés en homologie d'intersection*, Ann. of Math. (2) 141 (1995), no. 1, 147–179.
6. G. Barthel, J.-P. Brasselet, K.-H. Fieseler and L. Kaup, *Invariante Divisoren und Schnittho-mologie von torischen Varietäten.* Banach Center Publications, "Parameter Spaces", Vol. 36, Warszawa 1996, ed. P. Pragacz, pp. 9–23.

7. G. Barthel, J.-P. Brasselet, K.-H. Fieseler and L. Kaup, *Equivariant Intersection Cohomology of Toric Varieties*, Algebraic Geometry, Hirzebruch 70, 45–68, Contemp. Math.241, Amer. Math. Soc., Providence, R.I., 1999.
8. G. Barthel, J.-P. Brasselet, K.-H. Fieseler and L. Kaup, *Combinatorial Intersection Cohomology for Fans*, Tohoku Math. J. 54 (2002), 1–41.
9. G. Barthel, J.-P. Brasselet, K.-H. Fieseler and L. Kaup, *Combinatorial Duality and Intersection Product: A Direct Approach* Tohoku Math. J. (2) 57 (2003), no 2, 273–292.
10. A. Beilinson, J. Bernstein, O. Gabber and P. Deligne, *Faisceaux pervers*. Astérisque, 100, 1982.
11. B. Bendiffalah, *Shadow forms of Brasselet-Goresky-MacPherson* Transactions of the Amer. Math. Soc. Vol 347, N.12, Dec 1995, 4747–4764.
12. R. Benedetti, *Density of Morse functions on a complex space*, Math. Ann.229 (1977), pp. 135–139.
13. L. Billera and C. Lee, *Sufficiency of McMullen's condition for f-vectors of simplicial polytopes*, Bulletin AMS 2 (1980), 181–185.
14. C. Bisi, F. Bracci, T. Izawa and T. Suwa, *Localized intersection of currents and the Lefschetz coincidence point theorem*, Annali di Matematica Pura e Applicata 195 (2016) no 2, 601–621.
15. A. Borel, *$\mathcal{L}^2$-cohomology and intersection cohomology of certain arithmetic varieties*. In: Emmy Noether in Bryn Mawr, eds. B. Srinivasan and J. Sally, Springer, 1983, 119–131.
16. A. Borel and W. Casselman, *Cohomologie d'intersection et $\mathcal{L}^2$-cohomologie de variétés arithmétiques de rang rationnel 2*. CR Acad. Sci. Paris 301(1985), 369–373.
17. A. Borel and J.C. Moore, *Homology theory for locally compact spaces*, Michigan Math J. 7, 137–159 (1960).
18. A. Borel and al. *Intersection Cohomology*, Progress in Mathematics Vol 50, Birkhäuser, 1984.
19. W. Borho and R. MacPherson, *Représentations des groupes de Weyl et homologie d'intersection pour les variétés nilpotentes*. C.R. Acad. Sci. Paris, t. 292 Ser. I, 707–710 (1981).
20. R. Bott and L. Tu, *Differential forms in Algebraic Topolology*. Springer-Verlag, New-York, 1982.
21. T. Braden, *Remarks on the Combinatorial Intersection Cohomology of Fans*. Pure and Applied Mathematics Quarterly Volume 2, Number 4. 1149– 1186, 2006.
22. T. Braden and R.D. MacPherson, *Intersection homology of toric varieties and a conjecture of Kalai*, Comment. Math. Helv. 74 (1999), 442–455.
23. J.-P. Brasselet, *Définition combinatoire des homomorphismes de Poincaré, Alexander et Thom pour une pseudo-variété*, Astérisque no 82–83, (1981), p. 71–91.
24. J.-P. Brasselet, *de Rham theorems for singular varieties*, Contemp. Math. 161 (1994) 95–112.
25. J.-P. Brasselet K.-H. Fieseler and L. Kaup, *Caractérisation des variétés homologiques à l'aide des invariants de l'homologie d'intersection*, Banach Institut Varsovie, Vol. 33 (1996), p. 19–22.
26. J.-P. Brasselet et G.Gonzalez-Sprinberg, *Espaces de Thom et contre-exemples de J.L.Verdier et M.Goresky*, Bol. Soc. Bras. Mat. vol. 17 n°2 (1986).
27. J.-P. Brasselet et G.Gonzalez-Sprinberg, *Sur l'homologie d'intersection et les classes de Chern des variétés singulières*, Travaux en cours n°23 Hermann (1987), 5–11.
28. J.-P. Brasselet, M. Goresky and R. MacPherson, *Simplicial Differential Forms with Poles*, Amer. Journal of Maths., 113 (1991), 1019–1052.
29. J.-P. Brasselet, G. Hector et M. Saralegi, *Théorème de de Rham pour les variétés singulières*, Annals of Global Analysis and Geometry Vol. 9, no 3, 1991, 211–243.
30. J.-P. Brasselet, G. Hector et M. Saralegi, *$\mathcal{L}^2$-cohomologie des espaces stratifiés*, Manuscripta Mathematica. Vol. 76, 1992, 21–32.
31. J.-P. Brasselet and A. Legrand, *Differential forms on singular varieties and cyclic homology*, in: Singularity Theory, Liverpool 1996, in: London Math. Soc. Lecture Note Ser., vol. 263, Cambridge Univ. Press, 1999, pp. 175–187.
32. J.-P. Brasselet et M. Saralegui, *Variétés homologiques et homologie d'intersection*, C. R. Acad. Sci. Paris Sér. I, 314, (1992), 847–850.

33. J.-P. Brasselet and T. Suwa, *Local and global coincidence homology classes*, Journal of Fixed Point Theory and Applications, 23(2), 2021.
34. J.-P. Brasselet, A.K.M. Libardi, E.C. Rizziolli and M.J. Saia, *Cobordism of maps of locally orientable Witt spaces*, Publicationes Mathematicae Debrecen, 2019, 94 (3–4), pp. 299–317.
35. J.-P. Brasselet, A.K.M. Libardi, T.F.M. Monis, E.C. Rizziolli and M.J. Saia, *A Lefschetz coincidence theorem for singular varieties*. Singularities and Foliations. Geometry, Topology and Applications, pp. 513–535.
36. G. Bredon, *Sheaf theory.* New York: McGraw-Hill, 1967.
37. P. Bressler, *Intersection Cohomology on Nonrational Polytopes*. Compositio Mathematica volume 135, 245–278 (2003).
38. P. Bressler and V. Lunts, *Intersection cohomology on nonrational polytopes*, Compositio Mathematica 135(3), 2000.
39. P. Bressler and V. Lunts, *Hard Lefschetz theorem and Hodge-Riemann relations for intersection cohomology of nonrational polytopes*, Compositio Mathematica, volume 135, 245–278 (2003).
40. M. Brion, *Equivariant cohomology and equivariant intersection theory.* Feb. 2008. arXiv:math/9802063v1.
41. M. Brion and R. Joshua, *Vanishing of odd dimensional intersection cohomology II*, Mathematische Annalen volume 321, pages 399–437 (2001).
42. M. Brion and R. Joshua, *Intersection Cohomology of reductive varieties*, Journal of the European Mathematical Society, 6(4), 2003.
43. M. Brückner, *Die Elemente der vierdimensionalen Geometrie mit besonderer Beriicksichtigung der Polytope*, Jahresbericht des Vereins für Naturkunde zu Zwickau I.S. – 1893, 1–61.
44. M. Brückner, *Uber die Ableitung der allgemeinen Polytope und die nach Isomorphismus verschiedenen Typen der allgemeinen Achtzelle (Oktatope)*, Verhl. Nederl. Akad. Wetensch. Sect. I. 10, No. 1 (1909).
45. J.L. Brylinski, *(Co)-homologie d'intersection et faisceaux pervers,* Séminaire Bourbaki 589, Astérisque 92–93.
46. J.L. Brylinski, *Equivariant Intersection Cohomology,* IHES 1986 and in Kazhdan-Lusztig theory and related topics (Chicago, IL, 1989), 5–32, Contemp. Math., 139, Amer. Math. Soc., Providence, RI, 1992.
47. S.E. Cappell and J.L. Shaneson, *Singular spaces, characteristic classes, and intersection homology,* Annals of Mathematics134(1991), 325–374.
48. H. Cartan *Homologie et cohomologie singulière.*, Séminaire Henri Cartan, tome 1 (1948-1949), exposé no 5.
49. H. Cartan and S. Eilenberg, *Homological algebra,*, The Princeton University Press, Princeton, 1956.
50. W. Casselman, *Introduction to the $\mathcal{L}$-cohomology of arithmetic quotients of bounded symmetric domains.* In: Complex Analytic Singularities, Adv. Stud. Pure Math. 8 (1986), 69–93.
51. E. Cattani, A. Kaplan and W. Schmid , *$\mathcal{L}^2$ and intersection cohomologies for a polarizable variation of Hodge structure.* Inventiones mathematicae volume 87, pages 217–252 (1987).
52. D. Chataur, M. Saralegi-Aranguren and D. Tanré, *Blown-up intersection cohomology,* Contemp.Math., vol. 708, Amer. Math. Soc., Providence, RI, 2018, pp. 45–102.
53. D. Chataur, M. Saralegi-Aranguren and D. Tanré, *Intersection cohomology, simplicial blow-up and rational homotopy,* Mem. Amer. Math. Soc. 254 (2018), no. 1214, viii+108.
54. D. Chataur, M. Saralegi-Aranguren and D. Tanré, *Poincaré duality with cap products in intersection homology,* Adv. Math. 326 (2018), 314–351.
55. D. Chataur, M. Saralegi-Aranguren and D. Tanré, *Intersection homology: General perversities and topological invariance*, Ilinois J. Math. Volume 63, Number 1 (2019), 127–163.
56. D. Chataur, M. Saralegi-Aranguren and D. Tanré, *Steenrod squares on intersection cohomology and a conjecture of M Goresky and W Pardon,* Algebr. Geom. Topol.16 (2016), no. 4, p. 1851–1904.
57. D. Chataur and D. Tanré, *Natural operations in intersection homology*, arXiv:2005.04960v2.

58. J. Cheeger, *On the spectral geometry of spaces with cone-like singularities*. Proc Natl Acad Sci U S A. 1979 May; 76(5): 2103–2106.
59. J. Cheeger, *On the Hodge theory of Riemannian pseudomanifolds*. In: Geometry of the Laplace operator. Proceedings of Symposia in pure Mathematics 36, 91–146 (1980). Amer. Math. Soc. Providence, RI.
60. J. Cheeger, M. Goresky and R.MacPherson, *$\mathcal{L}^2$-cohomology and intersection cohomology of singular algebraic varieties*, in "Seminar on differential geometry" ed. by S. T. Yau, Princeton University Press, 1982, p. 303–340.
61. D.C. Cohen, M. Goresky and L. Ji, *On the Künneth Formula for Intersection Cohomology*, Trans. AMS 333 (1992), 63–69.
62. A. Connes *Noncommutative Differential Geometry*, Inst. Hautes Etudes Sci., Publ. Math. No. 62 (1985), 257–360.
63. M.A. de Cataldo and L. Migliorini, *The decomposition theorem, perverse sheaves and the topology of algebraic maps*. Bull. AMS Vol. 46, No. 4, 2009, 535–633.
64. J. Denef and F. Loeser, *Weights of exponential sums, intersection cohomology, and Newton polyhedra*. Invent. math. 106, 275–294 (1991).
65. M. Denh, *Die Eulersche Formel im Zusammenhang mit den Inhalt in der Nicht Euklidischen Geometrie*. Math. Ann. 61 (4), 561–586 (1905).
66. A. Dimca, *Sheaves in Topology*, Springer Science and Business Media, 2004.
67. T. Essig *About a De Rham Complex Describing Intersection Space Cohomology in a Three Strata Case*, in Ed. M. Banagl, Ruprecht-Karls-Universität Heidelberg, 2016. Also in arXiv:1804.06690v2.
68. K.H.Fieseler, *Rational Intersection Cohomology of Projective Toric Varieties*. Journal für die reine und angewandte Mathematik 413 (1991): 88–98.
69. K.H.Fieseler, *Towards a combinatorial intersection cohomology for fans*, C.R. Acad. Sci. Paris 300 (1999).
70. K.H.Fieseler, *Survey on combinatorial intersection homology* In: Math. Forschungsinstitut Oberwolfach, no. 20, p. 1071–1075. http://www2.math.uu.se/~khf/finis.pdf.
71. K.H.Fieseler and L. Kaup, *Singular Poincaré duality and intersection homology*, Proceedings of the 1984 Vancouver Conference in Algebraic Geometry. AMS, Providence, (1986), 113–161.
72. K. Fieseler and L. Kaup, *Theorems of Lefschetz type in intersection homology, I. The hyperplane section theorem*. Revue Roumaine de Mathématiques Pures et Appliquées 33, 175–195 (1988).
73. K. Fieseler and L. Kaup, *The vanishing invariants in intersection homology of complex spaces*, in : Singularities, Banach Center Publ. 20, 1988, 225–235.
74. G. Friedman, *Intersection homology Künneth theorems*. Mathematische Annalen volume 343, 371–395, (2009).
75. G. Friedman, *Superperverse intersection cohomology: stratification (in)dependence*, Math. Z. 252 (2006), no. 1, p. 49–70.
76. G. Friedman, *Singular chain intersection homology for traditional and super-perversities*, Trans. Amer. Math. Soc. 359 (2007), no. 5, p. 1977–2019.
77. G. Friedman, *Intersection homology with general perversities*, Geom. Dedicata 148 (2010), p. 103–135.
78. G. Friedman, *Intersection homology and Poincaré duality on homotopically stratified spaces*, Geom. Topol. Volume 13, Number 4 (2009), 2163–2204.
79. G. Friedman, *An introduction to intersection homology with general perversity functions*, in Topology of stratified spaces, Math. Sci. Res. Inst. Publ., vol. 58, Cambridge Univ. Press, Cambridge, 2011, p. 177–222.
80. G. Friedman and J. E. McClure, *Cup and cap product in intersection (co)homology*, Advances in mathematics, vol. 240, pp. 383–426, 2013.
81. W. Fulton, *Introduction to toric varieties*, Annals of Math. Studies, Princeton Univ. Press, 1993.

82. W. Fulton and R. MacPherson, *Categorical framework for the study of singular spaces.*, Memoirs of the Amer. Math. Soc. vol. 243, A.M.S., Providence, RI 1981.
83. W. Fulton and B. Sturmfels, *Intersection theory on toric varieties.* Topology, 36(2):335–353, 1997.
84. S.I. Gelfand, and R. MacPherson, *Verma modules and Schubert cells: a dictionary,* Séminaire d'Algèbre. Lecture Notes in Mathematics vol. 924. Berlin-Heidelberg-New York: Springer 1982.
85. S.I. Gelfand, and Y.I. Manin, *Methods of Homological Algebra,* Springer Monographs in Mathematics, Springer-Verlag, Berlin, New York: Springer-Verlag (2003).
86. R. Godement, *Topologie algébrique et théorie des faisceaux.* Actualites Scientifiques et Industrielles. 1252. Publications de l'Institut de Mathématique de l'Université de Strasbourg. XIII. Paris: Hermann VIII, 283 p. (1958).
87. M.Goresky, *Intersection homology operations,* Comment. Math. Helv. 59 (1984), 485–505.
88. M.Goresky, *$\mathcal{L}^2$-cohomology is intersection cohomology,* The zeta functions of Picard modular surfaces, 47–63, Univ. Montréal, Montréal, QC, 1992.
89. M.Goresky, *Course given at Brown University in the fall of 2016* https://www.math.ias.edu/~goresky/math2710/all.pdf.
90. M. Goresky, *Morse theory, stratifications and sheaves.* In: Cisneros-Molina, J.L., Lê, D.T., Seade, J. (eds.) Handbook of Geometry and Topology of Singularities I. Springer, Cham, 261–301.
91. M. Goresky et R. MacPherson, *La dualité de Poincaré pour les espaces singuliers,* C.R. Acad. Sci. Paris 284 (Janvier 1977), 1549–1551.
92. M. Goresky and R. MacPherson, *Intersection homology theory,* Topology 19 (1980), 135–162.
93. M. Goresky and R. MacPherson, *Morse Theory and Intersection Homology*, in Analyse et topologie sur les espaces singuliers, Astérisque 101–102, Soc. Math. France, 1982, pp. 135–192.
94. M. Goresky and R. MacPherson, *Intersection homology theory II,* Inv. Math. 71 (1983), 77–129.
95. M. Goresky and R. MacPherson, *Stratified Morse theory.* In Singularities, Part1 (Arcata, Calif., 1981), volume 40 of Proc. Sympos. Pure Math., pages 517–533. Amer. Math. Soc., Providence, R.I., 1983.
96. M. Goresky and R. MacPherson, *Stratified Morse theory,* Ergebnisseder Mathematik und ihrer Grenzgebiete volume 14, Springer-Verlag, Berlin, 1988.
97. M. Goresky and R. MacPherson, *Local contribution to the Lefschetz fixed point formula,* Inv. Math. 111 (1993), 1–33. 295, 296.
98. M. Goresky and R. MacPherson, *Letschetz fixed point theorem for intersection homology,* Comment. Math. Helvetici 60 (1985) 366–391.
99. M. Goresky and W. Pardon, *Wu Numbers of Singular Spaces,* Topology 28, (1989), 325–367.
100. A. Grothendieck, *Sur quelques points d'algèbre homologique,* Tôhoku Mathematical Journal, (2), 9: 119–221 (1957).
101. N. Habegger and L. Saper, *Intersection cohomology of cs-spaces and Zeeman's filtration,* Invent. Math. 105 (1991), 247–272.
102. H.A. Hamm and Lê D.T., *The Lefschetz Theorem for hyperplane sections.* In: Cisneros-Molina, J.L., Lê, D.T., Seade, J. (eds.) Handbook of Geometry and Topology of Singularities I. Springer, Cham, 467–514.
103. R. Hartshorne, *Residues and duality.* Lecture Notes in Mathematics vol. 20. Berlin-Heidelberg-New York: Springer 1966.
104. P. Heegaard, *Sur l'Analysis situs,* Bulletin de la Société Mathématique de France, vol. 44, 1916, p. 161–242 . French translation of the danish thesis : *Forstudier til en topologisk Teori for de algebraiske Fladers Sammenhæng* , Copenhague, Det Nordiske Forlag-Ernst Bojesen, 1898.
105. G. Hochschild, B. Kostant and A. Rosenberg *Differential Forms on regular affine Algebras,* Trans. AMS 102 (1962), 383–408.

106. C. Houzel, *Histoire de la théorie des faisceaux* Matériaux pour l'histoire des mathématiques au XXe siècle (Nice, 1996). Sémin. Congr., 3, Soc. Math. France, Paris, pp. 101–119, 1998. also in M. Kashiwara and P. Schapira, Sheaves on Manifolds: With a Short History "Les débuts de la théorie des faisceaux" by C. Houzel, Berlin/Heidelberg/Paris etc., Springer, 1990, 512 p.
107. W. C. Hsiang and V. Pati, *L^2-cohomology of normal algebraic surfaces I*, Invent. Math. 81 (1985), no. 3, 395–412.
108. R. Joshua, *Vanishing of odd-dimensional intersection cohomology*, Mathematische Zeitschrift 195(2) 239–253 (1987).
109. R. Joshua, *Equivariant intersection cohomology–a survey.* in Invariant Theory, Proc. of AMS Special Session, October 31–November 1, 1986, Contemporary Math. 88, Amer. Math. Soc., 1989, 25–31.
110. K. Karu, *Hard Lefschetz Theorem for Nonrational Polytopes*, (pr)e-printmath.AG/0112087(25 pages), 2002.
111. M. Kashiwara and T. Kawai, *The Poincaré lemma for variations of polarized Hodge structure.* Publ. RIMS 23 (1987), 345–407.
112. M. Kashiwara and P. Shapira, *Categories and Sheaves.* Grundlehren der Mathematischen Wissenschaften 332, Springer (2006).
113. H.C. King, *Topological invariance of intersection homology without sheaves,* Topology and its Applications, Volume 20, Issue 2, August 1985, Pages 149–160.
114. H.C. King, *Intersection homology and homology manifolds*, Topology 21 (1982), 229–234.
115. L. Kaup and K.-H. Fieseler, *Singular Poincarée duality and intersection homology,* in: Proc. 1984 Vancouver Conf. in Algebraic Geometry, Amer. Math. Soc., Providence, 1986, 113–161.
116. F. Kirwan and J. Woolf, *An introduction to intersection homology theory,* 2nd ed. Boca Raton, FL: Chapman Hall. 2006.
117. S. Kleiman, *The development of intersection homology theory.* A century of mathematics in America, Part II, Hist. Math. 2, Amer. Math. Soc., 1989, pp. 543–585.
118. B. Klinger, *Intersection homology and perverse sheaves.* http://webusers.imj-prg.fr/~bruno.klingler/cours/coursepervers.pdf.
119. F. Lazzeri, *Morse Theory on singular spaces*, in Singularités à Cargèse. Publ. Soc. Math. de France, 1973 (Astérisque 7-8).
120. S. Lefschetz, *Intersection of complexes on manifolds*, Proceedings of the National Academy of Sciences Vol. 11, 287–289, 1925.
121. S. Lefschetz, *Continuous transformations of manifolds*, Proceedings of the National Academy of Sciences Vol. 9, 90–93, 1923.
122. S. Lefschetz, *Intersections and transformations of manifolds*, Trans. Amer. Math. Soc. Vol. 28 No. 1. (1926), 1–49.
123. E. Looijenga, *$\mathcal{L}^2$–cohomology of locally symmetric varieties,* Compositio Math. 67 (1988), 3–20.
124. U. Ludwig, *A complex in Morse theory computing intersection homology*, Annales de l'Institut Fourier, Tome 67, no. 1 (2017), p. 197–236.
125. D. Massey, *Notes on perverse sheaves and vanishing cycles* http://arxiv.org/abs/math/9908107 http://www.massey.math.neu.edu/Massey/Massey_preprints/perverse_notes.pdf.
126. D. Massey, *Stratified Morse Theory: Past and Present* Pure and applied mathematics quarterly 2(4) October 2006. Also in http://www.massey.math.neu.edu/Massey/Massey_preprints/SMT_survey.pdf.
127. R. MacPherson, *Global questions in the topology of singular spaces,* Proceedings of the International Congress of Mathematicians, Vol. I, Warszawa, 1983. PWN– Polish Scientific Publishers, Warszawa, and Elsevier Science Publishers B.V., Amsterdam, (1984), 213–236.
128. R. MacPherson, *Intersection Homology and Perverse Sheaves,* Report December 15, 1990. Unpublished Notes: http://faculty.tcu.edu/gfriedman/notes/ih.pdf.
129. R. MacPherson and K. Vilonen, *Elementary construction of perverse sheaves*, Invent. Math. 84 (1986), 403–435.

130. L.G. Maxim, *Intersection Homology and Perverse Sheaves: with Applications to Singularities* Graduate Texts in Mathematics, 281. Springer.
131. C. Mc Crory, *Poincaré duality in spaces with singularities.* Thesis, Branseis University. May 1972. Revised version in : https://www.maths.ed.ac.uk/~v1ranick/papers/mccrory4.pdf.
132. C. Mc Crory, *Stratified general position.* Algebraic and Geometric Topology Lecture Notes in Mathematics Volume 664, 1978, pp 142–146.
133. C. Mc Crory and A. Parusiński, *Real intersection homology.* Topology and its Applications, Elsevier, 2018, 234, pp. 108–129.
134. P. McMullen, *The numbers of faces of simplicial polytopes*, Israel J. Math. 9 (1971), 559–570.
135. P. McMullen and D. W. Walkup, *A generalized lower-bound conjecture for simplicial polytopes.* Mathematika, Lond., 18: 264–273, 1971.
136. J. Milnor, *Morse Theory*, Ann. of Math. studies 51, Princeton Univ. Press, Princeton N.J. 1963.
137. M. Nagase, *$\mathcal{L}^2$-cohomology and intersection cohomology of stratified spaces,* Duke Math. J. 50(1983), 329–368.
138. M. Nagase, *Sheaf theoretic $\mathcal{L}^2$-cohomology,* Adv. Stud. Pure Math. $n°8$ (1987), 273–279, North Holland.
139. M.Nagase, *Remarks on the $\mathcal{L}^2$-cohomology of singular algebraic surfaces.* J. Math. Soc. Japan Vol. 41, No. 1, 1989, 97–116.
140. Thủy Nguyễn T. B., *A Remark on the Topology at Infinity of a Polynomial Mapping $F : \mathbb{C}^n \to \mathbb{C}^n$ via Intersection Homology.* Advances in Pure Mathematics. Vol. 6 No. 13, December 2016, 1037–1052.
141. Thủy Nguyễn T. B., *Etude de certains ensembles singuliers associés à une application polynomiale,* Thèse soutenue le 30 sept. 2013 Université de la Mediterranée, Marseille, France available on https://tel.archives-ouvertes.fr/tel-00875930/document.
142. Thủy Nguyễn T. B., *Geometry of singularities at infinity of the Pinchuk's map.* https://arxiv.org/pdf/1710.03318.pdf.
143. Thủy Nguyễn T. B., and M.A. Soares Ruas *On singular varieties associated to a polynomial mapping from $\mathbb{C}^n$ to $\mathbb{C}^{n-1}$.* Asian Journal of Mathematics Volume 22 (2018), Number 6, 1157–1172.
144. Thủy Nguyễn T. B., A. Valette and G. Valette, *On a Singular Variety Associated to a Polynomial Mapping.* Journal of Singularities, 7, 2013, 190–204.
145. T. Oda, *The intersection cohomology and toric varieties (Modern aspects of combinatorial structure on convex polytopes.* Kyoto university (1994).
146. T. Oda, *The intersection cohomology of toric varieties,* in Geometry of Toric Varieties and Convex Polytopes (T. Hibi, ed.), RIMS Kokyuroku 776, March 1992, 49–67.
147. T. Ohsawa, *Cheeger-Goresky-MacPherson conjecture for varieties with isolated singularities,* Math. Z. 206 (1991), 219–224.
148. T. Ohsawa, *On the $\mathcal{L}^2$-cohomology of complex spaces,* Math. Z. 209 (1992), 519–530.
149. G. Padilla, M. Saralegi-Aranguren *Intersection cohomology of the circle actions,* Topology and its Applications 154 (2007) 2764–2770.
150. W.L. Pardon and M. Stern *$\mathcal{L}^2 - \bar{\partial}$-cohomology of complex projective varieties.* J. Amer. Math. Soc. vol. 4, 1991, 603–621.
151. R. Pignoni, *Density and stability of Morse functions on a stratified space,* Ann. Scuola Nor. Sup. Pisa ser. IV vol. 6 (1979) 593–608.
152. H. Poincaré, *Sur la généralisation d'un théorème d'Euler relatif aux polyèdres,* Comptes rendus de l'Acad. des Sciences 117 (1893), 144-145. Œuvres de Poincaré XI, Gauthier-Villars Ed. Paris, 1953, 6–7.
153. H. Poincaré, *Analysis Situs,* Journal de l'Ecole Polytechnique ser 2, 1 (1895) pp. 1–123. (Reprinted in Œuvres de Poincaré VI, Gauthier-Villars Ed. Paris, 1953, VI, 193–288).
154. H. Poincaré, *Complément à l'Analysis Situs,* Rendiconti del Circolo matematico di Palermo, 13 (1899) pp. 285–343. (Reprinted in Œuvres de Poincaré VI, Gauthier-Villars Ed. Paris, 1953, VI, 290–337).

155. H. Poincaré, *Second complément à l'Analysis Situs,* Proceedings of the London Mathematical Society, 32 (1900), pp. 277–308. (Reprinted in Œuvres de Poincaré VI, Gauthier-Villars Ed. Paris, 1953, 338–370).
156. G. Pollini, *Intersection differential forms,* Rendiconti del Seminario Matematico della Università di Padova (2005) Volume: 113, page 71–97.
157. P. Popescu-Pampu, *La dualité de Poincaré,* Images des Mathématiques, CNRS, 2012 https://images.math.cnrs.fr/La-dualite-de-Poincare.html.
158. J. Prieto and M. Saralegi-Aranguren, *Equivariant intersection cohomology of the circle actions.* Revista de la Real Academia de Ciencias Exactas, Fisicas y Naturales. Serie A. Matematicas volume 108, 49–62 (2014).
159. K. Reidemeister *Topologie der Polyeder und kombinatorische Topologie der Komplexe,* Akademische Verlagsgesellschaft, Lpz., 1938-Mathematik und ihre Anwendungen in Monographien und Lehrbüchern, Band 17.
160. G. de Rham, *Sur l'analysis situs des variétés à n dimensions*, J. Math. Pures Appl. (9) 10 (1931), p. 115–200.
161. K. Rietsch, *An introduction to perverse sheaves.* arXiv:math/0307349v1. 26 Jul. 2003.
162. C. Rourke and B. Sanderson *Homology stratifications and intersection homology.* Geometry and Topology Monographs 2 (1999) 455–472.
163. L. Saper, $\mathcal{L}^2$*-cohomology and intersection homology of certain algebraic varieties with isolated singularities,* Invent. Math. 82 (1985), no. 2, 207–255.
164. L. Saper, $\mathcal{L}^2$*-cohomology of Kähler varieties with isolated singularities*, J. Differential Geometry, 36 (1992), 89–161.
165. L. Saper and M. Stern, $\mathcal{L}^2$*-cohomology of arithmetic varieties.* Annals of Mathematics. Second Series, Vol. 132, No. 1 (Jul., 1990), pp. 1–69.
166. M. Saralegi-Aranguren, *de Rham intersection cohomology for general perversities,* llinois J. Math. Volume 49, Number 3 (2005), 737–758.
167. J. Schürmann, *Topology of Singular Spaces and Constructible Sheaves*, Monografie Matematyczne 63, Birkhäuser Verlag, Basel, 2003.
168. J.P. Serre *Géomérie algébrique et géométrie analytique*, Annales de l'Institut Fourier, vol. 6,? 1956, p. 1–42.
169. P. Siegel, *Witt Spaces: a Geometric Cycle Theory for KO-Homology at odd primes*, Amer. J. Math., **105**, (1983), no. 5, 1067–1105.
170. D. Sommerville, *The relations connecting the angle sums and volume of a polytope in space of n dimensions.* Proceedings of the Royal Society London, Ser. A, Vol. 115, (1927), 103–119.
171. T. Springer, *Quelques applications de la cohomologie d'intersection,* Séminaire Bourbaki 589, Astérisque 92–93, Soc. Math. France, 1982, pp. 249–274.
172. R. Stanley, *Generalized h-vectors, intersection cohomology of toric varieties and related results*, M. Nagata, H. Matsumura, eds.,Commutative Algebra and Combinatorics, 187–213, Adv. Stud.Pure Math. 11, Kinokunia, Tokyo, and North Holland, Amsterdam/New York, 1987.
173. R. Stanley, *The number of faces of a simplicial convex polytope.* Adv. in Math. 35 (1980), 236–238.
174. N. Steenrod, *Homology with local coefficients.* Annals of Mathematics. 44 (4) 610–627, 1943.
175. N. Steenrod, *The Topology of Fibre Bundles* Princeton University Press, Princeton 1951. New edition 7th printing, Princeton landmarks in mathematics and physics, 1999.
176. N. Steenrod, *The Work and Influence of Professeur S. Lefschetz in Algebraic Topology.* in Algebraic Geometry and Topology: A Symposium in Honor of Solomon Lefschetz. Princeton Mathematical Series, Vol. 12. Princeton University Press, 1957.
177. T. Suwa, *Complex Analytic Geometry*, World Scientific, March 2021.
178. R. Thom, *Les singularités des applications différentiables*, Annales de l'Institut Fourier, vol. 6, 1956, p. 43–87.
179. D. Trotman, *Stratification theory* In: Cisneros-Molina, J.L., Lê, D.T., Seade, J. (eds.) Handbook of Geometry and Topology of Singularities I. Springer, Cham, 231–260.

180. A. Valette and G. Valette, *Geometry of polynomial mappings at infinity via intersection homology* Annales de l'Institut Fourier, vol. 74, 2014, p. 2147–2163.
181. G. Valette *A Lefschetz duality for intersection homology,* Geom. Dedicata 169 (2014), 283–299.
182. G. Valette $\mathcal{L}^\infty$*-cohomology is intersection cohomology*. Advances in Mathematics, Volume 231, Issues 3–4, 2012, 1818–1842.
183. J.L. Verdier, *Dualité dans la cohomologie des espaces localement compacts* Séminaire Bourbaki. Vol. 1965/1966. Exposé 300.
184. J.L. Verdier, *Catégories dérivés, état 0.* in Séminaire de Géométrie Algébrique du Bois-Marie, Cohomologie Étale, SGA 4 1/2 par P. Deligne. Lecture Notes in Mathematics vol. 569. Springer-Verlag 1977.
185. A. Weber, *A morphism of intersection homology and hard Lefschetz,* Algebraic geometry conference, 339–348, 1999.
186. H. Whitney *Geometric integration theory*, Princeton University Press, 1957.
187. B. Youssin, $\mathcal{L}^p$*-cohomology of cones and horns*, J. Differential Geom. 39 (1994), no. 3, 559–603.
188. S. Zucker, *Hodge theory with degeneration coefficients:* L_2*-cohomology in the Poincaré metric*, Ann. of Math. 109, 415–476 (1979).
189. S. Zucker, *Cohomology of warped products and arithmetic groups.* Inventiones Math. 70 (1982), 169–218.
190. S. Zucker, $\mathcal{L}^2$*-cohomology and intersection homology of locally symmetric varieties, II.* Compositio Math. 519 (1986), 339–398.

Chapter 6
Milnor's Fibration Theorem for Real and Complex Singularities

José Luis Cisneros-Molina and José Seade

Abstract Milnor's fibration theorem is a landmark in singularity theory; it allowed to deepen the study of the geometry and topology of analytic maps near their critical points. In this chapter we revisit the classical theory and we glance at some areas of current research. We start with a glimpse at the origin of the fibration theorem, which is motivated by the study of exotic spheres. We then discuss an elementary example where all the ingredients of the fibration theorem are described in simple terms, and we use this as a guideline all along the chapter. The first part concerns complex singularities, which is a fairly mature area of mathematics; we survey some of the main steps in this line of research and indicate a wide bibliography as well as relations with other chapters in this book. The second part concerns real singularities, a theory that still is in its youth, though it springs also from Milnor's seminal work.

Contents

J. L. Cisneros-Molina (✉)
Instituto de Matemáticas, Unidad Cuernavaca, Universidad Nacional Autónoma de México, Cuernavaca, Mexico
e-mail: jlcisneros@im.unam.mx

J. L. Cisneros-Molina · J. Seade
Laboratorio Solomon Lefschetz, International Research Laboratry CNRS, Cuernavaca, Mexico
e-mail: jseade@im.unam.mx

J. Seade
Instituto de Matemáticas, Universidad Nacional Autónoma de México, Mexico City, Mexico

J. L. Cisneros-Molina et al. (eds.), *Handbook of Geometry and Topology of Singularities II*, https://doi.org/10.1007/978-3-030-78024-1_6

6.1 Introduction

Milnor's fibration theorem [110] is a landmark in singularity theory, it allowed to deepen the study of the geometry and topology of analytic maps near their critical points. To each singular point of a complex hypersurface it associates a fibre bundle, known as the Milnor Fibration of the singularity.

The classification of singularities from a topological viewpoint started with the work of Brauner [14], he considered singular points of plane curves C in $\mathbb{C}^2$ and analyses the intersection $L = C \cap \mathbb{S}^3_\varepsilon$ of C with a small sphere $\mathbb{S}^3_\varepsilon$ of radius ε centred in the singular point. The intersection L is a knot or link in $\mathbb{S}^3_\varepsilon$ and the pair $(\mathbb{S}^3_\varepsilon, L)$ determines the local topology of the curve near the singular point (see Sect. 6.4). Further work in this direction was also done in [19, 79, 147, 165]. For normal surfaces, Mumford [114] proved that if V has a singularity at P then its link $L_V = V \cap \mathbb{S}^5_\varepsilon$ is never simply connected, hence, it cannot have the homotopy type of a 3-sphere $\mathbb{S}^3$. Later, Brieskorn proved [15] that in higher dimensions there is no analogue to Mumford's theorem, by showing examples of singularities of n-dimensional hypersurfaces whose link is homeomorphic to the $2n-1$-sphere, in some of these examples the link L of the singularity is diffeomorphic to the standard $2n-1$-sphere, while in other cases L is an *exotic* sphere [16, 70]. Motivated by Brieskorn's result, Milnor determined when the link L_V of a complex hypersurface V is a homology sphere [110, Theorem 8.5], a key ingredient for this result is Milnor's fibration theorem (see Sect. 6.2).

Let $f\colon (\mathbb{C}^{n+1}, \mathbf{0}) \to (\mathbb{C}, 0)$ be a holomorphic map taking the origin into the origin, with an isolated critical point at $\mathbf{0}$. Since f is analytic, there exists $r > 0$ sufficiently small so that $0 \in \mathbb{C}$ is the only critical value of the restriction $f|_{\mathbb{B}_r}$, where $\mathbb{B}_r$ is the open ball of radius r and center at $\mathbf{0}$. Set

$$V := f^{-1}(0) \quad \text{and} \quad V^* := V \setminus \{\mathbf{0}\}.$$

So V^* is an n-dimensional complex manifold. We know (see Sect. 6.4) that V^* meets transversely every sufficiently small sphere $\mathbb{S}_\varepsilon$ in $\mathbb{C}^{n+1}$ centered at $\mathbf{0}$ and contained in $\mathbb{B}_r$. The manifold $L_V := V \cap \mathbb{S}_\varepsilon$ is called the link of the singularity and its diffeomorphism type does not depend on the choice of the sphere. Then Milnor's fibration theorem [110, Theorem 4.8] says that for every such sphere $\mathbb{S}_\varepsilon$ we have a smooth fiber bundle (see Sect. 6.5)

$$\varphi := \frac{f}{|f|} : \mathbb{S}_\varepsilon \setminus L_V \longrightarrow \mathbb{S}^1 . \tag{6.1}$$

The fiber is diffeomorphic to a $2n$-ball to which one attaches handles of middle index. The number of such handles is now called the Milnor number of the singularity. In fact f can have non-isolated critical points and we still have fibration (6.1). In this case, the link L_V is not longer a smooth manifold and the Milnor fiber is diffeomorphic to a ball to which we must attach handles of various indices. The precise number of handles of each index is prescribed by the Lê numbers of the singularity, a concept introduced by Massey [99, 100] (see Sect. 6.6.2).

The fibers F_f are diffeomorphic to the complex manifolds obtained by considering a regular value t sufficiently near $0 \in \mathbb{C}$ and looking at the piece of $f^{-1}(t)$ contained within the open ball $\mathbb{B}_\varepsilon$ bounded by $\mathbb{S}_\varepsilon$ (see Fig. 6.1).

Milnor also proved a fibration theorem for real singularities [109, Theorem 2] or [110, Theorem 11.2]. Let $f : (\mathbb{R}^n, \mathbf{0}) \to (\mathbb{R}^p, 0)$, $n \geq p \geq 1$, be a real analytic map with an isolated critical point, one has a fiber bundle:

$$\varphi : \mathbb{S}_\varepsilon^{n-1} \setminus L_V \longrightarrow \mathbb{S}^{p-1} . \tag{6.2}$$

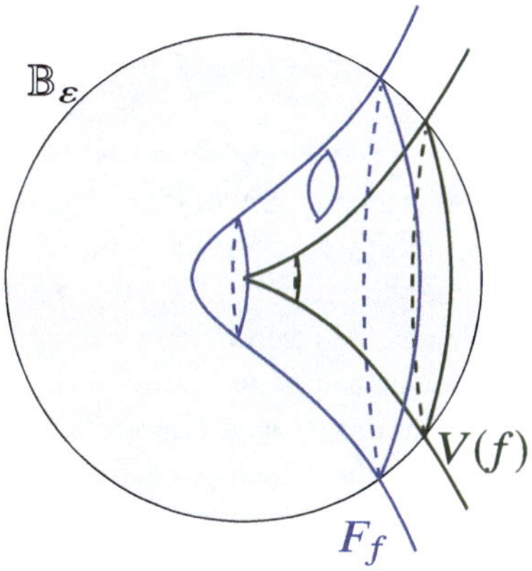

Fig. 6.1 The Milnor fiber F_f

In general the projection map φ can only be taken as $f/\|f\|$ in a neighborhood of the link L_V. Also, as pointed out [110, Sect. 11], the condition of having an isolated critical point is very restrictive. Generically the set of critical values has positive dimension, and even if the only critical value is 0, it is fairly stringent to ask for having an isolated critical point (see for instance [41]).

From his appearance, Milnor's fibration theorem has opened the way to countless insights and new understandings. It is a beautiful piece of mathematics, where many different branches, aspects and ideas, come together. In this chapter we revisit the classical theory in both the complex and real settings, we present various extensions of it, we glance at some areas of current research and we indicate some connections with other topics treated in this volume.

This chapter consists of two parts, the first one comprises Sects. 6.2 to 6.7 and is devoted to complex singularities. In Sect. 6.2 we explain how the search of exotic spheres in links of complex hypersurface lead to Milnor's fibration theorem. Sect. 6.3 presents the particular case of Brieskorn-Pham polynomials, these are polynomials studied by Pham [135] and by Brieskorn [16], where one can see in an elementary way all the properties and ingredients involved in Milnor's fibration theorem. This model example will be the guiding theme and we will be referring to it through the chapter. Sect. 6.4 explains the local conical structure of real and complex analytic sets. In Sect. 6.5 we present the classical fibration theorems: the fibration on a sphere (6.1), the fibration on a tube and their equivalence. Section 6.6 consists of three parts. The first one focus on the topology and geometry of the link of isolated singularities. The second part deals with the topology and geometry of the fibre, it starts introducing the Milnor number $\mu(f)$ of a holomorphic map with isolated critical point: the fibre of the Milnor fibration has the homotopy type of a wedge of spheres of middle dimension and $\mu(f)$ is the number of spheres. We present two ways to compute it, one topological and the other algebraic. We also present Lê numbers, which are a generalization of the Milnor number introduced by Massey for holomorphic maps with non-isolated critical point that we mentioned above. In the third part we mention relations of Milnor's fibration with other subjects: smoothings of singularities; vanishing cycles and the Milnor lattice; monodromy, open book decompositions and contact structures. Section 6.7 presents several extensions of Milnor's fibration. Firstly, we describe the fibration for an isolated complete intersection singularity (ICIS). Then, we present the Milnor-Lê fibration, a fibration on a tube for a holomorphic map germ $f\colon (X,\mathbf{0}) \to (\mathbb{C},0)$ on an analytic subset X of $\mathbb{C}^n$, and we introduce the canonical pencil o ff, which gives a fibration on a ball $\mathbb{B}_\varepsilon$ minus the variety $V = f^{-1}(0)$.

The second part of the chapter deals with real singularities and consists of Sects. 6.8 to 6.11. In Sect. 6.8 we present the classical Milnor fibration (6.2) for real analytic maps $f\colon (\mathbb{R}^n,\mathbf{0}) \to (\mathbb{R}^p,0)$, $n \geq p \geq 1$ with an isolated critical point. We discuss its main weaknesses compared with is complex counterpart, in particular the fact that in general, the projection is not given by $f/\|f\|$. At this respect, we introduce the strong Milnor condition and some results related to it. Section 6.9 is devoted to the existence of fibrations for real analytic maps with non-isolated critical point. First we consider the case of maps with isolated critical value, these include polar weighted homogeneous polynomials, an interesting family of real analytic polynomials which generalize complex weighted homogeneous polynomials and which behave much like them but also new structures can appear, for instance, they can have open book decompositions that cannot appear in complex singularities. In Sect. 6.10 we discuss a regularity condition that is necessary and sufficient to assure that if we have the fibration (6.2), we can take the projection φ to be $f/|f|$ everywhere. This is called d-regularity. We finish with Sect. 6.11 which is about critical points of analytic functions in the complex variables $z_1, \ldots, z_n$ and their conjugates. We encounter some examples of these functions already in previous sections. This type of functions were called mixed functions by Oka [119].

Complex Singularities

6.2 Exotic Spheres and the Birth of Milnor's Fibration

In 1956 Milnor [108] proved the astonishing result that topological manifolds can have inequivalent differentiable structures constructing the first "exotic spheres": smooth 7-manifolds homeomorphic to $\mathbb{S}^7$ but with non-equivalent differentiable structures.

The set $\mathcal{S}_n$ of equivalence classes of smooth structures on the n-sphere $\mathbb{S}^n$ is a monoid, with operation the connected sum and the identity element the standard sphere $\mathbb{S}^n$. Smale [154], Stallings [156] and Zeeman [166] proved that every homotopy n-sphere, $n \neq 3, 4$, is homeomorphic to the standard n-sphere $\mathbb{S}^n$. Also Smale [155] proved that two homotopy n-spheres, $n \neq 3, 4$, are h-cobordant if and only if they are diffeomorphic. Thus, for $n \neq 4$, the monoid $\mathcal{S}_n$ is isomorphic to the monoid Θ_n of all h-cobordism classes of homotopy n-spheres. The case $n = 3$ follows by Perelman's proof of Poincaré's conjecture.

Kervaire and Milnor [81] studied Θ_n and proved that this monoid is actually a finite abelian group. They noticed that Θ_n contains a "preferred subgroup" $bP_{n+1} \leq \Theta_n$, of those homotopy spheres that bound a parallelizable manifold, i.e., a manifold with trivial tangent bundle. This is a finite cyclic group which has finite index in Θ_n. This cyclic group is trivial for n even and has order 1 or 2 for $n \equiv 1 \pmod 4$, being generated by the Kervaire sphere [80], the boundary of the manifold obtained by plumbing two copies of the tangent disk bundle of $\mathbb{S}^n$. By [18] $bP_{n+1} \cong \mathbb{Z}_2$ (or equivalently, Kervaire's sphere is exotic) if $n \equiv 1 \pmod 8$. For $n \equiv 3 \pmod 4$, ($n + 1 = 4m$ for some $m > 1$) its order $|bP_{4m}|$ grows more than exponentially:

$$|bP_{4m}| = \left[2^{2m-2}\left(2^{2m-1} - 1\right)\right] \cdot \left[\text{numerator of}\left(\frac{4\mathcal{B}_m}{m}\right)\right], \tag{6.3}$$

where the $\mathcal{B}_m$ are the Bernoulli numbers. Thus for instance (see [69, 81]), for $n = 7, 11, 15$ or 19 there are, respectively, $|bP_{n+1}| = 28,\ 992,\ 8128$ and 130816

non-equivalent differentiable structures on the n-sphere that bound a parallelizable manifold. The structure of $\mathcal{S}_4$ is still an open problem (see [111, 112]).

6.2.1 Singularities and Exotic Spheres

Let (V, P) be a complex analytic variety of complex dimension n in some affine space $\mathbb{C}^N$, with a unique singular point at P. Then $V^* = V \setminus \{P\}$ is a complex n-manifold.

Proposition 6.2.1 *There exists $\varepsilon > 0$ sufficiently small, so that every sphere $\mathbb{S}_r$ in $\mathbb{C}^N$ of radius $r \leq \varepsilon$ and center at P meets V^* transversely.*

This is proved in [110, Chap. 2] when V is algebraic and it is a particular case of a general theorem about the local conical structure of analytic sets, see Sect. 6.4. It follows that if (V, P) is as above, then its *link* $L_V := \mathbb{S}_\varepsilon \cap V$ is a smooth real analytic manifold of dimension $2n - 1$.

For $n = 1$, L_V is a union of circles, one for each branch of V. For $n = 2$, in 1961 Mumford [114] proved that if V has a normal singularity at P then its link L_V is never simply connected, hence, it cannot have the homotopy type of a 3-sphere $\mathbb{S}^3$. In 1966 Brieskorn [15] proved that if V is given by the equation

$$z_0^3 + z_1^2 + \cdots + z_n^2 = 0, \qquad n \geq 3 \text{ odd},$$

which has an isolated singularity at the origin, then L_V is homeomorphic to the sphere $\mathbb{S}^{2n-1}$, showing that for dimensions higher than 2, there is not an analogous of Mumford's theorem. Brieskorn's result motivated the search of exotic spheres in the links L_V of complex hypersurfaces V, i.e., defined by one single equation, with isolated singularities. So one has to answer the following question:

Question 6.2.2 Can we know when L_V is a homotopy sphere?, and if so, can we determine which element in Θ_n it represents?

Question 6.2.2 was answered by the work of various people in the 1960s, most notably by E. Brieskorn, F. Hirzebruch and J. Milnor, see [44, 62] for more details of this interesting story.

Combining the result by Brieskorn with results of Jänich [78] (Independently proved in [71]), Hirzebruch [70] proved that the link $\Sigma(d, 2, \ldots, 2)$ of the singularity given by the equation

$$z_0^d + z_1^2 + \cdots + z_n^2 = 0, \qquad \text{with both } n \textit{ and } d \text{ odd},$$

is a homotopy sphere. In particular, $\Sigma(3, 2, 2, 2, 2, 2)$ is the 9-dimensional exotic Kervaire sphere. Also inspired by Brieskorn's result, Milnor, in a letter to J. Nash, considered in more generality singularities of hypersurfaces of the form

$$z_0^{a_0} + z_1^{a_1} + \cdots + z_n^{a_n} = 0, \quad a_i \geq 2\,, \tag{6.4}$$

and conjectured which of them have links which are spheres.

Pham [135] motivated by applications to the theory of elementary particles studied the smooth complex hypersurface $z_0^{a_0} + z_1^{a_1} + \cdots + z_n^{a_n} = 1$, he computed its homotopy type (see Property 6.3.8 in Sect. 6.3), its intersection pairing and its monodromy. Pham's computations were the ingredients that allowed Brieskorn [16] to prove Milnor's conjecture. One of these results is the following remarkable theorem of Brieskorn [16, Korollar 2].

Theorem 6.2.3 *Every exotic sphere of dimension $m = 2n - 1 > 6$ that bounds a parallelizable manifold is the link of some hypersurface singularity of the form*

$$z_0^{a_0} + z_1^{a_1} + \cdots + z_n^{a_n} = 0\,,$$

for some appropriate integers $a_i \geq 2$, $i = 0, 1, \ldots, n$.

As a particular case Brieskorn proved that the link of the singularity

$$z_0^3 + z_1^{6k-1} + z_2^2 + \cdots + z_{2m}^2 = 0\,; \qquad k \geq 1\,,\ m \geq 2\,.$$

is a $(4m-1)$-sphere in bP_{4m}. For $m = 2$ and $k = 1, \ldots, 28$ we get the 28 classes of 7-spheres, and for $m = 3$ and $k = 1, \ldots, 992$, we get the 992 classes of 11-spheres.

Afterwards, Milnor [110] answered Question 6.2.2 for general complex hypersurfaces with an isolated singularity using his fibration theorem.

Let V be the zero-locus of an analytic map $f : (\mathbb{C}^{n+1}, \mathbf{0}) \to (\mathbb{C}, 0)$ with an isolated critical point at $\mathbf{0}$. Equip its link L_V^{2n-1} with its natural differentiable structure as the transverse intersection $L_V = \mathbb{S}_\varepsilon \cap V$ of two smooth submanifolds of $\mathbb{C}^{n+1}$. One has a map:

$$\varphi := \frac{f}{|f|} : \mathbb{S}_\varepsilon \setminus L_V \longrightarrow \mathbb{S}^1\,, \tag{6.5}$$

and Milnor's fibration theorem says that this is a smooth fiber bundle.

Milnor also proves that the fiber F_t is diffeomorphic to the portion of a non-critical level $f^{-1}(t)$ contained within the ball $\mathbb{B}_\varepsilon$ bounded by $\mathbb{S}_\varepsilon$ (see Fig. 6.1). This implies that the normal bundle of F_t is trivial, being the inverse image of a regular value. Hence the tangent bundle TF_t is stably trivial, i.e., F_t is stably parallelizable, and we know from [81] that for compact connected manifolds with non-empty boundary, stably-parallelizable implies parallelizable. Thus we get:

Proposition 6.2.4 *The link L_V of every complex hypersurface isolated singularity bounds the fibers F_t, which are parallelizable manifolds.*

The point is to know when L_V is a homotopy sphere, and when this happens, which element it represents in bP_{2n}. Milnor proved that the link of every isolated hypersurface singularity in $\mathbb{C}^{n+1}$ is $(n-2)$-connected [110, Theorem 5.2] and the fiber F_t has the homotopy of a bouquet $\bigvee \mathbb{S}^n$ of spheres of middle dimension [110,

Theorems 6.5]. The number of spheres in the bouquet $\bigvee \mathbb{S}^n$ is strictly positive, unless V has no singularity. This is now called the Milnor number μ of f (see Sect. 6.6).

For $n > 2$ the link is simply connected and therefore the Hurewicz isomorphism implies that the homology of L_V also vanishes in dimensions $i = 1, \ldots, n-2$. Since L_V is always orientable, by the Poincaré duality isomorphism its homology vanishes in dimensions $n+i, i = 1, \ldots, n-2$ as well. Thus the only possibly non-zero groups are in dimensions $i = n, n-1$ and of course $i = 0, 2n-1$ where they are isomorphic to the group of the integers (or the corresponding ring of coefficients).

If $H_{n-1}(L_V)$ vanishes then $H_n(L_V)$ also vanishes, by duality, and L_V is a homology sphere. If $n \geq 2$, then L_V is simply connected by [110]. Hence, if $n > 2$ then Smale's theorem in [155] implies (Poincaré's Conjecture in dimensions ≥ 5) that L_V is actually homeomorphic to $\mathbb{S}^{2n-1}$. So the question is to decide when $H_{n-1}(L_V)$ vanishes, and it is here that the fibration theorem enters the scene.

Fix a fiber F_0 and consider the monodromy $h \colon F_0 \to F_0$ (a first return map, see Sect. 6.6.3) of the bundle (6.5) and the induced representation in the middle homology of the fiber

$$h_* \colon H_n(F_0) \to H_n(F_0) \,.$$

Let $\Delta(t) = \det(h_* - tI_*)$ be the characteristic polynomial of the monodromy. Corresponding to Milnor's fibration there is a Wang sequence (see p. 67 in [110]):

$$H_n(F_0) \xrightarrow{h_*-I_*} H_n(F_0) \longrightarrow H_n(\mathbb{S}_\varepsilon - L_V) \longrightarrow 0 \,.$$

By Alexander and Poincaré dualities we have

$$H_n(\mathbb{S}_\varepsilon - L_V) \cong H^n(L_V) \cong H_{n-1}(L_V)$$

and one arrives a t Mlnor's theorem [110, Theorem 8.5],

Theorem 6.2.5 *For $n > 2$ the link L_V is a topological sphere if and only if the determinant $\Delta(1)$ of $h_* - I_*$ is ± 1.*

If n is odd, $\dim L_V \equiv 1 \pmod 4$. The diffeomorphism class of L_V in bP_{2n} is determined by the Kervaire invariant $c(F_0) \in \mathbb{Z}_2$, and a theorem of Levine [92] asserts that in this case the Kervaire invariant is given by

$$c(F) = 0 if \Delta(-1) \equiv \pm 1 \pmod 8, \qquad c(F) = 1 if \Delta(-1) \equiv \pm 3 \pmod 8.$$

If n is even, $\dim L_V \equiv 3 \pmod 4$ and $2n = 4m$ for some $m > 1$. The diffeomorphism class of L_V in bP_{4m} is determined by the signature of the intersection form of the Milnor fibre

$$H_n(F_0) \otimes H_n(F_0) \to \mathbb{Z}.$$

Consider the collection of all $4m$ manifolds M which are stably parallelizable and are bounded by the $(4m-1)$-sphere. The corresponding signatures $\sigma(M) \in \mathbb{Z}$ form a group under addition. Let $\sigma_m > 0$ denote the generator of this group. Kervaire and Milnor proved [81, Theorem 7.5] that if Σ_1 and Σ_2 are homotopy $(4m-1)$-spheres, $m > 1$, which bound stably parallelizable manifolds M_1 and M_2 respectively, then Σ_1 and Σ_2 are h-cobordant, and therefore diffeomorphic, if and only if $\sigma(M_1) \equiv \sigma(M_2) \pmod{\sigma_m}$. On the other hand, an integer σ occurs as $\sigma(M)$ for some stably parallelizable manifold M bounded by a homotopy sphere if and only if $\sigma \equiv 0 \pmod 8$. Therefore $|bP_{4m}| = \sigma_m/8$ and it is given by the multiple of the numerator of $\frac{4\mathcal{B}_m}{m}$ given in (6.3), where $\mathcal{B}_m$ is the m-th Bernoulli number.

6.2.2 Open Questions

One may consider singularities which are not hypersurfaces, and try to produce other elements in the homotopy of spheres. To our knowledge, little is known about this problem. If we consider complex isolated complete intersection singularities, one always has a Milnor fibration and the fibers can be regarded as being the interior of compact parallelizable manifolds with boundary the link, by [65]. So in these cases, if the link is an exotic sphere, this is in $bP_{2n} \leq \Theta_{2n-1}$, which is the simplest and best understood part of Θ_{2n-1}.

The second author thanks Patrick Popescu-Pampu for bringing to his attention the following interesting question posed by Durfee [162, Problem H, p. 252]:

Question 6.2.6 Does every exotic sphere occur as the link of an isolated complex singularity?

A step for answering Question 6.2.6 is the question that Popescu-Pampu originally asked:

Question 6.2.7 Does there exist a complex isolated singularity whose link is a homotopy sphere that does not bound a parallelizable manifold?

Such examples, if they exist, would produce elements in the most mysterious part of the groups Θ_n.

6.3 Model Example: the Brieskorn-Pham Singularities

This section presents the particular case of Brieskorn-Pham polynomials, studied by Pham [135] and by Brieskorn [16], where one can see in an elementary way all the properties and ingredients involved in Milnor's fibration theorem.

Definition 6.3.1 A *Brieskorn-Pham polynomial* is a map $f : \mathbb{C}^{n+1} \to \mathbb{C}$ of the form:

$$z_0^{a_0} + \dots + z_n^{a_n}, \quad a_i \geq 2\,. \tag{6.6}$$

A simple computation shows that the origin $\mathbf{0} \in \mathbb{C}^{n+1}$ is its only critical point, so $V := f^{-1}(0)$ is a complex hypersurface with an isolated singularity at $\mathbf{0}$. Let d be the lowest common multiple of the a_i and for each $i = 0, \dots, n$ set $d_i = d/a_i$. Then one has a $\mathbb{C}^*$-action on $\mathbb{C}^{n+1}$ determined for every non-zero complex number $\lambda \in \mathbb{C}^*$ by

$$\lambda \cdot (z_0, \dots, z_n) \mapsto (\lambda^{d_0} z_0, \dots, \lambda^{d_n} z_n) \tag{6.7}$$

Observe that one has:

$$f(\lambda^{d_0} z_0, \dots, \lambda^{d_n} z_n) = \lambda^d f(z_1, \dots, z_n) \tag{6.8}$$

so f is *weighted homogeneous* (see Definition 6.3.9 below). This $\mathbb{C}^*$-action has $\mathbf{0}$ as its only fixed point and V is an invariant set, union of $\mathbb{C}^*$-orbits. This has the following properties:

Property 6.3.2 (*Conical structure*) Restricting the action (6.7) to the positive real numbers $t \in \mathbb{R}^+$, we get a flow such that:

- each orbit converges to $\mathbf{0}$ as t tends to 0, and it goes to infinity as t tends to ∞;
- each orbit is transverse to all spheres centered at $\mathbf{0}$. Hence V intersects transversely every $(2n+1)$-sphere $\mathbb{S}_r$ centered at $\mathbf{0}$, so $K_r := V \cap \mathbb{S}_r$ is a real codimension 2 smooth submanifold of $\mathbb{S}_r$;
- Given arbitrary spheres $\mathbb{S}_r, \mathbb{S}_{r'}$ centered at $\mathbf{0}$, the flow gives a diffeomorphism from $\mathbb{S}_r$ into $\mathbb{S}_{r'}$ taking K_r into $K_{r'}$. Moreover, the flow determines a 1-parameter group of diffeomorphisms that exhibits the pair $(\mathbb{C}^{n+1}, V)$ as the cone over $(\mathbb{S}_r, K_r)$. We denote the manifold K_r by L_f and call it *the link* (see [43, 110]).

Property 6.3.3 (*Constant argument*) The argument of the complex number $f(z)$ is constant on each orbit of the above $\mathbb{R}^+$-action, i.e., $f(z)/|f(z)| = f(tz)/|f(tz)|$ for all $t \in \mathbb{R}^+$.

Property 6.3.4 (*Fibration on tubes*)

- The restriction of the $\mathbb{C}^*$-action to $\mathbb{S}^1$ leaves invariant every sphere around $\mathbf{0}$.
- By (6.8), multiplication by $e^{i\theta}$ in $\mathbb{C}^n$ transports each fiber $f^{-1}(\zeta)$ into the fiber $f^{-1}(e^{i\theta d} \cdot \zeta)$. Hence $\mathbb{S}^1$ acts on each *tube* $N(\delta) := f^{-1}(\partial\mathbb{D}_\delta)$, carrying fibres of f into fibres of f, where $\partial\mathbb{D}_\delta \cong \mathbb{S}^1$ is the boundary of the disc in $\mathbb{C}$ of radius $\delta > 0$ and centered at 0. A direct computation shows that the orbits of this action are transverse to the fibers of f. So we have a smooth fiber bundle:

$$f \colon N(\delta) \to \partial\mathbb{D}_\delta. \tag{6.9}$$

- By Property 6.3.2 V is transverse to the unit sphere $\mathbb{S}^{2n+1}$ and since each point in $V \setminus \{\mathbf{0}\}$ is regular, each fiber $f^{-1}(t)$ with $|t|$ sufficiently small is also transverse to $\mathbb{S}^{2n+1}$.
- Hence, if we set $N(1, \delta) := N(\delta) \cap \mathbb{B}^{2n+2}$, where the 1 means that the ball $\mathbb{B}^{2n+2}$ has radius 1, we have that the fiber bundle (6.9) determines a fiber bundle:

$$f\colon N(1,\delta) \to \partial\mathbb{D}_\delta \cong \mathbb{S}^1\,. \tag{6.10}$$

This is the second classical version of Milnor's fibration for the map f, known as *Milnor-Lê fibration*.

Property 6.3.5 (*Fibration on the ball*) Observe that for each line $\mathcal{L}_\theta$ through the origin in $\mathbb{C}$, we may consider the set

$$X_\theta := \{z \in \mathbb{C}^n \mid f(z) \in \mathcal{L}_\theta\}\,.$$

Each X_θ is a real analytic hypersurface with an isolated singularity at $\mathbf{0}$, their union fills the entire $\mathbb{C}^{n+1}$ and their intersection is V. By Property 6.3.3, the orbits of the $\mathbb{R}^+$action are contained in the X_θ's. By Property 6.3.2, each X_θ is transverse to all the spheres, and by Property 6.3.4, the $\mathbb{S}^1$-action permutes these hypersurfaces. Thus, one has that these varieties define a pencil in $\mathbb{C}^{n+1}$, a sort of open-book where the binding is now the singular variety V, and each of these varieties is transverse to every sphere around $\mathbf{0}$. If we remove V from $\mathbb{C}^n$, for every ball around $\mathbf{0}$ we get a fiber bundle

$$\varphi = \frac{f}{|f|}\colon \mathbb{B}^{2n} \setminus V \longrightarrow \mathbb{S}^1\,. \tag{6.11}$$

The fiber over a point $e^{i\theta}$ is a connected component of $X_\theta \setminus V$. The other component is $f^{-1}(e^{-i\theta})$.

Property 6.3.6 (*Fibration on the sphere*) We now focus our attention near the origin, say restricted to the unit ball $\mathbb{B}^{2n+2}$ in $\mathbb{C}^{n+1}$. Since each X_θ meets transversely the sphere $\mathbb{S}^{2n+1} = \partial\mathbb{B}^{2n+2}$, the intersection is a smooth codimension 1 submanifold of the sphere, containing the link $L_f = V \cap \mathbb{S}^{2n+1}$. And since the orbits of the $\mathbb{S}^1$-action preserve the sphere $\mathbb{S}^{2n+1}$, the restriction of φ to $\mathbb{S}^{2n+1}$ defines the classical *Milnor fibration*:

$$\varphi = \frac{f}{|f|}\colon \mathbb{S}^{2n+1} \setminus L_f \longrightarrow \mathbb{S}^1\,. \tag{6.12}$$

Property 6.3.7 (*Equivalence of fibrations*) Each $\mathbb{R}^+$-orbit is everywhere transverse to the tube $N(\delta)$, by Property 6.3.2 it is transverse to the sphere $\mathbb{S}^{2n+1}$, and by Property 6.3.3, the complex numbers $f(z)$ have constant argument along each orbit. Thence the integral lines of this action determine a diffeomorphism between $N(1,d)$ and $\mathbb{S}^{2n+1}$ minus the part of the sphere contained inside the open solid tube $f^{-1}(\mathring{\mathbb{D}}_\delta)$. This determines the classical equivalence between the Milnor fibration in the sphere (6.12) and the Milnor-Lê fibration in the tube (6.10).

Property 6.3.8 (*Join*) Consider the fibre $F = f^{-1}(1)$. Pham [135] gave an explicit construction of a vanishing polyhedron $\mathcal{P}$ which is a deformation retract of F. The polyhedron $\mathcal{P}$ is homeomorphic to the join $G_{a_0} * \cdots * G_{a_n}$ where G_{a_j} is a set with a_j points. Hence, F has the homotopy type of a wedge $\bigvee S^n$ of n-spheres, where the number of spheres is $(a_0 - 1)(a_2 - 1)\dots(a_n - 1)$. This number is now called the Milnor number of the singularity.

6.3.1 Weighted Homogeneous Singularities

We now remark that everything we said above works in exactly the same way when V is defined by the following larger class of polynomials.

Definition 6.3.9 A complex polynomial $f\colon (\mathbb{C}^{n+1}, \mathbf{0}) \to (\mathbb{C}, 0)$ is *weighted homogeneous* of type $\{d; d_0, \dots, d_n\}$, where d and d_i are positive integers, if there is a $\mathbb{C}^*$ action on $\mathbb{C}^n$ of the form

$$\lambda \cdot (z_0, \dots, z_n) = (\lambda^{d_0} z_0, \dots, \lambda^{d_n} z_n)\,,$$

satisfying that for all $\lambda \in \mathbb{C}^*$ and for all $z \in \mathbb{C}^n$ one has:

$$f(\lambda \cdot z) = \lambda^d \cdot f(z)\,.$$

These all have the same Properties 6.3.2 to 6.3.7, while Property 6.3.8 generalizes to the following theorem by Oka [116, Theorem 1]:

Theorem 6.3.10 (Join theorem) *Let f be a polynomial in $\mathbb{C}^n \times \mathbb{C}^m$ such that $f(\mathbf{z}, \mathbf{w}) = g(\mathbf{z}) + g(\mathbf{w})$ for each $(\mathbf{z}, \mathbf{w}) \in \mathbb{C}^n \times \mathbb{C}^m$, where g and h are weighted homogeneous in $\mathbb{C}^n$ and $\mathbb{C}^m$ respectively. Let $X = f^{-1}(1) \subset \mathbb{C}^n \times \mathbb{C}^m$, $Y = g^{-1}(1) \subset \mathbb{C}^n$ and $Z = h^{-1}(1) \subset \mathbb{C}^m$. Then there is a natural homotopy equivalence between X and the join $Y * Z$.*

If V is the zero-locus of an analytic map $f\colon (\mathbb{C}^{n+1}, \mathbf{0}) \to (\mathbb{C}, 0)$ with an isolated critical point at $\mathbf{0}$, it has the same Properties 6.3.2 to 6.3.8, but instead of having a $\mathbb{C}^*$-action, there are appropriate vector fields whose flows have the desired properties.

6.3.2 Real Analytic Singularities

As we will see in the sequel, all real analytic isolated singularities can be equipped with flows that satisfy properties analogous to Property 6.3.2, and Properties 6.3.4 to 6.3.6, but not always Property 6.3.3. In analogy with Property 6.3.7, this implies that we have a fibration as in (6.10) and it can be carried to a fibration on the sphere as in (6.12) but the projection map φ may not always be taken to be $f/|f|$ away from a neighborhood of the link. Having also Property 6.3.3 grants that φ can be taken as $f/|f|$ everywhere, and this is equivalent to the map-germ being d-regular, a concept that we discuss in Sect. 6.10.

6.4 Local Conical Structure of Analytic Sets

In this section we present the generalization of Property 6.3.2 in Sect. 6.3. Consider a reduced, equidimensional real analytic space V of dimension n, defined in an open ball $\mathbb{B}_r(\mathbf{0}) \subset \mathbb{R}^N$ around the origin. Assume V contains the origin $\mathbf{0}$ and $V^* := V - \{\mathbf{0}\}$ is a real analytic manifold of dimension $n > 0$. The following is proved in [110] and it can be deduced from [95].

Theorem 6.4.1 *(Milnor 1968) There exists $\varepsilon > 0$ sufficiently small, so that every sphere in $\mathbb{R}^N$ centered at $\mathbf{0}$ and with radius $\leq \varepsilon$ intersects V^* transversely. Moreover, there is a smooth 1-parameter family of diffeomorphisms $\{\gamma_t\}$, $t \in [0, \varepsilon)$, such that γ_0 is the identity and if $\mathbb{S}_{\varepsilon-t}$ denotes the sphere of radius $\varepsilon - t$, then each γ_t carries the pair $(\mathbb{S}_\varepsilon, \mathbb{S}_\varepsilon \cap V)$ into $(\mathbb{S}_{\varepsilon-t}, \mathbb{S}_{\varepsilon-t} \cap V)$. The pair $(\mathbb{B}_\varepsilon, \mathbb{B}_\varepsilon \cap V)$ is homeomorphic to the cone over $(\mathbb{S}_\varepsilon, \mathbb{S}_\varepsilon \cap V)$.*

The idea of the proof is simple: consider the function $d\colon \mathbb{R}^N \to \mathbb{R}$ given by $d(x_1, ..., x_N) = x_1^2 + \cdots x_N^2$, so that d is the square of the function "distance to $\mathbf{0}$". The solutions of its gradient vector field ∇d are the straight rays that emanate from the origin. Let us adapt this vector field to V. For this, take the restriction d_V of d to V. At each point $x \in V^*$ the gradient vector $\nabla d_V(x)$ is obtained by projecting $\nabla d(x)$ to $T_x V^*$, the tangent space of V^* at x, so $\nabla d_V(x)$ vanishes if and only if $T_x V^* \subset T\mathbb{S}_x$. This means that a point $x \in V^*$ is a critical point of d_V if and only if V^* is tangent at x to the sphere passing through x and centered at $\mathbf{0}$. Just as in [110, Corollary 2.8], one has that d_V has at most a finite number of critical values corresponding to points in V^*, since it is the restriction of an analytic function on $\mathbb{B}_r(\mathbf{0})$. Hence V^* meets transversely all sufficiently small spheres around the origin in $\mathbb{R}^N$. The gradient vector field of d_V is now everywhere transversal to the spheres around $\mathbf{0}$, and it can be assumed to be integrable. Hence it defines a 1-parameter family of local diffeomorphisms of V^* taking each link into "smaller" links, proving Theorem 6.4.1.

Theorem 6.4.1 was extended in [20] to varieties with arbitrary singular locus using Whitney stratifications, we refer to [33, 59] for background material on stratifications and to chapter [160] by Trotman in Volume I of this Handbook for a survey in stratification theory. A more refined argument due to Durfee [43] (see also [91]) and based on the "Curve Selection Lemma" of [110], shows that in fact the diffeomorphism type of the manifold $V \cap \mathbb{S}_\varepsilon$ is also independent of the choice of the embedding of V in $\mathbb{R}^N$. One has.

Theorem 6.4.2 *Let V be a real or complex analytic set in $\mathbb{R}^m$ and P a singular point in V. Then there exists a Whitney stratification of $\mathbb{R}^m$ for which V is a union of strata, P is a point stratum, and one has the following (Fig. 6.2):*

1. *There exists $\varepsilon > 0$ sufficiently small, so that every sphere $\mathbb{S}_r$ in $\mathbb{R}^m$ of radius $r \leq \varepsilon$ and center at P meets transversely every stratum in V.*
2. *One has a homeomorphism of pairs: $(\mathbb{B}_\varepsilon, \mathbb{B}_\varepsilon \cap V) \cong \mathrm{Cone}\,(\mathbb{S}_\varepsilon, \mathbb{S}_\varepsilon \cap V)$.*
3. *The homeomorphism type of $L_V := \mathbb{S}_\varepsilon \cap V$ is independent of the choice of the defining equations for V.*

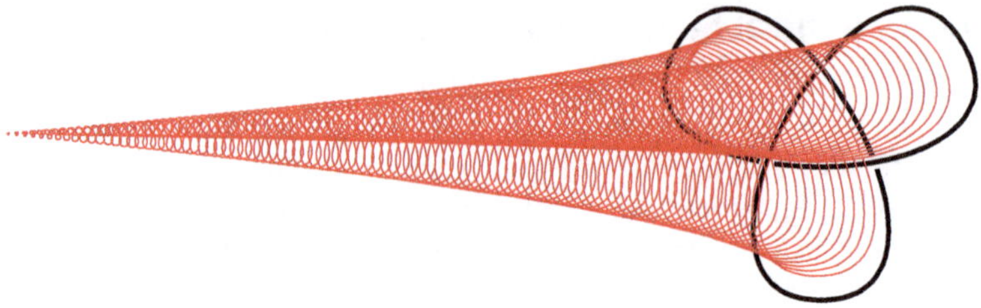

Fig. 6.2 The link of the singularity determines the topological type

Definition 6.4.3 The manifold $L_V := V \cap \mathbb{S}_\varepsilon$ is called the *link* of V at 0, and a sphere $\mathbb{S}_\varepsilon$ as in Theorem 6.4.2 is called a *Milnor sphere* for V and ε is called a *Milnor radius* for V. We also denote L_V by L_f when we want to emphasize the function rather than the space V.

6.5 The Classical Fibration Theorems for Complex Singularities

Properties 6.3.4 and 6.3.6 in Sect. 6.3 generalize respectively to the two versions of the classical Milnor's fibration theorem for complex singularities.

Theorem 6.5.1 (Fibration Theorem on the sphere) *Let U be an open neighborhood of the origin $\mathbf{0} \in \mathbb{C}^{n+1}$ and $f : (U, \mathbf{0}) \to (\mathbb{C}, 0)$ a complex analytic map. Set $V := f^{-1}(0)$ and $L_V := V \cap \mathbb{S}_\varepsilon$ where $\mathbb{S}_\varepsilon$ is a sufficiently small sphere in U centered at $\mathbf{0}$. Then,*

$$\varphi := \frac{f}{|f|} : \mathbb{S}_\varepsilon \setminus L_V \longrightarrow \mathbb{S}^1, \tag{6.13}$$

is a C^∞ fiber bundle.

The proof by Milnor [110] uses the Curve Selection Lemma, first to show that the map φ has no critical points and then to construct a complete vector field w on $\mathbb{S}_\varepsilon \setminus L_V$ that projects under φ onto the unit vector field tangent to $\mathbb{S}^1$. Its flow defines a 1-parameter subgroup of diffeomorphisms

$$h_t : \mathbb{S}_\varepsilon \setminus L_V \to \mathbb{S}_\varepsilon \setminus L_V \tag{6.14}$$

such that

$$h_t(F_\theta) = F_{\theta+t}. \tag{6.15}$$

This shows the local triviality of φ, i.e., that each fiber of φ has a neighborhood which is a product (Fig. 6.3).

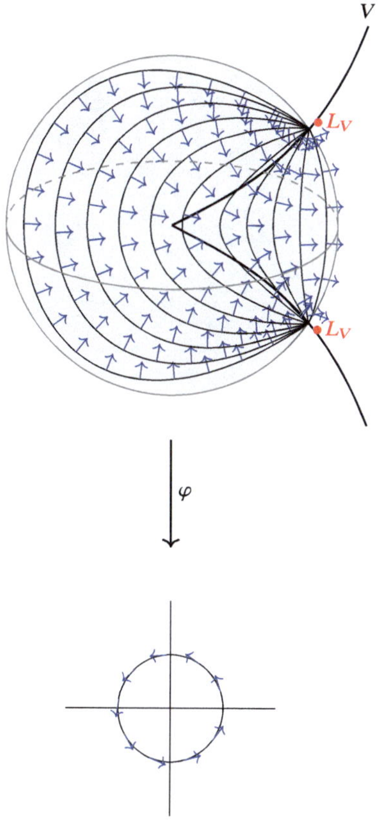

Fig. 6.3 Fibres of φ and trivializing vector field w on $\mathbb{S}_\varepsilon \setminus L_V$

Nowadays, the most common proof of Theorem 6.5.1 follows the original approach sketched by Milnor himself in a previous unpublished article [109]. The starting point is the following:

Theorem 6.5.2 (Fibration Theorem on the tube) *With the above hypothesis and notation, let $\delta > 0$ be sufficiently small with respect to ε, so that for every $t \in \mathbb{C}$ with $|t| \leq \delta$ the fiber $f^{-1}(t)$ meets the sphere $\mathbb{S}_\varepsilon$ transversely. Let $\mathbb{D}_\delta$ be the disc in $\mathbb{C}$ of radius δ and center at 0; let $\partial\mathbb{D}_\delta \cong \mathbb{S}^1$ be its boundary and set $N(\varepsilon, \delta) := f^{-1}(\partial\mathbb{D}_\delta) \cap \mathbb{B}_\varepsilon$, where $\mathbb{B}_\varepsilon$ is the open ball in $\mathbb{C}^{n+1}$ bounded by $\mathbb{S}_\varepsilon$. Then,*

$$f_{|N(\varepsilon,\delta)} : N(\varepsilon, \delta) \longrightarrow \partial\mathbb{D}_\delta, \tag{6.16}$$

is a C^∞ fiber bundle, (essentially) isomorphic to that in Theorem 6.5.1.

Definition 6.5.3 The manifold with boundary $N(\varepsilon, \delta)$ is called a *Milnor tube* (Fig. 6.4).

The word "essentially" in the last statement is because the fibers of (6.16) are compact, while those of (6.13) are open manifolds. To have an actual isomorphism of the two fibrations one must restrict the fibration (6.16) to the open ball.

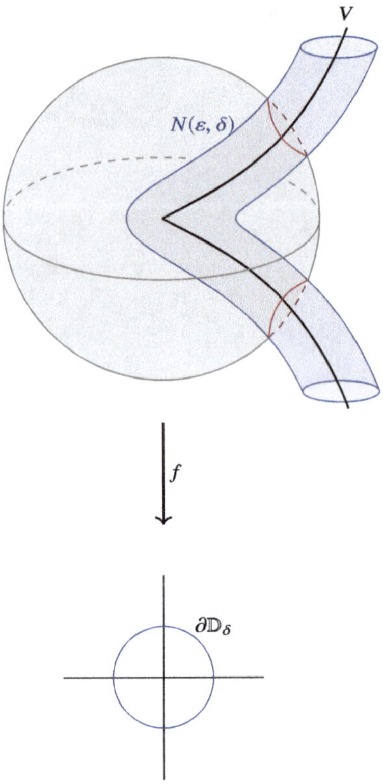

Fig. 6.4 Fibration on the Milnor tube $N(\varepsilon, \delta)$

Milnor proved this theorem in [109] when the map-germ f has an isolated critical point. In the general case, Milnor proved in [110] that the fibers of (6.16) (restricted to the open ball) are diffeomorphic to those of (6.13). In order to prove that one actually has a fiber bundle in Theorem 6.5.2 one must grant that given $\varepsilon > 0$ as above, there exists a δ as stated, such that all fiber $f^{-1}(t)$ with $|t| \leq \delta$ meet the sphere $\mathbb{S}_\varepsilon$ transversely. This was not known till 1977 when Hironaka proved in [68] that all complex valued holomorphic maps have a Thom Stratification. Actually in [66, Théorème 1.2.4], Hamm and Lê gave the existence of a *bonne stratification* which is equivalent to Thom's a_f condition. Using Hironaka's result, Lê Dũng Tráng [86] proved Theorem 6.5.2 in a more general context (see Theorem 6.7.1), so nowadays fibration (6.16) is called the *Milnor-Lê fibration*.

To prove Theorem 6.5.2 we restrict f to a sufficiently small open ball $\mathbb{B}_r$ around $\mathbf{0}$ so that $0 \in \mathbb{C}$ is its only critical value. We equip $\mathbb{B}_r$ with a Thom stratification [68], so that V is a union of strata and we assume that $\mathbf{0}$ itself is a stratum. Now let $\mathbb{S}_\varepsilon \subset \mathbb{B}_r$ be a Milnor sphere for f as defined in Definition 6.4.3, so that every sphere of radius $\leq r$ meets transversely each stratum in V; this is possible by Theorem 6.4.2. By compactness, this implies that there exists $\delta > 0$ such that for each $t \in \mathbb{C}$ with $0 < |t| \leq \delta$, the fiber $f^{-1}(t)$ meets $\mathbb{S}_\varepsilon$ transversely. Hence all fibers of (6.16) in

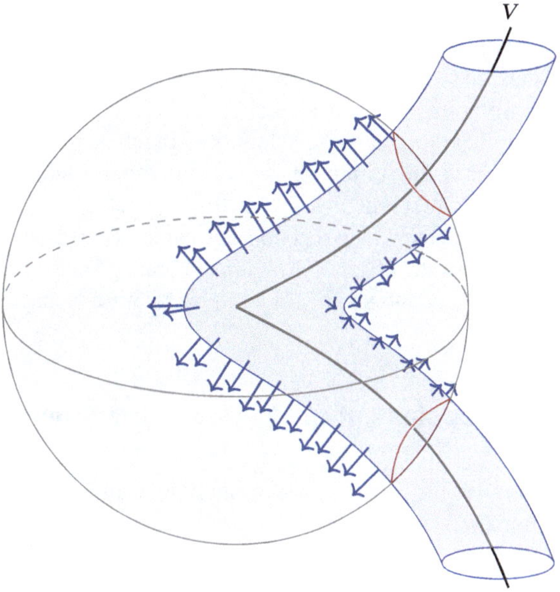

Fig. 6.5 The vector field ξ that carries a Milnor tube into the sphere

Theorem 6.5.2 are compact smooth manifolds with boundary. The proof of the local triviality is like in the usual proof of Ehresmann's fibration lemma, by lifting via the Jacobian of f appropriate vector fields in $\mathbb{C}$ to vector fields in $\mathbb{B}_\varepsilon$ which are normal to the fiber. The only additional thing is that we must choose the liftings so that the vector fields are also tangent to $\mathbb{S}_\varepsilon$, which is possible because the fiber is transverse to the sphere (see [150, 164]).

The next step to prove Theorem 6.5.1 is analogous to Property 6.3.7 in Sect. 6.3 and it is implicit in [110]: we inflate the Milnor tube, carrying the fibration in the tube into the fibration in the sphere as stated. This relies on the Curve Selection Lemma. The key for this is constructing a vector field as stated in the following lemma (Fig. 6.5).

Lemma 6.5.4 *There exists an integrable vector field ξ on $\mathbb{B}_\varepsilon \setminus V$ such that:*

1. *Its integral lines are transverse to all Milnor tubes $f^{-1}(\mathbb{S}^1_r)$;*
2. *Its integral lines are transverse to all spheres centered at* **0***;*
3. *Its integral lines travel along points where f has constant argument. That is, if z, w are points in $\mathbb{B}_\varepsilon \setminus V$ which are in the same integral line of ξ, then $f(z)/|f(z)| = f(w)/|f(w)|$.*

This allows one to inflate the tube to the sphere so that we get a homeomorphism

$$h\colon N(\varepsilon,\eta) \longrightarrow \overline{\mathbb{S}_\varepsilon \setminus N(\varepsilon,\eta)}\,,$$

in the obvious way: for each $z \in N(\varepsilon,\eta)$ we consider the unique integral line of ξ passing by z; we then travel along this integral line till it hits the sphere $\mathbb{S}_\varepsilon$. We thus get a fiber bundle $\varphi\colon \mathbb{S}_\varepsilon \setminus L_V \to \mathbb{S}^1$ with projection map $\varphi := f \circ h^{-1}$. The hard part is having one such vector field that further satisfies the third condition. This grants that the projection map in Theorem 6.5.1 can be taken as $\varphi = f/|f|$ and the two fibrations are equivalent.

Having these two equivalent fiber bundles associated to a map-germ brings great richness. The first fibration is interesting for topology and differential geometry. This has important relations with knot theory, open-book decompositions, contact and symplectic geometry. The second fibration lends itself more naturally to generalizations, and this has strong relations with algebraic geometry, as it exhibits the special fiber V as the limit of a flat family of complex manifolds that degenerate to V.

6.6 Topology of the Link and the Fiber

When V is the zero-locus of an analytic map $f\colon (\mathbb{C}^{n+1}, \mathbf{0}) \to (\mathbb{C}, 0)$ with an isolated singularity at the origin $\mathbf{0}$, its link is a closed, oriented, real analytic manifold of dimension $2n-1$ and it can be regarded as being the boundary of the Milnor fiber. Milnor [110, Sects. 5 and 6] and others have studied the topology of the link and the fibre of Milnor's fibration. Here we only say a few words about these important subjects.

6.6.1 The Link

When $n = 1$, the defining function f has an essentially unique decomposition into irreducible factors, $f = f_1^{a_1} \cdots f_r^{a_r}$. This determines a decomposition of its zero-locus V into irreducible components, the branches of V, so the link L_V is a knot or link in $\mathbb{S}^3$ with r components, one for each branch. We refer to [17] for a clear account on this subject.

For instance, if f is the homogeneous polynomial $z_1^k + z_2^k$, then L_V has k components and each is a torus knot of type $(1, 1)$, which means that it is a circle embedded in a torus $\mathbb{S}^1 \times \mathbb{S}^1$ in such a way that it goes once around a parallel and once around a meridian. If f is now the weighted homogeneous polynomial $z_1^2 + z_2^3$, then L_V has only one component, which is the trefoil knot depicted in Fig. 6.2. More generally, if f is $z_1^p + z_2^q$ for some $p, q \geq 2$, and k is the largest common divisor of p, q, then L_V has k components, and Brauner [14] proved that each is a torus knot of type (p', q') where $p' = p/k$ and $q' = q/k$.

In general, every knot defined by an analytic map-germ $\mathbb{C}^2 \to \mathbb{C}$ is an iterated torus knot determined by the Puiseux pairs of a Puiseux parameterization of its zero-locus (see [17] for details). Chapter [56] by García-Barroso et al. in Volume I of this Handbook describes how to encode the combinatorial information of a plane curve which determines its embedded topological type in a two-dimensional simplicial complex.

The case $n = 2$ goes back to Felix Klein [83] where he studies the complex surfaces obtained as quotients of $\mathbb{C}^2$ by a finite subgroup. His work implies that every manifold of the form $\mathbb{S}^3/\Gamma$ where Γ is a finite subgroup of SU(2) is the link of a complex surface in $\mathbb{C}^3$. The corresponding singularities are the rational double points; they have been studied by many authors, see for instance [42] or [150, Chap. 3]. For instance the link of the function:

$$f(z_1, z_2, z_3) = z_1^2 + z_2^3 + z_3^5 ,$$

is the celebrated Poincaré's homology 3-sphere, which is diffeomorphic to SO(3) divided by the group of rotations in $\mathbb{R}^3$ that preserve an icosahedron, with SO(3) being the group of rotations of the Euclidean 3-space and it is isomorphic to SU(2) divided by its center.

In general, given a holomorphic map-germ $f : (\mathbb{C}^3, \mathbf{0}) \to (\mathbb{C}, 0)$, the link L_f is always a connected Waldhausen 3-manifold with negative definite intersection form (see [115] for details). If f further is weighted homogeneous, then its zero-locus V is an invariant set of the restriction of the corresponding $\mathbb{C}^*$-action to $\mathbb{S}^1$, and as in Property 6.3.4 in Sect. 6.3. This $\mathbb{S}^1$-action also leaves invariant every sphere centered at the origin, therefore the link L_V actually is a Seifert 3-manifold. These manifolds were studied by Seifert [153] and are determined by a set of numerical invariants called the *Seifert invariants* (see also [133]). In [134] Orlik and Wegreich proved that if V is a surface in $\mathbb{C}^3$ given by a weighted homogeneous polynomial $f(z_1, z_2, z_3)$, there is an equivariant analytic deformation of V into a surface defined by one of five classes of polynomials, which induces an equivariant diffeomorphism of the corresponding links, so it is enough to compute the Seifert invariants of the links of these five classes of polynomials. The polynomials $z_1^p + z_2^q + z_3^r$, as in the example above, are one of those five types of weighted homogeneous singularities in $\mathbb{C}^3$. Chapter [107] by Michel in Volume I of this Handbook describes the topology of a surface singularity.

The following remains being an open question. This was studied by St. Yau, A. Durfee and others in the 1970s and 1980s with interesting partial results, but to our knowledge, there have not been significant improvements in several decades.

Problem 6.6.1 Characterize up to orientation preserving homeomorphism, the 3-manifolds that arise as links of isolated complex surface singularities in $\mathbb{C}^3$.

If in the problem above we erase the condition of being defined in $\mathbb{C}^3$ then the answer is known; this follows essentially by deep results by Grauert, Mumford (and independently Du Val) and Neumann (see [115]). The problem is determining which closed oriented Waldhausen manifolds with negative definite intersection form arise as the link of a hypersurface in $\mathbb{C}^3$.

For a general holomorphic map-germ $f : (\mathbb{C}^{n+1}, \mathbf{0}) \to (\mathbb{C}, 0)$ we know from [110, Theorem 5.2] that the link L_V is an orientable $(2n-2)$-connected smooth manifold of dimension $2n-1$. There are cases where the link can be described explicitly as for instance in Theorem 6.2.3 in Sect. 6.2 where we saw that every exotic sphere of dimension greater than 6 that bounds a parallelizable manifold, is the link of some Brieskorn-Pham polynomial.

Another interesting aspect is to study how the link L_V is embedded in the sphere. Such a pair $(\mathbb{S}_\varepsilon, L_V)$ is called an algebraic knot, a terminology introduce by Lê [86]. We briefly discussed above the case $n = 1$. For $n > 1$ very little is known. Yet, Milnor's fibration theorem says that these are all fibred knots (or perhaps links if $n = 1$), and if f has an isolated critical point, then one gets interesting open-book decompositions, a topic briefly discussed below. We refer to [139] for a thorough discussion of this interesting subject (see also [151, Section 11.2]).

6.6.2 The Fiber

Concerning the topology of the fiber, Milnor in [110, Theorems 5.1, 6.5] proved:

Theorem 6.6.2 *If V is the zero-locus of an analytic map $f : (\mathbb{C}^{n+1}, \mathbf{0}) \to (\mathbb{C}, 0)$ with an isolated singularity at the origin, then the fiber F of the Milnor fibration of V is a smooth parallelizable manifold of dimension $2n$ which has the homotopy type of a bouquet $\bigvee \mathbb{S}^n$ of spheres of dimension n. The number μ of spheres in the bouquet is strictly positive, unless V is smooth.*

Definition 6.6.3 The number $\mu = \mu(f)$ is called the *Milnor number* of f.

The number $\mu(f)$ can be computed topologically as the *Poincaré-Hopf index* of its gradient vector field $\nabla f(z) = \big(\frac{\partial f}{\partial z_1}(z), \dots, \frac{\partial f}{\partial z_{n+1}}(z)\big)$ [110, §7], and algebraically as the complex dimension of the *Jacobian algebra*

$$\mu(f) = \dim_{\mathbb{C}} \mathbb{C}\{z_1, \dots, z_{n+1}\} \Big/ \Big\langle \frac{\partial f}{\partial z_1}, \dots, \frac{\partial f}{\partial z_{n+1}} \Big\rangle .$$

This generalizes Property 6.3.8 in Sect. 6.3, and Pham's result says that the Milnor number of a Brieskorn-Pham polynomial is $(a_1 - 1)(a_2 - 1) \dots (a_n - 1)$. More generally, for f a weighted homogeneous polynomial of type $\{d; d_1, \dots, d_n\}$ with an isolated critical point at the origin, Milnor and Orlik [113] computed its Milnor number to be $\mu = (d/d_1 - 1) \cdots (d/d_n - 1)$.

Property 6.3.8 in Sect. 6.3 also generalizes to join theorems analogous to Theorem 6.3.10 by Oka. Let $g : (\mathbb{C}^n, 0) \to (\mathbb{C}, 0)$ and $h : (\mathbb{C}^m, 0) \to (\mathbb{C}, 0)$ be holomorphic maps with 0 an isolated singularity. Let Y and Z be the Milnor fibres of g and h respectively. Let $f(\mathbf{z}, \mathbf{w}) = g(\mathbf{z}) + h(\mathbf{w})$ for each $(\mathbf{z}, \mathbf{w}) \in \mathbb{C}^n \times \mathbb{C}^m$ and let X be the Milnor fibre of f. Sebastiani and Thom [152] proved that X is homotopically equivalent to the join $Y * Z$ of Y and Z. Sakamoto in [146] proved the general result when g and h are holomorphic functions with arbitrary singularities.

Theorem 6.6.2 was generalized by Hamm [65] to complex isolated complete intersection singularity germs (ICIS) (see Sect. 6.7). The Milnor fiber F is the piece of a non-critical level of the defining function that is contained within a small ball around the singularity. Hamm's theorem states that as in the hypersurface case, F has the homotopy type of a bouquet of spheres of middle dimension. The number of such spheres is called the Milnor number of the ICIS germ. One also has in this setting an algebraic formula for the Milnor number, called the *Lê-Greuel formula.* This says:

Theorem 6.6.4 *If $f_1, \ldots, f_k$ and g are holomorphic map germs $(\mathbb{C}^{n+k}, \mathbf{0}) \to (\mathbb{C}, 0)$ such that $f = (f_1, \ldots, f_k)$ and (f, g) define isolated complete intersection germs, then their Milnor numbers are related by:*

$$\mu(f) + \mu(f, g) = \dim_{\mathbb{C}} \frac{\mathcal{O}_{n+k,\mathbf{0}}}{(f, Jac_{k+1}(f, g))},$$

where $Jac_{k+1}(f, g)$ denotes the ideal generated by the determinants of all $(k+1)$ minors of the corresponding Jacobian matrix.

Since all the homology groups of the Milnor fiber of an ICIS vanish except in dimension 0 and in middle dimension, the above formula actually measures the change in the Euler characteristic of the Milnor fibers corresponding to f and to (f, g). This is the way how this formula generalizes to other settings (see Theorem 6.9.11 below).

In the case of hypersurfaces, Theorem 6.6.2 was actually refined by Milnor himself [110] in all dimensions except $n = 2$, and then completed by Lê-Perron [90] for $n = 2$ with a proof that works in all dimensions:

Theorem 6.6.5 *The Milnor fiber F of a hypersurface $V \subset \mathbb{C}^{n+1}$ with an isolated singularity is diffeomorphic to a $2n$-ball to which one attaches μ handles of middle index n, where μ is the Milnor number.*

For hypersurfaces of complex dimension n with non-isolated singularities, one has from [110] that the Milnor fiber F is a CW-complex of dimension n. This is also an immediate consequence of Andreotti-Frankel's theorem in [6] for Stein-manifolds. Massey's remarkable theorem in [101, 101] extends Theorem 6.6.5 to this setting.

Theorem 6.6.6 *Let $f: (\mathbb{C}^{n+1}, \mathbf{0}) \to (\mathbb{C}, 0)$ be a holomorphic map-germ and let F be its Milnor fiber. Assume the defining function f has a non-isolated critical point and let s be the complex dimension of its critical set. Then the germ of f has associated (generic) Lê numbers λ_i, $i = 0, \ldots, s$, and one has:*

- *If $s \leq n-2$, then F is obtained up to diffeomorphism from a $2n$-ball by successively attaching $\lambda^{n-k}_{f,\ell}(\mathbf{0})$ k-handles, where $n - s \leq k \leq n$ and $\lambda^{n-k}_{f,\ell}(\mathbf{0})$ is the $(n-k)^{th}$ Lê number.*
- *If the complex dimension of its critical set is $s = n-1$, then F is obtained up to diffeomorphism, from a real $2n$-manifold with the homotopy type of a bouquet of $\lambda^{n-1}_{f,\ell}(\mathbf{0})$ circles, by successively attaching $\lambda^{n-k}_{f,\ell}(\mathbf{0})$ k-handles, where $2 \leq k \leq n$.*

We refer to chapter [89] by Lê et al. in Volume I of this Handbook and chapter [102] by Massey in this volume for an account on this subject.

6.6.3 Vanishing Cycles, Open-Books and the Monodromy

Given a holomorphic map-germ $f\colon (\mathbb{C}^{n+1}, \mathbf{0}) \to (\mathbb{C}, 0)$, the Fibration Theorem 6.5.2 tells us that the Milnor fibers can be regarded as a 1-parameter flat family $\{F_t\}$ of complex manifolds that degenerate to the special fiber $F_0 = V := f^{-1}(0)$. This is the prototype of what is called a smoothing of the singularity, which means a flat deformation where all other fibers are smooth; we refer to Greuel's article [61] in Volume I of this Handbook for an account on that subject. Since the germ of V at $\mathbf{0}$ is a topological cone, by Theorem 6.4.2, this means that all the homology groups of F vanish in the limit. In particular, if the critical point of f at $\mathbf{0}$ is isolated, then Theorem 6.6.2 says that the only interesting homology group of the Milnor fiber F is in dimension n and it is generated by $\mu(f)$ cycles of dimension n, which are called *the vanishing cycles*. This group,

$$\mathcal{L}(f) := H_n(F, \mathbb{Z}) \cong \mathbb{Z}^{\mu(f)}\,,$$

is naturally equipped with a $(-1)^n$-symmetric bilinear form $\langle\ ,\ \rangle$ coming from the intersection of cycles. The group $\mathcal{L}(f)$, together with this additional structure is called the *Milnor lattice* of the singularity. The literature about it is vast; we refer to Ebeling's article [50] in Volume I of this Handbook or to Dimca's book [33] for accounts of that subject.

This can be seen in a different way that brings us to the theory of open-books. Given f as above, the first version of the Milnor fibration theorem says that

$$\varphi := \frac{f}{|f|}\colon \mathbb{S}_\varepsilon \setminus L_V \longrightarrow \mathbb{S}^1$$

is a fiber bundle. Consider the diffeomorphism $h_{2\pi}\colon \mathbb{S}_\varepsilon \setminus L_V \to \mathbb{S}_\varepsilon \setminus L_V$ in (6.14) given by the vector field which shows the local triviality of the Milnor fibration (6.13). This is the first return map of the flow defined by a lifting of the vector field on $\mathbb{S}^1$ given by multiplying the radial vector field by the complex number i. By (6.15) the fibre $F := F_0 = F_{2\pi}$ is invariant under $h_{2\pi}$, so the *geometric monodromy on the fibre* $h\colon F \to F$ is defined by the restriction $h = h_{2\pi}|_F$. It depends on the choice of the horizontal vector field w but its isotopy class does not depend on this choice. Hence, it induces a well-defined isomorphism $h_{*,i}\colon H_i(F; A) \to H_i(F; A)$. The coefficients are usually $\mathbb{Z}$, $\mathbb{Q}$, $\mathbb{R}$ or $\mathbb{C}$ depending on the situation. The isomorphism $h_{*,i}$ is called the i-th *monodromy isomorphism.*

If f has an isolated critical point at $\mathbf{0}$, then we have that $\tilde{H}_n(F; \mathbb{Z}) \cong \mathbb{Z}^\mu$ is generated by the vanishing cycles. One way to study the Milnor lattice of f and the monodromy isomorphism is to consider *distinguished basis* of vanishing cycles, for this we refer to Ebeling's survey articles [49, 50] and the references in there.

On the other hand, Milnor's Fibration (6.13) together with the link L_V give an open-book structure to the sphere $\mathbb{S}_\varepsilon$. Open-books were introduced by Winkelnkemper [163] and these have become an important concept in topology (see for example

[140]). Open-books allow to describe an arbitrary closed manifold in terms of lower dimensional ones. The work of E. Giroux and others shows closed connections between open-books and contact-geometry (see [139]).

Definition 6.6.7 Let M be a smooth closed n-manifold and let N be a codimension 2 submanifold of M with trivial normal bundle. An *open-book decomposition of* M consists of N together with a map $\pi : M \setminus N \to \mathbb{S}^1$ such that

- π is a locally trivial fibration and
- there exists a tubular neighbourhood of N diffeomorphic to $N \times \mathbb{D}^2$ such that the restriction of π to $N \times (\mathbb{D}^2 \backslash \{0\})$ is the map $(x, y) \mapsto y/||y||$.

The map π is called an *open-book fibration* of M, N is called the *binding* and the fibres of π are called the *pages*

It follows that the pages are all diffeomorphic and each page can be compactified by attaching the binding N as its boundary, getting a compact manifold with boundary.

In the original definition of open-books, Winkelnkemper starts with a compact manifold $\bar{F}$ with non-empty boundary $\partial \bar{F}$, together with a diffeomorphism h of $\bar{F}$ which is the identity on $\partial \bar{F}$. Now form the mapping cylinder $\bar{F}_h$ of h, which is a manifold with boundary $(\partial \bar{F}) \times \mathbb{S}^1$. Identifying (x, t) with (x, t') for each $x \in \partial \bar{F}$ and $t, t' \in \mathbb{S}^1$, we obtain a closed, differentiable manifold M. The fibres $\bar{F} \times t$ are the pages of the open-book, and their common boundary $N = \partial \bar{F} \times t$ is the binding. Notice that in this definition the pages are already compact manifolds with boundary, and their interiors are the pages above.

In the case we envisage here, the manifold M is the sphere $\mathbb{S}^{2n-1}$, the binding is the link L_V, the pages are the Milnor fibers and the diffeomorphism h is the geometric monodromy. We refer to [139] for a discussion on the canonical contact structure carried by this open-book decomposition.

6.7 Extensions and Refinements of Milnor's Fibration Theorem

As mentioned above, Hamm [65] (see also [94]) proved an extension of Milnor's fibration theorem for ICIS, isolated complete intersection singularity germs,

$$f = (f_1, \ldots, f_k) \colon \mathbb{C}^{n+k} \to \mathbb{C}^k .$$

We know from [60, Lemma 1.10] (see also [94]) that given such an f, one can always find good representatives so that the first $k - 1$ equations define an ICIS $\mathcal{W}$ of one dimension more, and the last equation defines an isolated singularity hypersurface germ in $\mathcal{W}$. Therefore we may see this as a special case of the following theorem by Lê Dũng Tráng [86] (see [29] for the last statement in Theorem 6.7.1, concerning the isomorphism of the two fibrations).

Theorem 6.7.1 (Milnor-Lê fibration) *Let X be an analytic subset of an open neighborhood U of the origin $\mathbf{0}$ in $\mathbb{C}^n$. Given $f\colon (X,\mathbf{0}) \to (\mathbb{C},0)$ holomorphic with a critical point at $\mathbf{0} \in X$* (in the stratified sense [59]), *let $V := f^{-1}(0)$, $\mathbb{B}_\varepsilon$ a closed ball of sufficiently small radius ε around $\mathbf{0} \in \mathbb{C}^n$ and $\mathbb{S}_\varepsilon$ its boundary. Then:*

1. *If $L_X = X \cap \mathbb{S}_\varepsilon$ is the link of X and $L_V = f^{-1}(0) \cap \mathbb{S}_\varepsilon \subset L_X$ is the link of V, one has a fiber bundle:*

$$\varphi = \frac{f}{|f|}\colon L_X \setminus L_V \longrightarrow \mathbb{S}^1. \tag{6.17}$$

2. *Now choose $\varepsilon >> \delta > 0$ sufficiently small and consider the Milnor tube*

$$N(\varepsilon,\delta) = X \cap \mathbb{B}_\varepsilon \cap f^{-1}(\partial\mathbb{D}_\delta)\,,$$

where $\mathbb{D}_\delta \subset \mathbb{C}$ is the disc of radius δ around $0 \in \mathbb{C}$. Then

$$f\colon N(\varepsilon,\delta) \longrightarrow \partial\mathbb{D}_\delta\,, \tag{6.18}$$

is a fiber bundle, C^∞-isomorphic to the previous bundle.

Notice that the fibers in (6.17) are subsets of the link $L_X := X \cap \mathbb{S}_\varepsilon$ while the fibers in (6.18) are contained in the interior of $X \cap \mathbb{B}_\varepsilon$, in analogy with the classical Milnor fibrations (6.13) and (6.16). As in Property 6.3.5 of Sect. 6.3, these statements can be refined by giving a fibration on the whole ball $\mathbb{B}_\varepsilon$ minus the variety $V := f^{-1}(0)$ which has the two fibrations in Theorem 6.7.1 as subfibrations. For this we need.

Theorem 6.7.2 (The Canonical Pencil) *For each $\theta \in [0,\pi)$, let $\mathcal{L}_\theta$ be the line through 0 in $\mathbb{R}^2$ with an angle θ with respect to the positive orientation of x-axis. Set $V = f^{-1}(0)$ and $X_\theta = f^{-1}(\mathcal{L}_\theta)$. Then:*

(i) The X_θ are all homeomorphic real analytic hypersurfaces of X with singular set $\mathrm{Sing}(V) \cup (X_\theta \cap \mathrm{Sing}(X))$. Their union is the whole space X and they all meet at V, which splits each X_θ in two homeomorphic halves.

(ii) If $\{S_\alpha\}$ is a Whitney stratification of X adapted to V, then the intersection of the strata with each X_θ determines a Whitney stratification of X_θ, and for each stratum S_α and each X_θ, the intersection $S_\alpha \cap X_\theta$ meets transversely every sphere in $\mathbb{B}_\varepsilon$ centered at $\mathbf{0}$.

(iii) There is a uniform conical structure for all X_θ, i.e., there is a homeomorphism

$$h\colon (X \cap \mathbb{B}_\varepsilon, V \cap \mathbb{B}_\varepsilon) \to \big(Cone(L_X), Cone(L_f)\big),$$

which restricted to each X_θ gives a homeomorphism $(X_\theta \cap \mathbb{B}_\varepsilon) \cong Cone(X_\theta \cap \mathbb{S}_\varepsilon)$.

The next theorem implies that the fibrations over the circle in Milnor's theorem actually are liftings of fibrations over $\mathbb{RP}^1$:

Theorem 6.7.3 (Fibration Theorem) *One has a commutative diagram of fiber bundles*

$$\begin{array}{ccc} (X \cap \mathbb{B}_\varepsilon) \setminus V & \xrightarrow{\ \varphi\ } & \mathbb{S}^1 \\ & \searrow^{\Psi} & \downarrow{\scriptstyle \pi'} \\ & & \mathbb{RP}^1 \end{array}$$

where $\Psi(x) = (Re(f(x)) : Im(f(x)))$ with fiber $(X_\theta \cap \mathbb{B}_\varepsilon) \setminus V$, π is the natural two-fold covering and $\varphi(x) = \frac{f(x)}{|f(x)|}$. The restriction of φ to the link $L_X \setminus L_f$ is the usual Milnor fibration (6.17), while the restriction to the Milnor tube $f^{-1}(\partial\mathbb{D}_\eta) \cap \mathbb{B}_\varepsilon$ is the fibration (6.18) (up to multiplication by a constant), and these two fibrations are equivalent.

The proof of Theorem 6.7.3 follows the same line as in the case when X is non-singular. The key point is constructing an appropriate integrable vector field in the vein of Lemma 6.5.4 above. When the ambient space X is singular we must consider stratified vector fields and use either Mather's controlled vector fields [104], or Verdier's rugose vector fields [161], which are all continuous and integrable. The proof in [29] of Theorem 6.7.3 also shows:

Corollary 6.7.4 *Let $f : (X, \mathbf{0}) \to (\mathbb{C}, 0)$ be as above, a holomorphic map with a critical point at $\mathbf{0} \in X$, and consider its Milnor fibration*

$$\varphi = \frac{f}{|f|} : L_X \setminus L_f \longrightarrow \mathbb{S}^1 .$$

If the germ $(X, \mathbf{0})$ is irreducible, then every pair of fibers of φ over antipodal points of $\mathbb{S}^1$ are glued together along the link L_f producing the link of a real analytic hypersurface X_θ, which is homeomorphic to the link of $\{Re\ f = 0\}$. Moreover, if both X and f have an isolated singularity at $\mathbf{0}$, then this homeomorphism is in fact a diffeomorphism and the link of each X_θ is diffeomorphic to the double of the Milnor fiber of f regarded as a smooth manifold with boundary L_f.

In [87], Lê sketched a proof of the fact that the Milnor fiber X_t of fibration (6.18) over a point $t \in \partial\mathbb{D}_\delta$, has as a deformation retract a vanishing polyhedron $\mathcal{P}_t \subset X_t$ and described the degeneration of X_t onto the special fiber X_0. Recently, a complete and detailed proof of this result was given in [88]. It is a far-reaching generalization of Pham's result in [135] for Brieskorn-Pham polynomials (see Property 6.3.8).

Real Singularities

We now look at Milnor fibrations for real analytic singularities. This emerged too from Milnor's seminal work in [109, 110]. We also remark that much of the discussion below goes through for meromorphic functions (see [12, 13, 63, 64]) and for semi-algebraic and subanalytic maps (see [45, 46, 98]). We also refer to the [34] for other aspects the theory that complement what we explain in this presentation.

6.8 Milnor Fibration for Real Analytic Maps

The following was stated as Hypothesis 11.1 in [110].

Condition 6.8.1 A real analytic map-germ $f\colon (\mathbb{R}^n, \mathbf{0}) \to (\mathbb{R}^p, 0)$, $n \geq p > 0$ *satisfies the Milnor condition at* $\mathbf{0}$ if the derivative $Df(x)$ has rank p at every point $x \in U \setminus \mathbf{0}$, where U is some open neighborhood of $\mathbf{0} \in \mathbb{R}^n$.

The following extends the fibration theorem to the real setting:

Theorem 6.8.2 (Milnor) *Let f satisfy Milnor condition at* $\mathbf{0}$. *For every $\varepsilon > 0$ sufficiently small, let $\delta > 0$ be sufficiently small with respect to ε and consider the Milnor tube $N(\varepsilon, \delta) := f^{-1}(\partial\mathbb{D}_\delta) \cap \mathbb{B}_\varepsilon$, where $\mathbb{D}_\delta$ is the disc in $\mathbb{R}^p$ of radius δ and center at 0, $\partial\mathbb{D}_\delta$ is its boundary and $\mathbb{B}_\varepsilon$ is the closed ball in $\mathbb{R}^n$ of radius ε and center* $\mathbf{0}$. *Then*

$$f|_{N(\varepsilon,\delta)} : N(\varepsilon, \delta) \longrightarrow \partial\mathbb{D}_\delta \ , \tag{6.19}$$

is a fiber bundle. Moreover, the tube $N(\varepsilon, \delta)$ is diffeomorphic to $\mathbb{S}_\varepsilon^{n-1} \setminus \big(f^{-1}(\mathring{\mathbb{D}}_\delta) \cap \mathbb{S}_\varepsilon^{n-1}\big)$, where $\mathring{\mathbb{D}}_\delta$ is the interior, and (6.19) *determines an equivalent fiber bundle:*

$$\varphi\colon \mathbb{S}_\varepsilon^{n-1} \setminus L_f \to \mathbb{S}^{p-1}\,, \tag{6.20}$$

where $L_f = f^{-1}(0) \cap \mathbb{S}_\varepsilon^{n-1}$ is the link. The projection φ is $f/\|f\|$ in a tubular neighborhood of L_f.

The statement that $\varphi = f/\|f\|$ in a tubular neighborhood of the link L_f is implicit in Milnor's book and it was made explicit in [30, 138]. The proof that (6.19) is a fibre bundle is an easy extension of the proof of Ehresmann's Fibration Lemma. As in the complex case, one then constructs an integrable vector field v in the ball $\bar{\mathbb{B}}_\varepsilon$, which is transverse to all spheres in this ball centered at $\mathbf{0}$, and transverse to all Milnor tubes. The integral curves of v allow us to carry $N(\varepsilon, \delta)$ diffeomorphically into the complement of $f^{-1}(\mathbb{D}_\delta) \cap \mathbb{S}_\varepsilon^{n-1}$ in the sphere $\mathbb{S}_\varepsilon^{n-1}$, keeping its boundary fixed (see Fig. 6.6), and one extends the induced fibration to all of $\mathbb{S}_\varepsilon^{n-1} \setminus L_f$ using for instance that the normal bundle of the link is trivial.

Yet, we cannot in general inflate the tube in such a way that the projection φ is $f/\|f\|$ everywhere. In fact this theorem has two weaknesses:

(1) It is much too stringent: map-germs satisfying Condition 6.8.1 are highly non-generic (see[41]).

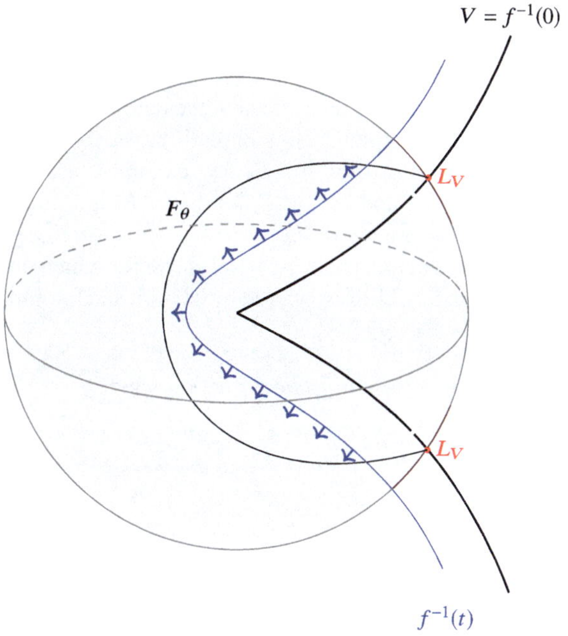

Fig. 6.6 Pushing the fiber from the interior of the ball toward the sphere

(2) One has no control over the projection map φ outside a neighborhood of the link.

Of course, every complex valued holomorphic function with an isolated critical point satisfies Condition 6.8.1, and so does if we compose such a function with a real analytic local diffeomorphism of either the target or the source. The interesting point is finding examples which are honestly real analytic. Milnor exhibited the following examples in his book [110], suggested to him by N. Kuiper. Let A denote either the complex numbers, the quaternions or the Cayley numbers, and define

$$h\colon A \times A \to A \times \mathbb{R}\,,$$

by $h(x, y) = (2x\bar{y}, |y|^2 - |x|^2)$. Milnor first proves (see [110, Lemma 11.6]) that this mapping carries the unit sphere of $A \times A$ to the unit sphere of $A \times \mathbb{R}$ by a *Hopf fibration*. Then he defines, more generally,

$$f\colon A^n \times A^n \to A \times \mathbb{R}\,,$$

by

$$f(x, y) = (2\langle x, y\rangle, \|y\|^2 - \|x\|^2)\,, \tag{6.21}$$

where $\langle\cdot,\cdot\rangle$ is the Hermitian inner product in A. This map is a local submersion on a punctured neighborhood of $(0, 0) \in A^n \times A^n$. The link of the corresponding singularity is the Stiefel manifold of 2-frames in A^n and the Milnor fibre is a disc bundle over the unit sphere of A^n.

For $p = 1$, Condition 6.8.1 is always satisfied (see for instance [161]). For maps into $\mathbb{R}^2$, generically the critical values are real curves converging to $(0, 0)$, though there are several families of singularities satisfying Condition 6.8.1, see for instance Sect. 6.8.1 below. For $p > 2$, few examples are known of map-germs satisfying Condition 6.8.1 and having a "non-trivial Milnor fibration", where non-trivial means that the fibers are not discs.

There are in fact pairs (n, p) as above for which no such examples exist, as stated in Theorem 6.8.3 below, proved by Church and Lamotke in [21], completing previous work by Looijenga [93].

Theorem 6.8.3 *Let n, p be positive integers.*

1. *If $2 \geq n - p \geq 0$, then such examples exist for the pairs $\{(2, 2), (4, 3), (4, 2)\}$.*
2. *If $n - p = 3$, non-trivial examples exist for $(5, 2)$ and $(8, 5)$, and perhaps for $(6, 3)$.*
3. *If $n - p \geq 4$, then such examples exist for all (n, p).*

In particular, if $p = 2$, such examples exist for all $n \geq 4$. The case $(6, 3)$ was left open and it was recently settled affirmatively in [37].

The proof in [21] follows the line in [93] and consists of an inductive process to decide for which pairs (n, p) there exists a codimension p submanifold K of the sphere $\mathbb{S}^{n-1}$ with a tubular neighborhood N which is a product $N \cong K \times D^p$, such that the natural projection $K \times (D^p \setminus \{0\}) \to \mathbb{S}^{p-1}$ given by $(x, y) \mapsto y/\|y\|$ extends to a smooth fiber bundle projection $\mathbb{S}_{n-1} \setminus K \to \mathbb{S}^{p-1}$. No explicit singularities satisfying the Milnor condition 6.8.1 were given.

The first explicit non-trivial example of a real analytic singularity with target $\mathbb{R}^2$ satisfying Condition 6.8.1, other than those in [110], was given by A'Campo [1]. This is the map $\mathbb{C}^{m+2} \to \mathbb{C}$ defined by

$$(u, v, z_1, \ldots, z_m) \longmapsto uv(\bar{u} + \bar{v}) + z_1^2 + \ldots + z_m^2\,. \qquad (6.22)$$

In the recent article [5] is proved that after attaching handles to the Milnor fibre, this becomes contractible; each handle corresponds to a critical point of an $\mathbb{R}$-morsification; in particular one recovers the formula of [82, Theorem 2.3] for the Euler characteristic of the Milnor fibres.

6.8.1 Strong Milnor Condition

The following notion was introduced in [144]:

Condition 6.8.4 Let $f\colon (\mathbb{R}^n, \mathbf{0}) \to (\mathbb{R}^p, 0)$, $n > p \geq 2$, be analytic and satisfy the Milnor condition at $\mathbf{0}$, and let L_f be its link. We say that f satisfies *the strong Milnor condition at* $\mathbf{0}$ if for every sufficiently small sphere $\mathbb{S}_\varepsilon$ around $\mathbf{0}$,

$$\frac{f}{|f|}\colon \mathbb{S}_\varepsilon \setminus L_f \to \mathbb{S}^{p-1} \tag{6.23}$$

is a fiber bundle.

Jacquemard [77] gives two conditions that insure that maps into $\mathbb{R}^2$ satisfying Condition 6.8.1 actually satisfy the strong Milnor condition. These conditions are sufficient but not necessary. The first of these is geometric, the second is algebraic. These are:

Condition A: There exists a neighborhood U of the origin in $\mathbb{R}^n$ and a real number $0 < \rho < 1$ such that for all $x \in U - 0$ one has:

$$\frac{|\langle \text{grad } f_1(x), \text{grad } f_2(x)\rangle|}{||\text{ grad } f_1(x)|| \cdot ||\text{ grad } f_2(x)||} \leq 1 - \rho\,,$$

where $\langle\cdot,\cdot\rangle$ is the usual inner product in $\mathbb{R}^n$.

Condition B: If ε_n denotes the local ring of analytic map-germs at the origin in $\mathbb{R}^n$, then the integral closures in ε_n of the ideals generated by the partial derivatives

$$\left(\frac{\partial f_1}{\partial x_1}, \frac{\partial f_1}{\partial x_2}, \ldots, \frac{\partial f_1}{\partial x_n}\right) \text{ and } \left(\frac{\partial f_2}{\partial x_1}, \frac{\partial f_2}{\partial x_2}, \ldots, \frac{\partial f_2}{\partial x_n}\right) \tag{6.24}$$

coincide, where f_1, f_2 are the components of f.

One has:

Theorem 6.8.5 (Jacquemard) *Let $f : (\mathbb{R}^n, \mathbf{0}) \to (\mathbb{R}^2, 0)$, $n > 2$, be an analytic map-germ. If the component functions f_1 and f_2 of f satisfy the previous two conditions A and B, then for every sphere $\mathbb{S}^{n-1}_\varepsilon$ of radius $\varepsilon > 0$ sufficiently small and centered at $\mathbf{0}$, one has that the projection map in* (6.20),

$$\varphi\colon \mathbb{S}^{n-1}_\varepsilon \setminus L_f \to \mathbb{S}^1\,,$$

can be taken to be $f/\|f\|$ everywhere.

In [148, 149] it is presented a method to construct an infinite family of singularities which satisfies the strong Milnor condition. This family was later studied in detail in [144].

Definition 6.8.6 A *twisted Brieskorn-Pham polynomial* of class $\{a_1, \ldots, a_n; \sigma\}$ is a polynomial $f\colon \mathbb{C}^n \to \mathbb{C}$ of the form

$$z_1^{a_1}\bar{z}_{\sigma(1)} + \cdots + z_n^{a_n}\bar{z}_{\sigma(n)},$$

where each $a_j \geq 2$, $j = 1, \ldots, n$ and σ is a permutation of the set $\{1, \ldots, n\}$ called the *twisting*.

In [144, 149] it was noticed that there exists a smooth action of $\mathbb{S}^1 \times \mathbb{R}^+$ on $\mathbb{C}^n$ of the form

$$(\lambda, r) \cdot (z_1, \dots, z_n) = (\lambda^{d_1} r^{p_1} z_1, \dots, \lambda^{d_n} r^{p_n} z_n), \quad \lambda \in \mathbb{S}^1, r \in \mathbb{R}^+, \tag{6.25}$$

where the d_j, p_j are positive integers such that $\gcd(d_1, \dots, d_n) = \gcd(p_1, \dots, p_n) = 1$, such that

$$f\big((\lambda, r) \cdot (z_1, \dots, z_n)\big) = \lambda^d r^p f(z_1, \dots, z_n), \tag{6.26}$$

for some positive integers d and p. So these polynomials are analogous to complex weighted homogeneous polynomials. In [144, 149] it was proved that twisted Brieskorn-Pham polynomials have an isolated critical point at $\mathbf{0}$ and the aforementioned actions imply that one has Properties 6.3.2 to 6.3.7 in Sect. 6.3, hence they satisfy the strong Milnor condition.

They are called twisted Brieskorn-Pham polynomials for their similarity with the classical Brieskorn-Pham polynomials (6.6) and the fact proved in [121, 144] (see also [28, 76]), that if the twisting σ is the identity, the corresponding open-book decompositions are equivalent to those of classical Brieskorn-Pham singularities. On the other hand, it was proved in [137] that the link of the twisted Brieskorn-Pham polynomial $z_1^{a_1}\bar{z}_2 + z_2^{a_2}\bar{z}_1$ is isotopic to the link of the complex singularity given by $z_1 z_2(z_1^{a_1} + z_2^{a_2})$, but their corresponding open-book decompositions are different. Similar statements hold in 3-variables: in [2–4] there are families of real analytic singularities with the strong Milnor condition, such that their open-books do not appear in complex singularities.

Problem 6.8.7 What is the equivalent of Theorem 6.8.3 for the strong Milnor condition? That is, for which pairs (n, p) there exists an analytic map-germ $f : (\mathbb{R}^n, \mathbf{0}) \to (\mathbb{R}^p, 0)$, $n > p \geq 2$, satisfying the strong Milnor condition at $\mathbf{0}$?

When $p = 2$, such maps exist for all $n \geq 4$. There are also several examples with $p = 3$ in [77].

It was noted in [144] that the above Condition B can be relaxed and still have sufficient conditions to guarantee the strong Milnor condition. For this we recall the notion of the real integral closure of an ideal as given in [55]:

Definition 6.8.8 Let I be an ideal in the ring ε_m. The *real integral closure* of I, denoted by $\overline{I}_{\mathbb{R}}$, is the set of $h \in \varepsilon_m$ such that for all analytic $\varphi : (\mathbb{R}, 0) \to (\mathbb{R}^m, 0)$, we have $h \circ \varphi \in (\varphi^*(I))\varepsilon_1$.

Given $f : (\mathbb{R}^n, 0) \to (\mathbb{R}^2, 0)$ as above, let us set:

Condition $\mathrm{B}_{\mathbb{R}}$: The real integral closures of the Jacobian ideals in (6.24) coincide.

For complex analytic germs conditions B and $\mathrm{B}_{\mathbb{R}}$ are equivalent (see [55, 159]). As pointed out in [144], essentially the same proof of Jacquemard in [77] gives:

Theorem 6.8.9 *Let $f : (\mathbb{R}^n, 0) \to (\mathbb{R}^2, 0)$ be an analytic map-germ that satisfies the Milnor condition. If its components f_1, f_2 satisfy the condition A above and condition $\mathrm{B}_{\mathbb{R}}$, then f satisfies the strong Milnor condition.*

This improvement of Theorem 6.8.5 was used in [144] to prove a stability theorem for real singularities with the strong Milnor condition. This was also used in [40] to find a theorem in the vein of Theorm 6.8.9 but using "regularity conditions" instead of Jacquemard's conditions. This inspired [144, 149] and the use of the canonical pencil described in Sect. 6.10. We refer for instance to [36, 37, 46, 48, 118, 119, 121, 131, 137] for recent work on the topology of the Milnor fibers.

Massey [103] improved Theorem 6.8.9 using a different viewpoint, via a generalized Łojasiewicz inequality (see the next section). Massey's viewpoint applies in a larger setting, not requiring f to have an isolated critical point, and it relaxes significantly condition $B_{\mathbb{R}}$.

6.8.2 Model Singularities

In [67] the authors study some families of isolated singularities, sharpening and extending results by Arnol'd ([7, 9]) and Kuiper [84] and J. Seade and others [144, 149]. They consider first a class of weighted homogeneous polynomials which they call *model polynomials*. These were considered by Arnol'd who used them as a first but main part of his far-reaching classification of singularities of functions. They characterize the model polynomials that have an isolated singularity and extend some of Arnol'd's techniques and results related to monomial bases of the algebra of a singularity, to polynomials with any number of variables.

Given a function $\phi\colon \{1, ..., n\} \to \{1, ..., n\}$, integers $p_i \geq 2$, and complex numbers $\lambda_i \neq 0$, one has a *model polynomial*:

$$f(z) = \lambda_1 z_1^{p_1} z_{\phi(1)} + \lambda_2 z_2^{p_2} z_{\phi(2)} + \ldots + \lambda_n z_n^{p_n} z_{\phi(n)} .$$

and the *twisted model polynomial*:

$$f(z) = \lambda_1 z_1^{p_1} \bar{z}_{\phi(1)} + \lambda_2 z_2^{p_2} \bar{z}_{\phi(2)} + \ldots + \lambda_n z_n^{p_n} \bar{z}_{\phi(n)} .$$

Given such a polynomial, one can associate to the corresponding function ϕ a graph defined as follows: take one vertex for each integer from 1 to n and put an arrow from the vertex i to the vertex $\phi(i)$. For instance, to the model polinomials

$$z_1^{p_1} z_2 + z_2^{p_2} z_3 + \ldots + z_{n-1}^{p_{n-1}} z_n + z_n^{p_n} ,$$
$$z_1^{p_1} z_2 + z_2^{p_2} z_3 + \ldots + z_n^{p_n} z_1 ,$$

correspond, respectively, the graphs shown in Fig. 6.7, which are called the *n-bamboo* and *n-cycle*, and similarly for the twisted ones.

It is then proved that a model (and a twisted model) polynomial has an isolated critical point at $\underline{0}$ if, and only if, every component of its graph is a bamboo or a cycle. The twisted Pham-Brieskorn polynomials in Definition 6.8.6 are a special case of

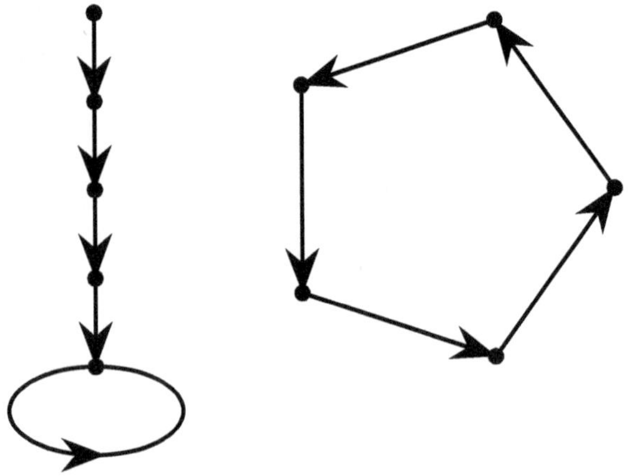

Fig. 6.7 Graphs 5-bamboo and 5-cycle

the twisted models. As pointed out in [67], it would be interesting to characterize the twisted model polynomials which have an isolated critical value. Since they are all quasihomogeneous over $\mathbb{R}$, all of these will have a Milnor-Lê fibration.

6.9 On Functions with a Non-isolated Critical Point

As noted before, considering map-germs $(\mathbb{R}^n, \mathbf{0}) \to (\mathbb{R}^p, 0)$ with an isolated critical point is very stringent. We now discuss the general case of arbitrary critical locus, starting with the slightly more general case of functions with an isolated critical value.

6.9.1 Functions with an Isolated Critical Value

Every holomorphic map-germ $(\mathbb{C}^m, \mathbf{0}) \to (\mathbb{C}, 0)$ with a critical point at $\mathbf{0}$ has an isolated critical value, and the fibration Theorems 6.5.1 and 6.5.2 hold in this setting. It is thus natural to look for extensions of Milnor's Theorem 6.8.2 for analytic map-germs $f := (f_1, \dots, f_p) : (\mathbb{R}^n, \mathbf{0}) \to (\mathbb{R}^p, 0)$ with a possibly non-isolated critical point at $\mathbf{0}$, such that $0 \in \mathbb{R}^p$ is the only critical value, i.e., all critical points of f are in the special fiber $V := f^{-1}(0)$. This was first done in [138].

Definition 6.9.1 If f admits a fibration in tubes of the type (6.19), then we call this the (local) *Milnor-Lê fibration* of f (or the Milnor fibration in tubes). If it admits a fibration on the spheres of the type (6.23), then we call this the (local) *Milnor fibration* of f (or the Milnor fibration on spheres).

Let $f, g\colon \mathbb{C}^n \to \mathbb{C}$ be holomorphic map-germs, in [138] the authors study conditions for functions of the form $f\bar{g}\colon \mathbb{C}^n \to \mathbb{C}$ with isolated critical value and meromorphic functions of the form $f/g\colon U \to \mathbb{CP}^1$, to have Milnor and Milnor-Lê fibrations.

Given a real analytic map-germ $f\colon (\mathbb{R}^n, \mathbf{0}) \to (\mathbb{R}^p, 0)$, $n > p \geq 1$, with an isolated critical value at $0 \in \mathbb{R}^p$, we want to know when is there a local Milnor-Lê fibration. That is, we want conditions to insure that given a ball $\mathbb{B}_\varepsilon$ bounded by a Milnor sphere $\mathbb{S}_\varepsilon$ for f (see Definition 6.4.3), there exists a ball $\mathbb{D}_\delta$ of some radius δ in $\mathbb{R}^p$, centered at 0, such that if we set $N(\varepsilon, \delta) := f^{-1}(\partial\mathbb{D}_\delta) \cap \mathbb{B}_\varepsilon$, then

$$f|_{N(\varepsilon,\delta)} : N(\varepsilon, \delta) \longrightarrow \partial\mathbb{D}_\delta$$

is a C^∞ fiber bundle.

We know from [110] that this is always satisfied when f has an isolated critical point (that is immediate from the implicit function theorem). Yet, when the critical point is not isolated the situation is more delicate. In [138] it was noticed that if the map-germ f is such that $V(f) = f^{-1}(0)$ has dimension > 0 and f has the Thom a_f-property, then f has a local Milnor-Lê fibration.

The study of Milnor fibrations for real analytic map-germs was also addressed by Massey [103]. Recall that in [66] Hamm and Lê used the complex analytic Łojasiewicz inequality to show that Thom stratifications exist. Massey gives the appropriate generalization for the real analytic setting:

Definition 6.9.2 An analytic germ $f\colon (\mathbb{R}^n, \mathbf{0}) \to (\mathbb{R}^2, 0)$ satisfies the *strong Łojasiewicz inequality* at $\mathbf{0}$ if there exists a neighborhood $\mathcal{W}$ of $\mathbf{0}$ and constants $c, \theta \in \mathbb{R}$ with $c > 0$, $0 < \theta < 1$, such that for all $x \in \mathcal{W}$ one has:

$$\|f(x)\|^\theta \leq c \min_{|(a,b)|=1} |a\nabla g(x) + b\nabla h(x)| \,.$$

In this case the germ f is said to be *Ł-analytic*.

The main theorem in [103] says.

Theorem 6.9.3 (Massey) *If f is Ł-analytic, then for every Milnor sphere $\mathbb{S}_\varepsilon$ there is a Milnor tube $N_f(\varepsilon, \delta)$ where f is a proper stratified submersion and the projection of a C^∞ fiber bundle. That is, Ł-analytic maps have Milnor-Lê fibrations.*

Now we need the following definition from [32].

Definition 6.9.4 Let $f\colon (\mathbb{R}^n, \mathbf{0}) \to (\mathbb{R}^p, 0)$, $n > p \geq 1$, have an isolated critical value at $0 \in \mathbb{R}^p$. We say that f has *the transversality property* if for every sufficiently small sphere $\mathbb{S}_\varepsilon$ around the origin in $\mathbb{R}^n$, there exists $\delta > 0$ such that all the fibers $f^{-1}(t)$ with $||t|| \leq \delta$ meet transversely the sphere $\mathbb{S}_\varepsilon$.

The transversality property appears already in [66], and in [131] this is called the Hamm-Lê condition. Maps with the Thom a_f-property and non-empty link have the transversality condition, but not conversely: there are examples by Oka [130] of maps with the transversality property which do not have the Thom a_f-property (see also [106]) and Ribeiro [141, 142] gives infinite families of maps with the transversality condition but without the Thom a_f-property.

The theorem below is Theorem 3.4 in [31]. This improves [138, Theorem 1.3].

Theorem 6.9.5 *Let $f\colon (\mathbb{R}^n, \mathbf{0}) \to (\mathbb{R}^p, 0)$, $n > p \geq 1$, have an isolated critical value at $0 \in \mathbb{R}^p$. Assume further that f has the transversality property and $V(f) := f^{-1}(0)$ has dimension greater than 0. Then f has a local Milnor-Lê fibration:*

$$f|_{N(\varepsilon,\delta)} : N(\varepsilon, \delta) \longrightarrow \partial\mathbb{D}_\delta\,, \tag{6.27}$$

with $N(\varepsilon, \delta) := f^{-1}(\partial\mathbb{D}_\delta) \cap \mathbb{B}_\varepsilon$ for some ball $\mathbb{D}_\delta \subset \mathbb{R}^p$, $0 < \delta \ll \varepsilon$. This determines an equivalent fiber bundle:

$$\varphi : \mathbb{S}_\varepsilon \setminus L_f \longrightarrow \mathbb{S}^{p-1}\,, \tag{6.28}$$

where the projection map φ is $f/\|f\|$ restricted to $[\mathbb{S}_\varepsilon \cap N(\varepsilon, \delta)]$.

The way to pass from the fibrations in tubes to that on the sphere is as before: one constructs a smooth vector field ζ in the ball $\mathbb{B}_\varepsilon$ minus V, satisfying:

- Each integral line is transversal to all spheres in $\mathbb{B}_\varepsilon$ centered at $\mathbf{0}$.
- Each integral line is transversal to all tubes $f^{-1}(\partial\mathbb{D}_\delta)$ contained in $\mathbb{B}_\varepsilon$.

The difference with the holomorphic setting is that we cannot guarantee now a third condition: that the vectors $f(z)$ are collinear for all points in each integral line (cf. Theorem 6.10.3 below). We discuss this in Sect. 6.10.

For maps of the type $f\bar{g}$ that we envisage above, there is a simple criterium in [54] and [106, Proposition 3.5] to decide whether or not the map has the transversality property. This is called CT-regularity in [106]. The advantage of this criterium is that it is easy to use in practice.

6.9.2 Polar Weighted Singularities

Polar weighted homogeneous maps are polynomial maps $f\colon \mathbb{R}^{2n} \to \mathbb{R}^2$ with $n > 1$ with isolated critical value which behave very much as complex homogeneous polynomials in the sense that there is a weighted action of $\mathbb{C}^* \equiv \mathbb{S}^1 \times \mathbb{R}^*$ on $\mathbb{R}^{2n}$ for which f brings out scalars. The difference with the complex case is that now scalars in $\mathbb{S}^1$ and those in $\mathbb{R}^*$ may act with different weights.

These were defined in [22] inspired by the definition and properties (6.25) and (6.26) of twisted Brieskorn-Pham polynomials given in [144, 149]. In particular these satisfy the transversality property, so they have a local Milnor-Lê fibration (6.27) and as in the complex case, they have a Milnor fibration (6.28).

We identify $\mathbb{R}^{2n}$ with $\mathbb{C}^n$ and $\mathbb{R}^2$ with $\mathbb{C}$ in the usual way.

Definition 6.9.6 A map $f\colon \mathbb{C}^n \to \mathbb{C}$ is *polar weighted homogeneous* if there exists a smooth action of $\mathbb{S}^1 \times \mathbb{R}^+$ on $\mathbb{C}^n$ of the form

$$(\lambda, r) \cdot (z_1, \dots, z_n) = (\lambda^{d_1} r^{p_1} z_1, \dots, \lambda^{d_n} r^{p_n} z_n), \quad \lambda \in \mathbb{S}^1, r \in \mathbb{R}^+,$$

where the d_j, p_j are positive integers such that $\gcd(d_1, \dots, d_n) = \gcd(p_1, \dots, p_n) = 1$, such that

$$f\big((\lambda, r) \cdot (z_1, \dots, z_n)\big) = \lambda^d r^p f(z_1, \dots, z_n),$$

for some integers d and p. The $\mathbb{S}^1$-action is called the *polar action*, while the $\mathbb{R}^+$-action is called the *radial action*.

Examples of this kind of maps include complex weighted homogeneous polynomials (where $d_i = p_i$) and twisted Brieskorn-Pham polynomials. In [22, 118] it is proved that this type of polynomials have isolated critical value and that the actions satisfy Properties 6.3.2 to 6.3.7. By Property 6.3.2 they satisfy the transversality property, and thus, they have a Milnor-Lê fibration (6.27); and by Properties 6.3.3, 6.3.5 and 6.3.7 they have a Milnor fibration given by $f/\|f\|$ which is equivalent to the Milnor-Lê fibration. With respect to Property 6.3.8, there is a Join Theorem for polar weighted homogeneous polynomials proved in [22], which generalizes Theorem 6.3.10 by Oka [116, Theorem 1] for complex weighted homogeneous polynomials.

In [27] a classification of polar weighted homogeneous polynomials with isolated critical point in three variables is given. It is proved that every polar weighted homogeneous polynomial with isolated critical point in three variables must contain certain polynomials which belong to one of five families, such families generalize the families given by Orlik and Wagreich mentioned in Sect. 6.6.1. It is also proved that the diffeomorphism type of the link does not change under small perturbation of the coefficients of the polynomial. The proof follows the one of Orlik and Wagreich's result [134]: given a family of polar weighted homogeneous polynomials with isolated critical point $f_{\mathbf{w}}\colon \mathbb{C}^3 \to \mathbb{C}$, where the parameter $\mathbf{w}$ is in $\mathbb{C}^r$ with $r > 3$, one constructs a manifold M with an $\mathbb{S}^1$-action, an open set $U \subset \mathbb{C}^r$ and a map $\phi\colon M \to U$, such that the action leaves $\phi^{-1}(\mathbf{w})$ invariant for all $\mathbf{w} \in U$, $\phi^{-1}(\mathbf{w}) \cong L_{\mathbf{w}}$ equivariantly, where $L_{\mathbf{w}}$ is the link of the polynomial $f_{\mathbf{w}}$, and ϕ is a locally trivial fibration. Unlike the case of complex weighted homogeneous polynomials, the set U can be disconnected and the topology of $L_{\mathbf{w}}$ can be different for $\mathbf{w}$ in different connected components. One example in two variables of this phenomenon was shown by Oka [119].

Problem 6.9.7 Determine how the topology of the link and the Milnor fibre change when the parameter $\mathbf{w}$ changes connected component of U.

Polar weighted homogeneous polynomials have been intensively studied by Oka [118–120, 122, 123] and other authors [11, 67, 72–76, 76].

6.9.3 Functions with Arbitrary Discriminant

We now consider the general setting and study Milnor fibrations for map-germs $(\mathbb{R}^n, \mathbf{0}) \to (\mathbb{R}^p, 0)$ with arbitrary critical points, following [23, 32]. We start with an example studied in [97] by López de Medrano.

Example 6.9.8 Consider maps $(f, g) : \mathbb{R}^n \to \mathbb{R}^2$ the form:

$$(f, g) = \left(\sum_{i=1}^{n} a_i x_i^2 \,,\, \sum_{i=1}^{n} b_i x_i^2 \right),$$

where the a_i and b_i are real constants in generic position in the Poincaré domain. This means that the origin is in the convex hull of the points $\lambda_i := (a_i, b_i)$ and no two λ_i are linearly dependent.

A simple calculation shows that (f, g) is a complete intersection and the corresponding link is a smooth non-empty manifold of real codimension 2 in the sphere. The critical points Σ of (f, g) are the coordinate axis of $\mathbb{R}^n$ and the set $\Delta(f, g)$ of critical values is the union of the n line-segments in $\mathbb{R}^2$ joining the origin to the points λ_i. Hence $\mathbb{R}^p$ splits into various connected components, and it is proved in [97] that the topology of the fibers over points in different components changes. Yet, we know from [32] that these map-germs have the transversality property given in Definition 6.9.4, and away from the critical set they have a Milnor-Lê fibration. In fact these maps are d-regular too, a concept that we discuss in Sect. 6.10 and implies that away from the discriminant, they have also a Milnor fibration on small spheres with projection map $(f, g)/\|(f, g)\|$.

More generally, consider now an open neighbourhood U of $\mathbf{0} \in \mathbb{R}^n$ and a C^ℓ map $f \colon (U, \mathbf{0}) \to (\mathbb{R}^p, 0)$, $n > p \geq 2$, $\ell \geq 1$, with a critical point at $\mathbf{0}$. Denote by $\mathcal{C}_f$ the set of critical points of f in $\mathbb{B}_\varepsilon$ and let Δ_f be the image $f(\mathcal{C}_f)$. These are the critical values of f; we call Δ_f the discriminant of f.

Definition 6.9.9 We say that the map-germ f has the *transversality property* at 0 if there exists a real number $\varepsilon_0 > 0$ such that, for every ε with $0 < \varepsilon \leq \varepsilon_0$, there exists a real number δ, with $0 < \delta \ll \varepsilon$, such that for every $t \in \mathbb{B}_\delta^k \setminus \Delta_f$ one has that either $f^{-1}(t)$ does not intersect the sphere $\mathbb{S}_\varepsilon^{n-1}$ or $f^{-1}(t)$ intersects $\mathbb{S}_\varepsilon^{n-1}$ transversely in $\mathbb{R}^n$.

The transversality condition of the fibers with small spheres ensures having a Milnor-Lê fibration, even for C^ℓ maps with non isolated critical values. Of course that as in Example 6.9.8, if the base of the fibration has several connected components (sectors), then the topology of the fibers can change from one sector to another. We have the following result from [32].

Proposition 6.9.10 *Let $f \colon (\mathbb{R}^n, \mathbf{0}) \to (\mathbb{R}^p, 0)$, $n \geq p \geq 2$, be a map-germ of class C^ℓ with $\ell \geq 1$. If f has the transversality property, then the restrictions:*

$$f|: \mathbb{B}^n_\varepsilon \cap f^{-1}(\mathring{\mathbb{D}}^p_\delta \setminus \Delta_f) \longrightarrow (\mathring{\mathbb{D}}^p_\delta \setminus \Delta_f) \cap Im(f),$$
$$f|: \mathbb{B}^n_\varepsilon \cap f^{-1}(\mathbb{S}^{p-1}_\delta \setminus \Delta_f) \longrightarrow (\mathbb{S}^{p-1}_\delta \setminus \Delta_f) \cap Im(f),$$

are C^ℓ fiber bundles, where ε and δ are small enough positive real numbers, $\mathbb{B}^n_\varepsilon \subset \mathbb{R}^n$ and $\mathbb{D}^p_\delta \subset \mathbb{R}^p$ are the closed balls of radius ε and δ centered at $\mathbf{0}$ and 0, respectively, $\mathring{\mathbb{D}}^p_\delta$ is the interior of the ball $\mathbb{D}^p_\delta$ and $\mathbb{S}^{p-1}_\delta$ is its boundary. If f is analytic, then the fibrations above are C^∞.

Ribeiro [141] calls this fibration Milnor-Hamm fibration and gives necessary and sufficient conditions for its existence as well as classes of maps which satisfy them. He also considers a "stratified transversality condition" which generalizes Definition 6.9.4.

In [23] the authors study the topology of the fibres of real analytic maps $\mathbb{R}^n \to \mathbb{R}^{p+k}$, $n > p + k$, inspired by the classical Lê-Greuel formula for the Milnor number of isolated, complex, complete intersection germs. The idea is that if the map germ is defined by analytic functions $(f_1, ..., f_p, g)$, then we can study the topology of its fibres by comparing it with the topology of the germ we get by dropping down g. We require for this that the map $f := (f_1, ..., f_p)$ actually satisfies the Thom a_f-property with respect to some Whitney stratification $\{S_\alpha\}$, and that its zero-set $V(f)$ has dimension ≥ 2 and it is union of strata. The map-germ $(\mathbb{R}^n, \mathbf{0}) \xrightarrow{g} (\mathbb{R}^k, 0)$ is assumed to have an isolated critical point in $\mathbb{R}^n$ with respect to the stratification $\{S_\alpha\}$. By Proposition 6.9.10 the map-germs f and (f, g) have associated local Milnor-Lê fibrations. Then one has the corresponding Lê-Greuel formula [23, Theorem 1]:

Theorem 6.9.11 *Let F_f and $F_{f,g}$ be Milnor fibres of f and (f, g) (any Milnor fibres, regardless of the fact that the topology of the fibers may depend on the connected component of the base once we remove the discriminant). Then one has:*

$$\chi(F_f) = \chi(F_{f,g}) + \mathrm{Ind}_{\mathrm{PH}} \nabla \tilde{g}|_{F_f \cap \mathbb{B}_{\varepsilon'}},$$

where $\tilde{g}: \mathbb{R}^n \to \mathbb{R}$ is given by $\tilde{g}(x) = \|g(x) - t_0\|^2$ with $t_0 \in \mathbb{R}^k$ such that $F_{f,g} = g|_{F_f}^{-1}(t_0)$ and $\mathbb{B}_{\varepsilon'}$ is a small ball in $\mathbb{R}^n$ centered at the origin.

The term $\mathrm{Ind}_{\mathrm{PH}} \nabla \tilde{g}|_{F_f \cap \mathbb{B}_{\varepsilon'}}$ on the right, which by definition is the total Poincaré-Hopf index in F_f of the vector field $\nabla \tilde{g}|_{F_f}$, can be expressed also in the following equivalent ways:

1. *As the Euler class of the tangent bundle of F_f relative to the vector field $\nabla \tilde{g}|_{F_f \cap \mathbb{B}_{\varepsilon'}}$ on its boundary;*
2. *As a sum of polar multiplicities relative to $\tilde{g}$ on $F_f \cap \mathbb{B}_{\varepsilon'}$.*
3. *As the index of the gradient vector field of a map $\tilde{g}$ on F_f associated to g;*
4. *As the number of critical points of $\tilde{g}$ on F_f;*
5. *When $p = 1 = k$, this invariant can also be expressed algebraically, as the signature of a certain bilinear form that originates from [8, 51, 52, 57, 58].*

When $n = 2m$, $p = 2q$, $k = 2$ and $(f, g): \mathbb{C}^m \to \mathbb{C}^{q+1}$ is holomorphic, this is a reformulation of the classical Lê-Greuel formula [60, 85], Theorem 6.6.4 above.

We remark that when $k = 1$ and the germs of f and (f, g) are both isolated complete intersection germs, there is a Lê-Greuel type formula in [47] expressed in terms of normal data of f with respect to an appropriate Whitney stratification. See also [105] for refinements of the above discussion.

6.10 Milnor Fibrations and d-Regularity

The concept of d-regularity introduced in [30] is inspired by [10, 40, 144, 149] and it is a key for understanding the difference between real and complex singularities concerning Milnor fibrations.

6.10.1 The Case of an Isolated Critical Value

Let U be an open neighborhood of the origin $\mathbf{0} \in \mathbb{R}^n$, and consider a real analytic germ $f\colon (U, \mathbf{0}) \to (\mathbb{R}^p, 0)$ which is a submersion for each $x \notin V := f^{-1}(0)$ and has a critical point at $\mathbf{0}$.

Definition 6.10.1 The *canonical pencil* of f is a family $\{X_\ell\}$ of real analytic spaces parameterized by $\mathbb{RP}^{p-1}$, defined as follows: for each $\ell \in \mathbb{RP}^{p-1}$, consider the line $\mathcal{L}_\ell \subset \mathbb{R}^p$ that determines ℓ, and set

$$X_\ell = \{x \in U \mid f(x) \in \mathcal{L}_\ell\}\,.$$

Note that every two distinct elements of the pencil X_ℓ and $X_{\ell'}$ satisfy

$$X_\ell \cap X_{\ell'} = V\,.$$

Each X_ℓ has dimension $n - p + 1$, is non-singular outside V and their union covers all of U.

Each line $\mathcal{L}$ intersects the sphere $\mathbb{S}^{p-1}$ in two antipodal points θ^- and θ^+. We decompose the line $\mathcal{L}$ into the corresponding half lines accordingly:

$$\mathcal{L} = \mathcal{L}^- \cup \{0\} \cup \mathcal{L}^+.$$

If we define $E_\ell^\mp$ to be the inverse image $f^{-1}(\mathcal{L}^\mp)$, respectively, then we can express each element of the canonical pencil as the following union (see Fig. 6.8):

$$X_\ell = E_\ell^- \cup V \cup E_\ell^+. \tag{6.29}$$

If L_V is the link of f, we can describe the fibers of the map $\varphi = f/\|f\|\colon \mathbb{S}_\varepsilon^{n-1} \setminus L_V \to \mathbb{S}^{p-1}$ as :

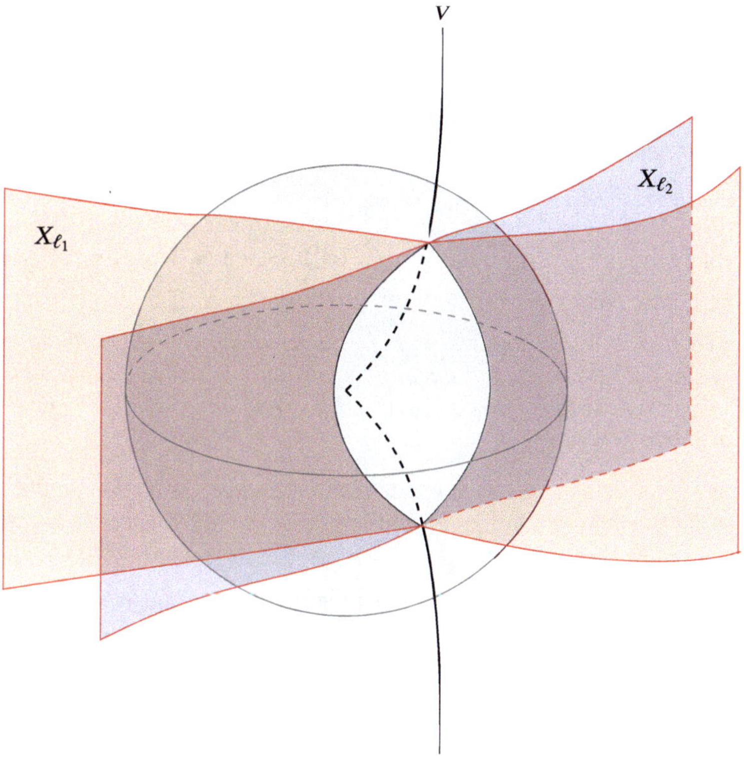

Fig. 6.8 Canonical pencil

$$\varphi^{-1}(\theta^-) = E_\ell^- \cap \mathbb{S}_\varepsilon^{n-1}, \qquad \varphi^{-1}(\theta^+) = E_\ell^+ \cap \mathbb{S}_\varepsilon^{n-1}.$$

Then we can write:

$$X_{\mathcal{L}} \cap \mathbb{S}_\varepsilon^{n-1} = (E_\ell^- \cap \mathbb{S}_\varepsilon^{n-1}) \cup L_V \cup (E_\ell^+ \cap \mathbb{S}_\varepsilon^{n-1}) = \varphi^{-1}(\theta^-) \cup L_V \cup \varphi^{-1}(\theta^+).$$

We now assume that $f : U \to \mathbb{R}^p$ is real analytic, with an isolated critical value at $\mathbf{0}$ and it is *locally surjective*, i.e. the restriction of f to every neighborhood of $\mathbf{0} \in U$ covers a neighborhood of $0 \in \mathbb{R}^p$.

Definition 6.10.2 We say that f is *d-regular* if there exists $\varepsilon_0 > 0$ such that for every $\varepsilon \leq \varepsilon_0$ and for every line $\mathcal{L}$ through the origin in $\mathbb{R}^p$, the sphere $\mathbb{S}_\varepsilon^{n-1}$ and the manifold $X_{\mathcal{L}} \setminus V$ are transverse.

As examples of d-regular maps one has:

- All holomorphic maps $\mathbb{C}^n \to \mathbb{C}$, all polar weighted homogeneous polynomials and real weighted homogeneous maps with an isolated critical value, are d-regular maps.

- If f and g are holomorphic maps $\mathbb{C}^2 \to \mathbb{C}$ such that the product $f\bar{g}$ has an isolated critical value at the origin, then the map $f\bar{g}$ is d-regular, by [138].
- The strongly non-degenerate mixed functions in [119] are all d-regular, by [30, 119] (see Sect. 6.11).
- Direct sums of d-regular maps. That is, if f is d-regular in the variables $(x_1, \ldots, x_n)$ and g is d-regular in the variables $(y_1, \ldots, y_m)$, then $f + g$ is d-regular in the variables $(x_1, \ldots, x_n, y_1, \ldots, y_m)$, by [30].

The following is a fundamental property of d-regularity. We refer to [30] for its proof.

Theorem 6.10.3 *The real analytic map f is d-regular if and only if there exists a smooth vector field ζ such that its integral lines are transverse to all spheres around $\mathbf{0}$, transverse to all Milnor tubes $f^{-1}(\partial\mathbb{D}_\eta) \cap \mathbb{B}_\varepsilon$, and tangent to each element $X_{\mathcal{L}}$ of the canonical pencil.*

Such a vector field allows us to inflate the tube and get a fibration on the sphere minus the link, granting that the projection map is $f/\|f\|$. Hence we get a slight refinement of [30, Theorem 1]:

Theorem 6.10.4 *Let $f := (f_1, \ldots, f_p)\colon (\mathbb{R}^n, \mathbf{0}) \to (\mathbb{R}^p, 0)$ be a locally surjective real analytic map with an isolated critical value at $0 \in \mathbb{R}^p$ and assume $V = f^{-1}(0)$ has dimension > 0. Then f admits a Milnor-Lê fibration if and only if it has the transversality property. If this is so, then f is d-regular at 0 if and only if one has a commutative diagram of smooth fibre bundles:*

$$\begin{array}{ccc} \mathbb{S}_\varepsilon^{n-1} \setminus L_f & \xrightarrow{\ \phi\ } & \mathbb{S}^{p-1} \\ & \underset{\psi}{\searrow} & \downarrow{\scriptstyle \pi'} \\ & & \mathbb{RP}^{p-1} \end{array}$$

where L_f is the link, $\psi = (f_1(x) : \cdots : f_p(x))$ and $\phi = \frac{f}{\|f\|} : \mathbb{S}_\varepsilon^{n-1} \setminus L_f \to \mathbb{S}^{p-1}$ is the Milnor fibration. Furthermore, if the two fibrations exist (one on a Milnor tube, another on the sphere minus the link), then these fibrations are smoothly equivalent. That is, there exists a diffeomorphism between their corresponding total spaces, carrying fibers into fibers.

This answers affirmatively a question raised by dos Santos [35], where the author proved it for $p = 2$ and f weighted homogeneous. The proof in [30] of the equivalence of the two fibrations has a small gap that has been filled in [24, 25] where the theorem is proved in the more general setting of real analytic maps with arbitrary linear discriminant (see Definition 6.10.8). In [30, 31] there are other criteria to determine d-regularity which can be useful in practice.

The following corollary is an immediate consequence of the previous theorem:

Remark 6.10.5 In [38, 39] the vector field of Theorem 6.10.3 is called a Milnor vector field and sufficient condition are given for its existence. In [24, 25] it is proved that such a vector field exists if and only if the map is d regular.

Corollary 6.10.6 *Given $f\colon (\mathbb{R}^n, \mathbf{0}) \to (\mathbb{R}^p, 0)$ as in the theorem above, consider its Milnor fibration*

$$\phi = \frac{f}{\|f\|} : \mathbb{S}_\eta^{n-1} \setminus L_f \to \mathbb{S}^{p-1}.$$

Then the union of the link L_f and each pair of fibres of ϕ over antipodal points of $\mathbb{S}^{p-1}$ corresponding to the line $\mathcal{L}_\ell$, is the link of the real analytic variety X_θ.

For instance, if $f : (\mathbb{C}^n, 0) \to (\mathbb{C}, 0)$ is holomorphic and it has an isolated critical point at 0, then $\{Re\, f = 0\}$ is a real hypersurface and its link is the double of the Milnor fiber of f with the link L_f as an equator. If $n = 2$, then the link of $Re\,(f) = 0$ is a compact Riemann surface of genus $2g_f + r - 1$ where g_f is the genus of the Milnor fibre of f and r the number of connected components of the link of f. Thus for instance, we know from [29] that for the map $(z_1, z_2) \overset{f}{\mapsto} z_1^2 + z_2^q$ one gets that the link of $Re\, f$ is a closed oriented surface in the 3-sphere, union of the Milnor fibres over the points $\pm i$; an easy computation shows that it has genus $q - 1$. This provides an explicit way to determine closed surfaces of all genera ≥ 1 in the 3-sphere by a single analytic equation.

It would be interesting to study the geometry and topology of the 4-manifolds one gets in this way, by considering the link of the hypersurface in $\mathbb{C}^3$ defined by the real part of a holomorphic function with an isolated critical point. For example, for the map $(z_1, z_2, z_3) \overset{f}{\mapsto} z_1^2 + z_2^3 + z_3^5$, the corresponding 4-manifold is the double of the E_8 manifold, whose boundary is Poincaré's homology 3-sphere.

The following related result was proved by Menegon and Seade [105] answering a question in [23].

Theorem 6.10.7 *Let $f = (f_1, \ldots, f_n)\colon (\mathbb{R}^m, 0) \to (\mathbb{R}^n, 0)$, with $1 < n < m$, be a map germ with an isolated critical point. For each set of indices $\{i_1, \ldots, i_k\}$ set*

$$V_{i_1,\ldots,i_k} := f_{i_1,\ldots,i_k}^{-1}(0) \cap \mathbb{B}_\varepsilon^m,$$

where $f_{i_1,\ldots,i_k} = (f_{i_1}, \ldots, f_{i_k})\colon (\mathbb{R}^m, 0) \to (\mathbb{R}^k, 0)$ and ε is a Milnor radius for f. The topology of $V_{i_1,\ldots,i_k}$ is independent of the choice of the indices $i_1, \ldots, i_k$. Moreover, the diffeomorphism type of its link it is also independent of the choice of the indices $i_1, \ldots, i_k$ and it is diffeomorphic to the boundary of the product of the Milnor fibre F of f and a closed $(n-k)$-disk. In the particular case when $k = n - 1$, that is precisely the double of F.

6.10.2 The General Case

In [32] the authors continue the work begun in [23] and extend the above discussion on d-regularity to differentiable functions $(\mathbb{R}^n, 0) \overset{f}{\to} (\mathbb{R}^p, 0)$ of class $\mathbb{C}^\ell$, $\ell \geq 1$,

$n \geq p \geq 2$, with a critical point at $\mathbf{0} \in \mathbb{R}^n$, arbitrary critical values Δ_f and non-empty link L_V. This is immediate when the discriminant Δ_f is linear:

Definition 6.10.8 The map-germ $f\colon (\mathbb{R}^n, 0) \to (\mathbb{R}^p, 0)$ has *linear discriminant* if for some representative f there exists $\eta = \eta(f) > 0$ such that the intersection of Δ_f with the closed ball $\mathbb{D}^p_\eta$ is a union of line-segments, i.e.:

$$\Delta_f \cap \mathbb{D}^p_\eta = \mathrm{Cone}\big(\Delta_f \cap \mathbb{S}^{p-1}_\eta\big)\,.$$

We call η a *linearity radius* for Δ_f. (The case when f has $0 \in \mathbb{R}^p$ as isolated critical value is considered to have linear discriminant with arbitrary linearity radius.)

Let $f\colon (\mathbb{R}^n, 0) \to (\mathbb{R}^p, 0)$ be as above, with linear discriminant, and consider a representative f with linearity radius $\eta > 0$. Set $\partial\Delta_\eta := \Delta_f \cap \mathbb{S}^{p-1}_\eta$. For each point $\theta \in \mathbb{S}^{p-1}_\eta$, let $\mathcal{L}_\theta \subset \mathbb{R}^p$ be the open segment of line that starts at the origin and ends at the point θ (but not containing these two points). Set $E_\theta := f^{-1}(\mathcal{L}_\theta)$, so each E_θ is a manifold of class C^ℓ for every θ in $\mathbb{S}^{p-1}_\eta \setminus \partial\Delta_\eta$.

Definition 6.10.9 Let $f\colon (\mathbb{R}^n, 0) \to (\mathbb{R}^p, 0)$ be a map-germ of class C^ℓ with $\ell \geq 1$ and linear discriminant. We say that f is *d-regular* at 0 if for some representative f there exists $\varepsilon_0 > 0$ small enough such that $f(\mathbb{B}^n_{\varepsilon_0}) \subset \mathring{\mathbb{D}}^p_\eta$, where η is a linearity radius for Δ_f, and such that every E_θ intersects the sphere $\mathbb{S}^{n-1}_\varepsilon$ transversely in $\mathbb{R}^n$, for every ε with $0 < \varepsilon \leq \varepsilon_0$ and for all $\theta \in \mathbb{S}^{p-1}_\eta \setminus \partial\Delta_\eta$ such that the intersection is not empty.

Example 6.10.10 Let $\mathbb{K}$ be either $\mathbb{R}$ or $\mathbb{C}$. Let $(f, g) : \mathbb{K}^n \to \mathbb{K}^2$ be a $\mathbb{K}$-analytic map of the form:

$$(f, g) = \left(\sum_{i=1}^{n} a_i x_i^q\,,\ \sum_{i=1}^{n} b_i x_i^q\right),$$

where $(a_i, b_i) \in \mathbb{K}$ are constants in generic position as in Example 6.9.8, and $q \geq 2$ is an integer. By [32] the discriminant Δ is linear and (f, g) is d-regular.

It is proved in [32] that the fibration theorems 6.10.3 and 6.10.4 extend to this general setting of C^ℓ maps with linear discriminant which have the transversality property and are d-regular. Also, there are in [32] examples of non-analytic maps for which these fibration theorems apply. In [141] M. Ribeiro gives some sufficient conditions for maps with linear discriminant to be d-regular (called ρ-regularity there) which ensure the existence of the fibration on the sphere.

Consider the following example that generalizes Example 6.10.10.

Example 6.10.11 Let $(f, g)\colon \mathbb{R}^n \to \mathbb{R}^2$ be:

$$(f, g) = \left(\sum_{i=1}^{n} a_i x_i^p\,,\ \sum_{i=1}^{n} b_i x_i^q\right),$$

with $p, q \geq 2$ integers and the (a_i, b_i) as above. If $p \neq q$, the discriminant $\Delta_{(f,g)}$ is not linear. Yet we can always linearize it with a homeomorphism h in $\mathbb{R}^2$. Moreover, these maps have the transversality property and they are d_h-regular in an appropriate sense that depends on the homeomorphism h. The fibration theorems in [32] extend to this setting, and in fact to all C^ℓ-maps that admit an appropriate "conic modification", a condition that seems to be always satisfied.

6.11 Singularities of Mixed Functions

A mixed function is a real analytic function $\mathbb{C}^n \to \mathbb{C}$ in the complex variables $z = (z_1, \ldots, z_n)$ and their conjugates $\bar{z} = (\bar{z}_1, \ldots, \bar{z}_n)$. Early appearance of this type of functions in singularity theory are, for instance, Kuiper's examples (6.21) in [110, Chap. 11], example (6.22) by A'Campo [1], as well as in the work of Rudolph [145]. The modern study of mixed functions in singularity theory springs from [22, 144, 150]: the twisted Brieskorn-Pham polynomials (Definition 6.8.6) and, more generally, polar weighted homogeneous polynomials (Sect. 6.9.2) are naturally mixed functions. So are the singularities studied by López de Medrano (see for instance [96]) that spring from the study of holomorphic linear actions of $\mathbb{C}^m$ in $\mathbb{C}^n$, $0 < m \ll n$ (see [151, Sect. 16]). Other examples of mixed functions are functions of the form $f\bar{g}\colon \mathbb{C}^n \to \mathbb{C}$ with f and g holomorphic studied in [136, 138]. This type of functions yield to a rich interplay between complex geometry and the theory of real analytic singularities, and that is the topic of Chaps. VI–VIII in [150].

The terms "mixed function and mixed singularity" were coined by Oka [119]. Oka's systematic study of mixed functions has turned this subject into a whole new line of research. In particular, inspired by the theory of complex singularities [117], Oka [119] introduces for mixed functions the useful notion of non-degeneracy with respect to a naturally defined Newton boundary. He uses this to prove a fibration theorem for strongly non-degenerate convenient mixed functions, and to study their topology. We give the necessary definitions to state this theorem and we refer to Oka's chapter [132] in this volume for more on the subject.

A *mixed analytic function* is a function $f\colon \mathbb{C}^n \to \mathbb{C}$ of the form

$$f(z) = \sum_{\nu,\mu} c_{\nu,\mu} z^\nu \bar{z}^\mu \,,$$

Assume for simplicity $c_{0,0} = 0$, so that $\mathbf{0} \in V(f) := f^{-1}(0)$. Following Oka, we call $V(f)$ a *mixed hypersurface*, though in general it has real codimension 2. The *(radial) Newton polygon (at the origin)* $\Gamma_+(f)$ is defined in the usual way: it is the convex hull of:

$$\bigcup_{c_{\nu,\mu}\neq 0} (\nu + \mu) + \mathbb{R}^{+n} \,,$$

where $\nu + \mu$ is the sum of the multi-indices of $z^\nu \bar{z}^\mu$, i.e., $\nu + \mu = (\nu_1 + \mu_1, \ldots, \nu_n + \mu_n)$.

In analogy with complex polynomials, define the *Newton boundary* $\Gamma(f)$ as the union o f the compact faces of $\Gamma_+(f)$. To every given positive integer (weight) vector $P = (p_1, \ldots, p_n)$ we associate a linear function ℓ_P on the Newton boundary $\Gamma(f)$ defined by:

$$\ell_P(\nu) = \sum_{j=1}^{n} p_j \nu_j$$

for $\nu \in \Gamma(f)$. Let $\Delta(P, f) = \Delta(P)$ be the face where ℓ_P attains its minimal value. Then, for a positive weight P, define the face function $f_P(z)$ by:

$$f_P(z) = \sum_{\nu+\mu \in \Delta(P)} c_{\nu,\mu} z^\nu \bar{z}^\mu .$$

We need some notation to introduce the main definitions. For a subset $J \subset \{1, 2, \ldots, n\}$, we define the subspace $\mathbb{C}^J = \{\mathbf{z} = (z_1, \ldots, z_n) \in \mathbb{C}^n \mid z_j = 0, j \notin J\}$, the subset $\mathbb{C}^{*J} = \{\mathbf{z} = (z_1, \ldots, z_n) \in \mathbb{C}^n \mid z_j = 0 \Leftrightarrow j \notin J\}$ abbreviating $\mathbb{C}^{*n} = \mathbb{C}^{*J}$ for $J = \{1, 2, \ldots, n\}$.

Definition 6.11.1 Let P be a strictly positive weight vector. We say that $f(z)$ is *non-degenerate* for P if the fiber $f^{-1}(0) \cap \mathbb{C}^{*n}$ contains no critical point of the map $f_P\colon \mathbb{C}^{*n} \to \mathbb{C}$. The map f is *strongly non-degenerate* for P if the mapping $f_P\colon \mathbb{C}^{*n} \to \mathbb{C}$ has no critical points at all, $\dim \Delta(P) \geq 1$ and $f_P\colon \mathbb{C}^{*n} \to \mathbb{C}$ is surjective. The function $f(z)$ is called non-degenerate (respectively strongly non-degenerate) if it is non-degenerate (respectively strongly non-degenerate) for every strictly positive weight vector P.

For $J \subset \{1, 2, \ldots, n\}$, consider the restriction $f^J := f|_{\mathbb{C}^J}$ and define the set

$$\mathcal{NV}(f) = \{I \subset \{1, \ldots, n\} \mid f^I \neq 0\}.$$

We call $\mathcal{NV}(f)$ the *set of non-vanishing coordinate subspaces* for f.

Definition 6.11.2 We say that f is *k-convenient* if $J \in \mathcal{NV}(f)$ for every $J \subset \{1, \ldots, n\}$ with $|J| = n - k$. We say that f is *convenient* if f is $(n-1)$-convenient.

We may now state the main results in [119, Theorems 29, 33, 36] concerning Milnor fibrations combined into a single statement.

Theorem 6.11.3 *Assume the mixed polynomial $f(z)$ is convenient and strongly non-degenerate. Then one has a fibration of the Milnor-Lê type in a Milnor tube as in* (6.16), *as well as a Milnor fibration on every sufficiently small sphere with projection map $f/\|f\|$, as in* (6.13), *and the two fibrations are smoothly equivalent.*

In [119] Oka also uses toric geometry to get a resolution of the corresponding singularity, in analogy with the complex case (see for instance [117]). He then uses this to study the topology of the links, as well as the topology of the Milnor fibers.

Oka has published several other articles studying different aspects of mixed functions, including [53, 121, 124–130] with important results that cover a wide spectrum of topics, from intersection theory to contact structures or even studying roots of equations applied in astronomy.

Remark 6.11.4 Inspired in [88], in [26] for a family of mixed functions a vanishing polyhedron is contructed and with it, a join theorem is proved. One can also consider *mixed maps* $f\colon \mathbb{C}^n \to \mathbb{C}^k$, for this we refer to [143] and the references in there.

Acknowledgements The first author is a Regular Associate of the Abdus Salam International Centre for Theoretical Physics, Trieste, Italy. He was supported by project CONACYT 253506 and a UNAM-DGAPA-PASPA sabbatical scholarship. The second named author acknowledges support from CONACYT, project 282937, and UNAM-PAPIIT, project IN 110517.

References

1. N. A'Campo. Le nombre de Lefschetz d'une monodromie. Indag. Math. **35**, 113–118 (1973)
2. H. Aguilar-Cabrera. New open-book decompositions in singularity theory. Geom. Dedicata **158**, 87–108 (2012)
3. H. Aguilar-Cabrera. Open-book decompositions of $\mathbb{S}^5$ and real singularities. Int. J. Math. **25** (2014)
4. H. Aguilar-Cabrera. The topology of real suspension singularities of type $f\bar{g} + z^n$. Math. Z. **277**, 209–240 (2014)
5. L. Andersen. On isolated real singularities I, October 2021. Preprint: arXiv:2110.04407v1 [math.AG]
6. A. Andreotti, T. Frankel. The Lefschetz theorem on hyperplane sections. Ann. Math. (2) **69**, 713–717 (1959)
7. V. I. Arnold. Normal forms of functions near degenerate critical points, the Weyl groups A_k, D_k, E_k and Lagrangian singularities. *Funkcional. Anal. i Priložen.*, 6(4):3–25, 1972
8. V. I. Arnold. The index of a singular point of a vector field, the Petrovskiĭ-Oleĭnik inequalities, and mixed Hodge structures. Funkcional. Anal. i Priložen., **12** (1), 1–14 (1978)
9. V. I. Arnold, S. M. Guseĭn-Zade, and A. N. Varchenko. *Singularities of differentiable maps. Vol. I*, volume 82 of *Monographs in Mathematics*. Birkhäuser Boston, Inc., Boston, MA, 1985. The classification of critical points, caustics and wave fronts, Translated from the Russian by Ian Porteous and Mark Reynolds
10. K. Bekka. *(C)-Régularité et Trivialité Topologique*. Lecture Notes in Mathematics 1462, Springer-Verlag (1991), 42–62
11. V. Blanlœil and M. Oka. Topology of strongly polar weighted homogeneous links. SUT J. Math., (1), 119–128 (2015)
12. A. Bodin, A. Pichon. Meromorphic functions, bifurcation sets and fiberd links. Math. Res. Lett. **14**, 413–422 (2007)
13. A. Bodin, A. Pichon, J. Seade. Milnor fibrations of meromorphic functions. Journal London Math. Soc. **80**, No. 2, 311–325 (2009)
14. K. Brauner. Zur Geometrie der Funktionen zweier komplexer Veränderlicher. II: Das Verhalten der Funktionen in der Umgebung ihrer Verzweigungsstellen. III: Klassifikation der Singularitäten algebroider Kurven. IV: Die Verzweigungsgruppen. Abhandlungen Hamburg, **6**, 1–55 (1928)
15. E. Brieskorn. Examples of singular normal complex spaces which are topological manifolds. Proc. Nat. Acad. Sci. U.S.A., **55**, 1395–1397 (1966)

16. E. Brieskorn. Beispiele zur Differentialtopologie von Singularitäten. Invent. Math, **2**, 1–14 (1966)
17. E. Brieskorn, H. Knörrer. *Plane algebraic curves*. Transl. from the German by John Stillwell. Reprint of the hardback ed. 1986. Modern Birkhäuser Classics, (2012)
18. E. H. Brown, Jr., F. P. Peterson. The Kervaire invariant of $(8k+2)$-manifolds. Amer. J. Math., **88**, 815–826 (1966)
19. W. Burau. Kennzeichnung der Schlauchknoten. Abh. Math. Sem. Univ. Hamburg, **9**(1), 125–133 (1933)
20. D. Burghelea, A. Verona. Local homological properties of analytic sets. Manuscripta Math. **7**, 55–66 (1972)
21. P. T. Church, K. Lamotke. Non-trivial polynomial isolated singularities. Nederl. Akad. Wetensch. Proc. Ser. A **78**, 149–154 (1975)
22. J. L. Cisneros-Molina. Join theorem for polar weighted homogeneous singularities. In *Singularities II*. Contemp. Math. 475, p. 43–59. A. M. S., Providence, RI, 2008. Proc. Int. School and Workshop on the Geometry and Topology of Singularities in honor of the 60th birthday of Lê D. T. Cuernavaca, Mexico, 2007. Eds. J.-P. Brasselet *et al*
23. J. L. Cisneros-Molina, N. G. Grulha, J. Seade. On the topology of real analytic maps. International Journal of Mathematics **25**, No. 7 (2014)
24. J. L. Cisneros-Molina, A. Menegon. Equivalence of Milnor and Milnor-Lê fibrations for real analytic maps. International Journal of Mathematics **30**, No. 14 (2019)
25. J. L. Cisneros-Molina, A. Menegon. Errata to equivalence of Milnor and Milnor-Lê fibrations for real analytic maps. https://doi.org/10.1142/S0129167X21500701
26. J. L. Cisneros-Molina and A. Menegon. Lê's vanishing polyhedron for a family of mixed functions. *Bulletin of the London Mathematical Society*, 51(6):1022–1038, 2019
27. J. L. Cisneros-Molina, A. Romano-Velázquez. Classification of Isolated Polar Weighted Homogeneous Singularities. In *Singularities in Geometry, Topology, Foliations and Dynamics*. p. 77–104, Birhauser Verlag, Trends in Mathematics 2017. Eds. J. L. Cisneros-Molina *et al*
28. J. L. Cisneros-Molina and A. Romano-Velázquez. The real embedding method to study mixed functions. In *A panorama of singularities*, volume 742 of *Contemp. Math.*, pages 13–30. Amer. Math. Soc., Providence, RI, 2020
29. J. L. Cisneros-Molina, J. Seade, J. Snoussi. Refinements of Milnor's fibration theorem for complex singularities. Adv. Math., **222**, 937–970 (2009)
30. J. L. Cisneros-Molina, J. Seade, J. Snoussi. Milnor Fibrations and d-regularity for real analytic Singularities. International Journal of Mathematics **21**, No. 4, 419-434 (2010)
31. J. L. Cisneros-Molina, J. Seade, J. Snoussi. Milnor Fibrations and the concept of d-regularity for analytic map germs. In *Real and complex singularities*, AMS Contemporary Mathematics 569, p 1–28, A. M. S., Providence, RI, 2012. Eds. V. Goryunov *et al*
32. J. L. Cisneros-Molina, A. Menegon, J. Seade, and J. Snoussi. Fibration theorems à la Milnor for differentiable maps with non-isolated singularities, February 2020. Preprint : arXiv:2002.07120 [math.AG]
33. A. Dimca. *Singularities and Topology of Hypersurfaces*. Universitext, Springer, 1992
34. A. A. Do Espirito Santo, D. Dreibelbis, M. F. Ribeiro, and R. N. A. Dos Santos. A quick trip through fibration structures. *Journal of Singularities*, 22:134–158, 2020. Proceedings of the 15th International Workshop on Real and Complex Singularities, 22-28th July, 2018, São Carlos, Brazil
35. R. N. A. dos Santos. Equivalence of real Milnor's fibrations for quasi homogeneous singularities. Rocky Mt. J. Math. **42**, No. 2, 439–449 (2012)
36. R. N. A. dos Santos, D. Dreibelbis, N. Dutertre. *Topology of the real Milnor fiber for isolated singularities.* In *Real and complex singularities*, AMS Contemporary Mathematics 569, p 1–28, A. M. S., Providence, RI, 2012. Eds. V. Goryunov *et al*
37. R. N. A. dos Santos, M. A. B. Hohlenwerger, O. Saeki, T. O. Souza. New examples of Neuwirth-Stallings pairs and non-trivial real Milnor fibrations. Annales de l'Institut Fourier, **66** no. 1, 83–104 (2016)

38. R. N. A. dos Santos, M. F. Ribeiro. Geometrical Conditions for the Existence of a Milnor Vector Field. Bull. Braz. Math. Soc., New Series (2020). https://doi.org/10.1007/s00574-020-00230-9
39. R. N. A. dos Santos, M. F. Ribeiro, M. Tibâr. Milnor–Hamm sphere fibrations and the equivalence problem. J. Math. Soc. Japan **72**(3), 945–957, July, 2020. https://doi.org/10.2969/jmsj/82278227
40. R. N. A. dos Santos, M. A. Ruas. Real Milnor fibrations and (c)-regularity. Manuscripta Math., **117**(2), 207–218 (2005)
41. A. A. du Plessis, C. T. C. Wall. Discriminants, vector fields and singular hypersurfaces. In *New developments in singularity theory*, D. Siersma et al. (ed.), Kluwer Acad. Publ. NATO Sci. Ser. II, Math. Phys. Chem. 21, 321-377, 2001. Proc. NATO Advanced Study Institute on new developments in singularity theory, Cambridge, UK, 2000
42. A. Durfee. Fifteen characterizations of rational double points and simple critical points. Enseign. Math., **25**, 131–163 (1979)
43. A. Durfee. Neighborhoods of algebraic sets. Trans. Amer. Math. Soc, **276**, 517–530 (1983)
44. A. Durfee. Singularities. In *History of topology*, North-Holland, p. 417–434. Amsterdam, 1999
45. N. Dutertre. On the topology of semi-algebraic functions on closed semi-algebraic sets. Manuscr. Math. **139**, 415–441 (2012)
46. N. Dutertre, R. N. A. dos Santos, Y. Chen, A. do Espirito Santo. Open book structures on semi-algebraic manifolds. Manuscr. Math. **149**, 205–222 (2016)
47. N. Dutertre, N. G. Grulha Jr. Lê-Greuel type formula for the Euler obstruction and applications. Adv. Math. **251**, 127–146 (2014)
48. N. Dutertre, R. N. A. dos Santos, Y. Chen, A. do Espirito Santo. Fibration structures and formulae for the Euler characteristics of Milnor fibers. RIMS Kokyuroku Bessatsu **B55**, 41–65 (2016)
49. W. Ebeling. Monodromy. In *Singularities and computer algebra*. Cambridge University Press. London Mathematical Society Lecture Note Series 324, 129-155, 2006. Selected papers of the conference on the occasion of Gert-Martin Greuel's 60th birthday. Eds. Ch. Lossen *et al*
50. W. Ebeling. Distinguished bases and monodromy of complex hypersurface singularities. Handbook of Geometry and Topology of Singularities, Volume I, 2020
51. W. Ebeling, S. Gusein-Zade, On the index of a vector field at an isolated singularity. Fields Inst. Commun., 24, 141-152, Amer. Math. Soc., Providence, RI, 1999. The Arnoldfest (Toronto, ON, 1997)
52. D. Eisenbud, H. I. Levine, An algebraic formula for the degree of a C^∞ map germ. With an appendix by B. Teissier: Sur une inegalité à la Minkowski pour les multiplicités. Ann. Math. **106**, 19–44 (1977)
53. C. Eyral, M. Oka. Whitney regularity and Thom condition for families of non-isolated mixed singularities. J. Math. Soc. Japan, **70**(4), 1305–1336 (2018)
54. J. Fernández de Bobadilla, A. Menegon Neto. The boundary of the Milnor fiber of complex and real analytic non-isolated singularities. Geom. Dedicata **173**(1), 143–162 (2014)
55. T. Gaffney. The integral closure of modules and Whitney equisingularity. Inv. Math. **102**, 301–322 (1992)
56. E. R. García-Barroso, P. D. González-Pérez, P. Popescu-Pampu. The combinatorics of plane curve singularities. How Newton polygons blossom into lotuses. Handbook of Geometry and Topology of Singularities, Volume I, 2020
57. X. Gómez-Mont, P. Mardesić, The index of a vector field tangent to a hypersurface and the signature of the relative Jacobian determinant. Ann. Inst. Fourier **47**, no. 5, 1523–1539 (1997)
58. X. Gómez-Mont, P. Mardesić, The index of a vector field tangent to an odd-dimensional hypersurface, and the signature of the relative Hessian. Funktsional. Anal. i Prilozhen. **33**, no. 1, 96, 1–13 (1999)
59. M. Goresky, R. MacPherson. *Stratified Morse theory*. Springer-Verlag, Berlin, 1988
60. G.-M. Greuel. Der Gauss-Manin Zusammenhang isolierter Singularitäten von vollständigen Durchschnitten. Dissertation. Göttingen, 1973. Math. Ann. **214** 235–266 (1975)

61. G.-M. Greuel. Deformation and smoothing of singularities. Handbook of Geometry and Topology of Singularities, Volume I, 2020
62. G.-M. Greuel, W. Purket. Life and work of Egbert Brieskorn (1936-2013). Journal of Singularities **18**, 7-34 (2018). Special volume in honor of E. Brieskorn. (See also arXiv:1711.09600)
63. S. M. Gusein-Zade, I. Luengo, A. Melle-Hernández. *Zeta functions of germs of meromorphic functions, and the Newton diagram,* Funct. Anal. Appl. **32**, 93–99 (1998)
64. S. M. Gusein-Zade, I. Luengo, A. Melle-Hernández. *On the topology of germs of meromorphic functions and its applications,* St. Petersburg Math. J. **11**, 775–780 (2000)
65. H. Hamm. Lokale topologische Eigenschaften komplexer Räume. Math. Ann., **191**, 235–252 (1971)
66. H. Hamm, D. T. Lê. Un théorème de Zariski du type de Lefschetz. Ann. Sci. Ec. Norm. Sup., **6** (series 4), 317–366 (1973)
67. L. Hernández de la Cruz, S. López de Medrano. Some families of isolated singularities. In *Singularities II.* Contemp. Math. 475, p. 79–87. A. M. S., Providence, RI, 2008. Proc. Int. School and Workshop on the Geometry and Topology of Singularities in honor of the 60th birthday of Lê D. T. Cuernavaca, Mexico, 2007. Eds. J.-P. Brasselet *et al*
68. H. Hironaka. *Stratification and flatness.* In *Real and complex singularities* (Proc. Ninth Nordic Summer School/NAVF Sympos. Math., Oslo, 1976), p. 199–265. Sijthoff and Noordhoff, 1977
69. F. Hirzebruch. *Topological Methods in algebraic geometry*. Springer Verlag, 1956
70. F. Hirzebruch. *Singularities and exotic spheres*. Sem. Bourbaki 1966/67, No.314, 20 p. 1968
71. W. C. Hsiang, W. Y. Hsiang. Some results on differentiable actions. Bull. Amer. Math. Soc., **72**, 134–138 (1966)
72. K. Inaba. On the enhancement to the Milnor number of a class of mixed polynomials. J. Math. Soc. Japan, **66**(1), 25–36 (2014)
73. K. Inaba. On fibered links of singularities of polar weighted homogeneous mixed polynomials. In *Singularities in geometry and topology 2011*, volume 66 of Adv. Stud. Pure Math., pages 81–92. Math. Soc. Japan, Tokyo, 2015
74. K. Inaba. On deformations of isolated singularities of polar weighted homogeneous mixed polynomials. Osaka J. Math., **53**(3), 813–842 (2016)
75. K. Inaba. Topology of the Milnor fibrations of polar weighted homogeneous polynomials. Manuscripta Math., **157**(3-4), 411–424 (2018)
76. K. Inaba, M. Kawashima, and M. Oka. Topology of mixed hypersurfaces of cyclic type. J. Math. Soc. Japan, **70**(1), 387–402 (2018)
77. A. Jacquemard, Fibrations de Milnor pour des applications réelles. Boll. Un. Mat. Ital. **B** (7), 591–600 (1989)
78. K. Jänich. Differenzierbare Mannigfaltigkeiten mit Rand als Orbiträume differenzierbarer G-Mannigfaltigkeiten ohne Rand. Topology, **5**, 301–320 (1966)
79. E. Kähler. Über die Verzweigung einer algebraischen Funktion zweier Veränderlichen in der Umgebung einer singulären Stelle. Math. Z., **30**(1), 188–204 (1929)
80. M. A. Kervaire. A manifold which does not admit any differentiable structure. Comment. Math. Helv., **34**, 257–270 (1960)
81. M. A. Kervaire, J. W. Milnor. Groups of homotopy spheres: I. Annals of Mathematics. Princeton University Press. **77**, 504–537 (1963)
82. G. M. Khimshiashvili. On the local degree of a smooth map. Sakharth. SSR Mecn. Akad. Moambe, **SSR85**(2), 309-312 (1977)
83. F. Klein. *Lectures on the icosahedron and the solution of equations of the fifth degree*. Dover, 1956
84. N. H. Kuiper. C^1-equivalence of functions near isolated critical points. In *Symposium on Infinite-Dimensional Topology (Louisiana State Univ., Baton Rouge, La., 1967)*, pages 199–218. Ann. of Math. Studies, No. 69. 1972
85. D. T. Lê. Computation of the Milnor number of an isolated singularity of a complete intersection. Funct. Anal. Appl. **8**, 127–131 (1974)
86. D. T. Lê. *Some remarks on relative monodromy*. In P. Holm, editor, in *Real and complex singularities* (Proc. Ninth Nordic Summer School/NAVF Sympos. Math., Oslo, 1976), pages 397–403. Sijthoff and Noordhoff, Alphen aan den Rijn, 1977

87. D.T. Lê, *Polyèdres évanescents et effondrements*, A fête of topology, 293-329, Academic Press, Boston, MA, 1988
88. D.T. Lê and A. Menegon Neto, *Vanishing polyhedron and collapsing map*, Math. Zeitschrift, v. 286 (2017), p. 1003–1040
89. D. T. Lê, J. J. Nuño-Ballesteros, J. Seade. The topology of the Milnor fiber. Handbook of Geometry and Topology of Singularities, Volume I, 2020
90. D. T. Lê, B. Perron. Sur la fiber de Milnor d'une singularité isolée en dimension complexe trois. C. R. Acad. Sci. Paris Sé r. A-B **289**, no. 2, A115–A118 (1979)
91. D. T. Lê, B. Teissier. Cycles évanescents et conditions de Whitney. In Proc. Symp. Pure Math, **40** (Part 2), 65–103 (1983)
92. J. Levine. Polynomial invariants of knots of codimension two. *Ann. of Math. (2)*, 84:537–554, 1966
93. E. Looijenga. A note on polynomial isolated singularities. Indag. Math., **33**, 418–421 (1971)
94. E. Looijenga. *Isolated Singular Points on Complete Intersections*. Cambridge Univ. Press, Cambridge, London, New York, New Rochelle, Melbourne, Sydney, 1984
95. S. Łojasiewicz. Triangulation of semi-analytic sets. Annali Sc. Norm. Sup. de Pisa, **18**, 449–474 (1964)
96. S. López de Medrano. Topology of the intersection of quadrics in $\mathbb{S}^n$. In *Algebraic topology* Springer Verlag Lecture Notes in Math., 1370, 280–292, 1989. (Arcata, CA, 1986), Eds. Carlsson *et al*
97. S. López de Medrano. Singularities of homogeneous quadratic mappings. Rev. R. Acad. Cienc. Exactas Fís. Nat. Ser. A Math. RACSAM, **108**, 95–112 (2014)
98. R. Martins, A. Menegon. Milnor-Lê type fibrations for subanalytic maps. Preprint 2018, arXiv:1806.05349
99. D. B. Massey. The Lê varieties. I. Invent. Math. **99**, 357–376 (1990)
100. D. B. Massey. The Lê varieties II. Invent. Math. **104**, 113–148 (1991)
101. D. B. Massey. *Lê Cycles and Hypersurface Singularities*. Lecture Notes in Mathematics 1615, Springer-Verlag (1995)
102. D. B. Massey. Lê Cycles and Numbers of hypersurface singularities. This volume, 2021
103. D. B. Massey. Real analytic Milnor fibrations and a strong Lojasiewicz inequality. In *Real and complex singularities*; Camb. Univ. Press. L. M. S. Lecture Note Series 380, 268-292, 2010. 10th International Workshop on Real and Complex Singularities São Carlos, Brazil, 2008. (Ed.) M. Manoel *et al*
104. J. Mather. Stratifications and mappings. In *Dynamical Systems*, Proc. Sympos. Univ. Bahia, Salvador 1971, 195–232, 1973. Ed. M. Peixoto
105. A. Menegon, J. Seade. On the Lê-Milnor fibration for real analytic maps. Math. Nachr. **290**, No. 2–3, 382–392 (2017)
106. A. Menegon, J. Seade. Vanishing zones and the topology of non-isolated singularities. Geom. Dedicata **202**, 321–335 (2019)
107. F. Michel. The Topology of Surface Singularities. Handbook of Geometry and Topology of Singularities, Volume I, 2020
108. J. W. Milnor. On manifolds homeomorphic to the 7-sphere. Annals of Mathematics. Princeton University Press. **64**, 399–405 (1956)
109. J. W. Milnor. On isolated singularities of hypersurfaces. Preprint June 1966. Unpublished
110. J. W. Milnor. *Singular points of complex hypersurfaces*. Annals of Mathematics Studies, 61, Princeton University Press, Princeton, N.J., 1968
111. J. W. Milnor. Differential topology forty-six years later. Notices Amer. Math. Soc., **58**(6), 804–809 (2011)
112. J. W. Milnor. Topology through the centuries: low dimensional manifolds. Bull. Amer. Math. Soc. (N.S.), **52**(4), 545–584 (2015)
113. J. W. Milnor, P. Orlik. Isolated singularities defined by weighted homogeneous polynomials. Topology, **9**, 385–393 (1970)
114. D. Mumford. The topology of normal singularities of an algebraic surface and a criterion for simplicity. Publ. Math. I.H.E.S., **9**, 5–22 (1961)

115. W. D. Neumann. A calculus for plumbing applied to the topology of complex surface singularities and degenerating complex curves. Trans. Amer. Math. Soc., **268**, 299–344 (1981)
116. M. Oka. On the homotopy types of hypersurfaces defined by weighted homogeneous polynomials. Topology, **12**, 19–32 (1973)
117. M. Oka. *Non-degenerate complete intersection singularity*. Actualités Mathématiques. Paris: Hermann. (1997)
118. M. Oka. Topology of polar weighted homogeneous hypersurfaces. Kodai Math. J., **31**, 163–182 (2008)
119. M. Oka. Non-degenerate mixed functions. Kodai Math. J. **33**, No. 1, 1–62 (2010)
120. M. Oka. On mixed plane curves of polar degree 1. In *The Japanese-Australian Workshop on Real and Complex Singularities—JARCS III*, volume 43 of Proc. Centre Math. Appl. Austral. Nat. Univ., pages 67–74. Austral. Nat. Univ., Canberra, 2010
121. M. Oka. On Mixed Brieskorn Variety. in *Topology of algebraic varieties and singularities*. AMS Contemporary Mathematics 538, 389–399, 2011. Eds. J. I. Cogolludo *et al*
122. M. Oka. On mixed projective curves. In *Singularities in geometry and topology*, volume 20 of IRMA Lect. Math. Theor. Phys., pages 133–147. Eur. Math. Soc., Zürich, 2012
123. M. Oka. Mixed functions of strongly polar weighted homogeneous face type. In *Singularities in geometry and topology 2011*, volume 66 of Adv. Stud. Pure Math., pages 173–202. Math. Soc. Japan, Tokyo, 2015
124. M. Oka. Intersection theory on mixed curves. Kodai Math. J., **35**(2), 248–267 (2012)
125. M. Oka. Contact structure on mixed links. Vietnam J. Math., **42**(3), 249–271 (2014)
126. M. Oka. Remark on the roots of generalized lens equations. SUT J. Math., **53**(2), 127–134 (2017)
127. M. Oka. Łojasiewicz exponents of non-degenerate holomorohic and mixed functions. Kodai Math. J., **41**(3), 620–651 (2018)
128. M. Oka. On the roots of an extended Lens equation and an application. In *Singularities and foliations. geometry, topology and applications*, volume 222 of Springer Proc. Math. Stat., p. 489–511. Springer, Cham, 2018
129. M. Oka. Smooth mixed projective curves and a conjecture. J. Singul., **18**, 329–349 (2018)
130. M. Oka. On Milnor fibrations of mixed functions, a_f-condition and boundary stability. Kodai Math. J. **38**, No. 3, 581–603 (2015)
131. M. Oka. On the connectivity of Milnor fiber for mixed functions. Preprint 2018
132. M. Oka. Introduction to mixed hypersurface singularity. This Volume, 2021
133. P. Orlik. *Seifert manifolds*, volume 291 of Lecture Notes in Mathematics. Springer-Verlag, Berlin, 1972
134. P. Orlik, Ph. Wagreich. Isolated singularities of algebraic surfaces with C^*-action. Ann. of Math., **93**, 205–228 (1971)
135. F. Pham. Formules de Picard-Lefschetz généralisées et ramification des intégrales. Bull. Soc. Math. France, **93**, 333–367 (1965)
136. A. Pichon. Real analytic germs $f\bar{g}$ and open-book decompositions of the 3-sphere, Int. J. Math. **16**, 1–12 (2005)
137. A. Pichon, J. Seade. Real singularities and open-book decompositions of the 3-sphere. Ann. Fac. des Sciences de Toulouse **12**, No. 2, 245–265 (2003)
138. A. Pichon, J. Seade. Fibered multilinks and real singularities $f\bar{g}$. Math. Annalen **342** (3), 487–514 (2008)
139. P. Popescu-Pampu. *Complex singularities and contact topology*. Winter Braids Lecture Notes Vol. 3, 1–74, 2016. Course no III
140. A. Ranicki. *High-dimensional knot theory*. Springer Monographs in Mathematics. Springer-Verlag, New York, 1998. Algebraic surgery in codimension 2, With an appendix by Elmar Winkelnkemper
141. M. F. S. Ribeiro. *Singular Milnor Fibrations*. PhD thesis, Universidade de São Paulo, April 2018
142. M. F. S. Ribeiro. New Classes of Mixed Functions Without Thom Regularity. Bulletin of the Brazilian Mathematical Society, New Series, Jul 2019

143. Maico F. Ribeiro, Antonio A. Espírito do Santo, and Fernando P. P. Reis. Milnor–Hamm Fibration for Mixed Maps. *Bulletin of the Brazilian Mathematical Society, New Series*, 2020
144. M. A. Ruas, J. Seade, A. Verjovsky, On real singularities with a Milnor fibration, in *Trends in singularities*, Birkhäuser, Basel, 191–213, 2002. Eds. A. Libgober *et al*
145. L. Rudolph. Isolated critical points of mappings from $\mathbb{R}^4 \to \mathbb{R}^2$ and a natural splitting of the Milnor number of a classical fibered link. I. basic theory; examples. Comment. Math. Helv., **62**, 630-645 (1987)
146. K. Sakamoto. Milnor fiberings and their characteristic maps. In *Manifolds—Tokyo 1973*. Univ. Tokyo Press, Tokyo, p. 145–150, 1975. (Proc. Internat. Conf., Tokyo, 1973)
147. O. Schreier. Über die gruppen $A^a B^b = 1$. Abh. Math. Sem. Univ. Hamburg, **3**(1), 167–169 (1924)
148. J. Seade. Fibred links and a construction of real singularities via complex geometry. Bol. Soc. Brasil. Mat. (N.S.), **27**(2), 199–215 (1996)
149. J. Seade. Open book decompositions associated to holomorphic vector fields. Bol. Soc. Mat. Mex. **3**, 323–335 (1997)
150. J. Seade. *On the topology of isolated singularities in analytic spaces*. Progress in Mathematics 241. Birkhäuser, 2006
151. J. Seade. On Milnor's fibration theorem and its offspring after 50 years. Bull. Amer. Math. Soc. (N.S.), **56**(2), 281–348 (2019)
152. M. Sebastiani, R. Thom. Un résultat sur la monodromie. Invent. Math. **13**, 90–96 (1971)
153. H. Seifert. Topologie dreidimensionaler gefaserter Räume. Acta Math., **60**, 147–238 (1933)
154. S. Smale. Generalized Poincaré's conjecture in dimensions greater than four. Ann. of Math. **74**(2), 391–406 (1961)
155. S. Smale. On the structure of manifolds. Amer. J. Math., **84**, 387–399 (1962)
156. J. R. Stallings. Polyhedral homotopy-spheres. Bull. Amer. Math. Soc., **66**, 485–488 (1960)
157. I. Tamura, Spinnable structures on differentiable manifolds. Proc. Japan Acad. **48**, 293–296 (1972). http://projecteuclid.org/euclid.pja/1195519667
158. I. Tamura, Foliations and spinnable structures on manifolds. Ann. Inst. Fourier (Grenoble) **23**(2), 197–214 (1973). http://www.numdam.org/item?id=AIF_1973__23_2_197_0
159. B. Teissier, Cycles évanescents, sections planes et conditions de Whitney. In *Singularités à Cargése*. Astérisque, Nos. 7 et 8, Soc. Math. France, Paris, p. 285–362, 1973
160. D. Trotman. Stratification theory. Handbook of Geometry and Topology of Singularities, Volume I, 2020
161. J.-L. Verdier. Stratifications de Whitney et théorème de Bertini-Sard. Invent. Math., **36**, 295–312 (1976)
162. C. Weber (ed.) *Noeuds, tresses et singularités*. Comptes rendus du séminaire tenu aux Plans-sur-Bex (Suisse) en mars 1982. Monographie No.31 de L'Enseignement Mathématique (1983)
163. H. E. Winkelnkemper. Manifolds as open books. Bull. Amer. Math. Soc., **79**, 45–51 (1973)
164. J. Wolf. Differentiable fibre spaces and mappings compatible with Riemannian metrics. Michigan Math. J., **11**, 65–70 (1964)
165. O. Zariski. On the Topology of Algebroid Singularities. Amer. J. Math., **54**(3), 453–465 (1932)
166. E. C. Zeeman. The generalised Poincaré conjecture. Bull. Amer. Math. Soc., **67**, 270, 1961

Chapter 7
Lê Cycles and Numbers of Hypersurface Singularities

David B. Massey

Abstract The Milnor number is the most important number associated to an isolated hypersurface singularity. It is an invariant of the ambient topological-type of the hypersurface, it is effectively algebraically calculable, it determines the homotopy-type of the Milnor fiber, and its constancy in a family implies that Thom's A_f condition is satisfied and that the ambient topological-type of the hypersurface is constant (outside of possibly one dimension). In this survey, we will review results on the Lê cycles and Lê numbers—results which tell us the extent to which the Lê numbers of a non-isolated hypersurface singularity are a good generalization of the Milnor number from the isolated case.

Contents

7.1 Introduction and Earlier Results

By now, the Milnor Fibration, introduced by Milnor in 1968 in "Singular Points of Complex Hypersurfaces" [42] is considered the most fundamental object in the study of the local topology of complex hypersurfaces. This fibration is discussed at length

D. B. Massey (✉)
Department of Mathematics, Northeastern University, Boston, MA 02115, USA
e-mail: d.massey@northeastern.edu

J. L. Cisneros-Molina et al. (eds.), *Handbook of Geometry and Topology of Singularities II*, https://doi.org/10.1007/978-3-030-78024-1_7

in the article in these volumes "The Topology of the Milnor Fibration", by Lê, Nuño Ballesteros, and Seade [28].

We wish to recall some definitions and results here, particularly in the case of isolated hypersurface singularities to lead to our discussion of Lê cycles and numbers, which we developed in [33, 34, 36], to deal with **non-isolated** hypersurface singularities.

Let $\mathcal{U}$ be a non-empty open subset $\mathbb{C}^{n+1}$, $n \geq 1$, and let $f : \mathcal{U} \to \mathbb{C}$ be an analytic function which is nowhere locally constant. Let $V(f) := f^{-1}(0)$ be the hypersurface defined by f. Let $\mathbf{z} := (z_0, \dots, z_n)$ be the standard analytic coordinates on $\mathcal{U}$, and let Σf denote the critical locus of f, i.e.,

$$\Sigma f := V\left(\frac{\partial f}{\partial z_0}, \cdots, \frac{\partial f}{\partial z_n}\right).$$

By a standard argument using the Curve Selection Lemma, if $\mathbf{p} \in \Sigma f$, then, in an open neighborhood $\mathcal{W}$ of $\mathbf{p}$, $\mathcal{W} \cap \Sigma f \subseteq \mathcal{W} \cap V(f - f(\mathbf{p}))$; one says that *the critical values of f are locally isolated.*

By the *singular set of* $V(f)$, we mean the points of $V(f)$ where $V(f)$ is **not** an analytic submanifold of $\mathcal{U}$; we denote this singular set by $\Sigma V(f)$. If $\mathbf{p} \in \Sigma V(f)$ and f is reduced at $\mathbf{p}$, then the previous paragraph implies that, in an open neighborhood $\mathcal{W}$ of $\mathbf{p}$, $\mathcal{W} \cap \Sigma V(f) = \mathcal{W} \cap \Sigma f$. So, for local questions for a reduced f, one frequently sees Σf and $\Sigma V(f)$ used interchangeably.

The function f being reduced at $\mathbf{p}$ is equivalent to $\dim_{\mathbf{p}} \Sigma f \leq n-1$. In particular, since $n \geq 1$, if $\mathbf{p} \in V(f)$ and $\dim_{\mathbf{p}} \Sigma f = 0$, then f is reduced at $\mathbf{p}$ and $\Sigma V(f) = \Sigma f$ near $\mathbf{p}$.

Milnor worked in the complex algebraic category; however, essentially all of his results in [42] still hold, with the same proofs, in the complex analytic category. Some of his results that we wish to recall are:

1. Milnor proved that the object that is now called the Milnor fibration over a circle is a smooth, locally trivial fibration. In fact, Milnor mentions two diffeomorphic versions of this fibration: one on a small sphere and one inside a small ball. It is most convenient for us to use the fibration inside a ball.
   So, for us here, the Milnor fibration of f at a point $\mathbf{p} \in V(f)$ is given by the restriction of f, $f : B^\circ_\epsilon(\mathbf{p}) \cap f^{-1}(\partial \mathbb{D}_\eta) \to \partial \mathbb{D}_\eta$ for $0 < \eta \ll \epsilon \ll 1$, where $B^\circ_\epsilon(\mathbf{p})$ is an open ball of radius ϵ centered at $\mathbf{p}$ and $\partial \mathbb{D}_\eta$ is the boundary circle of a disk of radius η centered at the origin in $\mathbb{C}$. The diffeomorphism-type of this fibration is independent of the choice of the sufficiently small η and sufficiently smaller η.
   The fiber is the **Milnor fiber** and, at $\mathbf{p} \in V(f)$, is denoted by $F_{f,\mathbf{p}}$. The action on the homology/cohomology of $F_{f,\mathbf{p}}$ induced by traveling counteclockwise once around the base circle in the Milnor fibration is called the **Milnor monodromy**.
2. Milnor proved that, if $\mathbf{p} \in V(f)$, then the Milnor fiber, $F_{f,\mathbf{p}}$, has the homotopy-type of a finite n-dimensional CW-complex. This implies that all of the homology groups are finitely-generated, are zero above dimension n, and that $H_n(F_{f,\mathbf{0}})$ is free Abelian.

3. Milnor proved that if f has an isolated critical point at $\mathbf{p}$, i.e., $\dim_{\mathbf{p}} \Sigma f = 0$, then $F_{f,\mathbf{p}}$ is $(n-1)$-connected. Combining this with the previous result, it follows that, in the case of an isolated critical point, the Milnor fiber has the homotopy-type of a finite bouquet (one-point union) of n-spheres; the number of spheres in this bouquet is the *Milnor number* and is denoted by μ (or $\mu_f(\mathbf{p})$ or $\mu_{\mathbf{p}}(f)$, or some other such variant). In particular, the reduced homology is trivial except in dimension n where the homology group is $\mathbb{Z}^{\mu}$.
4. Let $\mathbf{p} := (p_0, \dots, p_n) \in V(f)$. Milnor showed that if $\dim_{\mathbf{p}} \Sigma f = 0$, then the Milnor number of f at $\mathbf{p}$ is the degree of the normalized jacobian map from a small sphere around $\mathbf{p}$ to a sphere around the origin. As observed by Palamodov [46], this implies that the Milnor number $\mu_{\mathbf{p}}(f)$ can be calculated algebraically by taking the dimension as a complex vector space of the jacobian algebra
$$\frac{\mathbb{C}\{z_0 - p_0, \dots, z_n - p_n\}}{\langle \frac{\partial f}{\partial z_0}, \dots, \frac{\partial f}{\partial z_n} \rangle},$$
where $\mathbb{C}\{z_0 - p_0, \dots, z_n - p_0\}$ denotes the ring of analytic germs, convergent power series, at $\mathbf{p}$.
5. There is one final result of Milnor's that we wish to mention here. Suppose that f is a weighted homogeneous polynomial (i.e., there exist positive integers $r_0, \dots, r_n$ such that $f\,(z_0{}^{r_0}, \dots, z_n{}^{r_n})$ is a homogeneous polynomial). Then, the Milnor fiber, $F_{f,\mathbf{0}}$, is diffeomorphic to $f^{-1}(1)$.

One of the most fundamental properties of the Milnor fiber is often stated without reference: for reduced hypersurfaces, the homotopy-type of the Milnor fiber is an invariant of the local, ambient topological-type of the hypersurface. For an isolated critical point, this statement is equivalent to saying that the Milnor number is an invariant of the local, ambient topological-type of the reduced hypersurface; in this case, this result appears in a remark of Teissier in [52] in 1972 and in [53] in 1973. The general result, with a monodromy statement, is due to Lê in [22, 23], which both appeared in 1973.

Theorem 7.1.1 (Teissier, Lê) *Let $f : (\mathcal{U}, \mathbf{0}) \to (\mathbb{C}, 0)$ and $g : (\mathcal{U}, \mathbf{0}) \to (\mathbb{C}, 0)$ be reduced complex analytic functions which define hypersurfaces with the same ambient topological-type at the origin, i.e., such that there exist open neighborhoods $\mathcal{W}$ and $\mathcal{V}$ of the origin and a homeomorphism $h : \mathcal{W} \to \mathcal{V}$ such that $h(\mathbf{0}) = \mathbf{0}$ and $h(\mathcal{W} \cap V(f)) = \mathcal{V} \cap V(g)$.*

Then, there exists a homotopy-equivalence $\alpha : F_{f,\mathbf{0}} \to F_{g,\mathbf{0}}$ such that the induced isomorphism on homology commutes with the respective Milnor monodromy automorphisms.

Example 7.1.2 Consider $f : \mathbb{C}^2 \to \mathbb{C}$ given by $f(x, y) = y^2 - x^3$ and $g : \mathbb{C}^2 \to \mathbb{C}$ given by $g(x, y) = y^2 - x^3 - x^2$. The spaces $V(f)$ and $V(g)$ are respectively referred to as a **cusp** and a **node**. Both functions have isolated critical points at the origin.

We leave it as an exercise for the reader to verify that $\mu_{\mathbf{0}}(f) = 2$, while $\mu_{\mathbf{0}}(g) = 1$ and so, by Lê and Teissier's result above, $V(f)$ and $V(g)$ are not ambiently homeomorphic at the respective origins.

In fact, by considering the number of irreducible components of $V(f)$ and $V(g)$ at $\mathbf{0}$, which corresponds to the number of connected components of $V(f) - \{\mathbf{0}\}$ and $V(g) - \{\mathbf{0}\}$ in small neighborhoods of the origin, one sees that $V(f)$ and $V(g)$ are not even homeomorphic at their origins without worrying about the ambient space.

Note that non-homeomorphic hypersurfaces with isolated singularities can easily have the same Milnor numbers. Again, we leave it as an exercise for the reader to see that this happens if we take $h : \mathbb{C}^2 \to \mathbb{C}$ given by $h(x, y) = y^3 - x^3$ and $k : \mathbb{C}^2 \to \mathbb{C}$ given by $k(x, y) = y^2 - x^5$.

The **Sebastiani-Thom Theorem** was proved in various forms in [43, 47, 49]. This result states that:

Theorem 7.1.3 *If we have analytic functions $f : \mathcal{U} \to \mathbb{C}$ and $g : \mathcal{U}' \to \mathbb{C}$, and points $\mathbf{a} \in V(f)$, $\mathbf{b} \in V(g)$, then the Milnor fiber at $(\mathbf{a}, \mathbf{b})$ of the function $h : \mathcal{U} \times \mathcal{U}' \to \mathbb{C}$ defined by $h(\mathbf{w}, \mathbf{z}) := f(\mathbf{w}) + g(\mathbf{z})$ is homotopy-equivalent to the join, $F_{f,\mathbf{a}} * F_{g,\mathbf{b}}$, of the Milnor fibers of f and g.*

This determines the homology of $F_{h,\mathbf{0}}$ in a simple way, since the reduced integral homology of the join of two spaces X and Y is given by

$$\widetilde{H}_{j+1}(X * Y) = \sum_{k+l=j} \widetilde{H}_k(X) \otimes \widetilde{H}_l(Y) \oplus \sum_{k+l=j-1} \operatorname{Tor}\left(\widetilde{H}_k(X), \widetilde{H}_l(Y)\right).$$

In particular, if one takes g above to be the function from $\mathbb{C}$ to $\mathbb{C}$ given by squaring, then the function h above, $h(\mathbf{w}, z) := f(\mathbf{w}) + z^2$, is referred to as the **suspension of** f, since the Sebastiani-Thom Theorem tells one that the Milnor fiber of h is homotopy-equivalent to the suspension of the Milnor fiber of f.

There is another important result about the homotopy-type of Milnor fiber. The very nice general result of Kato and Matsumoto [21] is:

Theorem 7.1.4 *If $\mathbf{p} \in V(f)$ and $s := \dim_{\mathbf{p}} \Sigma f$, then $F_{f,\mathbf{p}}$ is $(n - s - 1)$-connected; in particular, when $s = 0$, one recovers the result of Milnor.*

In fact, the connectivity result above is the best possible general result of this type, as the following example shows.

Example 7.1.5 Consider $g := z_0 z_1 \cdots z_{s+1} + z_{s+2}^2 + \cdots + z_n^2$. We leave it as an exercise for the reader to verify, using results above, that this g has an s-dimensional critical locus at the origin and $F_{g,\mathbf{0}}$ has non-trivial homology in dimension $n - s$.

We wish to consider another classic non-isolated hypersurface singularity:

Example 7.1.6 The Whitney umbrella is the hypersurface in $\mathbb{C}^3$ is defined by the vanishing of $f = y^2 - ux^2$. After an analytic change of coordinates, given by $u =$

$x + t$, we obtain that $f = y^2 - x^3 - tx^2$, which presents the Whitney umbrella as a family, f_t, of nodes degenerating to a cusp.

Now, $F_{y^2+ux^2,\mathbf{0}}$ is homotopy-equivalent to the suspension of $F_{ux^2,\mathbf{0}}$. But, as ux^2 is homogeneous,

$$F_{ux^2,\mathbf{0}} \cong \left\{(u,x)|ux^2 = 1\right\} = \left\{\left(\frac{1}{x^2}, x\right)|x \neq 0\right\} \cong \mathbb{C}^*.$$

Thus, $F_{y^2+ux^2,\mathbf{0}}$ is homotopy-equivalent to the suspension of a circle, i.e., $F_{y^2+ux^2,\mathbf{0}}$ is homotopy-equivalent to a 2-sphere.

That the Milnor number, the number of n-spheres in the homotopy-type of the Milnor fiber, at an isolated hypersurface singularity $\mathbf{p} \in V(f)$ is calculated as the complex vector space dimension of the jacobian algebra implies that, if $\mathbf{p} \in \Sigma f$, then the Milnor fiber does **not** have the homology of a point. But what about the case of non-isolated hypersurface singularities? The following **result of A'Campo** [1] implies that, again, at even a non-isolated critical point, the Milnor fiber cannot have the homology of a point.

Theorem 7.1.7 *Suppose that* $\mathbf{p} \in V(f) \cap \Sigma f$. *Then the Lefschetz number of the Milnor monodromy of* f *at* $\mathbf{p}$ *is zero.*

As the homotopy-type of the Milnor fiber is an invariant of the local, ambient topological-type of the hypersurface at the origin, if one has a family of hypersurfaces with isolated singularities in which the local, ambient, topological-type is constant, then the Milnor number must remain constant in the family. In 1976, Lê and Ramanujam proved the converse of this; we describe their result now.

Let $\mathbb{D}^\circ$ be an open disk about the origin in $\mathbb{C}$, let $\mathcal{U}$ be an open neighborhood of the origin in $\mathbb{C}^{n+1}$, and let $f : (\mathbb{D}^\circ \times \mathcal{U}, \mathbb{D}^\circ \times \mathbf{0}) \to (\mathbb{C}, 0)$ be an analytic function; we write f_t for the function defined by $f_t(\mathbf{z}) := f(t, \mathbf{z})$. **Lê and Ramanujam** proved in [30] the following stunning result:

Theorem 7.1.8 *Suppose that, for all small* t, $\dim_{\mathbf{0}} \Sigma f_t = 0$ *and that the Milnor number of* f_t *at the origin is independent of* t. *Then, for all small* t,

1. *the fiber-homotopy type of the Milnor fibrations of* f_t *at the origin is independent of* t;
   and, if $n \neq 2$,
2. *the diffeomorphism-type of the Milnor fibrations of* f_t *at the origin are independent of* t, *and*
3. *the local, ambient, topological-type of* $V(f_t)$ *at the origin are independent of* t.

There is another important result about μ-constant families: the result of **Lê and K. Saito**. We continue with $f : (\mathbb{D}^\circ \times \mathcal{U}, \mathbb{D}^\circ \times \mathbf{0}) \to (\mathbb{C}, 0)$ as above. The result of [31] tells one how limiting tangent spaces to nearby level hypersurfaces of f approach the singularity.

Theorem 7.1.9 *Suppose that, for all small t, $\dim_{\mathbf{0}}\Sigma f_t = 0$ and that the Milnor number of f_t at the origin is independent of t. Then, $\mathbb{D}^\circ \times \mathbf{0}$ satisfies Thom's A_f condition at the origin with respect to the ambient stratum, i.e., if $\mathbf{p}_i$ is a sequence of points in $\mathbb{D}^\circ \times \mathcal{U} - \Sigma f$ such that $\mathbf{p}_i \to \mathbf{0}$ and such that the tangent planes $T_{\mathbf{p}_i} V(f - f(\mathbf{p}_i))$ converge to some $\mathcal{T}$, then $\mathbb{C} \times \mathbf{0} = T_{\mathbf{0}}(\mathbb{D}^\circ \times \mathbf{0}) \subseteq \mathcal{T}$.*

The result of Kato and Matsumoto can be obtained from a more general result of Lê, a result which is one of few which allows calculations concerning the homology of the Milnor fiber for an arbitrary hypersurface singularity.

Let $\mathcal{U}$ be an open neighborhood of the origin in $\mathbb{C}^{n+1}$ and let $f : (\mathcal{U}, \mathbf{0}) \to (\mathbb{C}, 0)$ be an analytic function. Let $L : \mathbb{C}^{n+1} \to \mathbb{C}$ be a generic linear form. Then, it is easy to see that if $\dim_{\mathbf{0}}\Sigma f \geqslant 1$, then $\dim_{\mathbf{0}}\Sigma(f_{|V(L)}) = (\dim_{\mathbf{0}}\Sigma f) - 1$.

Now, **Lê's Attaching Result**, the main result of [22], which we sketch the proof of in Sect. 7.3, is:

Theorem 7.1.10 *The Milnor fiber $F_{f,\mathbf{0}}$ is obtained from the Milnor fiber $F_{f_{|V(L)},\mathbf{0}}$ by attaching a certain number of n-handles (n-cells on the homotopy level); this number of attached n-handles is given by the intersection number $\left(\Gamma^1_{f,L} \cdot V(f)\right)_{\mathbf{0}}$, where $\Gamma^1_{f,L}$ denotes the relative polar curve of f with respect to L.*

We will define the relative polar curve and discuss how to calculate intersection numbers in later sections, but we can already see that Kato and Matsumoto's result follows inductively from this since we already know Milnor's result for isolated singularities and because attaching handles of index k does not affect the connectivity in dimensions $\leqslant k - 2$.

Not only does Lê's result imply Kato and Matsumoto's, but—assuming that $\left(\Gamma^1_{f,L} \cdot V(f)\right)_{\mathbf{0}}$ is effectively calculable—Lê's result enables the calculation of the Euler characteristic of the Milnor fiber, together with some Morse-type inequalities on the Betti numbers of the Milnor fiber; for instance, the n-th Betti number, $b_n\left(F_{f,\mathbf{0}}\right)$, is less than or equal to $\left(\Gamma^1_{f,L} \cdot V(f)\right)_{\mathbf{0}}$. However, these inequalities are usually far from being equalities.

So What Do We Want from Lê Numbers?

Suppose we have an analytic function, $f : \mathcal{U} \to \mathbb{C}$ with a critical locus of arbitrary dimension $s := \dim_{\mathbf{p}} \Sigma f$ at some point $\mathbf{p} \in V(f)$. What properties would we want Lê numbers, generalized Milnor numbers, of f at $\mathbf{p}$ to have and what are Lê **cycles**?

Let us answer the last question first: Lê cycles are effective analytic cycles, formal sums of non-negative integers times analytic subspaces, which we define first, so that we can take intersection numbers with affine linear subspaces in order to produce the Lê numbers.

For the Lê numbers associated to f, we want there to be $s + 1$ numbers which are effectively calculable; call the numbers $\lambda^0_f, \dots, \lambda^s_f$. In the case of an isolated singularity, we want λ^0_f to be the Milnor number of f and all other λ^i_f to be zero.

For arbitrary s, we would like to generalize Milnor's result for isolated singularities and show that the Milnor fiber of f at the origin has a handle decomposition in which the number of attached handles of each index are given by the appropriate λ^i_f.

Finally, we would like to have generalizations of the results of Lê and Ramanujam and Lê and Saito to families of hypersurface singularities of arbitrary dimension.

As we showed 25 years ago, the Lê numbers succeed at these goals to a great degree. We have also included in this survey some results that have been proved since then, for instance, Theorem 7.3.10 of Bobadilla, Remark 7.6.10, and almost all of Sect. 7.7.

7.2 Definitions and Basic Properties of Lê Cycles and Numbers

As in the introduction, we let $\mathcal{U}$ be a non-empty open subset $\mathbb{C}^{n+1}$, $n \geq 1$, and $f : \mathcal{U} \to \mathbb{C}$ an analytic function which is nowhere locally constant. We make a linear choice of coordinates $\mathbf{z} := (z_0, \ldots, z_n)$ on $\mathcal{U}$.

We are going to need some minimal intersection theory, just proper intersections (so the codimensions of the intersecting cycles add) inside a complex manifold, namely U. Such intersections produce well-defined intersection **cycles**, as opposed to rational equivalent **classes**. In the analytic setting, these cycles are locally finite (as opposed to finite sums in the algebraic setting) formal sums of integers times irreducible analytic sets. The reader is directed to [10], Chap. 7 and Sect. 8.2. If C and D are properly intersecting cycles, we write $C \cdot D$ for their intersection product. When a cycle C is 0-dimensional, so a collection of points with integer coefficients, we write $C_{\mathbf{p}}$ for the coefficient of $\mathbf{p}$ in C. Occasionally, to emphasize that we are considering $V(g_1, \ldots, g_k)$ as a cycle, we shall write $\left[V(g_1, \ldots, g_k)\right]$.

Definition 7.2.1 For $1 \leq k \leq n+1$, we say that the **Lê cycles**, $\left\{\Lambda^k_{f,\mathbf{z}}\right\}_k$, (of f with respect to $\mathbf{z}$) **exist** provided that all of the intersections of cycles below are proper.

We let $\Gamma^{n+1}_{f,\mathbf{z}} := U$ as a cycle. Then, for $1 \leq k \leq n+1$, we define the $(k-1)$-dimensional relative polar cycle $\Gamma^{k-1}_{f,\mathbf{z}}$ and the $(k-1)$-dimensional Lê cycle $\Lambda^{k-1}_{f,\mathbf{z}}$ by downward induction by:

$$\Gamma^k_{f,\mathbf{z}} \cdot V\left(\frac{\partial f}{\partial z_{k-1}}\right) = \Gamma^{k-1}_{f,\mathbf{z}} + \Lambda^{k-1}_{f,\mathbf{z}},$$

where $\Gamma^{k-1}_{f,\mathbf{z}}$ consists of the sum of those components of the intersection product which are not contained in Σf and $\Lambda^{k-1}_{f,\mathbf{z}}$ consists of the sum of those components of the intersection which are contained in Σf.

Note that, in a small neighborhood of a point $\mathbf{p} \in \Sigma f$, $\Gamma^0_{f,\mathbf{z}} = 0$, i.e.,

$$\Lambda^0_{f,\mathbf{z}} = \Gamma^1_{f,\mathbf{z}} \cdot V\left(\frac{\partial f}{\partial z_0}\right).$$

We wish to emphasize that the Lê and relative polar cycles depend on the choice the coordinates $\mathbf{z}$.

Remark 7.2.2 The $\left\{\Gamma^k_{f,\mathbf{z}}\right\}_k$ above are a version of the relative polar varieties of Hamm, Lê, and Teissier from [12, 53]; the only difference is that we do not require the coordinates to be so generic that the cycles are reduced. We could also say above that the "relative polar cycles exist", but it is too cumbersome to always have as an hypothesis that both the Lê cycles and relative polar cycles exist; so we just say the Lê cycles exist and the reader should understand that that also implies the relative polar cycles exist.

Note that the intersections above could fail to be proper on all of $\mathcal{U}$, but may be proper on a smaller open set $\mathcal{U}'$ (frequently a small open neighborhood of a point $\mathbf{p} \in \Sigma f$). In such a case, in the definition, we would replace the $\mathcal{U}$ with $\mathcal{U}'$.

From the definition, we immediately conclude;

Proposition 7.2.3 *Suppose the Lê cycles exist. Then, for all k such that $0 \leq k \leq n$:*

1. $\left|\Lambda^k_{f,\mathbf{z}}\right| \subseteq \Sigma f$, *where the vertical bars denote the underlying set. Consequently, locally near a point* $\mathbf{p}$, $\left|\Lambda^k_{f,\mathbf{z}}\right| \subseteq V(f - f(\mathbf{p}))$.
2. $\Lambda^k_{f,\mathbf{z}}$ *and* $\Gamma^k_{f,\mathbf{z}}$ *are effective (i.e., non-negative) and purely k-dimensional (which vacuously includes the case where the cycles are zero, i.e., the underlying sets are empty).*
3. $\left|\Gamma^{k+1}_{f,\mathbf{z}}\right| \cap \Sigma f = \bigcup_{j\leq k} \left|\Lambda^j_{f,\mathbf{z}}\right|$; *in particular,* $\Sigma f = \bigcup_{j\leq n} \left|\Lambda^j_{f,\mathbf{z}}\right|$ *Furthermore,*
4. *if* $s := \dim_{\mathbf{p}} \Sigma f$, *then in a neighborhood of* $\mathbf{p}$, $\Lambda^k_{f,\mathbf{z}} = 0$ *for* $k > s$ *and one may start the inductive process with* $\Gamma^{s+1}_{f,\mathbf{z}} = \left[V\left(\frac{\partial f}{\partial z_{s+1}}, \dots, \frac{\partial f}{\partial z_n}\right)\right]$. *In particular, if* $\mathbf{p} \in V(f)$ *and* $\dim_{\mathbf{p}} \Sigma f = 0$, *then, near* $\mathbf{p}$, *the only non-zero Lê cycle is* $\Lambda^0_{f,\mathbf{z}}$ *and*

$$\left(\Lambda^0_{f,\mathbf{z}}\right)_{\mathbf{p}} = \left[V\left(\frac{\partial f}{\partial z_0}, \dots, \frac{\partial f}{\partial z_n}\right)\right]_{\mathbf{p}} = \dim_{\mathbb{C}} \frac{\mathbb{C}\{z_0 - p_0, \dots, z_n - p_n\}}{\langle \frac{\partial f}{\partial z_0}, \dots, \frac{\partial f}{\partial z_n} \rangle} = \mu_{\mathbf{p}}(f),$$

the Milnor number of f at $\mathbf{p}$.

Before we give an example, we want to first define the Lê **numbers**.

Definition 7.2.4 Suppose that the Lê cycles of f with respect to $\mathbf{z}$ exist.

Then, for all $\mathbf{p} \in U$, for all k such that $0 \leq k \leq n$, we say that the **k-dimensional Lê number, $\lambda^k_{f,\mathbf{z}}(\mathbf{p})$ (respectively, k-dimensional relative polar number, $\gamma^k_{f,\mathbf{z}}(\mathbf{p})$), of f with respect to $\mathbf{z}$ at $\mathbf{p}$ exists** if and only if $\Lambda^k_{f,\mathbf{z}}$ (respectively, $\Gamma^k_{f,\mathbf{z}}$) properly intersects the affine linear subspace $V(z_0 - p_0, \dots, z_{k-1} - p_{k-1})$ at $\mathbf{p}$, in other words, if $\mathbf{p}$ is an isolated point in the intersection.

If it exists, then naturally we define $\lambda^k_{f,\mathbf{z}}(\mathbf{p})$ (respectively, $\gamma^k_{f,\mathbf{z}}(\mathbf{p})$) to be the corresponding intersection number, i.e.,

$$\lambda^k_{f,\mathbf{z}}(\mathbf{p}) := \left(\Lambda^k_{f,\mathbf{z}} \cdot V(z_0 - p_0, \dots, z_{k-1} - p_{k-1})\right)_{\mathbf{p}}$$

(respectively,

$$\gamma^k_{f,\mathbf{z}}(\mathbf{p}) := \left(\Gamma^k_{f,\mathbf{z}} \cdot V(z_0 - p_0, \ldots, z_{k-1} - p_{k-1})\right)_{\mathbf{p}}.)$$

When $k = 0$, we mean that $\lambda^0_{f,\mathbf{z}}(\mathbf{p}) := \left(\Lambda^0_{f,\mathbf{z}}\right)_{\mathbf{p}}$, i.e., the coefficient of $\mathbf{p}$ in the 0-dimensional cycle $\Lambda^0_{f,\mathbf{z}}$.

Note that if the Lê numbers (respectively, relative polar numbers) exist at $\mathbf{p}$, they exist at all points near $\mathbf{p}$.

Example 7.2.5 Let $f : \mathbb{C}^3 \to \mathbb{C}$ be given by $f(t, x, y) = y^2 - x^4 - tx^3$, and use coordinates (t, x, y) (in that order). We shall suppress the reference to these fixed coordinates throughout the remainder of this example.

First we find $\Sigma f = V(-x^3, -4x^3 - 3tx^2, 2y) = V(x, y)$, i.e., the t-axis. Since $\dim \Sigma f = 1$, we may begin with

$$\Gamma^2_f = V\left(\frac{\partial f}{\partial y}\right) = V(2y) = V(y).$$

$$\Gamma^2_f \cdot V\left(\frac{\partial f}{\partial x}\right) = V(y) \cdot V(-4x^3 - 3tx^2) = V(y) \cdot V\left((4x + 3t)x^2\right) =$$

$$V(y) \cdot \left[V(4x + 3t) + 2V(x)\right] = V(y, 4x + 3t) + 2V(y, x).$$

So, $\Gamma^1_f = V(y, 4x + 3t)$ and $\Lambda^1_f = 2V(y, x)$.

Thus,

$$\Gamma^1_f \cdot V\left(\frac{\partial f}{\partial t}\right) = V(y, 4x + 3t) \cdot V(-x^3) = V(y, 4x + 3t) \cdot 3V(x) = 3[\mathbf{0}] = \Lambda^0_f.$$

And, of course, since $\mathbf{0}$ is a critical point, $\Gamma^0_f = 0$.

Thus, the Lê cycles are defined. Furthermore, the Lê numbers at the origin are defined and we calculate them:

$$\lambda^1_f(\mathbf{0}) = \left(2V(y, x) \cdot V(t)\right)_{\mathbf{0}} = 2 \qquad \text{and} \qquad \lambda^0_f(\mathbf{0}) = 3.$$

In the next section, we will return to this example and explain what it tells us about the homology/cohomology of the Milnor fiber of f at $\mathbf{0}$.

Example 7.2.6 Let $h : \mathbb{C}^3 \to \mathbb{C}$ be given by $h(t, x, y) = y^2 + (t - x^2)^2$. We find that

$$\Sigma h = V(2(t - x^2),\ 2(t - x^2)(-2x),\ 2y) \ = \ V(t - x^2, y),$$

which is 1-dimensional.

We wish to look at this with two different coordinate systems, and see that we obtain different Lê numbers.

First, use coordinates (t, x, y) and suppress writing them throughout. We start with

$$\Gamma_h^2 = V\left(\frac{\partial h}{\partial y}\right) = V(2y) = V(y).$$

Then

$$\Gamma_h^2 \cdot V\left(\frac{\partial h}{\partial x}\right) = V(y) \cdot V\big(2(t - x^2)(-2x)\big) = V(y, x) + V(y, t - x^2).$$

So $\Gamma_h^1 = V(y, x)$ and $\Lambda_h^1 = V(y, t - x^2)$. Now,

$$\Gamma_h^1 \cdot V\left(\frac{\partial h}{\partial t}\right) = V(y, x) \cdot V\big(2(t - x^2)\big) = V(y, x, t) = \Lambda_h^0.$$

Thus we find $\lambda_h^1(\mathbf{0}) = \big(V(y, t - x^2) \cdot V(t)\big)_{\mathbf{0}} = 2$ and $\lambda_h^0(\mathbf{0}) = 1$ with respect to the coordinates (t, x, y).

Now use coordinates (x, t, y) and suppress writing them throughout. We again start with

$$\Gamma_h^2 = V\left(\frac{\partial h}{\partial y}\right) = V(2y) = V(y).$$

Then

$$\Gamma_h^2 \cdot V\left(\frac{\partial h}{\partial t}\right) = V(y) \cdot V\big(2(t - x^2)\big) = V(y, t - x^2).$$

So $\Gamma_h^1 = 0$ and $\Lambda_h^1 = V(y, t - x^2)$. Thus we find $\lambda_h^1(\mathbf{0}) = \big(V(y, t - x^2) \cdot V(x)\big)_{\mathbf{0}} = 1$, and $\lambda_h^0(\mathbf{0}) = 0$ with respect to the coordinates (x, t, y).

In the previous example, the Lê cycles and Lê numbers changed when we changed coordinate systems; however, the reader is invited to check that, for a generic coordinate choice, one always obtains $\Lambda_h^1 = V(t - x^2, y)$, $\Lambda_f^0 = 0$, $\lambda_h^1(\mathbf{0}) = 1$ and $\lambda_h^0(\mathbf{0}) = 0$.

However, we will see that, in the next example, the Lê **cycles** are not fixed even for generic coordinates (but the Lê numbers are).

Example 7.2.7 Let $h = y^2 - x^3 - (u^2 + v^2 + w^2)x^2$ and fix the coordinates (u, v, w, x, y).

Then,

$$\Sigma h = V(-2ux^2,\ -2vx^2,\ -2wx^2,\ -3x^2 - 2x(u^2 + v^2 + w^2),\ 2y) = V(x, y).$$

As Σh is 3-dimensional, we begin our calculation with Γ_h^4.

$$\Gamma_h^4 = V(-2y) = V(y).$$

$$\Gamma_h^4 \cdot V\left(\frac{\partial h}{\partial x}\right) = V(y) \cdot V(-3x^2 - 2x(u^2 + v^2 + w^2)) =$$

$$V(-3x - 2(u^2 + v^2 + w^2), y) + V(x, y) = \Gamma_h^3 + \Lambda_h^3.$$

$$\Gamma_h^3 \cdot V\left(\frac{\partial h}{\partial w}\right) = V(-3x - 2(u^2 + v^2 + w^2), y) \cdot V(-2wx^2) =$$

$$V(-3x - 2(u^2 + v^2), w, y) + 2V(u^2 + v^2 + w^2, x, y) = \Gamma_h^2 + \Lambda_h^2.$$

$$\Gamma_h^2 \cdot V\left(\frac{\partial h}{\partial v}\right) = V(-3x - 2(u^2 + v^2), w, y) \cdot V(-2vx^2) =$$

$$V(-3x - 2u^2, v, w, y) + 2V(u^2 + v^2, w, x, y) = \Gamma_h^1 + \Lambda_h^1.$$

$$\Gamma_h^1 \cdot V\left(\frac{\partial h}{\partial u}\right) = V(-3x - 2u^2, v, w, y) \cdot V(-2ux^2) =$$

$$V(u, v, w, x, y) + 2V(u^2, v, w, x, y) = 5[\mathbf{0}] = \Lambda_h^0.$$

Hence, $\Lambda_h^3 = V(x, y)$, $\Lambda_h^2 = 2V(u^2 + v^2 + w^2, x, y) =$ a cone (as a set),

$$\Lambda_h^1 = 2V(u^2 + v^2, w, x, y),$$

and $\Lambda_h^0 = 5[\mathbf{0}]$. Thus, at the origin, $\lambda_h^3 = 1$, $\lambda_h^2 = 4$, $\lambda_h^1 = 4$, and $\lambda_h^0 = 5$.

Note that Λ_h^1 depends on the choice of coordinates—for, by symmetry, i f we re-ordered u, v, and w, then Λ_h^1 would change correspondingly. Moreover, one can check that this is a generic problem.

Such "non-fixed" Lê cycles arise from the absolute polar varieties (see [32, 54]) of the higher-dimensional Lê cycles (we shall see this in Theorem 7.6.6). For instance, in the present case, Λ_h^2 is a cone, and its 1-dimensional polar variety varies with the choice of coordinates, but generically always consists of two lines; this is the case for Λ_h^1 as well. Though the Lê cycles are not even generically fixed, the Lê numbers turn out to be generically independent of the coordinates (see Corollary 7.6.7).

We usually refer to the following proposition as the *Teissier trick*, since it was first proved by Teissier in [53] in the case of isolated critical points, by parameterizing the irreducible components of the relative polar curve and using the Chain Rule from Calculus; the proof is the same for critical loci of arbitrary dimension. Of course, we present the formula from the trick using our notation, not Teissier's. We will use this trick, together with Lê's attaching theorem, Theorem 7.1.10, in a crucial way in Sect. 7.3.

Proposition 7.2.8 *Suppose that the Lê numbers and relative polar numbers of f with respect to $\mathbf{z}$ at $\mathbf{p} \in V(f)$ exist.*

Then, $\mathbf{p} \notin \Gamma^1_{f,\mathbf{z}}$ or $\dim_{\mathbf{p}} \Gamma^1_{f,\mathbf{z}} \cap V(f) = 0$, and

$$\left(\Gamma^1_{f,\mathbf{z}} \cdot V(f)\right)_{\mathbf{p}} = \lambda^0_{f,\mathbf{z}}(\mathbf{p}) + \gamma^1_{f,\mathbf{z}}(\mathbf{p}).$$

Now we wish to know how Lê cycles and numbers behave when one takes a hyperplane slice. This involves how relative polar cycles and numbers behave under slicing.

The following proposition is part of Proposition 1.21 of [36].

Proposition 7.2.9 *Let $\mathbf{p} := (p_0, \ldots, p_n) \in \Sigma f$, use coordinates $\mathbf{z} := (z_0, \ldots, z_n)$ on $\mathcal{U}$, and coordinates $\tilde{\mathbf{z}} := (z_1, \ldots, z_n)$ on $V(z_0 - p_0)$. Suppose that all of the Lê and relative polar numbers $\lambda^*_{f,\mathbf{z}}(\mathbf{p})$ and $\gamma^*_{f,\mathbf{z}}(\mathbf{p})$ exist.*

Then,

1. *all of the Lê and relative polar numbers $\lambda^*_{f_{|V(z_0-p_0)},\tilde{\mathbf{z}}}(\mathbf{p})$ and $\gamma^*_{f_{|V(z_0-p_0)},\tilde{\mathbf{z}}}(\mathbf{p})$ exist,*
2. $$\lambda^0_{f_{|V(z_0-p_0)},\tilde{\mathbf{z}}}(\mathbf{p}) = \lambda^1_{f,\mathbf{z}}(\mathbf{p}) + \gamma^1_{f,\mathbf{z}}(\mathbf{p}),$$
3. *for all k such that $1 \le k \le n-1$ and all j such that $1 \le j \le n$,*
$$\lambda^k_{f_{|V(z_0-p_0)},\tilde{\mathbf{z}}}(\mathbf{p}) = \lambda^{k+1}_{f,\mathbf{z}}(\mathbf{p}) \quad \text{and} \quad \gamma^j_{f_{|V(z_0-p_0)},\tilde{\mathbf{z}}}(\mathbf{p}) = \gamma^{j+1}_{f,\mathbf{z}}(\mathbf{p}),$$
and
4. *near $\mathbf{p}$, for all k such that $1 \le k \le n-1$ and all j such that $1 \le j \le n$,*
$$\Lambda^k_{f_{|V(z_0-p_0)},\tilde{\mathbf{z}}} = \Lambda^{k+1}_{f,\mathbf{z}} \cdot V(z_0 - p_0) \quad \text{and} \quad \Gamma^j_{f_{|V(z_0-p_0)},\tilde{\mathbf{z}}} = \Gamma^{j+1}_{f,\mathbf{z}} \cdot V(z_0 - p_0).$$

Example 7.2.10 Let us look at the function and first set of coordinates (t, x, y) from Example 7.2.6. Below, we will dispense with referencing the coordinates until it is crucial at the end.

We had (or easily calculate now) $h : \mathbb{C}^3 \to \mathbb{C}$ given by $h(t, x, y) = y^2 + (t - x^2)^2$, $\Sigma h = V(t - x^2, y)$, $\Gamma^3_h = \mathbb{C}^3$, $\Gamma^2_h = V(y)$, $\Gamma^1_h = V(y, x)$, $\Lambda^1_h = V(y, t - x^2)$, $\gamma^3_h(\mathbf{0}) = 1$, $\gamma^2_h(\mathbf{0}) = 1$, $\gamma^1_h(\mathbf{0}) = 1$, $\lambda^1_h(\mathbf{0}) = 2$, and $\lambda^0_h(\mathbf{0}) = 1$ with respect to the coordinates (t, x, y).

Now $\Sigma(h_{|V(t)}) = \Sigma(y^2 + x^4, t) = V(t) \cap \Sigma h$, so we may apply Proposition 7.2.9, to conclude that

$$\lambda^0_{h_{|V(t)},(x,y)}(\mathbf{0}) = \lambda^1_{h,(t,x,y)}(\mathbf{0}) + \gamma^1_{h,(t,x,y)}(\mathbf{0}) = 2 + 1 = 3.$$

As $h_{|_{V(t)}}$ has an isolated critical point at $\mathbf{0}$, $\lambda^0_{h_{|_{V(t)}},(x,y)}(\mathbf{0})$ equals $\mu_{\mathbf{0}}(h_{|_{V(t)}})$, and our calculation above agrees with

$$\mu_{\mathbf{0}}(h_{|_{V(t)}}) = \dim_{\mathbb{C}} \frac{\mathbb{C}\{x, y\}}{\langle 4x^3, 2y\rangle} = 3.$$

The following is an interesting separate result; it is Proposition 1.31 of [36].

Proposition 7.2.11 *Let $k \geqslant 1$. Suppose that the Lê cycles exist, and suppose, for all pairs of distinct irreducible germs, V and W, of Σf through $\mathbf{p}$, that*

$$\dim_{\mathbf{p}}(V \cap W) \leqslant k - 1.$$

Suppose that $\lambda^k_{f,\mathbf{z}}(\mathbf{p}) = 0$. Then, $\lambda^j_{f,\mathbf{z}}(\mathbf{p}) = 0$ for all $j \leqslant k$.

Finally, in this section, we need to discuss how generically one needs the coordinates to be in order for all of the Lê numbers and relative polar numbers to exist, and for us to know that the topological implications that we want later hold.

First, we need to recall a version of an A_f stratification, a weaker version than that described by Lê in [28]. The version we present was called a *good stratification* by Hamm and Lê in [12]. We continue with $f : \mathcal{U} \to \mathbb{C}$ being an analytic function which is nowhere locally constant.

Definition 7.2.12 If X is analytic space, an *analytic stratification* of X is a locally finite partition, $\{S_\alpha\}$, of X into analytic submanifolds—*the strata*—such the closure of each stratum is also analytic, and such that $\{S_\alpha\}$ satisfies the condition of the frontier, i.e., the closure of each stratum is a union of strata.

Suppose that we have an analytic function $f : \mathcal{U} \to \mathbb{C}$. A **good stratification** for f at a point $\mathbf{p} \in V(f)$ is an analytic stratification, $\mathfrak{G}$, of the hypersurface $V(f)$ in a neighborhood, $\mathcal{U}'$, of $\mathbf{p}$ such that the smooth part of $V(f)$ is a stratum and so that the stratification satisfies Thom's A_f condition with respect to $\mathcal{U}' - V(f)$. That is, if $\mathbf{q}_i$ is a sequence of points in $\mathcal{U}' - V(f)$ such that $\mathbf{q}_i \to \mathbf{q} \in S \in \mathfrak{G}$ and $T_{\mathbf{q}_i} V(f - f(\mathbf{q}_i))$ converges to some hyperplane $\mathcal{T}$, then $T_q S \subseteq \mathcal{T}$.

In [12], Hamm and Lê give a proof that good stratifications exist; of course, the existence also follows from Hironaka's result as stated in Theorem 3.5 of [28].

The notion defined below, that of *prepolar coordinates*, is crucial throughout the remainder of this article. It provides a generic condition on linear choices of coordinates which implies that all the Lê numbers and polar numbers are defined. Moreover, prepolarity seems to be the right condition to obtain many topological results. The importance of this definition cannot be overstated.

Definition 7.2.13 Suppose that $\{S_\alpha\}$ is a good stratification for f in a neighborhood, $\mathcal{U}$, of the origin. Let $\mathbf{p} \in V(f)$. Then, a hyperplane, H, in $\mathbb{C}^{n+1}$ through $\mathbf{p}$ is a **prepolar slice** for f at $\mathbf{p}$ with respect to $\{S_\alpha\}$ provided that H transversely intersects all the strata of $\{S_\alpha\}$—except perhaps the stratum $\{\mathbf{p}\}$ itself – in a neighborhood of $\mathbf{p}$.

If H is a prepolar slice for f at $\mathbf{p}$ with respect to $\{S_\alpha\}$, then, as germs of sets at $\mathbf{p}$, $\Sigma(f_{|_H}) = (\Sigma f) \cap H$ and $\dim_{\mathbf{p}} \Sigma(f_{|_H}) = (\dim_{\mathbf{p}} \Sigma f) - 1$ provided $\dim_{\mathbf{p}} \Sigma f \geqslant 1$; moreover, $\{H \cap S_\alpha\}$ is a good stratification for $f_{|_H}$ at $\mathbf{p}$ (see [12]).

By 2.1.3 of [12], for a fixed good stratification for f, prepolar slices are generic.

We say simply that H is a *prepolar slice* for f at $\mathbf{p}$ provided that there exists a good stratification with respect to which H is a prepolar slice.

Let $(z_0, \dots, z_n)$ be a linear choice of coordinates for $\mathbb{C}^{n+1}$, let $\mathbf{p} \in V(f)$, and let $\{S_\alpha\}$ be a good stratification for h at $\mathbf{p}$.

For $0 \leqslant i \leqslant n$, $(z_0, \dots, z_i)$ is a **prepolar-tuple for f at p with respect to** $\{S_\alpha\}$ if and only if $V(z_0 - p_0)$ is a prepolar slice for f at $\mathbf{p}$ with respect to $\{S_\alpha\}$ and for all j such that $1 \leqslant j \leqslant i$, $V(z_j - p_j)$ is a prepolar slice for $f_{|V(z_0-p_0,\dots,z_{j-1}-p_{j-1})}$ at $\mathbf{p}$ with respect to the good stratification

$$\{S_\alpha \cap V(z_0 - p_0, \dots, z_{j-1} - p_{j-1})\}.$$

As prepolar slices are generic, so are prepolar-tuples.

Naturally, we say that $(z_0, \dots, z_i)$ is a **prepolar-tuple for f at p** provided that there exists a good stratification for f at $\mathbf{p}$ with respect to which $(z_0, \dots, z_i)$ is a prepolar-tuple.

Note that, if $\mathbf{p} \in V(f)$ and $\dim_{\mathbf{p}} \Sigma f = 1$, then H is a prepolar slice for f at $\mathbf{p}$ if and only if $\dim_{\mathbf{p}} \Sigma(f_{|_H}) = 0$.

Remark 7.2.14 Note that in all of our discussion and definitions above of a good stratification and prepolar coordinates, we did **not** actually need a "stratification", that is, we did not use the condition of the frontier that the closure of each stratum must be a union of strata. An analytic partition into analytic submanifolds satisfying the A_f with respect to the ambient stratum would have sufficed.

The following is Theorem 1.28 of [36].

Theorem 7.2.15 *Suppose that $\mathbf{p} \in V(f) \cap \Sigma f$, let $s := \dim_{\mathbf{p}} \Sigma f$, and suppose that $(z_0, \dots, z_{s-1})$ is a prepolar-tuple at $\mathbf{p}$ (when $s = 0$, there is no condition on the coordinates). Then, for all k such that $0 \leq k \leq s$, $\lambda^k_{f,\mathbf{z}}(\mathbf{p})$ and $\gamma^k_{f,\mathbf{z}}(\mathbf{p})$ exist.*

Furthermore, if $(z_0, \dots, z_{n-1})$ is a prepolar-tuple at $\mathbf{x} \in V(f)$, then all of the Lê numbers and relative polar numbers of f with respect to $\mathbf{z}$ exist at $\mathbf{x}$.

In particular, for all $\mathbf{x} \in V(f)$, for a generic choice of coordinates $\mathbf{z}$, there is an open neighborhood of $\mathbf{x}$ in which all of the Lê cycles, relative polar cycles, Lê numbers, and relative polar numbers of f with respect to $\mathbf{z}$ exist.

7.3 Lê Numbers and the Topology of the Milnor Fiber

As in the previous sections, we let $\mathcal{U}$ be a non-empty open subset $\mathbb{C}^{n+1}$, $n \geq 1$, and $f : \mathcal{U} \to \mathbb{C}$ an analytic function which is nowhere locally constant. We make a linear choice of coordinates $\mathbf{z} := (z_0, \dots, z_n)$ on $\mathcal{U}$.

In this section, we give a handle decomposition, in terms of Lê numbers, of the Milnor fiber of f, where f may have a critical locus of arbitrary dimension. This decomposition is more refined than that obtained by iteratively applying Lê's attaching result from [22].

However, first, we wish to discuss three prior results of others which led to our general formulation in terms of Lê numbers. And, we should clarify a point that we have been sloppy about twice in the past; our result is about the **compact** Milnor fiber (see Proposition 1.9 of [28]). If $\mathbf{p} \in V(f)$, we denote the compact Milnor fiber of f at $\mathbf{p}$ by $\overline{F}_{f,\mathbf{p}}$. Thus,

$$\overline{F}_{f,\mathbf{p}} := B_\epsilon(\mathbf{p}) \cap f^{-1}(t),$$

where $0 < \epsilon \ll 1$, $B_\epsilon(\mathbf{p})$ is a **closed** ball of radius ϵ centered at $\mathbf{p}$, and $t \in \mathbb{D}^\circ_\eta$, where $\mathbb{D}^\circ_\eta$ is an open disk in $\mathbb{C}$ centered at the origin, and $0 < \eta \ll \epsilon$.

We also remind the reader that, if $\dim_{\mathbf{p}} \Sigma f = 0$, then the Milnor number of f at $\mathbf{p}$, $\mu_f(\mathbf{p})$, is equal to $\lambda^0_{f,\mathbf{z}}(\mathbf{p})$ (regardless of the choice of the coordinates $\mathbf{z}$).

Now we combine the result of Milnor in Theorem 6.6 of [42] for $n \neq 2$, and of Lê and Perron in Proposition 0 of [29] in the case where $n = 2$, to obtain:

Theorem 7.3.1 *Suppose that* $\dim_{\mathbf{p}} \Sigma f = 0$. *Then the compact Milnor fiber* $\overline{F}_{f,\mathbf{p}}$ *is obtained up to diffeomorphism from a real* $2n$*-ball by attaching* $\mu_f(\mathbf{p})$ n*-handles.*

Translating the result of Vannier in Proposition 1 of [57] into our notation, we have:

Theorem 7.3.2 (*Vannier*) *Suppose that* $\dim_{\mathbf{p}} \Sigma f = 1$, $n \geq 3$, *and* z_0 *is a generic linear form (prepolar is enough). Then the compact Milnor fiber* $\overline{F}_{f,\mathbf{p}}$ *is obtained up to diffeomorphism from a real* $2n$*-ball by attaching* $\lambda^1_{f,\mathbf{z}}(\mathbf{p})$ $(n-1)$*-handles and then attaching* $\lambda^0_{f,\mathbf{z}}(\mathbf{p})$ n*-handles.*

Remark 7.3.3 We wish to sketch the proofs of these previous two theorems so that the reader can easily see how to prove our generalization to functions with critical loci of arbitrary dimension. The reader should also consult Sect. 6 of [28].

We will assume that the point $\mathbf{p} \in V(f)$ is the origin for convenience. We will omit many technical details, but hopefully include enough so that all of the ideas are clear. Even without some technical details, this sketch is lengthy.

In what follows, we will frequently encounter the product of two smooth manifolds with boundary, producing a manifold with boundary and "corners". As is common, we "straighten the angles" or "smooth these corners" as discussed by Smale on page 396 of [51], which actually refers to Milnor's 1959 notes [41]. Throughout the remainder of this sketch, whenever we have the product of two smooth manifolds with boundary, we will assume, frequently without additional comment, that the corners have been smoothed to produce a new manifold with boundary.

We begin by describing Lê's proof of his attaching theorem, Theorem 7.1.10, where L in that statement would be our z_0, which we assume to be prepolar, and so $\Gamma^1_{f,L} = \Gamma^1_{f,z_0} = \Gamma^1_{f,\mathbf{z}}$.

For $t \in \mathbb{C}$, let $f_t := f_{|V(z_0 - t_0)}$. Choose $\epsilon > 0$ small enough so that the closed ball of radius ϵ centered at the origin in $V(z_0)$, $\{0\} \times B_\epsilon^{2n}$, is a Milnor ball for f_0 at the origin (see [28]). One can now choose τ and δ, $0 < \tau \ll \delta \ll \epsilon$, so that

- $\Gamma^1_{f,z_0} \cap \partial\big(\mathbb{D}^\circ_\delta \times B_\epsilon^{2n}\big) = \emptyset$,
- $N := \big(\mathbb{D}_\delta \times B_\epsilon^{2n}\big) \cap f^{-1}(\mathbb{D}^*_\tau)$ is the total space of a Milnor tube fibration (with corners) for f (as in Theorem 3.1 of [28]),
- $N \cap V(z_0)$ is the total space for a Milnor tube fibration for f_0, and
- $f^{-1}(\mathbb{D}_\tau) \cap \Gamma^1_{f,z_0} \cap \partial(\mathbb{D}_\delta \times B_\epsilon^{2n}) = \emptyset$.

Finally, we may assume that $(\{0\} \times B_\epsilon) \cap f_0^{-1}(\mathbb{D}_\tau)$ is diffeomorphic to the closed $2n$-ball $\{0\} \times B_\epsilon$ (by the argument of Milnor in Lemmas 5.9, 5.10, and Theorem 5.11 of [42]).

Let v be in the interior of $\mathbb{D}^*_\tau$. Lê's proof of his attaching result is via Morse Theory. He considers the function r given by restricting $|z_0|^2$ to the compact Milnor of f at the origin,

$$\overline{F}_{f,\mathbf{0}} := \big(\mathbb{D}_\delta \times B_\epsilon^{2n}\big) \cap f^{-1}(v)$$

(a technical argument is required to prove that this is diffeomorphic—after smoothing corners—to the standard compact Milnor fiber $B_\epsilon^{2n+2} \cap f^{-1}(v)$). It is easy to show that, for $0 < \omega \ll |v|$, $\big(\mathbb{D}_\omega \times B_\epsilon^{2n}\big) \cap f^{-1}(v)$ is diffeomorphic to the product of $\overline{F}_{f_0,\mathbf{0}}$ with $\mathbb{D}_\omega$. Lê then lets the value of r grow, starting from the small positive value ω^2 to δ^2.

For $r > \omega^2$, the function r has critical points precisely where the Milnor fiber intersects Γ^1_{f,z_0}. In Lê's case, z_0 is generic enough so that Γ^1_{f,z_0} is reduced, which implies that the critical points of z_0 restricted to $\overline{F}_{f,\mathbf{0}}$ are complex non-degenerate; this implies that each critical point of r has index n. With the assumption that z_0 is merely prepolar, $\Gamma^1_{f,\mathbf{z}}$ need not be reduced, but then one perturbs to the non-degenerate case; this splitting of the degenerate critical points is automatically counted correctly by the intersection number $\left(\Gamma^1_{f,z_0} \cdot V(f)\right)_{\mathbf{0}}$. Hence, for ease, throughout the remainder of this sketch, we will assume that Γ^1_{f,z_0} is reduced.

In terms of handles, what the argument above shows is that the compact Milnor fiber $\overline{F}_{f,\mathbf{0}}$ is obtained from the product of the the compact Milnor fiber $\overline{F}_{f_0,\mathbf{0}}$ with a disk and then attaching $\left(\Gamma^1_{f,z_0} \cdot V(f)\right)_{\mathbf{0}}$ n-handles.

We wish to describe the above attaching in a different way, though it is not truly a "different" way; it is still Morse theoretic, just more topological and less analytically rigorous since the Morse function(s) involved are not given analytically. This more topological viewpoint will be crucial for one's understanding of what we do later.

Consider the map T which is the restriction of (f, z_0) to the closure

$$\overline{N} := \big(\mathbb{D}_\delta \times B_\epsilon^{2n}\big) \cap f^{-1}(\mathbb{D}_\tau)$$

of N above, i.e., we are using $\mathbb{D}_\tau$ here, not $\mathbb{D}^*_\tau$. As Lê demonstrates, for generic z_0, the restriction of T to $\big(\mathbb{D}_\delta \times B_\epsilon^{2n}\big) \cap \Gamma^1_{f,z_0}$ is one-to-one; so we assume that z_0 is this generic. Thus we have $T : \overline{N} \to \mathbb{D}_\tau \times \mathbb{D}_\delta$ and the *Cerf diagram* C is the curve $T(\Gamma^1_{f,z_0})$.

The intersection $(\{v\} \times \mathbb{D}_\delta) \cap C$ consists of $\left(\Gamma^1_{f,z_0} \cdot V(f)\right)_{\mathbf{0}}$ points, which we refer to as the *Cerf points for* $f = v$. Consider a set $K \subseteq \{v\} \times \mathbb{D}_\delta$ formed as follows: take a small disk $\mathbb{D}_0$ around the origin in $\mathbb{D}_\delta$, connect $\{v\} \times \mathbb{D}_0$ to a collection of small disks (the Cerf disks) around the Cerf points for $f = v$ by any choice of smooth non-intersecting, non-self-intersecting paths in $\{v\} \times \mathbb{D}_\delta$, thicken and smooth the paths; this defines K. Let $\widehat{K} := T^{-1}(K)$.

Then $\widehat{K}$ is diffeomorphic to the entire Milnor fiber $F_{f,\mathbf{0}}$, because one encounters no more points in the Cerf diagram as one follows a flow taking the boundary of K to all of $\{v\} \times \mathbb{D}_\delta$ and this flow lifts by T to take $\widehat{K}$ to $F_{f,\mathbf{0}}$. In this description, each thickened, smoothed path and Cerf disk corresponds to the attaching of one n-handle. On the level of homology, this follows from excision and using deformation retracts; on the level of Morse theory and handles, one would have to produce a Morse function which increases along the pre-images of the thickened, smoothed paths of K, and has no critical points other than a single complex non-degenerate critical point above each Cerf point for $f = v$. It is topologically clear that such Morse functions exist but, without analytic descriptions of the thickened, smoothed paths, one has no hope of writing these Morse functions explicitly. Nonetheless, this description and viewpoint is fundamental to the remainder of the our argument, which proceeds by describing regions and an isotopy in $\mathbb{D}_\tau \times \mathbb{D}_\delta$, and lifting via T to regions in $\overline{N}$.

Recall from Proposition 7.2.8,

$$\left(\Gamma^1_{f,\mathbf{z}} \cdot V(f)\right)_{\mathbf{0}} = \lambda^0_{f,\mathbf{z}}(\mathbf{0}) + \gamma^1_{f,\mathbf{z}}(\mathbf{0}).$$

Note that, if $\dim_{\mathbf{0}} \Sigma f = 0$, then $\lambda^0_{f,\mathbf{z}}(\mathbf{0}) = \mu_f(\mathbf{0})$ and $\gamma^1_{f,\mathbf{z}}(\mathbf{0}) = \mu_{f_{|V(z_0)}}(\mathbf{0})$.

What the proofs o f L ê ad Perron, Vannier, and our own show is that, during the handle attaching which we described above, $\gamma^1_{f,\mathbf{z}}(\mathbf{0})$ of the attached n-handles cancel with $\gamma^1_{f,\mathbf{z}}(\mathbf{0})$ $(n-1)$-handles in a handle decomposition of the product of $\overline{F}_{f_{|V(z_0)},\mathbf{0}}$ with a disk.

Recall that $(\{0\} \times B_\epsilon) \cap f_0^{-1}(\mathbb{D}_\tau)$ is diffeomorphic to the closed $2n$-ball $\{0\} \times B_\epsilon$. It is easy to show that, for small $\omega > 0$,

$$\Gamma^1_{f,z_0} \cap \partial\big((\mathbb{D}^\circ_\omega \times B_\epsilon) \cap f^{-1}(\mathbb{D}_\tau)\big) = \emptyset$$

and the projection

$$\big(\mathbb{D}^\circ_\omega \times B_\epsilon\big) \cap f^{-1}(\mathbb{D}_\tau) \to \mathbb{D}^\circ_\omega$$

is a smooth trivial fibration (with boundary and corners). Thus, letting $b \in (\mathbb{D}^*_\omega)^\circ$, we have that $\overline{N}_b := (\{b\} \times B_\epsilon) \cap f_b^{-1}(\mathbb{D}_\tau)$ is diffeomorphic to the closed $2n$-ball $\{b\} \times B_\epsilon$.

Now consider the set $Z \subseteq \mathbb{D}_\tau \times \{b\}$ formed by taking a small closed disk $\mathbb{D}$ centered at (v, b), connecting it by non-intersecting paths to small non-intersecting closed disks centered at $(0, b)$ and disks, the Cerf disks for $z_0 = b$, at the $\gamma^1_{f,\mathbf{z}}(\mathbf{0}) = \left(\Gamma^1_{f,\mathbf{z}} \cdot V(z_0)\right)_{\mathbf{0}}$ points in $(\mathbb{D}_\tau \times \{b\}) \cap C$, and then thicken the paths and smooth the corners. Then, $T^{-1}(Z)$ is diffeomorphic to the entire space $\overline{N}_b$, because there are no more critical points of f_b as one flows outward from $T^{-1}(Z)$ to fill up all of $\overline{N}_b$; this, in turn, is diffeomorphic to the closed $2n$-ball $\{b\} \times B_\epsilon$. It is important to note that, if $\dim_{\mathbf{0}} \Sigma f = 0$, then 0 is not a critical value of f_b and so $T^{-1}(A)$ is diffeomorphic to $T^{-1}(Z)$, which is diffeomorphic to a closed $2n$-ball.

Let A be the subset of Z consisting of the disk around (v, b) and the Cerf disks, together with the corresponding thickened, smoothed paths. Let Y be the subset of Z consisting of the disk around (v, b) and the disk around $(0, b)$, together with the corresponding thickened, smoothed path. Let $\mathcal{U}$ and $\mathcal{W}$ be small open neighborhoods of $T^{-1}(Y)$ and $T^{-1}(A)$, respectively, which are homotopy-equivalent to $T^{-1}(Y)$ and $T^{-1}(A)$, and whose intersection is homotopy-equivalent to the pre-image to the small disk around (v, b) from above.

Then $\mathcal{U} \cup \mathcal{W}$ is contractible, $\mathcal{U} \cap \mathcal{W}$ is homotopy-equivalent to the Milnor fiber $F_{f_0,\mathbf{0}}$, and $\mathcal{W}$ is homotopy-equivalent to $T^{-1}(A)$, which is ĭs a smooth manifold with boundary obtained from the product of $\overline{F}_{f_0,\mathbf{0}}$ with a disk to which one attaches $\gamma^1_{f,\mathbf{z}}(\mathbf{0})$ n-handles, which come from the complex non-degenerate critical points of f_b, which correspond to the points in $(\mathbb{D}_\tau \times \{b\}) \cap C$. An easy Mayer-Vietoris argument (a slight generalization of that used by Vannier in [57]), as given in the proofs of Theorem 4.3 of [33] and Theorem 3.3 of [36], shows that $T^{-1}(A)$ has the homology of (even the homotopy-type of) a finite $(n-1)$-dimensional CW complex; in particular, $H_n(T^{-1}(A)) = 0$ and $H_{n-1}(T^{-1}(A))$ is free abelian. This means that, on the level of homology, the attached $\gamma^1_{f,\mathbf{z}}(\mathbf{0})$ n-handles cancel with $\gamma^1_{f,\mathbf{z}}(\mathbf{0})$ $(n-1)$-handles. Hence, if $n \geq 3$, then the results of Smale in [51] (see also Cerf's Lemme fundamental of Sect. III of [4]) tell us that the attached $\gamma^1_{f,\mathbf{z}}(\mathbf{0})$ n-handles cancel with $(n-1)$-handles in a handle decomposition of the product of $\overline{F}_{f_{|V(z_0)},\mathbf{0}}$ with a disk.

Now, as described in [29], one can "swing" the region A in $\mathbb{D}_\tau \times \mathbb{D}_\delta$ isotopically, by rotating the small disk centered at (v, b) and "sliding" the Cerf disks along the components of the Cerf diagram to obtain a subset $E \subseteq \{v\} \times \mathbb{D}_\delta$ which consists of a small closed disk, connected by non-intersecting paths to small non-intersecting closed disks centered at $\gamma^1_{f,\mathbf{z}}(\mathbf{0})$ of the $\left(\Gamma^1_{f,\mathbf{z}} \cdot V(f)\right)_{\mathbf{0}}$ points in $(\{v\} \times \mathbb{D}_\delta) \cap C$, and then thickening the paths and smoothing any corners.

This isotopy lifts to an isotopy from $T^{-1}(A)$ to $T^{-1}(E)$, and so $T^{-1}(E)$ is obtained from the product o f$\overline{F}_{f_{|V(z_0)},\mathbf{0}}$ with a disk by canceling $\gamma^1_{f,\mathbf{z}}(\mathbf{0})$ $(n-1)$-handles.

Note that the isotopy presents $T^{-1}(E)$ as the pre-image of a small disk in $\{v\} \times \mathbb{D}_\delta$, connected by thickened, smoothed paths to $\gamma^1_{f,\mathbf{z}}(\mathbf{0})$ of the Cerf disks for $f = v$. Now the entire Milnor fiber $\overline{F}_{f,\mathbf{0}}$ is obtained from $T^{-1}(E)$ by attaching the remaining n-handles from Lê's Attaching Theorem, by taking the pre-image under T of the region E with the remaining

$$\left(\Gamma^1_{f,\mathbf{z}} \cdot V(f)\right)_{\mathbf{0}} - \gamma^1_{f,\mathbf{z}}(\mathbf{0}) = \lambda^0_{f,\mathbf{z}}(\mathbf{0})$$

Cerf disks for $f = v$ attached via thickened, smoothed paths from E.

Recall from Proposition 7.2.9 that

$$\lambda^0_{f_{|V(z_0)}, \tilde{\mathbf{z}}}(\mathbf{0}) = \lambda^1_{f,\mathbf{z}}(\mathbf{0}) + \gamma^1_{f,\mathbf{z}}(\mathbf{0}).$$

If $\dim_{\mathbf{0}} \Sigma f \leq 1$, then this formula becomes $\mu_{f_{|V(z_0)}}(\mathbf{0}) = \lambda^1_{f,\mathbf{z}}(\mathbf{0}) + \gamma^1_{f,\mathbf{z}}(\mathbf{0})$, where $\lambda^1_{f,\mathbf{z}}(\mathbf{0}) = 0$ if $\dim_{\mathbf{0}} \Sigma f = 0$. The results of Lê and Perron and Vannier follow.

Using our notation and conventions from the remark above, what we have shown is:

Proposition 7.3.4 *Suppose that $V(z_0)$ is a prepolar slice for f at $\mathbf{0}$, and $n \neq 2$, then the Milnor fiber of f at $\mathbf{0}$ is obtained—up to diffeomorphism—from the product of a closed disk $\mathbb{D}$ with the Milnor fiber of f_0 at $\mathbf{0}$ by first attaching $\gamma^1_{f,\mathbf{z}}(\mathbf{0})$ n-handles, which cancel against $\gamma^1_{f,\mathbf{z}}(\mathbf{0})$ $(n-1)$-handles of $\mathbb{D} \times F_{f_0,\mathbf{0}}$, and then attaching $\lambda^0_{f,\mathbf{z}}(\mathbf{0})$ more n-handles.*

If $n = 2$, we have the same conclusion except that the canceling is only up to homotopy.

The following is Theorem 4.3 of [33] and Theorem 3.3 of [36]. The proof by induction is immediate from Proposition 7.3.4, together with the formulas for the Lê numbers of a hyperplane slice from Proposition 7.2.9.

Theorem 7.3.5 *Let $\mathcal{U}$ be an open subset of $\mathbb{C}^{n+1}$, let $f : \mathcal{U} \to \mathbb{C}$ be an analytic map, let $\mathbf{p} \in V(f)$, let s denote* $\dim_{\mathbf{p}} \Sigma f$, *and let $\mathbf{z} = (z_0, \dots, z_{s-1})$ be prepolar for f at* $\mathbf{p}$.

If $s \leqslant n-2$, then $F_{f,\mathbf{p}}$ is obtained up to diffeomorphism from a real $2n$-ball by successively attaching $\lambda^{n-k}_{f,\mathbf{z}}(\mathbf{p})$ k-handles, where $n - s \leqslant k \leqslant n$;

if $s = n-1$, then $F_{f,\mathbf{p}}$ is obtained up to diffeomorphism from a real $2n$-manifold with the homotopy-type of a bouquet of $\lambda^{n-1}_{f,\mathbf{z}}(\mathbf{p})$ circles by successively attaching $\lambda^{n-k}_{f,\mathbf{z}}(\mathbf{p})$ k-handles, where $2 \leqslant k \leqslant n$.

Remark 7.3.6 The reader should understand that we do not claim in Theorem 7.3.5 that there is no further cancellation of higher-dimensional attached handles with lower-dimensional handles; it is absolutely **not** true in general that the Lê numbers are equal to the Betti numbers of the Milnor fiber.

In the following corollary, we let $\tilde{b}_i(f, \mathbf{p})$ denote the degree i reduced Betti number of $F_{f,\mathbf{p}}$, and note that the Universal Coefficient Theorem implies that homology and cohomology yield the same Betti numbers.

Corollary 7.3.7 *Let $\mathcal{U}$ be an open subset of $\mathbb{C}^{n+1}$, let $f : \mathcal{U} \to \mathbb{C}$ be an analytic map, let $\mathbf{p} \in V(f)$, let s denote* $\dim_{\mathbf{p}} \Sigma f$, *and let $\mathbf{z} = (z_0, \dots, z_{s-1})$ be prepolar for f at* $\mathbf{p}$. *We will suppress the references to the coordinates below.*

Then there is a chain complex (the **Lê complex***)*

$$0 \to \mathbb{Z}^{\lambda^s_f(\mathbf{p})} \to \mathbb{Z}^{\lambda^{s-1}_f(\mathbf{p})} \to \cdots \to \mathbb{Z}^{\lambda^1_f(\mathbf{p})} \to \mathbb{Z}^{\lambda^0_f(\mathbf{p})} \to 0$$

such that the cohomology of the complex at the $\lambda^k_f(\mathbf{p})$ *term is isomorphic to* $\widetilde{H}^{n-k}(F_{f,\mathbf{p}}; \mathbb{Z})$.

Hence, the reduced Euler characteristic of the Milnor fiber of f *at* $\mathbf{p}$ *is given by*

$$\widetilde{\chi}(F_{f,\mathbf{p}}) = \sum_{i=0}^{s}(-1)^{n-i}\lambda^i_f(\mathbf{p})$$

and the reduced Betti numbers, $\tilde{b}_i(f; \mathbf{p})$, *satisfy Morse inequalities with respect to the Lê numbers, i.e., for all* k *with* $n - s \leqslant k \leqslant n$,

$$(-1)^k \sum_{i=n-s}^{k} (-1)^i \tilde{b}_i(f, \mathbf{p}) \leqslant (-1)^k \sum_{i=n-s}^{k} (-1)^i \lambda^{n-i}_f(\mathbf{p})$$

and

$$(-1)^k \sum_{i=k}^{n}(-1)^i \tilde{b}_i(f, \mathbf{p}) \leqslant (-1)^k \sum_{i=k}^{n}(-1)^i \lambda^{n-i}_f(\mathbf{p}).$$

In particular, $\tilde{b}_n(f, \mathbf{p}) \leq \lambda^0_f(\mathbf{p})$ *and* $\widetilde{H}^{n-s}(F_{f,\mathbf{p}}; \mathbb{Z})$ *is free abelian of rank* $\tilde{b}_{n-s}(f, \mathbf{p}) \leq \lambda^s_f(\mathbf{p})$.

Now we wish to give a generalization, on the cohomological level, of the main result of Siersma in [50]. Phrased in terms of Lê numbers, Siersma's result is that, if $\mathbf{p} \in V(f)$, $\dim_{\mathbf{p}} \Sigma f = 1$, and $\lambda^1_f(\mathbf{p}) = 1$, then either f defines a one-parameter constant Milnor number family or $F_{f,\mathbf{p}}$ has the homotopy-type of a bouquet of $(\lambda^0_f(\mathbf{p}) - 1)$ n-spheres.

Cohomologically, what this says is that, if $\lambda^1_f(\mathbf{p}) = 1$ and rank $H^{n-1}(F_{f,\mathbf{p}}; \mathbb{Z}) = 1$, then $\lambda^0_f(\mathbf{p}) = 0$. Together with Lê, we generalized this cohomological statement in Theorem 5.3 of [26] (here, with a slightly different phrasing):

Theorem 7.3.8 *Let* $\mathcal{U}$ *be an open subset of* $\mathbb{C}^{n+1}$, *let* $f : \mathcal{U} \to \mathbb{C}$ *be an analytic map, let* $\mathbf{p} \in V(f)$, *let* s *denote* $\dim_{\mathbf{p}}\Sigma f$, *and let* $\mathbf{z} = (z_0, \ldots, z_{s-1})$ *be prepolar for* f *at* $\mathbf{p}$. *Suppose that* $\lambda^s_f(\mathbf{p}) = \operatorname{rank} \widetilde{H}^{n-s}(F_{f,\mathbf{p}}; \mathbb{Z})$.

Then, Σf *is smooth at* $\mathbf{p}$ *and, for all* k *such that* $0 \leq k \leq s - 1$, $\lambda^k_f(\mathbf{p}) = 0$. *Hence, the reduced cohomology of* $F_{f,\mathbf{p}}$ *is zero outside of degree* $n - s$, *where it is isomorphic to* $\mathbb{Z}^{\lambda^s_f(\mathbf{p})}$.

We now wish to describe how Lê numbers provide a way to generalize Items (1) and (2) of Theorem 7.1.8, the result of Lê and Ramanujam, to families of non-isolated hypersurface singularities.

Our main result in [34] (also Theorem 9.4 of [36]) is:

Theorem 7.3.9 *Suppose that $\mathcal{U}$ is an open neighborhood of the origin in $\mathbb{C}^{n+1}$. Let $f_t : (\mathcal{U}, \mathbf{0}) \to (\mathbb{C}, 0)$ be a one-parameter complex analytic family in the variables $\mathbf{z} = (z_0, \dots, z_n)$.*

Let $s := \dim_{\mathbf{0}} \Sigma f_0$. Suppose that, for all t small, $(z_0, \dots, z_{s-1})$ is prepolar for f_t at $\mathbf{0}$ and that the Lê numbers, $\lambda^i_{f_t,\mathbf{z}}(\mathbf{0})$, are independent of t for each i with $0 \leqslant i \leqslant s$. Then,

1. *the homology of the Milnor fiber of f_t at the origin is independent of t if $|t|$ is sufficiently small;*

   if $s \leqslant n - 2$,

2. *the fiber homotopy-type of the Milnor fibrations of f_t at the origin is independent of t if $|t|$ is sufficiently small;*

   and, if $s \leqslant n - 3$,

3. *the diffeomorphism-type of the Milnor fibrations of f_t at the origin is independent of t if $|t|$ is sufficiently small.*

For many years, it was an open question whether or not the constancy of the Lê numbers in a family actually implied the constancy of the ambient topological-type. Then, in 2005, Bobadilla answered several questions along these lines in [5, 6].

What Bobadilla proves is:

Theorem 7.3.10 *Suppose that $\mathcal{U}$ is an open neighborhood of the origin in $\mathbb{C}^{n+1}$, where $n \geq 4$. Let $f_t : (\mathcal{U}, \mathbf{0}) \to (\mathbb{C}, 0)$ be a one-parameter complex analytic family in the variables $\mathbf{z} = (z_0, \dots, z_n)$.*

Let $s := \dim_{\mathbf{0}} \Sigma f_0$. Suppose that, for all t small, $(z_0, \dots, z_{s-1})$ is prepolar for f_t at $\mathbf{0}$ and that the Lê numbers, $\lambda^i_{f_t,\mathbf{z}}(\mathbf{0})$, are independent of t for each i with $0 \leqslant i \leqslant s$.

1. *The homotopy-type of the real link of $V(f_t)$ at the origin is independent of t if $|t|$ is sufficiently small.*
2. *Suppose that $s = 1$. Then the ambient topological-type of $V(f_t)$ at the origin is independent of t if $|t|$ is sufficiently small.*
3. *There is an example where $n = 4$ and $s = 3$ for which the ambient topological-type of $V(f_t)$ is* **not** *independent of t for $|t|$ sufficiently small. (See Example 9 of [5].)*

7.4 Lê-Iomdine Formulas and Thom's A_f Condition

In this section, we generalize formulas produced and used by Iomdine [17] and Lê [25] in the case of a 1-dimensional critical locus. We then look at applications of these formulas to the upper-semicontinuity of the Lê numbers and to Thom's A_f condition.

In much of this section, we will, for convenience, consider the case where the point under consideration is the origin; the generalizations to arbitrary points are obvious. So we assume that $\mathcal{U}$ is an open neighborhood of $\mathbf{0}$ in $\mathbb{C}^{n+1}$, $n \geq 1$, and $f : (\mathcal{U}, \mathbf{0}) \to (\mathbb{C}, 0)$ is an analytic function which is nowhere locally constant.

First, we need a definition of *polar ratios*; we follow [39, 50] (see also [36], Definition 4.1).

Definition 7.4.1 Suppose that Γ^1_{f,z_0} is one-dimensional at the origin.

- Let η be an irreducible component of Γ^1_{f,z_0} (with its reduced structure). Then $\eta \cap V(z_0)$ is zero-dimensional at the origin if and only if $\eta \cap V(f)$ is zero-dimensional at the origin, and in this case the **polar ratio** of η (for f at $\mathbf{0}$ with respect to z_0) is

$$\frac{(\eta \cdot V(f))_{\mathbf{0}}}{(\eta \cdot V(z_0))_{\mathbf{0}}} = \frac{\left(\eta \cdot V\left(\frac{\partial f}{\partial z_0}\right)\right)_{\mathbf{0}} + (\eta \cdot V(z_0))_{\mathbf{0}}}{(\eta \cdot V(z_0))_{\mathbf{0}}} = \frac{\left(\eta \cdot V\left(\frac{\partial f}{\partial z_0}\right)\right)_{\mathbf{0}}}{(\eta \cdot V(z_0))_{\mathbf{0}}} + 1.$$

- Let η be an irreducible component of Γ^1_{f,z_0} such that $\eta \cap V(z_0)$ is one-dimensional at the origin, i.e., contained in $V(z_0)$ near the origin. Then we define the polar ratio of η to be 1.

We will be interested in the **maximum polar ratio** over all of the irreducible components η of Γ^1_{f,z_0} at the origin.

If the set Γ^1_{f,z_0} is empty at the origin (i.e., $\mathbf{0} \notin \Gamma^1_{f,z_0}$), then we define this maximum polar ratio to be 1.

Remark 7.4.2 The case where h is a homogeneous polynomial of degree d is particularly easy to analyze. Provided that Γ^1_{h,z_0} is one-dimensional at the origin, each component of the polar curve is a line, and so the polar ratios are all 1 or d.

We are going to consider functions of the form $f + az_0^j$, where a is a non-zero complex number and j is suitably large. Clearly, however, the coordinate z_0 is extremely non-generic for $f + az_0^j$. Hence, if we are using the coordinates $(z_0, z_1, \dots, z_n)$ for f, we use the coordinates $(z_1, z_2, \dots, z_n, z_0)$ for $f + az_0^j$. The purpose of this "rotation" of the coordinate system is merely to get the z_0 coordinate out of the way. Typically, if f has an s-dimensional critical locus at the origin, then $f + az_0^j$ will have an $(s-1)$-dimensional critical locus at the origin; thus, it is only the choice of the coordinates $z_0, \dots, z_{s-1}$ that we care about for f, and the coordinates $z_1, \dots, z_{s-1}$ for $f + az_0^j$.

The following **Lê-Iomdine Formulas** are Theorem 4.5 of [36].

Theorem 7.4.3 *Let $j \geqslant 2$, let $f : (\mathcal{U}, \mathbf{0}) \to (\mathbb{C}, 0)$ be an analytic function, let s denote $\dim_{\mathbf{0}} \Sigma f$, and assume that $s \geqslant 1$. Let $\mathbf{z} = (z_0, \dots, z_n)$ be a linear choice of coordinates such that the Lê numbers of f at the origin are defined. Let a be a non-zero complex number, and use the coordinates $\tilde{\mathbf{z}} = (z_1, \dots, z_n, z_0)$ for $f + az_0^j$.*

If j is greater than or equal to the maximum polar ratio for f then, for all but a finite number of complex a, $\Sigma(f + az_0^j) = \Sigma f \cap V(z_0)$ as germs of sets at $\mathbf{0}$, $\dim_{\mathbf{0}} \Sigma(f + az_0^j) = s - 1$, the Lê numbers of $f + az_0^j$ at the origin exist, and

$$\lambda^0_{f+az_0^j,\tilde{\mathbf{z}}}(\mathbf{0}) = \lambda^0_{f,\mathbf{z}}(\mathbf{0}) + (j-1)\lambda^1_{f,\mathbf{z}}(\mathbf{0}),$$

and, for $1 \leqslant i \leqslant s-1$,

$$\lambda^i_{f+az_0^j,\tilde{\mathbf{z}}}(\mathbf{0}) = (j-1)\lambda^{i+1}_{f,\mathbf{z}}(\mathbf{0}).$$

Moreover, if we have the strict inequality that j is greater than the maximum polar ratio for f, then the above equalities hold for all non-zero a; in particular, this is the case if $j \geqslant 2 + \lambda^0_{f,\mathbf{z}}(\mathbf{0})$.

Corollary 7.4.4 *Let $f : (\mathcal{U}, \mathbf{0}) \to (\mathbb{C}, 0)$ be an analytic function, let s denote $\dim_{\mathbf{0}} \Sigma f$, and let $\mathbf{z} = (z_0, \ldots, z_n)$ be a linear choice of coordinates such that the Lê numbers of f at the origin are defined. Then, for $0 \ll j_0 \ll j_1 \ll \cdots \ll j_{s-1}$,*

$$f + z_0^{j_0} + z_1^{j_1} + \cdots + z_{s-1}^{j_{s-1}}$$

has an isolated singularity at the origin, and its Milnor number is given by

$$\mu(f + z_0^{j_0} + z_1^{j_1} + \cdots + z_{s-1}^{j_{s-1}}) = \sum_{i=0}^{s} \left(\lambda^i_{f,\mathbf{z}}(\mathbf{0}) \prod_{k=0}^{i-1} (j_k - 1) \right) =$$

$$\lambda^0_{f,\mathbf{z}}(\mathbf{0}) + (j_0 - 1)\lambda^1_{f,\mathbf{z}}(\mathbf{0}) + (j_1 - 1)(j_0 - 1)\lambda^2_{f,\mathbf{z}}(\mathbf{0}) + \ldots$$

$$+ (j_{s-1} - 1) \ldots (j_1 - 1)(j_0 - 1)\lambda^s_{f,\mathbf{z}}(\mathbf{0}).$$

If we apply Remark 7.4.2 and use the well-known fact the the Milnor number at the origin of an isolated critical point of a homogeneous polynomial in $n+1$ variables of degree d is $(d-1)^{n+1}$, then the above corollary immediately implies:

Corollary 7.4.5 *Let h be a homogeneous polynomial of degree d in $n+1$ variables, let $s = \dim_{\mathbf{0}} \Sigma h$, and suppose that $\lambda^i_{h,\mathbf{z}}(\mathbf{0})$ exists for all $i \leqslant s$. Then,*

$$\sum_{i=0}^{s} (d-1)^i \lambda^i_{h,\mathbf{z}}(\mathbf{0}) = (d-1)^{n+1}.$$

There are the following **uniform Lê-Iomdine formulas** from Theorem 4.15 of [36]; the point is that, in one-parameter families, one can pick an exponent j which works for all f_t where $|t|$ is small.

Theorem 7.4.6 *Let $f_t : (\mathcal{U}, \mathbf{0}) \to (\mathbb{C}, 0)$ be a one-parameter complex analytic family in the variables $\mathbf{z} = (z_0, \dots, z_n)$, where $\mathcal{U}$ is an open neighborhood of the origin in $\mathbb{C}^{n+1}$.*

As before, we use the coordinates $\mathbf{z} = (z_0, \dots, z_n)$ for f_t, we and use the rotated coordinates $\tilde{\mathbf{z}} = (z_1, z_2, \dots, z_n, z_0)$ for $f_t + z_0^j$.

Let $s := \dim_{\mathbf{0}} \Sigma f_0$, and suppose that $s \geqslant 1$. Suppose that the Lê numbers $\lambda^*_{f_t,\mathbf{z}}(\mathbf{0})$ are defined for all small t. Then, there exist $\tau > 0$ and j_0 such that, for all $j \geqslant j_0$ and for all t such that $0 < |t| < \tau$, $\dim_{\mathbf{0}} \Sigma(f_0 + z_0^j) = s - 1$, the Lê numbers of $f_t + z_0^j$ are defined, and

(i) $\lambda^0_{f_t+z_0^j,\tilde{\mathbf{z}}}(\mathbf{0}) = \lambda^0_{f_t,\mathbf{z}}(\mathbf{0}) + (j-1)\lambda^1_{f_t,\mathbf{z}}(\mathbf{0})$;

(ii) $\lambda^i_{f_t+z_0^j,\tilde{\mathbf{z}}}(\mathbf{0}) = (j-1)\lambda^{i+1}_{f_t,\mathbf{z}}(\mathbf{0})$, *for* $1 \leqslant i \leqslant s-1$; *and*

(iii) $\Sigma(f_t + z_0^j) = \Sigma f_t \cap V(z_0)$ *near* $\mathbf{0}$.

As the Milnor number in a family is upper-semicontinuous, Corollary 7.4.4 and Theorem 7.4.6 immediately imply:

Theorem 7.4.7 *Using the notation of the previous theorem, the tuple of Lê numbers*

$$\left(\lambda^s_{f_t,\mathbf{z}}(\mathbf{0}), \lambda^{s-1}_{f_t,\mathbf{z}}(\mathbf{0}), \dots, \lambda^0_{f_t,\mathbf{z}}(\mathbf{0})\right)$$

is lexicographically upper-semicontinuous in the t variable, i.e., for all t of small magnitude, either

$$\lambda^s_{f_0,\mathbf{z}}(\mathbf{0}) > \lambda^s_{f_t,\mathbf{z}}(\mathbf{0})$$

or

$$\lambda^s_{f_0,\mathbf{z}}(\mathbf{0}) = \lambda^s_{f_t,\mathbf{z}}(\mathbf{0}) \ \ \text{and} \ \ \lambda^{s-1}_{f_0,\mathbf{z}}(\mathbf{0}) > \lambda^{s-1}_{f_t,\mathbf{z}}(\mathbf{0})$$

or

$$\vdots$$

or

$$\lambda^s_{f_0,\mathbf{z}}(\mathbf{0}) = \lambda^s_{f_t,\mathbf{z}}(\mathbf{0}),\ \lambda^{s-1}_{f_0,\mathbf{z}}(\mathbf{0}) = \lambda^{s-1}_{f_t,\mathbf{z}}(\mathbf{0}),\ \dots,\ \lambda^1_{f_0,\mathbf{z}}(\mathbf{0}) = \lambda^1_{f_t,\mathbf{z}}(\mathbf{0}),$$

$$\text{and}\ \lambda^0_{f_0,\mathbf{z}}(\mathbf{0}) \geqslant \lambda^0_{f_t,\mathbf{z}}(\mathbf{0}).$$

The uniform Lê-Iomdine formulas enable us to apply the main result of Lê and Saito [31] about Thom's A_f condition to the case of families of non-isolated hypersurface singularities.

In all of the results we give below, it is extremely important that **our assumptions on the genericity of the coordinate system will be solely that the Lê numbers**

exist. This is a dimensional requirement which is very easy to check. This should be contrasted with the results of [14, 15].

Recall now the main result of Lê and Saito in [31]:

Theorem 7.4.8 *Let* $\mathbb{D}^\circ$ *be an open disk about the origin in* $\mathbb{C}$, *let* $\mathcal{U}$ *be an open neighborhood of the origin in* $\mathbb{C}^{n+1}$, *and let* $f : (\mathbb{D}^\circ \times \mathcal{U}, \mathbb{D}^\circ \times \mathbf{0}) \to (\mathbb{C}, 0)$ *be an analytic function; for all* $t \in \mathbb{D}^\circ$, *we write* f_t *for the function defined by* $f_t(\mathbf{z}) := f(t, \mathbf{z})$, *where* $f_t(\mathbf{0}) = 0$.

Suppose that $\dim_{\mathbf{0}} \Sigma f_0 = 0$ *and that, for all small* t, *the Milnor number of* f_t *at the origin is independent of* t. *Then,* $\mathbb{D}^\circ \times \{\mathbf{0}\}$ *satisfies Thom's* A_f *condition at the origin with respect to the ambient stratum, i.e., if* $\mathbf{p}_i$ *is a sequence of points in* $\mathbb{D}^\circ \times \mathcal{U} - \Sigma f$ *such that* $\mathbf{p}_i \to \mathbf{0}$ *and such that* $T_{\mathbf{p}_i} V(f - f(\mathbf{p}_i))$ *converges to some* $\mathcal{T}$, *then* $\mathbb{C} \times \mathbf{0} = T_{\mathbf{0}}(\mathbb{D}^\circ \times \{\mathbf{0}\}) \subseteq \mathcal{T}$.

Our first generalization of the result of Lê and Saito is Theorem 6.5 of [36].

Theorem 7.4.9 *Let* $\mathbb{D}^\circ$ *be an open disk about the origin in* $\mathbb{C}$, *let* $\mathcal{U}$ *be an open neighborhood of the origin in* $\mathbb{C}^{n+1}$, *and let* $f : (\mathbb{D}^\circ \times \mathcal{U}, \mathbb{D}^\circ \times \mathbf{0}) \to (\mathbb{C}, 0)$ *be an analytic function; we write* f_t *for the function defined by* $f_t(\mathbf{z}) := f(t, \mathbf{z})$.

Let $s = \dim_{\mathbf{0}} \Sigma f_0$. *Suppose that, for all small* t, *the Lê numbers* $\lambda^*_{f_t, \mathbf{z}}(\mathbf{0})$ *are defined and independent of* t. *Then,* $\mathbb{D}^\circ \times \mathbf{0}$ *satisfies Thom's* A_f *condition at the origin with respect to the ambient stratum, i.e., if* $\mathbf{p}_i$ *is a sequence of points in* $\mathbb{D}^\circ \times \mathcal{U} - \Sigma f$ *such that* $\mathbf{p}_i \to \mathbf{0}$ *and such that* $T_{\mathbf{p}_i} V(f - f(\mathbf{p}_i))$ *converges to some* $\mathcal{T}$, *then* $\mathbb{C} \times \mathbf{0} = T_{\mathbf{0}}(\mathbb{D}^\circ \times \mathbf{0}) \subseteq \mathcal{T}$.

From the above theorem, we obtain the following corollary, which is Corollary 6.6 of [36].

Corollary 7.4.10 *Let* $f : \mathcal{U} \to \mathbb{C}$ *be an analytic function on an open subset of* $\mathbb{C}^{n+1}$, *let* $\mathbf{z} = (z_0, \dots, z_n)$ *be a linear choice of coordinates for* $\mathbb{C}^{n+1}$, *let* M *be an analytic submanifold of* $V(f)$, *let* $\mathbf{q} \in M$, *and let* s *denote* $\dim_{\mathbf{q}} \Sigma f$.

If the Lê numbers $\lambda^*_{f, \mathbf{z}}(\mathbf{p})$ *are defined and independent of* $\mathbf{p}$, *for all* $\mathbf{p} \in M$ *near* $\mathbf{q}$, *then* M *satisfies Thom's* A_f *condition at* $\mathbf{q}$ *with respect to the ambient stratum; that is, if* $\mathbf{q}_i$ *is a sequence of points in* $\mathcal{U} - \Sigma f$ *such that* $\mathbf{q}_i \to \mathbf{q}$ *and such that* $T_{\mathbf{q}_i} V(f - f(\mathbf{q}_i))$ *converges to some* $\mathcal{T}$, *then* $T_{\mathbf{q}} M \subseteq \mathcal{T}$.

Remark 7.4.11 It is important to note that, in Corollary 7.4.10, we only require that the coordinates are generic enough so that the Lê numbers are defined; we are not requiring that the coordinates are prepolar.

On the other hand, Corollary 7.4.10 tells us how we can obtain good stratifications: if we have an analytic stratification of $V(h)$ such that the Lê numbers are defined and constant along the strata, then the stratification is actually a good stratification. However, there is no guarantee that the coordinates used to define the Lê numbers are prepolar with respect to this good stratification.

The following multi-parameter version of the result of Lê and Saito is Theorem 6.8 of [36].

Theorem 7.4.12 *Let M be an open neighborhood of the origin in $\mathbb{C}^k$, let $\mathcal{U}$ be an open neighborhood of the origin in $\mathbb{C}^{n+1}$, and let $f : (M \times \mathcal{U}, M \times \mathbf{0}) \to (\mathbb{C}, 0)$ be an analytic function; we write $f_{\mathbf{t}}$ for the function defined by $f_{\mathbf{t}}(\mathbf{z}) := f(\mathbf{t}, \mathbf{z})$, where $\mathbf{t} \in M$ and $\mathbf{z} \in \mathcal{U}$.*

Suppose that $\dim_{\mathbf{0}} \Sigma f_{\mathbf{0}} = 0$ *and that, for all* $\mathbf{t}$ *near the origin, the Milnor number of $f_{\mathbf{t}}$ at the origin is independent of* $\mathbf{t}$. *Then, $M \times \mathbf{0}$ satisfies Thom's A_f condition at the origin with respect to the ambient stratum, i.e. if $\mathbf{p}_i$ is a sequence of points in $M \times \mathcal{U} - \Sigma f$ such that $\mathbf{p}_i \to \mathbf{0}$ and such that $T_{\mathbf{p}_i} V(f - f(\mathbf{p}_i))$ converges to some $\mathcal{T}$, then $T_{\mathbf{0}}(M \times \mathbf{0}) \subseteq \mathcal{T}$.*

Finally, we have the multi-parameter generalization of the result of Lê and Saito, where the critical loci may have arbitrary dimension; this is Theorem 6.9 of [36].

Theorem 7.4.13 *Let M be an open neighborhood of the origin in $\mathbb{C}^k$, let $\mathcal{U}$ be an open neighborhood of the origin in $\mathbb{C}^{n+1}$, and let $f : (M \times \mathcal{U}, M \times \mathbf{0}) \to (\mathbb{C}, 0)$ be an analytic function; we write $f_{\mathbf{t}}$ for the function defined by $f_{\mathbf{t}}(\mathbf{z}) := f(\mathbf{t}, \mathbf{z})$, where $\mathbf{t} \in M$ and $\mathbf{z} \in \mathcal{U}$.*

Let $s = \dim_{\mathbf{0}} \Sigma f_{\mathbf{0}}$. Suppose that, for all small $\mathbf{t}$, *the Lê numbers $\lambda^*_{f_{\mathbf{t}},\mathbf{z}}(\mathbf{0})$ are defined and independent of* $\mathbf{t}$. *Then, $M \times \mathbf{0}$ satisfies Thom's A_f condition at the origin with respect to the ambient stratum, i.e. if $\mathbf{p}_i$ is a sequence of points in $M \times \mathcal{U} - \Sigma f$ such that $\mathbf{p}_i \to \mathbf{0}$ and such that $T_{\mathbf{p}_i} V(f - f(\mathbf{p}_i))$ converges to some $\mathcal{T}$, then $T_{\mathbf{0}}(M \times \mathbf{0}) \subseteq \mathcal{T}$.*

7.5 Aligned Singularities and Hyperplane Arrangements

Hyperplane arrangements are a much-studied area of mathematics. However, from the point of view of analytic/algebraic hypersurfaces, hyperplane arrangements are somewhat unnatural since they are not preserved by analytic or algebraic changes of coordinates. However, *hypersurfaces with aligned singularities* are more general than hyperplane arrangements, but have some similar, desirable properties with respect to Lê cycles and numbers.

All of this material on aligned singularities is taken from Chap. 7 of [36].

For convenience, throughout this section, we concentrate our attention on hypersurface germs at the origin. So we again assume that $\mathcal{U}$ is an open neighborhood of $\mathbf{0}$ in $\mathbb{C}^{n+1}$, $n \geq 1$, and $f : (\mathcal{U}, \mathbf{0}) \to (\mathbb{C}, 0)$ is an analytic function which is nowhere locally constant.

Recall the definition of a *good stratification* in Definition 7.2.12.

Definition 7.5.1 If $f : (\mathcal{U}, \mathbf{0}) \to (\mathbb{C}, 0)$ is an analytic function, then an **aligned good stratification** for f at the origin is a good stratification for f at the origin in which the closure of each stratum of the singular set is smooth at the origin.

If such an aligned good stratification exists, we say that f has an **aligned singularity** at the origin.

If $\{S_\alpha\}$ is an aligned good stratification for f at the origin, then we say that a linear choice of coordinates, $\mathbf{z}$, is **an aligning set of coordinates** for $\{S_\alpha\}$ provided that for each i, $V(z_0, \dots, z_{i-1})$ transversely intersects the closure of each stratum of dimension $\geqslant i$ of $\{S_\alpha\}$ at the origin. Naturally, we say simply that a set of coordinates, $\mathbf{z}$, is **aligning for** f at the origin provided that there exists an aligned good stratification for f at the origin with respect to which $\mathbf{z}$ is aligning.

Note that, given an aligned singularity, aligning sets of coordinates are generic and prepolar.

Closely related to this notion is:

Definition 7.5.2 If $f : (\mathcal{U}, \mathbf{0}) \to (\mathbb{C}, 0)$ is an analytic function on an open subset of $\mathbb{C}^{n+1}$, then a linear choice of coordinates, $\mathbf{z}$, for $\mathbb{C}^{n+1}$ is *pre-aligning for* f at the origin provided that for each Lê cycle, $\Lambda^i_{f,\mathbf{z}}$, and for each irreducible component, C, of $\Lambda^i_{f,\mathbf{z}}$ passing through the origin, the following conditions are satisfied:

1. $\dim_{\mathbf{0}} C = i$;
2. C is smooth at the origin;
3. $V(z_0, z_1, \dots, z_{i-1})$ transversely intersects C at the origin.

Proposition 7.5.3 *If f has an aligned singularity at the origin, then for a generic linear choice of coordinates $\mathbf{z}$, $\mathbf{z}$ is prepolar for f at each point, $\mathbf{p}$, near the origin and, hence, for each such $\mathbf{p}$, the reduced Euler characteristic of the Milnor fiber of f at $\mathbf{p}$ is given by*

$$\tilde{\chi}(F_{f,\mathbf{p}}) = \sum_{i=0}^{s} (-1)^{n-i} \lambda^i_{f,\mathbf{z}}(\mathbf{p}).$$

Proposition 7.5.4 *Suppose that $\{S_\alpha\}$ is an aligned good stratification for f at the origin and that $\mathbf{z}$ is an aligning set of coordinates for $\{S_\alpha\}$ at the origin. Then, as germs of sets at the origin, for all i,*

$$\Lambda^i_{f,\mathbf{z}} \subseteq \bigcup_{\dim_{\mathbf{0}} S_\alpha = i} \overline{S}_\alpha.$$

Hence, $\mathbf{z}$ is a pre-aligning set of coordinates for f at the origin.

Our main interest in aligned singularities is due to:

Proposition 7.5.5 *Suppose that f has an aligned s-dimensional singularity at the origin and that the coordinates $\mathbf{z}$ are aligning. Then, the Lê cycles and Lê numbers can be characterized topologically in the following inductive manner:*

As a set, $\Lambda^s_{f,\mathbf{z}}$ *equals the union of the s-dimensional components of the singular set of* f*. To determine the Lê cycle, to each s-dimensional component,* C *of* Σf*, we assign the multiplicity* $m_C = (-1)^{n-s}\tilde{\chi}(F_{f,\mathbf{p}})$ *for generic* $\mathbf{p} \in C$*, where* $F_{f,\mathbf{p}}$ *denotes the Milnor fiber of* f *at* $\mathbf{p}$ *and* $\tilde{\chi}$ *is the reduced Euler characteristic. Moreover, for all* $\mathbf{p} \in |\Lambda^s_{f,\mathbf{z}}|$, $\lambda^s_{f,\mathbf{z}}(\mathbf{p}) = \sum_{\mathbf{p}\in C} m_C$.

Now, suppose that we have defined the Lê numbers, $\Lambda^i_{f,\mathbf{z}}(\mathbf{p})$ *for all* $i \geqslant k+1$ *and for all* $\mathbf{p}$ *near the origin.*

Then, as a set, $\Lambda^k_{f,\mathbf{z}}$ *equals the closure of the k-dimensional components of the set of points* $\mathbf{p} \in V(f)$*, where*

$$\tilde{\chi}(F_{f,\mathbf{p}}) \neq \sum_{i=k+1}^{s} (-1)^{n-i}\lambda^i_{f,\mathbf{z}}(\mathbf{p}).$$

The Lê cycle is defined by assigning to each irreducible component C *of this set the multiplicity*

$$m_C = (-1)^{n-k}\left(\tilde{\chi}(F_{f,\mathbf{p}}) - \sum_{i=k+1}^{s} (-1)^{n-i}\lambda^i_{f,\mathbf{z}}(\mathbf{p})\right),$$

for generic $\mathbf{p} \in C$*. Finally, for all* $\mathbf{p} \in |\Lambda^k_{f,\mathbf{z}}|$*, we have* $\lambda^k_{f,\mathbf{z}}(\mathbf{p}) = \sum_{\mathbf{p}\in C} m_C$.

The following two corollaries are immediate:

Corollary 7.5.6 *If* f *has an aligned singularity at the origin, then all aligning coordinates* $\mathbf{z}$ *determine the same Lê cycles and Lê numbers.*

Corollary 7.5.7 *Let* f *and* g *be reduced, analytic germs with aligned singularities at the origin in* $\mathbb{C}^{n+1}$*. Let* $\mathbf{z}$ *and* $\tilde{\mathbf{z}}$ *be aligning sets of coordinates for* f *and* g*, respectively. If* H *is a local, ambient homeomorphism from the germ of* $V(f)$ *at the origin to the germ of* $V(g)$ *at the origin, then as germs of sets at the origin,*

$$H(\Lambda^i_{f,\mathbf{z}}) = \Lambda^i_{g,\tilde{\mathbf{z}}},$$

for all i*, and for all* $\mathbf{p}$ *near the origin in* $\mathbb{C}^{n+1}$*,*

$$\lambda^i_{f,\mathbf{z}}(\mathbf{p}) = \lambda^i_{g,\tilde{\mathbf{z}}}(H(\mathbf{p})),$$

for all i*.*

Hyperplane Arrangements

The remainder of this section is devoted to hyperplane arrangements, and is taken from Chap. 5 of [36]. The study of hyperplane arrangements is quite complex and touches on many areas of mathematics (see, for instance, [44, 45]).

A **central hyperplane arrangement in** $\mathbb{C}^{n+1}$ is the zero-locus of an analytic function $h : \mathbb{C}^{n+1} \to \mathbb{C}$ where h is a product of d linear forms on $\mathbb{C}^{n+1}$ (here, we are not necessarily assuming that the forms are distinct).

Recall that Corollary 7.4.5 states: let h be a homogeneous polynomial of degree d in $n+1$ variables, let $s = \dim_{\mathbf{0}} \Sigma h$, and suppose that the Lê numbers of h exist at the origin. Then,

$$\sum_{i=0}^{s}(d-1)^i \lambda^i_{h,\mathbf{z}}(\mathbf{0}) = (d-1)^{n+1}.$$

This formula allows us to calculate the Lê numbers for a central hyperplane arrangement in a purely combinatorial manner from the lattice of flats of the arrangement (see [45] and below). It was experimentally observed by D. Welsh and G. Ziegler that there was a fairly trivial relationship between the Lê numbers of the arrangement and the Möbius function (defined later in this section). This relationship generalizes to matroid-based polynomial identities (see [40]).

In the remainder of this section, we give the combinatorial characterization of the Lê numbers for central hyperplane arrangements and describe the relation between the Lê numbers and the Möbius function.

Example 7.5.8 Suppose we have such an h. In this case, $V(h)$ equals the union of hyperplanes, $\{H_i\}_{i\in I}$, where I is the indexing set $\{1, \ldots, d'\}$, each H_i occurs with some multiplicity $m_i :=$ mult H_i, and $\sum m_i = d$ (in particular, if h is reduced, then each $m_i = 1$ and $d' = d$).

There is an obvious good Whitney stratification of $V(h)$ obtained from the "flats" of the hyperplane arrangement; the collection of flats is given by $\{w_{_J}\}_{J\subseteq I}$, where

$$w_{_J} := \bigcap_{i\in J} H_i.$$

Note that the entire affine space is $\mathbb{C}^{n+1}$ is considered to be a flat by considering the empty indexing set J.

If we now take the stratification $\{S_{_J}\}_{J\subseteq I}$, where

$$S_{_J} = w_{_J} - \bigcup_{J \subsetneq K} w_{_K},$$

then clearly h is analytically trivial along the strata, and therefore one has trivially a Whitney stratification. In words, the strata are intersections of the hyperplanes minus smaller intersections of hyperplanes.

We wish to calculate the Lê numbers of h at the origin with respect to generic coordinates $\mathbf{z}$. As h is analytically trivial along the strata, it is easy to see that, as sets, the Lê cycles are given by the unions of the flats of correct dimension. Hence, as cycles, for all k,

$$\Lambda^k_{h,\mathbf{z}} = \sum_{\dim S_{_J}=k} a_{_J}[w_{_J}]$$

for some a_J. By Proposition 7.2.9, a_J may be calculated by taking any $\mathbf{p} \in S_J$ and a normal slice N to S_J in $\mathbb{C}^{n+1}$ at $\mathbf{p}$, and then $a_J = \lambda^0_{h_{|N}}(\mathbf{p})$, where we use generic coordinates. After a translation to make the point $\mathbf{p}$ the origin, we see that $h_{|N}$ at $\mathbf{p}$ is again (up to multiplication by units) a product of linear forms of degree $e_J := \sum_{i \in J} m_i$.

Therefore, we may use Corollary 7.4.5 to calculate the Lê numbers of h at the origin by a downward induction on the dimension of the flats. (In the following, it looks nicer if we suppress the subscripts.) We denote a hyperplane in the arrangement by H, a flat by w or v, and define

$$e(w) := \sum_{w \subseteq H} \text{mult } H.$$

Next, we define the *vanishing Möbius function*, η, by downward induction on the dimension of the flats. For a hyperplane, H, in the arrangement, define

$$\eta(H) := \text{ mult } H - 1;$$

for a smaller dimensional flat, w, Corollary 7.4.5 tells us that we need to require

$$\eta(w) := (e(w) - 1)^{n+1-\dim w} - \sum_{v \supsetneq w} \eta(v) \cdot (e(w) - 1)^{\operatorname{codim}_v w}.$$

This equality is equivalent to

$$\sum_{v \supseteq w} (e(w) - 1)^{\dim v} \eta(v) = (e(w) - 1)^{n+1}.$$

Finally, having calculated the vanishing Möbius function, one has that, for all i,

$$\lambda^i_{h,\mathbf{z}}(\mathbf{0}) = \sum_{\dim w = i} \eta(w).$$

By Theorem 7.3.5, knowing the Lê numbers of the hyperplane arrangement gives us the Euler characteristic of the Milnor fiber together with Morse inequalities on the Betti numbers. (Another method for computing the Euler characteristic of the Milnor fiber from the data provided by the containment relations among the flats, i.e. by knowing the *intersection lattice*, is given in [45].)

Now, we wish to describe the relation between the Lê numbers of a central arrangement and the Möbius function—this is the result which is generalized in [40].

Let h be the product of d distinct linear forms on $\mathbb{C}^{n+1}$, so that each hyperplane in the arrangement $V(h)$ occurs with multiplicity 1. Let $\mathcal{A}$ denote the collection of hyperplanes which are components of $V(h)$. We use the variable H to denote hyperplanes in $\mathcal{A}$. We use the letters v and w to denote flats of arbitrary dimension.

Finally, in agreement with our notation in Example 7.5.8, let $e_{\mathcal{A}}(v)$ = the number of hyperplanes of $\mathcal{A}$ which contain the flat v.

As we saw above, the Lê numbers of a central hyperplane arrangement can be described in terms of a function $\eta_{\mathcal{A}}$ defined inductively on the flats by: for all $H \in \mathcal{A}$, $\eta_{\mathcal{A}}(H) = 0$, and for all flats w,

$$\sum_{w \subseteq v} (e_{\mathcal{A}}(w) - 1)^{\dim v} \eta_{\mathcal{A}}(v) = (e_{\mathcal{A}}(w) - 1)^{n+1}.$$

The *Möbius function*, $\mu_{\mathcal{A}}$, on $\mathcal{A}$ is defined inductively on the flats by: $\mu_{\mathcal{A}}(\mathbb{C}^{n+1}) = 1$ and for all flats $v \subsetneq w$,

$$\sum_{\text{flats } u, v \subseteq u \subseteq w} \mu_{\mathcal{A}}(u) = 0.$$

Here, we subscript by η, e, and μ by $\mathcal{A}$ because our proof is by induction on the ambient dimension, and the inductive step requires slicing $\mathcal{A}$ by hyperplanes, N, **not** contained in $\mathcal{A}$. This will produce new arrangements inside the ambient space N. So it is important that we indicate which arrangement is under consideration.

More notation now, related to the slicing. We will be taking two kinds of hyperplane slices. N will denote a prepolar hyperplane slice through the origin in $\mathbb{C}^{n+1}$, i.e. a hyperplane slice which contains no flats of $\mathcal{A}$ other than the origin. We will also use normal slices to the one-flats; if v is a one-dimensional flat and $p_v \in v - 0$, N_v will denote a normal slice to v at p_v—that is, N_v is a hyperplane in $\mathbb{C}^{n+1}$ which transversely intersects v at p_v. We use $\mathcal{A} \cap N$ to denote the obvious induced arrangement in N (which is identified with $\mathbb{C}^n$). The arrangement $\mathcal{A} \cap N_v$ is considered as a central arrangement where p_v becomes the origin and all hyperplanes not containing p_v are ignored. Note that the number of hyperplanes in the arrangement $\mathcal{A} \cap N_v$ is $e_{\mathcal{A}}(v)$.

An arrangement is *essential* provided that the origin is a flat of the arrangement (hence, the arrangement is not trivially a product).

The following is Theorem 5.6 of [36], and is the motivating basic result for all of the further results in [40].

Theorem 7.5.9 *If $\mathcal{A}$ is a an essential, central hyperplane arrangement consisting of d hyperplanes in $\mathbb{C}^{n+1}$, then*

$$\eta_{\mathcal{A}}(\mathbf{0}) = (d-1)(-1)^{n+1}\mu_{\mathcal{A}}(\mathbf{0}) = (d-1)|\mu_{\mathcal{A}}(\mathbf{0})|.$$

7.6 Other Characterizations of the Lê Cycles

In this section, we will discuss four other characterizations of Lê cycles: one involving the blow-up of the jacobian ideal, one involving the characteristic cycle of the vanishing cycles, one involving a general process for perverse sheaves applied to the complex of vanishing cycles, and one involving a characterization in terms of a constructible function.

Throughout this section, we assume that $\mathcal{U}$ is a non-empty open subset of $\mathbb{C}^{n+1}$, $n \geq 1$, $\mathbf{z} = (z_0, \ldots, z_n)$ is a coordinate system for $\mathbb{C}^{n+1}$, and $f : \mathcal{U} \to \mathbb{C}$ is an analytic function which is nowhere locally constant.

Let $j(f)$ denote the jacobian ideal of f, i.e.,

$$j(f) := \left\langle \frac{\partial f}{\partial z_0}, \ldots, \frac{\partial f}{\partial z_n} \right\rangle.$$

We consider the blow-up, $\mathrm{Bl}_{j(f)}\mathcal{U}$, of $\mathcal{U}$ along $j(f)$ inside $\mathcal{U} \times \mathbb{P}^n$, and let E denote the exceptional divisor. We let $\pi : \mathcal{U} \times \mathbb{P}^n \to \mathcal{U}$ be the projection. We use $[w_0 : \ldots w_n]$ for homogeneous coordinates on $\mathbb{P}^n$ and consider distinguished copies of $\mathbb{P}^j$ inside of $\mathbb{P}^n$; we write $\mathbb{P}^j \times \{\mathbf{0}\}$ for $\{[w_0 : \cdots : w_j : 0 : \cdots : 0] \in \mathbb{P}^n\}$.

The following is Theorem II.1.26 of [37]; this characterization of the Lê cycles was suggested to us by T. Gaffney and can also be derived from the work of van Gastel in [56]. In more generality, we prove this in Corollary I.2.22 of [37].

Theorem 7.6.1 *Let* $\mathbf{p} \in V(f) \cap \Sigma f$ *and suppose that* $\mathbf{z} := (z_0, \ldots, z_n)$ *is a prepolar for* f *at* $\mathbf{p}$.

Then,

1. *the Lê numbers and polar numbers* $\lambda^*_{f,\mathbf{z}}(\mathbf{p})$ *and* $\gamma^*_{f,\mathbf{z}}(\mathbf{p})$ *exist,*

   and there exists a neighborhood Ω *of* $\mathbf{p}$ *such that*

2. *for all* j *such that* $0 \leqslant j \leqslant k$*, the exceptional divisor* E *properly intersects* $\Omega \times \mathbb{P}^j \times \{\mathbf{0}\}$ *in* $\mathcal{U} \times \mathbb{P}^n$*, and*

3. $$\Gamma^{j+1}_{f,\mathbf{z}} = \pi_*(\mathrm{Bl}_{j(f)}\,\mathcal{U} \cdot (\Omega \times \mathbb{P}^j \times \{\mathbf{0}\}))$$

   and

   $$\Lambda^j_{f,\mathbf{z}} = \pi_*(E \cdot (\Omega \times \mathbb{P}^j \times \{\mathbf{0}\})),$$

   where the intersection takes place in $\mathcal{U} \times \mathbb{P}^n$ *and* π_* *denotes the proper push-forward.*

To present our next characterization of the Lê numbers, we will use the derived category of bounded, constructible complexes of sheaves and characteristic cycles. We will give a quick treatment of characteristic cycles, but must direct the reader to other sources for the basics of the derived category; we recommend the books [7, 20, 48]. We will also follow Appendix 6.A of [11] for our discussion of Morse groups/modules.

We should mention that the vanishing cycles of Kashiwara and Schapira are shifted by one from what essentially all other sources use, and we will **not** use their shift. Also, as there are various conventions on the signs in the characteristic cycle, we point out that we use the convention which makes all of the coefficients in the characteristic cycle of a perverse sheaf non-negative.

Throughout this discussion, since we are interested in integral cohomology, we will fix the base ring $\mathbb{Z}$ for our complexes of sheaves of modules (however, we could use other base rings, and it is common to use the fields $\mathbb{Q}$ or $\mathbb{C}$).

We let $\mathcal{U}$ be an open neighborhood of the origin of $\mathbb{C}^{n+1}$, and let X be a closed, analytic subset of $\mathcal{U}$. We let $\mathbf{z} := (z_0, \ldots, z_n)$ be coordinates on $\mathcal{U}$.

Recall that the *complex link*, $\mathbb{L}_{X,\mathbf{p}}$, of X at $\mathbf{p}$ is the Milnor fiber of a generic affine form, restricted to X, at $\mathbf{p}$. That is, the complex link is

$$\mathbb{L}_{X,\mathbf{p}} := B_\epsilon^\circ(\mathbf{p}) \cap X \cap V(L - b),$$

where $B_\epsilon^\circ(\mathbf{p})$ is an open ball in $\mathcal{U}$ of radius ϵ, where $0 < \epsilon \ll 1$, centered at $\mathbf{p}$, L is a generic affine form which is zero at $\mathbf{p}$, and b is a complex number such that $0 < |b| \ll \epsilon$. The homotopy-type of the complex link is an analytic invariant of the germ of X at $\mathbf{p}$.

Let $\mathfrak{S}$ be a complex analytic Whitney stratification of X, with connected strata. Let $\mathbf{F}^\bullet$ be a bounded complex of sheaves of $\mathbb{Z}$-modules on X, which is constructible with respect to $\mathfrak{S}$. For each $S \in \mathfrak{S}$, we let $d_S := \dim S$, and let $(\mathbb{N}_{X,S}, \mathbb{L}_{X,S})$ denote *complex Morse data for S in X*, consisting of a normal slice and complex link of S in X. Recall that, if $\mathbf{p} \in S$, then $\mathbb{L}_{X,S}$ is the complex link of the normal slice to S at $\mathbf{p}$, i.e., $\mathbb{L}_{X,S} = \mathbb{L}_{\mathbb{N}_{X,S},\mathbf{p}}$. The homeomorphism-type of the pair $(\mathbb{N}_{X,S}, \mathbb{L}_{X,S})$ is independent of the choices.

Definition 7.6.2 For each $S \in \mathfrak{S}$ and each integer k, the isomorphism-type of the $\mathbb{Z}$-module $m_S^k(\mathbf{F}^\bullet) := \mathbb{H}^{k-d_S}(\mathbb{N}_{X,S}, \mathbb{L}_{X,S}; \mathbf{F}^\bullet)$ is independent of the choice of $(\mathbb{N}_{X,S}, \mathbb{L}_{X,S})$; we refer to $m_S^k(\mathbf{F}^\bullet)$ as the *degree k Morse module of S with respect to* $\mathbf{F}^\bullet$.

Remark 7.6.3 The shift by d_S above is present so that perverse sheaves can have non-zero Morse modules in only degree 0; in fact, by 9.5.2 of [19] (or Corollary 4.27 of [38]), having possibly non-zero Morse modules only in degree zero is equivalent to being perverse.

Definition 7.6.4 Define $c_S(\mathbf{F}^\bullet) := \sum_{k\in\mathbb{Z}}(-1)^k \operatorname{rank}(m_S^k(\mathbf{F}^\bullet))$, and define the **characteristic cycle of $\mathbf{F}^\bullet$** (in $T^*\mathcal{U}$) to be the analytic cycle

$$\mathrm{CC}(\mathbf{F}^\bullet) := \sum_{S\in\mathfrak{S}} c_S(\mathbf{F}^\bullet)\left[\overline{T_S^*\mathcal{U}}\right].$$

Note that, by Remark 7.6.3, if $\mathbf{F}^\bullet$ is a perverse sheaf, then

$$\mathrm{CC}(\mathbf{F}^\bullet) := \sum_{S\in\mathfrak{S}} \operatorname{rank}(m_S^0(\mathbf{F}^\bullet))\left[\overline{T_S^*\mathcal{U}}\right].$$

Consider the shifted constant sheaf $\mathbb{Z}^\bullet_{\mathcal{U}}[n+1]$ on $\mathcal{U}$; this is a perverse sheaf (see, for instance, Sect. 5 of [7]). Then the shifted complex of vanishing cycles $\phi_f[-1]\mathbb{Z}^\bullet_{\mathcal{U}}[n+1]$ is a perverse sheaf on $V(f)$. This is a complex of sheaves such that, for all $\mathbf{p} \in V(f)$, the stalk cohomology $H^k(\phi_f[-1]\mathbb{Z}^\bullet_{\mathcal{U}}[n+1])_{\mathbf{p}}$ is isomorphic to the reduced cohomology of the Milnor fiber $\widetilde{H}^{k+n}(F_{f,\mathbf{p}}; \mathbb{Z})$.

More generally, suppose that X is an analytic space (embedded in affine space so that the open balls we are about to use make sense), that $g : X \to C$ is an analytic map, and that $\mathbf{A}^\bullet$ is a bounded, constructible complex of sheaves of $\mathbb{Z}$-modules on X. For all $\mathbf{x} \in V(g)$, there is still a well-defined Milnor fibration and Milnor fiber $F_{g,\mathbf{x}}$; see [24].

Then the **nearby cycles**, $\psi_g\mathbf{A}^\bullet$, and **vanishing cycles**, $\phi_g\mathbf{A}^\bullet$, are bounded, constructible complexes of sheaves of $\mathbb{Z}$-modules on $V(g)$ such that the stalk cohomology at $\mathbf{x} \in V(g)$ is isomorphic to the hypercohomology (respectively, relative hypercohomology) of $F_{g,\mathbf{x}}$, i.e., for all $\mathbf{x} \in V(g)$,

$$H^k(\psi_g\mathbf{A}^\bullet)_{\mathbf{x}} \cong \mathbb{H}^k(F_{g,\mathbf{x}};\ \mathbf{A}^\bullet) \quad \text{and} \quad H^k(\phi_g\mathbf{A}^\bullet)_{\mathbf{x}} \cong \mathbb{H}^{k+1}(B^\circ_\epsilon(\mathbf{x}) \cap X,\ F_{g,\mathbf{x}};\ \mathbf{A}^\bullet),$$

where $B^\circ_\epsilon(\mathbf{x})$ is a sufficiently small open ball of radius $\epsilon > 0$, centered at $\mathbf{x}$, in a local embedding of X into affine space.

If $\mathbf{P}^\bullet$ is a perverse sheaf on X, then the shifted nearby and vanishing cycles $\psi_g[-1]\mathbf{P}^\bullet := (\psi_g\mathbf{P}^\bullet)[-1]$ and $\phi_g[-1]\mathbf{P}^\bullet := (\phi_g\mathbf{P}^\bullet)[-1]$ are perverse sheaves on $V(g)$.

Now, identifying the cotangent space $T^*\mathcal{U}$ of $\mathcal{U}$ with $\mathcal{U} \times \mathbb{C}^{n+1}$ and its projectivization with $\mathcal{U} \times \mathbb{P}^n$, the result of Kashiwara et al. [18], Lê and Mebkhout [27] is:

Theorem 7.6.5 *The exceptional divisor E of the blow-up of $\mathcal{U}$ along the jacobian ideal (from above) is equal to $\mathbb{P}(\mathrm{CC}(\phi_f[-1]\mathbb{Z}^\bullet_{\mathcal{U}}[n+1]))$, the projectivized characteristic cycle of the shifted vanishing cycles.*

Thus, replacing E in Theorem 7.6.1 with $\mathbb{P}(\mathrm{CC}(\phi_f[-1]\mathbb{Z}^\bullet_{\mathcal{U}}[n+1]))$ and using the same notation, we obtain Theorem 10.14 and Corollary 10.15 of [36]:

Theorem 7.6.6 *Suppose that the coordinates $\mathbf{z} := (z_0, \ldots, z_n)$ are prepolar for f at a point $\mathbf{p} \in V(f)$ and write*

$$\mathbb{P}(\mathrm{CC}(\phi_f[-1]\mathbb{Z}^\bullet_{\mathcal{U}}[n+1])) = \sum_{S \in \mathfrak{S}} m_S\, \mathbb{P}\big(\overline{T^*_S\mathcal{U}}\big).$$

Then, there exists a neighborhood Ω of $\mathbf{p}$ in which, for all j,

$$\Lambda^j_{f,\mathbf{z}} = \pi_*\left(\mathbb{P}(\mathrm{CC}(\phi_f[-1]\mathbb{Z}^\bullet_{\mathcal{U}}[n+1])) \cdot (\Omega \times \mathbb{P}^j \times \{\mathbf{0}\})\right) =$$

$$\sum_{S \in \mathfrak{S}} m_S\, \pi_*\big(\mathbb{P}\big(\overline{T^*_S\mathcal{U}}\big) \cdot (\Omega \times \mathbb{P}^j \times \{\mathbf{0}\})\big) = \sum_{S \in \mathfrak{S}} m_S\, \Gamma^j_{\mathbf{z}}(\overline{S}),$$

where $\Gamma^j_{\mathbf{z}}(\overline{S})$ is the (possibly non-reduced) j-dimensional absolute polar variety (see [32]) of $\overline{S}$ with respect to the affine forms (or corresponding flag)

$$(z_0 - p_0, \ldots, z_j - p_j).$$

As the absolute polar multiplicities are generically independent of the coordinates (see [55]), we conclude:

Corollary 7.6.7 *For a generic choice of the coordinates at a point, the Lê numbers are independent of the coordinate choice.*

Before we give another characterization of the Lê numbers, we first need the following proposition, which is Proposition 10.2 of [36]; we need this proposition so that the theorem that we state is not vacuously true.

Proposition 7.6.8 *Suppose that $\mathbf{p} \in V(f)$. Then, for a generic linear choice of coordinates $\mathbf{z} := (z_0, \ldots, z_n)$, there exists open neighborhood Ω of $\mathbf{p}$ such that $\mathbf{z}$ is prepolar at $\mathbf{x}$ for all $\mathbf{x} \in \Omega$.*

The following theorem, which gives another characterization of the Lê numbers, is a combination of Theorem 7.6.6 above, Corollary 4.20, Theorem 4.24 and Theorem 6.4 of [38] (also using Definition 4.1 of [38]). In the theorem, we will not distinguish in the notation between an affine form and its restriction to a subspace.

Theorem 7.6.9 *Let $\mathbf{P}^\bullet := \phi_f[-1]\mathbb{Z}^\bullet_{\mathcal{U}}[n+1]$. Suppose that $\mathbf{p} \in V(f)$, that Ω is an open neighborhood of $\mathbf{p}$ and that, for all $\mathbf{x} \in \Omega \cap V(f)$, the coordinate system $\mathbf{z}$ is prepolar at $\mathbf{x}$. Then, for all k and for all $\mathbf{x} \in \Omega$, $\mathbf{x}$ is an isolated point in the support of the perverse sheaf*

$$\phi_{z_k - x_k}[-1]\psi_{z_{k-1}-x_{k-1}}[-1] \ldots \psi_{z_0 - x_0}[-1]\mathbf{P}^\bullet$$

and, hence, the stalk cohomology at $\mathbf{x}$ is (possibly) non-zero only in degree zero and, for all k,

$$H^0(\phi_{z_k - x_k}[-1]\psi_{z_{k-1}-x_{k-1}}[-1] \ldots \psi_{z_0 - x_0}[-1]\mathbf{P}^\bullet)_{\mathbf{x}} \cong \mathbb{Z}^{\lambda^k_{f,\mathbf{z}}(\mathbf{x})},$$

where, of course, $\lambda^k_{f,\mathbf{z}}(\mathbf{x})$ is the k-dimensional Lê number of f at $\mathbf{x}$.

Remark 7.6.10 As we saw in Corollary 7.3.7, there is a chain complex

$$0 \to \mathbb{Z}^{\lambda^s_{f,\mathbf{z}}(\mathbf{x})} \to \mathbb{Z}^{\lambda^{s-1}_{f,\mathbf{z}}(\mathbf{x})} \to \cdots \to \mathbb{Z}^{\lambda^1_{f,\mathbf{z}}(\mathbf{x})} \to \mathbb{Z}^{\lambda^0_{f,\mathbf{z}}(\mathbf{x})} \to 0$$

such that the cohomology of the complex at the $\lambda^k_{f,\mathbf{z}}(\mathbf{x})$ term is isomorphic to $\widetilde{H}^{n-k}(F_{f,\mathbf{x}}; \mathbb{Z})$.

It is true more generally that, if $\mathbf{P}^\bullet$ is a perverse sheaf, then there is a chain complex with terms

$$H^0(\phi_{z_k-x_k}[-1]\psi_{z_{k-1}-x_{k-1}}[-1]\dots\psi_{z_0-x_0}[-1]\mathbf{P}^\bullet)_{\mathbf{x}},$$

whose cohomology is isomorphic to the stalk cohomology of $\mathbf{P}^\bullet$ at x, provided that for all k and for all $\mathbf{x} \in \Omega$, $\mathbf{x}$ is an isolated point in the support of the iterated nearby and vanishing cycles; see Theorem 5.3 of [35] and Theorem 4.16 of [38]. In addition, a repeated application of Theorem 3.3 of [38], together with Theorem 3.4 of [38], tells us that, if the Morse modules of $\mathbf{P}^\bullet$ are free abelian, then all of the modules

$$H^0(\phi_{z_k-x_k}[-1]\psi_{z_{k-1}-x_{k-1}}[-1]\dots\psi_{z_0-x_0}[-1]\mathbf{P}^\bullet)_{\mathbf{x}}$$

are also free abelian (of finite rank).

Our final characterization of the Lê cycles and numbers is formal, and is in terms of the function given by the reduced Euler characteristic of the Milnor fiber. In fact, given Proposition 7.6.8 and Corollary 7.3.7, it is easy to conclude Proposition/Definition 10.6 and Remark 10.7 of [36]:

Theorem 7.6.11 *Suppose that* $\mathbf{p} \in V(f)$, *that* Ω *is an open neighborhood of* $\mathbf{p}$ *and that, for all* $\mathbf{x} \in \Omega \cap V(f)$, *the coordinate system* $\mathbf{z}$ *is prepolar at* $\mathbf{x}$.

Then, the Lê cycles $\Lambda^k_{f,\mathbf{z}}$ *are the unique cycles in* $\Omega \cap V(f)$ *such that each* $\Lambda^k_{f,\mathbf{z}}$ *is purely k-dimensional, properly intersects* $V(z_0 - x_0, \dots, z_{k-1} - x_{k-1})$ *at each* $\mathbf{x} \in \Lambda^k_{f,\mathbf{z}}$ *and, for all* $\mathbf{x} \in V(f)$,

$$\tilde{\chi}(F_{f,\mathbf{x}}) = \sum_{k=0}^{n} (-1)^{n-k} \left(\Lambda^k_{f,\mathbf{z}} \cdot V(z_0 - x_0, \dots, z_{k-1} - x_{k-1})\right)_{\mathbf{x}}.$$

Remark 7.6.12 The reader should appreciate the "strange" implication of Theorem 7.6.11. Using essentially nothing other than the reduced Euler characteristics of the Milnor fibers, one can produce the Lê numbers which then yield seemingly strictly more data, such as the Morse inequalities of Corollary 7.3.7. However, as discussed in [35] (and related to the results above), this is an implication of the fact that the shifted vanishing cycles are a perverse sheaf.

7.7 Projective Lê Cycles

Throughout this section, $h : \mathbb{C}^{n+1} \to \mathbb{C}$ will denote a (non-zero) homogeneous polynomial of degree d, and we continue to let $\mathbf{z} = (z_0, \dots .z_n)$ be coordinates for $\mathbb{C}^{n+1}$, and assume that the Lê and relative polar numbers of h at $\mathbf{0}$ with respect to $\mathbf{z}$ exist. We assume throughout that we are not in the trivial case where $d = 1$, i.e., where h is linear; hence, $\mathbf{0} \in \Sigma h$.

As all of the partial derivatives of h are homogeneous (of degree $(d-1)$), all of the positive-dimensional relative polar cycles and Lê cycles of h are conic ($\mathbb{C}^*$-conic), and so projectivize to yield cycles in $\mathbb{P}^n$. For $k \geq 0$, define the k**-dimensional**

projective Lê cycle of h and the k**-dimensional projective relative polar cycle of** h by

$$(\mathbb{P}\Lambda_{h,\mathbf{z}})^k := \mathbb{P}(\Lambda^{k+1}_{h,\mathbf{z}}) \quad \text{and} \quad (\mathbb{P}\Gamma_{h,\mathbf{z}})^k := \mathbb{P}(\Gamma^{k+1}_{h,\mathbf{z}}).$$

As we wrote in Remark 4.13 of [36], it seems reasonable to define the **global projective Lê numbers** in terms of the Lê numbers of h at the origin (with respect to coordinates $\mathbf{z}$ such that the Lê numbers exist): for $k \geq 1$,

$$(\mathbb{P}\lambda_{h,\mathbf{z}})^k := \lambda^{k+1}_{h,\mathbf{z}}(\mathbf{0}).$$

We mentioned there that the data from $\lambda^0_{h,\mathbf{z}}(\mathbf{0})$ is not lost, since Corollary 7.4.5 tells us that

$$\lambda^0_{h,\mathbf{z}}(\mathbf{0}) = (d-1)^{n+1} - \sum_{i=1}^{n}(d-1)^i\lambda^i_{h,\mathbf{z}}(\mathbf{0})$$

or, in our new terminology and notation:

Proposition 7.7.1 *The 0-dimensional Lê number of h at the origin can be calculated using the global projective Lê numbers of h:*

$$\frac{\lambda^0_{h,\mathbf{z}}(\mathbf{0})}{d-1} = (d-1)^n - \sum_{i=0}^{n-1}(d-1)^i(\mathbb{P}\lambda_{h,\mathbf{z}})^i.$$

In the remainder of this section, we wish to take this projective case further. Everything here was produced in collaboration with Paolo Aluffi during our visit to Florida State University in early 2020.

While we are interested in the projective setting, most of the results seem to be best stated and proved in the affine situation.

Theorem 7.7.2 *The following formulas for the Lê numbers and relative polar numbers at the origin hold:*

1. $\gamma^{n+1}_{h,\mathbf{z}}(\mathbf{0}) = 1$, $\lambda^{n+1}_{h,\mathbf{z}}(\mathbf{0}) = 0$, *and* $\gamma^0_{h,\mathbf{z}}(\mathbf{0}) = 0$.
2. *For* $0 \leq k \leq n$,

$$\gamma^k_{h,\mathbf{z}}(\mathbf{0}) + \lambda^k_{h,\mathbf{z}}(\mathbf{0}) = (d-1)\gamma^{k+1}_{h,\mathbf{z}}(\mathbf{0}).$$

In particular, $\lambda^0_{h,\mathbf{z}}(\mathbf{0}) = (d-1)\gamma^1_{h,\mathbf{z}}(\mathbf{0})$.

Proof The proof of Item (1) is trivial. We will prove Item (2).

Consider $\Gamma^1_{h,\mathbf{z}}$, which is conic and 1-dimensional; thus, as a set, $\Gamma^1_{h,\mathbf{z}}$ is a collection of lines. Consequently, as $\partial h/\partial z_0$ is homogeneous of degree $(d-1)$,

$$\lambda^0_{h,\mathbf{z}}(\mathbf{0}) = \left(\Gamma^1_{h,\mathbf{z}} \cdot V\left(\frac{\partial h}{\partial z_0}\right)\right)_{\mathbf{0}} = (d-1)\left(\Gamma^1_{h,\mathbf{z}} \cdot V(z_0)\right)_{\mathbf{0}} = (d-1)\gamma^1_{h,\mathbf{z}}(\mathbf{0}).$$

Now we use hyperplane slicing from Proposition 7.2.9, to find

$$\gamma^k_{h,\mathbf{z}}(\mathbf{0}) + \lambda^k_{h,\mathbf{z}}(\mathbf{0}) = \left(\Gamma^{k+1}_{h,\mathbf{z}} \cdot V\left(\frac{\partial h}{\partial z_k}\right) \cdot V(z_0, \ldots, z_{k-1})\right)_{\mathbf{0}} =$$

$$\left(\Gamma^1_{h_{|V(z_0,\ldots,z_{k-1})},(z_k,\ldots,z_n)} \cdot V\left(\frac{\partial h}{\partial z_k}\right)\right)_{\mathbf{0}} = (d-1)\left(\Gamma^1_{h_{|V(z_0,\ldots,z_{k-1})},(z_k,\ldots,z_n)} \cdot V(z_k)\right)_{\mathbf{0}} =$$

$$(d-1)\left(\Gamma^{k+1}_{h,\mathbf{z}} \cdot V(z_0, \ldots, z_{k-1}, z_k)\right)_{\mathbf{0}} = (d-1)\gamma^{k+1}_{h,\mathbf{z}}(\mathbf{0}).$$

From this theorem, one can use induction arguments to easily conclude:

Corollary 7.7.3 *For all k such that $1 \leq k \leq n+1$,*

$$\gamma^k_{h,\mathbf{z}}(\mathbf{0}) = \sum_{j=0}^{k-1} \frac{\lambda^j_{h,\mathbf{z}}(\mathbf{0})}{(d-1)^{k-j}} = (d-1)^{n-k+1} - \sum_{m=0}^{n-k} (d-1)^m \lambda^{m+k}_{h,\mathbf{z}}(\mathbf{0}).$$

Now we wish to projectivize the base space in Theorem 7.6.1 and consider the blow-up

$$\eta : \mathrm{Bl}_{j(h)}\mathbb{P}^n \to \mathbb{P}^n$$

of $\mathbb{P}^n$ along the jacobian ideal $j(h)$. We view $\mathrm{Bl}_{j(h)}\mathbb{P}^n$ as a subset of $\mathbb{P}^n \times \mathbb{P}^n$, and continue to denote the exceptional divisor by E.

If we projectivize the relative polar cycle portion of Theorem 7.6.1, we obtain:

Theorem 7.7.4 *Suppose that $(z_0, \ldots, z_n)$ are prepolar coordinates for h at $\mathbf{0}$. Then,*

1. *the Lê numbers and polar numbers $\lambda^*_{h,\mathbf{z}}(\mathbf{0})$ and $\gamma^*_{f,\mathbf{z}}(\mathbf{0})$ exist,*
2. *for all j such that $0 \leqslant j \leqslant k$, the exceptional divisor E properly intersects $\mathbb{P}^n \times \mathbb{P}^j \times \{\mathbf{0}\}$ in $\mathbb{P}^n \times \mathbb{P}^n$,*
3. $$(\mathbb{P}\Gamma_{h,\mathbf{z}})^j = \eta_*(\mathrm{Bl}_{j(h)}\,\mathbb{P}^n \cdot (\mathbb{P}^n \times \mathbb{P}^j \times \{\mathbf{0}\})),$$
   where the intersection takes place in $\mathbb{P}^n \times \mathbb{P}^n$ and η_* denotes the proper push-forward, and
4. *$\{\mathbf{0}\} \times \mathbb{P}^{n-j}$ properly intersects $(\mathbb{P}\Gamma_{h,\mathbf{z}})^j$ in a finite collection of points (possibly with multiplicity) in $\mathbb{P}^n$ and*
   $$\gamma^{j+1}_{h,\mathbf{z}}(\mathbf{0}) = \int \left(\{\mathbf{0}\} \times \mathbb{P}^{n-j}\right) \cdot (\mathbb{P}\Gamma_{h,\mathbf{z}})^j = \text{degree}\left((\mathbb{P}\Gamma_{h,\mathbf{z}})^j\right).$$

As all choices of prepolar coordinates produce the product intersections in Item (2) above, as the degree does not change under rational equivalence, and using Item (2) of Theorem 7.7.2, we conclude:

Corollary 7.7.5 *The values of the Lê numbers and relative polar numbers of h at the origin are independent of the choice of prepolar coordinates for h at* $\mathbf{0}$.

Theorem 7.7.4 identifies the relative polar numbers as the **projective degrees of the polar/gradient map**. See [8, 9], Example 19.4 of [13], and Sect. 3 of [3]. Following the notation and terminology of this last reference, we conclude:

Corollary 7.7.6 *Suppose that $(z_0, \dots, z_k)$ is a prepolar tuple for h at* $\mathbf{0}$.

Then, for $0 \le i \le n$, the i-th polar degree, g_i, of h is equal to the $(n-i+1)$-dimensional relative polar multiplicity of h at $\mathbf{0}$, i.e., $g_i = \gamma_{h,\mathbf{z}}^{n-i+1}(\mathbf{0})$.

Now that Corollary 7.7.6 identifies the relative polar multiplicities as the projective degrees of the polar/gradient map, and as we can write the relative polar multiplicities in terms of Lê numbers via Corollary 7.7.3, there are many new formulas which hold involving relative polar multiplicities and Lê numbers which follow immediately from formulas involving the projective degrees of the polar/gradient map.

For instance, we conclude immediately from the work of June Huh in [16]:

Corollary 7.7.7 *Suppose that $(z_0, \dots, z_k)$ is a prepolar tuple for h at* $\mathbf{0}$.

Then the sequence of relative polar multiplicities is log-concave, i.e., for all k where $0 \le k \le n-1$,

$$\left(\gamma_{h,\mathbf{z}}^{k+1}(\mathbf{0})\right)^2 \ge \gamma_{h,\mathbf{z}}^{k}(\mathbf{0}) \cdot \gamma_{h,\mathbf{z}}^{k+2}(\mathbf{0}).$$

Using Aluffi's formula from Theorem 2.1 of [2], we can write the Schwartz-MacPherson-Chern class of the projective hypersurface $\mathbb{P}(V(h))$ in terms of relative polar multiplicities and/or Lê numbers. In particular, reading off the degree zero part of this formula, we obtain the Euler characteristic of $\mathbb{P}(V(h))$:

Corollary 7.7.8 *Suppose that $(z_0, \dots, z_k)$ is a prepolar tuple for h at* $\mathbf{0}$.

Then, the Euler characteristic of $\mathbb{P}(V(h))$ is given by

$$\chi\left(\mathbb{P}(V(h))\right) = n+1+(-1)^n \sum_{m=1}^{n+1} (-1)^m \gamma_{h,\mathbf{z}}^{m}(\mathbf{0}) =$$

$$n+1+\frac{(1-d)^{n+1}-1}{d} + (-1)^n \sum_{j=1}^{n} \frac{(d-1)^j-(-1)^j}{d} \lambda_{h,\mathbf{z}}^{j}(\mathbf{0}) =$$

$$n+1+\frac{(1-d)^{n+1}-1}{d} + (-1)^n \sum_{j=1}^{n} \frac{(d-1)^j-(-1)^j}{d} (\mathbb{P}\lambda_{h,\mathbf{z}})^{j-1}.$$

Remark 7.7.9 The reader may wonder why one would be interested in the formulas for the Euler characteristic of $\mathbb{P}(V(h))$ in terms of Lê or projective Lê numbers when the formula in terms of the relative polar multiplicities is so simple (because the formula in terms of projective degrees of the polar/gradient map is so simple).

The point is that, if the dimension of the critical locus of h is fairly small compared to n, then there are far fewer non-zero Lê numbers than relative polar multiplicities.

References

1. A'Campo, N.: Le nombre de Lefschetz d'une monodromie. Proc. Kon. Ned. Akad. Wet., Series A **76**, 113–118 (1973).
2. Aluffi, P.: Computing characteristic classes of projective schemes. J. Symbolic Comput. **35**, 3–19 (2003).
3. Aluffi, P.: Euler characteristics of general linear sections and polynomial Chern classes. Rend. Circ. Mat. Palermo **62**, 3–26 (2013).
4. Cerf, J.: La théorie de Smale sur le h-cobordisme des variétés. Séminaire Henri Cartan **14, no. 11-13**, 1–23 (1961-1962).
5. de Bobadilla, J. F.: Answers to some equisingularity questions. Invent. Math. **161**, 657–675 (2005).
6. de Bobadilla, J. F.: Topological equisingularity of hypersurfaces with 1-dimensional critical set. Advan. Math. **248**, 1199–1253 (2013).
7. Dimca, A.: Sheaves in Topology. Universitext. Springer-Verlag (2004).
8. Dimca, A., Papadima, S.: Hypersurface complements, Milnor fibers and higher homotopy groups of arrangments. Annal of Math. **158 (2)**, 473–507 (2003).
9. Dolgachev, I.: Polar Cremona transformations. Michigan Math. Journ. **48**, 191–202 (200).
10. Fulton, W.: Intersection Theory, *Ergeb. Math.*, vol. 2. Springer-Verlag (1984).
11. Goresky, M., MacPherson, R.: Stratified Morse Theory, *Ergeb. der Math.*, vol. 14. Springer-Verlag (1988)
12. Hamm, H., Lê D. T.: Un théorème de Zariski du type de Lefschetz. Ann. Sci. Éc. Norm. Sup. **6 (series 4)**, 317–366 (1973).
13. Harris, J.: Algebraic Geometry, *Graduate Texts in Mathematics.*, vol. 133. Springer-Verlag (1992).
14. Henry J.-P., Merle, M.: Conditions de régularité et éclatements. Ann. Inst. Fourier **37**, 159–190 (1987).
15. Henry, J.-P., Merle, M., and Sabbah, C. : Sur la condition de Thom stricte pour un morphisme analytique. Ann. Sci. L'Ecole Norm. Sup. **17**, 227–268 (1984).
16. Huh, J.: Milnor numbers of projective hypersurfaces and the chromatic polynomial of graphs. Journ. Amer. Math. Soc. **25**, 907–927 (2021).
17. I. N. Iomdin: Variétés complexes avec singularités de dimension un. Sibirsk. Mat. Z. **15**, 1061–1082 (1974).
18. Kashiwara, M.: Systèmes d'equations micro-differentielles. Dépt. de Math., Univ. Paris-Nord 8 (1978).
19. Kashiwara, M., Schapira, P.: Microlocal Study of Sheaves, *Astérisque*, vol. 128. Soc. Math. France (1985).
20. Kashiwara, M., Schapira, P.: Sheaves on Manifolds, *Grund. math. Wissen.*, vol. 292. Springer-Verlag (1990).
21. Kato, M., Matsumoto, Y.: On the connectivity of the Milnor fibre of a holomorphic function at a critical point. Proc. of 1973 Tokyo manifolds conf. pp. 131–136 (1973).
22. Lê, D. T.: Calcul du Nombre de Cycles Évanouissants d'une Hypersurface Complexe. Ann. Inst. Fourier, Grenoble **23**, 261–270 (1973).
23. Lê, D. T.: Topologie des Singularites des Hypersurfaces Complexes, *Astérisque*, vol. 7-8, pp. 171–182. Soc. Math. France (1973).
24. Lê, D. T.: Some remarks on Relative Monodromy. In: P. Holm (ed.) Real and Complex Singularities, Oslo 1976, pp. 397–404. Nordic Summer School/NAVF (1977).
25. Lê, D. T.: Ensembles analytiques complexes avec lieu singulier de dimension un (d'après I.N. Iomdin). Séminaire sur les Singularités (Paris, 1976–1977) Publ. Math. Univ. Paris VII pp. 87–95 (1980).
26. Lê, D. T., Massey, D.: Hypersurface Singularities and Milnor Equisingularity. Pure and Appl. Math. Quart., special issue in honor of Robert MacPherson's 60th birthday **2, no.3**, 893–914 (2006).

27. Lê, D. T., Mebkhout, Z.: Variétés caractéristiques et variétés polaires. C.R. Acad. Sc. Paris **296**, 129–132 (1983).
28. Lê, D. T., Nuño Ballesteros, J. J., Seade, J.: The Topology of the Milnor fibration (2021).
29. Lê, D. T., Perron, B.: Sur la fibre de Milnor d'une singularité isolée en dimension complexe trois. C. R. Acad. Sci. Pairs Sér. A **289**, 115–118 (1979).
30. Lê, D. T., Ramanujam, C. P.: The Invariance of Milnor's Number implies the Invariance of the Topological Type. Amer. Journ. Math. **98**, 67–78 (1976).
31. Lê, D. T., Saito, K.: La constance du nombre de Milnor donne des bonnes stratifications. C.R. Acad. Sci. **277**, 793–795 (1973).
32. Lê, D. T., Teissier, B.: Variétés polaires locales et classes de Chern des variétés singulières. Annals of Math. **114**, 457–491 (1981).
33. Massey, D.: The Lê Varieties, I. Invent. Math. **99**, 357–376 (1990).
34. Massey, D.: The Lê Varieties, II. Invent. Math. **104**, 113–148 (1991).
35. Massey, D.: Numerical Invariants of Perverse Sheaves. Duke Math. J. **73**(2), 307–370 (1994).
36. Massey, D.: Lê Cycles and Hypersurface Singularities, *Lecture Notes in Math.*, vol. 1615. Springer-Verlag (1995).
37. Massey, D.: Numerical Control over Complex Analytic Singularities, *Memoirs of the AMS*, vol. 778. AMS (2003)
38. Massey, D.: Singularities and Enriched Cycles. Pacific J. Math. **215, no. 1**, 35–84 (2004).
39. Massey, D., Siersma, D.: Deformations of Polar Methods. Ann. Inst. Fourier **42**, 737–778 (1992).
40. Massey, D., Simion, R., Stanley, R., Vertigan, D., Welsh, D., Ziegler, G.: Lê Numbers, Matroid Identities, and the Tutte Polynomial. J. Combinatorial Theory **70**, 118–133 (1997).
41. Milnor, J.: Differentiable manifolds which are homotopy spheres (1959). Notes, Princeton.
42. Milnor, J.: Singular Points of Complex Hypersurfaces, *Annals of Math. Studies*, vol. 77. Princeton Univ. Press (1968).
43. Oka, M.: On the homotopy type of hypersurfaces defined by weighted homogeneous polynomials. Topology **12**, 19–32 (1973).
44. Orlik, P., Solomon, L.: Combinatorics and Topology of Complements of Hyperplanes. Invent. Math. **56**, 167–189 (1980).
45. Orlik, P., Terao, H.: Arrangements of Hyperplanes, *Grund. Math. Wissen*, vol. 300. Springer-Verlag (1992).
46. Palamodov, V.P.: Multiplicity of holomorphic mappings. Funct. Anal. Appl. **1**, 218–226 (1967).
47. Sakamoto, K.: The Seifert matrices of Milnor fiberings defined by holomorphic functions. J. Math. Soc. Japan **26 (4)**, 714–721 (1974).
48. Schürmann, J.: Topology of Singular Spaces and Constructible Sheaves, *Monografie Matematyczne*, vol. 63. Birkhäuser (2004).
49. Sebastiani, M. , Thom, R.: Un résultat sur la monodromie. Invent. Math. **13**, 90–96 (1971).
50. Siersma, D.: Isolated line singularities. Proc. Symp. Pure Math. **35, part 2, Arcata Singularities Conf.**, 485–496 (1983).
51. Smale, S.: Generalized Poincaré's Conjecture in Dimensions Greater than Four. Ann. of Math. (2) **74**, 391–406 (1961).
52. Teissier, B.: Déformations à type topologique constant II. Séminaire Douady-Verdier, Secrétariat E.N.S. Paris (1972).
53. Teissier, B.: Cycles évanescents, sections planes et conditions de Whitney. Astérisque **7-8**, 285–362 (1973).
54. Teissier, B.: Variétés polaires I: Invariants polaires des singularités d'hypersurfaces. Invent. Math. **40 (3)**, 267–292 (1977).
55. Teissier, B.: Variétés polaires II: Multiplicités polaires, sections planes, et conditions de Whitney, Proc. of the Conf. on Algebraic Geometry, La Rabida 1981. Springer Lect. Notes **961**, 314–491 (1982).
56. van Gastel, L.: Excess Intersections and a Correspondence Principle. Invent. Math. **103 (1)**, 197–222 (1991).
57. Vannier, J. P.: Familles à un paramètre de fonctions analytiques à lieu singulier de dimension un. C. R. Acad. Sc. Paris **303, Série I, no. 8**, 367–370 (1986).

Chapter 8
Introduction to Mixed Hypersurface Singularity

Mutsuo Oka

Abstract In 1968, Milnor introduced the fibration structure $\varphi : \mathbb{S}_\varepsilon^{2n-1} \setminus K \to \mathbb{S}^1$ for a given holomorphic function $f : (\mathbb{C}^n, \mathbf{0}) \to (\mathbb{C}, 0)$ where $\varphi = f/|f|$ and ε is chosen small enough and $K = f^{-1}(0) \cap \mathbb{S}_\varepsilon^{2n-1}$ [24]. From a viewpoint of algebraic geometry, it is more convenient to study the tubular fibration $f : E(\varepsilon, \delta)^* \to \mathbb{D}_\delta^*$ where $E(\varepsilon, \delta)^* = \{\mathbf{z} \in \mathbb{B}_\varepsilon^{2n} \mid 0 < |f(\mathbf{z})| \le \delta\}$ with $\delta \ll \varepsilon$ and $\mathbb{D}_\varepsilon^* := \{t \in \mathbb{C} \mid |t| \le \varepsilon\} \setminus \{0\}$ [15, 22]. After this fundamental result, many researches have been carried out in various related directions. Among them, the generalization of the fibration structure and related geometry to the situation $f : \mathbb{R}^m \to \mathbb{R}^p$ attract many researchers which has led to many researches even today. Milnor also proved the existence of the spherical fibration $\varphi : \mathbb{S}_\varepsilon^{m-1} \setminus K \to \mathbb{S}^{p-1}$ for sufficiently small ε under the assumption of isolated critical points at the origin (Theorem 11.2, [24]). The problem here is that φ is not necessarily the canonical one $f/|f|$. J. Seade studied when φ can be the natural map $f/|f|$ in [53]. For further information, see the references in [53] and also [7–9]. In this survey, we concentrate on the case $m = 2n$ and $p = 2$, namely $f : \mathbb{C}^n \to \mathbb{C}$ where the mapping is considered as a mixed function. It turns out that this class of functions produces a rich class of links which are fibered over a circle. We will try to give a survey of the basics to study mixed hypersurface singularities using the method of the non-degenerate Newton boundary (and also toric modification) which is a very powerful tool for the study of the complex analytic singularity theory.

Contents

M. Oka (✉)
Tokyo University of Science, Kagurazaka 1-3, Shinjuku-ku, Tokyo 162-8601, Japan
e-mail: oka@rs.kagu.tus.ac.jp

J. L. Cisneros-Molina et al. (eds.), *Handbook of Geometry and Topology of Singularities II*, https://doi.org/10.1007/978-3-030-78024-1_8

8.1 A Quick Trip to the Complex Hypersurface Singularity Theory

8.1.1 Milnor Fibration

Let U be an open neighborhood of the origin $\mathbf{0} \in \mathbb{C}^n$ and let $f : (U, \mathbf{0}) \to (\mathbb{C}, 0)$ be a holomorphic function. $\mathbf{0}$ is called *a critical point* or *a singular point* of the hypersurface $V := \{\mathbf{z} \mid f(\mathbf{z}) = 0\}$ if $\partial f := (\frac{\partial f}{\partial z_1}, \dots, \frac{\partial f}{\partial z_n})$ vanishes at $\mathbf{0}$. Milnor proved the following fundamental result.

Theorem 8.1.1 ([24]) *There exists a positive number ε such that*

$$\varphi : \mathbb{S}_\varepsilon^{2n-1} \setminus K_\varepsilon \to \mathbb{S}^1, \quad \varphi(\mathbf{z}) = f(\mathbf{z})/|f(\mathbf{z})|$$

is a locally trivial fibration where $\mathbb{S}_\varepsilon^{2n-1}$ is the sphere of radius ε and $K_\varepsilon = V \cap \mathbb{S}_\varepsilon^{2n-1}$. The equivalence class of this fibration does not depend on the choice of a small enough ε.

Recall that two fibration $p : E \to \mathbb{S}^1$ and $p' : E' \to \mathbb{S}^1$ is C^∞-equivalent (or C^0-equivalent) if there is a diffeomorphism (or homeomorphism) $\varphi : E \to E'$ such that $p = p' \circ \varphi : E \to \mathbb{S}^1$. In this survey, we always consider C^∞-equivalence. K_ε is called *a link* of the hypersurface. Brieskorn found an exotic sphere as a link of a certain polynomial. For example, $f(\mathbf{z}) = z_1^3 + z_2^{6k-1} + z_3^2 + \cdots + z_n^2$ with n odd gives exotic spheres for any $k \geq 1$. His discovery of exotic spheres as links of complex hypersurfaces defined by so called Brieskorn polynomials pushed Milnor to start a systematic study of links of hypersurfaces. There is a nice survey about these topics in [9]. The fiber $\varphi^{-1}(1)$ is usually denoted as F and it is called a Milnor fiber. A basic result on the topology of links is.

Theorem 8.1.2 ([24]) *The link K_ε is $(n-3)$-connected and F has a homotopy type of (at most) an $(n-1)$-dimensional CW-complex. If further $\mathbf{0}$ is an isolated singularity of V, F is $(n-2)$-connected and homotopic to a bouquet of $(n-1)$-spheres.*

In the case of isolated singularity, the Betti-number $b_{n-1}(F)$ is usually called *the Milnor number* of f at the origin and is denoted by $\mu(f)$. Kato-Matsumoto generalized this connectivity assertion as follows.

Theorem 8.1.3 (Kato-Matsumoto [19]) *Let Σ be the singular locus of f and let $s = \dim_0 \Sigma$. Then the Milnor fiber is $(n-2-s)$-connected.*

Monodromy of the Fibration

Consider the spherical Milnor fibration $\varphi : \mathbb{S}_\varepsilon^{2n-1} \setminus K_\varepsilon \to \mathbb{S}^1$. Construct a vector field $\mathcal{X}$ on $\mathbb{S}_\varepsilon^{2n-1} \setminus K_\varepsilon$ so that $d\varphi(\mathcal{X}(\mathbf{z})) = \frac{\partial}{\partial\theta}(\varphi(\mathbf{z}))$ where $\frac{\partial}{\partial\theta}$ is the unit angular vector field on $\mathbb{S}^1$. We can construct $\mathcal{X}$ so that it is integrable over any finite time interval. For this purpose, we can use Lemma 8.1.5 in Sect. 8.1.2 to assume the additional condition: $\mathcal{X}(\mathbf{z})$ is tangent to $\{\mathbf{z} \mid |f(\mathbf{z})| = \text{const.}\}$ for $\mathbf{z} \in \mathbb{S}_\varepsilon^{2n-1} \setminus K_\varepsilon$ where $|f(\mathbf{z})| \leq \delta$ with a mall positive δ. Let $F = \varphi^{-1}(1)$ be the Milnor fiber. By the integration over $[0, \theta]$, we get a family of diffeomorphisms $h_\theta : F \to F_\theta := \varphi^{-1}(e^{i\theta})$. Then $h := h_{2\pi} : F \to F$ and h is called *a geometric monodromy* of the Milnor fibration. The geometric monodromy h depends on the choice of $\mathcal{X}$ but the isotopy class of h does not depend on $\mathcal{X}$ and it induces a well-defined isomorphism on the homology group $h_{*j} : H_j(F) \to H_j(F)$ (or on cohomology group $h^{*j} : H^j(F) \to H^j(F)$). For a tubular Milnor fibration, the monodromy is defined similarly. See also Cisneros-Seade's chapter, Vol. 2 in this handbook [9].

Characteristic Polynomials and Zeta Function

Let $h : F \to F$ be a geometric monodromy map. Then the complement of the link $E = \mathbb{S}_\varepsilon \setminus K_\varepsilon$ is obtained by glueing $F \times \{1\}$ and $F \times \{0\}$ by h on the product $F \times [0, 1]$. *The j-th characteristic polynomial $P_j(t)$* is defined by the characteristic polynomial of j-th monodromy homomorphism $h_{*j} : H_j(F; \mathbb{R}) \to H_j(F; \mathbb{R})$. The zeta function of the monodromy is defined by the alternating product [24]:

$$\zeta(t) = P_0(t)^{-1} P_1(t) \cdots P_{n-1}(t)^{(-1)^{n-2}}.$$

In the case of an isolated singularity, $\zeta(t) = P_{n-1}(t)^{(-1)^{n-2}}(t-1)^{-1}$ and $\deg P_{n-1}(t) = \mu(f)$. Therefore

$$\mu(f) = (-1)^n(\deg \zeta(t) + 1).$$

Recall that the degree of a rational function $p(t)/q(t)$ is defined by $\deg\ p - \deg\ q$. Using the Wang exact sequence of the spherical Milnor fibration (see [58], or Lemma 8.4, [24]), the following criterion is well-known (Theorem 8.5, [24]).

Proposition 8.1.4 (Milnor [24]) *Assume that f has an isolated singularity at the origin. K_ε is a homology sphere (a homotopy sphere for $n \geq 4$) if and only if $P_{n-1}(1) = \pm 1$.*

8.1.2 The Hamm-Lê lemma and a Tubular Milnor Fibration

To study the Milnor fibration from the view point of algebraic geometry, it is more convenient to use the so called *tubular Milnor fibration*. This fibration was systematically studied by Hamm-Lê to prove Zariski's hyperplane section theorem [15]. Refer also to the paper of A'Campo [1] where a tubular fibration and resolution of the function f is used for the calculation of the zeta function.

Lemma 8.1.5 (Hamm-Lê, Lemme (2.1.4), [15]) *Let f be a holomorphic function defined in a neighborhood U of the origin in $\mathbb{C}^n$. Then there exists a positive number r_0 satisfying the following property. Take any positive number $r_1 \leq r_0$. There exists a positive number $\delta(r_1)$ such that*

(SN) (Smoothness of the nearby fibers) For any non-zero η, $|\eta| \leq \delta(r_1)$, the fiber $f^{-1}(\eta) \cap \mathbb{B}^{2n}_{r_0}$ is non-singular.
(ST) (Strong transversality) For any $r_1 \leq r \leq r_0$, the sphere $\mathbb{S}^{2n-1}_r$ and the fiber $f^{-1}(\eta)$ with η, $|\eta| \leq \delta(r_1)$ intersect transversely.

This lemma is important when f has a non-isolated singularity at the origin. Take any positive number $r \leq r_0$ and $\delta \leq \delta(r)$ and consider the mapping $f : E(r,\delta)^* \to \mathbb{D}^*_\delta$ where

$$E(r,\delta)^* := \{\mathbf{z} \in \mathbb{B}^{2n}_r \,|\, 0 \neq |f(\mathbf{z})| \leq \delta\}, \quad \mathbb{D}^*_\delta := \{\eta \in \mathbb{C} \,|\, 0 \neq |\eta| \leq \delta\}.$$

Then by the Ehresmann's fibration theorem [60],

Theorem 8.1.6 *$f : E(r,\delta)^* \to \mathbb{D}^*_\delta$ is a locally trivial fibration. Its restriction over $\mathbb{S}^1_\delta$ is equivalent to the spherical Milnor fibration.*

We call this fibration *a tubular Milnor fibration.* In the comparison with this fibration, *a spherical Milnor fibration* is the original fibration on the complement of a link described in Theorem 8.1.1. For the proof of the equivalence of two Milnor fibrations, see for example Theorem (2.2), [29].

Resolution and A'Campo Formula

Let $f : (U, \mathbf{0}) \to (\mathbb{C}, 0)$ be a holomorphic function where U is an open neighborhood of the origin in $\mathbb{C}^n$ and put $V = f^{-1}(0)$. A holomorphic mapping $\pi : X \to U$ is called *a good resolution of* f if the following conditions are satisfied.

1. X is an n-dimensional complex manifold.
2. π is a proper mapping and the restriction $\pi : X \setminus \pi^{-1}(V) \to U \setminus V$ is biholomorphic.
3. Put $\pi^{-1}(V) = (\pi^* f)^{-1}(0) = \tilde{V} + \sum_{j=1}^{r} m_j E_j$ and put $D = \pi^{-1}(\mathbf{0})$. Here $\tilde{V}$ is the strict transformation of V. Here m_j is the multiplicity of $\pi^* f$ along E_j. Then $\tilde{V}$ and each divisors E_j $(j = 1, \dots, r)$ are non-singular and the reduced divisor of $(\pi^* f)^{-1}(0)$ has only normal crossing singularities.

Put $E_0 = \tilde{V}$ and let $E'_j = E_j \cap D \setminus \bigcup_{k \neq j} E_k$. Then the formula of A'Campo is given as follows.

Theorem 8.1.7 (A'Campo [1], Theorem (5.2), [29]) *The zeta function of the monodromy of the Milnor fibration is given by the following.*

$$\zeta(t) = \prod_{j=1}^{r} (1 - t^{m_j})^{-\chi(E'_j)}.$$

In the case of an isolated singularity and $\pi : X \setminus D \to U \setminus \{\mathbf{0}\}$ is biholomorphic, E_j is included in $\pi^{-1}(\mathbf{0})$. So the calculation of $\zeta(t)$ is reduced to the computation of the Euler characteristic $\chi(E'_j)$. Let $\tilde{E}(r, \delta)^* := \pi^{-1}(E(r, \delta)^*)$. Then the tubular Milnor fibration can be understood as the fibration $\pi^* f : \tilde{E}(r, \delta)^* \to \mathbb{D}^*_\delta$. Here is the point where the Newton boundary and the non-degeneracy condition come in.

8.1.3 Weighted Homogeneous Polynomials

Let $f(\mathbf{z}) = \sum_\nu c_\nu \mathbf{z}^\nu$ be a polynomial (or a Laurent polynomial) and let $P = {}^t(p_1, \dots, p_n)$ be an integer vector. f is called *a weighted homogeneous polynomial (or Laurent polynomial) of degree dwith the weight vector* P if $\sum_{j=1}^{n} p_j \nu_j = d$ for any ν with $c_\nu \neq 0$. Here d is assumed to be a non-zero integer and $\nu = (\nu_1, \dots, \nu_n)$. We say that P is *a strictly positive weight vector* (respectively *a non-negative weight vector*) if $p_i > 0$ for any i (resp. $p_i \geq 0$ *for any* i). When a weighted homogeneous polynomial is given, there is an associated $\mathbb{C}^*$-action on $\mathbb{C}^n$ which is defined by $t \circ \mathbf{z} = (z_1 t^{p_1}, \dots, z_n t^{p_n})$. Here $\mathbb{C}^* = \mathbb{C} \setminus \{0\}$. Then f satisfies the equality $f(t \circ \mathbf{z}) = t^d f(\mathbf{z})$. Taking a differential in t and putting $t = 1$, we get the Euler equality

$$d \cdot f(\mathbf{z}) = \sum_{j=1}^{n} p_j z_j \frac{\partial f}{\partial z_j}(\mathbf{z}). \tag{8.1}$$

This equality implies the critical value of f is only 0 or empty. The importance of weighted homogeneous polynomials is the fact that it appears canonically as a face function of the Newton boundary.

8.1.4 Newton Boundary and Non-degeneracy

(a) (Newton polyhedron of a germ) Let $f(\mathbf{z}) = \sum c_\nu \mathbf{z}^\nu$ be a germ of a holomorphic function expanded at $\mathbf{z} = \mathbf{0}$. The Newton polyhedron $\Gamma_+(f)$ for a germ of function f is defined by the convex hull of the union $\bigcup_{\nu, c_\nu \neq 0}(\nu + \mathbb{R}^n_+)$. It is a non-compact polyhedron and we define the Newton boundary $\Gamma(f)$ as the union of compact faces of $\Gamma_+(f)$. The Newton boundary $\Gamma(f)$ is useful for the study of the local geometry of $f^{-1}(0)$ at the origin. Unless otherwise stated, we consider this situation in this survey.

(b) (For global geometry) Let $h(\mathbf{z}) = \sum_{i=1}^{m} c_i \mathbf{z}^{\nu_i}$ be a Laurent polynomial. The Newton polygon $\Delta(h)$ is defined as the convex hull of $\{\nu_i \mid c_{\nu_i} \neq 0\}$. Note that in this case, we do not add the upper right quadrant $\mathbb{R}^n_+$ and thus $\Delta(f)$ is a compact polyhedron.

In this survey, we consider the local germ case, unless otherwise stated. Let Ξ be a face of $\Gamma(f)$ (of any dimension). The face function of f for Ξ is defined by

$$f_\Xi(\mathbf{z}) := \sum_{\nu \in \Xi} c_\nu \mathbf{z}^\nu.$$

We say that f is (Newton) *non-degenerate on* Ξ if $f_\Xi : \mathbb{C}^{*n} \to \mathbb{C}$ has no critical point. Here $\mathbb{C}^{*n}$ is the maximal torus:

$$\mathbb{C}^{*n} = \{\mathbf{z} = (z_1, \ldots, z_n) \mid z_j \neq 0, \ \forall j\}.$$

We say that f is *non-degenerate* if f is non-degenerate on every face Ξ of $\Gamma(f)$. For Laurent polynomials, the non-degeneracy is defined as follows. Let $h(\mathbf{z})$ be a Laurent polynomial. For a face Ξ of $\Delta(h)$, we say that h is (Newton) *non-degenerate on* Ξ if $h_\Xi : \mathbb{C}^{*n} \to \mathbb{C}$ has no critical point. We say h is *non-degenerate* if $h : \mathbb{C}^{*n} \to \mathbb{C}$ has no critical point and h is non-degenerate on every face Ξ of $\Delta(h)$. We only use the non-degeneracy of Laurent polynomial in Lemma 8.3.40.

Dual Newton Diagram and Associated Toric Modification

Consider the space of non-negative rational weight vectors $N_{\mathbb{Q}}^{+} \subset \mathbb{Z}^n$. Take a weight vector $P = {}^t(p_1, \ldots, p_n) \in N_{\mathbb{Q}}^{+}$. To emphasize the weight vectors, we denote weight vectors by column vectors. P defines a linear function ℓ_P on $\Gamma_+(f)$ by $\ell_P(\nu) = \sum_{i=1}^{n} p_i \nu_i$. For simplicity, we write $\ell_P(\nu) = P(\nu)$ from now on. Let $d(P, f)$ be the minimal value of $\ell_P|_{\Gamma_+(f)}$ and $\Delta(P)$ be the face where ℓ_P takes its minimal value. We also write $d(P)$ for $d(P, f)$ for simplicity. As $P \in N_{\mathbb{Q}}^{+}$, we see that $d(P)$ exists and it is a no-negative integer. Define $f_P(\mathbf{z}) := f_{\Delta(P)}(\mathbf{z})$. Note that f_P is a weighted homogeneous function of degree $d(P)$ with the weight vector P. If P is strictly positive, f_P is a polynomial. We introduce an equivalence relation in $N_{\mathbb{Q}}^{+}$. We say P, Q are equivalent and denote it as $P \sim Q$, if $\Delta(P) = \Delta(Q)$. This gives a polyhedral cone subdivision of $N_{\mathbb{Q}}^{+}$ which we call *the dual Newton diagram* of f and denote it as $\Gamma^*(f)$. Here a cone subdivision is a subdivision which satisfies (1) $P \sim rP$ for any $r > 0$ and (2) if $P \sim Q$, any point R on the line segment $\overline{PQ}$ is also equivalent to P.

For a polyhedral cone, the closure of an equivalence class $[P]$ can be written as a closed cone

$$\sigma = \mathrm{Cone}(P_1, \ldots, P_k) := \left\{ Q = \sum_{i=1}^{k} t_i P_i \mid t_i \geq 0 \right\}$$

and $[P]$ includes the open cone Int σ. In this survey, a cone is always a closed cone. Here we assume that $\{P_1, \ldots, P_k\}$ is the minimal set of generators of σ. The vectors $P_1, \ldots, P_k$ are called *the vertices of* σ. We can take P_i to be a primitive integer vector. A polyhedral cone subdivision Σ^* of $N_{\mathbb{Q}}^{+}$ is called *a regular simplicial cone subdivision, or a regular fan* if each maximal dimensional cone $\tau = \mathrm{Cone}(P_1, \ldots, P_n)$ in Σ^* is generated by n primitive integer vectors so that $\{P_1, \ldots, P_n\}$ is a $\mathbb{Z}$-basis of $\mathbb{Z}^n$. Equivalently $\det(p_{ji})_{1 \leq i,j \leq n} = \pm 1$ where $P_i = {}^t(p_{1i}, \ldots, p_{ni})$. We say Σ^* is *a regular simplicial cone subdivision of* $\Gamma^*(f)$ if Σ^* is a regular fan which is also a subdivision of $\Gamma^*(f)$.

Remark 8.1.8 Recall that $f(\mathbf{z})$ is *convenient* if the Newton boundary $\Gamma(f)$ intersect with every coordinate axis. Equivalently $f(\mathbf{z})$ contains a monomial $c_j z_j^{a_j}$ with $c_j \neq 0$ for each $j = 1, \ldots, n$. Suppose that f is convenient. Then $(n-1)$-dimensional cone Cone $(E_1, \ldots, \overset{\vee}{E_i}, \ldots, E_n)$ is an equivalence class. Here $E_1, \ldots, E_n$ are canonical generators. Namely $E_i = {}^t(0, \ldots, \underset{i}{\underbrace{1}}, \ldots, 0)$.

Suppose Σ^* is a regular simplicial cone subdivision of $\Gamma^*(f)$. Let $\mathcal{S}$ be the set of n-dimensional cones in Σ^*. Take $\sigma \in \mathcal{S}$. Choose primitive integer vectors $P_1, \ldots, P_n$ so that $\sigma = \mathrm{Cone}(P_1, \ldots, P_n)$. We identify σ with the unimodular matrix $(P_1, \ldots, P_n) = (p_{ij})_{1 \leq i,j \leq n}$ and we consider an affine space $\mathbb{C}_\sigma^n$ labeled with σ and the coordinates $\mathbf{u}_\sigma = (u_{\sigma,1}, \ldots, u_{\sigma,n})$ and the holomorphic mapping $\pi_\sigma : \mathbb{C}_\sigma^n \to \mathbb{C}^n$ which is defined by

$$\pi_\sigma(\mathbf{u}_\sigma) = \mathbf{z}, \quad z_j = u_{\sigma,1}^{p_{j1}} \cdots u_{\sigma,n}^{p_{jn}},\ j = 1, \ldots, n.$$

Note that the restriction of π_σ to the maximal torus $\mathbb{C}^{*n}$ is a group isomorphism of $\mathbb{C}^{*n}$. In the same way, the homomorphism π_A is defined for any unimodular matrix A. It satisfies the canonical composition rule $\pi_A \circ \pi_B = \pi_{AB}$. Therefore π_σ is a birational map and the inverse is given by $\pi_{\sigma^{-1}}$ which is at least well-defined on the torus $\mathbb{C}^{*n} \subset \mathbb{C}^n$. We construct n-dimensional complex manifold X, first taking the disjoint union $\amalg_{\sigma \in \mathcal{S}} \mathbb{C}^n_\sigma$ and then glue $\mathbb{C}^n_\sigma$ and $\mathbb{C}^n_\tau$ by $\mathbf{u}_\tau = \pi_{\tau^{-1}\sigma}(\mathbf{u}_\sigma)$ wherever $\pi_{\tau^{-1}\sigma} : \mathbb{C}^n_\sigma \to \mathbb{C}^n_\tau$ is well-defined. By the definition, there is a birational holomorphic mapping $\hat{\pi} : X \to \mathbb{C}^n$ which is called *the toric modification associated with* Σ^*. By the construction, $\{(\mathbb{C}^n_\sigma, \mathbf{u}_\sigma),\ \sigma \in \mathcal{S}\}$ give canonical charts of X. Note that the restriction of $\hat{\pi}$ to $\mathbb{C}^n_\sigma$ is nothing but π_σ. The main property of this modification is:

Theorem 8.1.9 *Assume that f is non-degenerate. Then $\hat{\pi} : X \to \mathbb{C}^n$ is a proper biholomorphic mapping which gives a good resolution of f.*

Assume that $f(\mathbf{z})$ is convenient. Then we may assume that the vertices of Σ^* other than the canonical one E_i, $i = 1, \ldots, n$ are strictly positive. See Theorem (3.4), [29] for further details. For a coordinate chart $\mathbb{C}^n_\sigma$ with $\sigma = \mathrm{Cone}(P_1, \ldots, P_n)$, the divisor $\hat{E}(P_i)$ is defined by the closure of $\{u_{\sigma,i} = 0\}$. If P_i is strictly positive, $\hat{E}(P_i)$ is an exceptional divisor i.e. $\hat{\pi}(\hat{E}(P_i)) = \mathbf{0}$.

The Further Basic Properties of $\hat{\pi} : X \to \mathbb{C}^n$

1. The pull-back of a monomial is given by $\hat{\pi}^*\mathbf{z}^\nu = u_{\sigma,1}^{P_1(\nu)} \cdots u_{\sigma,n}^{P_n(\nu)}$ in the coordinate chart $\mathbb{C}^n_\sigma$ with $\sigma = \mathrm{Cone}(P_1, \ldots, P_n)$. Thus we have

$$\hat{\pi}^* f(\mathbf{u}_\sigma) = f_\sigma(\mathbf{u}_\sigma) \prod_{i=1}^n u_{\sigma,i}^{d(P_i)}$$

$$\text{where } f_\sigma(\mathbf{u}_\sigma) := \sum_\nu c_\nu \prod_{j=1}^n u_{\sigma,j}^{P_j(\nu)-d(P_j)}$$

The intersection of $\tilde{V}$ with the exceptional divisor $\hat{E}(P_i)$ is defined as

$$\hat{E}(P_i) \cap \tilde{V} = \{\mathbf{u}_\sigma \mid u_{\sigma,i} = f_{\sigma,P_i}(\mathbf{u}_\sigma) = 0\},$$

$$f_{\sigma,P_i}(\mathbf{u}_\sigma) := \pi_\sigma^{-1} * f_{P_i}(\mathbf{u}_\sigma) / \prod_{j=1}^n u_{\sigma,j}^{d(P_j)}$$

$$= \sum_{\nu \in \Delta(P_i)} c_\nu \prod_{j=1}^n u_{\sigma,j}^{P_j(\nu)-d(P_j)}$$

Here $\tilde{V}$ is the strict transform of V. Thus f_{σ,P_i} is described by the pull-back of the face function f_{P_i} divided by $\prod_{j=1}^{n} u_{\sigma,j}^{d(P_j)}$. Note that f_{σ,P_i} does not contain the variable $u_{\sigma,i}$.

2. For two vertices P, Q, the intersection $\hat{E}(P) \cap \hat{E}(Q)$ is non-empty if and only if there is a maximal cone $\sigma = \mathrm{Cone}(P_1, \ldots, P_n)$ such that $P = P_1$, $Q = P_2$.
3. Put $E(P) := \hat{E}(P) \cap \tilde{V}$. Then $E(P) \neq \emptyset$ if and only if $\dim \Delta(P) \geq 1$.

For the calculation of the zeta function through A'Campo's formula, we need only consider the maximal dimensional faces of each coordinate subspaces $\mathbb{C}^I$ and $\chi(\hat{E}(P)')$ can be computed combinatorially (see Lemma 8.3.40). The zeta function is given combinatorially by Varchenko [57] as follows.

$$\zeta(t) = \prod_I \zeta_I(t), \ \zeta_I = \prod_{Q \in S_I} (1 - t^{d(Q, f^I)})^{-\chi(Q)}$$

$$\chi(Q) = (-1)^{|I|-1} |I|! \mathrm{Vol}_{|I|} C(\Delta(Q, f^I), \mathbf{0}) / d(Q, f^I).$$

Here Vol_m is the m-dimensional Euclidean volume and $\mathcal{S}_I$ is the set of weight vectors associated to the maximal dimensional faces of $\Gamma(f^I)$. $C(\Delta, \mathbf{0})$ is the cone of Δ with the origin, i.e. $C(\Delta, \mathbf{0}) = \{r\nu \mid 0 \leq r \leq 1, \ \nu \in \Delta\}$. See Theorem (5.3), [29] for details.

Example 8.1.10 Let Σ^* be the regular simplicial cone subdivision with vertices $\{E_1, \ldots, E_n, P\}$, $P = {}^t(1, \ldots, 1)$. Then the associated toric modification $\hat{\pi} : X \to \mathbb{C}^n$ is nothing but the ordinary blowing up at the origin and X has n coordinate charts

$\sigma_i = \mathrm{Cone}\,(E_1, \ldots, \underset{\smile}{\overset{i}{P}}, \ldots, E_n), i = 1, \ldots, n.$

Example 8.1.11 1. $f(z_1, z_2) = z_1^2 - z_2^3$. The dual Newton diagram $\Gamma^*(f)$ is the left side of Fig. 8.1. It has three vertices $E_1, E_2, P = {}^t(3, 2)$. Taking a section with the dotted line, we denote it as the right side graph with black vertices. The dual Newton diagram is not regular. We add two vertices $S = {}^t(1, 1), T = {}^t(2, 1)$ to make it regular. Consider the cone $\sigma = \mathrm{Cone}(S, P)$. Then $\pi_\sigma^* f(\mathbf{u}_\sigma) = u_{\sigma,1}^2 u_{\sigma,2}^6 (1 - u_{\sigma,1})$. $\tilde{V}$ is defined by $u_{\sigma,1} - 1 = 0$ and intersects transversely with $\hat{E}(P)$ which is defined by $u_{\sigma,2} = 0$. The multiplicities of $\hat{\pi}^* f$ one $\hat{E}(S)$ and $\hat{E}(P)$ are 2 and 6 respectively.
2. Let $f(z_1, z_2, z_3) = z_1^d + z_2^d + z_3^d$. Then the dual Newton diagram is given by 4 vertices E_1, E_2, E_3 and $P = {}^t(1, 1, 1)$. $\Gamma^*(f)$ is already regular and the corresponding toric modification is nothing but the blowing up at the origin.

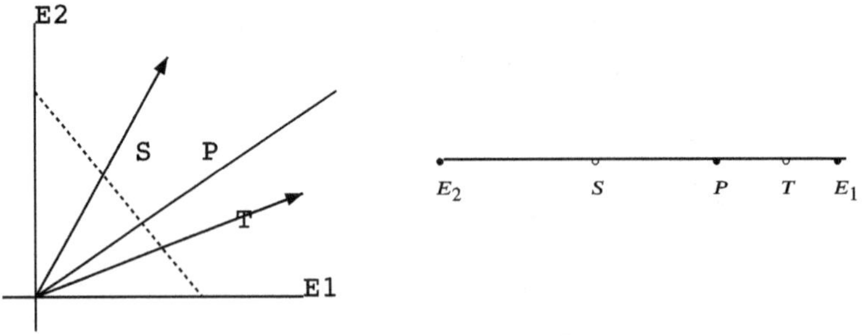

Fig. 8.1 Left:$\Gamma^*(f)$, right: section

8.2 Mixed Hypersurface Singularities

8.2.1 Mixed Analytic Functions

A complex valued real analytic function $f(\mathbf{z}, \bar{\mathbf{z}})$ defined on an open set U of $\mathbb{C}^n$ is called *a mixed function* if there is an analytic function of $2n$-variables $f(\mathbf{z}, \mathbf{w})$ defined on $U \times \check{U}$ such that $f(\mathbf{z}, \bar{\mathbf{z}})$ is obtained by substituting $w_i = \bar{z}_i$, $i = 1, \ldots, n$. Here $\check{U} = \{\bar{\mathbf{z}} \,|\, \mathbf{z} \in U\}$. Assuming $\mathbf{0} \in U$ for simplicity, $f(\mathbf{z})$ can be expanded as $f(\mathbf{z}, \bar{\mathbf{z}}) = \sum_{\nu,\mu} c_{\nu,\mu} \mathbf{z}^\nu \bar{\mathbf{z}}^\mu$. Here $\mathbf{z} = (z_1, \ldots, z_n)$, $\bar{\mathbf{z}} = (\bar{z}_1, \ldots, \bar{z}_n)$, $\nu = (\nu_1, \ldots, \nu_n)$, $\mu = (\mu_1, \ldots, \mu_n)$ and $\mathbf{z}^\nu = z_1^{\nu_1} \cdots z_n^{\nu_n}$, $\bar{\mathbf{z}}^\mu = \bar{z}_1^{\mu_1} \cdots \bar{z}_n^{\mu_n}$. $\mathbf{z}^\nu \bar{\mathbf{z}}^\mu$ is called *a mixed monomial*.

We consider the germ of a real analytic variety $(V, \mathbf{0})$, $V = f^{-1}(0) \subset \mathbb{C}^n$ and its local geometry unless otherwise stated. Writing $z_j = x_j + iy_j$, $\bar{z}_j = x_j - iy_j$, f can be written as $f(\mathbf{x}, \mathbf{y}) = g(\mathbf{x}, \mathbf{y}) + ih(\mathbf{x}, \mathbf{y})$ where $g(\mathbf{x}, \mathbf{y}) = \Re f(\mathbf{x}, \mathbf{y})$, $h(\mathbf{x}, \mathbf{y}) = \Im f(\mathbf{x}, \mathbf{y})$ and $V = \{(\mathbf{x}, \mathbf{y}) \,|\, g(\mathbf{x}, \mathbf{y}) = h(\mathbf{x}, \mathbf{y}) = 0\}$. Thus the mixed hypersurface $V = f^{-1}(0)$ can be equivalently described as a complete intersection variety of real codimension 2 from the analogy of complex analytic hypersurfaces. Conversely, if a complete intersection variety of real codimension 2 in $\mathbb{R}^{2n}$ is given as

$$V = \{(\mathbf{x}, \mathbf{y}) \in \mathbb{R}^{2n} \,|\, g(\mathbf{x}, \mathbf{y}) = h(\mathbf{x}, \mathbf{y}) = 0\},$$

take complex variables $z_1, \ldots, z_n$ $(z_j = x_j + iy_j)$ and substitute

$$x_j = (z_j + \bar{z}_j)/2, \quad y_j = (z_j - \bar{z}_j)/2i$$

in g, h and define $f(\mathbf{z}, \bar{\mathbf{z}}) = g(\mathbf{z}, \bar{\mathbf{z}}) + ih(\mathbf{z}, \bar{\mathbf{z}})$. Then V can be described as a mixed hypersurface defined by the mixed function f. The difference is that when we use a mixed function description, we fix the ordering of g, h so that it gives an orientation of the normal bundle of the hypersurface V. We define several gradient vectors.

$$dg = (g_{x_1}, g_{y_1}, \dots, g_{x_n}, g_{y_n}), \text{ where } g_{x_j} = \frac{\partial g}{\partial x_j}, g_{y_j} = \frac{\partial g}{\partial y_j},$$

$$\partial f = (f_{z_1}, \dots, f_{z_n}), \ \bar{\partial} f = (f_{\bar{z}_1}, \dots, f_{\bar{z}_n}), \text{ where } f_{z_j} = \frac{\partial f}{\partial z_j}, f_{\bar{z}_j} = \frac{\partial f}{\partial \bar{z}_j}.$$

Proposition 8.2.1 *Let $k : \mathbb{C}^n \to \mathbb{R}$ be a real valued mixed function. Then $\overline{\partial k} = \bar{\partial} k$ and $dk = 2\bar{\partial} k$.*

Here $(x_1, y_1, \dots, x_n, y_n) \in \mathbb{R}^{2n}$ is canonically identified with $(x_1 + iy_1, \dots, x_n + iy_n) \in \mathbb{C}^n$.

8.2.2 Mixed Singularities

Consider a mixed analytic function $f(\mathbf{z}, \bar{\mathbf{z}})$ defined on an open set $U \subset \mathbb{C}^n$. A point $\mathbf{a} = (\alpha_1, \dots, \alpha_n) \in U$ is called *a critical point of f* or *a mixed singular point* if the tangential mapping $df_{\mathbf{a}} : T_{\mathbf{a}}\mathbb{C}^n \to T_{f(\mathbf{a})}\mathbb{C} \cong T_{f(\mathbf{a})}\mathbb{R}^2$ is not surjective. If $df_{\mathbf{a}}$ is surjective, $\mathbf{a}$ is called *a mixed non-singular* (or *a regular*) point of f.

Proposition 8.2.2 (Proposition 1, [30]) *The following conditions are equivalent.*

1. *$\mathbf{a} = (a_1, \dots, a_n)$ is a critical point of f.*
2. *There exists a complex number α, $|\alpha| = 1$ such that $\overline{\partial f(\mathbf{a}, \bar{\mathbf{a}})} = \alpha \bar{\partial} f(\mathbf{a}, \bar{\mathbf{a}})$.*

For brevity, we often denote $f(\mathbf{a}, \bar{\mathbf{a}})$ by $f(\mathbf{a})$, $\overline{\partial f(\mathbf{a}, \bar{\mathbf{a}})}$ by $\overline{\partial f}(\mathbf{a})$.

Lemma 8.2.3 ([5], Lemma 2, [38]) *Assume that $V = f^{-1}(0)$ is mixed non-singular at $\mathbf{p} \in \mathbb{S}_r^{2n-1} \cap V$. Then the following conditions are equivalent.*

1. *The intersection of the sphere $\mathbb{S}_r^{2n-1}$ and V at $\mathbf{p}$ is not transverse.*
2. *There exists a complex number $\alpha \in \mathbb{C}^*$ such that $\mathbf{p} = \alpha \overline{\partial f}(\mathbf{p}) + \bar{\alpha} \bar{\partial} f(\mathbf{p})$.*
3. *There exist real numbers c, c' such that*

$$\mathbf{p} = c\bar{\partial} g(\mathbf{p}) + c'\bar{\partial} h(\mathbf{p}).$$

Remark 8.2.4 For a mixed hypersurface $V = f^{-1}(0)$, $\mathbf{a} \in V$ can be a non-singular point as a real algebraic variety, even when $\mathbf{a}$ is a mixed singular point. For example, if $f = |z_1|^2 + \cdots + |z_n|^2 - 1$, V is a $(2n-1)$-dimensional sphere and is smooth everywhere, but it is also mixed singular everywhere. "Mixed-non-singular at $\mathbf{a}$" implies V is smooth and of real codimension 2 at $\mathbf{a}$.

8.2.3 A Tubular Milnor Fibration of a Real Analytic Mapping

Consider a real analytic mapping $f : U \to \mathbb{C}$ where U is an open neighborhood of the origin of $\mathbb{C}^n$ and assume that f satisfies the following conditions.

(SN) (Smoothness of the nearby fibers) There exists an open neighborhood W of $\mathbf{0} \in \mathbb{C}^n$ and 0 is the unique critical value of the restriction $f|_W$.

(T) (Transversality) There exist positive numbers r_0 and $\delta \ll r_0$ such that $\mathbb{B}_{r_0}^{2n} \subset W$ and for any $\eta \in \mathbb{D}_\delta^*$, $f^{-1}(\eta)$ and the sphere $\mathbb{S}_{r_0}^{2n-1}$ intersect transversely.

Here $\mathbb{D}_\delta := \{\eta \in \mathbb{C} \mid |\eta| \le \delta\}$ and $\mathbb{D}_\delta^* := \{\eta \in \mathbb{D}_\delta \mid \eta \ne 0\}$. Then by the Ehresmann fibration theorem (see [60]), we have

Proposition 8.2.5 *$f : E(r_0, \delta)^* \to \mathbb{D}_\delta^*$ is a locally trivial fibration where*

$$E(r_0, \delta)^* := \{\mathbf{z} \in \mathbb{B}_{r_0}^{2n} \mid 0 \ne |f(\mathbf{z})| \le \delta\}.$$

Furthermore we consider the stronger transversality condition:

(ST) (Stronger transversality) For any positive $r \le r_0$, there exists a positive number $\delta(r)$ such that for any non-zero $\eta \in \mathbb{D}_{\delta(r)}^*$ and any ρ, $r \le \rho \le r_0$, $f^{-1}(\eta)$ and the sphere $\mathbb{S}_\rho^{2n-1}$ intersects transversely.

Proposition 8.2.6 *Assume that f satisfy the conditions (SN) and (ST). The above fibration $f : E(r_0, \delta)^* \to \mathbb{D}_\delta^*$ does not depend on the choice of r_0 and $\delta \ll r_0$.*

The above fibration is called *the tubular Milnor fibration of f*.

8.2.4 Stratification and Thom's a_f-Regularity

Stratification

In this section, we recall basic definitions about stratification. The stratification theory is introduced by Whitney [59] and Thom [55]. Let X be a real smooth manifold of dimension n and let V be a closed subset.

Let $\mathcal{S} = \{M_\alpha; \alpha \in A\}$ be a family of mutually disjoint submanifolds of X where M_α is a subset of V. The partition $\mathcal{S}$ is called *a smooth stratification of V* if the following conditions are satisfied:

(0) (Partition of V) V is the union of M_α for $\alpha \in A$.

(i) (Locally closedness) Each element of $\mathcal{S}$ is a locally closed smooth submanifold of X.

(ii) (Locally finiteness) For each point $x \in V$, there exists an open neighborhood U of x in X such that $\{\alpha \in A; U \cap M_\alpha \ne \emptyset\}$ is finite.

(iii) (Frontier condition) $\overline{M_\alpha} \cap M_\beta \ne \emptyset$ implies $\overline{M_\alpha} \setminus M_\alpha \supset M_\beta$. Here $\overline{M}$ is the closure of M in X.

An element M_α in $\mathcal{S}$ is called a stratum. $\mathcal{S}$ is called *Whitney a-regular* if the following condition is satisfied for any M_α, M_β with $M_\beta \subset \overline{M_\alpha} \setminus M_\alpha$. Take a point $p_\infty \in M_\beta$ and a sequence $p_\nu \in M_\alpha$ which converges to p_∞. Taking a subsequence if necessary, we assume that $\lim_{\nu\to\infty} T_{p_\nu} M_\alpha = \tau$ in a suitable Grassmannian variety. Then $\tau \supset T_{p_\infty} M_\beta$. $\mathcal{S}$ satisfies b-regularity if it satisfies the following. Take another sequence q_ν on M_β which converges to p_∞ and assume that the direction $\overline{p_\nu q_\nu}$ converges to ℓ. Then $\ell \in \tau$. It is known that b-regularity implies a-regularity. $\mathcal{S}$ is called Whitney regular if it satisfied b-regularity. For further details, see Mather [23] and Trotman's chapter in this Handbook [56].

Thom's a_f-Regularity

Consider a real analytic mapping $f : U \to \mathbb{C}$ as above. We say that f satisfies a_f-regularity of Thom if there exists a positive number r_0 such that in $\mathbb{B}^{2n}_{r_0} \setminus f^{-1}(0)$, f has no critical point and there exists a stratification $\mathcal{S}$ of $f^{-1}(0) \cap \mathbb{B}^{2n}_{r_0}$ with finite strata satisfying the following condition. For any sequence $p_\nu \in \mathbb{B}^{2n}_{r_0} \setminus f^{-1}(0)$ such that $T_{p_\nu} f^{-1}(f(p_\nu)) \to \tau$ in the real Grassmannian space and $p_\nu \to p_\infty$ with $p_\infty \in M \in \mathcal{S}$. Then $\tau \supset T_{p_\infty} M$. Then the following is well-known.

Lemma 8.2.7 (Proposition 1, [38]) *Suppose that f satisfies a_f-regularity for some stratification $\mathcal{S}$ of $f^{-1}(0)$. Assume that there exists $r_0 > 0$ such that the sphere $\mathbb{S}^{2n-1}_r$, $r \le r_0$ and each stratum $M \in \mathcal{S}$ of a positive dimension intersect transversely. Then the condition (ST) is satisfied.*

Proof Let $V(\mathbf{p}) := f^{-1}(f(\mathbf{p}))$. Assume that (ST) does not hold for some r_1, $0 < r_1 < r_0$. Then there exists a sequence $\mathbf{p}_\nu \to \mathbf{p}_\infty \in M$ with $\|\mathbf{p}_\infty\| \ge r_1$ such that the sphere $S^{2n-1}_{\|\mathbf{p}_\nu\|}$ and $V(\mathbf{p}_\nu)$ is not transversal i.e., $T_{\mathbf{p}_\nu} V(\mathbf{p}_\nu) \subset T_{\mathbf{p}_\nu} S^{2n-1}_{\|\mathbf{p}_\nu\|}$. Taking a subsequence if necessary, we assume that $\lim_{\nu\to\infty} T_{\mathbf{p}_\nu} V(\mathbf{p}_\nu) = \tau$. By continuity, $\tau \subset T_{\mathbf{p}_\infty} S^{2n-1}_{\|\mathbf{p}_\infty\|}$. On the other hand, $T_{\mathbf{p}_\infty} M \subset \tau$ by the a_f-regularity assumption. This is a contradiction as $T_{\mathbf{p}_\infty} M \not\subset T_{\mathbf{p}_\infty} S^{2n-1}_{\|\mathbf{p}_\infty\|}$ by the transversality assumption of M and $\mathbb{S}^{2n-1}_{\|\mathbf{p}_\infty\|}$ at p_∞. □

8.3 Milnor Fibrations for Mixed Functions

8.3.1 Mixed Functions and Newton Boundary

For a given germ of a mixed analytic function $f(\mathbf{z}, \bar{\mathbf{z}}) = \sum_{\nu,\mu} c_{\nu,\mu} \mathbf{z}^\nu \bar{\mathbf{z}}^\mu$, we will define the Newton boundary $\Gamma(f)$, following the definition of the Newton boundary of a complex analytic functions.

First we define *the Newton polyhedron* $\Gamma_+(f)$ of f as

$$\Gamma_+(f) = \text{convex hull of} \left\{ \bigcup_{c_{\nu,\mu} \neq 0} (\nu + \mu + \mathbb{R}^n_+) \right\}.$$

$\Gamma_+(f)$ has a canonical polyhedron structure. *The Newton boundary* $\Gamma(f)$ is defined by the union of compact faces of $\Gamma_+(f, \mathbf{z})$. Let $\mathbb{Z}^n_{\ge 0} = \{{}^t(p_1, \dots, p_n) \in \mathbb{Z}^n \mid p_i \in$

$\mathbb{Z}_{\geq 0}\}$ where $\mathbb{Z}_{\geq 0}$ is the set of non-negative integers. A vector P defines a linear function ℓ_P on $\Gamma_+(f)$ by

$$\Gamma_+(f) \ni \eta = (\eta_1, \dots, \eta_n) \mapsto \ell_P(\eta) := \eta_1 p_1 + \cdots + \eta_n p_n.$$

A vector $P \in \mathbb{Z}^n_{\geq 0}$ is considered as a weight vector by $\deg_P z_i = \deg_P \bar{z}_i = p_i$. To distinguish from the polar weight which is defined later, we call this weight *the radial weight* and we denote it as $\mathrm{rdeg}_P\, \mathbf{z}^\nu \bar{\mathbf{z}}^\mu = \sum_{i=1}^n p_i(\nu_i + \mu_i)$. Put $d(P) = d(P, f)$ as the minimal value of $\ell_P : \Gamma_+(f) \to \mathbb{R}_{\geq 0}$ and put $\Delta(P) = \{\eta \in \Gamma_+(f) \mid \ell_P(\eta) = d(P)\}$. If P is strictly positive (i.e. $p_i > 0$, $\forall i$), $\Delta(P)$ is a compact face of $\Gamma_+(f)$, i.e. $\Delta(P) \subset \Gamma(f)$. For a given face $\Delta \subset \Gamma_+(f)$ or a given weight vector $P = {}^t(p_1, \dots, p_n) \in \mathbb{Z}^n_{\geq 0}$, we associate the face functions f_Δ and $f_P = f_{\Delta(P)}$respectively by

$$f_\Delta(\mathbf{z}, \bar{\mathbf{z}}) = \sum_{\nu+\mu\in\Delta} c_{\nu,\mu} \mathbf{z}^\nu \bar{\mathbf{z}}^\mu$$

$$f_P(\mathbf{z}, \bar{\mathbf{z}}) = \sum_{\ell_P(\mu+\nu)=d(P)} c_{\nu,\mu} \mathbf{z}^\nu \bar{\mathbf{z}}^\mu = \sum_{\nu+\mu\in\Delta(P)} c_{\nu,\mu} \mathbf{z}^\nu \bar{\mathbf{z}}^\mu.$$

f_Δ and f_P are called *the face function of the face* Δ or *the face function of the weight* P *respectively.*

8.3.2 *Non-degeneracy of Mixed Functions*

In the study of a complex analytic hypersurface singularity, the concept of non-degeneracy of Newton boundary plays an important role. We introduce the same concept for a mixed hypersurface. Consider a mixed function germ $f(\mathbf{z}, \bar{\mathbf{z}}) = \sum_{\nu,\mu} a_{\nu,\mu} \mathbf{z}^\nu \bar{\mathbf{z}}^\mu$ and its Newton boundary. Take a face $\Delta \in \Gamma(f)$. We say that f is *non-degenerate on* Δ if 0 is not a critical value of the mapping $f_\Delta : \mathbb{C}^{*n} \to \mathbb{C}$. If the inverse image $f^{-1}(0)$ is empty, it is considered to be non-degenerate. f is *strongly non-degenerate on* Δ if $f_\Delta : \mathbb{C}^{*n} \to \mathbb{C}$ has no critical point. We assume also f_Δ is surjective for $\dim \Delta \geq 1$. Here $\mathbb{C}^{*n}$ is the maximal torus:

$$\mathbb{C}^{*n} := \{\mathbf{z} = (z_1, \dots, z_n) \in \mathbb{C}^n \mid z_j \neq 0,\ j = 1, \dots, n\}.$$

We say that f is *non-degenerate* (respectively *strongly non-degenerate*) if f is non-degenerate (resp. strongly non-degenerate) on every face $\Delta \in \Gamma(f)$.

8.3.3 Mixed Functions of one Variable ($n = 1$)

To get a better understanding of the complexity of the strong non-degeneracy, we consider first mixed polynomials of one variable. The Newton boundary is a single point d which is the minimal radial degree of the monomials in f. Let f_d be the face function. In general, it is a finite linear sum of monomials of type $z^a\bar{z}^{d-a}$. Consider the simplest case: $f_d = z^a\bar{z}^b$, $a+b=d$. It is easy to see that

Lemma 8.3.1 *Assume that $f_d(z) = z^a\bar{z}^b$. Then $f_d : \mathbb{C}^* \to \mathbb{C}^*$ is $|a-b|$-fold covering map for $a \neq b$. If $a = b$, the image of f_d is the positive half line of the real axis.*

Recall $\mathbb{C}^* := \mathbb{C} \setminus \{0\}$. For the case where f_d is a linear sum of several mixed monomials of the same radial degree, the situation is more complicated. A loop $\sigma : [0, 2\pi] \to \mathbb{C}^*$ is called *a positive monotone loop* or *a negative monotone loop* if the argument of $\sigma(\theta)$ is a monotone increasing or monotone decreasing respectively. If σ is a monotone loop of rotation number m, there are exactly m solutions $0 \le \theta_1 < \cdots < \theta_m < 2\pi$ of the equation $\arg \sigma(\theta) = \tau$ for any τ, $0 \le \tau \le 2\pi$.

First let us consider the behavior of a linear form. A linear form $az + b\bar{z}$ is called *positive* (respectively *critical* or *negative*) if $|a| > |b|$ (resp. $|a| = |b|$ or $|a| < |b|$). Then the basic property of linear forms is

Lemma 8.3.2 *Consider a linear form $f_d = z + \lambda\bar{z}$ for simplicity.*

1. *Assume that $|\lambda| = 1$ and put $\lambda = e^{2i\theta}$. Then the image of $f_d : \mathbb{C}^* \to \mathbb{C}$ is the line which is the union of the two half lines $L_\theta := \{\eta \in \mathbb{C}^* \mid \arg(\eta) = \theta, \text{ or } \theta + \pi\}$ and $\{0\}$.*
2. *Assume that $|\lambda| < 1$. Then the restriction of f_d to the unit disk, $\theta \mapsto f_d(e^{i\theta})$ is a monotone increasing loop and $f_d : \mathbb{C}^* \to \mathbb{C}^*$ is a diffeomorphism.*
3. *Assume $|\lambda| > 1$. Then the restriction of f_d to the unit disk $|z| = 1$ is a monotone decreasing loop and $f_d : \mathbb{C}^* \to \mathbb{C}^*$ is a diffeomorphism.*

Proof (1) If $\lambda = 1$, the assertion is obvious. For $\lambda = e^{2i\theta}$, use the change of the variable $w = ze^{-i\theta}$ and the equality $z + e^{2i\eta}\bar{z} = e^{i\theta}(w + \bar{w})$.
For the assertions (2) and (3), assume first that λ is a positive real number. Put $z = e^{i\theta}$ and $f_d(z) = re^{i\xi}$. Then we have

$$re^{i\xi} = \begin{cases} e^{i\theta}(1 + \lambda e^{-2i\theta}), & 0 < \lambda < 1 \\ \lambda e^{-i\theta}(1 + \lambda^{-1}e^{2i\theta}), & 1 < \lambda. \end{cases}$$

Thus we have

$$\xi = \begin{cases} \theta + \Re(-i\log(1 + \lambda e^{-2i\theta})), & 0 < \lambda < 1 \\ -\theta + \Re(-i\log((1 + \lambda^{-1}e^{2i\theta})), & \lambda > 1. \end{cases}$$

For $0 < \lambda < 1$, we have

$$\begin{aligned}\frac{d\xi}{d\theta} &= 1 + \Re\frac{-2\lambda e^{-2i\theta}}{1+\lambda e^{-2i\theta}}\\ &= 1 + \frac{-2\lambda^2 - 2\lambda\cos(2\theta)}{1+\lambda^2+2\lambda\cos(2\theta)} = \frac{1-\lambda^2}{1+\lambda^2+2\lambda\cos(2\theta)} > 0.\end{aligned}$$

For $\lambda > 1$,

$$\begin{aligned}\frac{d\xi}{d\theta} &= -1 + \Re\frac{2\lambda^{-1}e^{2i\theta}}{1+\lambda^{-1}e^{2i\theta}}\\ &= -1 + \frac{2\lambda^{-2} + 2\lambda^{-1}\cos(2\theta)}{1+\lambda^{-2}+2\lambda^{-1}\cos(2\theta)} = \frac{-1+\lambda^{-2}}{1+\lambda^{-2}+2\lambda^{-1}\cos(2\theta)} < 0.\end{aligned}$$

Thus the image of the circle $|z| = 1$ is a monotone increasing loop for $0 < \lambda < 1$ and a monotone decreasing loop for $\lambda > 1$. As $f_d(rz) = r^d f_d(z)$ by homogeneity, the image $f_d(\mathbb{C}^*)$ is equal to $\mathbb{C}^*$ and $f_d|_{\mathbb{C}^*}$ has no critical points.

If λ is not a positive real number, put $\lambda = re^{2i\tau}$, $r > 0$ and change the coordinate by $w = ze^{-i\tau}$ and put $w = e^{i\xi}$, then $\xi = \theta - \tau$ where $z = e^{i\theta}$. Use the equality $z + \lambda\bar{z} = e^{i\tau}(w + r\bar{w})$, to get

$$\begin{aligned}\arg(z+\lambda\bar{z}) &= \tau + \Re(-i\log(w+r\bar{w}))\\ \frac{d\arg(z+\lambda\bar{z})}{d\theta} &= \frac{d\arg(w+\lambda\bar{w})}{d\xi}\\ &= \begin{cases} 1 + \Re\dfrac{-2re^{-2i\xi}}{1+re^{-2i\xi}}, & r < 1\\ -1 + \Re\dfrac{2r^{-1}e^{2i\xi}}{1+r^{-1}e^{2i\xi}}, & r > 1.\end{cases}\end{aligned} \tag{8.2}$$

□

Now we consider the general case. Assume that $f_d(z)$ is factored as

$$f_d(z) = cz^p\bar{z}^q\prod_{i=1}^{\ell}(z+\lambda_i\bar{z}),\ c \neq 0,\ d = p+q+\ell. \tag{8.3}$$

By Lemma 8.3.2, we have

Lemma 8.3.3 *If f_d satisfies (a) $p - q > 0$ and $|\lambda_i| < 1, \forall i$, or (b) $p - q < 0, |\lambda_i| > 1, \forall i$, then f_d defines a $|d'|$-fold covering map where $d' = p - q + \ell$ or $d' = p - q - \ell$ respectively.*

Proof Assume for example that (a) $p - q > 0$ and $|\lambda_i| < 1, \forall i$. Put $z = e^{i\theta}$, $\lambda_j = r_je^{i\tau_j}$ and $w_j = ze^{-i\tau_j} = e^{i\theta_j}$. Then $\theta_j = \theta - \tau_j$. Then by Lemma 3 and (8.2),

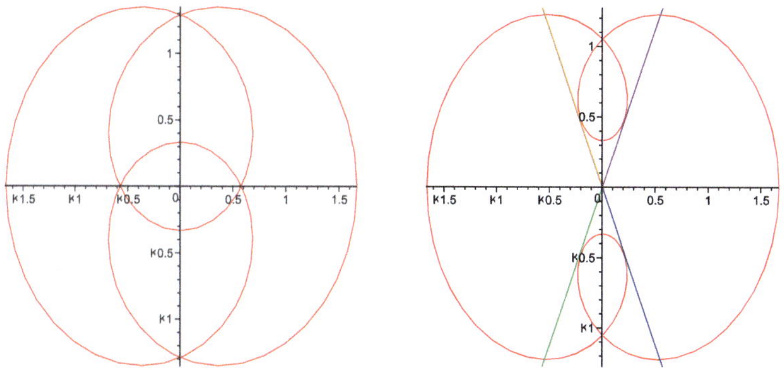

Fig. 8.2 Left:$\sigma(1, 2/3)$, right:$\sigma(2/3, 1)$

$$\frac{d\arg(f_d(z))}{d\theta} = (p-q) + \sum_{j=1}^{\ell} \frac{d\arg(w_j + \lambda_j \bar{w}_j)}{d\theta_j}$$

$$(p-q) + \sum_{j=1}^{\ell}\left(1 + \Re\frac{-2r_j e^{-2i\theta_j}}{1 + r_j e^{-2i\theta_j}}\right) \geq p - q > 0.$$

Thus the image of the unit circle is a monotone increasing loop and the rotation number of f_d is d' and the assertion follows. The case (b) is treated similarly. □

For f_d to be strongly non-degenerate at the origin, it is necessary that $f_d(\mathbb{C}^*) = \mathbb{C}^*$ and $f_d : \mathbb{C}^* \to \mathbb{C}^*$ is a covering map.

Corollary 8.3.4 *For $n = 1$, if the face function f_d satisfies one of the conditions in Lemma 8.3.3, $f(z)$ is strongly non-degenerate.*

Example 8.3.5 Consider the polynomial $f_d = z^2(az + b\bar{z})$ with $a, b \in \mathbb{R}_+$. Consider the loop $\sigma(a, b) := f_d|_{\mathbb{S}^1}$. Note that $\sigma(1, 2/3)$ is monotone increasing and the rotation number of $\sigma(1, 2/3)$ is 3. On the other hand, $\sigma(2/3, 1)$ has the rotation number 1 and it is not monotone. Note that in the latter case, there exist 4 directions $0 < \theta_1 < \theta_2 < \theta_3 < \theta_4 \leq 2\pi$ so that the half line $\{re^{i\theta_j} \mid r > 0\}$ is tangent to the graph, say at $r_j e^{i\theta_j}$ for $j = 1, \ldots, 4$. See Figure 8.2. This point $r_j e^{i\theta_j}$ corresponds to a critical point of $f_d : \mathbb{C}^* \to \mathbb{C}^*$.

Remark 8.3.6 1. If the factorization (8.3) does not have any critical factors $\forall\lambda_i$, $|\lambda_i| \neq 1$, the rotation number of $f_d/|f_d| : \mathbb{S}^1 \to \mathbb{S}^1$ is given as $p - q + \ell_+ - \ell_-$ where $\ell_\pm$ is the number of the positive factors (resp. the negative factors) in the product [35].

2. Consider a mixed function $f(z) = f_d(z) + R(z)$ where the monomials of $R(z)$ have lower radial degrees. Consider a root α of $f(z) = 0$. α is called *simple* if the Jacobian f at $z = \alpha$ is ± 1 for a sufficiently small ε. α is called a *positive* (respectively

a *negative*) if the rotation number is 1 (resp. -1). Assume that f has only simple zeros. Take a big disk $\mathbb{D}_S$ so that $\mathbb{D}_S$ contains all zeros of $f(z) = 0$. Then the rotation number of $f/|f| : \mathbb{S}^1_S \to \mathbb{S}^1$ is given as $p - q + \ell_+ - \ell_-$ and it is equal to $\rho_+ - \rho_-$ where $\rho_\pm$ is the number of positive and negative zeros respectively. (Theorem 17, [35]).

3. In particular, if $f(z) = z^{d_p+r}\bar{z}^r +$ (lower terms) and assume that $f = 0$ has only simple zeros, the number of zeros satisfies the inequality $\rho_+ + \rho_- \ge d_p = \rho_+ - \rho_-$.

8.3.4 Mixed Weighted Homogeneous Polynomials

Radial Weighted Homogeneous Polynomials and Polar Weighted Homogeneous Polynomials

A mixed polynomial $f = \sum_{\nu,\mu} a_{\nu,\mu}\mathbf{z}^\nu\bar{\mathbf{z}}^\mu$ is called *a radial weighted homogeneous polynomial* of degree d_r with respect to an integral weight vector $Q = {}^t(q_1, \dots, q_n)$ if it satisfies the condition

$$a_{\nu,\mu} \neq 0 \Longrightarrow \sum_{j=1}^n q_j(\nu_j + \mu_j) = d_r. \tag{8.4}$$

d_r is called *the radial degree* and we denote it as $d_r = \mathrm{rdeg}_Q\, f$. We usually assume that $q_i \ge 0$ and $d_r > 0$. $\mathbb{R}_+$-action on $\mathbb{C}^n$ is associated by $(\rho, \mathbf{z}) \mapsto \rho \circ \mathbf{z} := (\rho^{q_1} z_1, \dots, \rho^{q_n} z_n)$ and f satisfies the equality:

$$f(\rho \circ \mathbf{z}, \rho \circ \mathbf{z}) = \rho^{d_r} f(\mathbf{z}, \bar{\mathbf{z}}).$$

Taking the differential in ρ and putting $\rho = 1$, the Euler equality for the radial weight Q takes the form:

$$d_r\, f(\mathbf{z}, \bar{\mathbf{z}}) = \sum_{i=1}^n \left\{ q_i z_i \frac{\partial f}{\partial z_i}(\mathbf{z}, \bar{\mathbf{z}}) + q_i \bar{z}_i \frac{\partial f}{\partial \bar{z}_i}(\mathbf{z}, \bar{\mathbf{z}}) \right\}. \tag{8.5}$$

Proposition 8.3.7 *For a given germ of a mixed function $f(\mathbf{z}, \bar{\mathbf{z}})$ and a strictly positive weight vector $P = (p_1, \dots, p_n)$, the face function $f_P(\mathbf{z}, \bar{\mathbf{z}})$ is a radial weighted homogeneous polynomial with the weight vector P and the radial degree $d(P, f)$.*

We say that $f(\mathbf{z}, \bar{\mathbf{z}}) = \sum_{\nu,\mu} a_{\nu,\mu}\mathbf{z}^\nu\bar{\mathbf{z}}^\mu$ is *a polar weighted homogeneous polynomial* of degree d_p with respect to an integral weight vector $P = (p_1, \dots, p_n)$ if it satisfies the condition

$$a_{\nu,\mu} \neq 0 \Longrightarrow \sum_{j=1}^n p_j(\nu_j - \mu_j) = d_p. \tag{8.6}$$

d_p is called *the polar degree of* f and we denote it by $d_p = \mathrm{pdeg}_P\, f$. Here p_i can be negative but we assume that $d_p \neq 0$.

Assume that P is an integer vector and f is polar weighted with respect to P. There is an associated $\mathbb{S}^1$ action on $\mathbb{C}^n$ which satisfies the following equality.

$$e^{i\theta} \circ \mathbf{z} := (z_1 e^{ip_1\theta}, \dots, z_n e^{ip_n\theta}), \quad e^{i\theta} \in \mathbb{S}^1,$$
$$f(e^{i\theta} \circ \mathbf{z}) = e^{id_p\theta} f(\mathbf{z}).$$

Taking the differential in θ and substituting $\theta = 0$, we get the polar Euler equality:

$$d_p\, f(\mathbf{z}, \bar{\mathbf{z}}) = \sum_{i=1}^{n} \left\{ p_i z_i \frac{\partial f}{\partial z_i}(\mathbf{z}, \bar{\mathbf{z}}) - p_i \bar{z}_i \frac{\partial f}{\partial \bar{z}_i}(\mathbf{z}, \bar{\mathbf{z}}) \right\}. \tag{8.7}$$

Note that in both cases, we can take the weight vector to be a primitive integer vector. *The normalized radial weight vector* $\hat{Q}$ is defined by $\hat{Q} = (q_1/d_r, \dots, q_n/d_r)$. Similarly *the normalized polar weight vector* is defined by $\hat{P} = (p_1/d_p, \dots, p_n/d_p)$.

If f is polar weighted homogeneous with respect to a weight vector P and f is also radial weighted homogeneous with respect to a primitive vector Q, we say that f is *a mixed weighted homogeneous polynomial.* In [30, 31], we called such a polynomial *a polar weighted homogeneous polynomial.* We will change the terminology as *a mixed weighted homogeneous polynomial* to distinguish those polynomials which are only polar weighted and the other mixed polynomials which are also radial weighted homogeneous.

As an example, consider $f_1 := z_1^3\bar{z}_1^2 + z_1^2\bar{z}_1 + z_2^2\bar{z}_2$ and $f_2 = z_1^3\bar{z}_1^2 + z_2^2\bar{z}_2$. Both polynomials are polar weighted homogeneous with respect to $P = {}^t(1, 1)$. f_2 is radial weighted homogeneous of degree 15 with the weight vector $Q = (3, 5)$ but f_1 can not be radial weighted homogeneous.

For given primitive polar and radial weight vectors $P = {}^t(p_1, \dots, p_n)$, $Q = {}^t(q_1, \dots, q_n)$, we associate $\mathbb{R}_+ \times \mathbb{S}^1$-action in $\mathbb{C}^n$ as follows.

$$\rho e^{i\theta} \circ (z_1, \dots, z_n) = (\rho^{q_1} e^{ip_1\theta} z_1, \dots, \rho^{q_n} e^{ip_n\theta} z_n), \ \rho e^{i\theta} \in \mathbb{R}_+ \times \mathbb{S}^1.$$

Observe that f is a mixed weighted homogeneous polynomial of polar degree $d_p = \mathrm{pdeg}\, f$ and radial degree $d_r = \mathrm{rdeg}\, f$ with the respective weight vector P, Q if and only if it satisfies the following equality.

$$f(\rho e^{i\theta} \circ \mathbf{z}, \overline{\rho e^{i\theta} \circ \mathbf{z}}) = \rho^{d_r} e^{id_p\theta} f(\mathbf{z}, \bar{\mathbf{z}}). \tag{8.8}$$

The weight vector is not unique unless we assume that P is a primitive integral vector i.e. $\gcd(p_1, \dots, p_n) = 1$. To avoid this ambiguity, it is convenient to consider the normalized weight vector.

Strongly Mixed Weighted Homogeneous Polynomials

A mixed weighted homogeneous polynomial $h(\mathbf{z}, \bar{\mathbf{z}})$ is called *strongly mixed weighted homogeneous* if the polar weight vector is equal to the radial weight vector i.e. $P = Q$. In this case, the associated $\mathbb{R}_+ \times \mathbb{S}^1$-action is in fact $\mathbb{C}^*$-action:

$$t \circ \mathbf{z} = (z_1 t^{p_1}, \ldots, z_n t^{p_n}), \quad t \in \mathbb{C}^*.$$

If further $P = {}^t(1, \ldots, 1)$, f is called *a strongly mixed homogeneous polynomial.* In this case, the associated $\mathbb{C}^*$-action is simply the canonical action $(t, \mathbf{z}) \mapsto t\mathbf{z}$ and the above equality (8.8) says that $f(\mathbf{z}, \bar{\mathbf{z}}) = 0$ defines a well-defined projective hypersurface in $\mathbb{P}^{n-1}$.

The Global Milnor fibration of a Mixed Weighted Homogeneous Polynomial

Mixed weighted homogeneous polynomials play a similar fundamental role for the description of the Milnor fibration of a general mixed function.

Proposition 8.3.8 *Let $f(\mathbf{z}, \bar{\mathbf{z}})$ be a mixed weighted homogeneous polynomial with a polar weight vector $P = {}^t(p_1, \ldots, p_n)$ and a radial weight vector $Q = {}^t(q_1, \ldots, q_n)$. Let d_p and d_r be the polar and radial degrees respectively.*

1. *$f : \mathbb{C}^n \to \mathbb{C}$ has a unique critical value 0 and the restricted mappings*

$$f : \mathbb{C}^n \setminus f^{-1}(0) \to \mathbb{C}^*$$
$$f : \mathbb{C}^{*n} \setminus f^{-1}(0) \to \mathbb{C}^*$$

are locally trivial fibrations. Let F and F^* be the fibers. The monodromy mapping $h : F \to F$ or $h : F^* \to F^*$ is given by

$$h(\mathbf{z}) = e^{2\pi i/d_p} \circ \mathbf{z} = (e^{2\pi p_1 i/d_p} z_1, \ldots, e^{2\pi p_n i/d_p} z_n).$$

2. *Assume that a radial weight vector Q is strictly positive and V has an isolated singularity at the origin.*

   a. *For any $r > 0$,*

$$\varphi : \mathbb{S}_r^{2n-1} \setminus K_r \to \mathbb{S}^1, \quad \varphi(\mathbf{z}) = f(\mathbf{z})/|f(\mathbf{z})|$$

   is a locally trivial fibration (called a spherical Milnor fibration*) where $K_r = f^{-1}(0) \cap \mathbb{S}_r^{2n-1}$ and the isomorphism class does not depend on the choice of the radius r.*

   b. *The global Milnor fibration is homotopically equivalent to the restriction over $\mathbb{S}_\delta^1$*

$$f : f^{-1}(\mathbb{S}^1_\delta) \to \mathbb{S}^1_\delta$$

and its isomorphism class does not depend on $\delta > 0$ and it is isomorphic to the spherical Milnor fibration. Here $\mathbb{S}^1_\delta := \{w \in \mathbb{C} \mid |w| = \delta\}$.

We call F and F^* the Milnor fiber and the toric Milnor fiber respectively. For the proof, see Proposition 3 and Proposition 4, [30].

Corollary 8.3.9 *Let $f(\mathbf{z}, \bar{\mathbf{z}})$ be a mixed weighted homogeneous polynomial. Then $f : \mathbb{C}^{*n} \to \mathbb{C}$ has no critical point if and only if $f^{-1}(0) \subset \mathbb{C}^{*n}$ is mixed non-singular.*

Proof By the above Proposition, 0 is the only possible critical value of $f : \mathbb{C}^{*n} \to \mathbb{C}$. □

8.3.5 *Milnor Fibrations for Strongly Non-degenerate Mixed Functions*

We study Milnor fibrations for a general mixed function $f(\mathbf{z}, \bar{\mathbf{z}})$, which is not necessarily mixed weighted homogeneous.

Non-vanishing Coordinate Subspaces and Vanishing Coordinate Subspaces

For a subset $J \subset \{1, 2, \dots, n\}$, $\mathbb{C}^J$ is the subspace $\{\mathbf{z} \in \mathbb{C}^n \mid z_j = 0,\ j \notin J\}$ and the restriction is denoted as $f^J := f|_{\mathbb{C}^J}$. There are two classes of subspaces which are characterized as follows.

$$\mathcal{I}_{nv}(f) = \{I \subset \{1, \dots, n\} \mid f^I \not\equiv 0\}$$
$$\mathcal{I}_v(f) = \{I \subset \{1, \dots, I\} \mid f^I \equiv 0\}.$$

The subspace $\mathbb{C}^I$ with $I \in \mathcal{I}_{nv}(f)$ (respectively with $I \in \mathcal{I}_v(f)$) is called *a non-vanishing coordinate subspace of f* (resp. *a vanishing coordinate subspace*). Define subset $V^\sharp$ of V as

$$V^\sharp := \bigcup_{I \in \mathcal{I}_{nv}(f)} V \cap \mathbb{C}^{*I}.$$

Recall that f is convenient if and only if $\mathcal{I}_{nv}(f) = \{\forall I \subset \{1, \dots, n\} \mid I \neq \emptyset\}$. In particular, if f is convenient, $V^\sharp = V \setminus \{\mathbf{0}\}$.

Lemma 8.3.10 (Theorem 19, [31]) *Suppose that $f(\mathbf{z}, \bar{\mathbf{z}})$ is non-degenerate. Then there exists a positive number r_0 such that the following property holds.*

1. (Smoothness) $V^\sharp \cap \mathbb{B}^{2n}_{r_0}$ is non-singular. In particular, if f is convenient, $V \setminus \{\mathbf{0}\}$ is non-singular and $\mathbf{0}$ is an isolated singularity of $V = f^{-1}(0)$.

2. *(Transversality) The family of spheres* $\mathbb{S}_r^{2n-1}$, $0 < r \le r_0$ *intersect transversely with* $V^\sharp$.

This assertion is a generalization of the assertion for holomorphic functions (Corollary (2.3), [29]).

Proof (1) We denote the set of mixed singular points of V by $\Sigma(V)$. Assuming the assertion does not hold, we can find a real analytic curve $\mathbf{z}(t) \in \mathbb{C}^n$, $0 \le t \le 1$ by the Curve Selection Lemma [14, 24] such that $\mathbf{z}(t) \in \Sigma(V) \cap V^\sharp$ for $t > 0$ and $\mathbf{z}(0) = \mathbf{0}$. By Proposition 8.2.2, there exists an analytic curve $\lambda(t) \in \mathbb{S}^1 \subset \mathbb{C}$

$$\overline{\partial f}(\mathbf{z}(t), \bar{\mathbf{z}}(t)) = \lambda(t)\bar{\partial} f(\mathbf{z}(t), \bar{\mathbf{z}}(t)). \tag{8.9}$$

Put $I = \{j \mid z_j(t) \not\equiv 0\}$. As $\mathbf{z}(t) \in V^\sharp$ and $I \in \mathcal{I}_{nv}(f)$, $f^I \not\equiv 0$. Note that f^I is also non-degenerate. Assume $I = \{1, \dots, m\}$ for simplicity and consider the Taylor expansion of $f^I(\mathbf{z}(t), \bar{\mathbf{z}}(t))$:

$$z_i(t) = b_i t^{a_i} + (\text{higher terms}), \quad b_i \ne 0,\ i = 1, \dots, m$$
$$\lambda(t) = \lambda_0 + \lambda_1 t + (\text{higher terms}), \quad \lambda_0 \in \mathbb{S}^1 \subset \mathbb{C}.$$

Put $A = {}^t(a_1, \dots, a_m)$ and let f^I_A be the face function of $f^I(\mathbf{z}, \bar{\mathbf{z}})$, $d = d(A, f^I) > 0$ and $\mathbf{b} = (b_1, \dots, b_m) \in \mathbb{C}^{*m}$.

$$\frac{\partial f}{\partial z_j}(\mathbf{z}(t), \bar{\mathbf{z}}(t)) = \frac{\partial f^I_A}{\partial z_j}(\mathbf{b}, \bar{\mathbf{b}})t^{d-a_j} + (\text{higher terms}), \quad j = 1, \dots, m$$
$$\frac{\partial f}{\partial \bar{z}_j}(\mathbf{z}(t), \bar{\mathbf{z}}(t)) = \frac{\partial f^I_A}{\partial \bar{z}_j}(\mathbf{b}, \bar{\mathbf{b}})t^{d-a_j} + (\text{higher terms}), \quad j = 1, \dots, m,$$
$$\mathrm{ord}_t \frac{\partial f^I}{\partial z_j}(\mathbf{z}(t), \bar{\mathbf{z}}(t)) = \mathrm{ord}_t \frac{\partial f^I}{\partial \bar{z}_j}(\mathbf{z}(t), \bar{\mathbf{z}}(t)),\ j = 1, \dots, m \text{ by (8.9)}.$$

Combining this and (8.9), we get

$$\overline{\partial f^I_A}(\mathbf{b}) = \lambda_0 \bar{\partial} f^I_A(\mathbf{b}).$$

On the other hand, as $f^I(\mathbf{z}(t)) \equiv 0$, we have the equality $f^I_A(\mathbf{0})$. Then 0 becomes a critical value of $f^I_A : \mathbb{C}^{*I} \to \mathbb{C}$ but this is a contradiction to the strong non-degeneracy assumption as $\mathbf{b} \in \mathbb{C}^{*I}$.

The assertion (2) follows from the same argument as the proof of Corollary 2.9, [24]. We give another proof using non-degeneracy. Assume that the transversality assertion does not hold. Using the Curve Selection Lemma and Lemma 8.2.3, there exist real an analytic curve $\mathbf{z}(t) \in \mathbb{C}^n$ and an analytic scalar function $c(t)$, $0 \le t \le 1$ such that $\mathbf{z}(t) \in V^\sharp$ for $t > 0$, $\mathbf{z}(t) \ne \mathbf{0}$, $\mathbf{z}(0) = \mathbf{0}$ and $\mathbf{z}(t)$ satisfies the equality:

$$\mathbf{z}(t) = \bar{c}(t)\overline{\partial f}(\mathbf{z}(t), \bar{\mathbf{z}}(t)) + c(t)\bar{\partial} f(\mathbf{z}(t), \bar{\mathbf{z}}(t)). \tag{8.10}$$

Put $I = \{j \mid z_j(t) \not\equiv 0\}$. As $\mathbf{z}(t) \in V^\sharp$, $I \in \mathcal{I}_{nv}(f)$ and $f^I \not\equiv 0$. Assume $I = \{1, \dots, m\}$ and consider the Taylor expansions of $f^I(\mathbf{z}(t))$, $\mathbf{z}(t)$ and $c(t)$ as follows:

$$\begin{aligned} z_i(t) &= b_i t^{a_i} + (\text{higher terms}), \quad b_i \neq 0, \ i = 1, \dots, m \\ c(t) &= c_0 t^e + (\text{higher terms}). \end{aligned}$$

Put $A = (a_1, \dots, a_m)$, $d = d(A, f^I) > 0$, $\mathbf{b} = (b_1, \dots, b_m) \in \mathbb{C}^{*m}$ and $a_{\min} := \min\{a_i \mid i \in I\}$, $I_{\min} := \{i \mid a_i = a_{min}\}$. From (8.10), $2a_{min} = e + d$ and

$$\bar{c}_0 \bar{f}_{A,z_j}(\mathbf{b}) + c_0 f_{A,\bar{z}_j}(\mathbf{b}) = \begin{cases} 0, & j \notin I_{min} \\ b_j, & j \in I_{min}. \end{cases} \tag{8.11}$$

As $f(\mathbf{z}(t), \bar{\mathbf{z}}(t)) \equiv 0$, $f_A^I(\mathbf{b}) = 0$ and multiplying $\bar{b}_i a_i$ to (8.11) and taking the summation for $i = 1, \dots, m$, we get

$$\begin{aligned} 0 < \sum_{i=1}^m a_i b_i \bar{b}_i &= \sum_{i=1}^m (\bar{c}_0 \bar{f}_{A,z_j}(\mathbf{b}) + c_0 f_{A,\bar{z}_j}(\mathbf{b})) \bar{b}_i a_i \\ &= \sum_{i=1}^m \bar{c}_0 a_i \bar{b}_i \bar{f}_{A,z_j}(\mathbf{b}) + c_0 \sum_{i=1}^m a_i \bar{b}_i f_{A,\bar{z}_j}(\mathbf{b}) \\ &= \overline{\left(c_0 \sum_{i=1}^m a_i b_i f_{A,z_j}(\mathbf{b}) \right)} + c_0 \sum_{i=1}^m a_i \bar{b}_i f_{A,\bar{z}_j}(\mathbf{b}) \\ &\overset{\star}{=} \overline{\left(-c_0 \sum_{i=1}^m a_i \bar{b}_i f_{A,\bar{z}_j}(\mathbf{b}) \right)} + c_0 \sum_{i=1}^m a_i \bar{b}_i f_{A,\bar{z}_j}(\mathbf{b}) \\ &= 2i\Im(c_0 \sum_{i=1}^m a_i \bar{b}_i f_{A,\bar{z}_j}(\mathbf{b})) \in i\mathbb{R}. \end{aligned}$$

where the equality $\overset{\star}{=}$ follows from the radial Euler equality of $f_A(\mathbf{b})$ and the equality $f_A(\mathbf{b}) = 0$. This is an obvious contradiction, as the first term is a positive real number. □

8.3.6 *The Milnor Fibration for Convenient Mixed Functions*

We first show the existence of the Milnor fibration assuming f is convenient and strongly non-degenerate.

Lemma 8.3.11 *Assume that $f(\mathbf{z}, \bar{\mathbf{z}})$ is a strongly non-degenerate mixed function. Let r_0 be a small positive number as in Theorem 8.3.10. There exists a positive number δ_0 satisfying the following.*

1. *Nearby fiber* $V_\eta := f^{-1}(\eta)$ *is non-singular in the ball* $\mathbb{B}^{2n}_{r_0}$ *for any* $\eta \neq 0$, $|\eta| \le \delta_0$.
2. *Assume further that* f *is convenient. Taking* r_0 *sufficiently small if necessary, for any positive number* r_1, $r_1 \le r_0$, *there is a positive number* $\delta(r_1) \le \delta_0$ *and for any non-zero complex number* η, $|\eta| \le \delta(r_1)$, *and* r, $r_1 \le r \le r_0$, *the intersection* $V_\eta \cap \mathbb{S}^{2n-1}_r$ *is transversal.*

Proof We first show the assertion (1). For this assertion, the convenience is not necessary. Suppose the assertion does not hold and we use the Curve Selection Lemma [14, 24] to find an analytic curve $\mathbf{z}(t)$, $0 \le t \le 1$ so that $\mathbf{z}(0) = \mathbf{0}$ and $f(\mathbf{z}(t), \bar{\mathbf{z}}(t)) \neq 0$ for $t \neq 0$ and $\mathbf{z}(t)$ is a critical point of $f : \mathbb{C}^n \to \mathbb{C}$. By Proposition 8.2.2, there exists a real scalar function $\lambda(t) \in \mathbb{S}^1$ satisfying the following.

$$\overline{\partial f}(\mathbf{z}(t), \bar{\mathbf{z}}(t)) = \lambda(t)\bar{\partial} f(\mathbf{z}(t), \bar{\mathbf{z}}(t)). \tag{8.12}$$

The argument is similar as that of the proof of Lemma 8.3.10. Put $I := \{j \mid z_j(t) \not\equiv 0\}$. For simplicity, suppose $I = \{1, \dots, m\}$ and consider f^I. As $f(\mathbf{z}(t), \bar{\mathbf{z}}(t)) = f^I(\mathbf{z}(t), \bar{\mathbf{z}}(t)) \not\equiv 0$, we have $f^I \not\equiv 0$. We expand $\mathbf{z}(t)$ and $\lambda(t)$ in Taylor expansions.

$$z_i(t) = b_i t^{a_i} + (\text{higher terms}), \quad b_i \neq 0,\ i = 1, \dots, m$$
$$\lambda(t) = \lambda_0 + \lambda_1 t + (\text{higher terms}), \quad \lambda_0 \in \mathbb{S}^1 \subset \mathbb{C}.$$

Put $A = {}^t(a_1, \dots, a_m)$ and $\mathbf{b} = (b_1, \dots, b_m)$.

$$\frac{\partial f}{\partial z_j}(\mathbf{z}(t), \bar{\mathbf{z}}(t)) = \frac{\partial f^I_A}{\partial z_j}(\mathbf{b})t^{d-a_j} + (\text{higher terms}),$$
$$\frac{\partial f}{\partial \bar{z}_j}(\mathbf{z}(t), \bar{\mathbf{z}}(t)) = \frac{\partial f^I_A}{\partial \bar{z}_j}(\mathbf{b})t^{d-a_j} + (\text{higher terms}), \quad d = d(A, f^I).$$
$$\mathrm{ord}_t \frac{\partial f^I}{\partial z_j}(\mathbf{z}(t), \bar{\mathbf{z}}(t)) = \mathrm{ord}_t \frac{\partial f^I}{\partial \bar{z}_j}(\mathbf{z}(t), \bar{\mathbf{z}}(t)).$$

Thus by (8.12),

$$\overline{\partial f^I_A}(\mathbf{b}) = \lambda_0 \bar{\partial} f^I_A(\mathbf{b})$$

which implies $\mathbf{b}$ is a critical point of $f^I_A : \mathbb{C}^{*I} \to \mathbb{C}$. This is a contradiction to the strong non-degeneracy assumption.

(2) Take $r_0 > 0$ so that $\mathbb{S}^{2n-1}_r, 0 < r \le r_0$ intersect transversely with $V^\sharp = V \setminus \{\mathbf{0}\}$. Now fix a positive $r_1 \le r_0$. As $V \cap \mathbb{B}^{2n}_{r_1,r_0}$ is compact, we can take $\delta(r_1)$ so that $f^{-1}(\eta)$ and $\mathbb{S}^{2n-1}_r$ intersect transversely for any $r_1 \le r \le r_0$ and $|\eta| \le \delta(r_1)$. Here $\mathbb{B}^{2n}_{r_1,r_0} := \{\mathbf{z} \mid r_1 \le \|\mathbf{z}\| \le r_0\}$. □

Remark 8.3.12 Without assuming the convenience, V can have a non-isolated singularity and we can not use the compactness argument as above to prove the assertion

(2). Therefore for the proof of the existence of a Milnor fibration for non-convenient mixed functions, we need "local tameness" which we will introduce in Sect. 8.3.8.

Corollary 8.3.13 (Existence of Tubular Milnor fibration) *Suppose $f(\mathbf{z}, \bar{\mathbf{z}})$ is a strongly non-degenerate and convenient mixed function. Take $r_0 > 0$ and $\delta(r_0)$ as Lemma 8.3.11. Put $E(r_0, \delta_0)^* = \{\mathbf{z} \in \mathbb{C}^n \mid \|\mathbf{z}\| \le r_0,\ 0 < |f(\mathbf{z}, \bar{\mathbf{z}})| \le \delta_0\}$ and $\mathbb{D}^*_{\delta_0} := \{\eta \in \mathbb{C} \mid 0 < |\eta| \le \delta_0\}$. Then*

$$f : E(r_0, \delta_0)^* \to \mathbb{D}^*_{\delta_0} \tag{8.13}$$

is a locally trivial fibration and the isomorphism class does not depend on the choice of r_0 and δ_0.

8.3.7 The Spherical Milnor Fibration

We assume always that $f(\mathbf{z}, \bar{\mathbf{z}})$ is a strongly non-degenerate mixed function. We consider the spherical Milnor fibration problem. When does the mapping $f/|f| : \mathbb{S}_r^{2n-1} \setminus K_r \to \mathbb{S}^1$ give a fibration? We consider two vector fields on $\mathbb{C}^n \setminus V$.

$$\mathbf{v}_1(\mathbf{z}, \bar{\mathbf{z}}) = \overline{\partial \log f}(\mathbf{z}, \bar{\mathbf{z}}) + \bar{\partial} \log f(\mathbf{z}, \bar{\mathbf{z}}) \tag{8.14}$$

$$\mathbf{v}_2(\mathbf{z}, \bar{\mathbf{z}}) = i\left(\overline{\partial \log f}(\mathbf{z}, \bar{\mathbf{z}}) - \bar{\partial} \log f(\mathbf{z}, \bar{\mathbf{z}})\right) \tag{8.15}$$

Hereafter we denote $\mathbf{v}_i(\mathbf{z}, \bar{\mathbf{z}})$ simply by $\mathbf{v}_i(\mathbf{z})$. We will show that $\mathbf{v}_1, \mathbf{v}_2$ are gradient vectors of $\Re \log f(\mathbf{z}) = \log|f(\mathbf{z})|$ and $\Im \log f(\mathbf{z}) = \arg f(\mathbf{z})$ by the following calculation. We denote *the Hermitian inner product* in $\mathbb{C}^n$ by $(\mathbf{v}, \mathbf{w})$. Namely $(\mathbf{v}, \mathbf{w}) = \sum_{j=1}^n v_j \bar{w}_j$ for $\mathbf{v} = (v_1, \dots, v_n)$, $\mathbf{w} = (w_1, \dots, w_n) \in \mathbb{C}^n$. Note that $\Re(\mathbf{v}, \mathbf{w})$ is *the Euclidean inner product* in $\mathbb{R}^{2n}$ under the canonical identification

$$\mathbf{v} = (v_1, \dots, v_n) \in \mathbb{C}^n \iff (v_{11}, v_{12}, \dots, v_{n1}, v_{n2}) \in \mathbb{R}^{2n}$$

where $v_j = v_{j1} + i v_{j2}$.

Proposition 8.3.14 ((4), (5), Sect. 5.2 in [31]) *For a given analytic curve $\mathbf{z}(t)$, $\mathbf{z}(0) = \mathbf{w}$ with $\frac{d\mathbf{z}}{dt}(0) = \mathbf{v}$, the following equality holds.*

$$\left.\frac{d \log|f(\mathbf{z}(t))|}{dt}\right|_{t=0} = \left.\frac{d\Re \log f(\mathbf{z}(t))}{dt}\right|_{t=0} = \Re(\mathbf{v}, \mathbf{v}_1(\mathbf{w})) \tag{8.16}$$

$$\left.\frac{d \arg f(\mathbf{z}(t))}{dt}\right|_{t=0} = \left.\frac{\Im \log f(\mathbf{z}(t))}{dt}\right|_{t=0} = \Re(\mathbf{v}, \mathbf{v}_2(\mathbf{w})), \tag{8.17}$$

$$\left.\frac{d \log f(\mathbf{z}(t))}{dt}\right|_{t=0} = \Re(\mathbf{v}, \mathbf{v}_1(\mathbf{w})) + i\Re(\mathbf{v}, \mathbf{v}_2(\mathbf{w})). \tag{8.18}$$

Proof We use the equality $\log f(\mathbf{z}) = \log|f(\mathbf{z})| + i \arg f(z)$. Let $\mathbf{z}(t)$ be as above.

$$\left.\frac{d\log f(\mathbf{z}(t))}{dt}\right|_{t=0} = \left.\frac{d\Re(\log f)(\mathbf{z}(t))}{dt}\right|_{t=0} + i\left.\frac{d\Im\log f(\mathbf{z}(t))}{dt}\right|_{t=0}$$

$$\begin{aligned}\left.\frac{d\Re(\log f)(\mathbf{z}(t))}{dt}\right|_{t=0} &= \Re\left((\mathbf{v}, \overline{\partial\log f(\mathbf{w})}) + (\bar{\mathbf{v}}, \bar\partial\log(\mathbf{w}))\right)\\ &= \Re(\mathbf{v}, \mathbf{v}_1(\mathbf{w}))\\ \left.\frac{d\Re(\arg f)(\mathbf{z}(t))}{dt}\right|_{t=0} &= \Re\left((\mathbf{v}, -i\overline{\partial\log f}(\mathbf{w})) + (\bar{\mathbf{v}}, i\overline{\bar\partial\log f(\mathbf{w})})\right)\\ &= \Re(\mathbf{v}, \mathbf{v}_2(\mathbf{w})).\end{aligned}$$

□

Corollary 8.3.15 *Take a point $\mathbf{p} \in f^{-1}(\eta) \cap \mathbb{B}^{2n}_{r_0}$ with $0 \neq |\eta| \leq \delta_0$ as in Corollary 8.3.13. Then $T_{\mathbf{p}} f^{-1}(\eta) = \{\mathbf{w} \in \mathbb{C}^n \mid \Re(\mathbf{w}, \mathbf{v}_i(\mathbf{p})) = 0,\ i = 1, 2\}$.*

To prove the existence of a spherical Milnor fibration, we use the next lemmas.

Lemma 8.3.16 (Lemma 30, [31]) *Take a point $\mathbf{z} \in \mathbb{S}^{2n-1}_\varepsilon \setminus K_\varepsilon$. $\mathbf{z}$ is a critical point of $\varphi = f/|f|$ if and only if $\mathbf{v}_2(\mathbf{z})$ and $\mathbf{z}$ are linearly dependent over $\mathbb{R}$.*

Lemma 8.3.17 (Lemma 31, [31]) *Assume that $f(\mathbf{z}, \bar{\mathbf{z}})$ is strongly non-degenerate. There exists a positive number ε_1 so that for any $\mathbf{z} \in \mathbb{B}^{2n}_{\varepsilon_1} \setminus f^{-1}(0)$, $\mathbf{v}_2(\mathbf{z})$ and $\mathbf{z}$ are linearly independent over $\mathbb{R}$.*

Assume that f is a convenient, strongly non-degenerate mixed function. Using Lemma 8.3.16, Corollary 8.3.15, Lemmas 8.3.17 and 8.3.11, we have

Theorem 8.3.18 *For ε small enough, $\varphi : S_\varepsilon \setminus K_\varepsilon \to \mathbb{S}^1$ is a locally trivial fibration.*

Lemma 8.3.11 is used to construct a horizontal vector field $\mathcal{V}(\mathbf{z})$ of φ which is controlled near K_ε so that it can be integrated over any finite time interval. Here $\mathcal{V}(\mathbf{z})$ is horizontal if $d\varphi_{\mathbf{z}}(\mathcal{V}(\mathbf{z})) = \frac{\partial}{\partial\theta} \in T_{\varphi(\mathbf{z})}\mathbb{S}^1$ and controlled if $\mathcal{V}$ is tangent to $|f(\mathbf{z})| = const$. See Sect. 8.3.10 below for further detail.

8.3.8 Milnor Fibrations for Non-convenient Mixed Functions

In this section, we study non-degenerate mixed functions without assuming the convenience. This means that the singularity might be non-isolated. For a non-negative weight vector $P = {}^t(p_1, \ldots, p_n)$, put $I(P) := \{i \mid p_i = 0\}$. Note that $d(P) = 0$ if and only $\Gamma_+(f) \cap \mathbb{R}^{I(P)} \neq \emptyset$. Recall that $d(P)$ is the minimum value of the linear function ℓ_P restricted $\Gamma(f)$ (see Sect. 8.1.4). Therefore

$$d(P) > 0 \iff f^{I(P)} \equiv 0 \iff I(P) \in \mathcal{I}_v(f).$$

To study non-convenient mixed functions, we consider a slightly bigger boundary $\Gamma_{nc}(f)$ of $\Gamma_+(f)$. Namely $\Gamma_{nc}(f)$ is the union of $\Gamma(f)$ and faces $\Delta \subset \Gamma_+(f)$ which is not included in any proper coordinate subspace.

Fix $I \in \mathcal{I}_v(f)$. Consider the distance function on $\mathbb{C}^I$ defined by $\rho_I(\mathbf{z}) := \sqrt{\sum_{i\in I}|z_i|^2}$. We say that f is *locally tame along the vanishing coordinate subspace* $\mathbb{C}^I$ if there exists $r_I > 0$ so that for any $\mathbf{a}_I = (\alpha_i)_{i\in I} \in \mathbb{C}^{*I}$ with $\rho_I(\mathbf{a}_I) \le r_I$ and for any P with $I(P) = I$, $f_{P,\mathbf{a}_I} := f_P|_{z_I=\mathbf{a}_I}$ is strongly non-degenerate as a function of $\{z_j \mid j \in I^c\}$. f is called *locally tame* if f is locally tame for any vanishing coordinate subspace $\mathbb{C}^{*I}$ [38].

Remark 8.3.19 The locally tame strong non-degeneracy is weaker than the super strong non-degeneracy defined in [31] (Definition 49) (i.e., $r_I = \infty$) but it plays the same role for the existence of a local Milnor fibration.

We first observe the following property.

Proposition 8.3.20 (Proposition 7, [31]) *Suppose that $f(\mathbf{z},\bar{\mathbf{z}})$ is strongly non-degenerate and locally tame along vanishing coordinate subspaces. Then for $I \in \mathcal{I}_{nv}$, f^I is also strongly non-degenerate and locally tame along vanishing coordinate subspaces.*

Put $r_{nc} = \min\{r_I \mid I \in \mathcal{I}_v(f)\}$. Choose r_0 so that

$$\forall r \le r_0 \implies \mathbb{S}_r^{2|I|-1} \pitchfork V^{*I},\ \forall I \in \mathcal{I}_{nv}(f).$$

Here $V^{*I} = V \cap \mathbb{C}^{*I}$ and $\pitchfork$ implies the intersection is transverse. We can generalize the Hamm-Lê Lemma [15] for mixed functions as follows.

Lemma 8.3.21 (Lemma 7, [38]) *Suppose that $f(\mathbf{z},\bar{\mathbf{z}})$ is a strongly non-degenerate mixed function and f is locally tame along vanishing coordinate subspaces. Put $\hat{r}_0 = \min\{r_{nc}, r_0\}$. Then for any $0 < r_1 \le \hat{r}_0$, there exists a positive number $\delta(r_1)$ satisfying the following.*

1. *(Smoothness of the nearby fibers) The fiber $V_\eta := f^{-1}(\eta)$, $\eta \ne 0$, $|\eta| \le \delta(r_1)$ is non-singular in the ball $\mathbb{B}^{2n}_{\hat{r}_0}$, i.e. 0 is the unique critical value of f on $\mathbb{B}^{2n}_{\hat{r}_0} \cap f^{-1}(\mathbb{D}_{\delta(r_1)})$.*
2. *(Strong transversality) For any non-zero η, $|\eta| \le \delta(r_1)$ and any r with $r_1 \le r \le \hat{r}_0$, the sphere $\mathbb{S}_r^{2n-1}$ and the hypersurface V_η intersect transversely.*

Corollary 8.3.22 (Theorem 9, [38]) *Assume that $f(\mathbf{z},\bar{\mathbf{z}})$ is strongly non-degenerate and locally tame along vanishing coordinate subspaces. Take $\hat{r}_0 > 0$ as in Lemma 8.3.21. For any $r_1 \le \hat{r}_0$, let $\delta(r_1)$ be as in Lemma 8.3.21. Then*

$$f : E(r_1, \delta(r_1))^* \to \mathbb{D}^*_{\delta(r_1)}$$

is a locally trivial fibration and the equivalence class of this fibration does not depend on the choice of r_1 and $\delta(r_1)$.

We call this fibration *the tubular Milnor fibration at the origin.* Take ε_1 as Lemma 8.3.16 and take $r_1 \le \min\{\hat{r}_0, \varepsilon_1\}$. The following is the existence assertion of the spherical fibration for non-convenient case.

Theorem 8.3.23 *Assume that $f(\mathbf{z}, \bar{\mathbf{z}})$ is strongly non-degenerate and locally tame along vanishing coordinate subspaces. Take $\hat{r}_0$ as in Lemma 8.3.21 and take $r_1 \le \hat{r}_0$, $\delta(r_1)$ as in Corollary 8.3.22. Then*

$$f/|f| : \mathbb{S}_{r_1}^{2n-1} \setminus K \to \mathbb{S}^1, \quad \textit{where } K = f^{-1}(0) \cap \mathbb{S}_{r_1}^{2n-1}$$

is a locally trivial fibration.

We call this fibration *the spherical Milnor fibration at the origin.*

Proof By Lemma 8.3.11, $\{\mathbf{z}, \bar{\partial} g(\mathbf{z}), \bar{\partial} h(\mathbf{z})\}$ are linearly independent over $\mathbb{R}$ for $\mathbf{z}$ with $0 \ne |f(\mathbf{z})| \le \delta(r_1)$. On the other hand, $\{\mathbf{v}_1(\mathbf{z}), \mathbf{v}_2(\mathbf{z})\}$ and $\{\bar{\partial} g(\mathbf{z}), \bar{\partial} h(\mathbf{z})\}$ span the same subspace of real dimension two (Lemma 2, [38]). Therefore we can construct on $\mathbb{S}_{r_1}^{2n-1} \setminus K$ a vector field $\mathcal{V}$ so that $\Re(\mathcal{V}(\mathbf{z}), \mathbf{v}_2(\mathbf{z})) = 1$ and in the neighborhood of K, it also satisfies

$\Re(\mathcal{V}(\mathbf{z}), \mathbf{v}_1(\mathbf{z})) = 0$. This implies along the integral curves, the absolute value of f does not change in the neighborhood of K, as long as $\Re(\mathcal{V}(\mathbf{z}), \mathbf{v}_1(\mathbf{z})) = 0$. Thus this gives the structure of a locally trivial fibration on $f/|f| : \mathbb{S}_{r_1}^{2n-1} \setminus K \to \mathbb{S}^1$, where $K = f^{-1}(0) \cap \mathbb{S}_{r_1}^{2n-1}$. □

The transversality (2) of Lemma 8.3.21 of the nearby fibers also follows from the Thom's a_f-regularity (see next lemma) and Lemma 8.2.7.

Lemma 8.3.24 (Theorem 20, [38], Theorem 3.14, [11]) *Suppose that $f(\mathbf{z}, \bar{\mathbf{z}})$ is strongly non-degenerate and tame along vanishing coordinate subspaces. Take $\hat{r}_0 > 0$ as in Lemma 8.3.21 and we consider in $\mathbb{B}_{\hat{r}_0}^{2n}$. Consider the canonical stratification $\mathcal{S}_{can}$ where*

$$\mathcal{S}_{can} := \{V^{*I}, \ \mathbb{C}^{*I} \setminus V^{*I}, \ I \in \mathcal{I}_{nv}\} \cup \{\mathbb{C}^{*I}, \ I \in \mathcal{I}_v\}.$$

Then f satisfies Thom a_f-regularity with respect to $\mathcal{S}_{can}$.

8.3.9 Topological Stability

Consider an analytic family of mixed functions $f_t(\mathbf{z}, \bar{\mathbf{z}})$, $0 \le t \le 1$ such that $\Gamma_{nc}(f_t)$ is constant. This implies $\mathcal{I}_{nv}(f_t)$, $\mathcal{I}_v(f_t)$ are also constant. Suppose also that f_t is locally tame and strongly non-degenerate for any t. There is a canonical stratification of $\mathbb{C}^n \times [0, 1]$ which is given as follows.

$$\mathcal{S}_{can} := \{\mathcal{V}^{*I}, \ \mathbb{C}^{*I} \times [0,1] \setminus \mathcal{V}^{*I}, \ I \in \mathcal{I}_{nv}\} \cup \{\mathbb{C}^{*I} \times [0,1], \ I \in \mathcal{I}_v\} \text{ where}$$
$$\mathcal{V}^{*I} := \{(\mathbf{z}, t) \in \mathbb{C}^{*I} \times [0,1] \mid f_t(\mathbf{z}) = 0\}, \quad I \in \mathcal{I}_{nv}(f_t).$$

Then we have

Theorem 8.3.25 (Theorem 3.14, [11]) *Assume that $f_t(\mathbf{z}, \bar{\mathbf{z}})$ is an analytic family satisfying the above condition. Then f_t, $0 \le t \le 1$ satisfies Whitney regularity with respect to the canonical stratification and the topological type of $(V(f_t), \mathbf{0})$ is constant for t and their tubular Milnor fibrations are equivalent. In particular, if f_t are convenient with the constant Newton boundary $\Gamma(f_t)$, the same assertion holds.*

Example

Consider a mixed function $f(z, \bar{z})$ of one variable z whose Newton boundary is $\{d\}$ and assume that the face function is given as

$$f_d(z) = z^p \bar{z}^q \prod_{j=1}^{\ell}(z + \lambda_j \bar{z}), \; d = p + q + \ell. \tag{8.19}$$

Assume that $|\lambda_j| < 1, \forall j$ and $p - q \ge 0$ and $p - q + \ell \ge 1$. Consider the family $f_{d,t}(z) := z^p \bar{z}^q \prod_{j=1}^{\ell}(z + t\lambda_j \bar{z})$. Then $f_{d,1} = f_d$ and $f_{d,0}(z) = z^{p+\ell}\bar{z}^q$. By Lemma 8.3.3, f_t is strongly non-degenerate for any $0 \le t \le 1$ and it satisfies the above assumption. Thus we have

Lemma 8.3.26 *$f_d(z, \bar{z})$ is equivalent to the monomial function $z^{p+\ell}\bar{z}^q$.*

Let $f_k(z_k)$ be a mixed function of one variable z_k and assume that the lowest term has degree d_k and the face function is written as

$$f_{k,d_k}(z_k) = z_k^{p_k} \bar{z}_k^{q_k} \prod_{j=1}^{\ell_k}(z_k + \lambda_{kj} \bar{z}_k)$$

with $p_k - q_k \ge 0$, $p_k - q_k + \ell_k > 0$ and $|\lambda_{kj}| < 1$ for any $j = 1, \dots, \ell_k$. Consider a mixed function

$$f(\mathbf{z}, \bar{\mathbf{z}}) = f_1(z_1, \bar{z}_1) + \cdots + f_n(z_n, \bar{z}_n) \quad \text{and}$$

a mixed Brieskorn polynomial and a holomorphic Brieskorn polynomial:

$$\begin{aligned} B(\mathbf{z}, \bar{\mathbf{z}}) &= z_1^{p_1+\ell_1} \bar{z}_1^{q_1} + \cdots + z_n^{p_n+\ell_n} \bar{z}_n^{q_n} \\ b(z) &= z_1^{p_1-q_1+\ell_1} + \cdots + z_n^{p_n-q_n+\ell_n}. \end{aligned}$$

Theorem 8.3.27 *$(V(f), \mathbf{0})$ is equivalent to $(V(B), \mathbf{0})$. It is also equivalent to $(V(b), \mathbf{0})$.*

Proof First, by Theorem 8.3.25, f is equivalent to $\sum_{k=1}^{n} f_{k,d_k}(z_k, \bar{z}_k)$. Secondly each f_{k,d_k} is equivalent to $z_k^{p_k+\ell_k}\bar{z}_k^{q_k}$ by Lemma 8.3.26. Finally we use again Theorem 8.3.25 to get the assertion. For the equivalence with$(V(b), \mathbf{0})$, use Lemma 8.3.45 below. □

8.3.10 *Equivalence of Tubular and Spherical Milnor Fibrations*

Under the assumption of the strong non-degeneracy and the local tameness, the tubular and the spherical Milnor fibrations are equivalent. We use two gradient vectors $\mathbf{v}_1(\mathbf{z})$ and $\mathbf{v}_2$ defined in (8.14), (8.15). First we prepare the following.

Lemma 8.3.28 (Lemma 34, [31]) *Assume that f is a strongly non-degenerate locally tame mixed function. Then there exists a positive number r_0 so that for any $\mathbf{z}$ with $\|\mathbf{z}\| \le r_0$ and $f(\mathbf{z}, \bar{\mathbf{z}}) \neq 0$, three vectors*

$$\mathbf{z}, \quad \mathbf{v}_1(\mathbf{z}, \bar{\mathbf{z}}), \quad \mathbf{v}_2(\mathbf{z}, \bar{\mathbf{z}})$$

are either (i) linearly independent over $\mathbb{R}$ or (ii) they are linearly dependent over $\mathbb{R}$ and the relation can be written as

$$\mathbf{z} = a\,\mathbf{v}_1(\mathbf{z}, \bar{\mathbf{z}}) + b\,\mathbf{v}_2(\mathbf{z}, \bar{\mathbf{z}}),\ a, b \in \mathbb{R}. \tag{8.20}$$

and the coefficient a is positive.

In Lemma 34, [31], f is assumed to be convenient but the proof is exactly the same for a local tame mixed function. We consider two Milnor fibrations.

$$\begin{aligned} f :& \partial E(r_0, \delta_0)^* \to \mathbb{S}^1_{\delta_0}, & (8.21)\\ & \text{where } \partial E(r_0, \delta_0)^* = \{\mathbf{z} \in \mathbb{B}^{2n}_{r_0} \mid |f(\mathbf{z})| = \delta_0\} \\ f/|f| :& S^{2n-1}_{r_0} \setminus N(K) \to \mathbb{S}^1, & (8.22)\\ & \text{where } N(K) = \{\mathbf{z} \in \mathbb{S}^{2n-1}_{r_0} \mid |f(\mathbf{z})| < \delta_0\}. \end{aligned}$$

The first fibration is the restriction of the tubular Milnor fibration to the boundary and it is homotopically equivalent to the tubular Milnor fibration. The second fibration is equivalent to the spherical Milnor fibration.

Theorem 8.3.29 *The above two fibrations are equivalent.*

Proof Construct a vector field $\mathcal{V}(\mathbf{z})$ in $\mathbb{B}^{2n}_{r_0} \cap \{\mathbf{z} \mid |f(\mathbf{z}, \bar{\mathbf{z}})| \ge \delta\}$ so that the following conditions are satisfied.

$$(a)\ \Re(\mathcal{V}(\mathbf{z}), \mathbf{v}_2(\mathbf{z})) = 0,\quad (b)\ \Re(\mathcal{V}(\mathbf{z}), \mathbf{v}_1(\mathbf{z})) > 0,\quad (c)\ \Re(\mathcal{V}(\mathbf{z}), \mathbf{z}) > 0.$$

Construct $\mathcal{V}$ locally and glue together using a partition of the unity.
(i) For a point $\mathbf{p}$ where $\{\mathbf{p}, \mathbf{v}_1(\mathbf{p}), \mathbf{v}_2(\mathbf{p})\}$ are linearly independent, choose a small open ball neighborhood $U_{\mathbf{p}}$ so that three vectors are linearly independent on $U_{\mathbf{p}}$. We take $\mathcal{V}(\mathbf{z})$ so that (a), (b),(c) are satisfied.
(ii) For a point $\mathbf{p}$ where $\{\mathbf{p}, \mathbf{v}_1(\mathbf{p}), \mathbf{v}_2(\mathbf{p})\}$ are linearly dependent over $\mathbb{R}$. Then $\Re(\mathbf{p}, \mathbf{v}_1(\mathbf{p})) > 1$ by (8.20). We choose $\mathcal{V}(\mathbf{z}) = \mathbf{v}_1(\mathbf{z})$. Taking a sufficiently small open ball neighborhood $U_{\mathbf{p}}$ so that $\Re(\mathbf{w}, \mathbf{v}_1(\mathbf{w})) > 0$ for any $\mathbf{w} \in U_{\mathbf{p}}$. Note that $\Re(\mathbf{v}_1(\mathbf{w}), \mathbf{v}_2(\mathbf{w})) = 0$ by the definition of $\mathbf{v}_1$ and $\mathbf{v}_2$. As $\mathbb{B}^{2n}_{r_0} \cap \{\mathbf{z} \mid |f(\mathbf{z})| \geq \delta(r_0)\}$ is compact, we get a finite neighborhood $U_{\mathbf{p}_1}, \dots, U_{\mathbf{p}_\nu}$ which covers $\mathbb{B}^{2n}_{r_0} \cap \{\mathbf{z} \mid |f(\mathbf{z})| \geq \delta(r_0)\}$. Apply the partition of the unity method to this open covering, to construct a global vector field $\mathcal{V}$. It is easy to see that $\mathcal{V}$ satisfies the conditions $(a), (b), (c)$ everywhere. Then the integration of $\mathcal{V}$ gives a diffeomorphism $\psi : \partial E(r_0, \delta) \to \mathbb{S}^{2n-1}_{r_0} \setminus N(K)$. The condition (a) implies arg $f(\mathbf{z}(t))$ is constant along the integral curve and (b) implies that the integral curve $\mathbf{z}(t)$ starting at $\mathbf{z}(0) = \mathbf{p} \in \partial E(r_0, \delta_0)^*$ stays outside of the tube $\{|f| > \delta_0\}$ for $t > 0$. (c) imply $\|\mathbf{z}(t)\|$ are monotone increasing. Thus the integral curve $\mathbf{z}(t)$ arrives at $\mathbf{q} = \mathbf{z}(t_0) \in \mathbb{S}^{2n-1}_{r_0} \setminus N(K)$ after a finite time t_0 which depends on the initial point $\mathbf{p}$. Define ψ by $\psi(\mathbf{p}) = \mathbf{q}$ and ψ gives an equivalence diffeomorphism of two fibration. □

8.3.11 *Real Blowing Up and a Resolution of a Real Type*

Let $f(\mathbf{z}, \bar{\mathbf{z}})$ be a convenient non-degenerate mixed function and let $V = f^{-1}(0)$. Let $\Gamma^*(f)$ be the dual Newton diagram and let Σ^* be a regular simplicial cone subdivision and consider the associated toric modification $\hat{\pi} : X \to \mathbb{C}^n$. See Sect. 8.1.4 and Theorem 8.1.9 for the definition and the basic properties of this modification. Let $\tilde{V}$ be the strict transform of V in X. It turns out that $\tilde{V}$ still has some small singularity. To solve these singularities, we need real blowing ups along exceptional divisors.

Real Blowing-Up

Consider one dimensional complex line with the coordinate $z = x + iy$. Consider the imbedding $\iota : \mathbb{C} \setminus \{0\} \to \mathbb{C} \times \mathbb{RP}^1$ defined by $\iota(z) = (z, [x : y])$ where $\mathbb{RP}^1$ is the real projective space of dimension 1. The real blowing-up is the closure of the image of ι, say $X_{\mathbb{R}}$. There is a canonical projection $\pi_{\mathbb{R}} : X_{\mathbb{R}} \to \mathbb{C}$ so that $\pi_{\mathbb{R}} : X_{\mathbb{R}} \setminus \pi_{\mathbb{R}}^{-1}(0) \to \mathbb{C} \setminus \{0\}$ is a real analytic diffeomorphism and $E := \pi_{\mathbb{R}}^{-1}(0)$ is isomorphic to the real projective space $\mathbb{RP}^1$. We call E *the real exceptional divisor.* $X_{\mathbb{R}}$ has two charts $U_1 := \{x \neq 0\}$ and $U_2 = \{y \neq 0\}$. On U_1 and U_2, we can take the respective coordinates (x', s) and (y', t) where $s = y/x,\ t = x/y$ and $\pi_{\mathbb{R}}$ is defined by

$$U_1 \ni (x', s) \xrightarrow{\pi_{\mathbb{R}}} (x, y) = (x', x's) \in \mathbb{C} \ on\ U_1$$
$$U_2 \ni (y', t) \xrightarrow{\pi_{\mathbb{R}}} (x, y) = (y't, y') \in \mathbb{C} \ on\ U_2.$$

Note that the exceptional divisor is defined by $x' = 0$ in U_1 or $y' = 0$ in U_2.

Real Blowing Up Along Complex Divisors

Assume X is a complex manifold and D is a smooth complex divisor. Then we can take a real blowing up along the normal bundle of D, $\pi_{\mathbb{R}} : Y \to X$. Then the exceptional divisor E is a $\mathbb{RP}^1$-bundle over D. Let $\hat{\pi} : X \to \mathbb{C}^n$ be a toric modification associated with a regular simplicial cone subdivision of $\Gamma^*(f)$. We apply this real blowing up for the exceptional divisors $\hat{E}(P)$ of $\hat{\pi}$.

Lemma 8.3.30 *We assume that f is non-degenerate. Let $\sigma = \mathrm{Cone}(P_1, P_2, \dots, P_n)$ be a regular n-simplicial cone in Σ^*. Consider the face function $f_{P_1}(\mathbf{z}, \bar{\mathbf{z}})$ of $f(\mathbf{z}, \bar{\mathbf{z}}) = \sum_{\nu,\mu} c_{\nu,\mu} \mathbf{z}^\nu \bar{\mathbf{z}}^\mu$ which is a non-degenerate radially weighted homogenous polynomial of degree $d_r = d(P_1)$ with the weight vector P_1. Then the pull-back is given as*

$$\pi_\sigma^* f_{P_1}(\mathbf{u}_\sigma, \bar{\mathbf{u}_\sigma}) = \sum_{\nu+\mu \in \Delta(P_1)} c_{\nu,\mu} \prod_{i=1}^{n} u_{\sigma,i}^{P_i(\nu)} \bar{u}_{\sigma,i}^{P_i(\mu)}.$$

If further f is polar weighted of degree d_p with the same weight P_1, Then

$$\pi_\sigma^* f_{P_1}(\mathbf{u}_\sigma, \bar{\mathbf{u}}_\sigma) = u_{\sigma,1}^{\alpha} \bar{u}_{\sigma,1}^{\beta} g(\mathbf{u}_\sigma{}', \bar{\mathbf{u}}_\sigma'),$$
$$\text{where } \mathbf{u}_\sigma' = (u_{\sigma,2}, \dots, u_{\sigma,n}),\ \alpha = \frac{d_r + d_p}{2},\ \beta = \frac{d_r - d_p}{2}.$$

Proof The last equality follows from $P_i(\nu) + P_i(\mu) = d_r$ and $P_i(\nu) - P_i(\mu) = d_p$. □

Note that $\pi_\sigma^* f_{P_1}$ is not divisible by $u_{\sigma,1}^{d(P_i)}$ like in the holomorphic case.

Corollary 8.3.31 *Let $\pi_{\mathbb{R}} : Z \to \mathbb{C}_\sigma^n$ be the real blowing up along the divisor $\hat{E}(P_1) = \{u_{\sigma,1} = 0\}$. Let E be the exceptional divisor of $\pi_{\mathbb{R}}$ and let $\hat{V}$ be the strict transform of $V = f^{-1}(0)$ into X and let $\tilde{V}$ be the strict transform of $\hat{V}$ to Z. Then $\tilde{V}$ is non-singular over $\hat{E}(P_1)^* := \{0\} \times \mathbb{C}_\sigma^{*(n-1)}$ and intersects transversely with E.*

Proof Put $u_{\sigma,1} = x_1 + iy_1$ and put $s = y_1/x_1$, $t = x_1/y_1$. Consider the coordinate chart $y_1 \neq 0$ with coordinates $(y_1', t, \mathbf{u}_\sigma')$. In these coordinates, $\pi_{\mathbb{R}}(y_1', t, \mathbf{u}_\sigma') = (y_1' t, y_1', \mathbf{u}_\sigma')$ and E is defined by $y_1' = 0$. Take $\mathbf{q} \in E \cap \tilde{V} \cap \{t \neq 0\}$. Assume $\mathbf{q} = (0, t_0, \mathbf{v})$ in these coordinates. We assume $\mathbf{v} \in \mathbb{C}_\sigma^{*(n-1)}$ for simplicity. Then using the equalities $u_{\sigma,1} = y_1'(t+i)$, $\bar{u}_{\sigma,1} = y_1'(t-i)$, we can write

$$\pi_{\mathbb{R}}^* f_\sigma(y_1', t, \mathbf{u}_\sigma') = y_1'^{d_r} f_\sigma',$$
$$f_\sigma' = \sum_{\nu,\mu} c_{\nu,\mu} (t+i)^{P_1(\nu)} (t-i)^{P_1(\mu)} \prod_{i=2}^{n} u_{\sigma,i}^{P_i(\nu)} \bar{u}_{\sigma,i}^{P_i(\mu)}.$$

Thus we observe that $\tilde{V}$ is defined by $f'_\sigma = 0$ and f'_σ does not contain y'_1. As $\pi_{\mathbb{R}} : Z \setminus E \to \mathbb{C}^n_\sigma$ is a diffeomorphism and $\pi^*_\sigma f^{-1}(0) \cap \mathbb{C}^{*n}_\sigma$ is non-singular by the non-degeneracy, $\{f'_\sigma = 0\} \cap \mathbb{R}^* \times \mathbb{C}^{*(n-1)}_\sigma$ is non-singular at $\mathbf{q}' = (1, t_0, \mathbf{v})$. Thus the Jacobian matrix J of $\{\Re f'_\sigma, \Im f'_\sigma\}$ with respect to $(y'_1, t, x_2, \ldots, y_n)$ has rank two at $\mathbf{q}'$. Here $u_{\sigma,j} = (x_j + iy_j)$ for $j = 2, \ldots, n$. It takes the form

$$J = \begin{pmatrix} 0 & a_1 & \frac{\partial \Re f'_\sigma}{\partial x_2} & \cdots & \frac{\partial \Re f'_\sigma}{\partial y_n} \\ 0 & a_2 & \frac{\partial \Im f'_\sigma}{\partial x_2} & \cdots & \frac{\partial \Im f'_\sigma}{\partial y_n} \end{pmatrix}, \quad a_1, a_2 \in \mathbb{C}.$$

Note that the Jacobian of $\{\Re f'_\sigma, \Im f'_\sigma\}$ at $\mathbf{q}$ is the same with the Jacobian at $\mathbf{q}'$. On the other hand, E is defined by $y'_1 = 0$. Thus the Jacobian matrix of $\{y_1, \Re f'_\sigma, \Im f'_\sigma\}$ at $\mathbf{q}$ is written as

$$\begin{pmatrix} 1 & 0 \\ 0 & J \end{pmatrix}$$

This implies $\tilde{V}$ and E intersect transversely at $\mathbf{q}$. □

A Good Resolution of a Real Type

Let f be a convenient non-degenerate mixed function and let $V = f^{-1}(0)$. Let $\Phi : Y \to \mathbb{C}^n$ be a proper mapping where Y is a real analytic manifold of real dimension $2n$. We say that Φ is *a good resolution of a real type* of f if it satisfies the following properties.

1. $\Phi^{-1}(\mathbf{0}) = D,\ D = D_1 \cup \cdots \cup D_m$ where each D_j is a non-singular real divisor of real codimension 1.
2. $\Phi : Y \setminus \Phi^{-1}(\mathbf{0}) \to \mathbb{C}^n \setminus \{\mathbf{0}\}$ is a diffeomorphism.
3. Let $\tilde{V}$ be the closure of $\Phi^{-1}(V) \cap (Y - D)$ (we call it the strict transform of V). $\tilde{V}$ is non-singular and real dimension $2(n-1)$.
4. $\tilde{V} \cup_{i=1}^m D_i$ has only normal crossing singularities in the following sense. Take any $\mathbf{q} \in D_1 \cap \cdots \cap D_k \cap \tilde{V}$. Then there is a chart U with real analytic coordinates $(v_1, \ldots, v_{2n})$ in the neighborhood of $\mathbf{q}$ such that $D_i = \{v_i = 0\},\ 1 \le i \le k$ and $\tilde{V} = \{v_{k+1} = v_{k+2} = 0\}$. In this neighborhood U, $\Phi^* f(\mathbf{v}) = v_{k+1} + iv_{k+2}$.

The Construction of a Resolution of a Real Type

Let Σ^* be a regular simplicial cone subdivision of $\Gamma^*(f)$ and let $\hat{\pi} : X \to \mathbb{C}^n$ the the associated toric modification. We assume that the vertices of Σ^* are strictly positive except the canonical ones $E_1, \ldots, E_n$, as f is convenient. Thus $\hat{\pi} : X \setminus \hat{\pi}^{-1}(\mathbf{0}) \to \mathbb{C}^n \setminus \{\mathbf{0}\}$ is biholomorphic. Let $\mathcal{V}$ be the set of strictly positive vertices of Σ^* and let $\pi_{\mathbb{R}} : Y \to X$ be the real blowing ups along every exceptional divisor $\hat{E}(P),\ P \in \mathcal{V}$. Consider the composition $\Phi = \hat{\pi} \circ \pi_{\mathbb{R}} : Y \to \mathbb{C}^n$.

Theorem 8.3.32 (Theorem 24, [31]) *$\Phi : Y \to \mathbb{C}^n$ is a good resolution of real type of f.*

The proof follows from Lemma 8.3.30 and Corollary 8.3.31. For further details, see [31]. There is one class of mixed functions for which the toric modification is already a good resolution. A convenient mixed function $f(\mathbf{z}, \bar{\mathbf{z}}) = \sum_{\nu,\mu} c_{\nu,\mu}\mathbf{z}^{\nu}\bar{\mathbf{z}}^{\mu}$ is called *a mixed function of a strongly polar weighted homogeneous face type* [33], if it satisfies the next conditions.

1. For any maximal dimensional face $\Delta \subset \Gamma(f)$, $f_\Delta(\mathbf{z})$ is a strongly mixed weighted homogeneous polynomial with respect to the unique integer weight P, normal to the face Δ.
2. For any (ν, μ) with $c_{\nu,\mu} \neq 0$ and $\nu + \mu \notin \Delta(P_1)$,

$$P_1(\nu) > \frac{\mathrm{rdeg}_{P_1} f + \mathrm{pdeg}_{P_1} f}{2}, \quad P_1(\mu) > \frac{\mathrm{rdeg}_{P_1} f - \mathrm{pdeg}_{P_1} f}{2}.$$

Remark 8.3.33 In [33], the condition (2) is forgotten in the definition of a mixed function of a strongly polar weighted homogenous face type, though in the proof of Theorem 11, (2) is used. Thus the condition (2) must be added to the definition.

Example 8.3.34 1. If $f(\mathbf{z}, \bar{\mathbf{z}})$ is a strongly mixed homogeneous polynomial, f is a mixed function of a strongly polar weighted homogeneous face type.

2. Assume that $g(\mathbf{z}) = \sum_{\nu} c_\nu \mathbf{z}^\nu$ is a holomorphic function. Consider a branched covering $\varphi_{a,b} : \mathbb{C}^n \to \mathbb{C}^n$, $a \neq b, \varphi(z) = (z_1^a\bar{z}_1^b, \dots, z_n^a\bar{z}_n^b)$. Put $f(\mathbf{z}, \bar{\mathbf{z}}) := \varphi_{a,b}^* g(\mathbf{z}, \bar{\mathbf{z}})$. Then f is a mixed function of a strongly polar weighted homogeneous face type. In [37], it is proved that the link of f has a canonical contact structure.

Theorem 8.3.35 (Theorem 11, [33]) *Assume that $f(\mathbf{z}, \bar{\mathbf{z}})$ is a convenient strongly non-degenerate mixed function of strongly polar weighted homogeneous face type. Then the toric modification $\hat{\pi} : X \to \mathbb{C}^n$ is a good resolution of f and there is a formula of the Varchenko type.*

Here a good resolution means the following. For any coordinate chart $(\mathbb{C}_\sigma, \mathbf{u}_\sigma)$ with $\sigma = \mathrm{Cone}(P_1, \dots, P_n)$,

$$\pi_\sigma^* f = \prod_{i=1}^{n} u_{\sigma,i}^{a_i}\bar{u}_{\sigma,i}^{b_i} g(\mathbf{u}_\sigma, \bar{\mathbf{u}}_\sigma) \text{ where}$$

$$a_i = \frac{\mathrm{rdeg}_{P_i}(f) + \mathrm{pdeg}_{P_i}(f)}{2}, \quad b_i = \frac{\mathrm{rdeg}_{P_i}(f) - \mathrm{pdeg}_{P_i}(f)}{2}.$$

where g is a mixed function defining the strict transform $\tilde{V}$ and $\tilde{V}$ and $\{\hat{E}(P_i), i = 1, \dots, n\}$ intersect transversely if they intersect. Note that the topology of f is not uniquely determined by the combinatorics of $\Gamma(f)$ for mixed functions.

Example

1. Let $f_1(\mathbf{z}) = z_1^2\bar{z}_1 - z_2\bar{z}_2^2$ and $f_2(\mathbf{z}) = z_1^2\bar{z}_1 - z_2^2\bar{z}_2$ and let $C_1 = f_1^{-1}(0)$ and $C_2 = f_2^{-1}(0)$. They have the same dual Newton diagram which is generated by E_1, E_2, P where $P = {}^t(1,1)$. The associated toric modification is just a blowing up. Take the chart $\sigma = \mathrm{Cone}(P, E_2)$. $\pi_\sigma(u_{\sigma,1}, u_{\sigma,2}) = (z_1, z_2)$ with $z_1 = u_{\sigma,1}$, $z_2 = u_{\sigma,1}u_{\sigma,2}$. Thus

$$\pi_\sigma^* f_1 = u_{\sigma,1}^2\bar{u}_{\sigma,1} - u_{\sigma,1}u_{\sigma,2}\bar{u}_{\sigma,1}^2\bar{u}_{\sigma,2}^2 = u_{\sigma,1}\bar{u}_{\sigma,1}(u_{\sigma,1} - \bar{u}_{\sigma,1}u_{\sigma,2}\bar{u}_{\sigma,2}^2)$$
$$\pi_\sigma^* f_2 = u_{\sigma,1}^2\bar{u}_{\sigma,1} - u_{\sigma,1}^2u_{\sigma,2}^2\bar{u}_{\sigma,1}\bar{u}_{\sigma,2} = u_{\sigma,1}^2\bar{u}_{\sigma,1}(1 - u_{\sigma,2}^2\bar{u}_{\sigma,2}).$$

Now note that $\lim_{u_{\sigma,1}\to 0} |u_{\sigma,2}\bar{u}_{\sigma,2}^2| = \lim_{u_{\sigma,1}\to 0} |\frac{u_{\sigma,1}}{\bar{u}_{\sigma,1}}| = 1$. Thus we can see that $\tilde{C}_1 \cap \hat{E}(P) = \{(0, u_{\sigma,2}) \mid |u_{\sigma,2}| = 1\}$. While $\tilde{C}_2 \cap \hat{E}(P) = \{(0,1)\}$ and we can see that $\tilde{C}_2$ and $\hat{E}(P)$ intersect transversely. The reason is that f_2 is strongly mixed homogeneous. In particular, f_2 is a mixed function of a strongly polar weighted homogeneous face type.

Consider the real modification $\pi_{\mathbb{R}} : Y \to X$ on the normal bundle of $\hat{E}(P)$. Put $u_{\sigma,1} = x + yi$. Take the coordinate chart $\{y \neq 0\}$ and put $t = x/y$.

$$\begin{aligned}\pi_\sigma^* f_1 &= (t^2y^2 + y^2)\left((yt + iy) - (yt - iy)u_{\sigma,2}\bar{u}_{\sigma,2}^2\right)\\ &= (t^2y^2 + y^2)y\left((t+i) - (t-i)u_{\sigma,2}\bar{u}_{\sigma,2}^2\right).\end{aligned}$$

Namely in this chart, the strict transform of C_1 is defined by

$$u_{\sigma,2}\bar{u}_{\sigma,2}^2(t-i) - (t+i) = 0.$$

We can see that this is a non-singular real curve.

2. Let $g_s(\mathbf{z}, \bar{\mathbf{z}}) = \sum_{j=1}^n z_j^2\bar{z}_j + s\sum_{i=1}^{n-1} z_i^2\bar{z}_n$. Note that g_s is a strongly mixed homogeneous polynomial. $\Gamma^*(g_s)$ is the regular fan with vertices $\{E_, \dots, E_{n-1}, P\}$ with $P = {}^t(1,\dots,1)$. Thus $\Sigma^* = \Gamma^*(g_s)$ and the toric modification is the ordinary blowing up at the origin. Take the chart $\mathbb{C}_\sigma^n$ with $\sigma = \mathrm{Cone}\,(E_1, \dots, E_{n-1}, P)$ with coordinates $(u_1, \dots, u_n)$. Then we have

$$\hat{\pi}^* g_s(\mathbf{u}) = u_n^2\bar{u}_n\{u_1^2\bar{u}_1 + \cdots + u_{n-1}^2\bar{u}_{n-1} + 1 + s\left(u_1^2 + \cdots + u_{n-1}^2\right)\}$$

and $\hat{V}$ is defined by

$$\hat{g}_s := u_1^2\bar{u}_1 + \cdots + u_{n-1}^2\bar{u}_{n-1} + 1 + s(u_1^2 + \dots u_{n-1}^2) = 0$$

and non-singular for a generic s. For a sufficiently small s, g_s is isomorphic to g_0 which is a simplicial hyperplane polynomial and isotopic to the trivial holomorphic link of the hyperplane $z_1 + \cdots + z_n = 0$. See Lemma 8.3.45 below. For s large, the topology is different. Let us see this phenomenon more explicitly for $n = 2$ and

$s = 2$. Then V is defined by $z_1^2\bar{z}_1 + z_2^2\bar{z}_2 + z_1^2\bar{z}_2 = 0$ and the local equation of the projective curve defined by $g_2 = 0$, $\bar{V}$, is given as

$$\bar{g}_2 := u_1^2\bar{u}_1 + 1 + 2u_1^2 = 0, \quad u_1 = z_1/z_2, \text{ in } \mathbb{P}^1.$$

Put $u_1 = x + iy$. Then

$$\bar{g}_2 = (x^3 + xy^2 + 1 + 2x^2 - 2y^2) + i(x^2y + 4xy + y^3).$$

Solve the real solutions of $\Re\bar{g}_2 = \Im\bar{g}_2 = 0$. Thus $\hat{V} \cap \hat{E}(P) = \{(-1/8 \pm i\sqrt{31}/8, 0), (\beta, 0)\}$ where β is the real root of $x^3 + 2x^2 + 1 = 0$. This implies that $g_2 = 0$ has three branches at the origin.

8.3.12 *Simplicial Mixed Polynomials*

Consider a mixed polynomial $f(\mathbf{z}, \bar{\mathbf{z}}) = \sum_{j=1}^{s} c_j \mathbf{z}^{\mathbf{n}_j}\bar{\mathbf{z}}^{\mathbf{m}_j}$ where the coefficients $c_1, \dots, c_s$ are non-zero complex numbers. We associate the following Laurent polynomial $\hat{f}(\mathbf{w})$ with $f(\mathbf{z}, \bar{\mathbf{z}})$:

$$\hat{f}(\mathbf{w}) := \sum_{j=1}^{s} c_j \mathbf{w}^{\mathbf{n}_j - \mathbf{m}_j}, \quad \mathbf{w} = (w_1, \dots, w_n) \in \mathbb{C}^n.$$

Proposition 8.3.36 *Assume that $f(\mathbf{z}, \bar{\mathbf{z}})$ is a polar weighted homogeneous polynomial of degree d_p with respect to the polar weight $P = (p_1, \dots, p_n)$. Then $\hat{f}(\mathbf{w})$ is also a weighted homogeneous Laurent polynomial of degree d_p with respect to the same weight vector P.*

Put $\mathbf{n}_j = (n_{j,1}, \dots, n_{j,n})$, $\mathbf{m}_j = (m_{j,1}, \dots, m_{j,n}) \in \mathbb{N}^n$ for $j = 1, \dots, s$. $f(\mathbf{z}, \bar{\mathbf{z}})$ is called *simplicial* if two sets of s vectors $\{\mathbf{n}_j + \mathbf{m}_j \mid j = 1, \dots, s\}$ and $\{\mathbf{n}_j - \mathbf{m}_j \mid j = 1, \dots, s\}$ are both linearly independent in $\mathbb{R}^n$. Consider $n \times s$ matrices $N = (n_{i,j})$ and $M = (m_{i,j})$. If f is simplicial, rank $(N + M) =$ rank $(N - M) = s$ and it is clear that $s \leq n$. If $s = n$ we say f is *a full simplicial polynomial.* If $s = n$, $f(\mathbf{z}, \bar{\mathbf{z}})$ is simplicial if and only if $\det(N \pm M) \neq 0$.

Proposition 8.3.37 *Let $f(\mathbf{z}, \bar{\mathbf{z}}) = \sum_{j=1}^{s} c_j \mathbf{z}^{\mathbf{n}_j}\bar{\mathbf{z}}^{\mathbf{m}_j}$. If $f(\mathbf{z}, \bar{\mathbf{z}})$ is simplicial, $f(\mathbf{z}, \bar{\mathbf{z}})$ is a mixed weighted homogeneous polynomial under suitable polar and radial weight vectors.*

Proof Assume that $\hat{P}$ is a normalized polar weight vector. Then it must satisfy the equalities:

$$(\mathbf{n}_1 - \mathbf{m}_1, \hat{P}) = \cdots = (\mathbf{n}_s - \mathbf{m}_s, \hat{P}) = 1. \tag{8.23}$$

By the assumption, rank$(N - M) = s$. Therefore this has a non-trivial solution. Similarly a normalized radial weight vector $\hat{Q}$ must satisfy the equation

$$(\mathbf{n}_1 + \mathbf{m}_1, \hat{Q}) = \cdots = (\mathbf{n}_s + \mathbf{m}_s, \hat{Q}) = 1 \tag{8.24}$$

which has a solution as $\mathrm{rank}(N + M) = s$.

Remark 8.3.38 A polar weight vector or a radial weight vector which are solutions of (8.23) or (8.24) respectively need not be a non-negative vector.

Examples

Take integers $a_i \geq 1,\ b_i \geq 0,\ i = 1, \dots, n$ and consider the following polynomials which have isolated singularities at the origin.

$$\begin{aligned}
B_{\mathbf{a},\mathbf{b}}(\mathbf{z}, \bar{\mathbf{z}}) &= z_1^{a_1+b_1}\bar{z}_1^{b_1} + \cdots + z_n^{a_n+b_n}\bar{z}_n^{b_n} \text{ (mixed Brieskorn)}\\
f_I(\mathbf{z}, \bar{\mathbf{z}}) &= z_1^{a_1+b_1}\bar{z}_1^{b_1} z_2 + \cdots + z_{n-1}^{a_{n-1}+b_{n-1}}\bar{z}_{n-1}^{b_{n-1}} z_n + z_n^{a_n+b_n}\bar{z}_n^{b_n} \text{ (tree type)}\\
f_{II}(\mathbf{z}, \bar{\mathbf{z}}) &= z_1^{a_1+b_1}\bar{z}_1^{b_1} z_2 + \cdots + z_{n-1}^{a_{n-1}+b_{n-1}}\bar{z}_{n-1}^{b_{n-1}} z_n + z_n^{a_n+b_n}\bar{z}_n^{b_n} z_1 \text{ (cyclic type).}
\end{aligned}$$

The associated Laurent polynomials are given by the following polynomials:

$$\hat{B}_{\mathbf{a},\mathbf{b}}(\mathbf{w}) = w_1^{a_1} + \cdots + w_n^{a_n} \tag{8.25}$$
$$\hat{f}_I(\mathbf{w}) = w_1^{a_1} w_2 + \cdots + w_{n-1}^{a_n} w_n + w_n^{a_n} \tag{8.26}$$
$$\hat{f}_{II}(\mathbf{w}) = w_1^{a_1} w_2 + \cdots + w_{n-1}^{a_n} w_n + w_n^{a_n} w_1. \tag{8.27}$$

They are corresponding to the simplicial weighted homogeneous polynomials with an isolated singularity which are in the classification list of Orlik-Wagreich [44]. Mixed polynomials $B_{\mathbf{a},\mathbf{b}}, f_I, f_{II}$ are typical simplicial polynomials and their associated polynomials $\hat{B}_{\mathbf{a},\mathbf{b}}, \hat{f}_I, \hat{f}_{II}$ are weighted homogeneous polynomials with isolated singularity at the origin. See Orlik-Wagreich [44].

We assume that $f(\mathbf{z}, \bar{\mathbf{z}}) = \sum_{j=1}^{s} c_j \mathbf{z}^{\mathbf{n}_j} \bar{\mathbf{z}}^{\mathbf{m}_j}$ is a simplicial mixed polynomial and let $P = (p_1, \dots, p_n)$ be the polar weight vector and d_p be the polar degree. Put $F^*(a) = f^{-1}(a) \cap \mathbb{C}^{*n}$, $\hat{F}^*(a) := \hat{f}^{-1}(a) \subset \mathbb{C}^{*n}$ for $a \in \mathbb{C}^*$. They can be identified with the respective toric Milnor fibers of f and $\hat{f}$. Note that $F^*(a) \cong F^*(1)$ and $\hat{F}^*(a) \cong \hat{F}^*(1)$. The following theorem says that these toric Milnor fibrations are isomorphic.

Theorem 8.3.39 ([30]) *Let $f(\mathbf{z}, \bar{\mathbf{z}})$ be a simplicial mixed polynomial and let $\hat{f}(\mathbf{w})$ be the associated Laurent polynomial. Let d_p be the polar degree of $f(\mathbf{z})$. There exists a canonical diffeomorphism $\varphi : \mathbb{C}^{*n} \to \mathbb{C}^{*n}$ which maps $F(a)^*$ into $\hat{F}(a)^*$, $\forall a \in \mathbb{C}^*$ and makes the following diagram commutative. In particular, the toric Milnor fibrations of f and $\hat{f}$ are isomorphic.*

$$\begin{array}{ccc} \mathbb{C}^{*n} & \xrightarrow{\varphi} & \mathbb{C}^{*n} \\ \downarrow f & & \downarrow \hat{f} \\ \mathbb{C}^{*} & \xrightarrow{\mathrm{id}} & \mathbb{C}^{*} \end{array}$$

φ is also compatible with the monodromy maps h and $\hat{h}$ so that the diagram is commutative.

$$\begin{array}{ccc} F(a)^* & \xrightarrow{\varphi} & \hat{F}(a)^* \\ \downarrow h & & \downarrow \hat{h} \quad a \in \mathbb{C}^* \\ F(a)^* & \xrightarrow{\varphi} & \hat{F}(a)^* \end{array}$$

Here for $a \neq 0$, $F(a)^* = f^{-1}(a) \cap \mathbb{C}^{*n}$, $\hat{F}(a)^* = \hat{f}^{-1}(a) \cap \mathbb{C}^{*n}$.

Proof Recall that $f(\mathbf{z})$ and $\hat{f}(\mathbf{w})$ are polar weighted under the same weight vector $P = (p_1, \ldots, p_n)$. The $\mathbb{S}^1$ actions are defined by

$$\rho \circ \mathbf{z} = (\rho^{p_1} z_1, \ldots, \rho^{p_n} z_n), \quad \rho \circ \mathbf{w} = (\rho^{p_1} w_1, \ldots, \rho^{p_n} w_n)$$

where $\mathbf{z}$ and $\mathbf{w}$ are respective coordinates of f and $\hat{f}$. Their monodromy maps $h : F^* \to F^*$ and $\hat{h} : \hat{F}^* \to \hat{F}^*$ are defined by the action of $e^{2\pi i/d_p}$. Recall that

$$\hat{f}(\mathbf{w}) = \sum_{j=1}^{s} c_j \mathbf{w}^{\mathbf{n}_j - \mathbf{m}_j}.$$

Suppose we find a diffeomorphism $\varphi : \mathbb{C}^{*n} \to \mathbb{C}^{*n}$ so that $\varphi(\mathbf{z}) = \mathbf{w}$ which satisfies

$$\mathbf{w}^{\mathbf{n}_j - \mathbf{m}_j} = \mathbf{z}^{\mathbf{n}_j} \bar{\mathbf{z}}^{\mathbf{m}_j}, \forall z, \quad j = 1, \ldots, s. \tag{8.28}$$

Assuming this, observe that

$$\hat{f}(\mathbf{w}) = f(\mathbf{z}), \quad \varphi(e^{i\theta} \circ \mathbf{z}) = e^{i\theta} \circ \mathbf{w}$$

and thus φ is $\mathbb{S}^1$-equivariant and therefore commutes with the respective monodromies.

Now we will construct such φ as follows. Express z_j, w_j in polar coordinates as

$$z_j = \rho_j \exp(i\theta_j), \quad w_j = \xi_j \exp(i\eta_j), \quad j = 1, \ldots, n.$$

First put

$$\eta_j = \theta_j, \quad j = 1, \ldots, n.$$

As $\arg \mathbf{z}^{\mathbf{n}_j} \bar{\mathbf{z}}^{\mathbf{m}_j} = \arg \mathbf{w}^{\mathbf{n}_j - \mathbf{m}_j}$,

$$\begin{aligned}
&\mathbf{z}^{\mathbf{n}_j}\bar{\mathbf{z}}^{\mathbf{m}_j} = \mathbf{w}^{\mathbf{n}_j-\mathbf{m}_j} \\
&\iff |\mathbf{z}^{\mathbf{n}_j+\mathbf{m}_j}| = |\mathbf{w}^{\mathbf{n}_j-\mathbf{m}_j}| \\
&\iff (n_{j1}+m_{j1})\log\rho_1 + \cdots + (n_{jn}+m_{jn})\log\rho_n \\
&\qquad = (n_{j1}-m_{j1})\log\xi_1 + \cdots + (n_{jn}-m_{jn})\log\xi_n
\end{aligned}$$

Rewriting these equalities in the matrix expression:

$$(N+M)\begin{pmatrix}\log\rho_1\\ \vdots\\ \log\rho_n\end{pmatrix} = (N-M)\begin{pmatrix}\log\xi_1\\ \vdots\\ \log\xi_n\end{pmatrix}. \tag{8.29}$$

First we assume that $s = n$. Put $(N-M)^{-1}(N+M) = (\lambda_{ij}) \in \mathrm{GL}(n,\mathbb{Q})$. Now we define φ as follows.

$$\begin{aligned}
&\varphi : \mathbb{C}^{*n} \to \mathbb{C}^{*n}, \quad \mathbf{z} \mapsto \mathbf{w} \\
&\mathbf{z} = (\rho_1\exp(i\theta_1),\ldots,\rho_n\exp(i\theta_n)) \\
&\mathbf{w} = (\xi_1\exp(i\theta_1),\ldots,\xi_n\exp(i\theta_n)) \\
&\xi_j = \exp\Big(\sum_{i=1}^{n}\lambda_{ji}\log\rho_i\Big), \quad j = 1,\ldots,n.
\end{aligned}$$

It is clear that $\varphi : \mathbb{C}^{*n} \to \mathbb{C}^{*n}$ is a diffeomorphism.

Suppose $s < n$. We simply take a solution of (8.29) as a linear system of equations in $\log\xi_1,\ldots,\log\xi_n$.

Now the toric Milnor fibrations

$$f : \mathbb{C}^{*n}\setminus f^{-1}(0) \to \mathbb{C}^*, \quad \hat{f} : \mathbb{C}^{*n}\setminus\hat{f}^{-1}(0) \to \mathbb{C}^*$$

are defined using $\mathbb{S}^1$ actions under the same weight vector P and their monodromy maps h^* and $\hat{h}^*$ are defined using $\mathbb{S}^1$ action:

$$\begin{aligned}
\exp i\theta \circ \mathbf{z} &= (\exp(ip_1\theta)z_1,\ldots,\exp(ip_n\theta)z_n) \\
\exp i\theta \circ \mathbf{w} &= (\exp(ip_1\theta)w_1,\ldots,\exp(ip_n\theta)w_n)
\end{aligned}$$

as follows.

$$\begin{aligned}
h^* &: F^* \to F^*, \quad \mathbf{z} \mapsto \exp(2\pi i/d_P)\circ\mathbf{z} \\
\hat{h}^* &: \hat{F}^* \to \hat{F}^*, \quad \mathbf{w} \mapsto \exp(2\pi i/d_P)\circ\mathbf{w}
\end{aligned}$$

Thus it gives a commutative diagram.

$$\begin{array}{ccc} F(\alpha)^* & \xrightarrow{h^*} & F(\alpha)^* \\ \downarrow \varphi & & \downarrow \varphi \\ \hat{F}(\alpha)^* & \xrightarrow{\hat{h}^*} & \hat{F}(\alpha)^* \end{array}, \quad \forall \alpha \in \mathbb{C}^*.$$

□

For the topological information of $\hat{F}^*$, the following Lemmas are useful.

Lemma 8.3.40 (Kouchnirenko [21], Oka [28]) *Let $h(z_1, \ldots, z_n)$ be a non-degenerate Laurent polynomial and put $Z^* := \{\mathbf{z} \in \mathbb{C}^{*n} \mid h(\mathbf{z}) = 0\}$. Then the Euler characteristic of Z^* is given as*

$$\chi(F^*) = (-1)^{n-1} n! \mathrm{Vol}_n \, \Delta(h).$$

Assume $h(\mathbf{z}) = \sum_\nu c_\nu \mathbf{z}^\nu$. Recall that the Newton polyhedron of h, $\Delta(h)$ is the convex hull of $\{\nu \mid c_\nu \neq 0\}$.

Lemma 8.3.41 (Corollary (1.1.2) [28], Corollary (4.6.1) [29]) *Let*

$$W^* := \{\mathbf{z} \in \mathbb{C}^{*n} \mid h_1(\mathbf{z}) = \cdots = h_k(\mathbf{z}) = 0\}$$

be a full non-degenerate complete intersection variety. Then the inclusion map $\iota : W^* \to \mathbb{C}^{*n}$ is an $(n-k)$-equivalence.

Here W^* is called to be *full* if dim $\Delta(h_i) = n$ for any $i = 1, \ldots, k$. In the case $k = 1$, it reduces to

Corollary 8.3.42 *Let $\hat{F}^*$ be the toric Milnor fiber of $\hat{f}$ and assume that f is full simplicial. Then $\iota : \hat{F}^* \to \mathbb{C}^{*n}$ is an $(n-1)$-equivalence. In particular, if $n \geq 3$, $\pi_1(\hat{F}^*)$ is isomorphic to the free abelian group $\mathbb{Z}^n$.*

Corollary 8.3.43 *Let $\hat{F}^*$ be the toric Milnor fiber of $\hat{f}$ and assume that f is simplicial and $1 < s < n$. Then $\iota : \hat{F}^* \to \mathbb{C}^{*n}$ is $(s-1)$-equivalence. In particular, if $s \geq 3$, $\pi_1(\hat{F}^*)$ is isomorphic to the free abelian group $\mathbb{Z}^n$.*

Proof Assume that $\hat{f}(\mathbf{z}, \bar{\mathbf{z}}) = \sum_{j=1}^s \mathbf{z}^{\mathbf{n}_j - \mathbf{m}_j}$. Take a suitable unimodular matrix Ξ[1] so that $g(\mathbf{u}) := \pi_\Xi^* \hat{f}(\mathbf{u}) = \sum_{i=1}^s \mathbf{v}^{\xi_i}$ where $\mathbf{v} = (u_1, \ldots, u_s)$ and $\xi_1, \ldots, \xi_s \in {\mathbb{Z}_+}^s$. Put $F_g^* := \{\mathbf{v} = (u_1, \ldots, u_s) \in \mathbb{C}^{*s} \mid g(\mathbf{u}) = 1\}$. $\pi_\Xi^{-1}(\hat{F}^*) = F_g^* \times \mathbb{C}^{*(n-s)}$ where $\hat{F} = \hat{f}^{-1}(1)$. The inclusion $F_g^* \to \mathbb{C}^{*s}$ is a $(s-1)$-equivalence by Corollary 8.3.42. Thus the assertion follows immediately.

Remark 8.3.44 1. The numbers λ_{ij} need not to be positive and the diffeomorphism φ in Theorem 8.3.39 is only defined on $\mathbb{C}^{*n}$ in general.

[1] Consider the sublattice $A = (< \mathbf{n}_1 - \mathbf{m}_1, \ldots, \mathbf{n}_s - \mathbf{m}_s > \otimes \mathbf{Q}) \cap \mathbb{Z}^n$ and take a $\mathbb{Z}$-base $\{\mathbf{b}_1, \ldots, \mathbf{b}_s\}$ and extend it to a lattice base $B = \{\mathbf{b}_1, \ldots, \mathbf{b}_s, \ldots, \mathbf{b}_n\}$ and consider B as a unimodular matrix. Take $\Xi = B^{-1}$.

2. Theorem 8.3.39 is not true for non-simplicial mixed polynomials. As an example, consider a mixed homogeneous polynomial g: $g(\mathbf{z},\bar{\mathbf{z}}) = \sum_{j=1}^{n} z_j^{d+q}\bar{z}_j^q + s\mathbf{z}^\nu \prod_{j=1}^{q} z_j\bar{z}_j$ with $|\nu| = d$ and the associated homogeneous polynomial $\hat{g}(w) = \sum_{j=1}^{n} w_j^d + s\mathbf{w}^\nu$. For a generic s, g and $\hat{g}$ have isolated singularities at the origin. By Orlik-Milnor [26], $\hat{g}$ is equivalent to $\sum_{j=1}^{n} w_j^d$ but g and $\sum_{j=1}^{n} z_j^{d+q}\bar{z}_j^q$ are not generally equivalent for a big s. Thus g and $\hat{g}$ are not equivalent in general. See also Sect. 8.3.11, Example 2.

3. Ruas-Seade-Verjovsky studied a mixed Brieskorn polynomial $f(\mathbf{z},\bar{\mathbf{z}}) = z_1^{a_1+b_1}\bar{z}_1^{b_1} + \cdots + z_n^{a_n+b_n}\bar{z}_n^{b_n}$ in [50] In this case, $\varphi : \mathbb{C}^{*n} \to \mathbb{C}^{*n}$ is explicitly given as follows.

$$w_j^{a_j} = z_j^{a_j+b_j}\bar{z}_j^{b_j}, \quad j = 1, \ldots, n$$

which is equivalent to $w_j = z_j|z_j|^{2b_j/a_j}$ for $j = 1, \ldots, n$. This φ can be extended continuously to $\mathbb{C}^n$ but the differentiability fails on the coordinate subspaces. By normalizing φ, it gives a topological isomorphism of the Milnor fibrations of $B_{\mathbf{a},\mathbf{b}}$ and $\hat{B}_{\mathbf{a},\mathbf{b}}$.

Two links are isotopic. In fact, the following result is given in [17, 32].

Lemma 8.3.45 ([17, 32]) *The links of mixed hypersurfaces $B^{-1}(0)$, $f_I^{-1}(0)$, $f_{II}^{-1}(0)$ on the sphere $\mathbb{S}_r^{2n-1}$ are isotopic to the corresponding links of the hypersurfaces defined by the associated polynomial $\hat{B}(\mathbf{w})$, $\hat{f}_I(\mathbf{w})$, $\hat{f}_{II}(\mathbf{w})$.*

Proof Consider the family of mixed polynomials:

$$\begin{aligned} B_t(\mathbf{z},\bar{\mathbf{z}}) &= (1-t)B(\mathbf{z},\bar{\mathbf{z}}) + t\hat{B}(\mathbf{z}) \\ f_{I,t}(\mathbf{z},\bar{\mathbf{z}}) &= (1-t)f_I(\mathbf{z},\bar{\mathbf{z}}) + t\hat{f}_I(\mathbf{z}) \\ f_{II,t}(\mathbf{z},\bar{\mathbf{z}}) &= (1-t)f_{II}(\mathbf{z},\bar{\mathbf{z}}) + t\hat{f}_{II}(\mathbf{z}). \end{aligned}$$

Put $F_t(\mathbf{z},\bar{\mathbf{z}})$, $0 \le t \le 1$ be any one of the above family and put $V_t = \{\mathbf{z} \in \mathbb{C}^n \mid F_t(\mathbf{z},\bar{\mathbf{z}}) = 0\}$. Then the proof is reduced to the following Lemma and the Ehresmann theorem.

Lemma 8.3.46 *For any $0 \le t \le 1$, V_t is non-singular except at the origin and the intersection of $\mathbb{S}_r^{2n-1}$ and V_t is transversal for any $r > 0$.*

Proof We will give a proof for the mixed Brieskorn polynomial, following the proof of Lemma 1 and Lemma 2 of [32].

$$\begin{aligned} f(\mathbf{z},\bar{\mathbf{z}}) &= z_1^{a_1+b_1}\bar{z}_1^{\,b_1} + \cdots + z_n^{a_n+b_n}\bar{z}_n^{\,b_n}, \\ \hat{f}(\mathbf{z},\bar{\mathbf{z}}) &= z_1^{a_1} + \cdots + z_n^{a_n}, \\ f_t(\mathbf{z},\bar{\mathbf{z}}) &= (1-t)(z_1^{a_1+b_1}\bar{z}_1^{\,b_1} + \cdots + z_n^{a_n+b_n}\bar{z}_n^{\,b_n}) + t(z_1^{a_1} + \cdots + z_n^{a_n}) \\ &= \sum_{j=1}^{n} z_j^{a_j}(t + (1-t)|z_j|^{2b_j}). \end{aligned}$$

Consider the family of mixed hypersurfaces $V_t = f_t^{-1}(0)$, $0 \le t \le 1$.
First we will show that $V_t \setminus \{\mathbf{0}\}$ *is non-singular*. For $t = 0$ and $t = 1$, the assertion is obvious. So we assume that $0 < t < 1$. Suppose $0 \neq \mathbf{w} \in \mathbb{C}^n$ is a mixed singular point. Then by Proposition 8.2.2, there exists $\lambda \in \mathbb{S}^1$ such that

$$\overline{\partial f_t}(\mathbf{w}, \bar{\mathbf{w}}) = \lambda\, \bar{\partial} f_t(\mathbf{w}, \bar{\mathbf{w}}).$$

This is equivalent to

$$(a_j + b_j)\bar{w_j}^{a_j+b_j-1} w_j^{b_j}(1-t) + a_j \bar{w_j}^{a_j-1} t = b_j w_j^{a_j+b_j} \bar{w_j}^{b_j-1}(1-t)\lambda, \ \forall j.$$

Multiplying $\bar{w_j}$ to both sides and we get

$$\bar{w_j}^{a_j}\left\{(a_j+b_j)|w_j|^{2b_j}(1-t) + a_j t\right\} = b_j w_j^{a_j} |w_j|^{2b_j}(1-t)\lambda. \qquad (8.30)$$

Putting the left side and the right side of (8.30) L_j and R_j respectively, we get

$$|L_j| \ge |w_j|^{a_j+2b_j}(a_j+b_j)(1-t) \ge |w_j|^{a_j+2b_j} b_j(1-t) = |R_j|$$

and the equality holds only if $w_j = 0$. However as $\mathbf{w} \neq 0$, there exists j with $w_j \neq 0$ and the equality (8.30) does not hold for this j which gives a contradiction to (8.30). Thus V_t has a unique singularity at the origin.

Now we show the transversality of V_t and the sphere $\mathbb{S}^{2n-1}_{\|\mathbf{w}\|}$ at $\mathbf{w}$. For this assertion, it is enough to show the existence of a vector $\mathbf{v} \in T_{\mathbf{w}} V_t$ which is transversal to the sphere. Take an arbitrary point $\mathbf{w} \in V_t \cap \mathbb{S}^{2n-1}_{r_0}$. Recall that

$$f_t(\mathbf{z}, \bar{\mathbf{z}}) = \sum_{i=1}^n \psi_j(z_j, \bar{z_j}), \quad \psi_j(z_j, \bar{z_j}) := z_j^{a_j}\left(t + (1-t)|z_j|^{2b_j}\right).$$

We will construct real valued functions $\varphi_j(r)$, $j = 1, \dots, n$ such that

$$\varphi_j(1) = 1, \ \ \psi_j(\varphi_j(r) w_j, \varphi_j(r)\bar{w_j}) = r\psi_j(w_j, \bar{w_j}), \quad j = 1, \dots, n$$

and define an analytic curve into V_t by

$$\varphi : [1-\epsilon, 1+\epsilon] \to V_t, \quad \varphi(r) = (\varphi_1(r) w_1, \dots, \varphi_n(r) w_n).$$

As $f_t(\varphi(r), \overline{\varphi(r)}) = r\, f_t(\mathbf{w}, \overline{\mathbf{w}}) = 0$, the image of φ is in fact a real analytic curve in V_t, starting from $\mathbf{w}$ at $r = 1$. Define $\mathbf{v} := \frac{d\varphi}{dr}|_{r=1}$. As $\mathbf{v} \in T_{\mathbf{w}} V_t$, it is only necessary to show $\mathbf{v} \notin T_{\mathbf{w}} \mathbb{S}^{2n-1}_{r_0}$ which implies the transversality of V_t and $\mathbb{S}^{2n-1}_{r_0}$ at $\mathbf{w}$.
1. Construction of φ_j: If $w_j = 0$, put $\varphi_j \equiv 1$. Suppose $w_j \neq 0$. Note that $\psi_j(\varphi_j(r) w_j, \varphi_j(r)\bar{w_j}) = r\psi_j(w_j, \bar{w_j})$ is equivalent to

$$\varphi_j(r)^{a_j}(t + (1-t)|w_j|^{2b_j}\varphi_j(r)^{2b_j}) = r(t + (1-t)|w_j|^{2b_j}). \qquad (8.31)$$

Put $\hat{\psi}_j(s) := s^{a_j}(t+(1-t)|w_j|^{2b_j}s^{2b_j})$. The left side of (8.31) is equal to $\hat{\psi}_j(\varphi_j(r))$. As $\hat{\psi}_j(s)$ is a strictly monotone increasing in s, using the Implicit Function theorem, we solve

$$\hat{\psi}_j(s) - r\,|w_j|^{a_j}(t+(1-t)|w_j|^{2b_j}) = 0$$

in s with the initial condition $\varphi(1) = 1$ to obtain $\varphi_j(r)$.
2. $\mathbf{v} := (v_1, \ldots, v_n)$ is not zero: If $w_j \neq 0$, $v_j = \frac{d\varphi_j}{dr}(1)w_j$. Differentiating the equality

$$\hat{\psi}_j(\varphi_j(r)) - r\,|w_j|^{a_j}(t+(1-t)|w_j|^{2b_j}) \equiv 0$$

in r and putting $r = 1$, we get

$$\frac{d\hat{\psi}_j}{ds}(1)\frac{d\varphi_j}{dr}(1) = |w_j|^{a_j}(t+(1-t)|w_j|^{2b_j})$$

which implies $v_j \neq 0$.
3. $\mathbf{v} \notin T_{\mathbf{w}}\mathbb{S}^{2n-1}_{r_0}$: Consider the square of the norm function $\rho(\mathbf{z}) := \sum_{i=1}^n \|z_i\|^2$. As

$$\begin{aligned}\frac{d\rho(\varphi(r))}{dr}|_{r=1} &= \frac{d\sum_{j=1}^n |\varphi_j(r)^2|w_j|^2}{dr}|_{r=1}\\ &= 2\sum_{j=1}^n \frac{d\varphi_j}{dr}(1)\varphi_j(1)|w_j|^2 > 0\end{aligned}$$

$\mathbf{v}$ is not tangent to the sphere.

The proof for the cases F_I, F_{II} is bit more complicated and we refer to [32] for the tree case and to [17] for the cyclic case. □

8.3.13 The Join Theorem

Assume that $f_1(\mathbf{z}, \bar{\mathbf{z}})$ and $f_2(\mathbf{w}, \bar{\mathbf{w}})$ are mixed weighted homogeneous polynomials of variables $\mathbf{z} = (z_1, \ldots, z_n)$ and $\mathbf{w} = (w_1, \ldots, w_m)$ respectively. Let $\hat{P}_1 = {}^t(p_1, \ldots, p_n)$ and $\hat{Q}_1 = {}^t(q_1, \ldots, q_n)$ be the normalized polar and radial weight vectors of f_1 and let $\hat{P}_2 = (r_1, \ldots, r_m)$, $\hat{Q}_2 = {}^t(s_1, \ldots, s_m)$ be the normalized polar and radial weight vectors of f_2. Then the Join Theorem of holomorphic functions [27] can be generalized as follows.

Lemma 8.3.47 (Cisneros-Molina [6]) *Let f_1, f_2 be as above and put $\mathbf{u} = (\mathbf{z}, \mathbf{w}) \in \mathbb{C}^{n+m}$. Consider the mixed polynomial $f(\mathbf{u}, \bar{\mathbf{u}}) := f_1(\mathbf{z}, \bar{\mathbf{z}}) + f_2(\mathbf{w}, \bar{\mathbf{w}})$. Then we have*

1. *$f(\mathbf{u})$ is a mixed weighted homogeneous polynomial of normalized polar weight vector $\hat{P} = {}^t(p_1, \dots, p_n, r_1, \dots, r_m)$ and the normalized radial weight vector $\hat{Q} = {}^t(q_1, \dots, q_n, s_1, \dots, s_m)$.*
2. *Let F_1, F_2 and F be the respective Milnor fibers of the global Milnor fibrations of f_1, f_2 and f. Put h_1, h_2 and h be the respective monodromy mappings. Then there exists a natural homotopy equivalence $\iota : F \to F_1 * F_2$ and the following diagram is commutative.*

$$\begin{array}{ccc} F & \xrightarrow{h} & F \\ \downarrow \iota & & \downarrow \iota \\ F_1 * F_2 & \xrightarrow{h_1 * h_2} & F_1 * F_2 \end{array}$$

Here $F_1 * F_2$ is the join product of F_1 and F_2. For the definition of $F_1 * F_2$ and the basic properties, refer Milnor [25]. The proof is exactly the same as that of the Join Theorem for holomorphic functions [21, 27].

8.3.14 Topology of the Milnor Fiber

For the holomorphic function $f(\mathbf{z})$, the Milnor fiber F has a homotopy type of $(n-1)$-dimensional CW-complex. Moreover if $\mathbf{0}$ is an isolated singularity, F is homotopic to a bouquet of $(n-1)$-spheres by Milnor [24]. For the proof of the both assertions, the Morse function method and the complex structure of F play a key role.

For a mixed function, there does not exist any systematical result on either upper bound of CW-complex dimension or connectivity of the fiber assuming isolatedness of the singularity. The following are some results for special cases.

(1) Assume that $f(\mathbf{z}, \bar{\mathbf{z}})$ has a tubular Milnor fibration and assume that the locus of non-singular points of $V = f^{-1}(0)$ is at least one-dimensional. Then F is connected (Theorem 2.3, [43]).

(2) Let f be a mixed function of two variables with an isolated singularity at the origin and we assume that it has a tubular Milnor fibration. Then F is a connected open Riemann surface. Thus the homotopy dimension is 1.

(3) Assume that $f(\mathbf{z}, \bar{\mathbf{z}})$ is a full simplicial mixed polynomial (See Sect. 8.3.12). Let F^* be the toric Milnor fiber. It is diffeomorphic to the toric Milnor fiber of Laurent polynomial by Theorem 8.3.39 of Sect. 8.3.12. The inclusion map $\iota : \hat{F}^* \to \mathbb{C}^{*n}$ is $(n-1)$-equivalence ([28], Corollary (4.6.1) [29]). So $F^* \to \mathbb{C}^{*n}$ is also $(n-1)$-equivalence. If $s < n$, by change of toric coordinates, $\hat{f}$ can be a simplicial Laurent polynomial g of s variables and thus $\hat{F}_f^* \cong \mathbb{C}^{*(n-s)} \times F_g^*$ (Corollary 8.3.43) and $\chi(F_f^*) = 0$. In particular, if $s = n$ and $\Gamma(f)$ is 1-convenient, i.e. $F \cap \{z_i = 0\} \neq \emptyset$ for $i = 1, \dots, n$, F is simply connected. The Euler characteristic can be computed by the additive formula of Euler characteristics as the restriction f^I is also simplicial for any I. Thus using the decomposition $F = \amalg_I F^{*I}$ (here $F^{*I} = \mathbb{C}^{*I} \cap \{f^I = 1\}$, one can show that F has a homotopy type of $(n-1)$-dimensional CW-complex.

For the calculation of $\chi(F^*)$, use Lemma 8.3.40 in Sect. 8.3.12. Let us see some examples in the case of $n = 3$. Note that In this case, F is homotopic to a bouquet of 2-spheres and $\chi(F) = \mu(F) + 1$.

Here are some examples.

(i) $f = z_1^{a_1+b_1}\bar{z}_1^{b_1} + z_2^{a_2+b_2}\bar{z}_2^{b_2} + z_3^{a_3+b_3}\bar{z}_3^{b_3}$. This is equivalent to $z_1^{a_1} + z_2^{a_2} + z_3^{a_3}$. Thus its Milnor number is $(a_1 - 1)(a_2 - 1)(a_3 - 1)$.

(ii) $f = z_1^{a_1+b_1}\bar{z}_1^{b_1} z_2 + z_2^{a_2+b_2}\bar{z}_2^{b_2} z_3 + z_3^{a_3+b_3}\bar{z}_3^{b_3}$. Then f has the same decomposition as $z_1^{a_1} z_2 + z_2^{a_2} z_3 + z_3^{a_3}$ and $F = F^* \amalg F^{*\{2,3\}} \amalg F^{*\{3\}}$ and $\chi(F) = a_1a_2a_3 - a_2a_3 + a_3 = \mu + 1$.

(iii) $f = z_1^{a_1}\bar{z}_2 + z_2^{a_2}\bar{z}_3 + z_3^{a_3}$. Then $\chi(F) = a_1a_2a_3 - a_2a_3 + a_3$.

(iv) $f = z_1^{a_1+b_1}\bar{z}_1^{b_1} z_2 + z_2^{a_2+b_2}\bar{z}_2^{b_2} z_3 + z_3^{a_3+b_3}\bar{z}_3^{b_3} z_1$. Then it has the same decomposition as $z_1^{a_1} z_2 + z_2^{a_2} z_3 + z_3^{a_3} z_1$ and $F = F^* \amalg F^{*\{2,3\}} \amalg F^{*\{3,1\}} \amalg F^{*\{1,2\}}$ and $\chi(F) = a_1a_2a_3 + 1$. Note that $\chi(F^{*\{i,j\}}) = 0$ for $i \neq j$.

(v) $f = z_1^{a_1}\bar{z}_2 + z_2^{a_2}\bar{z}_3 + z_3^{a_3}\bar{z}_1$. Then $\chi(F) = a_1a_2a_3 - 1$.

(4) Let $f = f_1 + \cdots + f_k$ where f_j is a mixed weighted homogenous polynomial of one or two variables. Assume that the variables of f_j and f_i are all different for $i \neq j$ so that the variables of f are the disjoint sum of variables of f_j for $j = 1, \ldots, k$. We assume that the critical points of f_j are isolated. Let F_j be the Milnor fiber of f_j and let F be the Milnor fiber of f. Then by Join Theorem (Lemma 8.3.47), F is homotopic to the join of $F_1, \ldots, F_k$. Thus F is $(n-2)$-connected where n is the number of variables. Using the recent result of Inaba [16], the assertion holds without assuming that f_j, $j = 1, \ldots, k$ are mixed weighted homogeneous.

8.3.15 The Milnor Fibration for $f\bar{g}$

In this section, we consider a mixed function H which takes the form $H(\mathbf{z}, \bar{\mathbf{z}}) = f(\mathbf{z})\bar{g}(\mathbf{z})$ where f, g are holomorphic functions. Here we mean $\bar{g}(z) := \overline{g(z)}$. We consider hypersurfaces $V(f) := f^{-1}(0)$, $V(g) := g^{-1}(0)$, $V(H) := H^{-1}(0)$ and the intersection variety $V(f, g) = V(f) \cap V(g)$. As the points of the intersection $V(f) \cap V(g)$ are singular points of $V(H)$, $f\bar{g}$ can not be strongly non-degenerate for $n \geq 3$ by Lemma 8.3.10. However $H(\mathbf{z}, \bar{\mathbf{z}})$ is a very special type of mixed function, as it is defined by two holomorphic functions f, g. We consider the existence of the Milnor fibration for such a mixed function. Pichon and Seade have studied such functions, especially for the case $n = 2$ ([46–48]). There are also related works by Fernandez de Bobadilla and Menegon Neto [12] and also [2, 3, 18, 45] in a general setting. This section follows completely from [41, 42].

Basic Assumption

Recall that a real analytic mapping $\varphi : \mathbb{R}^{2n} \to \mathbb{R}^2$ has a 1-dimensional critical values in general. Thus without any assumption on f, g, there are many examples which do not have a Milnor fibration. So we need some assumptions which are easy to check and under which H can not have a non-isolated critical value locus. More precisely we look for the conditions so that the conditions (SN) and (ST) of Sect. 8.2.3 will be satisfied. For simplicity, we assume that

(I1) f and g are holomorphic functions such that $V(f)$ and $V(g)$ have isolated singularities at the origin.
(I2) The intersection variety $V(f, g) := \{f = g = 0\}$ is a complete intersection variety with an isolated singularity at the origin.

We fix a positive number $r_0 > 0$ so that $V(f), V(g), V(f, g)$ are only singular at the origin in the ball $\mathbb{B}^{2n}_{r_0}$, f, g has only $\mathbf{0}$ as a critical value on $\mathbb{B}^{2n}_{r_0}$ and for any r with $0 < r \le r_0$, the sphere $\mathbb{S}^{2n-1}_r$ intersects transversely with these varieties. We consider the canonical stratification of $V(H) = V(f) \cup V(g)$, $\mathcal{S} := \{V(f)^*, V(g)^*, V(f, g)^*, \{\mathbf{0}\}\}$ where $V(f)^* = V(f) \setminus V(f, g)$ and $V(f, g)^* = V(f, g) \setminus \{\mathbf{0}\}$.

The Multiplicity Condition

We wish to have a condition to get the isolatedness of the critical values for $H = f\bar{g}$. We say that H satisfies *the multiplicity-condition* if there exists a good resolution $\pi : X \to \mathbb{C}^n$ of the holomorphic function $h = fg$ such that

(i) $\pi : X \setminus \pi^{-1}(\mathbf{0}) \to \mathbb{C}^n \setminus \{\mathbf{0}\}$ is biholomorphic and the divisor defined by $\pi^*(fg) = 0$ has only normal crossing singularities and the respective strict transforms $\tilde{V}(f)$, $\tilde{V}(g)$ of $V(f)$ and $V(g)$ are smooth.
(ii) Put $\pi^{-1}(\mathbf{0}) = \cup_{j=1}^s D_j$ where $D_1, \ldots, D_s$ are smooth compact divisors in X. Denote the respective multiplicities of $\pi^* f$ and $\pi^* g$ along D_j by m_j and n_j. Then $m_j \ne n_j$ for $j = 1, \ldots, s$.

Fernandez de Bobadilla and Menegon Neto have considered the multiplicity condition for the case of plane curves [12].

Lemma 8.3.48 (Isolatedness of critical values, [41]) *Under the assumption (I1), (I2) and the multiplicity-condition, there exists a positive number r_1 so that H restricted on $\mathbb{B}^{2n}_{r_1}$ has the only critical value* $\{0\}$.

Proof We denote $\pi^{-1}(V(fg)) = D_1 \cup \cdots \cup D_{s+2}$ where $D_{s+1} = \tilde{V}(f)$ and $D_{s+2} = \tilde{V}(g)$. Put $D = \pi^{-1}(\mathbf{0})$. For simplicity, we put $D_{s+1} = \widetilde{V}(f)$ and $D_{s+2} = \widetilde{V}(g)$. In this notation, we put $m_{s+1} = 1, m_{s+2} = 0$ and $n_{s+1} = 0, n_{s+2} = 1$. Take an arbitrary point $p \in D$ and assume that $p \in \bigcap_{j\in J} D_j \setminus \bigcup_{j\notin J} D_j$ where $J \subset \{1, \ldots, s+2\}$. By the assumption (i), $|J| \le n$. Then there is a local holomorphic chart U_d with coordinates $(u_1, \ldots, u_n)$ and an injective map $\tau : J \to \{1, \ldots, n\}$ so that $u_{\tau(j)} = 0$ defines D_j in U_d and by the multiplicity assumption (i) and (ii), we can write

$$\pi^* f = k_f \prod_{j\in J} u_{\tau(j)}^{m_j}, \quad \pi^* g = k_g \prod_{j\in J} u_{\tau(j)}^{n_j}. \tag{8.32}$$

where k_f, k_g are units on U_d. We choose U_d small enough so that $U_d \cap \bigcup_{j \notin J} D_j = \emptyset$. Consider the pull-back $\tilde{H} := \pi^* H$. By the assumption, we can write $\tilde{H}$ in U_d as

$$\tilde{H} = k_f \bar{k}_g \prod_{j \in J} u_{\tau(j)}^{m_j} \bar{u}_{\tau(j)}^{n_j}.$$

Note that $J \cap \{1, \dots, s\} \neq \emptyset$ as $p \in D$. By Proposition 8.2.2, if $\mathbf{u} \in U_d$ is a critical point of $\tilde{H}$, we must have $|\frac{\partial \tilde{H}}{\partial u_j}|/|\frac{\partial \tilde{H}}{\partial \bar{u}_j}| = 1$. However the above expression says that

$$\left| \frac{\partial \tilde{H}}{\partial u_j} / \frac{\partial \tilde{H}}{\partial \bar{u}_j} \right| = \left| \left(m_j + u_{\tau(j)} \frac{\partial k_f}{\partial u_{\tau(j)}} \bar{k}_g \right) / \left(n_j + \bar{u}_{\tau(j)} \frac{\partial \bar{k}_g}{\partial \bar{u}_{\tau(j)}} k_f \right) \right|$$
$$\to \frac{m_j}{n_j} \neq 1, \ \mathbf{u} \to p.$$

Put

$$H_{\tau(j)} := \frac{\partial \tilde{H}}{\partial u_j}, \quad H'_{\tau(j)} := \frac{\partial \tilde{H}}{\partial \bar{u}_j}.$$

Thus we can take a smaller neighborhood U'_d if necessary and we may assume that $|H_{\tau(j)}| \neq |H'_{\tau(j)}|$ for any $\mathbf{u} \in U'_d \setminus D \cup D_{s+1} \cup D_{s+2}$. We do this operation for any $p \in D$. As D is compact, we find finite points $p_1, \dots, p_\mu$ such that $\cup_{i=1}^{\mu} U'_{p_i} \supset D$. Put $W = \cup_{i=1}^{\mu} U'_{p_i}$. W is an open set containing D so that $\tilde{H} : W \setminus (D \cup D_{s+1} \cup D_{s+2}) \to \mathbb{C}^*$ has no critical point. Put $W' = \pi(W)$. W' is an open neighborhood of the origin in $\mathbb{C}^n$ and $\pi : W \setminus \pi^{-1}(V(H)) \to W' \setminus V(H)$ is biholomorphic. This implies that $H : W' \setminus H^{-1}(0) \to \mathbb{C}^*$ has no critical point. This proves the assertion.

In general, the tangent space of a mixed hypersurface does not have a complex structure. However in our case, we have the following assertion which is also observed in the proof of Theorem 3.1, [45].

Proposition 8.3.49 (Proposition 15, [41]) *Let $H = f\bar{g}$ be as in the basic assumption and assume that $V_\eta := H^{-1}(\eta) \cap \mathbb{B}_r^{2n}$ is mixed non-singular. For any point $\mathbf{p} \in V_\eta$, $T_{\mathbf{p}} V_\eta$ contains a complex subspace $T_{\mathbf{p}} f^{-1}(f(\mathbf{p})) \cap T_{\mathbf{p}} g^{-1}(g(\mathbf{p}))$.*

The above observation implies the following assertion.

Lemma 8.3.50 (Theorem 3.1, [45], Lemma 5, [41]) *Let $\mathcal{S}$ be the stratification of $V(H)$ and let $r_1 \leq r_0$ be the positive number as in Lemma 8.3.48. We assume also that each strata in $\mathcal{S}$ is smooth in $\mathbb{B}_{r_1}^{2n}$. Then H satisfies Thom's a_f-condition.*

By Lemmas 8.3.48 and 8.3.50 and 8.2.7, we have the following.

Corollary 8.3.51 *Assume (I1), (I2) and the multiplicity condition. Then f has a tubular Milnor fibration.*

The Spherical Milnor Fibration for $f\bar{g}$

We now consider the spherical Milnor mapping $\varphi : \mathbb{S}_r^{2n-1} \setminus K \to \mathbb{S}^1$ defined by $\varphi(\mathbf{z}) := H(\mathbf{z})/|H(\mathbf{z})|$ where $K = V(H) \cap \mathbb{S}_r^{2n-1}$. For the existence of a spherical fibration, we need a stronger assumption than the basic assumption. We assume in this subsection that

(n1) $f(\mathbf{z})$ and $g(\mathbf{z})$ are convenient non-degenerate holomorphic functions in the neighborhood of the origin with respect to the Newton boundaries.
(n2) $V(f, g) = \{\mathbf{z} \in \mathbb{C}^n \mid f(\mathbf{z}) = g(\mathbf{z}) = 0\}$ is a non-degenerate complete intersection variety[2] in the sense of Newton boundary [29].

We call (n1) and (n2) *the Newton non-degeneracy condition.*

The hypersurfaces $V(f)$ and $V(g)$ have isolated singularities at the origin by the convenience and non-degeneracy assumption (n1). The intersection variety $V(f, g)$ also has an isolated singularity at the origin and the intersections of $V(f)$, $V(g)$ are transverse outside of the origin by (n2). See Lemma (2.2) [29].

We consider a little stronger assumption than the multiplicity condition. We say that H satisfies *the Newton multiplicity condition* if for any strictly positive weight vector P, the weighted degrees of f and g under P are not equal, i.e. $d(P, f) \neq d(P, g)$.

The Newton multiplicity condition can be checked by the Newton boundaries $\Gamma(f)$ and $\Gamma(g)$ as follows.

Proposition 8.3.52 (Proposition 9, [41]) *Assume that f, g have convenient Newton boundaries. Then H satisfies the Newton multiplicity condition if and only if $\Gamma(f) \cap \Gamma(g) = \emptyset$.*

Taking an admissible toric modification $\hat{\pi} : X \to \mathbb{C}^n$ for the dual Newton diagram $\Gamma^*(fg)$, as a good resolution, the following is obvious.

Proposition 8.3.53 *Assume that the Newton non-degeneracy condition (n1), (n2) and the Newton multiplicity condition. Then (I1), (I2) and the multiplicity condition are satisfied with the toric modification $\hat{\pi} : X \to \mathbb{C}^n$.*

Lemma 8.3.54 (Lemma 11, [41]) *We assume (n1), (n2) and the Newton multiplicity condition. There exists a positive number r_3 so that $\varphi : \mathbb{S}_r^{2n-1} \setminus K \to \mathbb{S}^1$ has no critical points for any r, $0 < r \leq r_3$.*

Combining with the transversality (T), we get the following.

Corollary 8.3.55 (Spherical Milnor fibration) *Assuming (n1), (n2) and the Newton multiplicity condition, $\varphi : \mathbb{S}_r^{2n-1} \setminus K \to \mathbb{S}^1$ gives a local trivial fibration for any $r \leq \min\{r_3, r_2\}$ where r_2 is a positive number in Lemma 8.3.21. Here $K = H^{-1}(0) \cap \mathbb{S}_r^{2n-1}$.*

[2] This means that for any strictly positive weight vector P, $f_P = g_P = 0$ is a non-singular complete intersection variety in $\mathbb{C}^{*n}$.

8.3.16 Mixed Projective Hypersurfaces

Recall that a mixed polynomial $f(\mathbf{z},\bar{\mathbf{z}}) = \sum_{j=1}^{\ell} c_j \mathbf{z}^{\nu_j}\bar{\mathbf{z}}^{\mu_j}$ is *a strongly mixed weighted homogeneous polynomial* if there exists a strictly positive weight vector $P = {}^t(p_1,\dots,p_n)$ so that f is mixed weighted homogeneous under the same weight P with $\mathrm{pdeg}_P\, f = d_p$ and $\mathrm{rdeg}_P = d_r$. Then the associated $\mathbb{R}^+ \times \mathbb{S}^1$ action on $\mathbb{C}^n$ is in fact the $\mathbb{C}^*$ action which is defined by

$$\tau \circ \mathbf{z} = (z_1\tau^{p_1},\dots,z_n\tau^{p_n}),\ \ \tau \in \mathbb{C}^*.$$

In particular, we say $f(\mathbf{z},\bar{\mathbf{z}})$ is *a strongly mixed homogeneous polynomial* if further $P = {}^t(1,\dots,1)$.

Assume that $f(\mathbf{z},\bar{\mathbf{z}})$ is a strongly mixed weighted homogeneous polynomial with radial degree d_r and polar degree d_p respectively and let $P = (p_1,\dots,p_n)$ be the weight vector. Then $f(\mathbf{z},\bar{\mathbf{z}})$ satisfies the equality:

$$f((t,\rho)\circ \mathbf{z}) = t^{d_r}\rho^{d_p} f(\mathbf{z},\bar{\mathbf{z}}),\quad (t,\rho) \in \mathbb{R}^+ \times \mathbb{S}^1. \tag{8.33}$$

Let $\tilde{V}$ be the mixed affine hypersurface

$$\tilde{V} = f^{-1}(0) = \{\mathbf{z} \in \mathbb{C}^n \mid f(\mathbf{z},\bar{\mathbf{z}}) = 0\}.$$

We consider the global fibration $f : \mathbb{C}^n \setminus \tilde{V} \to \mathbb{C}^*$. Then the Milnor fiber F is defined by the hypersurface $f^{-1}(1)$. The monodromy map $h : F \to F$ is defined by

$$h(\mathbf{z}) = \exp(\frac{2\pi i}{d_p}) \circ \mathbf{z} = \left(\exp(\frac{2p_1\pi\, i}{d_p})z_1,\dots,\exp(\frac{2p_n\pi\, i}{d_p})z_n\right).$$

We also consider the weighted projective hypersurface V defined by

$$V = \{(z_1 : z_2 : \cdots : z_n) \in \mathbb{P}(P)^{n-1} \mid f(\mathbf{z},\bar{\mathbf{z}}) = 0\}$$

where $\mathbb{P}(P)^{n-1}$ is the weighted projective space defined by the equivalence induced by the above $\mathbb{C}^*$-action:

$$\mathbf{z} \sim \mathbf{w} \iff \exists \tau \in \mathbb{C}^*,\ \mathbf{w} = \tau \circ \mathbf{z}.$$

It is well-known that $\mathbb{P}(P)^{n-1}$ is an orbifold with at most cyclic quotient singularities. See [10, 52]. By (8.33), the hypersurface $V = \{[\mathbf{z}] \in \mathbb{P}^{n-1}(P) | f(\mathbf{z}) = 0\}$ is well-defined. Consider the quotient map $\pi : \mathbb{C}^n \setminus \{O\} \to \mathbb{P}(P)^{n-1}$ and the restrictions $\pi|_F : F \to \mathbb{P}(P)^{n-1} \setminus V$. This is a branched cyclic covering of order d_p whose branching locus is the union of coordinate hyperplanes $H_j = \{z_j = 0\}$. The zeta function of the monodromy can be computed by the Milnor method Theorem 9.6, [24].

A Canonical Orientation

It is well known that a complex analytic smooth variety has a canonical orientation which comes from the complex structure (see for example p. 18, [13]). Let $\tilde{V} = f^{-1}(0)$ be a mixed hypersurface.

Proposition 8.3.56 *There is a canonical orientation on the smooth part of a mixed hypersurface.*

Proof Take a mixed regular point $\mathbf{a} \in \tilde{V}$. The normal bundle $\mathcal{N}$ of $\tilde{V} \subset \mathbb{C}^n$ has a canonical orientation so that $df_{\mathbf{a}} : \mathcal{N}_{\mathbf{a}} \to T_{f(\mathbf{a},\bar{\mathbf{a}})}\mathbb{C}$ is an orientation preserving isomorphism. This gives a canonical orientation on $\tilde{V}$ so that the ordered union of the oriented frames $\{v_1, \dots, v_{2n-2}, n_1, n_2\}$ of $T_{\mathbf{a}}\mathbb{C}^n$ is the orientation of $\mathbb{C}^n$ if and only if $\{v_1, \dots, v_{2n-2}\}$ is an oriented frame of $T_{\mathbf{a}}\tilde{V}$ where $\{n_1, n_2\}$ is an oriented frame of the normal bundle $\mathcal{N}_{\mathbf{a}}$. □

For brevity, we now concentrate on a strongly mixed homogeneous polynomial $f(\mathbf{z})$. Let $\tilde{V} = f^{-1}(0)$ and let $V = \{f = 0\} \subset \mathbb{P}^{n-1}$ be the corresponding mixed projective hypersurface for simplicity. V also has a canonical orientation.

The Milnor Fiber

Consider the Hopf fibration $\pi : \mathbb{C}^n \setminus \{\mathbf{0}\} \to \mathbb{P}^{n-1}$ and its restriction to the Milnor fiber F. As f is a strongly mixed homogeneous polynomial, it is easy to see that $\pi : F \to \mathbb{P}^{n-1} \setminus V$ is a cyclic covering of order d_p where d_p is the polar degree of f and the group of the covering transformation is generated by the monodromy map

$$h : F \to F, \quad \mathbf{z} \mapsto \exp(\frac{2\pi i}{d_p}) \circ \mathbf{z} = \exp(\frac{2\pi i}{d_p})\mathbf{z}$$

and the action is free. Thus we have

Proposition 8.3.57 *1.* $\chi(F) = d_p \chi(\mathbb{P}^{n-1} \setminus V)$.
2. *The following sequence is exact.*

$$1 \to \pi_1(F) \xrightarrow{\pi_\sharp} \pi_1(\mathbb{P}^{n-1} \setminus V) \to \mathbb{Z}/d_p\mathbb{Z} \to 1.$$

Corollary 8.3.58 *If $d_p = 1$, the projection $\pi : F \to \mathbb{P}^{n-1} \setminus V$ is a diffeomorphism.*

The monodromy map $h : F \to F$ gives a free $\mathbb{Z}/d_p\mathbb{Z}$ action on F. Thus using the periodic monodromy argument Theorem 9.6 in [24], we get

Proposition 8.3.59 *The zeta function of $h : F \to F$ is given by*

$$\zeta(t) = (1 - t^{d_p})^{-\chi(F)/d_p}.$$

In particular, if $d_p = 1$, $h = \mathrm{id}_F$ and $\zeta(t) = (1 - t)^{-\chi(F)}$.

The Degree of A Mixed Projective Hypersurface

Let $\mathcal{M}(q+2r,q,n)$ be the space of non-degenerate, convenient strongly mixed homogeneous polynomials of n variables $\mathbf{z}=(z_1,\dots,z_n)$ with the radial degree $q+2r$ and the polar degree q. By the definition, $f\in\mathcal{M}(q+2r,q,n)$ defines a mixed affine hypersurface in $\mathbb{C}^n$ with an isolated singularity at the origin. Suppose $f(\mathbf{z},\bar{\mathbf{z}})\in\mathcal{M}(q+2r,q,n)$ and let

$$V=\{[\mathbf{z}]\in\mathbb{P}^{n-1}\mid f(\mathbf{z},\bar{\mathbf{z}})=0\}.$$

By the non-degeneracy assumption, V is non-singular. It has a fundamental class $[V]\in H_{2n-4}(V,\mathbb{Z})$. The topological degree of V is the integer d so that $\iota_*[V]=d[\mathbb{P}^{n-2}]$ where $\iota: V\to\mathbb{P}^{n-1}$ is the inclusion map and $[\mathbb{P}^{n-2}]$ is the homology class of a canonical hyperplane $\mathbb{P}^{n-2}\subset\mathbb{P}^{n-1}$.

Theorem 8.3.60 *The topological degree of V is equal to the polar degree q. Namely the fundamental class $[V]$ corresponds to $q[\mathbb{P}^{n-2}]\in H_{2(n-2)}(\mathbb{P}^{n-1})$ by the homomorphism ι_* induced by the inclusion map.*

Proof Suppose that f is a non-degenerate mixed polynomial in $\mathcal{M}(q+2r,q,n)$. Take a generic 1-dimensional complex line L which is isomorphic to $\mathbb{P}^1$. Then the degree is given by the intersection number $[V]\cdot[L]$. Now, changing the coordinates if necessary, we may assume that

$$L: z_j=a_{j1}z_1+a_{j2}z_2,\ \ j=3,\dots,n. \tag{8.34}$$

Note that a linear change of coordinates does not violate the mixed strong homogeneity of f. Substituting (8.34) in $f(\mathbf{z},\bar{\mathbf{z}})$ to eliminate the variables $z_3,\dots,z_n$, we see that the intersection $V\cap L$ is described by

$$g(z_1,z_2,\bar z_1,\bar z_2)=0,\quad [z_1:z_2]\in L=\mathbb{P}^1$$

where g is the mixed polynomial after substituting $z_j=a_{j1}z_1+a_{j2}z_2$ and $\bar z_j=\bar a_{j1}\bar z_1+\bar a_{j2}\bar z_2$ for $j\ge 3$. As g is still a strongly mixed homogeneous polynomial in z_1,z_2 under the restriction to L, g is written as

$$g(\mathbf{z},\bar{\mathbf{z}})=f(\mathbf{z},\bar{\mathbf{z}})|_L=\sum_{\nu,\mu}c_{\nu,\mu}z_1^{\nu_1}z_2^{\nu_2}\bar z_1^{\mu_1}\bar z_2^{\mu_2},\ \mathbf{z}=(z_1,z_2),$$

where the summation are for the multi-integers $\nu=(\nu_1,\nu_2)$, $\mu=(\mu_1,\mu_2)$ such that

$$|\nu|+|\mu|=q+2r,\quad |\nu|-|\mu|=q.$$

Thus the polynomial $g(z_1,z_2,\bar z_1,\bar z_2)$ is a strongly mixed homogeneous polynomial of radial degree $q+2r$ and of polar degree q. Taking a linear change of coordinates if necessary, we may assume that the intersections are in the affine space

$z_2 \neq 0$. This implies that g has a monomial $z_1^{q+r}\bar{z}_1^r$ with a non-zero coefficient. Use the affine coordinate $z = z_1/z_2$ for the affine coordinate chart $\{z_2 \neq 0\}$. Then g takes the form:

$$g(z, \bar{z}) = \sum_{i,j} c_{i,j} z^i \bar{z}^j, \quad i \leq q + r,\ j \leq r,\ i + j \leq q + 2r$$

where $c_{q+r,r} \neq 0$. Let $\{\alpha_1, \ldots, \alpha_m\}$ be the root of $=g(z, \bar{z}) = 0$. We can see easily that the local intersection number at α_j is given as

$$I(V, L, \alpha_j) = \frac{1}{2\pi} \int_{|z-\alpha_j|=\varepsilon} \mathrm{Gauss}(g) d\theta$$

where $z - \alpha_j = \varepsilon \exp(i\theta)$ and $\mathrm{Gauss}(g)(z, \bar{z}) = \arg\, g(z, \bar{z})$ and ε is a sufficiently small positive number. In fact, the orientation of V is defined so that a frame $\{v_1, \ldots, v_{2n-4}\}$ at α_j is positive if and only if $\{v_1, \ldots, v_{2n-4}, n_1, n_2\}$ are positive where n_1, n_2 are frames of the normal bundle of V oriented by f. On the other hand, $\{\frac{\partial}{\partial x}, \frac{\partial}{\partial y}\}$ is also a frame of the normal bundle where $w = x + iy$. The orientations $\{n_1, n_2\}$ and $\{\frac{\partial}{\partial x}, \frac{\partial}{\partial y}\}$ are compatible if and only if the Gauss map at α_j has the positive rotation. A root (or a zero point) α of $g(z, \bar{z}) = 0$ is called *simple* if α is a regular point of $g : \mathbb{C} \to \mathbb{C}$. It is called *positive* or *negative* if the Gauss map at $z = \alpha$ has the rotation number 1 or -1 respectively. Taking line L generically, we may assume that the roots of g is all simple. Topologically the local intersection number of V and L is the mapping degree of the Gauss mapping at a root $z = \alpha_j$:

$$\mathrm{Gauss}(g) : \mathbb{S}^1_\varepsilon(\alpha_j) \cong \mathbb{S}^1 \to \mathbb{S}^1.$$

Take a sufficiently large positive number R. Consider the region $D := \mathbb{D}_R \setminus \cup_{j=1}^m \mathrm{Int}\, \mathbb{D}_\varepsilon(\alpha_j)$. Here $\mathbb{D}_\varepsilon(\alpha_j) := \{z \in \mathbb{C} \mid |z - \alpha_j| \leq \varepsilon\}$ and $\mathbb{S}^1_\varepsilon(\alpha_j) = \partial\mathbb{D}_\varepsilon(\alpha_j)$. The Gauss map extends to D and as $[\partial D] = \sum_{i=1}^m [\partial\mathbb{D}_\varepsilon(\alpha_j)]$ in $H_2(D)$, we see that

$$\sum_{j=1}^m \frac{1}{2\pi} \int_{|z-\alpha_j|=\varepsilon} \mathrm{Gauss}(g) d\theta = \frac{1}{2\pi} \int_{|z|=R} \mathrm{Gauss}(g) d\theta.$$

The right hand side is equal to the mapping degree of

$$\mathrm{Gauss(g)} : \mathbb{S}^1_R \to \mathbb{S}^1$$

which is equal to q, as the highest degree part of g is $c_{q+r,r} z^{q+r} \bar{z}^r$.

Moduli of Strongly Mixed Homogeneous Polynomials and Zero Points of Mixed Polynomials

We consider the moduli space of convenient non-degenerate mixed homogeneous polynomials of two variables $\mathcal{M}(n+m, n-m, 2)$. Recall $f \in \mathcal{M}(n+m, n-m, 2)$ defines a finite number of points in $\mathbb{P}^1$ as zero points which are all simple zeros of $f = 0$ in $\mathbb{P}^1$. Consider the subset of mixed polynomials of 1-variable which is defined as follows.

$$M(n+m, n, m) := \{\check{f}(z, \bar{z}) \in \mathbb{C}[z, \bar{z}] \mid (\star 1),\ (\star 2)\}$$

where $(\star 1)$: $\deg \check{f} = n + m$, $\deg_z \check{f} = n$, $\deg_{\bar{z}} \check{f} = m$, and $(\star 2)$: $f = 0$ has only simple zeros.

We consider also a subspace of $M(n+m, n, m)$ defined by

$$L(n+m, n, m) := \{f \in M(n+m, n, m) \mid f(z, \bar{z}) = \bar{z}^m q(z) - p(z),\ (\sharp)\}$$
$$\text{where } (\sharp) : \deg_z q(z) = n,\ \deg_z p(z) \le n$$

and $p(z), q(z) \in \mathbb{C}[z], r(\bar{z}) \in \mathbb{C}[\bar{z}]$ and the coefficient of z^n in q is non-zero. We have the canonical inclusion:

$$L(n+m, n, m) \subset M(n+m, n, m).$$

The class $L(n+m, n, m)$ comes from the set of harmonic functions $\bar{z}^m - \frac{p(z)}{q(z)}$ as the numerators. Especially $L(n+1, n, 1)$ corresponds to *the lens equation* $\bar{z} = \frac{p(z)}{q(z)}$ which has been studied by some astronomers (see [49]). We call $\bar{z}^m - \frac{p(z)}{q(z)} = 0$ *a generalized lens equation*. The corresponding numerators are called *a generalized lens polynomial*. There is a canonical correspondence

$$\psi : \mathcal{M}(n+m, n-m, 2) \to M(n+m, n, m),\ f(\mathbf{z}, \bar{\mathbf{z}}) \mapsto \check{f}(z, \bar{z})$$

where $\check{f}(z, \bar{z}) = f(\mathbf{z}, \bar{\mathbf{z}})/(z_2^n \bar{z}_2^m)$ and $z = z_1/z_2$, $\bar{z} = \bar{z}_1/\bar{z}_2$. The polynomial $\check{f}(z, \bar{z})$ is nothing but the affine equation of $f = 0$ in $\mathbb{P}^1$ with respect to the affine chart $\{z_2 \neq 0\}$ and the coordinate $z = z_1/z_2$. In a connected component U of $\mathcal{M}(n+m, n-m, 2)$, the Milnor fibrations of $f \in U$ does not depend on f and thus the Euler numbers $\chi(f^{-1}(1))$, $f \in U$ are constant. Let $\rho(\check{f})$ be the number of zeros of $\check{f} = 0$ in $\mathbb{P}^1$ which is equal to the number of lines in the affine cone $f = 0$ in $\mathbb{C}^2$. Put $R(\hat{\rho}) := \{\hat{\rho}(\check{f}) \mid \check{f} \in M(n+m, n, m)\}$. Let $\rho_\pm$ be the number of positive and negative roots. Then we have the equality $n - m = \rho_+ - \rho_-$ and $\rho(\check{f}) = \rho_+ + \rho_-$. Recall

$$\chi(F) = (n-m)\chi(\mathbb{P}^1 - V(\check{f})) = (n-m)(2 - \rho(\check{f})).$$

Therefore if $\rho(\check{f}) \neq \rho(\check{g})$, polynomials $f, g \in \mathcal{M}(n+m, n-m, 2)$ belong to different components. Thus the number of connected components of $\mathcal{M}(n+m, n-m, 2)$ is not smaller than the number of $R(\hat{\rho})$. For $M(n+1, n, 1)$, it is known that

$R(\hat{\rho})$ is precisely $\{n-1, n+1, \ldots, 5n-5\}$ by Bleher, Homma, Ji and Roeder [4]. Rhie constructed an explicit polynomial with $\rho(\check{f}) = 5n-5$. See [39, 49]. About the number of connected components of $M(n+m, n, m)$, $m \geq 2$, an estimation is given in [40]

Corollary 8.3.61 (Corollary 5 [40]) *$\rho(\check{f})$ for $f \in L(n+m, n, m)$ can take at least the values $\{n+m-2, n+m, \ldots, 5n+m-6\}$.*

In $M(n+m, n, m)$, the value can take $\{n-m, n-m+2, \ldots, n+m-4\}$ as well and thus the value set covers $\{n-m, n-m+2, \ldots, 5n+m-6\}$. We do not know if $5n+m-6$ is the optimal upper bound or not.

Projective Mixed Curves

We consider projective curves of polar degree q:

$$C = \{[\mathbf{z}] = [z_1 : z_2 : z_3] \in \mathbb{P}^2 \mid f(\mathbf{z}, \bar{\mathbf{z}}) = 0\}$$

where f is a strongly mixed homogeneous polynomial with pdeg $f = q$ and $f \in \mathcal{M}(p, q, 3)$. We have seen that the topological degree (=embedding degree) of C is q by Theorem 8.3.60. The genus g of C is not an invariant of q, under a fixed p, q. Recall that for a differentiable curve C of genus g, embedded in $\mathbb{P}^2$, with the topological degree q, we have the following Thom's inequality, which was conjectured by Thom, [54] and proved by several people, for example, Kronheimer-Mrowka [20]:

$$g \geq \frac{(q-1)(q-2)}{2}$$

where the right side number is the genus of algebraic curves of degree q, given by the Plücker formula. The mixed projective curve has much flexibility.

As an example, we consider the mixed polynomial

$$h_{q,r,j}(\mathbf{w}, \bar{\mathbf{w}}) = (w_1^{q+j}\bar{w}_1^j + w_2^{q+j}\bar{w}_2^j)(w_1^{r-j} - \alpha w_2^{r-j})(\bar{w}_1^{r-j} - \beta \bar{w}_2^{r-j}), \quad r \geq j \geq 0$$

with $\alpha, \beta \in \mathbb{C}^*$ generic [34, 36]. Note that $h_{q,r,j}$ is a strongly polar homogeneous polynomial with a radial degree $q+2r$ and a polar degree q respectively i.e., $h_{q,r,j} \in \mathcal{M}(q+2r, q, 2)$. Let $H_{q,r,j} := h_{q,r,j}^{-1}(1)$ the Milnor fiber of $h_{q,r,j}$. Note that $\rho(\check{h}_{q,r,j}) = q + 2(r-j)$ and $\chi(H_{q,r,j}) = -(q+2(r-j))q + 2q$. We consider the following join type polynomial.

$$f_{q,r,j}(\mathbf{z}, \bar{\mathbf{z}}) = h_{q,r,j}(\mathbf{w}, \bar{\mathbf{w}}) + z_3^{q+r}\bar{z}_3^r \in \mathcal{M}(q+2r, q, 2), \quad \mathbf{w} = (z_1, z_2).$$

Let $F_{q,r,j} = f_{q,r,j}^{-1}(1)$ be the Milnor fiber. By the Join Theorem (Cisneros-Molina [6]), $\chi(F_{q,r,j}) = q(q-1)(q-2) + 2q(q-1)(r-j) + q$. Let $C_{q,r,j}$ be the projective curve of degree q defined by $\{f_{q,r,j}(\mathbf{z},\bar{\mathbf{z}})=0\}$ in $\mathbb{P}^2$. Then the genus $g(C_{q,r,j})$ of $C_{q,r,j}$ is given by

$$g(C_{q,r,j})=\frac{(q-1)(q-2)}{2}+(q-1)(r-j)\geq\frac{(q-1)(q-2)}{2}.$$

For $q=2$, we get

$$g(C_{2,r,j})=(r-j)\geq 0.$$

Thus changing r, j, we have degree 2 curves of arbitrary genus. To get a mixed curve of degree 1 for a given genus g, we consider a twisted join type polynomial:

$$\begin{cases} h(\mathbf{w},\bar{\mathbf{w}}) & := (z_1+z_2)(z_1^r-\alpha z_2^r)(\bar{z}_1^r-\beta\bar{z}_2^r),\ |\alpha|\neq|\beta|, \\ f_r(\mathbf{z},\bar{\mathbf{z}}) & := h(\mathbf{w},\bar{\mathbf{w}})+\bar{z}_2 z_3^{r+1}\bar{z}_3^{r-1}. \end{cases}$$

Let $S_r : f_r=0$. Then the topological degree of S_r is 1 and the genus of S_r is r (Corollary 10, [34]).

8.3.17 Remarks and Problems

(a) About $f\bar{g}$. (1) The equivalence of tubular and spherical Milnor fibrations can be proved in the same way as Lemma 8.3.28 (see Theorem 16, [42]).
(2) To treat the case where $V(f)$, $V(g)$ have non-isolated singularities, the convenience assumption in Corollary 8.3.55 can be replaced by the following assumption (see [42]).
1. $f(\mathbf{z})$ and $g(\mathbf{z})$ is locally tame and Newton non-degenerate.
2. The variety $f=g=0$ is non-degenerate and locally tame in the sense that if $\mathbb{C}^J\in\mathcal{V}_f\cap\mathcal{V}_g$ and for any P with $I(P)=J$, there exists a positive number r_J such that $f_P=g_P=0$ is non-degenerate as a variety of variables $\{z_i | i\notin J\}$ fixing $\mathbf{z}_J$ with $\|\mathbf{z}_J\|\leq r_J$. Here $\mathcal{V}_f$, $\mathcal{V}_g$ are vanishing coordinate subspaces of f and g.

(b) (Further Join Theorem) Suppose that $g(\mathbf{z},\bar{\mathbf{z}})$ and $h(\mathbf{w},\bar{\mathbf{w}})$ are mixed functions of n-variables $\mathbf{z}=(z_1,\dots,z_n)$ and m-variables $\mathbf{w}=(w_1,\dots,w_m)$ which have tubular Milnor fibrations at the origin but we do not assume that f, g are mixed weighted homogeneous polynomials. Consider the function $f(\mathbf{z},\mathbf{w}):=f(\mathbf{z})+g(\mathbf{w})$ of $n+m$ variables. Show that f has a tubular Milnor fibration and the Milnor fiber is homotopic to the join of the respective Milnor fibers $F_f * F_g$. For holomorphic functions, this is proved by Sakamoto [51]. For mixed weighted homogeneous polynomials, the assertion is proved by Cisneros-Molina [6]. For general pair of mixed functions g, h which are not mixed weighted homogeneous, the assertion is proved by Inaba [16].

(c) Let $f(z)=\sum_{i=0}^{d} c_i z^{d-i}\bar{z}^i$ be a mixed polynomial of one variable z. Let $U=\{f \mid f:\mathbb{C}^*\to\mathbb{C}$ has no critical point$\}$. Characterize the necessary and sufficient condition for f to be in U. This is equivalent to $f(\mathbb{C}^*)=\mathbb{C}^*$ and $f:\mathbb{C}^*\to\mathbb{C}^*$ is a finite covering map. A sufficient condition is given in Lemma 8.3.3. Give a good estimation of the number o f connected components of U.

(d) Let $f(\mathbf{z},\bar{\mathbf{z}})$ be a mixed weighted homogeneous polynomial with radial degree d_r and polar degree d_p. Let $Q:=(q_1,\dots,q_n)$ and $P=(p_1,\dots,p_n)$ be the radial and polar weight vectors respectively. We have seen in Proposition 8.3.8 that it has a spherical Milnor fibration and a global Milnor fibration and they are equivalent. If $V(f)$ has an isolated singularity at the origin, by the compactness argument, it satisfies (ST) condition in Sect. 8.2.3.
Problem. Suppose that $f(\mathbf{z},\bar{\mathbf{z}})$ is a mixed weighted homogeneous polynomial with a non-isolated singularity. Does f satisfy the (ST)-condition in general so that a tubular Milnor fibration exists and it is equivalent to the spherical Milnor fibration and also to the global Milnor fibration?

(e) Let $f(\mathbf{z},\bar{\mathbf{z}})$ be a convenient strongly mixed homogeneous polynomial of polar degree d and let $F=f^{-1}(1)$ be the Milnor fiber and $V\subset\mathbb{P}^{n-1}$ be the corresponding projective hypersurface. Their Euler characteristics are related as $\chi(F)=d(n-\chi(V))$. In the case $n=2$, $\chi(F)=d(2-\rho(V))$. Thus $-\chi(F)\ge d^2-2d=(d-1)^2-1$ or $\rho(V)\ge d$ as $\rho(V)=\rho_+ + \rho_- \ge d$ (see Sect. 8.3.16). In the case $n=3$, Thom's inequality gives

$$\chi(F)\ge d(3-3d+d^2),\ n=3.$$

The equalities in both cases are taken if f is in the same connected component of the moduli space as the mixed Brieskorn polynomials $z_1^{q+d}\bar{z}_1^q+z_2^{q+d}\bar{z}_2^q$ for $n=2$ or $z_1^{d+q}\bar{z}_1^q+z_2^{d+q}\bar{z}_2^q+z_3^{d+q}\bar{z}_3^q$ for $n=3$. So for higher dimensions, we propose the conjecture:

$$Conjecture:\ (-1)^{n-1}\chi(F)\ge(d-1)^n+(-1)^{n-1},\ \forall n.$$

and the equality is taken for the component which contains the mixed Brieskorn polynomial. Or if we assume further that F is homotopic to a bouquet of $(n-1)$-spheres, the stronger conjecture is

$$\mu(f)\ge(d-1)^n.$$

A similar conjecture can be given for mixed homogeneous polynomials.

References

1. N. A'Campo. *La fonction zeta d'une monodromie.* Commentarii Mathematici Helvetici, 50, 233–248, 1975

2. R. N. Araujo dos Santos, M. Ribeiro and M. Tibar. *Fibrations of highly singular map germs*, Bull. Sci. Math. 55, (2019), 92–111
3. R. N. Araujo dos Santos, M. Ribeiro and M. Tibar. *Milnor-Hamm sphere fibrations and the equivalence problem*, arXiv:1810.05158
4. P. Bleher, Y. Homma, L. Ji, P. Roeder. *Counting zeros of harmonic rational functions and its application to gravitational lensing*, Int. Math. Res. Not. IMRN, 8, 2014, 2245–2264
5. Y. Chen. *Ensembles de bifurcation des polynômes mixtes et polyèdres de Newton*, Thèse, Université de Lille I, 2012
6. J. L. Cisneros-Molina. *Join Theorem for polar weighted homogeneous singularities*. In Singularities II. Contemp. Math. 475, p. 43–59. A. M. S., Providence, RI, 2008. Proc. Int. School and Workshop on the Geometry and Topology of Singularities in honor of the 60th birthday of Lê D. T. Cuernavaca, Mexico, 2007. Eds. J.-P. Brasselet *et al*
7. J. L. Cisneros-Molina, N. G. Grulha, J. Seade. *On the topology of real analytic maps*. Int. J. Maths. **25** (2014), 1450069, 30 pp.
8. J. L. Cisneros-Molina, A. Menegón, J. Seade, J. Snoussi. *Fibration theorems and d-regularity for differentiable map-germs with non-isolated critical value*. Preprint 2017
9. J. L. Cisneros-Molina and J. Seade. *Milnor's fibration theorem for real and complex singularities*, in Volume 2 in this Handbook
10. I. Dolgachev. *Weighted projective varieties*, in Group actions and vector fields (Vancouver, B.C., 1981), Lecture Notes in Math., 956, 34–71, Springer, Berlin, 1982
11. C. Eyral and M. Oka. *Whitney regularity and Thom condition for families of non-isolated mixed singularities*. J. Math. Soc. Japan 70 (2018), no. 4, 1305–1336
12. J. Fernandez de Bobadilla and A. Menegon Neto. *The boundary of the Milnor fibre of complex and real analytic non-isolated singularities*. Geom Dedicata, 173, (2014), 143–162
13. P. Griffiths and J. Harris. *Principles of algebraic geometry*. Wiley Classics Library. John Wiley & Sons Inc., New York, 1994. Reprint of the 1978 original
14. H. Hamm. *Lokale topologische Eigenschaften komplexer Räume*. Math. Ann., 191:235–252, 1971
15. H. A. Hamm and D. T. Lê. *Un théorème de Zariski du type de Lefschetz*. Ann. Sci. École Norm. Sup. (4), 6:317–355, 1973
16. K. Inaba, *Join theorem for real analytic singularities*, arXiv:2002.06549v1, February, 2020
17. K. Inaba, M. Kawashima and M. Oka. *Topology of mixed hypersurfaces of cyclic type*. J. Math. Soc. Japan 70 (2018), no 1, 387–402
18. C. Joita and M. Tibar. Images of analytic map germs, arXiv:1904.07571v4, April, 2020
19. M. Kato and Y. Matsumoto. *On the connectivity of the Milnor fiber of a holomorphic function at a critical point*. in Manifolds—Tokyo 1973 (Proc. Internat. Conf., Tokyo, 1973), 131–136, 1975
20. P. B. Kronheimer and T. S. Mrowka. *The genus of embedded surfaces in the projective plane*. Math. Res. Lett., 1(6):797–808, 1994
21. A. G. Kouchnirenko, *Polyèdres de Newton et nombres de Milnor*. Invent. Math. 32, 1–31, 1976
22. D. T. Lê. *Some remarks on relative monodromy*, In Real and complex singularities (Proc. Ninth Nordic Summer School/NAVF Sympos. Math., Oslo, 1976), 397–403, 1977
23. J. Mather. *Notes on topological stability*. Bull. Amer. Math. Soc. (N.S.), 49(4):475–506, 2012
24. J. Milnor. *Singular points of complex hypersurfaces*. Annals of Mathematics Studies, 61. Princeton University Press, Princeton, N.J., 1968
25. J. Milnor *Construction of universal bundles, II* , Annals of Math. 63, 1956, 430–436
26. J. Milnor and P. Orlik. *Isolated singularities defined by weighted homogeneous polynomials*. Topology, 9:385–393, 1970
27. M. Oka. *On the homotopy types of hypersurfaces defined by weighted homogeneous polynomials*. Topology, 12:19–32, 1973
28. M. Oka. *On the topology of full non-degenerate complete intersection variety*. Nagoya J. Math. 121 (1991), 137–148
29. M. Oka. *Non-degenerate complete intersection singularity*. Hermann, Paris, 1997

30. M. Oka. *Topology of polar weighted homogeneous hypersurfaces*. Kodai Math. J., 2008 , Vol.31, No.2, 163–182
31. M. Oka. *Non-degenerate mixed functions*. Kodai Math. J., 33(1):1–62, 2010
32. M. Oka. *On mixed Brieskorn variety.* Topology of algebraic varieties and singularities. 389-399, Contemp. Math., 538, Amer. Math. Soc., Providence, RI, 2011
33. M. Oka. *Mixed functions of strongly polar weighted homogeneous face type*. Singularities in Geometry and Topology 2011, Advanced Study in Pure Math. 66, 173-202, 2015
34. M. Oka. *On mixed plane curves of polar degree 1*, in The Japanese-Australian Workshop on Real and Complex Singularities—JARCS III, Proc. Centre Math. Appl. Austral. Nat. Univ., 43, 67–74. 2010
35. M. Oka. *Intersection theory on mixed curves*, Kodai Math. J., 35, 2012, no. 2, 248–267
36. M. Oka. *On mixed projective curves,* Singularities in Geometry and Topology, IRMA Lect. Math. Theor. Phys. Eur. Math. Soc., Zürich., 20, (2012), 133–147
37. M. Oka. *Contact structure on mixed links*, Vietnam J. Math., 42, 2014, NO. 3, 249–271
38. M. Oka. *On Milnor fibrations of mixed functions, a_f-condition and boundary stability*. Kodai J. Math., 38, (2015), 581–603
39. M. Oka. *On the roots of an extended Lens equation and an application* in Singularities and foliations. geometry, topology and applications, Springer Proc. Math. Stat., 222, 489–511, 2018
40. M. Oka. *Remark on the roots of generalized lens equations*, SUT J. Math., 53, 2017, no. 2, 127–134
41. M. Oka. *On the Milnor fibration for* $f(\mathbf{z})g(\bar{\mathbf{z}})$, arXiv: 1812.10909v3, available in European Journal of Math. (web), vol.1, 1, 2020
42. M. Oka. *On the Milnor fibration for* $f(\mathbf{z})g(\bar{\mathbf{z}})$*II*, arXiv:1909.01168v, September, 2019
43. M. Oka. *On the connectivity of Milnor fiber for mixed functions.* Contemp. Math., 742, Amer. Math. Soc., Providence, RI, 2020, 181–187
44. P. Orlik and P. Wagreich. *Singularities of algebraic surfaces with* C^* *action*. Math. Ann., 193, 1971, 121–135
45. A.J. Parameswaran and M. Tibar. *Corrigendum to "Thom irregularity and Milnor tube fibrations"*, Bull. Sci. Math., 153, (2019), 120–123
46. A. Pichon and J. Seade. *Real singularities and open-book decompositions of the 3-sphere*, Ann. Fac. Sci. Toulouse Math. (6), 12, (2003), 2, 245–265
47. A. Pichon and J. Seade. *Fibred multilinks and singularities* $f\bar{g}$, Math. Ann., 342, (2008), 3, 487–514
48. A. Pichon and J. Seade. *Milnor fibrations and the Thom property for maps* $f\bar{g}$, Journal of Singularities, 3, (2011), 144–150
49. S. H. Rhie. *n-point Gravitational Lenses with* $5(n-1)$ *Images*. arXiv:astro-ph/0305166, May 2003
50. M. A. S. Ruas, J. Seade, and A. Verjovsky. *On real singularities with a Milnor fibration.* In Trends in singularities, Trends Math., pages 191-213. Birkhäuser, Basel, 2002
51. K. Sakamoto. *Milnor fiberings and their characteristic maps.* In Manifolds-Tokyo 1973 (Proc. Internat. Conf., Tokyo, 1973), 145-150. Univ. Tokyo Press, Tokyo, 1975
52. I. Satake. *On a generalization of the notion of manifold*, Proc. Nat. Acad. Sci. U.S.A., 42, 1956, 359–363
53. J. Seade. *On the topology of isolated singularities in analytic spaces*. Progress in Mathematics 241. Birkhäuser, 2006
54. R. Thom. *Sur l'homologie des variétés algébriques réelles.* Differential and Combinatorial Topology (A Symposium in Honor of Marston Morse), Princeton Univ. Press, Princeton, N.J., 1965, pp. 255–265
55. R. Thom *Ensembles et morphismes stratifiés*, Bull. Amer. Math. Soc. 75, 1969, 240–284
56. D. Trotman. *Stratification theorey*, Chapter 4, in this Handbook Vol. 1
57. A. N. Varchenko, *Zeta-function of monodromy and Newton's diagram*, Invent. Math., 37, 253–262, 1976
58. H. C. Wang, *The homology groups of the fibre bundles over a sphere*, Duke J. Math. 16, 1949, 33–38

59. H. Whitney. *Elementary structure of real algebraic varieties.* Ann. of Math. (2), 66:545–556, 1957
60. J. A. Wolf. *Differentiable fibre spaces and mappings compatible with Riemannian metrics.* Michigan Math. J., 11:65–70, 1964

Chapter 9
From Singularities to Polyhedral Products

Santiago López de Medrano

Abstract The topology of intersections of concentric ellipsoids in $\mathbb{R}^n$ have been studied for some time under the different names of *links of pencils of quadrics*, *intersections of quadrics in* $\mathbb{R}^n$ and *real moment-angle manifolds*. During this time, different authors were working independently and with different aims without knowing about the others' work. When the relations were stablished, the interchange of ideas was useful and in some cases there were fruitful collaborations between members of different schools. We will tell this story, ending with some recent unpublished work on singular intersections that extends previous results.

Contents

S. López de Medrano (✉)
Instituto de Matemáticas, Universidad Nacional Autónoma de México, Mexico City, Mexico
e-mail: santiago@matem.unam.mx

J. L. Cisneros-Molina et al. (eds.), *Handbook of Geometry and Topology of Singularities II*, https://doi.org/10.1007/978-3-030-78024-1_9

9.1 Introduction

Euler's differential equations represent the motion of a free rigid body in three space. Invariants of this motion are given by the kinetic energy and the magnitude of the angular momentum, both being quadratic forms in the appropriate coordinates. Jacques Philippe Marie Binet (1786–1856) and Louis Poinsot (1777–1859) gave a geometrical description of the motion with the use of an ellipsoid in $\mathbb{R}^3$ representing a fixed level set of one of the invariants and drawing on it the possible curves of intersection with the ellipsoids for different levels of the other invariant. Since then, this figure has been used in many books and courses of Mechanics and solid models have been made[1]:

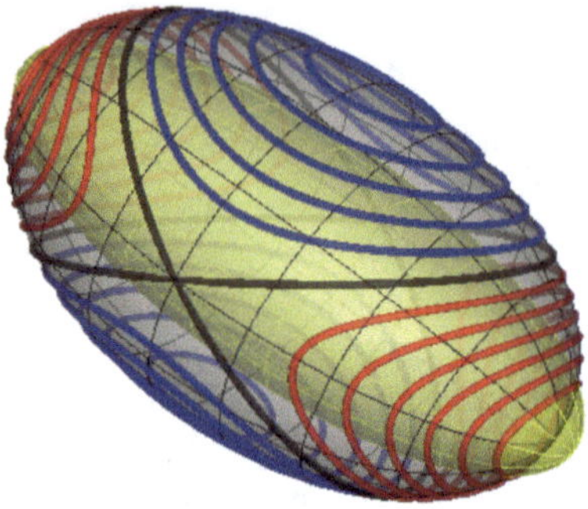

Today we can see in this figure the topology of the different possible intersections of two concentric ellipsoids in $\mathbb{R}^3$: it can be empty or homeomorphic to a pair of disjoint circles, if the ellipsoids are transverse, two points or two circles crossing at two points if they have generic tangencies (the more degenerate cases, when the ellipsoids have a common equator or when they are equal, do not appear). It does not seem that the geometry of these intersections has been studied, their curvature and torsion might be interesting.

However, the intersections of two or more ellipsoids in $\mathbb{R}^n$ for higher n does not seem to have been considered. The rigid body in Relativity Theory or Quantum Mechanics does not seem to exist and, although there are studies of the rigid body in classical mechanics in higher dimension [57, 62] or [68] with many invariants, the geometry of the intersections of their level sets does not seem to have been considered.

As far as we know it was in the period 1980–1991 that different papers were published studying intersections of concentric ellipsoids in $\mathbb{R}^n$ as well as abstract versions of them (within a very general theory). They emerged independently from three areas of Mathematics (Singularity Theory, Dynamical Systems and Algebraic Topology) and their interconnections were discovered only after the theories had been somewhat developed. We will try to describe these different sources and some consequences.

[1] Figure PoinsotDrehimpulsEllipsoid.png taken from [76] under Creative Commons licence CC BY-SA 4.0.

9.2 Singularity Theory

The first published paper I know where the topology of the intersections of concentric ellipsoids is studied is the remarkable 1980 Presidential Address to the London Mathematical Society *Stability, pencils and polytopes* by Wall [74].

The paper's aim is the understanding of the first instances of topological instability of differentiable maps $f : N \to P$ between differentiable manifolds. We can only try, in what follows, to give a rough idea of the appearance of intersections of quadrics in the realm of topological stability.

The first part of the paper is a delightful introduction to the theory of C^∞ stability of such maps which starts by recalling that a map f as above is called stable if any small enough perturbation g of f is equivalent to f through diffeomorphisms h and k of N and P (by composition $k \circ g = f \circ h$) which are perturbations of the identity maps. Then he recalls part of the important results of Morse, Whitney, Thom and Mather. In particular, he recalls the first example (due to René Thom) of the fact that the set of C^∞ stable maps is not dense in some dimensions: essentially it is the map $f : \mathbb{R}^8 \to \mathbb{R}^6$ given by

$$f(x_1, x_2, x_3, x_4, x_5, x_6, x_7, x_8) =$$

$$= (x_1, x_2, x_3, x_4, q_1(x_5, x_6, x_7, x_8), q_2(x_5, x_6, x_7, x_8))$$

where q_1 and q_2 is a generic pair of quadratic forms.

This mapping has rank 4 at the origin and its zero set is the quadratic cone given by the zeros of q_1 and q_2 in the $x_1 = x_2 = x_3 = x_4 = 0$ subspace. The set $M(6, 8; 4)$ of matrices 6×8 of rank 4 has codimension 8 inside the space of all 6×8 matrices, so if the differential of f at the origin is transversal to $M(6, 8; 4)$, all the mappings in a neighborhood U of f in the space of mappings $\mathbb{R}^8 \to \mathbb{R}^6$ will have an isolated point of rank 4 and would be locally equivalent to a mapping of the same form as f, but with different quadratic terms. If two of these maps were C^∞ equivalent, the differentials of the diffeomorphisms would be linear maps giving an equivalence between the quadratic terms. But not all pairs of quadratic forms in 4 variables are linearly equivalent (there is a real invariant) so in U the C^∞ type of the maps will vary and none of them will be C^∞ stable.

Among the many results by John Mather are the characterization of the dimensions for which the stable maps are dense, and the fact that C^∞ topological stable maps are dense. The question addressed by Wall now is to characterize those that are not topologically stable. We can not go into the technical details about the varieties in the space of jets involved, but the main idea is that certain mappings have deformations that involve two quadratic forms, and now the question is whether two such pairs are or not *topologically equivalent*. So Wall attacks with force this question, which occupies 10 pages of the article.

(1) Normal forms: So we have now a pair q_1, q_2 of quadratic forms in any number of variables (or the *pencil of quadrics* $\lambda q_1 + \mu q_2$) and their set of common zeros is a

cone that one naturally intersects with the unit sphere obtaining a link L. For a large dense open set of such pencils (those for which at least one of them is non-degenerate, say q_1) he produces a *linear normal form*:

$$q_1 = \Sigma_1^r a_i x_i^2 + \Sigma_1^s 2u_j v_j$$

$$q_2 = \Sigma_1^r b_i x_i^2 + \Sigma_1^s \left(\beta_j (u_j^2 - v_j^2) + 2\alpha_j u_j v_j\right)$$

where (x_i, y_i, u_j, v_j) are coordinates in the space $\mathbb{R}^n$ and b_i/a_i and $\alpha_j \pm i\beta_j$ are the real and complex eigenvalues of the pair q_1, q_2, that is, the eigenvalues of the matrix $A_1^{-1}A_2$, where A_i is the matrix of q_i in some coordinate system.

After observing that all the complex eigenvalues, being non-zero, can be moved around without changing the topological type, he proceeds to the study of the *diagonalizable real cases*.

(2) For the real cases

$$q_1 = \Sigma_1^r a_i x_i^2 = 0$$

$$q_2 = \Sigma_1^r b_i x_i^2 = 0$$

Wall shows that one can assume the pairs (a_i, b_i) lay in the unit circle and proves that the link L is non-singular if, and only if, no pair of these points is antipodal. Then he moves them into the least number of *blocks* of coincident points without breaking that condition and defines the *characteristic* to be the (cyclic) list of the sizes of those blocks $(m_1, m_2, \ldots, m_{2g+1})$, where g is receives the name *genus* of the link. And he shows why the characteristic determines the topology of the link.

In the next section he constructs a convex polytope Π associated to the link: it is the quotient by the action of the group generated by reflections in the coordinate planes which can be identified with the polytope:

$$\Sigma_1^r a_i t_i = 0$$

$$\Sigma_1^r b_i t_i = 0$$

$$\Sigma_1^r t_i = 1$$

$$t_i \geq 0$$

from which the link can be recovered. A deep study of the polytope leads him to another description that should be called the *Wall diagram*, which he had introduced in 1961 for another (failed) purpose.

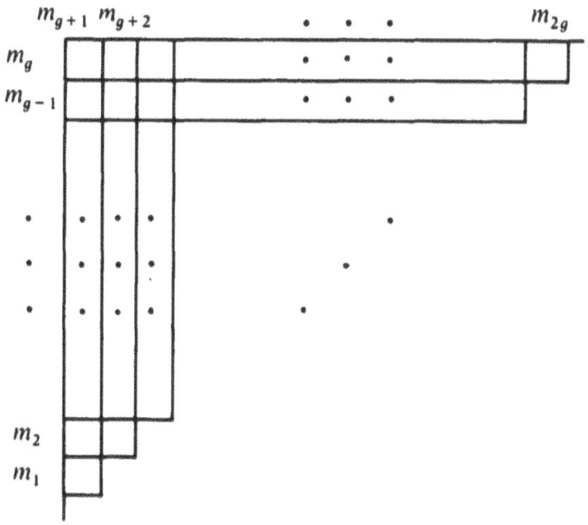

From this he is able to prove easily that, when $g = 1$, L is a product of 3 spheres ($S^{m_1-1} \times S^{m_2-1} \times S^{m_3-1}$) and when $g > 1$ to compute the homology of L, seen as a *frustum* of a product of simplices, through an impressive induction feat (guided by the *Wall diagram*) involving iterated Mayer-Vietoris sequences.

This complicated homology is sufficient for constructing many examples of non-topologically stable germs from deformations of pairs of quadratic forms.

There are, however, other cases missing, in particular those corresponding to the non-diagonalizable pairs which involve complex eigenvalues, for which Wall says that he has not computed the homology groups and that even that alone will not suffice. To fill this gap, Wall develops other tools involving deformations of higher degree. After several more pages he manages to attain his goal.

It is implicit in this work a topological normal form for the non-diagonalizable pairs of quadratic forms . (This form would be very useful to us many years later in the description of their topology, [41]). Also, it is a curious fact that there is a link (of a non-quadratic singularity) that is the connected sum of two products of spheres, due to James Damon, while all known quadratic connected sums of that type known so far have always an odd number of terms, a fact that deserves more attention. Wall states that there is a perfect correspondence between the configuration of the coefficients and the polytope that is valid for any number of quadrics (and is known as the *Gale diagram* in polytope theory) and gives also the abstract definition of L as a quotient of $\Pi \times \mathbb{Z}_2^n$ by the natural equivalence relation dictated by the facets of Π. Many more geometric ideas that appear in this paper will certainly open ways to other developments.

His computation of the homology of the diagonal case should certainly be studied and contrasted with other independent computations that were later discovered which give the splitting of the homology groups (Sect. 9.3.1), in principle for any number of quadrics, including the most general one so far, based on a stable homotopical splitting of polyhedral products [6], a vast generalization of these objects, that gives a splitting for any generalized homology theory. (See Sect. 9.5.2).

It must be said that the study of topological (un)stability of maps has been continued for decades by Wall and Andrew Du Plessis. It is condensed in the 1995 book *The Geometry of Topological Stability* [33] with almost 600 pages. It would be a hard task to follow the line that joins the results in *Stability, pencils and polytopes* to those in the book.

9.3 Dynamical Systems

Intersections of quadrics have appeared several times in dynamical systems. We recall here some of those appearances.

9.3.1 Complex Differential Equations

Camacho, Kuiper and Palis raised in 1978 a question about the topology of the intersection of two quadratic cones that appeared in the study of Complex Dynamical Systems [21]. Marc Chaperon generalized those dynamical systems into actions of more general groups and also raised the same type of topological questions [22].

Unaware of the work by Wall (and also of Chaperon's developments) I started to look at that question in January 1984, following a suggestion by Alberto Verjovsky.

Camacho-Kuiper-Palis developed a theory of topological equivalence of linear complex differential equations which then they applied to general non-linear systems with critical points. We will recall only the linear case:

Consider a system of linear, complex differential equations in $\mathbb{C}^n$:

$$\dot{z} = \Lambda z$$

where $z \in \mathbb{C}^n$ and Λ an $n \times n$ non-singular matrix of complex numbers.

If any pair of eigenvalues of Λ is linearly independent over $\mathbb{R}$, then the system is called *hyperbolic* and since this implies that they are all different, the system can be diagonalized:

$$\dot{z_i} = \lambda_i z_i \quad i = 1, \dots n$$

The solutions are obviously exponentials involving a complex variable τ.

$$z_i(\tau) = z_i(0)exp(\lambda_i \tau)$$

And it seems that there is not much more to say. Nevertheless, there is much hidden topology in this simple system.

The solutions (other than the solution $z = 0$) are parametrized by τ so they are 1-dimensional complex manifolds. There are two possibilities:

(1) When the origin is *not* in the convex hull of the coefficients λ_i, then all the non-zero solutions have the origin in their closure and one says that the system is in the *Poincaré* domain. Topologically, the system is just like the one corresponding to the identity matrix, whose solutions are the origin and all the lines through the origin with the origin removed.

(2) When the origin is in the convex hull of the coefficients λ_i, then most of the solutions keep their distance from the origin and one says that those solutions are *Siegel leaves* and that the system is in the *Siegel domain* . Some solutions have the origin in their closure and are called *Poincaré leaves*. We can only make a sketch of the picture by drawing the real analog in two dimensions:

There are three types of orbits: the origin, the straight half lines that approach the origin and the curved ones that are the Siegel leaves.

The origin and the straight lines are points where the space of orbits is not Hausdorff: any neighborhood of the origin in the quotient space contains the classes of the straight lines and any neighborhood of the class of a vertical line intersects every neighborhood of the class of a horizontal one! The actual complex version is more complicated: the Siegel leaves approach a straight leaf spiraling around it at the same time. In fact, the velocity of their rolling is determined by the eigenvalues and is a *topological invariant*, so these systems are not topologically stable!

Restricting to the Siegel leaves we have a civilized quotient: Each orbit has a unique point that is closest to the origin, so the space of Siegel leaves can be identified with the space of those closest points which is a subset of $\mathbb{C}^n$: it is a Hausdorff space!

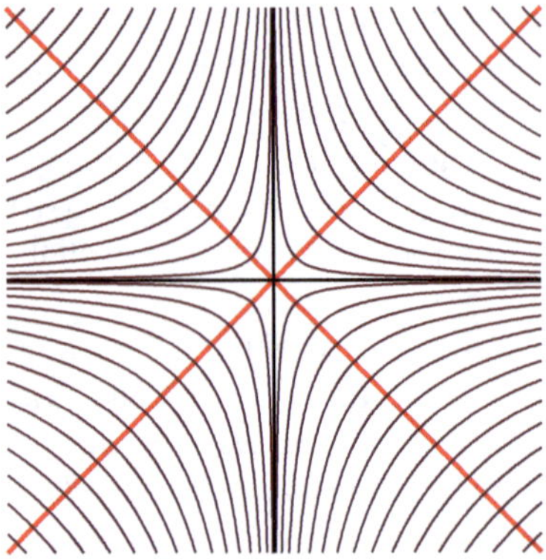

Not only that, they proved that those points are the non-zero solutions of equations

$$\Sigma \lambda_i \, |z_i|^2 = 0$$

It turns out that it is a smooth manifold as a consequence of the hyperbolicity assumption. But it is a hybrid real-complex object: a real manifold, not a holomorphic submanifold of the space in which it lives, but if we add to it the leaves themselves which are holomorphic submanifolds it becomes holomorphic (an open set of $\mathbb{C}^n$!) We will see later that the fact of being hybrid is not a sin but a blessing...

For the moment, to understand its topology we can observe that, because it is conical, it is the product of the real line times its intersection with the unit sphere, which we will denote by $\mathcal{Z}$:

$$\Sigma \lambda_i \, |z_i|^2 = 0$$

$$\Sigma \, |z_i|^2 = 1$$

In real terms, it is the intersection of three quadrics in $\mathbb{R}^{2n}$.

$$\Sigma a_i (x_i^2 + y_i^2) = 0$$

$$\Sigma b_i (x_i^2 + y_i^2) = 0$$

$$\Sigma (x_i^2 + y_i^2) = 1$$

which can also be thought of, by adding large enough multiples of the last equation to the first ones, as an intersection of three ellipsoids with the same center and axes, that is an intersection of coaxial ellipsoids.

At first thought one could bet that someone should have described its topology, sometime in the first half of the XXth century... Having found nothing in the literature I proceeded to study them in several steps [49, 50]:

(1) Characterizing and classifying the regular cases: drawing the coefficients λ_i in the complex plane, it is easy to see that the system is regular if, and only if, the

origin is not in one of the segments joining the coefficients. This is obviously implied by the hyperbolicity condition but is weaker, and this will be fundamental for the classification and topological description of the varieties. It is natural to call this condition *weak hyperbolicity, WH*, but I did not use this name until I learned that Marc Chaperon had already discovered, used and named this property around 1979.

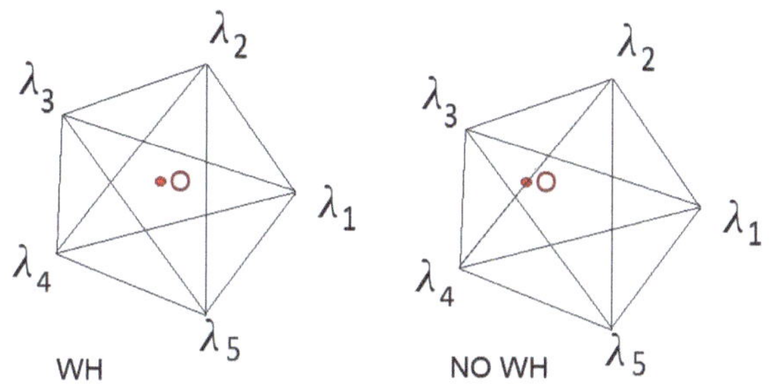

Then I showed that any such configuration of coefficients can be deformed, preserving WH (and therefore the topology of $\mathcal{Z}$), into a normal form consisting of the vertices of a regular polygon with an odd number of sides, where the ith vertex appears with a certain multiplicity $n_i \geq 1$: this gives a *cyclic partition.*

$$n = n_1 + n_2 + \cdots + n_{2\ell+1}$$

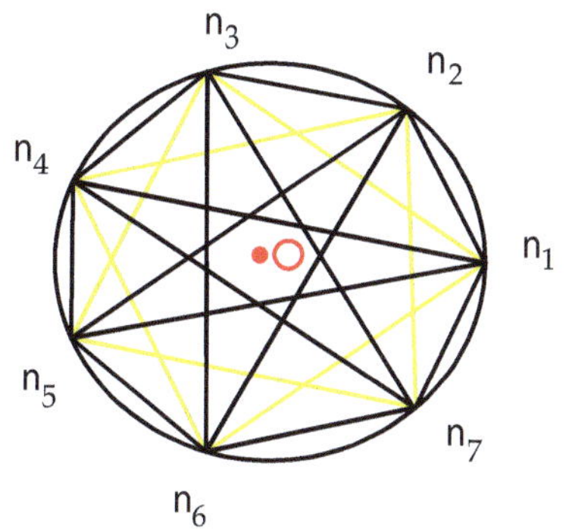

$$n = n_1 + n_2 + n_3 + n_4 + n_5 + n_6 + n_7$$

This is enough to describe the simplest cases: if $\ell = 0$ then $\mathcal{Z}$ is empty and if $\ell = 1$ then we can consider the example $\lambda_1 = 1$, $\lambda_2 = 1 + i$, $\lambda_3 = 1 - i$ with multiplicities $n = n_1 + n_2 + n_3$. Then the three real equations can be easily manipulated to obtain the equations of three spheres with separate variables so that

$$\mathcal{Z} = S^{2n_1-1} \times S^{2n_2-1} \times S^{2n_3-1}.$$

For $\ell > 1$ it is impossible to separate the variables and one has to go deeper.

(2) The construction of the associated convex polytope P :

$$\Sigma_{i=1}^{n}\lambda_i r_i = 0, \quad \Sigma_{i=1}^{n} r_i = 1, \quad r_i \geq 0$$

which can be identified with the quotient of $\mathcal{Z}$ by the natural action of the torus $(S^1)^n$ given by the action of each factor on the corresponding variable z_i.

The combinatorics of P in fact determines completely the manifold and the action, since they can be recovered abstractly from $P \times (S^1)^n$ by taking a natural quotient determined by the faces of P. This gives a cell decomposition of $\mathcal{Z}$ where each cell is a product of a face of P with a coordinate sub-torus of $(S^1)^n$ and shows immediately that $\mathcal{Z}$ is always connected.

(3) To read in the polytope the connectivity of $\mathcal{Z}$ and a splitting of its homology: After writing the chains of the cell complex and their boundaries in some examples, a pattern emerges that gives a decomposition of the homology of $\mathcal{Z}$ into a direct sum of small pieces, one for each union of facets of P.

$$H_i(\mathcal{Z}) = \oplus_{I \subset [n]} H_{i-|I|}(P, P_I)$$

$$P_I = \bigcup_{i \in I} P_i \text{ and } P_i = P \cap \{r_i = 0\}$$

The rough idea is that a relative cycle of (P, P_I) of dimension r, when rotated by an adequate sub-torus of dimension s produces a cycle of dimension $r + s$ of $\mathcal{Z}$ and this is compatible with boundary operators. And then one has to do the concrete computation from the normal form $n = n_1 + n_2 + \cdots + n_r$ of the above decomposition of the homology, which turns out to be without torsion.

One can see easily that $\mathcal{Z}$ is simply connected if, and only if, P intersects all coordinate subspaces $z_i = 0$, which is the case for any $\ell > 1$, and we will assume this from now on. But the homology of a manifold, even if simply-connected, does not determine its differentiable type. There is an unwritten principle in differential topology that states: if you want to know a compact manifold, try to find and understand a compact manifold it bounds. (That is how one understands the exotic spheres discovered by Milnor: in the sphere itself there are no non-trivial homotopy invariants, only when one looks at the homotopy invariants of a manifold it bounds is that one sees something unusual).

(4) The construction of a cobordism $\mathcal{Q}$, a compact manifold whose boundary is Z: Add one more variable x_0 to form the space $\mathbb{R} \times \mathbb{C}^n$ and assign to x_0^2 one of the coefficients in the equations of $\mathcal{Z}$ to build a manifold $\mathcal{Z}'$ given by equations

$$\lambda_1 x_0^2 + \Sigma_{i=1}^{n}\lambda_i |z_i|^2 = 0$$

$$\Sigma_{i=1}^{n} x_i^2 = 1$$

And then take half of it: let Q be its intersection with $\{x_0 \geq 0\}$. Then clearly $\mathcal{Z}$ is the boundary of Q. We need now to understand Q, which is a kind of intersection of quadrics... but not of the kind we have been studying!

First of all, it does not lay in a complex space: it involves one real variable x_0. And secondly, it is only one half of an intersection.

Well, that means that we have to start over again and do steps (1), (2), (3), (4) for the same type of intersections of quadrics, but now with real variables. And then do steps (1), (2), (3) for the halves of these quadrics to see what happens.

So we now have to consider all real intersections of quadrics Z:

$$\Sigma_{i=1}^{n} A_i x_i^2 = 0$$

$$\Sigma_{i=1}^{n} x_i^2 = 1$$

with $A_i \in \mathbb{R}^2$ and also Z_+, its intersection with $\{x_1 \geq 0\}$ which are again intersections of coaxial ellipsoids.

Part (1) is identical as before for both Z and Z_+: weak hyperbolicity of the configuration of coefficients A_i is equivalent to regularity and the normal form is again a regular polygon with multiplicities given by a partition of n. So is part (2) with polytope P:

$$\Sigma_{i=1}^{n} A_i r_i = 0, \quad \Sigma_{i=1}^{n} r_i = 1, \quad r_i \geq 0$$

only now it is the quotient of Z by the natural action of $\mathbb{Z}_2^n$ given by reflection on the coordinate hyperplanes of $\mathbb{R}^n$.

The cell decomposition of Z consists of P and all its images by those reflections. This lets us describe completely one case: for the partition $5 = 1 + 1 + 1 + 1 + 1$ it is easy to see that P is a pentagon with one edge in every coordinate hyperplane.

Then Z is given by iterated reflections of P on all 5 coordinate hyperplanes of $\mathbb{R}^5$, so it is formed by 32 pentagons, 80 edges and 40 vertices, from which we obtain its Euler characteristic and conclude that Z is the surface of genus 5.

Reflecting in only 4 of the 5 coordinate hyperplanes of $\mathbb{R}^5$ we obtain Z_+, which is easy to see that it is a torus minus 4 open disks.

Part (3) is more complicated because a face of P times a subgroup of $\mathbb{Z}_2^n$ is not a cell. But at the level of chains the process can be mimicked algebraically to obtain the following splitting:

$$H_i(Z) = \oplus_{I \subset \{1,\ldots,n\}} H_i(P, P_I)$$

$$P_I = \bigcup_{i \in I} P_i \text{ and } P_i = P \cap \{r_i = 0\}$$

and the concrete computation of the above summands from the normal form $n = n_1 + n_2 + \cdots + n_r$ follows as in the case of the homology of $\mathcal{Z}$.

The same computation works for the homology of any half of Z of the form $Z_+ = Z \cap \{x_j \geq 0\}$, only excluding the index j from I.

Now Z can be disconnected if P is non-empty and does not intersect a coordinate hyperplane $x_i = 0$. We can assume from now on that this does not happen (otherwise each component can be treated separately). It is simply connected if, and only if, any pair of facets of P has a non-empty intersection. To prove this last statement I used the dual of the cell structure of Z obtained by doing all the iterated reflections of P : Take a point y in the interior of P and reflect it on all the copies of P, thus obtaining the vertices of an n-cube C, one for each reflected copy of P, that is for every $(n-3)$ cell of Z. Then add n edges joining y to its reflections on the coordinate hyperplanes and reflect those edges in all possible ways, thus obtaining the 1-skeleton of C. These edges correspond to the $n-4$ cells of Z which are the reflections of the facets of P and for each pair of facets we have constructed an empty square and all its reflections. Whenever two facets of P intersect fill up those squares. Continuing this way until there are no more non-empty intersections of facets we get a polyhedron Z^* which has one cell of dimension $d-i$ (where $d = n-3$ is the dimension of Z) for every cell of dimension i of Z and the correspondence inverts the adjacency relation. Thus Z^* is combinatorially equivalent to the dual of the cell complex of Z defined by the copies of P so Z^* is homeomorphic to Z. Therefore, Z is simply connected when the 2-skeleton of Z^* equals the 2-skeleton of the cube, that is, when every pair of facets intersects. The same argument shows that Z is k-connected if, and only if, any collection of $k+1$ facets of P intersect, but that follows also from the computation of the homology. Much later, Z^* turned out to be a special case of the polyhedral product construction (Sect. 9.5.2).

(4) The construction of a cobordism: Let $Z' \subset \mathbb{R}^{n+1}$ be given by equations

$$A_1 x_0^2 + \Sigma_{i=1}^{n} A_i x_i^2 = 0$$

$$\Sigma_{i=1}^{n} x_i^2 = 1$$

And we take half of it:

$$Z'_+ = Z' \cap \{x_0 \geq 0\}.$$

(5) The topological description of the cobordism: it can be shown with some work that Z'_+ is a handlebody: a connected sum along the boundary of products of spheres with disks: $S^a \times D^b$. This requires the proof of several lemmas: that all the homology classes of Z'_+ can be represented by embedded spheres with trivial normal bundle, coming from the boundary, which joined by thin tubes form a handlebody H and that the space between the boundary of H and Z is an h-cobordism. The h-cobordism theorem implies that Z is diffeomorphic to the boundary of H. But the h-cobordism theorem requires some hypotheses: the cobordism should be simply-connected and of dimension at least 6 so the proof works only in with some restrictions. The result was proved in all cases with $\ell \geq 4$, for all but one of the cases for $\ell = 3$ and for most cases if $\ell = 2$. Fortunately, this includes all the original cases $\mathcal{Z}$ that motivated this study. By the end of 1984 I had the result, but it took me sometime to fill up the

details. The result is actually true with no exceptions, as we always believed, but it took several years and a change of point of view to complete:

For $\ell = 0$, Z is empty.
For $\ell = 1$, $Z = S^{n_1-1} \times S^{n_2-1} \times S^{n_3-1}$.
For $\ell > 1$, $Z = \#\left(S^{d_i-1} \times S^{n-d_i-2}\right)$, where $d_i = n_i + \ldots + n_{i+\ell-1}$.

To get the topology of the manifold $\mathcal{Z}$ corresponding to the same partition one just has to multiply all n_i by 2 in the above formulas. (That is why I never published the homology decomposition of $\mathcal{Z}$).

In late 1986 I received a letter from Terry Wall, including a copy of *Stability, pencils and polytopes*, where I learned that he had already done parts (1), (2) above, essentially in the same way, and about part (3), he had computed the homology of Z in a way I still have to study and understand some day. This was hard to swallow but, on the other hand, the splitting of the homology (that was immediately generalizable for the smooth intersection of any number of quadrics and now we know is valid for more general spaces and has a geometric foundation), the characterization of simple connectedness through the construction of the dual complex Z^* and the good parts (4) and (5) were not in Wall's paper.

9.3.2 Higher Dimensional Group Actions

Similar actions of $\mathbb{R}^m$ on $\mathbb{R}^n$ and $\mathbb{C}^m$ on $\mathbb{C}^n$ for $m > 1$ can be studied in the same way. In the work of Chaperon, the notion of weak hyperbolicity plays an essential role in the study of those non-linear actions.

The linear actions of $\mathbb{C}^m$ on $\mathbb{C}^n$ are given by m commuting linear vector fields that generate a foliation in C^n with generic leaves of complex dimension m and again, in some cases, there are leaves that do not have the origin in its closure called again Siegel leaves. Their union is an open set of $\mathbb{C}^n$ and in each Siegel leaf there is a unique point closest to the origin. The union of those points (which we will denote by $\mathcal{V}$) is the space of Siegel leaves of the system. It is invariant under the action of $\mathbb{C}^*$ and is given by the non-zero solutions of the equations

$$\Sigma \Lambda_i \left|z_i\right|^2 = 0$$

with $\Lambda_i \in \mathbb{C}^m$. To study this space it one intersects it with the unit sphere $\Sigma \left|z_i\right|^2 = 1$ to obtain, generically, a smooth compact manifold, which, we will denote again by by $\mathcal{Z}$.

This is clearly the quotient of $\mathcal{V}$ under the action of $\mathbb{R}_+$. We can take also the action of $\mathbb{C}^*$ and get a manifold in complex projective space, but that is another story: their relation with the complex flow make them a source of important examples in the theory of Complex Manifolds (see Sect. 9.4.1).

From the point of view of topology, this can be seen as an intersection of real quadrics. So is the case for certain intermediate objects defined by Laurent Meersseman, whose equations have complex variables, but real coefficients:

$$\Sigma A_i \, |z_i|^2 = 0$$

$$\Sigma \, |z_i|^2 = 1$$

with $A_i \in \mathbb{R}^m$, which we will denote by $Z^{\mathbb{C}}$. They are now known as *moment-angle* manifolds and we will have a lot to say about them in this and later sections in relation with dynamical systems, geometry and their role in Toric Topology and many generalizations.

So we know that it is enough to deal with the topology of the intersections of quadrics that appear in the real case (coming from linear actions of $\mathbb{R}^m$ on $\mathbb{R}^n$). We will see that many of the constructions for the case $m = 2$ can be generalized, but the classification of the regular cases turns out to be practically impossible for $m \geq 3$.

So we take all real intersections of quadrics Z:

$$\Sigma A_i x_i^2 = 0$$

$$\Sigma x_i^2 = 1$$

with $A_i \in \mathbb{R}^m$, that are known now known also as real *moment-angle* manifolds or as *intersections of coaxial ellipsoids in* $\mathbb{R}^n$. We try to follow the same steps as in the case $m = 2$:

For parts (2), (3) and (4) we have exactly the same general results with the same formulas as for the case $m = 2$ considered in the previous section.

As for part (1), characterizing the regular cases is easy: Weak hyperbolicity means that the origin is not in the convex hull of any set of m or less coefficients, and is equivalent to the fact that the system of equations is regular, so under this hypothesis Z is a smooth manifold of dimension $d = n - m - 1$.

But classifying them is a very different story. It is still true that, for all m, any WH configuration can be assumed to be in the unit sphere of $\mathbb{R}^m$ and then deformed into a primitive one with multiplicities, but there are just too many primitive configurations. Just image, for $m = 3$, any generic triangulation of S^2 and take the configuration formed by its vertices, which we can assume that satisfies WH. If in the interior of a triangle there is the antipode of some other vertex, then no two of the vertices of that triangle can be put together without breaking WH. If there is no such vertex, well, just add to the configuration a generic point that does have its antipode in that triangle. The result will be an arbitrarily complicated configuration that can not be reduced to a simpler one with multiplicities.

With the failure of step (1) we cannot expect to describe all possible topological types of Z, it seems that we can only look at the simplest examples. But, still, some families of them have been detected and studied:

An interesting family with $m > 2$ was introduced to us by Hirzebruch in 1986: in our terms, they are obtained by taking P to be the n-gon embedded in $\mathbb{R}^n_+$ with an edge in each coordinate hyperplane. Then the associated intersection of quadrics is formed by 2^n faces, $n\,2^{n-1}$ edges and $n\,2^{n-2}$ vertices, so it is the surface of Euler characteristic $2^{n-2}(4-2n+n) = 2^{n-2}(4-n)$ and genus $2^{n-3}(n-4)+1$. Some other surfaces, with non-diagonal quadratic equations were also found in joint work with Vinicio Gómez Gutiérrez [40].

The varieties Z we had described for $m = 2$ have also a rather simple topology, no much more complicated than that of a higher dimensional analog of a compact orientable surface. One could also take products of the known ones any number of times, to obtain more examples for $m > 2$ and for sometime it looked as if they could be all like this, or at least that they would all have free homology groups. For sometime, nothing moved.

Around the turn of the century the paper by Frédéric Bosio and Laurent Meersseman, *Real quadrics in* $\mathbb{C}^n$, *complex manifolds and convex polytopes* [15], in the context of moment-angle manifolds $Z^{\mathbb{C}}$, showed that this was not true: they showed that for any simplicial complex K one can find a simple polytope P and a set of its facets I such that the quotient P/P_I has the homotopy type of K. This implies in particular that the associated $Z^{\mathbb{C}}$ can have any amount of torsion in its homology groups. This is true also for the associated Z in the corresponding summand $H_i(P, P_I)$ of its homology group.

But the paper [15] not only proved that the problem was more complicated than we had thought, it also opened many doors to attack it: a formula for the product in the cohomology ring of $Z^{\mathbb{C}}$, a very beautiful theory of cobordism of simple polytopes to describe the passage from one type of simple polytope to another one, including the description of the simplest changes of the topological type of $Z^{\mathbb{C}}$ (*wall-crossing*) when the polytope suffers an elementary transformation between two different combinatorial types (one of such transformations is the truncation of a vertex of the polytope), as well as many more ideas and questions. Among them, they brought attention to a special family of polytopes, called dual-neighbourly: a simple polytope P of even dimension $d = 2p$ is called *dual-neighbourly* if every collection of p facets of P has a non-empty intersection (Cf. [17, p. 92] and [15, p. 114]). They are dual to the much studied *neighbourly* ones. Studying the cohomology ring of $Z^{\mathbb{C}}$ for those polytopes, they conjectured that they were connected sums of sphere products, just like in the case $m = 2$. But it took sometime and a change of point of view to be able to use the ideas and to attack the questions contained in this paper.

What we knew so far was enough for doing some more work in dynamical systems as we will see now, and in Complex Geometry as we will see in the following section.

9.3.3 Generalized Hopf Bifurcations

In the classical Poincaré-Andronov-Hopf bifurcation, an attracting fixed point of a flow in $\mathbb{R}^2$ bifurcates into an attracting cycle. Answering a question by Jean Diebolt, Marc Chaperon and his Ph.D. student Mathilde Kammerer-Colin de Verdière showed that it was possible in higher dimensional spaces to obtain attracting spheres from a fixed point and, they also obtained products of any number of spheres. "Spheres and products of spheres" sounded familiar to me. After I suggested that there could be other moment-angle manifolds playing the same role I was invited to join them. First we obtained some, and finally we obtained all of them [24].

A brief idea of the existence of such a bifurcation is explained as follows: Take a non-empty intersection of quadrics Z:

$$\Sigma_{i=1}^{n} A_i x_i^2 = 0$$

$$\Sigma_{i=1}^{n} x_i^2 - 1 = 0$$

with $A_i \in \mathbb{R}^m$ satisfying the weak hyperbolicity condition.

Then the vector field

$$X = \Sigma_{i=1}^{n} x_i \left(1 - \Sigma_{j=1}^{n} x_j^2 - \Sigma_{j=1}^{n} A_i \cdot A_j x_j^2\right) \frac{\partial}{\partial x_i}$$

has Z as an attracting normally hyperbolic point-wise invariant manifold.

It took us some time to realize that $X = -1/2\nabla\Phi$ where Φ is the sum of the squares of the left sides of the equations of Z above. This simplified the proof of the previous statement considerably [24].

Now consider a given a family of vector fields $\xi_u(z)$ on $\mathbb{C}^n$ with parameter $u \in \mathbb{R}^q$ such that $\xi_0(0) = 0$ and the eigenvalues of $D\xi_0(0)$ are all purely imaginary, different and non-zero. Then we can assume by changes of coordinates in C^n depending on the parameter u that $\xi_u(0) = 0$ and that $D\xi_0(0)$ is a linear diagonal vector field $L(x)$ whose diagonal terms are the eigenvalues. In other words, that it represents a collection of uncoupled linear oscillators.

Applying the theory of normal forms one can assume now that, for all j, the jth coordinate of ξ_u is, up to terms of order greater than 3, of the form

$$z_j \left(\lambda_j(u) + i\mu_j(u) - \Sigma_{k=1}^{n}(a_{jk}(u) + ib_{jk}(u))|z_k|^2\right)$$

where $\lambda_j(u)$, $\mu_j(u)$, $a_{jk}(u)$ and $b_{jk}(u)$ are smooth functions such that $a_{jk}(0) = 1 + A_j \cdot A_k$. Assume also that for some vector v_0, we have $D\lambda_j(0)v_0 = 1$.

A deep theorem by Chaperon [23] implies the following: Assuming some simple non-resonance conditions on the eigenvalues, that $q \geq n$ and some generic conditions on the λ_j, a_{jk}, then there is an open set of values of u that has 0 in its closure and contains an open cone containing v_0 such that ξ_u admits an invariant, attracting, normally hyperbolic manifold, diffeomorphic to the moment-angle manifold:

$$\Sigma_{i=1}^{n} A_i |z_i|^2 = 0$$

$$\Sigma_{i=1}^{n} |z_i|^2 = 1$$

For the proof in a more general setting, including the version of the above result in the case of transformations instead of flows, as well as many other results about the birth of invariant manifolds from fixed points, the reader is referred to [23].

With some of the ideas above, Genaro de la Vega and I constructed a theoretical dynamic model generalizing the famous system by May and Leonard to any number of species in competition [30]. In the May-Leonard system each of three competing species is dominant for some time and after a while its population decreases, letting another species be the dominant one until it comes the turn of the third one to dominate, and that process repeats cyclically, again and again. The process can be represented in the first octant as a dynamical system where all trajectories within its interior tend towards the unit simplex and those in the simplex itself spiral towards its border.

As in the case $5 = 1 + 1 + 1 + 1 + 1$, whose polytope is a regular pentagon in $\mathbb{R}^5_+$, we constructed an embedding of the regular n-gon in $\mathbb{R}^n_+$ with the same properties and then define the dynamical system which has it as an attracting point-wise invariant normally hyperbolic manifold. Then, adding a (small) vector field with a repelling fixed point in the center of the n-gon from which all the other trajectories escape spiraling toward the borders, one obtains a system with a slightly curved invariant n-gon (due to the normal hyperbolicity) on which the dynamics is of the same type, thus making each species in his turn to be dominant over the other ones.

The intersection of quadrics corresponding to this regular n-gon is invariant under the dihedral group. These surfaces are well-known in the theory of Riemann surfaces as certain complete intersections of complex quadrics. Our version led to the question of finding higher dimensional ones with this symmetry and from there to an interesting development with unexpected connections with other fields of pure and applied Mathematics associated to the Discrete Fourier Transform matrix ([54]).

It would be interesting to produce some simple examples of bifurcations where Z is a sphere to illustrate the possibilities of this type of bifurcation in applications.

9.4 Geometry

Following ideas from Borcea, Haefliger and Loeb-Nicolau, Alberto Verjovsky and I constructed from the intersections of quadrics $\mathcal{Z}$ related to the Siegel leaves of an action of $\mathbb{C}$ a family of compact, complex manifolds that in most cases are not symplectic [56]. A generalization by Laurent Meersseman [59], now called LV-M manifolds, that includes all even-dimensional moment-angle manifolds, has been much studied and has many implications that go from the classical Complex Manifold Theory to a recently developed Quantum Toric Geometry. The reader will find an excellent and more detailed survey by Alberto Verjovsky in [73].

The odd-dimensional moment-angle-manifolds admit a contact structure, and so do other non-diagonal intersections of quadrics constructed by Barreto and Verjovsky. Also, the moment-angle-manifolds appear in Symplectic Geometry as Lagrangian submanifolds with special properties.

9.4.1 Complex Geometry

The Classical Examples and Their Deformations

The first example of a compact complex manifold which is not algebraic was given by Hopf in 1948. It is constructed as follows: Given a real number $r > 1$, we can take in $\mathbb{C}^n \backslash 0$ the action of the infinite cyclic group given by $m \cdot z = r^m z$. The quotient manifold is diffeomorphic to $S^{2n-1} \times S^1$. Since the action is holomorphic and totally discontinuous, this manifold inherits a natural complex structure. To see it is not symplectic, and therefore cannot be algebraic, we can use the following well known facts:

- *On every symplectic manifold M of real dimension $2n$ there exists an element $x \in H^2(M)$ such that $x^n \neq 0$ in $H^{2n}(M)$.*
- *Every projective, algebraic manifold admits a Kähler structure and, in particular, it admits a symplectic structure.*

For the Hopf manifolds, since $H^2(S^{2n-1} \times S^1) = 0$ for $n > 1$, they cannot be symplectic.

Calabi and Eckmann [20] generalized Hopf's construction to give complex structures on $S^{2p-1} \times S^{2q-1}$, which are again non-symplectic for the same reason. They used the holomorphic bundle with fibre an elliptic curve

$$S^{2p-1} \times S^{2q-1} \to \mathbb{C}P^{p-1} \times \mathbb{C}P^{q-1}$$

The Hopf manifold was constructed using a dynamical system. But nobody used this fact until Haefliger, inspired by Arnol'd [2] and completing results by C.Borcea, generalized the construction by taking the quotient of $\mathbb{C}^n \backslash 0$ by any action of the infinite cyclic group which is holomorphic and totally discontinuous. For all these, the quotient is again topologically $S^{2n-1} \times S^1$. Haefliger [43], using the Poincaré-Dulac Theorem, obtained a complete description of these manifolds and their small deformations, showing in particular that they come either from linear actions, like

$$m \cdot (z_1, \ldots, z_n) = (\alpha_1^m z_1, \ldots, \alpha_n^m z_n)$$

(where α_i are complex numbers with $|\alpha_i| > 1$), or from certain non linear deformations of them when there are resonances among the α_i.

Jean Jacques Loeb and Marcel Nicolau extended Haefliger's results to the Calabi–Eckmann situation, thus obtaining a very complete description of a large class of complex structures on $S^{2n-1} \times S^{2m-1}$ together with their deformations [47]. This was based on the study of dynamical systems in the Poincaré domain which are transversal to that product of spheres and give different complex structures on it when parameters vary.

The *LV-M* Manifolds

When we learned in a conference about the work of Loeb-Nicolau, Alberto Verjovsky immediately suggested the construction of complex manifolds from dynamical systems in the Siegel domain and soon we were able to do that for $m = 1$, obtaining a class of non-Kähler compact, complex manifolds, called now *LV manifolds*, that included all the classical examples and many, many more. Soon after, Loeb and Nicolau studied the complex geometry of them [48].

Then Laurent Meersseman extended the construction for all m producing what are now known as *LV-M manifolds*, and went a lot deeper in the analysis of the complex geometry: he studied the existence of meromorphic functions and 1-forms, holomorphic vector fields, transverse Kähler foliations, analytic subsets and holomorphic submanifolds, Hodge numbers,...

Later, in joint work with Alberto [60] he showed that, under a rationality condition, the leaves of the transverse Kähler foliations are compact and are actually the fibers of holomorphic bundles over toric varieties, actually over *any* toric variety with at most certain simple singularities. This generalizes enormously the classical construction of Calabi–Eckmann. All this experience has even led Meersseman to obtain new general results in the classical theory of deformations of complex structures.

Recalling that $\mathcal{V}$ is the space of Siegel leaves of a linear action of $\mathbb{C}^m$ on $\mathbb{C}^n$ and that it is transversal to the foliation by them on an open set, we can use an observation by Haefliger stating that this implies that it has a complex structure. To get a compact, complex manifold we can take its quotient by the action of multiplying by a real number bigger than one, obtaining a complex structure on $\mathcal{Z} \times S^1$ (as in the Hopf manifolds) or, more interestingly, by taking the quotient of $\mathcal{V}$ by the scalar action of $\mathbb{C}^*$, i.e., by projectivizing it. So we obtain a *projective moment-angle manifold*:

$$\mathcal{PZ} = \mathcal{V}/\mathbb{C}^* = \mathcal{Z}/S^1 \subset \mathbb{C}P^{n-1}$$

We can also view the complex structure of $\mathcal{PZ}$ as follows: the foliation by Siegel leaves is invariant under the action of $\mathbb{C}^*$ so defines a holomorphic foliation of an open set in $\mathbb{C}P^{n-1}$ to which $\mathcal{PZ}$ is transversal.

$\mathcal{PZ}$ is given by homogeneous equations in $\mathbb{C}P^{n-1}$:

$$\Sigma \Lambda_i \, |z_i|^2 = 0$$

with $\Lambda_i \in \mathbb{C}^m$, but, again, it is not a holomorphic submanifold of $\mathbb{C}P^{n-1}$.

These are the *LV-M manifolds*.

They include all the linear classical examples mentioned above (the Hopf and Calabi-Eckman manifolds and their linear deformations) which we know are not symplectic.

Another source of known LV-M manifolds is the following: when a moment-angle manifold $Z^{\mathbb{C}}$ is odd dimensional, it can be considered (in many ways) as the real version of one with complex coefficients and quotient an LV-M manifold. But, when it is even dimensional, Meersseman observed that $Z^{\mathbb{C}} \times S^1$ is odd dimensional, so its projectivization is an LV-M manifold. But its projectivization is $Z^{\mathbb{C}}$ itself! So we have many examples of LV-M manifolds, in particular, many that are 2-connected, so they are compact complex manifolds that are obviously not symplectic.

The LV-M manifolds also include all elliptic curves and all the complex tori $(S^1)^{2k}$. All of them are complex manifolds that are Kähler and therefore symplectic, but appear here embedded in $\mathbb{C}P^{n-1}$ in a non-holomorphic way. They are just real algebraic smooth subvarieties. It can be proved that these are the only LV-M manifolds that admit a symplectic structure.

The topology of LV-M manifolds, other than the ones that are moment-angle manifolds, is much more complicated than those we studied above. In a few cases they are connected sums of some sphere products plus a sphere bundle. But it seems that most of them have a cohomology ring that is not that of a connected sum.

Many developments associated with LV-M manifolds followed, among others:

A generalization of LV-M manifolds by Frédéric Bosio, now called LV-M-B manifolds [14].

The interpretation of the LV-M-B manifolds in the context of Geometric Invariant Theory (GIT) quotients by Cupit-Foutou and Zaffran [27].

The work on almost complex structures by Demailly and Gaussier [31], where LV-M-B manifolds play an important role.

The tranversally Kähler foliation on the LV-M manifolds when the rationality condition does not apply has created a lot of interest among geometers, generating new geometric objects like quasifolds by Prato [67], via symplectic geometry, and Kähler quasifolds by Battaglia and Prato [11]. Recent work by Battaglia and Zaffran appears in [12]. These foliations, understood as *non-commutative* toric varieties, are the basis of the new Theory of Quantum Toric Geometry by Katzarkov et al. [45, 46].

9.4.2 Contact and Symplectic Geometry

If the even dimensional moment-angle manifolds do not admit symplectic structures (except for a few, well-determined cases), all the odd-dimensional ones (and large families of intersections of ellipsoids) admit contact structures. Additionally, there are contact structures in a family of concentric intersections of ellipsoids. And then moment-angle manifolds do appear in symplectic geometry as lagrangian submanifolds of certain types.

All Odd-Dimensional Moment-Angle Manifolds Admit Contact Structures

If $Z^{\mathbb{C}}$ is odd-dimensional then it admits a contact structure [8].

This is proved by showing that $Z^{\mathbb{C}}$ is an almost-contact manifold. A recent theorem by Borman et al. ([13]) implies the result. (Recall that a $(2n+1)$-dimensional manifold M is called *almost contact* if its tangent bundle admits a reduction to $SU(n) \times \mathbb{R}$). The proof uses the S^1-bundle map $Z^{\mathbb{C}} \mapsto Z^{\mathbb{C}}/S^1 = \mathcal{PZ}$ and the fact that this quotient is a LV-M and therefore admits a complex structure. Therefore, $Z^{\mathbb{C}}$ has an atlas modeled on $\mathbb{C}^{n-2} \times \mathbb{R}$. Computing the changes of coordinates of the charts one sees that their differentials lie in a subgroup of $GL(2n-3, \mathbb{R})$ that retracts by Gram-Schmidt onto $SU(n-2) \times \mathbb{R}$.

Large Families of Odd-Dimensional Coaxial Intersections of Ellipsoids Admit Contact Structures

First consider the odd-dimensional coaxial intersections of ellipsoids that are connected sums of spheres products:

An odd dimensional product $S^m \times S^n$ of two spheres admits a contact structure by results of Eliashberg [34] and Giroux [38]. And it was shown by Meckert [58] and more generally by Weinstein [75] (see also [34]) that the connected sum of contact manifolds of the same dimension is a contact manifold. Therefore all odd dimensional connected sums of sphere products admit contact structures.

Additionally, it was proved by Bourgeois [16] (see also Theorem 10 in [38]) that if a closed manifold M admits a contact structure, then so does $M \times T^{2m}$. Therefore, all intersections of ellipsoids $Z \times T^{2m}$, where Z is a connected sum of sphere products, admit contact structures.

As we shall see in Sect. 9.5.2, there are very many intersections of coaxial ellipsoids are connected sums of sphere products so all the odd-dimensional ones admit contact structures and so do their products with even dimensional tori [8].

A New Family of Odd-Dimensional Concentric Intersections of Ellipsoids That Admit Contact Structures

In [9] there is a different construction of contact structures on certain odd-dimensional concentric but not coaxial intersections o f ellipsoids.

$$w_1^2 + \Sigma_{j=1}^n \lambda_j^1 |z_j|^2 = 0$$

$$\dots$$

$$w_m^2 + \Sigma_{j=1}^n \lambda_j^m |z_j|^2 = 0$$

$$\Sigma_{k=1}^m |w_k|^2 + \Sigma_{j=1}^n |z_j|^2 = 1$$

The proof uses a geometric heat flow due to Altschuler and Wu [1] to deform a 1-form into a contact structure. The topology of these new objects has not yet been studied.

Intersections of Quadrics as Lagrangian Submanifolds

Intersections of ellipsoids do play a role in Symplectic Geometry as special types of Lagrangian submanifolds or constitutive elements o f them:

(a) A Lagrangian submanifold $L \subset \mathbb{C}^n$ is called H-minimal if its volume is critical under Hamiltonian deformations. They are the analogs of minimal submanifolds in Differential Geometry. Mironov and Panov [61] constructed embeddings of many intersections of quadrics as Lagrangian H-minimal submanifolds.

(b) Another type of Lagrangian submanifold called monotone (see [63] for a definition) have been constructed by Vardan Oganesyan with the help of intersections of ellipsoids. He constructs interesting examples of such submanifolds that are fibre bundles over tori of different dimensions, whose fibers are products of spheres, connected sums of products of spheres like $\#_5(S^{2p-1} \times S^{n-2p-2})$, or the surfaces of genus 5 or 17. Some of these manifolds have different embeddings as monotone Lagrangians with different minimal Maslov index. See also [64].

9.5 To the Polyhedral Product Functor

9.5.1 *Coxeter Groups, Small Covers and Toric Manifolds*

Unaware of the work described in the previous sections, Michael Davis and Tadeusz Januszkiewicz, in the very important article *Convex polytopes, Coxeter orbifolds and torus actions* [29], give the abstract construction of the intersections of coaxial ellipsoids and moment angle manifolds that was mentioned in the previous sections, only in a more general setting.

The main objective of the article is the study of certain similar spaces constructed from simple polytopes called *small covers* and *toric manifolds* which are seen as real versions of the classical toric varieties of Complex Algebraic Geometry. They are defined as quotients of products of the polytope times a power of $\mathbb{Z}_2$ or a torus. These constructions are extended in Chap. 2 of the paper to *simple polyhedral complexes* (not necessarily convex) seen as duals of simplicial complexes and these sections cover the first half of the paper.

It is only in Sect. 4 that spaces that include the ones we have denoted Z and $Z^{\mathbb{C}}$ appear in Sect. 4.1, as quotients of the product of a simple polyhedral complex P times a group that can be a power of $\mathbb{Z}_2$ or a torus and are denoted generically by $\mathscr{Z}$. The space corresponding to the group $\mathbb{Z}_2$ is identified, in the case that P is a simple polytope, as the *universal abelian cover of the polytope seen as a right-angled orbifold*. Otherwise, these spaces receive no name. (It was later that the now generally accepted name *moment-angle complexes* was given to them and the letter Z with more or less ornaments is used for them and their variants). These spaces are not mentioned in the very long and detailed introduction to the article, their properties are not much studied and no examples are discussed. One gets the impression that they are conceived mainly as tools for constructing other very interesting spaces and proving some very deep results about them, such as the computation of their cohomology groups and rings, in particular the analog of the Danilov-Jurkiewicz Theorem about toric varieties. The article also connects these spaces with Coxeter groups and with symplectic manifolds constructed by Delzant.

This article is usually called *seminal* for a very good reason: it has generated an enormous quantity of very important work on different areas of Algebraic Topology and other fields. The amount of this work generated by the slightly mentioned objects $\mathscr{X}$ is comparable to the one generated by the main objects of study in the article (small covers and toric manifolds). It would be difficult to determine which of the two has had a greater offspring. A few years later, in a classic on the subject, the book by Buchstaber and Panov [18], we see already the two types of objects in equal terms. Later classics like [6, 15, 32] deal only with moment-angle manifolds or moment-angle complexes.

The relation between these objects and the intersection of quadrics was first mentioned in two papers, coming from the two different sides: the published version of the paper by Bosio and Meersseman [15] and the paper by Denham and Suciu [32], always in the context of moment-angled manifolds $Z^{\mathbb{C}}$.

One can also observe that the paper by Wall has had repercussions in Singularity Theory, Algebra and Algebraic Geometry, while the papers on intersections of quadrics have had repercussions in Dynamical Systems, Complex Geometry and Singularity Theory. But for a long time neither of them attracted the attention of algebraic topologists. Only decades after their first appearance, when the relation between the two points of view was stablished, have they been recognized as precursors of the algebraic topology of polyhedral products .

9.5.2 The Polyhedral Product Functor

We cannot give a complete history of the development of the subject after [29], we refer the reader to the books by Buchstaber and Panov [18, 19], to the article [6] by Anthony Bahri, Martin Bendersky, Fred Cohen and Samuel Gitler about *polyhedral products*, which are generalized moment-angle complexes and the recent survey [7] and its bibliography.

Various precursors of the construction of moment-angle complexes and their splittings converge in [6] to the *polyhedral product functor construction*,[2] whose definition we quote:

Let $\underline{X} = \{(X_i, A_i)_{i=1}^m\}$ denote a set of pairs of CW–complexes.

Let K denote an abstract simplicial complex with m vertices labeled by the set $[m] = \{1, 2, \ldots, m\}$.

For every simplex σ in K, let $X^\sigma = \prod_{i=1}^m Y_i \subset \prod_{i=1}^m X_i$ where

[2] Another precursor discovered later was Coxeter [26] who built surfaces by essentially the same procedure.

$$Y_i = \begin{cases} X_i \text{ if } i \in \sigma \\ A_i \text{ if } i \in [m] \setminus \{\sigma\}. \end{cases}$$

The polyhedral product is

$$Z(K; \underline{X}) = \bigcup_{\sigma \in K} X^\sigma$$

In other words, $Z(K; \underline{X})$ is the set of points $(x_1, x_2, \ldots, x_m) \in \prod_{i=1}^m X_i$ such that the set of i such that x_i is not in A_i is a simplex of K. (A point in X^σ may have more points in A_i for $i \notin \sigma$, but that only means that the set of points not in A_i is smaller than σ and therefore a simplex of K). The formal definition above is important for considering $Z(K; \underline{X})$ as an homotopy colimit.

This is a functor from the category simplicial complexes with morphisms simplicial embeddings to the category of CW-complexes and continuous maps.

This definition includes all the spaces we have defined above (Z, $Z^{\mathbb{C}}$, Z_+) and many more that might appear in the future, as well as the spaces $\mathscr{L}$ and many others like, for example, the complexes $Z(K; (X, A))$ constructed in [32], which is the case where all the spaces (X_i, A_i) are equal.

To see this in the case of an intersection Z with an associated polytope P of dimension d with n facets, one on each coordinate hyperplane, we take K to be the dual of the boundary of P. Then Z is homeomorphic to $Z(K; (D^1, S^0))$: the above construction coincides with the that of the dual cell complex Z^* of Sect. 9.3.1. Also one has

$$Z^{\mathbb{C}} = Z(K; (D^2, S^1))$$

$$Z_+ = Z(K; ((D^1, \{1\}), (D^1, S^0), \ldots, (D^1, S^0))$$

So these complexes include also the half-intersections Z_+ that are not included when all the (X_i, A_i) are the same. But in the same way they include fourths, eights, etc., and any manifolds with corners that may appear.

But here come the best part of all: in [6] there is a *geometrical splitting of the suspension of* $Z(K; \underline{X})$ valid in all cases, in terms of the suspensions of simpler spaces! This is a fantastic confirmation and generalization of the splittings of homology we mentioned in Sects. 9.3.1 and 9.3.2. The splitting, being geometrical, extends to any generalized homology or cohomology theory, is compatible with stable cohomology operations o f any kind or order, etc.

For our case, where K is the dual of the boundary of the simple polytope P the result of the splitting is that there are homotopy equivalences:

$$\Sigma Z = \Sigma(Z(K; (D^1, S^0))) \to \bigvee_{I \notin K} \Sigma^2 |K_I|.$$

$$\Sigma Z^{\mathbb{C}} = \Sigma(Z(K;(D^2,S^1))) \to \bigvee_{I \notin K} \Sigma^{2+|I|}|K_I|.$$

where $K_I \subseteq K$ is the *full sub-complex* of K consisting of all simplices of K which have all of their vertices in I.

Polyhedral products have had enormous impact in Algebraic Topology, especially in homotopy theory, which we cannot describe. We refer to the recent survey [7]. We will try to explain only some of the consequences in the study of the topology of intersections of quadrics.

The pair of articles [6, 15] gave a new life to the study of the topology of intersections of quadrics. Combined, they showed that the problem of describing the topology of them was unthinkable: by the realization theorem of [15] (see Sect. 9.3.2), and the homotopy splitting theorem of [6], any stable cohomological operation, primary, secondary or of any order can appear non-trivially in an intersection of ellipsoids.[3] But more than any of the deep technical results that they contained, these articles were for us an incredible source of questions and ideas, a way of looking not at each example one by one, but of looking at the whole of them and their connections.

Having discovered the connection between our objects of study, and after some time of adapting to each other's point of view [53], Samuel Gitler and I were able to prove the Bosio-Mersseman conjecture (see the end of Sect. 9.3.2) by starting with the real case: P is a dual-neighbourly polytope of dimension $2p$, or equivalently, $Z(P)$ is $(p-1)$-connected, so it has homology in dimension p only (besides the 0 and top dimensional ones) and it must be torsion-free. We proved in [37], adapting the techniques of the case $m = 2$ to this more abstract situation (constructing the cobordism, representing its homology by embedded spheres coming from the boundary and applying the h-cobordism theorem) that, if $p \neq 2$, then $Z(P)$ is diffeomorphic to a connected sum of copies of $S^p \times S^p$.

Then we showed that certain operations on P preserved the fact that the corresponding $Z(P)$ are connected sums of sphere products:

(a) The book construction $P \mapsto P'$ on any simple polytope P consists in taking the product $P \times [0, 1]$ and one of its facets $F_i = P \cap \{x_i = 0\}$ and, for each given point $u \in F_i$, identifying all points (u, t) for $t \in [0, 1]$ into a single point. In terms of the configuration of coefficients it corresponds to repeating the coefficient A_i as a coefficient of a new variable. The proof that this operation preserves connected sums follows the same techniques of case $m = 2$ and also requires dimension and connectivity hypotheses. So one can consider succesive book constructions along different facets which amounts to giving multiplicities to the A_i as before.

In a particular case, if P is dual-neighbourly of dimension $d = 2p$ and we give multiplicity 2 to the coefficients of Z we obtain $Z^{\mathbb{C}}$ that is therefore also a connected sum of sphere products and this proves the original Bosio–Meersseman conjecture for $p \geq 2$ (when $p = 2$ we can start from P'). But this is only one case of an infinite

[3] For sometime, it had been known that non-trivial triple Massey products can appear and are frequent [10, 32].

lattice of polytopes obtained by successive book constructions on different facets starting from P or P' whose associated manifolds are connected sums if $d \geq 4$.

(b) The truncation of a vertex of P gives a new polytope P_{v} and its corresponding manifold Z_{v}. Z_{v} is diffeomorphic to $Z\#Z\#(2^{m-d}-1)(S^1 \times S^{d-1})$.

(b') The truncation of a vertex of P followed by the book construction on the new facet produces a new polytope $(P_{\mathrm{v}})'$ and the corresponding manifold $(Z_{\mathrm{v}})'$, it preserves simply connectedness and we have a complete description of its topology in terms of that of Z similar to the one in (b).

(c) The truncation of an edge of P gives a new polytope P_{e} and its corresponding manifold Z_{e}. If Z is simply connected, then Z_{e} is diffeomorphic to $Z\#Z\#(2^{m-d-1})(S^2 \times S^{d-2})\#(2^{m-d-1}-1)(S^1 \times S^{d-1})$. This can be false if Z is not simply connected.

(c') The truncation of an edge of P followed by the book construction on the new facet produces a new polytope $(P_{\mathrm{e}})'$ and its corresponding manifold $(Z_{\mathrm{e}})'$. If Z is simply connected, then an explicit geometrical description like the one in c) can be given of $(Z_{\mathrm{e}})'$, which turns out to be again simply connected.

Starting from P or P', operations (b), (b'), (c) and (c') can be applied iteratively any number of times each with only one condition on its order: operations (c) and (c') can not longer be applied after one of the operations (b) or (c), because they make the manifold non-simply connected. The topology of the final result is a connected sum of sphere products that can be completely described in terms of that of Z.

Starting from P we can also apply iteratively operations (a), (b') and (c') in any order and we obtain always polytopes whose associated manifolds are connected sums. At the end one can also add one operation (c) and, after that, any number of operations (b).

But this is still not a complete description of the topology of those connected sums: we would need to know how many terms and which products appear.

Recently I proved that for an even dimensional dual-neighbourly polytope P of dimension $d = 2p$ at least 6 and and $n = d + m + 1$ facets, the number of terms in the connected sum, which we can naturally call the *genus* of $Z(P)$ and will denote by $g(p, m)$, can be expressed in any of the following equivalent forms [55]:

(i) $g(p, m) = \Sigma_{j=0}^{m-1} \binom{j+p}{p} 2^j$

(ii) $g(p, m)$, *as a sequence parametrized by* m, *has generating function*

$$\frac{z}{(1-z)(1-2z)^{p+1}}$$

For $p = 1$ this gives the gives the formula for the surfaces and for $p = 2$ we get the conjectured genus of the 4-dimensional ones, in which case it has not been proved (but must be true) that they are connected sums of sphere products. For small m these sequences appear in the Sloane Encyclopedia of Integer Sequences [72] with different combinatorial and geometrical interpretations.

Operation (a) does not change the genus and Z' is the connected sum of $g(p, m)$ copies of $S^p \times S^{p+1}$. But when applied more times and in different facets, or after other of the operations, it is difficult to have control of which sphere products appear because one has to understand the facets of P and even so the splitting of the homology may render the computation complicated. So we can only determine the genus of the result. The reader may try the case where P is the hexagon.

The good news is that there is a very large number of dual-neighbourly polytopes: experts in the field consider that most of the simple polytopes are neighbourly (which is verified by explicit computations) and it is a fact and their number grows very fast with their dimension. See [42, pp. 129, 129a and 129b], [77, p. 402, Sect. 4] and [66]. And these are only the roots of infinite lattices of non-neighbourly polytopes stemming from each of them by applying the book construction on the different facets. If we apply on them iteratively the truncation operations we will get a really enormous quantity of intersections that are connected sum of sphere products and for a large number of them we can specify completely their topology.

We also solved another question from [15], when we showed, computing the cohomology rings, that the manifolds Z and $Z^{\mathbb{C}}$ associated to the truncated cube are not connected sums and we showed that there is an essential difference between the two, thus contradicting a published result. At the end we announced, prematurely, a formula for the cohomology ring of any Z, but the proof ran into some technical problems and then the Bahri–Bendersky–Cohen–Gitler team obtained a version in their own language. Matthias Franz [35] has recently proved the announced formula. I still hope to complete one day its geometrical proof in the spirit of our old splitting of the homology.

With this new perspective, in joint work with Vinicio Gómez Gutiérrez, we were able to use our version of Wall's topological normal form

$$q_1 = \Sigma_1^r a_i x_i^2 + \Sigma_1^s 2u_j v_j$$

$$q_2 = \Sigma_1^r b_i x_i^2 + \Sigma_1^s (u_j^2 - v_j^2)$$

to describe the topology of all the diagonalizable and non-diagonalizable smooth intersections (i.e., intersections of concentric ellipsoids) for $m = 2$ [41].

They include the unit tangent bundle of the sphere, products of two spheres and connected sums as before, including cases with three terms, as well as the ones in the diagonal case with s added to the dimensions of the spheres in each term.

For the proof we used another operation consisting in adding new terms to the equations (passing from s to $s+1$) and the truncation method inherited from [15], but with our own touch, to deal with all the remaining cases, including the diagonal ones (intersections of coaxial ellipsoids) not proved in 1984. (While writing this I realize that this is a kind of geometric version of Wall's frustum method for computing the homology (see Sect. 9.2).

Other results obtained were the construction of examples for equivariant cohomology [36], the study of the smoothness in some intersections of quadrics with dihedral symmetry [54] originated in the work about the generalized May-Leonard system (9.3.3) and the study of the manifold associated to the dodecahedron [3]. A different approach [39], based on the theory of oriented matroids allows us, among other things, to deal with some 4-dimensional cases. There is still much work to do in the description of all those intersection of quadrics involved, some of which are non-diagonal.

9.6 Back to Singularity Theory

Recall the varieties $\mathcal{Z}$ of Sect. 9.3.1, whose equations can be written in the form $\Sigma\lambda_i z_i \bar{z_i} = 0$, $\Sigma z_i \bar{z_i} = 1$. The study of these varieties suggested several new developments in Singularity Theory involving polynomials in complex variables z_i and their conjugates. One of these developments was the important work of José Seade about singularities of varieties and of vector fields on them, in the spirit of the Milnor Fibration Theorem [69, 70] and the more recent [71]). This work (and the article [44] I wrote with my Ph.D. student Luis Hernández de la Cruz, more in the spirit of the classical classification of singularities), initiated a long list of articles, mainly by Seade himself, by José Luis Cisneros and by Mutsuo Oka (starting with [25] and [65]) and their collaborators and followers.

But the real varieties Z can also been studied in the spirit of Singularity Theory. The cones on the smooth ones can be seen as varieties with an isolated singularity which asks to be smoothed and one can also study the singular intersections of quadrics Z (for which there is also a formula for the splitting of its homology) and the smoothings of their simplest cases. The following generalizes to every m published results about the case $m = 2$ [51, 52]. Some of the those results will appear in [4].

9.6.1 *Quadratic Cones*

Now we consider a quadratic homogeneous mapping:

$$F : \mathbb{R}^n \to \mathbb{R}^m$$

$$F(x) = \sum_{i=1}^{n} A_i x_i^2$$

where $A_i \in \mathbb{R}^m$.

For $B \in \mathbb{R}^m$, let $V_B = F^{-1}(B)$ and $\varphi_B(x) = |x|^2$ restricted to V_B. So V_0 is a cone with a singularity at the origin, but it is an isolated one when (WH) is satisfied by the coefficients A_i.

We shall consider also $V_{B,R}$, the intersection of V_B with the ball $|x| \leq R$:

$$\Sigma_1^n A_i x_i^2 = B$$

$$\Sigma_1^n x_i^2 \leq R^2$$

F has a singularity at $x = 0$ which under (WH) is the only singularity in V. To describe the generic singularities of $\varphi_B(x)$ we need stronger conditions: not only hyperbolicity (meaning that no m of the A_i are linearly dependent, as in the first dynamical example Sect. 9.3.1) but *strong hyperbolicity* (SH) meaning that also no $m+1$ of the A_i are affinely dependent.

The singularities of φ_B on V_B correspond to the intersections of the ray through B with the simplices spanned by m or less of the A_i. Under (SH) they would be a finite number of non degenerate singularities.

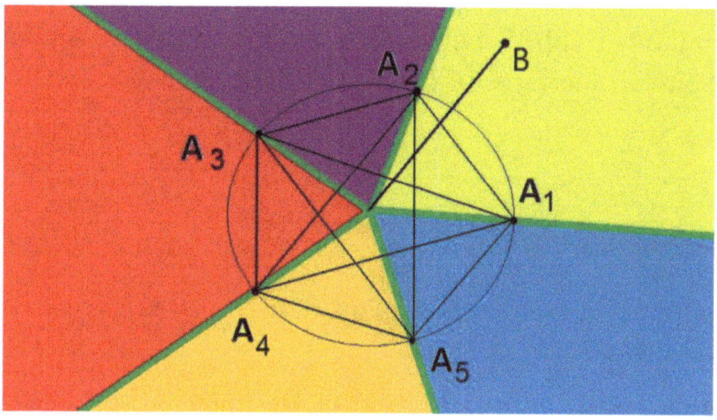

Rays and sectors of smoothness. Here $V_{B,R}$ is smooth and φ_B is Morse.

Then the diffeomorphism type of $V_{B,R}$ does not change with R for R sufficiently large and it would be that of a half intersection in $\mathbb{R}^{n+1}$:

$$\Sigma_0^n A_i x_i^2 = 0$$

$$\Sigma_0^n x_i^2 = 1$$

$$x_0 \geq 0$$

where $A_0 = -B/R^2$.

V_B will be diffeomorphic to the interior of this manifold with boundary. In general, it is an intersection of hyperboloids whose topology would depend on the region where B is in the complement of the union of all cones on the collections of m of the A_i, so when B crosses a positive cone generated by no more than m of the A_i it generally changes its topology. The generic transition would happen when B crosses through the interior of one of those cones.

For the case $m = 2$ we cannot use the normal form used to describe the topology of Z because it does not satisfy (SH) if some of the points have multiplicity $n_i > 1$. So we have to substitute it by another one where near the ith vertex of the regular polygon we place n_i different points λ_j in the unit circle, and then (SH) will be satisfied, any two being linearly independent and any three non-collinear.

In the case of a complex singularity all pre-images of points close to the origin have the same topology, according to Milnor's fibration theorem. In contrast, here for $m = 2$ we can have different pre-images as we go around the circle. Enrique Artal has proposed that it could be interesting to follow the changes of the homology of $V_{B,R}$ (or of its boundary) as we go around the origin to see if there is nevertheless some kind of holonomy.

9.6.2 *Singular Intersections and Smoothings*

We now consider intersections Z where (WH) is not satisfied . The intersection must be singular when the polytope is not simple, but also may be singular if P is simple, but not transverse to all the faces of the first orthant. The simplest example is this:

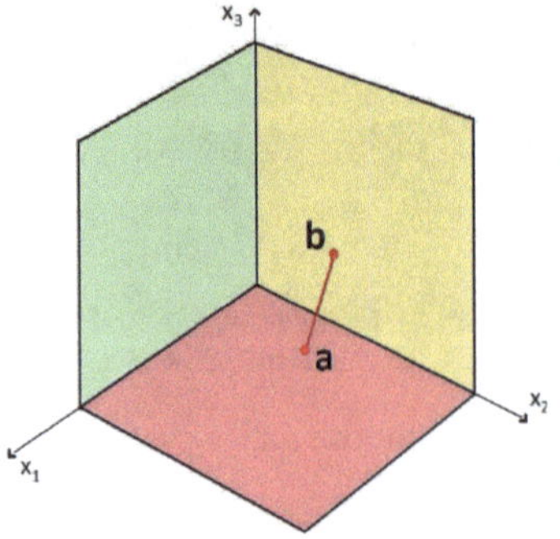

(1) A transverse interval.

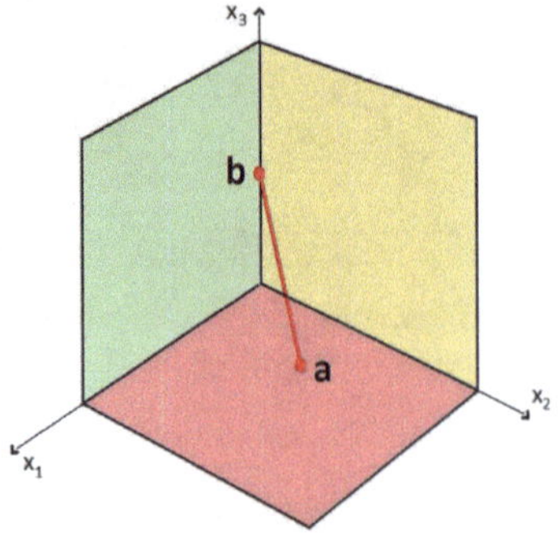

(2) A non-transverse interval.

(1) shows a transverse interval with equations $x_1 - x_2 + x_3 = 0$, $x_1 + x_2 + x_3 = 1$. Reflecting it on the $x_1 = 0$ and $x_3 = 0$ planes we get a (piecewise linear) S^1. Reflecting now on the $x_2 = 0$ plane gives a second copy. Z is diffeomorphic to $S^1 \times S^0$.

(2) shows a non-transverse interval with equations $x_1 - x_2 = 0, x_1 + x_2 + x_3 = 1$, joining $\mathbf{a} = (1/2, 1/2, 0)$ and $\mathbf{b} = (0, 0, 1)$. Reflecting it on the $x_1 = 0$ and $x_2 = 0$ planes we get four segments stemming from $\mathbf{b}$. Reflecting now on the $x_3 = 0$ gives the suspension of four points on that plane and Z is not a manifold.

This example generalizes as follows: for $n = p + q$ let P be an interval in $\mathbb{R}^n$ where one vertex has p coordinates equal to zero and the other one has the other q coordinates equal to zero. It is easy to show that Z is the *complete bipartite graph* $K_{2^p,2^q}$, in other words the join $[2^p] * [2^q] = (S^0)^p * (S^0)^q$.

For a transverse triangle in $\mathbb{R}^4_+$ we know that Z is diffeomorphic to $Z = S^2 \times S^0$.

For a simplest non-transverse triangle with equations

$$x_2 + x_3 - x_4 = 0, \quad x_1 + x_2 + x_3 + x_4 = 1.$$

and vertices: $(1, 0, 0, 0)$, $(0, 1/2, 0, 1/2)$, $(0, 0, 1/2, 1/2))$, $Z = \Sigma(S^1 \times S^0)$:

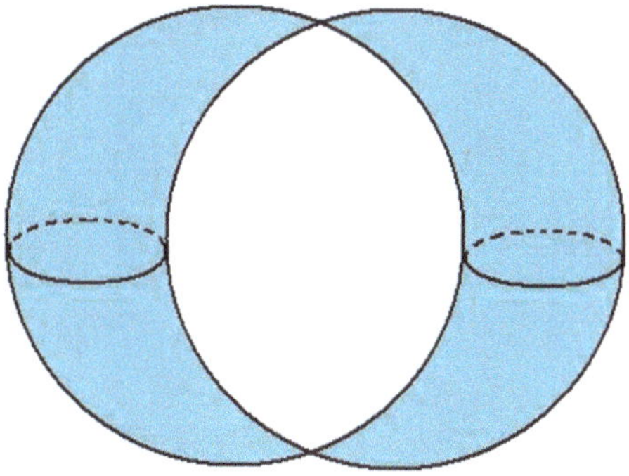

This could be the transition from a transverse triangle to a square, that is, from $Z = S^2 \times S^0$ to $Z = S^1 \times S^1$.

A more degenerate singularity would be a triangle with equations $x_3 - x_4 = 0,\ \ x_1 + x_2 + x_3 + x_4 = 1$ and vertices: $(1, 0, 0, 0),\ (0, 1, 0, 0),\ (0, 0, 1/2, 1/2))$.

Z is the union of two maximal 2-spheres in S^3 that intersect in a maximal S^1. A projection in $\mathbb{R}^3$ could look like the union of two spheres intersecting transversely in a small circle:

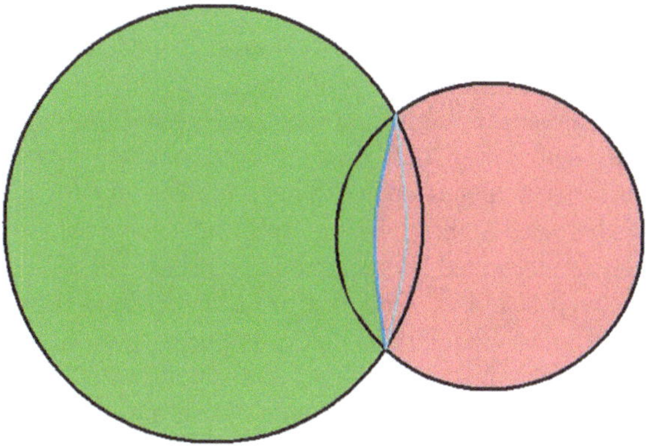

The last case would be a triangle with equations: $x_4 = 0\,,\ \ x_1 + x_2 + x_3 + x_4 = 1$, that is, the unit simplex in $\mathbb{R}^3_+$. Then Z is the 2-sphere in $\mathbb{R}^3$. It is a smooth manifold, but a singular variety in $\mathbb{R}^4$. It can be proved [52] that only in degenerate cases (when one of the variables or one of the equations plays no role) it may happen that (WH) is not satisfied but Z is a manifold, answering an old question by José Seade.

Other examples are the suspension and the join:

If to the equations of Z one adds one more variable x_{n+1} to obtain the system:

$$\Sigma_{i=1}^{n} A_i x_i^2 = 0, \ \ \Sigma_{i=1}^{n+1} x_i^2 = 1$$

the new intersection of quadrics is the suspension of Z and the new polytope is the pyramid on P.

More generally, if we have two varieties $Z(A)$ and $Z(B)$ given by

$$\Sigma_{i=1}^{n} A_i x_i^2 = 0, \;\; \Sigma_{i=1}^{n} x_i^2 = 1$$

$$\Sigma_{j=1}^{m} B_j y_j^2 = 0, \;\; \Sigma_{j=1}^{m} y_j^2 = 1$$

one can build up the system

$$\Sigma_{i=1}^{n} A_i x_i^2 = 0$$

$$\Sigma_{j=1}^{m} B_j y_j^2 = 0$$

$$\Sigma_{i=1}^{n} x_i^2 + \Sigma_{j=1}^{m} y_j^2 = 1$$

which clearly represents the join $P(A) * P(B)$ of the polytopes and the join of the varieties $Z(A) * Z(B)$. Compare with the equations of the product $Z(A) \times Z(B)$ obtained by adding to the above three the equation

$$\Sigma_{i=1}^{n} x_i^2 - \Sigma_{j=1}^{m} y_j^2 = 0$$

Other examples are the non-simple polytopes geometrically embedded. This means that each intersection with a coordinate hyperplane is a facet. The algebraic topology of the corresponding singular moment-angle manifolds has been studied by Ayzenberg and Buchstaber in [5]. They have proved that they are *homotopy equivalent* (but not necessarily homeomorphic) to polyhedral products.

The good news here is that the homology splitting is valid also for singular intersections adequately reinterpreted: P_i should be interpreted as $P \cap \{x_i = 0\}$, independently of the fact that it is a facet or not or if some $P_i = P_j$ as in our first example.

In general, the simplest singularities that can appear are those for which there is a unique collection S of at most m of the vectors A_i, such that the origin in $\mathbb{R}^m$ is a convex combination of them. This implies that they are exactly m and are linearly dependent, but they do not lay in an $(m-2)$ dimensional subspace of $\mathbb{R}^m$.

This means that we can assume they are the m vertices of an $m-1$ simplex in $\mathbb{R}^{m-1}$ which can be deformed into a standard one: its first $m-1$ elements being the standard basis of $\mathbb{R}^{m-1}$ and its last one the point with all coordinates equal to -1. Let us look at some examples:

Example 1: A regular polygon with an added point.

Take a regular polygon with an odd number of sides in the unit circle of $\mathbb{R}^2$ and add the antipodal point of one of the vertices. In this configuration the two antipodal points form the only pair that breaks (WH).

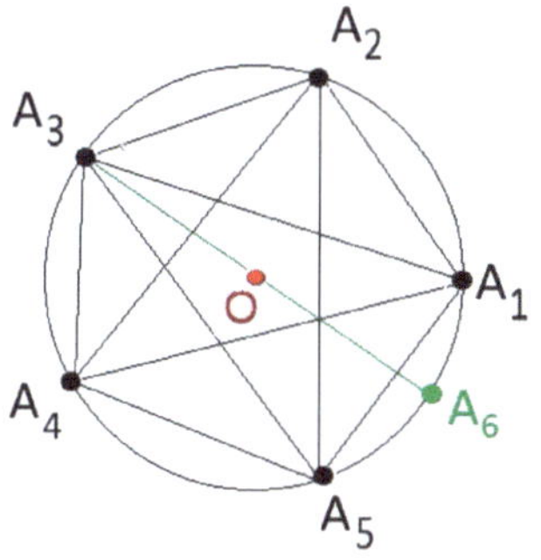

One can add multiplicities to all the points except to those in the antipodal pair to get more examples. These would give all types of configurations with codimension one singularities for $m = 2$. Now moving the new point slightly in both directions we would get in general two different partitions and two smoothings with different topology.

This example can be generalized: for any generic configuration of n points $A_i \in \mathbb{R}^m$ with non-empty associated polytope P that satisfies (WH) one can take $m-1$ that are linearly independent, which we can assume are $A_1, \ldots, A_{m-1}$ and let H be the hyperplane of $\mathbb{R}^m$ they generate. Then add to the configuration a new point A_0 which is in H (but not in any other hyperplane generated by other of the A_i) and such that the origin is in the interior of the convex hull of $A_0, A_1, \ldots, A_{m-1}$. The configuration $A_0, A_1, \ldots, A_n$ will give an intersection of ellipsoids with a codimension 1 singularity.

A smoothing of this singular intersection can be obtained simply by moving A_0 slightly out from H. Actually, we can obtain two smoothings, generally different, by moving that coefficient into each of the two half-spaces defined by H. The topology of the deformation in the neighborhood of the singular point can be described.

In the space of all possible configurations of n points in $\mathbb{R}^m$, the ones with a singularity form a stratified set of codimension 1 separating the open dense set of regular ones into chambers. All the configurations in a given chamber give diffeomorphic smooth intersections of ellipsoids and combinatorially equivalent polytopes. Those with a single singularity of the above type form an open dense stratum, and a generic deformation between two transverse intersections of ellipsoids can be approximated by one that only has codimension 1 singularities and is tranverse to their stratum. The reader is invited to follow how in the case $n = 5$, $m = 2$, Z can transit from being empty to being the surface of genus 5, passing through being four spheres and being two tori.

This is our version of *wall-crossing*. It differs from the version in [15] for moment-angle manifolds, mentioned at the end of Sect. 9.3.2, only in that it is presented in terms of configurations of coefficients instead of polytopes and the fact that in includes crossing walls where a singular intersection appears, but the combinatorial type of the polytope does not change. (Recall the first example and figure in this section).

Final remarks: In the story we have followed there is an ironic zig-zag between the "real" and "complex" cases (according to the variables involved): I started with the question of the space of complex Siegel leaves and their associated manifold $\mathcal{Z}$ (Sect. 9.3.1). To describe its topology I was forced to study the real case Z which I solved, but with certain exceptions. Nevertheless, this incomplete result was enough to describe *all* the complex cases $\mathcal{Z}$.

Then, for sometime the complex case was the one more studied (mainly for its relevance in Kähler Complex Geometry (Sect. 9.4.1)). In particular, Bosio and Meersseman in [15] conjectured that the moment-angle manifolds $Z^{\mathbb{C}}$ associated to dual-neighbourly even dimensional polytopes were connected sums of sphere products (recall (Sect. 9.3.2)). After some time, Sam Gitler and I (Sect. 9.5.2) were able to prove the B-M conjecture by proving again first the real version and deriving from it the original complex one. We also developed some real versions of their truncation operation to obtain more connected sums of sphere products in the real case. Only that part of our truncation results ran into troubles when we tried to apply them to the complex case. But another variation of the truncation idea coming from the complex case [15] was the clue for proving finally the remaining real cases mentioned in the first paragraph of this remark.

More recently, I proved a quantitative Bosio–Meersseman conjecture in the real case, giving a formula for the number of terms in the connected sum as a function of the dimension and the number of facets of the polytope (Sect. 9.5.2), which is enough to determine completely its topology. The formula works also for the complex case $Z^{\mathbb{C}}$ (the original Bosio–Meersseman conjecture), but now this is not enough to determine completely its topology, since in this case not all the terms are equal, and I can not tell, for the moment, how many of each type there are or if this depends only on the dimension and the number of facets of the polytope.

Acknowledgements Conversations with Marc Chaperon, Fred Cohen, José Seade and Sebastià Xambó contributed much to this paper. This work was partially supported by grant PAPIIT IN102918, UNAM.

References

1. Altschuler, S. J., Wu, L. F., *On deforming confoliations*. J. Diff. Geom. 54 (2000), pp.75–97.
2. Arnold, V. I. *Geometrical Methods in the Theory of Ordinary Differential Equations*. Springer-Verlag, 2008.
3. Artal, E., López de Medrano, S. y Lozano, M. T. *The dodecahedron: from intersections of quadrics to Borromean rings*, in A Mathematical Tribute to Professor José María Montesinos, Universidad Complutense de Madrid (2015), pp.61–79.
4. Artal, E., López de Medrano, S. y Lozano, M. T., *Singular intersections of quadrics II*, in process.
5. Ayzenberg, A., Buchstaber, V. M., *Moment-angle complexes and polyhedral products for convex polytopes*, arXiv:1010.1922, 2010.
6. Bahri, A., Bendersky, M., Cohen, F. and Gitler, S., *The polyhedral product functor: a method of computation for moment-angle complexes, arrangements and related spaces*, Advances in Mathematics 225 (2010), pp.1634–1668.

7. Bahri, A., Bendersky, M. and Cohen, F., *Polyhedral products and features of their homotopy theory* in Handbook of Homotopy Theory, Taylor and Francis, 2019.
8. Barreto, Y., López de Medrano, S. and Verjovsky, A., *Some open book and contact structures on moment-angle manifolds*. Bol. Soc. Mat. Mex. 23 (2017), pp.423–437.
9. Barreto, Y. and Verjovsky, A., *Moment-angle manifolds, intersection of quadrics and higher dimensional contact manifolds*, Moscow Mathematical Journal 14, 2014, pp. 669–696.
10. Baskakov, I. V., *Massey triple products in the cohomology of moment-angle complexes*. Russian Math. Surveys 58 (2003), pp.1039–1041.
11. Battaglia, F. and Prato, E. *Generalized toric varieties for simple nonrational convex polytopes*, Int. Math. Res. Not. 24 (2001) pp.1315–1337.
12. Battaglia, F. and D. Zaffran, D., *Foliations modeling nonrational simplicial toric varieties* Int. Math. Res. Not. IMRN 22 (2015), pp.11785–11815.
13. Borman, M. S., Eliashberg, Y. and Murphy, E., *Existence and classification of overtwisted contact structures in all dimensions*, Acta Mathematica, 215 (2015), pp.281–361.
14. Bosio, F., *Variétés complexes compactes: une généralisation de la construction de Meersseman et López de Medrano-Verjovsky*. Ann. Inst. Fourier (Grenoble), 51 (2001), pp.1259–1297.
15. Bosio, F., and Meersseman, L., *Real quadrics in $\mathbb{C}^n$, complex manifolds and convex polytopes*. Acta Math. 197 (2006), pp.53–127.
16. Bourgeois, F., *Odd dimensional tori are contact manifolds*, International Mathematics Research Notices, 30 (2002), pp.1571–1574.
17. Brøndsted, A., *An introduction to convex polytopes*, 1983, Springer-Verlag.
18. Buchstaber, V. M. and Panov, T. E., *Torus actions and their applications in Topology and Combinatorics*, University Lecture Seriers, AMS (2002).
19. Buchstaber, V. M. and Panov, T. E., *Toric Topology*, American Mathematical Society, 2015.
20. Calabi, E. and Eckmann, B., *A class of compact, complex manifolds which are not algebraic*. Ann. Math. 58 (1953), pp.494–500.
21. Camacho, C., Kuiper, N. and Palis, J., *The topology of holomorphic flows with singularities*, Publications Mathématiques I.H.E.S., 48 (1978), pp.5–38.
22. Chaperon, M., *Géométrie différentielle et singularités de systèmes dynamiques*. Astérisque 138–139 (1986).
23. Chaperon, M., *Generalized Hopf bifurcations: a birth lemma*, Moscow Mathematical Journal 11 (2011), in memory of V.I. Arnold, pp.413–438.
24. Chaperon, M. and López de Medrano, S., *Birth of attracting compact invariant submanifolds diffeomorphic to moment-angle manifolds in generic families of dynamics*, C. R. Acad. Sci. Paris, Ser. I 346 (2008), pp.1099–1102.
25. Cisneros-Molina, J. L., *Join theorem for polar weighted homogeneous singularities*, Contemporary Mathematics, 475 (2008), pp.79–87.
26. Coxeter, H., *Regular skew polyhedra in three and four dimension, and their topological analogues*, Proceedings of the London Mathematical Society, 2 (1938), pp.33–62.
27. Cupit-Foutu, S. and Zaffran, D., *Non-Kähler manifolds and GIT quotients*. Math. Z. 257 (2007), pp.783–797.
28. Davis M., *The geometry and Topology of Coxeter Groups*, Springer-Verlag, 2008.
29. Davis M. and Januszkiewicz, T., *Convex polytopes, Coxeter orbifolds and torus actions*, Duke Math. Journal 62 (1991), pp.417–451.
30. De la Vega, G., López de Medrano, S., *Generalizing the May-Leonard System to any number of species*, in: Progress and Challenges in Dynamical Systems. Springer Proceedings in Mathematics & Statistics 54 (2017), pp 395–407. Springer Verlag, Berlin, Heidelberg.
31. Demailly, J.-P. and Gaussier, H., *Algebraic embeddings of smooth almost complex structures*. J. Eur. Math. Soc. 19 (2017), pp.3391–3419.
32. Denham, G. and Suciu, A., *Moment-angle complexes, monomial ideals, and Massey products*, Pure and Applied Math. Quarterly 3 (2007), pp.25–60.
33. Du Plessis, A. and Wall, C. T. C., *The geometry of topological stability*, London Mathematical Society. Oxford, New York : Clarendon Press, 1994.

34. Eliashberg, Y., *Topological characterization of Stein manifolds of dimension* >2, International Journal of Mathematical 1 (1990), pp.29–46.
35. Franz, M., *Dga models for moment-angle complexes*, arXiv:2006.01571, 2020.
36. Franz, M., López de Medrano, S. and Malik, J., *Mutants of compactified representations revisited*, Boletín de la Sociedad Matemática Mexicana. Third Series 23 (2017), pp.511–526.
37. Gitler, S. and López de Medrano, S., *Intersections of quadrics, moment-angle manifolds and connected sums*, Geometry and Topology 17 (2013), pp.1497–1534.
38. Giroux, E., *Geometrie de contact: de la dimension trois vers les dimensions superieures*, ICM, Vol. II (2002), pp.405–414.
39. Gómez Gutiérrez, V., *Cyclic polytopes, oriented matroids and intersections of quadrics*, Boletín de la Sociedad Matemática Mexicana 23 (2017), pp.87–118.
40. Gómez Gutiérrez, V. and López de Medrano, S., *The topology of the intersections of quadrics II*, Boletín de la Sociedad Matemática Mexicana 20 (2014), pp.237–255.
41. Gómez Gutiérrez, V. and López de Medrano, S., *Surfaces as Complete Intersections*, in Riemann and Klein Surfaces, Automorphisms, Symmetries and Moduli Spaces, Contemporary Math. 629 (2014), pp171–180.
42. Grünbaum, B., *Convex Polytopes*, 2nd ed., Graduate Texts in Mathematics 221, Springer-Verlag, New York, 2003.
43. Haefliger A., *Deformations of transversely holomorphic flows on spheres and deformations of Hopf manifolds*, Compos. Math. 55 (1985), pp.241–251.
44. Hernández de la Cruz, L. and López de Medrano, S. *Some families of isolated singularities.* Contemporary Mathematics 475 (2008), pp.79–87.
45. Katzarkov, L., Lupercio, E., Meersseman, L. and Verjovsky, A., *The definition of a non-commutative toric variety*. Algebraic Topology: Appl. New Dir. 620 (2014), pp.223–250.
46. Katzarkov, L., Lupercio, E., Meersseman, L. and Verjovsky, A., *Quantum (Non-commutative) Toric Geometry: Foundations*, Advances in Mathematics 391(2):107945 (2021).
47. Loeb J. J. and Nicolau, M., *Holomorphic flows and complex structures on products of odd-dimensional spheres*, Math. Ann. 306 (1996), pp.781–817.
48. Loeb J. J. and Nicolau, M., *On the complex geometry of a class of non-Kaehlerian manifolds.* Israel J. Math. 110 (1999), pp.371–379.
49. López de Medrano, S., *The space of Siegel leaves of a holomorphic vector field*, in Holomorphic Dynamics (Mexico, 1986), Lecture Notes in Math. 1345, pp.233–245. Springer Verlag, Berlin, 1988.
50. López de Medrano, S., *Topology of the intersection of quadrics in* $\mathbb{R}^n$, in Algebraic Topology (Arcata Ca, 1986), Springer Verlag Lecture Notes in Math. 1370 (1989), pp.280–292, Springer Verlag.
51. López de Medrano, S., *Singularities of real homogeneous quadratic mappings*, Revista de la Real Academia de Ciencias Exactas, Fisicas y Naturales. Serie A. Matemáticas 108 (2014), pp.95–112.
52. López de Medrano, S., *Singular Intersections of Quadrics I*, in Singularities in Geometry, Topology, Foliations and Dynamics. Trends in Mathematics, 2017, pp.155–170. Birkhäuser.
53. López de Medrano, S., *Samuel Gitler and the topology of intersections of quadrics*, Boletín de la Sociedad Matemática Mexicana. Third Series 23 (2017) pp.5–21.
54. López de Medrano, S., *Smoothness in Some Varieties with Dihedral Symmetry and the DFT Matrix*, in Singularities, Algebraic Geometry, Commutative Algebra, and Related Topics, 2018, pp.391–409, Springer-Verlag.
55. López de Medrano, S., *On the genera of moment-angle manifolds associated to dual-neighborly polytopes, combinatorial formulas and sequences*, arXiv:2003.07508v.2.
56. López de Medrano, S. and Verjovsky, A., *A new family of complex, compact, nonsymplectic manifolds.* Bol. Soc. Brasil. Mat., 28 (1997), pp.253–269.
57. Manakov, S. V., *Note on the integration of Euler's equations of the dynamics of an n-dimensional rigid body*. Funct. Anal. and Its Appl. 10 (1976), pp.328–329.
58. Meckert, C., *Forme de contact sur la somme connexe de deux variétés de contact de dimension impaire*, Annales de L'Institut Fourier 32 (1982), pp.251–260.

59. L. Meersseman, L., *A new geometric construction of compact complex manifolds in any dimension*. Math. Ann., 317 (2000), pp.79–115.
60. Meersseman, L. and Verjovsky, A., *Holomorphic principal bundles over projective toric varieties*, J. Reine Angew. Math., 572 (2004), pp.57–96.
61. Mironov, A. E., Panov, T. E. *Intersections of quadrics, moment-angle manifolds, and Hamiltonian-minimal Lagrangian embeddings*, Funct. Anal. Its Appl. 47 (2013), pp.38–49.
62. Mishchenko, A. S., *Integral geodesics of a flow on Lie groups*, Functional Analysis and Its Applications 4 (1970), pp.232–235.
63. Oganesyan, V., *Monotone Lagrangian submanifolds of* $\mathbb{C}^n$ *and toric topology*, arXiv:1812.05007v5, 2020.
64. Oganesyan, V., Sun, Y., *Products and connected sums of spheres as monotone Lagrangian submanifolds*, Journal of Geometry and Physics 163 (2021), 104114.
65. Oka, M., *Topology of polar weighted homogeneous hypersurfaces*, Kodai Mathematical Journal 31 (2008), pp.163–182.
66. Padrol, A., *Neighborly and almost neighborly configurations, and their duals*, Doctoral Thesis, Universitat Politèctnica de Catalunya, 2013.
67. Prato, E., *Simple non-rational convex polytopes via symplectic geometry*. Topology 40 (2001), pp.961–975.
68. Ratiu, T., *The motion of the free n-dimensional rigid body*. Indiana University Mathematics Journal, 29 (1980), pp.609–629.
69. Seade, J., *Open book decompositions associated to holomorphic vector fields*, Boletín Sociedad Matemática Mexicana 3 (1997), pp.323–335.
70. Seade, J., *On the Topology of Isolated Singularities in Analytic Spaces*. Birkhäuser, 2005.
71. Seade, J., *On Milnor's fibration theorem and its offspring after 50 years*, Bull. Amer. Math. Soc. 56 (2019), pp.281–348.
72. Sloane, N. J. A., *On-Line Encyclopedia of Integer Sequences (OEIS)*, https://oeis.org.
73. Verjovsky, A., *Intersection of Quadrics in* $\mathbb{C}^n$, *Moment-Angle Manifolds, Complex Manifolds and Convex Polytopes* in Complex Non-Käler Geometry. Springer, 2019, pp.163–240.
74. Wall, C. T. C., Presidential address *Stability, Pencils and Polytopes*, Bulletin of the London Mathematical Society 12, 1980, pp.401–421.
75. Weinstein, A., *Contact surgery and symplectic handlebodies*, Hokkaido Math. J., Vol. 620 (1991), pp.241–251.
76. Wikipedia: *Mouvement á la Poinsot, Elipsoide de Poinsot* or *Poinsotsche Konstruktion*, https://es.wikipedia.org/wiki/Elipsoide_de_Poinsot.
77. Ziegler. G., *Recent Progress on Polytopes*, Contemporary Mathematics 223 (1999).

Chapter 10
Complements to Ample Divisors and Singularities

Anatoly Libgober

Abstract The paper reviews recent developments in the study of Alexander invariants of quasi-projective manifolds using methods of singularity theory. Several results in topology of the complements to singular plane curves and hypersurfaces in projective space extended to the case of curves on simply connected smooth projective surfaces.

Contents

A. Libgober (✉)
Department of Mathematics, University of Illinois, Chicago, IL 60607, USA
e-mail: libgober@uic.edu

J. L. Cisneros-Molina et al. (eds.), *Handbook of Geometry and Topology of Singularities II*, https://doi.org/10.1007/978-3-030-78024-1_10

10.1 Introduction

These notes review interactions between singularity theory and the study of fundamental groups and more generally the homotopy type of the complements to divisors on smooth projective varieties. The main question considered here is how the local topology of singularities as well as their global geometry affect the topology of the complement. Several surveys updating the state of the subject at respective points in time were written over the years (cf. [54]) but most often focusing on specific situations: complements to plane curves, arrangements of lines or hyperplanes etc. reflecting that earlier studies of the complements were mainly focused on the case of plane curves. Below we consider the complements to divisors D on smooth projective surfaces X and their fundamental groups, sometimes indicating how a generalization to the case of homotopy types of the complements in manifolds of dimension greater than two looks like, but mostly referring to other publications for additional details on homotopy invariants beyond fundamental groups. An earlier appearances of the studies of the complements in the context of general pairs (X, D) and their fundamental groups can be traced to the 80s. Some results on the topology of the complements in such set up did appear in [92, 104, 181]. A much earlier, beautiful results, especially those showing the role of the abelian varieties in the subject were obtained by Italian school (cf. [43] for a modern exposition).

The invariants of the fundamental groups with known strong relation to singularity theory are the Alexander type invariants, introduced in [132] and called their *characteristic varieties*. The connections besides singularity theory run through the knot theory, the Hodge theory of quasi-projective varieties, study of elliptic fibrations, symplectic geometry to mention a few.

There are three major approaches to the study of characteristic varieties of fundamental groups. One is topological, allowing their calculation in terms of a presentation of the fundamental group via generators and relations, obtained typically using braid monodromy. The other one is geometric, going through a study of homology of the abelian covers and eventually leading to determination of characteristic varieties in terms of local type of singularities and dimensions of the linear systems determined by the divisor and the local type of singularities. Finally, one can calculate the characteristic varieties using Deligne extensions of bundles endowed with a flat connection. Whole theory is a combination of methods and ideas from all these areas.

Many results related to the discussion of this paper are presented in volume [45] where for the most part the case of plane curves was considered. The exposition which follows, describes a generalization to the context of the complements to divisors on smooth simply connected surfaces. A very fruitful approach to a study of the complements to divisors is via resolutions of singularities, reducing the case when a divisor has arbitrary singularities to the case of divisors with normal crossings. In this way one replaces the complexity of the divisor by complexity of compactification and the complexity of individual components. The goal here is rather to study how complexity of singularities affects fundamental groups of the complements. Trying to make this paper more independent, we included some basic material which is scattered through existing literature and for which we could not find good references (e.g. theory of branched covers, the relation between quasi-adjunction and multiplier ideals etc.). We also survey several results on the fundamental groups which appear in the last 10–20 years providing an overview of the new results in this area. Several results here are new or did not appear in the literature: they include the divisibility of Alexander polynomials of complements on simply-connected surfaces, extending the case of plane curves (cf. Theorem 10.3.3), calculation of characteristic varieties in terms of classes of irreducible components in Picard group and invariants of quasi-adjunction of singularities (cf. Theorem 10.4.18) and others.

The content of the paper is as follows. In Sect. 10.2 we discuss an analog of classical method of Van Kampen (cf. [214]) to obtain presentations of the fundamental groups of the complements to divisors on smooth surfaces in terms of mapping class group valued monodromy associated to a divisor. We also review conditions on a divisor which allow to deduce that the fundamental group of the complement is abelian. In Sect. 10.3 we firstly extend the theory of Alexander invariants of plane algebraic curves (cf. [129, 143]) to the complements of curves on smooth projective surfaces (for an earlier work cf. [62]). In particular we obtain a result unifying the divisibility theorems in the case of plane curves, showing the divisibility of global Alexander polynomials respectively in terms of local Alexander polynomials and the Alexander polynomials at infinity. Many results depend on some sort of positivity assumptions of the components which suggest an interesting problem understanding the fundamental groups and its invariants when positivity is lacking. Theory of Alexander invariants is closely related to the study of homology of abelian covers. In Sect. 10.3.3 we present basic definitions and then describe approaches enumerating covers either in terms of subgroups of fundamental groups or in terms of eigensheaves of direct images of the structure sheaf. The most interesting results about Alexander invariants are obtained through interaction of topological and algebro-geometric view points. The last part of this section deals with multivariable Alexander invariants from topological view point. We included a brief discussion of multivariable Alexander invariants for quasi-projective invariants in higher dimensions including recent results on propagation (cf. [159] for another recent overview of this and related aspects). Section 10.4 discusses a calculation of characteristic varieties in terms of superabundances of the linear systems associated with a divisor on a smooth projective surface using ideals of quasi-adjunction of singularities of the divisor. The ideals of quasi-adjunction, defined in terms of branched covers of the germs of divisors, can be viewed as the multiplier ideals which received much attention over last 20–30 years. The role of these ideals in the study of the fundamental groups is to specify the linear system which dimensions determine the characteristic varieties and hence allow to give their geometric description. The section contains also another description of characteristic varieties using the Deligne's extension and ends with a brief review of the relations between the characteristic varieties and other invariants studied in Singularity theory, including Bernstein-Sato polynomials and Hodge decomposition of characteristic varieties. Section 10.5 mostly is based on recent preprint [47] which describes the results on distribution of Alexander type invariants when complexity (in appropriate sense) of the divisor increases. We describe the finiteness results when one searches for fundamental groups of the complements with large free quotients. The last section discusses several recent calculations of the fundamental groups of the complements. In the 80s scarcity of examples of quasi-projective groups and fundamental groups of the complements was viewed as impediment to development of general theory. In recent years this problem was amply addressed and we present some of the most consequential results.

In these notes, we tried at least to direct a reader to the most important recent developments but nevertheless several important topics were not covered here. Those missing include the relation between the Alexander invariants and the Mordell-Weil groups of isotrivial fibrations (cf. [146]), Chern numbers of algebraic surfaces and arrangements of curves (cf. [183]), free subgroups of the fundamental groups (cf. [72]), virtual nilpotence of virtually solvable quasi-projective groups (cf. [10]), singularities of varieties of representations of the fundamental groups (cf. [120]), the complements to symplectic curves (cf. [23, 24, 97]) among others.

The theory described below appears to be far from completion. Many interesting problems remain very much open (some are mentioned throughout the text) and a thorough understanding of the fundamental groups or homotopy type of quasi-projective varieties is still out of reach.

Finally, I want to thank Alex Degtyarev as well as the referee of this paper for reading the final version of the text and very helpful comments.

10.2 Braid Monodromy, Presentations of Fundamental Groups and Sufficient Conditions for Commutativity

10.2.1 Braid Monodromy Presentation of Fundamental Groups.

In the case of plane curves, Zariski-van Kampen method (cf. [214, 219]) is the oldest tool for finding presentations of the fundamental groups of the complements. A convenient way to state the theorem is in terms of braid monodromy. Its systematic use was initiated in [164] and in such form admits a natural generalization to the complements to divisors on arbitrary algebraic surface which we describe in this section. Braid monodromy became an important tool in symplectic geometry (cf.

[24]). A good exposition of braid monodromy of curves on ruled surfaces can be found in [57], Sect. 5.1.

Let X be a smooth projective surface and let D be a reduced divisor on X. To describe a presentation of $\pi_1(X \setminus D, p)$, $p \in X \setminus D$ we make several choices, on which the presentation will depend.

- Select a pencil[1] of hyperplane sections of $X \subset \mathbb{P}^N$, generic for the pair (X, D). Its base locus is a generic codimension 2 subspace $P \subset \mathbb{P}^N$ and we can consider the projection with the center at P, i.e. the map $\mathbb{P}^N \setminus P \to \mathbb{P}^1$ sending to $p \in \mathbb{P}^N \setminus P$ to the hyperplane containing P and p. Its restriction to X produces a regular map $\pi : X \setminus X \cap P \to \mathbb{P}^1$. Denoting by $\widetilde{X}$ the blow up of the surface X at the base locus $P \cap X$ of the pencil, we obtain a regular map $\widetilde{X} \to \mathbb{P}^1$. Assuming that P was selected so that $D \cap P = \emptyset$ and still denoting by D its preimage in $\widetilde{X}$ we obtain the map $\tilde{\pi} : \widetilde{X} \setminus D \to \mathbb{P}^1$. Seifert-van Kampen theorem implies that $\pi_1(X \setminus D) = \pi_1(\widetilde{X} \setminus D)$ and so we can do calculations on $\widetilde{X}$.
- Let $B = \{b_1, \ldots, b_k\} \subset \mathbb{P}^1$ be the set consisting of the critical values of $\tilde{\pi}$[2] and the images of the fibers of π, either containing a singular point of D or containing a point of D which is critical point of restriction $\pi|_D$.
- Let $\Omega \subset \mathbb{P}^1$ be a subset, containing B and isotopic to a disk in $\mathbb{P}^1$, and let $b_0 \in \partial\Omega$ be a point on the boundary of Ω.
- Let $\partial B_\epsilon(p)$ be the boundary of a small ball $B_\epsilon(p)$ in X[3] centered at a point $p \in X \cap P$ or, equivalently, the boundary of a small regular neighborhood of the exceptional curve E_p in $\widetilde{X}$ contracted to $p \in X$. The map $\tilde{\pi}$ restricted to $\partial B_\epsilon(p)$ is the Hopf fibration $\partial B_\epsilon = S^3 \to \mathbb{P}^1 = S^2$. Using its trivialization over Ω, we define a section over $\Omega \setminus B$: $s_p : \Omega \setminus B \to \tilde{\pi}^{-1}(\Omega \setminus B)$.
- Let $F_{b_i}, i = 0, 1, ..., k$ be the fiber of $\tilde{\pi}$ over b_i. The curves $F_{b_i}, i = 1, ..., k$ either have singularities at critical points of π or contain singular points of D or have non-transversal intersections with D, while F_{b_0} is smooth closed Riemann surface having genus $g = \frac{F_{b_0}(F_{b_0}+K)}{2} + 1$ where K is the canonical divisor of X.
- For any $p \in P \cap X$, let $\bar{F}^\circ_{b_0}$ be the surface with one connected boundary component obtained by removing from F_{b_0} its intersection with the above regular neighborhood of E_p. Denote by $\mathcal{M}(\bar{F}^\circ_g, [d]) = Diff^+(\overline{F_{b_0} \setminus (F_{b_0} \cap B_\epsilon(p))}, [F_{b_0} \cap D])$ the mapping class group of the Riemann surface with boundary with d marked points (cf. [88]) i.e. the group of isotopy classes of orientation preserving diffeomorphisms taking the subset $[d]$ of cardinality d into itself and constant on the boundary of the Riemann surface.

Definition 10.2.1 The braid monodromy of the pair (X, D) (for selected pencil on X) is the monodromy map

$$\mu : \pi_1(\Omega \setminus B, b_0) \to \mathcal{M}(\bar{F}^\circ_g, [d]) \tag{10.1}$$

---

[1] I.e. a family of divisors parametrized by $\mathbb{P}^1$.

[2] Those are absent in the classical case on pencils of lines $X = \mathbb{P}^2$ of Zariski-van Kampen theorem.

[3] We assume that there are no vanishing cycles corresponding to critical points of π and no points of D inside this ball.

obtained by

(a) selecting a loop (denoted in b) and c) below as γ) for each homotopy class in $\pi_1(\Omega \setminus B, b_0)$,

(b) a trivialization of the locally trivial fibration $\pi^{-1}(\gamma) \to \gamma$ i.e. a differentiable map $\pi^{-1}(b_0) \times [0,1] \to \pi^{-1}(\gamma)$ inducing a diffeomorphism of the fiber over $t \in [0,1]$ onto the fiber over the image of t in parametrization $[0,1] \to \gamma$ of the loop.

(c) assigning to γ the diffeomorphism of $\bar{F}_g^\circ = \pi^{-1}(b_0)$ sending a point $q \in \pi^{-1}(b_0)$ to the point $q' \in \pi^{-1}(b_0)$ to which the trivialization mentioned in b) takes the end point $q \times 1$ of the segment $q \times [0,1] \subset \pi^{-1}(b_0) \times [0,1]$ in $\pi^{-1}(\gamma)$.

One verifies that, though the diffeomorphism in (c) depends on both, the loop γ in (a) and the trivialization in (b), its class in the mapping class group does not depend on these choices.

Recall that the mapping class group $\mathcal{M}(\bar{F}_g^\circ, [d])$ acts on $\pi_1(\bar{F}_g^\circ \setminus [d], q)$ (here q is the base point which we assume is on the boundary of $\bar{F}_g^\circ$). For example in the case $g = 0$ the group $\mathcal{M}(\bar{F}_0^\circ, [d])$ is the Artin's braid group on d-strings i.e. the group of orientation preserving diffeomorphisms of a 2-disk Δ, constant on the boundary and taking into itself a given subset of Δ of cardinality d. It has a well known presentation:

$$< \sigma_1, ..., \sigma_{d-1}, \quad |\sigma_i\sigma_{i+1}\sigma_i = \sigma_{i+1}\sigma_i\sigma_{i+1} \quad \sigma_i\sigma_j = \sigma_j\sigma_i, 1 < |i-j| > \tag{10.2}$$

Note that the center of (10.2) is generated by $[\sigma_1(\sigma_2\sigma_1)(\sigma_3\sigma_2\sigma_1)....(\sigma_{d-1}...\sigma_1)]^2$ (cf. [98] Sect. 4.3). The action on the free group $\pi_1(\Delta \setminus [d], p)$ is given by

$$\sigma_i(t_i) = t_{i+1}, \sigma_i(t_{i+1}) = t_{i+1}^{-1}t_it_{i+1}, \sigma_i(t_j) = t_j, \quad j \neq i, i+1 \tag{10.3}$$

[4](which is the canonical action of the mapping class group on the fundamental group for appropriate choice of generators t_i of the latter). This way in the case of $X = \mathbb{P}^2$ one obtains the monodromy with the values in the Artin's braid group, the case described in [164]. The homomorphism (10.1) in [164] is described in a more combinatorial form, as a product of collection of braids. The ordered collection of factors in this product is the collection of braids corresponding to so called "good ordered system of generators" of the free group $\pi_1(\Omega \setminus B, b_0)$ (cf. [164] for details).

To define the final ingredient for our presentation of $\pi_1(X \setminus D)$, we consider the gluing map of the boundaries of $\pi^{-1}(\Omega)$ and $\pi^{-1}(\mathbb{P}^1 \setminus \Omega)$ which can be viewed as a map $\Phi : \pi^{-1}(\partial\Omega) \to \pi^{-1}(\partial(\mathbb{P}^1 \setminus \Omega))$, both spaces being locally trivial fibrations over $\partial\Omega = S^1$, preserving the set $D \cap \pi^{-1}(\partial\Omega)$ and commuting with projection onto S^1. Such map takes the loop $s_p(\partial\Omega)$ (as above, s_p is a section of restriction of the Hopf bundle over $\mathbb{P}^1$) to the loop $S^1 \to S^1 \times (F_{b_0} \setminus [d]) \to F_{b_0} \setminus [d]$ and hence determines a conjugacy class in the fundamental group of its target. We shall denote this class $\rho_{X,D}$. In the case of plane curve of degree d transversal to the line at infinity and pencil of lines, complement to the base point is the total space of line bundle $\mathcal{O}_{\mathbb{P}^1}(1)$, the gluing map Φ induced by positive generator of $\pi_1(GL_2(\mathbb{C}))$ which shows

[4] This implies that $\sigma_i^{-1}(t_i) = t_it_{i+1}t_i^{-1}, \sigma_i^{-1}(t_{i+1}) = t_i$.

that $\rho_{X,D} = \gamma_1 \cdot \ldots \cdot \gamma_d$ is the product of standard ordered system of generators of fundamental group of the complement in generic fiber to the intersection of this fiber with the curve.

The mapping class group valued monodromy determines the fundamental group as follows:

Theorem 10.2.2 *One has the isomorphism:*

$$\pi_1(X \setminus D) = \pi_1(C_0 \setminus C_0 \cap D)/\{(\mu(\gamma_j)\alpha_i)\alpha_i^{-1}, \rho_{X,D}\} \tag{10.4}$$

Theorem 10.2.2 reduces a calculation of the fundamental group to the calculation of the braid monodromy and the element $\rho_{X,D}$. The literature on calculations of braid monodromies of curves is very large and is very hard to review. We refer to [164, 166] where explicite expressions were obtained for the braid monodromy of smooth plane curves, branching curves of generic projections of smooth surfaces in $\mathbb{P}^3$, generic arrangements of lines and branching curves of generic projections of various embeddings of quadric. The survey [209] and the book [57] also are good references for more recent developments. We refer to the former for examples of calculations of fundamental groups using van Kampen method and references to other works on calculation of braid monodromy and the latter for computer use in calculations of braid monodromy and the fundamental groups.

Besides the fundamental group, the braid monodromy defines the homotopy type of the complement (cf. [135] for precise statement). It is however and open problem, if the homotopy type of the complement $X \setminus D$ is determined by the fundamental group and the topological Euler characteristic of the complement (cf. [135] for a discussion of this problem). Considering dependence of the braid monodromy on the curve and numerous choices made in its construction, in [16] the authors found conditions implying that the homeomorphism type of the triples $(\mathbb{P}^2, L, C)$, where C is a plane curve and L is one of the lines of the pencil used to construct the braid monodromy (the line at infinity), determines the braid monodromy. Braid monodromy is an essential tool in showing the existence of symplectic singular curves not isotopic to algebraic ones (cf. [24, 165]).

10.2.2 Abelian Fundamental Groups

The question "whether the fundamental group of the complement to a nodal curve is abelian" was known as "Zariski problem" since it was realized that Severi's proof of irreducibility of the family of plane curves with fixed degree and the number of nodes is incomplete (cf. [194]). Zariski derived commutativity of the fundamental groups of the complements to nodal curves using that irreducibility implies existence of degeneration of a nodal curve to a union of lines without points of multiplicity greater than two. Once one has degeneration, the relation between fundamental groups of the complements to a curve and to its degenerations (i.e. that given a degeneration

$C_0 = limC_t$ one has surjection $\pi_1(\mathbb{P}^2 \setminus C_0) \to \pi_1(\mathbb{P}^2 \setminus C_t)$ which is a consequence of definition of the braid monodromy and presentation (10.4)) implies the commutativity. Severi statements (and with it the Zariski proof [221]) was eventually validated (cf. [111]). A proof of commutativity, based on connectedness theorem, was found prior to this by Fulton [93] for algebraic fundamental group and by Deligne [66] in topological case.

The central result on commutativity of fundamental groups of complements to divisors is due to Nori with the key step being a generalization of Lefschetz hyperplane section theorem (cf. [174]). See [94, Chap. 5] and [127, Chap. 3].

Theorem 10.2.3 (Nori's weak Lefschetz theorem) *Let U be a connected complex manifold of dimension greater than one and let $i : H \to U$ be the embedding of a connected compact complex-analytic subspace defined by a locally principal sheaf of ideals. Let $q : U \to X$ be a locally invertible map to a smooth projective variety, $h = q \circ i$, and $R \subset X$ be a Zariski closed subset. Assume that $O_U(H)|H$ is ample. Then*

A: $G = \mathrm{Im}\pi_1(U \setminus q^{-1}(R)) \to \pi_1(X \setminus R)$ is a subgroup of a finite index.

B: If $q(H) \cap R = \emptyset$ then $\pi_1(H) \to \pi_1(X \setminus R)$ is a subgroup of a finite index.

C: If $\dim X = \dim U = 2$ then index of subgroup G of $\pi_1(X \setminus R)$ is at most $\frac{(Div(h))^2}{H^2}$ where the divisor in numerator is the first Chern class of $h_* O_H$, the Cartier divisor on X corresponding to the divisor H on U (cf. [174], 3.16 for details).

If q is embedding and H is reduced, this becomes Zariski-Lefschetz hyperplane section theorem (cf. [114, 220]). One has to note a subtlety in the finiteness of index in A and the index bound in C (cf. [110]). Typically $\pi_1(q(H))$ is much bigger than $\pi_1(H)$: for example if $H \to q(H)$ is normalization and H is rational and $q(H)$ nodal then $\pi_1(q(H))$ is free group with the rank equal to the number of nodes. Nevertheless the following is still open:

Problem 10.2.4 (*M. Nori*) Let D be an effective divisor of a surface X and $D^2 > 0$. Let N be a normal subgroup of $\pi_1(X)$ generated by the images of the fundamental groups of the normalizations of all irreducible components of D. Is the index of N in $\pi_1(X)$ finite? In particular, can a surface with infinite fundamental group contain a rational curve with positive self-intersection?

One of the main consequence of Theorem 10.2.3 is the following:

Corollary 10.2.5 *Let D and E be curves on smooth projective surface intersecting transversally and such that D has nodes as the only singularities. Assume that for each irreducible component C of D one has $C^2 > 2r(C)$. Then $N = Ker(\pi_1(X \setminus (D \cup E)) \to \pi_1(X \setminus E))$ is a finitely generated abelian group and the centralizer of N has a finite index in $\pi_1(X \setminus (D \cup E))$.*

This immediately implies that the fundamental group of a nodal curve in $\mathbb{P}^2$ is abelian (indeed, for irreducible curve of degree d the maximal number of nodes $r(C) = \frac{(d-1)(d-2)}{2}$ satisfies $2r(C) < d^2$). Moreover, for an irreducible plane curve with $r(C)$ nodes and $\kappa(C)$ cusps (with local equation $u^2 = v^3$) one obtains that

$\pi(\mathbb{P}^2 \setminus C)$ is abelian if $C^2 > 6\kappa(C) + 2r(C)$ (apply Corollary 10.2.5 to resolution of cusps only). On non-simply connected surfaces, the kernel $\pi_1(X \setminus C) \to \pi_1(X)$ belongs to the center if $C^2 > 4r(C)$ (though for $4r(C) \geq C^2 > 2r(C)$ the centralizer of this kernel still has a finite index, see [174] p. 324).

A result pointing out toward a positive answer to the Problem 10.2.4 appears in [126] and can be stated as follows. Let X be a smooth projective variety and Y be a subvariety such that $\pi_1(Y) \to \pi_1(X)$ is surjective. Let $f : Z \to Y$ be dominant morphism where all irreducible components of Z are normal. Let N be the normal subgroup of $\pi_1(X)$ generated by the images of irreducible components of Z. Then for any n, $\pi_1(X)/N$ has only finitely many n-dimensional complex representations, all of which are semi-simple (an obvious attribute a finite group).

Applications of Nori's results include [197, 198]. Paper [199] studies further exact sequence of the fiber spaces. Papers [211, 212] give conditions in opposite direction than the one considered by Nori, guaranteeing that the fundamental group of the complement is NON abelian. An important outcome of Nori's Weak Lefschetz theorem is that it provides an instance for the finiteness of the index of the image of the fundamental groups for compositions $H \to C \to X$ where as above H, X are smooth and $H \to C$ is dominant. This more general context was considered in [110] in the framework of the study of the representations of the fundamental groups of varieties dominating divisors in the moduli spaces of (pointed) curves (with level structure), under the heading of "non-abelian strictness theorems".

10.3 Alexander Invariants

10.3.1 Alexander Polynomials

Alexander polynomial of knots and links was introduced by James W. Alexander in 1928 (cf. [3]). In response to a question by D. Mumford (cf. [168]), who noticed its relation to a construction used by O. Zariski, the Alexander polynomials were put in [129] in the context of complements to plane algebraic curves. This extension blends the algebraic geometry and the methods introduced by Fox (cf. [90]) and Milnor (cf. [161]) for the study of knots. Various generalizations, in which (a zero set of) polynomial was replaced by a subvariety of a torus and involving germs of singularities (cf. [140]), extensions to higher dimensions (cf. [134]) and to curves in complex surfaces (cf. [62]), were considered as well. A twisted versions (cf. [44, 143, 158]) were studied more recently. Below we shall describe the Alexander polynomials in the context of divisors on simply connected surfaces and refer to [144] for the history of the subject and further references.

Let X be a smooth simply connected projective surface and let D be a divisor on X with irreducible components D_i. Let $\{[D_i]\} = H^2(D, \mathbb{Z}) = \oplus_i H^2(D_i, \mathbb{Z})$ denote a free abelian group generated by the cohomology classes corresponding to the irreducible components of D. For $\alpha \in H_2(X, \mathbb{Z})$, we put $D_\alpha = \sum_i (\alpha, [D_i])[D_i] \in$

$\{[D_i]\}$, where $[D_i] \in H_2(X, \mathbb{Z})$ is the fundamental class of the component D_i and denote by $\{D_\alpha\}$ the subgroup of $\{[D_i]\}$ generated by the classes $D_\alpha, \alpha \in H_2(X, \mathbb{Z})$. $\{D_\alpha\}$ is the image of the homomorphism $H_2(X, \mathbb{Z}) \to H^2(D, \mathbb{Z})$ obtained using the excision and duality isomorphisms giving $H_2(X, X - D, \mathbb{Z}) = H_2(T(D), \partial T(D), \mathbb{Z}) = H^2(D, \mathbb{Z})$ where $T(D)$ is a tubular neighborhood of D in X and $\partial T(D)$ is its boundary. From the exact sequence:

$$H_2(X, \mathbb{Z}) \to H^2(D, \mathbb{Z}) \to H_1(X \setminus D, \mathbb{Z}) \to H_1(X, \mathbb{Z}) = 0 \tag{10.5}$$

we deduce that

$$\{[D_i]\}/\{D_\alpha\} = H_1(X \setminus D, \mathbb{Z}). \tag{10.6}$$

For example for an irreducible projective (resp. affine) plane curve D of degree d we obtain $H_1(\mathbb{P}^2 \setminus D, \mathbb{Z}) = \mathbb{Z}/d\mathbb{Z}$ (resp. $H_1(\mathbb{C}^2 \setminus D) = \mathbb{Z}$).

Alexander polynomial is an invariant of the complement to a reduced divisor D and a surjection $\phi : \pi_1(X \setminus D) \to C$ where C is a cyclic group. We state the definition for a finite CW complex Y endowed with a surjection $\phi : \pi_1(Y) \to C$ such that $H_1(Y_\phi, \mathbb{Q})$ is finite dimensional where Y_ϕ is the covering space corresponding to the subgroup $Ker(\phi) \subset \pi_1(Y)$ (cf. [115] Sect. 1.3).[5]

If C is finite then the finiteness of the dimension of $H_1(Y_\phi, \mathbb{Q})$ is automatic.[6] If $H_1(Y, \mathbb{Z})$ is infinite cyclic then $H_1(Y_\phi, \mathbb{Q})$ also is finite-dimensional as follows for example from (10.8) below.

For the covering map $Y_\phi \to Y$, we have the exact compactly supported homology sequence corresponding to the sequence of chain complexes

$$0 \to C_*(Y_\phi, \mathbb{Q}) \overset{t-1}{\to} C_*(Y_\phi, \mathbb{Q}) \to C_*(Y, \mathbb{Q}) \to 0 \tag{10.7}$$

Here the first two terms are viewed as the modules over the group ring $\mathbb{Q}[\mathbb{Z}] = \mathbb{Q}[t, t^{-1}]$, where t denotes preferred generator of $C = \mathbb{Z}$ in multiplicative notations, and the left map being multiplication by $t - 1$. Hence

$$H_2(Y, \mathbb{Q}) \to H_1(Y_\phi, \mathbb{Q}) \overset{t-1}{\to} H_1(Y_\phi, \mathbb{Q}) \to H_1(Y, \mathbb{Q}) \tag{10.8}$$

$$\to H_0(Y_\phi, \mathbb{Q}) \overset{(t-1)}{\to} H_0(Y_\phi, \mathbb{Q})$$

Consider the cyclic decomposition of $H_1(Y_\phi, \mathbb{Q})$, viewed as a module over $\mathbb{Q}[t, t^{-1}]$,

[5] In this case we call Y_ϕ 1-finite.

[6] An example of infinite cyclic covers which is infinite in dimension 1 is given by the complement to a set [3] containing 3 points in $\mathbb{P}^1$. Let (a, b) be generators of the free group $\pi_1(\mathbb{P}^1 \setminus [3])$ and ϕ is the quotient of the normal subgroup generated by b. Then $\mathbb{P}^1 \setminus [3]$ is homotopy equivalent to a wedge of two circles and $(\mathbb{P}^1 \setminus [3])_\phi$ can be viewed as a real line with the circle attached at each integer point of this line with the covering group $\mathbb{Z}$ acting via translations. In particular $H_1((\mathbb{P}^1 \setminus [3])_\phi, \mathbb{Z})$ is a free abelian group with countably many generators.

$$H_1(Y_\phi, \mathbb{Q}) = \oplus \mathbb{Q}[t, t^{-1}]^{a_0} \oplus_p \mathbb{Q}[t, t^{-1}]/(p(t)) \tag{10.9}$$

where the summation is over a finite number of monic polynomials p.

One of immediate consequences is that if $rkH_1(Y, \mathbb{Q}) = 1$, then the multiplication by $t - 1$ in the top row in (10.8) is surjective (since clearly the multiplication by $t - 1$ is trivial on H_0) and hence in (10.9) $a_0 = 0$.[7] Moreover, $(t - 1)^\alpha$, $\alpha \in \mathbb{N}$ is not among the polynomials $p(t)$.

Definition 10.3.1 Let Y be a CW-complex as above.

If $a_0 = 0$ in the decomposition (10.9) one defines the Alexander polynomial $\Delta(t)$ of (Y, ϕ) as the order of the $\mathbb{Q}[t, t^{-1}]$-module $H_1(Y_\phi, \mathbb{Q})$ i.e. as the product.

$$\Delta(t) = \prod p(t) \tag{10.10}$$

In the case when X is a smooth projective surface and D is a reduced divisor, we call $\Delta(t)$, the *global* Alexander polynomial of $X \setminus D$ (and the surjection ϕ of its fundamental group).

$\Delta(t)$ has integer coefficients, is well defined up to $\pm t^i$, $i \in \mathbb{Z}$ and, it follows from (10.8) that, $rkH_1(Y, \mathbb{Q}) = 1$ implies $\Delta(1) \neq 0$. If the target of ϕ is a finite cyclic group then, since $\mathbb{Q}[\mathbb{Z}_n] = \mathbb{Q}[t, t^{-1}]/(t^n - 1)$, instead of (10.9) one has

$$H_1(Y_\phi) = \oplus[\mathbb{Q}[t, t^{-1}]/(t^{\mathrm{ord}C} - 1)]^{a_o} \oplus \mathbb{Q}[t, t^{-1}]/p(t) \tag{10.11}$$

and the Alexander polynomial defined to be the order (10.10) of this $\mathbb{Q}[t, t^{-1}]$-module.

This construction, when applied to the intersection of D with a small sphere about a singular point P of D and when ϕ is given by evaluation of the linking number in this sphere with D, yields the *local Alexander polynomial.* It is not hard to show the following (cf. [128]).

Proposition 10.3.2 *The local Alexander polynomial coincides with the characteristic polynomial of the local monodromy of the singularity of D at P.*

10.3.2 A Divisibility Theorem

This is the central result on the Alexander polynomials allowing to obtain information about $\Delta(t)$ in terms of geometry of D. In many cases it leads to its determination or makes possibilities for $\Delta(t)$ rather limited. The case of curves in $\mathbb{P}^2$ appears in [129].

Theorem 10.3.3 *Let $D = D_1 \bigcup D_2$ be a divisor on X such that D_1 is ample. Let $\phi_{X \setminus D} : \pi_1(X \setminus D) \to H_1(X \setminus D, \mathbb{Z}) \to C$ be a surjection onto a cyclic group C*

---

[7] The condition $a_0 = 0$ is equivalent to finite dimensionality of $H_1(Y_\phi, \mathbb{Q})$ over $\mathbb{Q}$.

(either infinite or finite) and $T(D_1)$ *denotes a small regular neighborhood of the divisor* D_1*. Assume also that* ϕ *maps the meridian*[8] *of each irreducible component of* D_1 *to the generator of* C *corresponding to the variable* t *of the Alexander polynomial. Then*

1. The cyclic cover $(X \setminus D)_\phi$ *is 1-finite and so is* $(T(D_1) \setminus D \cap T(D_1))_{\phi_T}$ *where* ϕ_T *is the composition* $\pi_1((T(D_1) \setminus D \cap T(D_1))) \to H_1((T(D_1) \setminus D \cap T(D_1))) \to H_1(X \setminus D) \to C$ *of the map induced be embedding and the surjection* $\phi_{X \setminus D}$*.*

2. Let $\Delta_{\phi_{X \setminus D}}, \Delta_{\phi_T}$ *be the Alexander polynomials of* $X \setminus D$ *and* $T(D_1) \setminus D \cap T(D_1)$ *corresponding to surjections* ϕ *and* ϕ_T *respectively. One has the following divisibility:*

$$\Delta_{\phi_{X \setminus D}}(t) | \Delta_{\phi_T} \tag{10.12}$$

3. Let $\{p_i\}$ *be the set consisting of singular points of* D_1 *and the points* $D_1 \cap D_2$*. For each* p_i *let* B_{p_i} *denotes a small ball in* X *centered at this point. Let* Δ_{p_i} *denotes the Alexander polynomial of* $B_{p_i} \setminus D \cap B_{p_i}$ *relative to the map* $\phi_i : H_1(B_{p_i} \setminus D \cap B_{p_i}) \to C$ *induced by embedding* $B_{p_i} \setminus D \cap B_{p_i} \to X \setminus D$*. Then*

$$\Delta_{\phi_T} = (t-1)^\alpha \prod \Delta_{p_i} \quad \alpha \in \mathbb{Z}. \tag{10.13}$$

In particular, the roots of the Alexander polynomials $\Delta_{\phi_{X \setminus D}}$ *and* Δ_{ϕ_T} *are roots of unity.*

Proof Ampleness of D_1 implies that for $n \gg 0$ there exist a smooth curve $\tilde{D}_1$ on X linearly equivalent to nD_1 and belonging to $T(D_1)$. Moreover, we can assume that $\tilde{D}_1$ is transversal to all components of D.

Weak Lefschetz theorem (cf. [114, 174]) implies that the composition in the middle row of the following diagram is a surjection:

$$\begin{array}{ccccc}
 & & Ker\phi_T & \rightarrow & Ker\phi_{X \setminus D} \\
 & & \downarrow & & \downarrow \\
\pi_1(\tilde{D}_1 \setminus \tilde{D}_1 \cap D) & \rightarrow & \pi_1(T(D_1) \setminus D \cap T(D_1)) & \rightarrow & \pi_1(X \setminus D) \\
 & & \downarrow & & \downarrow \\
 & & H_1(T(D_1) \setminus D \cap T(D_1), \mathbb{Z}) & & H_1(X \setminus D, \mathbb{Z}) \\
 & & & \searrow & \downarrow \\
 & & & & C
\end{array} \tag{10.14}$$

Therefore the right map in that row and hence also $Ker\phi_T \to Ker\phi_{X \setminus D}$ both are surjective. The condition that meridians are taken by ϕ_T to non-zero element of C

[8] I.e. a loop consisting of a path connecting the base point with a point in vicinity of the irreducible component of D, the oriented boundary of a small disk in X transversal to this component of D at its smooth point and not intersecting the other components of D, with the same path used to return back to the base point; orientation of the small disk must be positive i.e. such that its orientation will be compatible with the complex orientations of smooth locus of divisor and the ambient manifold. As an element of the fundamental group, only the conjugacy class of a meridian is well defined.

implies that the covering space $(T(D_1) \setminus D \cap T(D_1))_{\phi_T}$ is 1-finite (cf. [99]) and surjectivity of the maps of the kernels implies that so is $(X \setminus D)_{\phi}$. Since the map in the top row in (10.14) is surjective, $\mathbb{Q}[C]$-module $H_1((X \setminus D)_{\phi}, \mathbb{Q})$ is a quotient of $H_1((T(D_1) \setminus D \cap T(D_1))_{\phi_T}, \mathbb{Q})$ hence the divisibility relation (10.12) follows.

Finally, taking $T(D_1)$ sufficiently thin, $(T(D_1) \setminus D_1) \setminus \bigcup_i (B_{p_i} \setminus D \cap B_{p_i})$ can be assumed isotopic to the trivial C^∞-fibration $(T(D_1) \setminus D_1) \setminus \bigcup_i (B_{p_i} \setminus D \cap B_{p_i}) \to D_1 \setminus \{p_i\}$, with the fiber being isotopic to a punctured 2-disk. Due to assumption that meridians of all components are mapped to generator corresponding to t, the Alexander polynomial of $(T(D_1) \setminus D_1) \setminus \bigcup_i (B_{p_i} \setminus D \cap B_{p_i})$ is a power of $t-1$. The decomposition

$$T(D_1) \setminus D \cap T(D_1) = \left[(T(D_1) \setminus D_1) \setminus \bigcup_i (B_{p_i} \setminus D \cap B_{p_i}) \right] \cup \bigcup_i (B_{p_i} \setminus D \cap B_{p_i}) \tag{10.15}$$

induces decomposition of the cover $(T(D_1) \setminus D)_{\phi_T}$ of $T(D_1) \setminus D$ corresponding to subgroup $Ker\phi_T$ of $\pi_1(T(D_1) \setminus D \cap T(D_1))$ into a union of preimages of each subspace on the right in (10.15). Now the Mayer -Vietoris sequence implies the part 3 of the Theorem (also the 1-finiteness of $T(D_1) \setminus T(D_1) \cap D)$).

Corollary 10.3.4 ([129]) *Let C be an irreducible curve in $\mathbb{P}^2$ and L be the line at infinity. Then $H_1(\mathbb{P}^2 \setminus C \bigcup L, \mathbb{Z}) = \mathbb{Z}$ and the Alexander polynomial of $\mathbb{P}^2 \setminus C \bigcup L$ with respect to the abelianization, divides the product of the Alexander polynomials of links of all singularities of $C \bigcup L$. It also divides the Alexander polynomial of the link at infinity i.e. the Alexander polynomial of the complement $S_\infty \setminus C \cap S_\infty$ where S_∞ is the boundary of a small (in the metric on $\mathbb{P}^2$) regular neighborhood of $L \subset \mathbb{P}^2$.*

Proof It follows from (10.6) and Theorem 10.3.3 applied to C and L separately. More precisely, the part 2 (resp. part 2 and 3) of Theorem 10.3.3 show that the global Alexander polynomial divides the Alexander polynomial at infinity (resp. of the product of local Alexander polynomials).

Corollary 10.3.5 ([62]) *Let D be a divisor of a simply connected surface X. Let S be a subset of the set of singular points of D belonging to an irreducible component D' of D such that on log-resolution $\tilde{X}$ of singularities of D' outside of S, for proper preimage $\tilde{D}'$ one has $(\tilde{D}')^2 > 0$. Then one has divisibility:*

$$\Delta_{\phi_{X \setminus D}} | \prod_{p_i \in S} \Delta_{p_i} \tag{10.16}$$

Proof Condition on self-intersection implies that $\tilde{D}'$ is ample. Now the claim follows immediately from the Theorem 10.3.3 applied to the proper preimage of D on $\tilde{X}$ and its component $\tilde{D}'$ since $X \setminus D = \tilde{X} \setminus \tilde{D}$ because only points on deleted divisor D are blown up.

Example 10.3.6 *Milnor fibers of homogeneous polynomials and arrangements of lines* Let $\mathcal{A} \subset \mathbb{P}^2$ be an arrangement of lines given by equations $L_i(x, y, z) = 0, i =$

1, ..N. Milnor fiber $\prod L_i(x,y,z)=1$ of the cone $\prod L_i=0$ over this arrangement (denoted below $M_{\mathcal{A}_L}$) can be identified with the $\mathbb{Z}/N\mathbb{Z}$-cyclic cover of the complement $\mathbb{P}^2\setminus\mathcal{A}$. Theorem 10.3.3 gives restrictions on the degree of the characteristic polynomial of the monodromy operator acting on $H_1(M_{\mathcal{A}},\mathbb{Q})$ (which can be identified with the Alexander polynomial of $\mathbb{P}^2\setminus\mathcal{A}$) in terms of multiplicities of point of $\mathcal{A}$ along one of the lines (cf. [139]). For example, if $\mathcal{A}$ has only triple points along one of the lines, it follows that the characteristic polynomial of the monodromy of Milnor fiber has form $(t-1)^{N-1}(t^2+t+1)^{\kappa},\kappa\ge 0$. See [73, 74, 175] for other numerous applications.

10.3.3 Branched Covers

A branched cover of a complex space Y is a finite dominant morphism $f:X\to Y$. We will consider only the case when X is normal and Y is smooth. Ramification locus $R_f\subset X$ is the support of the quasi-coherent sheaf $\Omega_{X/Y}$ and the branch locus is $f(R_f)\subset Y$. It has codimension 1 (Nagata-Zariski purity of the branch locus cf. [222]).

Given an irreducible divisor $D\subset Y$ on a complex manifold Y one associates to (Y,D) a discrete valuation $\nu_D:\mathbb{C}(Y)\to\mathcal{N}_D$ of the field of meromorphic functions on Y given by $\nu_D(\phi))=ord_D(\phi),\phi\in\mathbb{C}(Y)$ (cf. [113], p.130). Here $\mathcal{N}_D$ is the subgroup of $\mathbb{Z}$ generated by the values of $\nu_D(\phi),\phi\in\mathbb{C}(Y)$. For a branched cover $X\to Y$ and a pair of irreducible divisors $D\subset Y,\Delta'\subset X$ where Δ' is a component of $f^*(D)^{red}$ the map $f^*:\mathbb{C}(Y)\to\mathbb{C}(X)$ induces the map $f^*_{\mathcal{N}}:\mathcal{N}_D\to\mathcal{N}_{\Delta'}$ .The index $[\mathcal{N}_{\Delta'}:f^*\mathcal{N}_D]$ (cf. [223] cf. Chap. 6, Sect. 12) is the ramification index e_D of f along the component Δ'. One has $e_{D'}=1$, unless D' is a component of R_f. Restriction of a branched cover $X\to Y$ onto the complement to the ramification divisor induces étale map $X\setminus R_f\to Y\setminus D$ where $D=f(R_f)$ is the branch locus. In particular given a branched cover $f:X\to Y$, selection of base point $p\in Y\setminus D$ allows to construct monodromy:

$$\pi_1(Y\setminus D)\to Sym(f^{-1}(p)) \tag{10.17}$$

into permutation group of points in the preimage of p assigning to each loop and a point $a\in f^{-1}(p)$ the end of the lift of the loop starting at a.

The set of equivalence classes of unramified covers $f_Z:Z\to Y\setminus D$, where f_{Z_1},f_{Z_2} are considered to be equivalent iff there exists biholomorphic isomorphism $h:Z_1\to Z_2$ such that $f_{Z_1}=f_{Z_2}\circ h$, is in one to one correspondence with the subgroups of $\pi_1(Y\setminus D,p)$ where $p\in Y\setminus D$ is a base point. The correspondence is given by assigning to f_Z the subgroup $(f_Z)_*\pi_1(Z,p')\subset\pi_1(Y\setminus D,p)$. This correspondence depends on a choice of a base point $p'\in f^{-1}(p)\subset X\setminus R_f$, but the subgroups corresponding to $p',p'',p'\ne p''$ are conjugate. A cover $Y\to X$ is called Galois if the corresponding subgroup is normal. The quotient of the fundamental group by this subgroup is the Galois group of the cover. This group is the image of the monodromy (10.17).

A branched cover is Galois if and only if the extension of the fields of meromorphic functions $\mathbb{C}(X)/\mathbb{C}(Y)$ is Galois and the Galois group of the cover is the Galois group of this field extension.

It follows from above discussion that the Galois group G acts freely on $X \setminus R_f$ with the quotient $Y \setminus D$ and one has the exact sequence $0 \to \pi_1(X \setminus R_f, p') \to \pi_1(Y \setminus D, p) \to G \to 0$. Vice versa, given an unramified cover $f : X \setminus R_f \to Y \setminus D$, it follows from Riemann Extension Theorem for normal spaces (cf. [101] Chap. 7, Sects. 4 and 2) that this action on $X \setminus R_f$ extends to the G-action on X via biregular automorphisms.

For an irreducible component $\Delta \subset R_f$ of the ramification divisor, the subgroup $I(\Delta)$ of G of automorphisms which fixes all $x \in \Delta$ is called the decomposition group of Δ or inertia group of Δ (cf. [104], Expose V, Sect. 2).[9] Action of inertia group on the tangent space at a smooth point $x \in \Delta$ which it fixes, induces the action on the normal space of Δ. The character ψ_I of this 1-dimensional representation of the cyclic group $I(\Delta)$ generates the group of characters $Char(I(\Delta))$. In particular one has a well defined map $Char I(\Delta) \to \mathbb{Z} : \chi \to i_\chi$ where $\chi = \psi^{i_\chi}, 0 \le i_\chi < \text{ord} I(x)$.

The extension is called abelian (resp. cyclic) if it is Galois and the Galois group is abelian (resp. cyclic). For a branched Galois cover $X \to Y$, the ramification index $e_{\Delta'}$ is the same for all irreducible components $\Delta' \subset X$, having the same image $D \subset Y$. Moreover, the order of the inertia group $|I(\Delta')| = e_{\Delta'}$. If r is the number of f-preimages of a generic point $D \subset Y$ then one has $|G| = re_{\Delta'}$.

The above correspondence between subgroups of the fundamental group and covers in Galois case becomes the correspondence (for fixed pair (Y, D)) between surjections $P : \pi_1(Y \setminus D) \to G$ and covers with Galois group G. Given a surjection, one can construct the corresponding cover, i.e. unique, up to homeomorphism over $Y \setminus D$, topological space Y' and the map $f : Y' \to Y \setminus D$ making Y' into unramified covering space with group G, as follows. This space Y' can be viewed as the quotient of the space of paths in $Y - D$ with a fixed initial points $p \in Y \setminus D$ with two paths γ_1, γ_2 being equivalent iff they have the same end point and the homotopy class of $\gamma_1^{-1} \circ \gamma_2 \in \pi_1(Y \setminus D, p)$ has trivial image in G. Y', being an unramified cover, inherits from $Y \setminus D$ the structure of complex manifold so that f' is étale. We have the following theorem.

Theorem 10.3.7 (Grauert-Remmert, cf. [101] Chap. 7, Grothendieck, [104], SGA1, Chap. XII, Sect. 5) *Let $f^* : X^* \to Y \setminus D$ be a finite unramified map of complex spaces where Y is a smooth complex manifold and D a divisor on Y. Then there i s a unique normal space X containing X^* as a dense subset and morphism $f : X \to Y$ such that $f^* = f|_{X^*}$.*

[9] Recall that the ground field here is $\mathbb{C}$. For varieties over non-algebraically closed fields, the inertia group $I(x)$ of $x \in \Delta$ (which is the subgroup of the decomposition group consisting of automorphisms inducing trivial automorphism of the extension of the residue fields of $f(x)$ by the residue field of x (cf. [104] Expose V, Sect. 2) is a proper subgroup of the decomposition group.

The inertia group of a component $\Delta' \subset X$ of ramification divisor in a Galois cover with a group G is the cyclic subgroup of G generated by the image in G of a representative g in the conjugacy class $\gamma \in \pi_1(Y \setminus D)$ of a meridian of D. If the order of this image is s, then g^s can be lifted to $X \setminus \Delta'$ as a closed path (i.e. g^s belongs to the kernel of surjection $\pi_1(Y \setminus D, p) \to G$) and is homotopic in the complement to the ramification divisor in X to a meridian of Δ'.

10.3.4 Abelian Covers

We discuss two ways to enumerate branched covers with abelian Galois group over a manifold with given branch locus. One is topological, which follows immediately from the discussion of previous section (cf. [137]) and another is algebro-geometric (cf. [4, 178]). A different important perspective, from a view point of root-stacks is discussed in [196].

Since by Hurewicz theorem $H_1(Y \setminus D, \mathbb{Z})$ is the abelianization of $\pi_1(Y \setminus D)$, any surjection of the fundamental group onto an abelian group G factors as $\pi_1(Y \setminus D) \to H_1(Y \setminus D, \mathbb{Z}) \to G$. Hence we have the following.

Corollary 10.3.8 *Equivalence class of a branched cover of a complex manifold Y and having a divisor $D = \bigcup D_i$ as its branch locus is determined by the surjection $H_1(Y \setminus D) \to G$ taking neither of meridians of D to identity. Vice versa, an abelian branched cover $f : X \to Y$ determines the branch divisor $D \subset Y$ and the above surjection of the homology group. Moreover, this correspondence induces the map from the set of irreducible components of the branch locus to the set of cyclic subgroups of the Galois group (inertia subgroups of the irreducible components of ramification divisor).*

An algebro-geometric description of abelian covers (cf. [178]) is given in terms of the collections of line bundles labeled by the characters of G. Given such a cover $f : X \to Y$, one obtains the decomposition into eigen-sheaves:

$$f_*(O_X) = \bigoplus_{\chi \in Char G} \mathcal{L}_\chi^{-1} \tag{10.18}$$

The left hand side has the structure of a sheaf of algebras and the work [178] describes the data specifying such structure on the right in (10.18). This is done in terms of classes of components of branch locus in $Pic(Y)$, the Galois group G, the collection of cyclic subgroups H and generators ψ_H of each group $Char H$ satisfying the following compatibility conditions. Once the O_Y-algebra structure, say $\mathcal{A}$, on the right in (10.18) is specified, the branched cover is just $Spec\mathcal{A} \to Y$.

For a pair (H, ψ), where $H \in Cyc(G)$ and ψ is a generator of $Char H$, let $D_{H,\psi}$ be the union of irreducible components D of the branch locus which have H as its inertia group and ψ as the character of representation of H on the normal space to a component of the ramification locus over D fixed by H. To a character $\chi \in CharG$ corresponds $r^{\chi}_{H,\psi} \in \mathbb{N}$ such that $\chi|_H = \psi^{r^{\chi}_{H,\psi}}, 0 \leq r^{\chi}_{H,\psi} < ordH$. For a pair $\chi_1, \chi_2 \in CharG$, let us set $\epsilon^{H,\psi}_{\chi_1,\chi_2}$ to be 0 (resp.1) if $\chi_1|_H = \psi^{i_{\chi_1}}, \chi_2|_H = \psi^{i_{\chi_2}}, 0 \leq i_{\chi_1}, i_{\chi_2} < CardH$ and $i_{\chi_1} + i_{\chi_2} < CardH$ (resp. $i_{\chi_1} + i_{\chi_2} \geq CardH$).

Then the bundles $\mathcal{L}_\chi$ in (10.18) satisfy the relations (cf. [178]):

$$\mathcal{L}_{\chi_1\chi_2} = \mathcal{L}_{\chi_1} \otimes \mathcal{L}_{\chi_2} \bigotimes_{H \in Cyc(G), \psi \in CharH} O(D_{H,\psi})^{\epsilon^{H,\psi}_{\chi_1,\chi_2}} \tag{10.19}$$

In fact, if $\chi_1, .., \chi_s$ are generators of a decomposition of Char(G) into a direct sum of cyclic subgroups and d_j is the order of χ_j, $j = 1, ..., s$ then:

$$d_\chi \mathcal{L}_\chi = \sum_{H,\psi} \frac{d_\chi r^{\chi}_{H,\psi}}{|H|} D_{H,\psi} \tag{10.20}$$

Vice versa (cf. [178]), given

(a) a finite abelian group G,
(b) a smooth compact complex manifold Y,
(c) a divisor D on Y with assignment to each irreducible component a cyclic subgroup H of G and a generator ψ_H of Char(H)
(d) collection of line bundles $\mathcal{L}_\chi$, $\chi \in$ Char(G) labeled by the characters of G

with (a), (b), (c), (d) satisfying the relations (10.19), there is abelian branched cover X of Y satisfying (10.18) (cf. [178]. The data (a), (b), (c), (d) subject to (10.19) called *the building data.*

We will show how to recover from Y, D and surjection $H_1(Y \setminus D, \mathbb{Z}) \to G$ the parts (c), (d) of the building data and vice versa, the building data determines the surjection onto the covering group.

Proposition 10.3.9 *Let Y be a smooth projective manifold and let $D = \bigcup_{i=1}^r D_i$ be a divisor with irreducible components D_i. The surjection $\pi : H_1(Y \setminus D) \to G$ onto an abelian group G determines for each character $\chi \in$ Char$G$ the bundle $\mathcal{L}_\chi$ so that the bundles $\mathcal{L}_\chi$, $\chi \in$ Char$G$ satisfy the relations (10.19). Moreover, the bundles $\mathcal{L}_\chi^{-1}$ are the eigenbundles of decomposition (10.18) for the covering corresponding to π (cf. Corollary 10.3.8). Vice versa, a building data determines the surjection $H_1(Y \setminus D, \mathbb{Z}) \to G$.*

Proof To a unitary character $\chi \in H^1(Y \setminus D, U(1))$ one associates the element in $Pic(X)$ as follows. One has the following high dimensional version of the exact sequence (10.5):

$$H_2(Y, \mathbb{Z}) \to H^{2\dim D}(D, \mathbb{Z}) \to H_1(Y \setminus D, \mathbb{Z}) \to H_1(Y, \mathbb{Z}) = 0 \tag{10.21}$$

Applying $Hom(\cdot, \mathbb{K})$, $\mathbb{K} = \mathbb{Z}, \mathbb{R}, U(1)$ to the terms of (10.21) we obtain:

$$\begin{array}{ccccc} 0 \to Hom(H_1(Y \setminus D, \mathbb{Z}), \mathbb{Z}) & \to & H_{2\dim D}(D, \mathbb{Z}) & \stackrel{\iota_{\mathbb{Z}}}{\to} & Pic(Y) \subset H^2(Y, \mathbb{Z}) \\ \downarrow & & \downarrow & & \downarrow \\ 0 \to \quad H^1(Y \setminus D, \mathbb{R}) & \to & H_{2\dim D}(D, \mathbb{R}) & \stackrel{\iota_{\mathbb{R}}}{\to} & Pic(Y) \otimes \mathbb{R} \subset H^2(Y, \mathbb{R}) \\ \downarrow & & \downarrow \exp & & \downarrow \exp \\ 0 \to \quad H^1(Y \setminus D, U(1)) & \stackrel{\varrho}{\to} & H_{2\dim D}(D, U(1)) & \stackrel{\iota_{U(1)}}{\to} & H^2(Y, \mathbb{Z}) \otimes U(1) \end{array} \tag{10.22}$$

Here ϱ is evaluation of a character on the meridian of the irreducible component,[10] the vertical arrows are induced by the exponentiation $exp : \mathbb{R} \to U(1), a \to e^{2\pi i a}$ and the map ι (for each choice of coefficients) assigns to a homology class, the class in $H^{2\dim D}(Y)$ which corresponds to the linear function on $H_{2\dim D}(X)$ given by the intersection index with this class. A lift $exp^{-1}(\varrho)$ of $\chi \in H^1(Y \setminus D, U(1))$ determines uniquely the element $\tilde{\chi}$ in the unit cube in $H_{2\dim D}(D, \mathbb{R})$, which is a fundamental domain for the action of the group $H_{2\dim D}(D, \mathbb{Z})$ on the latter and which has $H_{2\dim D}(D, U(1))$ as the quotient. Since $\iota_{U(1)}$ takes $\exp(\tilde{\chi})$ to the trivial class in $H^2(Y, \mathbb{Z}) \otimes U(1)$ we obtain that $\iota_{\mathbb{R}}(\tilde{\chi}) \in H^2(Y, \mathbb{R})$ is an integral class. Since it has the Hodge type $(1, 1)$, this class defines a line bundle. We shall denote it as $\mathcal{L}_\chi$.

Let $\chi(\gamma_{D_i}) = exp(2\pi i \alpha_i), \alpha_i \in \mathbb{Q}, 0 \le \alpha_i < 1$. If $ord\chi = d$ then $\alpha_i = \frac{\nu_i}{d}, 0 \le \nu_i < d, i \in \mathbb{N}$. It follows that $\iota_{\mathbb{R}}(exp^{-1}(\varrho(\chi))) = \sum \frac{\nu_i}{d}[D_i]$ defines an integral class in $H^2(Y, \mathbb{Z})$ and $\mathcal{L}_\chi$ is the bundle with the first Chern class corresponding to this integral class.[11] More directly, integrality can be seen as follows: since $\chi(D_\gamma) = 1, \forall \gamma \in H_2(Y, \mathbb{Z})$, it follows from (10.6) that one has $\prod_i exp(2\pi i \alpha_i)^{(\gamma, D_i)} = 1$. Hence $(\gamma, \sum_i \frac{\nu_i}{d}[D_i]) \in \mathbb{Z}$ for all $\gamma \in H_2(Y, \mathbb{Z})$ i.e. $\sum_i \frac{\nu_i}{d} D_i$ is an integral class. Let $\mathcal{L}_\chi$ be the bundle with the first Chern class corresponding to this class. The bundle $\mathcal{L}_\chi^d = \sum \nu_i \mathcal{O}(D_i)$ has a section and it follows from the calculation in [84] Sect. 3.6 (cf. also [127] Remarks 4.1.7 and 10.3.10 below) that for the cyclic cover $\pi_\chi : Y_\chi \to Y$ corresponding to $Ker(\chi)$ one has $\pi_\chi(\mathcal{O}) = \sum_0^{d-1} \mathcal{L}_\chi^{-k}$. Moreover, $\mathcal{L}_\chi^{-k}$ is the eigensheaf with corresponding character χ^k. Considering the full G-cover and factoring it through Y_χ, one sees that this is also the eigenbundle in the G-cover corresponding to π. Now the relations (10.19) follows from [178], Theorem 2.1.

Vice versa, the map Char$G = Hom(G, U(1)) \to H_{2\dim D}(D, U(1))$ sending a component with inertia group H to $exp(2\pi i \frac{r^\chi_{H,\psi}}{\text{ord}\, H})$ due to relations (10.19) lifts to the map to $H^1(Y \setminus D, U(1))$ and hence by duality induces the surjection $H_1(Y \setminus D, \mathbb{Z}) \to G$.

Example 10.3.10 (cf. [127], Sect. 4.1.B or [84]) Let Y be a smooth projective variety, $D \subset Y$ be a very ample divisor. Let $\mathcal{L}$ be a very ample line bundle such that $\mathcal{L}^d = \mathcal{O}(D)$. Clearly, the bundles $\mathcal{L}^i$ form a part of a building data for $G = \mathbb{Z}_d$. The corresponding cover can be obtained as follows. Let $s \in H^0(Y, \mathcal{O}(D))$ be a section with zero-scheme D and $\nu_d : [\mathcal{L}] \to [\mathcal{O}(D)]$ the map of the total spaces of the line bundles given by $v \in \mathcal{L} \to v^{\otimes d} \in \mathcal{L}^d$. Then $\nu_d^{-1}(s(Y))$ is a smooth subvariety Y_d of the total space of the line bundle $\mathcal{L}$ and its projection π onto the base endows Y_d

[10] Recall that this follows from identification $H_{2\dim D}(D, U(1)) = H^{2\dim D}(Y, Y \setminus D, U(1))$ obtained by excision and Lefschetz duality.

[11] Recall that Y is simply connected and hence $\mathcal{L}_\chi$ is well defined.

with the structure of the branched cover over Y with branch locus D. The divisibility of the fundamental class of D by d, implies that if $H_1(Y, \mathbb{Z}) = 0$, then there is well defined surjection $H_1(Y \setminus D, \mathbb{Z}) \to \mathbb{Z}_d$. It assignes to a 1-cycle δ representing a class in $H_1(Y \setminus D, \mathbb{Z})$, the modulo d intersection index of a 2-chain in Y having δ as its boundary. So Y_d is the cyclic cover of Y branched over D with Galois group $\mathbb{Z}_d$ and corresponding to this surjection of $H_1(Y \setminus D, \mathbb{Z})$. The inertia group of any point of D is $\mathbb{Z}_d$. On the other hand $\pi_*(O_{Y_d}) = \oplus_{i=0}^{i=d-1} \mathcal{L}^{-i}$ and $\mathcal{L}^{-i}$ is the eigen-bundle corresponding to the character of $\mathbb{Z}_d$ given by $\chi_i : j \to exp(\frac{2\pi\sqrt{-1}ij}{d})$. The relation (10.19) is immediate.

Vice versa, given the surjection $H_1(Y \setminus D, \mathbb{Z}) \to \mathbb{Z}_d$, the diagram (10.22) shows that the character χ_i, taking value $exp(\frac{2\pi\sqrt{-1}i}{d})$ on generator of $\mathbb{Z}/d\mathbb{Z}$, has as the lift $i_{\mathbb{R}}(exp^{-1}(\varrho(\chi_i)) = c_1(\mathcal{L}^i)$ and in this way producing a building data.

In the case $Y = \mathbb{P}^2$, D is an irreducible curve of degree d with equation $f(x_0, x_1, x_2) = 0$, one has $H_1(\mathbb{P}^2 \setminus D, \mathbb{Z}) = \mathbb{Z}_d$ and the cover corresponding to this isomorphism is biholomorphic to a hypersurface $V_f : u^d = f(x_0, x_1, x_2)$ in $\mathbb{P}^3$. Moreover the decomposition (10.18) becomes $f_*(O_{V_f}) = \oplus_{i=0}^{d-1} O_{\mathbb{P}^2}(-i)$.

Example 10.3.11 (cf. [118, 119]) Let $\mathcal{A}$ be an arrangement of r lines in $\mathbb{P}^2$. Then $H_1(\mathbb{P}^2 \setminus \mathcal{A}, \mathbb{Z}) = \mathbb{Z}^r/\{(1, ..., 1)\} = \mathbb{Z}^{r-1}$. Let $H_1(\mathbb{P}^2 \setminus \mathcal{A}, \mathbb{Z}) \to G = \mathbb{Z}_n^r/\{(1, ..., 1)\} = \mathbb{Z}_n^{r-1}$ sending the meridian the i-th line to $(0, ..., 0, 1, ...0) \mod n$. A character of G can be identify with a vector $(\frac{a_1}{n}, ..., \frac{a_r}{n})$, $0 \le a_i < n$, $\sum_1^r \frac{a_i}{n} \in \mathbb{Z}$. Let us denote this character $\chi_{a_1,...,a_r}$. The inertia group H_i of the i-th line is the subgroup of G isomorphic to $\mathbb{Z}_n$ and generated by $(0, ..., 0, 1, ...0) \mod n$ (all components except the i-th are zero) and the character ψ of H_i takes the value $exp\frac{2\pi i}{n}$ on the corresponding generator. It follows from discussion of Proposition 10.3.9 that

$$\mathcal{L}^{-1}_{\chi_{a_1,...,a_r}} = O_{\mathbb{P}^2}\left(-(\frac{\sum_1^r a_i}{n})\right)$$

See [119] for a direct calculation of the direct image of the structure sheaf using that this abelian cover is the restriction of the Kummer cover: $\mathbb{P}^{r-1} \to \mathbb{P}^{r-1}$ given by $(x_1, ..., x_r) \to (x_1^n, ..., x_r^n)$.

Example 10.3.12 Let D be the hypersurface in $\mathbb{C}^n$ given by $f_1(x_1, ..., x_n) \cdot ... \cdot f_r(x_1, ..., x_n) = 0$ where $f_i \in \mathbb{C}[x_1, .., x_n]$ are irreducible. Using a non-compact version of the calculation (10.6) one obtains $H_1(\mathbb{C}^n \setminus D) = \mathbb{Z}^r$. Let $p : H_1(\mathbb{C}^n \setminus D) \to G$ be a surjection onto an abelian group. Then to p corresponds the cover $P : V_{p,\bar{D}} \to \mathbb{P}^n$ branched over the projective closure $\bar{D}$ of D and possibly over the hyperplane at infinity with the following properties. The order r_i of $p(\gamma_i) \in G$ coincides with the ramification index of the branched cover P at $P^{-1}(s)$ where s is a generic point in D_i. At a generic point $s \in \mathbb{P}^{n-1}$ the ramification index at $P^{-1}(s)$ is the order in G of the class $p(\sum(degf_i)\gamma_i) \in H_1(\mathbb{C}^n \setminus D, \mathbb{Z})$. An explicit model of such covering can be obtained as the normalization of the projective closure of affine complete intersection in $\mathbb{C}^{n+r}$ given by equations:

$$z_i^{r_i} = f_i(x_1, ..., x_n) \quad i = 1, ..., r. \tag{10.23}$$

10.3.5 Characteristic Varieties

Alexander Invariants and Jumping Loci of Local Systems

A multivariable generalization of Alexander polynomials was proposed in [137] as follows. Let Y be a finite CW complex and $\phi : \pi_1(Y) \to A$ be a surjection onto a finitely generated abelian group. The unbranched abelian cover $\pi_\phi : Y_\phi \to Y$ corresponding to ϕ comes with a free action of A via cellular maps. Hence the compact supported homology $H_k(Y_\phi, \mathbb{C})$ and also its exterior powers

$$\Lambda^i H_k(Y_\phi, \mathbb{C}) \tag{10.24}$$

can be considered as the modules over the group algebra $\mathbb{C}[A]$ of A.

Definition 10.3.13 (*cf.* [132]) The affine subvariety $Char_i^k(Y, \phi)$ of the torus $Spec$ $\mathbb{C}[A]$ defined as support of the module $\Lambda^i H_k(Y_\phi, \mathbb{C})$ (cf. [81]) is called *the depth i characteristic variety of Y in dimension k (corresponding to surjection ϕ)*[12]

Standard results from commutative algebra (cf. [35] or [81] Chap. 20) show that $Char_i^k(Y)$ is the zero set of the i-th Fitting ideal of $\mathbb{C}[A]$-module $H_k(Y_\phi, \mathbb{C})$ i.e. the ideal generated by $(n - i + 1) \times (n - i + 1)$ minors of the matrix of a presentation of this module via n generators and m relations. Moreover, for $k = 1$, which unless otherwise stated will be our focus for the rest of this section, presentation of the module $H_1(Y_\phi, \mathbb{C})$ can be studied using the matrix of Fox derivatives giving presentation of $\mathbb{C}[A]$-module $H_1(Y_\phi, \tilde{p}, \mathbb{C})$ where $\tilde{p} = \pi_\phi^{-1}(p)$ is the preimage of $p \in Y$ (cf. [132]). As a consequence, this implies that for a CW complex having as its fundamental group a group with deficiency 1 and the homomorphism ϕ being the abelianization, the characteristic variety $Char_1^1(Y)$ of depth 1, has codimension 1 in $Spec\mathbb{C}[A]$. For example this is the case for $Y = S^3 \setminus L$ where L is a link. In fact, $Char_1^1(S^3 \setminus L)$ is the zero set of the multivariable Alexander polynomial of L [122]. In the case of algebraic curves in $\mathbb{C}^2$, the codimension of $Char_1^1$ is typically larger than 1 except for the case $A = \mathbb{Z}$ in which case it is the zero set of the 1-variable Alexander polynomial discussed in Sect. 10.3.1[13]

There is a different interpretation of these subvarieties of the complex tori $Spec\mathbb{C}[A]$.[14] Recall (cf. for example [49], Chap. 5) that a rank l local system on a CW complex Y is a l-dimensional linear representation of the fundamental group $\rho : \pi_1(Y) \to GL(l, \mathbb{C})$. (Co)homology of a local system are obtained as the cohomology of the chain complex:

$$\ldots \to C_i(\tilde{Y}, \mathbb{C}) \otimes_{\pi_1(Y)} \mathbb{C}^l \to \ldots \tag{10.25}$$

[12] The support is assumed to be a reduced variety.

[13] However $Char_1(\pi_1(\mathbb{P}^1 \setminus [3]), ab) = (\mathbb{C}^*)^2$ where [3] is a subset containing 3 points and ab is the abelianization of the free group.

[14] $Spec\mathbb{C}[A]$ is algebraic group with $CardTorA$ connected components with $(\mathbb{C}^*)^{rkA}$ being the component of identity.

where $C_i(Y, \mathbb{C})$ are the chains with compact support on the universal cover considered as a module over the group ring of the fundamental group.

An important feature of local systems is the following: in the case when Y is a smooth quasi-projective variety, the cohomology of local systems admit Hodge-deRham description given by Deligne (cf. [67]). First of all representations of fundamental groups can in interpreted as locally trivial vector bundles with constant transition functions (cf. [67], Corollary 1.4), which in turn, in the case when Y is a smooth manifold, can be interpreted as flat (integrable) connections $\nabla : \mathcal{V} \to \Omega^1_Y \otimes \mathcal{V}$ on a holomorphic vector bundle $\mathcal{V}$ (i.e. a $\mathbb{C}$-linear map satisfying Leibnitz rule). This differential operator can be extended to higher degree forms and lead to a twisted deRham complex:

$$... \to \Omega^p(Y) \otimes \mathcal{V} \xrightarrow{\nabla} \Omega^{p+1} \otimes \mathcal{V} \to ... \tag{10.26}$$

The (co)homology of the latter are identified with the (co)homology $H^*(Y, \rho)$ of the complex (10.25) since both are derived functors of the functor sending a representation to the subspace of invariants (cf. [67], Propositions 2.27 and 2.28).

Using (co)homology with twisted coefficients of rank one local systems, one can define *jumping loci*:

$$\mathcal{V}^k_i(Y) = \{\rho \in Hom(\pi_1(Y), \mathbb{C}) | H_k(Y, \rho) \geq i\} \tag{10.27}$$

There is the canonical identification of $Hom(H_1(Y, \mathbb{Z}), \mathbb{C}^*)$ and $Spec\mathbb{C}[H_1(Y, \mathbb{Z})]$ making two collections of subvarieties of both tori correspond to each other[15]:

$$Char^1_i(Y) \setminus \{1\} = \mathcal{V}^1_i(Y) \setminus \{1\} \tag{10.28}$$

This was shown to be the case for any finite CW complex Y in [117] for $k = 1$ and arbitrary i (cf. also [75, 137]) i.e. when one is interested in invariants of $\pi_1(Y)$ and for $i = 1$ but with arbitrary k (cf. [177]) when one considers invariants of the homotopy type.

Homology of Abelian Covers

Characteristic varieties determine the homology of covering spaces as follows.

Proposition 10.3.14 (cf. [70, 116, 132]) *Let $\phi : \pi_1(Y) \to A$ be a surjection onto a finite abelian group A and let Y_ϕ be the corresponding unbranched cover of Y with the Galois group A. Let $\nu(A)$ be the image of embedding $\phi^* : Spec\mathbb{C}[A] \to Spec\mathbb{C}[H_1(Y, \mathbb{Z})]$. Then:*

[15] The order of vanishing of the Fitting ideal of $H_1(Y_\phi, \mathbb{C})$ at (1, ..., 1) in general is different than $rkH_1(Y, \mathbb{C})$ which is the first Betti number of trivial local system.

$$rkH_k(Y_\phi, \mathbb{C}) = \sum_i Card(\nu(A) \cap \mathcal{V}_i^k(Y)) \tag{10.29}$$

This follows from the definition of jumping loci (10.27), the extension of classical Shapiro lemma (cf. [70]) from group cohomology to arbitrary spaces (cf. [49]) i.e. in our notations the identification $H_k(Y_\phi, \mathbb{C}) = H_k(Y, \mathbb{C}[A])$, and the decomposition $\mathbb{C}[A] = \oplus_{\chi \in Spec\mathbb{C}[A]} \mathbb{C}_\chi$ where $\mathbb{C}_\chi$ is the 1-dimensional representation of A given by the character χ.

Now consider the case of abelian branched covers which according to Corollary 10.3.8 are specified by the branching locus and the abelian quotient of the fundamental group of the complement to the latter. The proof below is a version of the argument due to M. Sakuma (cf. [188]).

Proposition 10.3.15 *Let X be a smooth simply-connected projective surface and $D = \bigcup D_i$ a reduced divisor. Let $\phi : H_1(X \setminus D, \mathbb{Z}) \to A$ be a surjection onto a finite abelian group. For a character $\chi \in A^*$ where $A^* = Hom(A, \mathbb{C}^*)$, let D^χ be the union of irreducible components D_i of D such that for the meridian $\delta_i \in H_1(X \setminus D, \mathbb{Z})$ of D_i one has $\chi(\delta_i) \neq 1$. Denote by $d(D^\chi, \chi)$[16] the maximum of the integers i such that $\chi \in Char_i^1(X \setminus D^\chi)$, where $Char_i^1(X \setminus D^\chi)$ is the depth i characteristic variety of the curve D^χ as defined in Definition 10.3.13. Let $\widetilde{\bar{X}_\phi}$ be a resolution of singularities of the branched cover $\bar{X}_\phi$ of X ramified along D corresponding to above surjection ϕ. Then*

$$rkH_1(\widetilde{\bar{X}_\phi}, \mathbb{C}) = \sum_{\chi \in A^*} d(D^\chi, \chi) \tag{10.30}$$

Proof Denote by $Sing\bar{X}_\phi$ the set of singularities of $\bar{X}_\phi$. This is a finite set mapped by the covering map into the set $Sing(D)$ of singularities of D. We will start by showing that

$$rkH_1(\widetilde{\bar{X}_\phi}, \mathbb{C}) = rkH_1(\bar{X}_\phi \setminus Sing(\bar{X}_\phi), \mathbb{C}) \tag{10.31}$$

Indeed, if E is the exceptional set of a resolution $\widetilde{\bar{X}_\phi} \to \bar{X}_\phi$, then

$$\widetilde{\bar{X}_\phi} \setminus E = \bar{X}_\phi \setminus Sing(\bar{X}_\phi) \tag{10.32}$$

On the other hand, the exact sequence of the pair $(\widetilde{\bar{X}_\phi}, \widetilde{\bar{X}_\phi} \setminus E)$ and the identification $H^i(\widetilde{\bar{X}_\phi}, \widetilde{\bar{X}_\phi} \setminus E) = H_{4-i}(E)$ yield:

$$0 \to H^1(\widetilde{\bar{X}_\phi}) \to H^1(\widetilde{\bar{X}_\phi} \setminus E) \to H_2(E) \to H^2(\widetilde{\bar{X}_\phi}) \tag{10.33}$$

[16] The integer $d(D^\chi, \chi)$ is called *the depth* of the character χ of the curve D^χ.

Together with injectivity of the right map in (10.33), which is a consequence of Mumford theorem on non-degeneracy of the intersection form on a resolution of a surface singularity (cf. [167]), we obtain (10.31).

By universal coefficients theorem, allowing to switch to cohomology, the identity (10.30) will follow from the following calculation of dimensions of χ-eigenspaces

$$dim H^1(\bar{X}_\phi \setminus Sing(\bar{X}_\phi))_\chi = d(D^\chi, \chi) \tag{10.34}$$

for all characters $\chi \in Char(A)$.

To show (10.34), note that the group A acts on $\bar{X}_\phi \setminus Sing(\bar{X}_\phi)$ with the quotient $X \setminus Sing(D)$. For any character χ of group A we consider the cyclic branched cover $(X \setminus Sing(D))_\chi$ of $X \setminus Sing(D)$ corresponding to composition $\pi_1(X \setminus D) \to A \to Im(\chi)$. One has the biregular isomorphism:

$$\bar{X}_\phi \setminus Sing(\bar{X}_\phi)/Ker\chi = (X \setminus Sing(D))_\chi \tag{10.35}$$

The group $A/Ker(\chi) = Im(\chi)$ acts on the left side of (10.35) and the identification (10.35) is $Im(\chi)$-equivariant. The transfer $H^*(\bar{X}_\phi \setminus Sing(\bar{X}_\phi)) \to H^*(\bar{X}_\phi \setminus Sing(\bar{X}_\phi)/Ker)$ (cf. [33], p.118) provides $Im(\chi)$-equivariant isomorphism $H^*(\bar{X}_\phi \setminus Sing(\bar{X}_\phi))^{Ker(\chi)} = H^*(\bar{X}_\phi \setminus Sing(\bar{X}_\phi)/Ker(\chi))$ which implies that

$$H^*(\bar{X}_\phi \setminus Sing(\bar{X}_\phi)_\chi = H^*(\bar{X}_\phi \setminus Sing(\bar{X}_\phi)_\chi^{Ker(\chi)} = H^*((X \setminus Sing(D))_\chi)_\chi \tag{10.36}$$

Equation (10.34) is obvious for trivial character and the isomorphism (10.36) shows that (10.30) follows from the cyclic case of the Proposition for non-trivial χ.[17]

Finally, the cyclic cover $(X \setminus Sing(D))_\chi$ is a totally ramified cover of $X \setminus Sing D$ branched over D^χ and the cyclic case with non-trivial χ follows from the calculation of homology of unbranched covers in Proposition 10.3.14 since the action of $Im(\chi)$ on kernel and cokernel of the map induced by the embedding of the cyclic unbranched cover $(X \setminus D^\chi)_\chi$ with Galois group $Im(\chi)$

$$H^1((X - Sing D)_\chi) \to H^1((X \setminus D^\chi)_\chi) \tag{10.37}$$

is trivial.

Structure of Characteristic Varieties

The central result on the structure of characteristic varieties of quasi-projective manifolds is obtained from their interpretation (cf. [117]) as the jumping loci of the cohomology of local systems, which allows to apply deep Hodge theoretical methods [9]. It asserts that the irreducible components of characteristic varieties are a

[17] We also use that removal 0-dimensional set $Sing D$ from a 4-dimensional manifold does not change the first Betti number.

finite order cosets of subtori of $Spec\mathbb{C}[H_1(X \setminus D, \mathbb{Z})]$ and that these components are the pull backs of the characteristic varieties of fundamental groups of curves via holomorphic maps. The origins of such correspondence are going back to Zariski [221], Beauville [30], Green-Lazarsfeld [103], Simpson [195] in projective case with quasi-projective case being addressed by Arapura [9]. In a very special case when the quasi-projective manifold is a complement to an irreducible plane singular curve the assertion of the finitness of the order of cosets becomes the cyclotomic property of the roots of Alexander polynomials ([129], cf. Theorem 10.3.3) and does not require Hodge theory (unlike the Theorem 10.3.16). We shall quote an orbifold version (cf. [19]) of this correspondence between the holomorphic maps and the components of characteristic varieties

Theorem 10.3.16 (cf. [9, 19]) *Let $\mathcal{V}_i^1(X \setminus D)^{irr}$ be an irreducible component of jumping locus of 1-dimensional cohomology of a smooth quasi-projective variety $X \setminus D$. Then $\mathcal{V}_i^1(X \setminus D)^{irr}$ is a coset of finite order of a subtorus of the commutative algebraic group $Spec\mathbb{C}[H_1(X \setminus D, \mathbb{Z})]$. Moreover, there exist an orbifold curve C^{orb}, an irreducible component $\mathcal{V}_i(\pi_1^{orb}(C^{orb}))^{irr}$ and holomorphic orbifold map $f : X \setminus D \to C^{orb}$ such $\mathcal{V}_i^1(X \setminus D)^{irr} = f^*(\mathcal{V}_i(\pi_1^{orb}(C^{orb}))^{irr})$.*

Corollary 10.3.17 *The array of Betti numbers of finite abelian covers of X branched over D determines the characteristic varieties of the fundamental group of the complement.*

Proof Translated subgroups are specified by the points of finite order on the torus which they contain. A point χ of the finite order belongs to the i-th characteristic variety if and only if the multiplicity of χ in the cyclic cover corresponding to the group $Im(\chi)$ is at least i. The claim follows.

A very important application of translated subgroup property is that it provides a necessary conditions on a group be quasi-projective i.e. to be a fundamental group of a smooth quasi-projective variety. For application of these and ideas from different ideas, not discussed here, to the problem of characterisation of quias-projective and quasi-Kahler groups (in particular the comparison with the fundamental groups of 3-manifolds) see: [10, 19, 32, 76, 77, 91, 124, 133]

10.3.6 Isolated Non-normal Crossings

A generalization of results on Alexander invariants from Sects. 10.3.1–10.3.5 providing invariants of the homotopy type beyond fundamental groups was proposed in [141]. The starting point is the following:

Theorem 10.3.18 (cf. [141], Theorem 2.1) *Let $X, dimX > 2$ be a smooth simply connected projective variety and let D be a divisor such that all its irreducible components are smooth and ample. Then $\pi_1(X \setminus D)$ is abelian and $\pi_i(X \setminus D) = 0$ for $2 \le i \le dimX - 1$.*

This theorem and Lefschetz hyperplane sections theorem have the following as an immediate corollary.

Corollary 10.3.19 ([141]) *Let X be as in Theorem 10.3.18, let $D = \bigcup_1^r D_i$ be a divisor with ample irreducible components and let $NNC(D)$ be the subvariety of X consisting of points $x \in X$ at which D fails to be a normal crossings divisor.*[18] *Let $s = dim NNC(D)$.*[19] *Then $\pi_i(X \setminus D) = 0$ for $i \le d - s - 2$. Moreover, if $H = \bigcap_1^s H_i$ is a sufficiently general intersection of very ample divisors on X then $D \cap H$ is a divisor on $X \cap H$ with isolated non normal crossings and*

$$\pi_i(X \setminus D) = \pi_i((X \setminus D) \cap H) \quad i \le d - s - 1 \tag{10.38}$$

In particular the first non-trivial homotopy group of the complement to a divisor with ample components can be calculated using the divisor $D \cap H$ with isolated non-normal crossings on H.

The high dimensional analogs of the results on the Alexander invariants of the complement to curves described in [141] give a similar description of the homotopy group $\pi_{d-1}(X \setminus D)$ where X as in Theorem 10.3.18, D is a divisor with ample components but now D is allowed to have *isolated* non normal crossings (INNC) i.e. $dim NNC(D) = 0$. The role of the first homology of the infinite abelian cover in the case of complements to curves is played by the first non-vanishing homotopy group $\pi_i(X \setminus D), i > 1$. In fact one has the identification:

$$H_{dim X-1}(\widetilde{X \setminus D}, \mathbb{Z}) = \pi_{dim X-1}(X \setminus D) \tag{10.39}$$

where $\widetilde{X \setminus D}$ is the universal (hence also universal abelian, cf. Theorem 10.3.18) cover of $X \setminus D$. The $\mathbb{Z}[\pi_1(X \setminus D)]$-module structure equivalently can be obtained using the Whitehead product (cf. [141]) and the characteristic variety of $X \setminus D$ in dimension $dim X - 1$ is the support of the module $\pi_{dim X-1}(X \setminus D) \otimes \mathbb{C}$.

In the case when a point in $NNC(D)$ belongs to only one component, failure to be normal crossing means that the point is just an isolated singularity of D. In this case the local information about the Alexander invariants is contained in the Milnor fiber of the singularities of D and includes the characteristic polynomial of the monodromy as well as some the Hodge theoretical invariants (cf. [141]). Reducible analog of isolated singularities are isolated non-normal crossings i.e. the intersections of hypersurfaces which are smooth and transversal everywhere except for a single point. It turns out that many features of isolated singularities described by Milnor [163] have counterparts in INNC case. They include the analogs of high connectivity of Milnor fibers, analogs of monodromy action on the cohomology, its cyclotomic properties and others.

---

[18] We call D a divisor with isolated non-normal crossings, if dim $NNC(D) = 0$.

[19] The convention is that if $x \notin D$ then D does have normal crossing at x and the dimension of empty set is -1.

To be specific, recall that if $f(x_0, ..., x_n) = 0$ is a germ of an isolated singularity, V_f is its zero set, B_ϵ is a small ball about the singular point of f, ∂B_ϵ is the boundary sphere then we have the following.[20]

Theorem 10.3.20 (cf. [163]) *(i) The complement $B_\epsilon \setminus (V_f \cap B_\epsilon)$ is homotopy equivalent to $\partial B_\epsilon \setminus (V_f \cap \partial B_\epsilon)$.*

(ii) There is a locally trivial fibration $\partial B_\epsilon \cap (V_f \cap \partial B_\epsilon) \to S^1$, with the fiber (i.e. the Milnor fiber) being homotopy equivalent to a wedge of spheres of dimension n.

Corollary 10.3.21 *If $f(x_0, ..., x_n)$ is a germ of isolated singularity, then the universal cyclic cover of $B_\epsilon \setminus (B_\epsilon \cap V_f)$ is homotopy equivalent to a finite wedge of spheres of dimension n. Moreover, the action of the deck transformation on the homology of the universal cyclic cover coincides with the action of the monodromy operator on the homology of the Milnor fiber. In particular, the characteristic polynomial of the monodromy coincides with the Alexander polynomial of $\partial B_\epsilon \setminus (V_f \cap \partial B_\epsilon)$.*

A generalization of these results to multi-component germs is as follows (which is the local counterpart of Corollary 10.3.19).

Theorem 10.3.22 (cf. [140]) *Let X_r be a germ $f_1(x_0, ..., x_n) \cdot ... \cdot f_r(x_0, ..., x_n) = 0$ of a hypersurface which is product of r irreducible germs. Then for $n > 1$*

$$\pi_1(\partial B_\epsilon \setminus (X_r \cap \partial B_\epsilon)) = \mathbb{Z}^r \quad \pi_i(\partial B_\epsilon \setminus (X_r \cap \partial B_\epsilon)) = 0 \ , 2 \le i \le n-1 \quad (10.40)$$

The role of the Milnor fiber is now played by the universal abelian[21] cover of the complement $\widetilde{\partial B_\epsilon \setminus (X_r \cap \partial B_\epsilon)}$ and the monodromy action is replaced by the action of fundamental group of the complement to INNC germ on the universal abelian cover via deck transformations. The homology $H_n(\widetilde{\partial B_\epsilon \setminus (X_r \cap \partial B_\epsilon)}, \mathbb{C})$ is equipped with the action of the group algebra $\mathbb{C}[H_1(\partial B_\epsilon \setminus (X_r \cap \partial B_\epsilon))]$. The $H_1(\partial B_\epsilon \setminus (X_r \cap \partial B_\epsilon), \mathbb{Z})$-action can be, as in global case, identified the Whitehead product of the elements of $\pi_n(\partial B_\epsilon \setminus (X_r \cap \partial B_\epsilon)$ with the elements of $\pi_1(\partial B_\epsilon \setminus (X_r \cap \partial B_\epsilon)) = H_1(\partial B_\epsilon \setminus (X_r \cap \partial B_\epsilon))$. The group ring of the latter is the ring of Laurent polynomials and the role of the characteristic polynomial of the monodromy is played by the subvariety of the torus $Spec\mathbb{C}[H_1(\partial B_\epsilon \setminus (X_r \cap \partial B_\epsilon))]$ which is the support of this module [81].

Example 10.3.23 (cf. [140]) Let $f_{d_i}(x_0, ..., x_n) = 0, i = 1, ..., r$ be the equations of smooth sufficiently general hypersurfaces in $\mathbb{P}^n$. Then the union in $\mathbb{C}^{n+1}$ of cones over these r hypersurfaces is isolated non-normal crossing. The support of this module is the zero set of $t_1^d \cdot ... \cdot t_r^{d_r} - 1 = 0$.

Example 10.3.24 Let $f_1(x, y)....f_r(x, y)$ be a germ of reducible curve in$\mathbb{C}^2$. The support of the universal abelian cover of the complement to the link coincides the zero set of the multivariable Alexander polynomial of the link. Further properties of this support and its Hodge theoretical properties are discussed in [39, 40].

---

[20] Only the last claim in (ii) requires singularity of f to be isolated.

[21] Which is also the universal cover for $n \ge 2$ since the fundamental group is abelian in this case.

We refer to the work [141] Sects. 5 and 6 for description of the structure of characteristic varieties of *global* INNC in terms of the homotopy groups of local INNC (of germs), as well as the relationship between the cohomology of local systems, the homology of branched cover and the translated subgroups of $Spec\mathbb{C}[\pi_1(X \setminus D)]$ which are the irreducible components of $Supp\pi_{dimX-1}(X \setminus D) \otimes \mathbb{C}$. The divisibility relations extending the divisibility theorem from Sect. 10.3.2 to the case of hypersurfaces in $\mathbb{P}^n$ with isolated singularities, the Thom Sebastiani theorems for the orders of the homotopy groups and other results on the topology of the complements to are discussed in [134]. For some results in non-isolated case, cf. Sect. 10.3.8.

10.3.7 Twisted Alexander Invariants

A generalization of Alexander polynomials was proposed in the context of knot theory which uses as an additional input a linear representation of the fundamental group (cf. [123] for a discussion of this generalization). A twisted version of Alexander polynomials of algebraic plane curves was considered in [44]. In [143] a multivariable extension of this construction called *a characteristic varieties of a CW complex twisted by a unitary representation* was defined as follows.

Let $\pi : \pi_1(X) \to U(V)$ be a unitary representation of the fundamental group of a CW complex such that $H_1(X, \mathbb{Z})$ is a free abelian group of a positive rank.[22] Here V is a complex vector space endowed with a Hermitian bilinear form and viewed as a left $\mathbb{C}[\pi_1(X)]$-module. Let $\tilde{X}$ be the universal cover of X. For a (left or right) module M over the algebra $\mathbb{C}[\pi_1(X)]$ (which is associative but possibly non-commutative), we denote by $M^\flat$ the module obtained by restriction of the coefficients to the group algebra $\mathbb{C}[\pi_1'(X)] \subset \mathbb{C}[\pi_1(X)]$ of the commutator subgroup $\pi_1'(X)$ of $\pi_1(X)$. Let $C_*(\tilde{X})$ denotes chain complex of $\tilde{X}$ endowed with the natural structure of (a right) $\mathbb{C}[\pi_1(X)]$-module. Consider the following complex of tensor products of $\mathbb{C}[\pi_1'(X)]$-modules.

$$C_*(\tilde{X})^\flat \otimes_{\mathbb{C}[\pi_1'(X)]} V^\flat : \quad g(c \otimes v) = cg^{-1} \otimes gv \tag{10.41}$$

The group $\pi_1(X)$ acts on the module (10.41) and the restriction of this action to the commutator $\pi_1'(X)$ is trivial. Hence (10.41) obtains the structure of $\mathbb{C}[H_1(X, \mathbb{Z})]$-module. It passes to the homology of the complex (10.41). We denote the resulting homology modules as $H_i(X_{ab}, V_{ab})$.

Definition 10.3.25 The support of $\mathbb{C}[H_1(X, \mathbb{Z})]$-module $\Lambda^l H_i(\tilde{X}_{ab}, V_{ab})$ we call *the ρ-twisted degree i, l-th characteristic variety of X* This is a subvariety of the torus $Spec\mathbb{C}[H_1(X, \mathbb{Z})]$ which we denote as $Ch_i^l(X, \rho)$.

In the case when $rkH_1(X, \mathbb{Z}) = 1$ the support is a finite subset of $\mathbb{C}^*$ and hence the zero set of a unique monic polynomial of degree $Card(Ch_i^l(X, \rho))$. More generally, a surjection $\epsilon : H_1(X, \mathbb{Z}) \to \mathbb{Z}$ defines the surjection of the group algebras and hence

[22] Torsion freeness condition of $H_1(X, \mathbb{Z})$ is introduced to simplify the exposition.

the embedding $\epsilon^* : \mathbb{C}^* = Spec\mathbb{C}[\mathbb{Z}] \to Spec\mathbb{C}[H_1(X, \mathbb{Z})]$. The Alexander polynomial $\Delta_n^l(X, \rho, \epsilon)$ is the unique monic polynomial of minimal degree having in $\mathbb{C}^*$ the roots $(Im\epsilon^*) \cap Ch_n^l(X, \rho)$.

We refer to [143] Sect. 5 for other results on the relation between the twisted Alexander polynomials and characteristic varieties in the context of complements to plane curves. For example the cyclotomic property of the roots of Alexander polynomials becomes the following: the roots of a ρ-twisted Alexander polynomial belong to a cyclotomic extension of the extension of $\mathbb{Q}$ generated by the eigenvalues of ρ (cf. [143], Theorem 5.4).

10.3.8 Alexander Invariants of the Complements Without Isolatedness Properties

Investigations of the Alexander invariants of the complements to hypersurfaces with isolated singularities which were discussed in Sects. 10.3.1, 10.3.2, 10.3.5 and 10.3.6 were extended to the case of hypersurfaces with non-isolated singularities and further to smooth quasi-projective varieties by L. Maxim and his collaborators (cf. [148, 151, 158]). The D-modules (cf. [186, 192]), the category of complexes of constructible sheaves, the perverse sheaves and peripheral complex (going back to [41] and first studied in this context in [158]) are the key technical tools used by these authors. We will review two most important outcomes of this approach: propagation of characteristic varieties and extension of divisibility theorem for Alexander polynomials. The propagation property of characteristic varieties was first noticed in the context of arrangements of hyperplanes (cf. [71, 82]) and extended further in [149, 150]. The key observation is pure topological and concerns the spaces satisfying a version of cohomological duality.

Definition 10.3.26 (*cf.* [31, 71, 149]) Let X be a finite CW complex and $G = \pi_1(X, x_0), x_0 \in X$. A topological space is called a duality space of dimension n if $H^p(X, \mathbb{Z}[G]) = 0,\ p \neq n$ and $H^n(X, \mathbb{Z}[G])$ is non-zero and torsion free. A spaces X is called *an abelian* duality space if for $A = H_1(X, \mathbb{Z})$ one has $H^p(X, A) = 0,\ p \neq n$ and $H^n(X, A) \neq 0$ is torsion free.

Theorem 10.3.27 (cf. [150], Theorem 3.16) *Let X be an abelian duality space of dimension n. Then the cohomology jumping loci of the characters of the fundamental group $\pi_1(X)$ satisfy the following properties (the so call propagation package [149, 150]:*

(i) Propagation: Subvarieties $\mathcal{V}^i(X)$ form descending chain:

$$\mathcal{V}^n(X) \supseteq \mathcal{V}^{n-1}(X) \supseteq .... \supseteq \mathcal{V}^0(X) \tag{10.42}$$

(ii) Codimension bound:

$$\mathrm{codim}\mathcal{V}^{n-i} \geq i$$

(iii) *Irreducible components: if V is an irreducible component of codimension d in $\mathcal{V}^n(X)$ then $V \subset \mathcal{V}^{n-d}$*
(iv) *Generic vanishing: for characters ρ of the fundamental group $\pi_1(X)$ in a Zariski open set in $Hom(\pi_1(X), \mathbb{C}^*)$ one has $H^i(X, L_\rho) = 0$ for $i \neq \dim X$.*
(v) *Signed Euler characteristic property: $(-1)^{dim X}\chi(X) > 0$*
(vi) *Betti numbers inequality: $b_i(X) > 0$ for $0 \le i \le n$ and $b_1(X) \ge n$.*

Use of jump loci of constructible complexes on semi-abelian varieties (which are Albanese varieties of X is appropriate sense) is the key step in the proof of this result.

This theorem suggests the problem of identifying the abelian duality spaces. In this direction one has the following.

Theorem 10.3.28 (cf. [149]) *(i) Let X a compact Kahler manifold which is abelian duality space. Then X is biholomorphic to an abelian variety.*

(ii) Let X be quasi-projective manifold, such the albanese map $X \to Alb(X)$ is proper. Then X is an abelian duality space. In particular a complement to a union of hypersurfaces in $\mathbb{C}^n$ satisfies the propagation package.

Next we will describe the identity having as a special case the divisibility relation between product of local Alexander polynomials of singularities and the global Alexander polynomial from Sect. 10.3.2 in the case of curves and in [134] in higher dimensions. Let $f(z_0, ..., z_{n+1}) = 0$ be a homogeneous polynomial of degree d having as its zero set the hypersurface $V_f \subset \mathbb{P}^{n+1}$ and let $f_d(z_1, .., z_d) = f(0, z_1, ..., z_{n+1})$ be the equation of the intersection of V_f with the hyperplane H_∞ at infinity. The map

$$f : \mathbb{P}^{n+1} \setminus (H_\infty \cup V_f) \to \mathbb{C}^* \tag{10.43}$$

which is the restriction of $f : \mathbb{C}^{n+1} = \mathbb{P}^{n+1} \setminus \mathbb{P}^n \to \mathbb{C}$ given by $(1, z_1, ..., z_{n+1}) \to f(1, z_1, ..., z_{n+1})$ allows to define the infinite cyclic cover $\widetilde{\mathbb{P}^{n+1} \setminus (H_\infty \cup V_f)}$ corresponding to the kernel of the map $\pi_1(\mathbb{P}^{n+1} \setminus (H_\infty \cup V_f) \to \pi_1(\mathbb{C}^*) = \mathbb{Z}$ induced by (10.43). For each $1 \le i \le n$ one has a well defined (up to a unit of $\mathbb{C}[t, t^{-1}]$) polynomial $\Delta_i(t)$which is the order of the $\mathbb{C}[t, t^{-1}]$-module $H_i(\widetilde{\mathbb{P}^{n+1} \setminus (H_\infty \cup V_f)})$ (cf. [158]). Let $\psi_f \mathbb{Q}_{\mathbb{C}^{n+1}}$ denotes Deligne's nearby cycles complex associated to f and let $\psi_i(t)$ be the order of $H^{2n+1-i}(V_f \cap \mathbb{C}^{n+1}, \psi_f \mathbb{Q}_{\mathbb{C}^{n+1}})$. Note that $\Delta_i(t) = \psi_i(t)\ for\ i < n$ and $\Delta_n(t)$ divides ψ_n. Let $h(t)$ be the characteristic polynomial of the Milnor fiber $f_d(z_1, .., z_{n+1}) = 1$. The final ingredient is the determinant $det\phi$ of the bilinear form:

$$H_{n+1}(\widetilde{\mathbb{P}^{n+1} \setminus (H_\infty \cup V_f)}, \mathbb{Q}(t)) \otimes H_{n+1}(\widetilde{\mathbb{P}^{n+1} \setminus (H_\infty \cup V_f)}, \mathbb{Q}(t)) \xrightarrow{\phi} \mathbb{Q}(t) \tag{10.44}$$

given by $(\alpha, \beta) \to \alpha \cdot i(\beta)$ where $i : \partial(\mathbb{P}^{n+1} \setminus (H_\infty \cup V_f))) \to \mathbb{P}^{n+1} \setminus (H_\infty \cup V_f)$ is the embedding of the boundary $\partial(\mathbb{P}^{n+1} \setminus (H_\infty \cup V_f))$ of a small regular neighborhood of $V_f \cup H_\infty$, and "$\cdot$" is the Poincare pairing in the homology of the infinite cyclic cover (cf. [162]) of the pair $((\mathbb{P}^{n+1} \setminus (H_\infty \cup V_f), (\partial\mathbb{P}^{n+1} \setminus (H_\infty \cup V_f)))$. With these definitions one can describe the relation between the global Alexander invariants and the data of the starta of a singular hypersurface as follows.

Theorem 10.3.29 (cf. [151]) *Let $f \in \mathbb{C}[x_1, ..., x_{n+1}]$ be defining polynomial of an affine hypersurface $F \subset \mathbb{C}^{n+1}$, f_d be the top degree form of homogenization of f with corresponding Milnor fiber F_h and $h_i(t)$ be the Alexander polynomial associated to $H_i(F_h)$. Let $\psi_n(t)$ denotes the Alexander module associated to $H_c^{2n-i}(V_0, \psi_f \mathbb{Q}_{\mathbb{C}^{n+1}})$. Finally, let ϕ be the intersection form on the infinite cover associated with f. Then*

$$h_n(t)\psi_n(t) = \delta_n(t)^2 det(\phi) \tag{10.45}$$

In the case when V_f is transversal to H_∞ and has only isolated singularities, this is translated into the following relation (cf. [151]):

$$(t-1)^{|e(\mathbb{P}^{n+1} \setminus (H_\infty \cup V_f))| + (-1)^{n+1}} (t^d - 1)^{\frac{1}{d}((d-1)^{n+1} + (-1)^n)} \prod_{p \in Sing(V_f)} \Delta_p = \Delta_n^2 det(\phi) \tag{10.46}$$

i.e. one obtains topological interpretation of terms converting divisibility into equality. In the case of plane curves one recovers the result in [44].

10.4 Ideals of Quasiadjunction and Multiplier Ideals

Now we will turn to calculation of components of characteristic varieties in terms of dimensions of linear systems determined by the singular points of the curve. For earlier expositions of this material in the case of plane curves cf. [142] or [18].

10.4.1 Ideals and Polytopes of Quasi-adjunction

Definition 10.4.1 Let X be a complex n-dimensional manifold, $P \in X$ be a point and let B_P be a small ball centered at P. Let D be a reduced divisor on X containing P, $f \in O_P$ be a reduced germ of a holomorphic function having D as its divisor and let $f(x_1, ..., x_n) = f_1(x_1, ..., x_n) \cdot ... \cdot f_r(x_1, ..., x_n)$ be its prime factorization. Let

$$(j_1, ..., j_r), (m_1, .., m_r), 0 \le j_i < m_i \tag{10.47}$$

be two arrays of integers. Consider abelian branched cover $V_{m_1,....,m_r}$ of B ramified over D and corresponding to the component-wise reduction $H_1(B_P \setminus D \cap B_P, \mathbb{Z}) = \mathbb{Z}^r \to \oplus_1^r \mathbb{Z}/m_i\mathbb{Z}$ (cf. Sect. 10.3.3). After selecting local coordinates near P, this cover can be viewed as a germ at the origin in $\mathbb{C}^{n+r}$ with coordinates $(z_1, ..., z_r, x_1, ..., x_n)$ given by the local equations:

$$z_1^{m_1} = f_1(x_1, ..., x_n), ..., z_r^{m_r} = f_r(x_1, ..., x_n), \tag{10.48}$$

The covering map of subvariety (10.48) onto B_P is given by projection

$$(z_1, ..., z_r, x_1, ..., x_n) \to (x_1, ..., x_n).$$

The ideal of quasi-adjunction $\mathcal{A}(f|j_1, ..., j_r|m_1, ..., m_r)$ of $f(x_1, ..., x_n)$, corresponding to the array (10.47), is the ideal of germs $\phi \in O_P$ in the local ring of P, such that the n-form

$$\omega_\phi = \frac{\phi(x_1, ..., x_n)dx_1 \wedge ... \wedge dx_n}{z_1^{m_1-j_1-1} .....z_r^{m_r-j_r-1}} \tag{10.49}$$

defined on the smooth locus of (10.48), can be extended over a log resolution of (10.48).

One shows that the ideal $\mathcal{A}(f|j_1, .., j_r|m_1, ..., m_r)$ is independent of a resolution of (10.48) and that it depends only on the vector:

$$\left(\frac{j_1+1}{m_1}, ..., \frac{j_r+1}{m_r}\right) \in [0, 1]^r \subset \mathbb{R}^r \tag{10.50}$$

rather than on specific values of j_i, m_i. To see that dependence is only on (10.50), let us consider the following resolution of the germ (10.48): select a log-resolution

$$\mu : (\tilde{X}, \tilde{D}) \to (B_P, B_P \cap D) \tag{10.51}$$

with the exceptional set $E = \bigcup_1^K E_k$ i.e. assume that $E \cup \mu^*(D)$ is a normal crossing divisor. Consider a resolution $\mathcal{X}_m$ of the *normalization* $\tilde{X}_{m_1,...,m_r}$ of the fiber product $\tilde{X} \times_X V_{m_1,..,m+r}$

$$\begin{array}{ccc} & \mathcal{X}_{m_1,...,m_r} & \\ & \downarrow & \\ & \tilde{X}_{m_1,...,m_r} & \\ & \downarrow & \\ \tilde{X} \times_X V_{m_1,..,m_r} & \to & V_{m_1,...,m_r} \\ \tilde{\pi}_{m_1,...,m_r} \downarrow & & \pi_{m_1,...,m_r} \downarrow \\ \tilde{X} & \stackrel{\mu}{\to} & \mathbb{C}^n \end{array} \tag{10.52}$$

The normalization $\tilde{X}_{m_1,..,m_r}$ has abelian quotient singularities and hence is $\mathbb{Q}$-Gorenstein Kawamata log-terminal (cf. [27]). In particular an n-form on the smooth locus of $\tilde{X}_{m_1,..,m_r}$ extends over exceptional locus of $\mathcal{X}_m \to X_{m_1,..,m_r}$ iff it is holomorphic on the smooth locus of $\tilde{X}_{m_1,..,m_r}$ (cf. [102]). Let $a_{k,i} = ord_{E_k}\mu^*(f_i)$, $c_k = ord_{E_k}\mu^*(dx_1 \wedge ... \wedge dx_n)$, $e_k(\phi) = ord_{E_k}\mu^*(\phi)$. We obtain that $\tilde{\pi}_{m_1,..,m_r}(\omega_\phi)$ extends over the smooth locus of normalization $\tilde{X}_{m_1,...,m_r}$ iff for all irreducible components $E_k \subset X$ one has:

$$\sum_i a_{k,i}\frac{j_i+1}{m_i} > \sum_i a_{k,i} - e_k - c_k - 1 \quad e_k = e_k(\phi) \tag{10.53}$$

Definition 10.4.2 For a choice of collection $\mathcal{E} = \{e_k\}, k = 1, ..., K$ of non-negative integers labeled by irreducible components of resolution (10.51) and such that there exist a germ $\phi \in O_P$ such that $e_k = e_k(\phi)$, the closure $\mathcal{P}(\mathcal{E})$ in the r-cube $[0, 1]^r$ of the set of solutions to inequalities (10.53) is called a *polytope of quasi-adjunction*[23] of the germ $f = f_1 \cdot .... \cdot f_r$.

Using polytopes $\mathcal{P}(\mathcal{E})$ or inequalities (10.53), one can describe necessary and sufficient conditions on (10.47) assuring extendability of the form ω_ϕ (cf. 10.49) on resolution of singularities of abelian covers in the tower (10.48). While the polytopes $\mathcal{P}(\mathcal{E})$ form a set which is partially ordered by inclusion and closely related to Alexander invariants, sometimes it is convenient also to use a partition of $[0, 1]^r$ into a union of (locally closed) *non-intersecting and possibly non-convex* polytopes Q compatible with polytopes $\mathcal{P}(\mathcal{E})$.

To describe these polytopes consider the hyperplanes in $\mathbb{R}^r$

$$\sum_i a_{k,i}(\gamma_i - 1) + c_k = \epsilon_k, \epsilon_k \in \mathbb{Z}_{<o} \tag{10.54}$$

labeled by the exceptional components $E_k, k = 1, ..., K$ of resolution μ (cf. (10.51)) where collections of integers ϵ_k are such that there exist $\phi \in O_P$ satisfying $\epsilon_k = e_k(\phi)$ and the remaining coefficients in (10.54) are defined just before (10.53). The hyperplanes (10.54) partition the unit cube $[0, 1]^r$ into a union of locally closed polytopes Q of various dimensions, such that $\upsilon = (..., \gamma_i, ..) = (..., \frac{j_i+1}{m_i}, ...)$ and $\upsilon' = (..., \gamma_i', ..) = (..., \frac{j_i'+1}{m_i'}, ...)$ belong to the same polytope iff υ, υ' satisfy the same sets of inequalities (10.53). Equivalently, each polytope Q coincides with the polytope formed by an equivalence class of points in $[0, 1]^r$ when one considers points equivalent if they have the same set of polytopes $\mathcal{P}(\mathcal{E})$ containing each. We wil call the polytopes Q, defining the decomposition of the unit cube into disjoint union, *the strict polytopes of quasi-adjunction* to distinguish them from the ordinary polytopes of quasi-adjunction $\mathcal{P}(\mathcal{E})$.

For a fixed vector υ with coordinates (10.50) the germs ϕ which satisfy the inequalities (10.53) form the ideal $\mathcal{A}_Q \subset O_P$ depending only on the polytope Q and not on a choice of $\upsilon \in Q$. In particular, the ideal of quasi-adjunction $\mathcal{A}(f|j_1, ..., j_r|m_1, ..., m_r)$ is the ideal $\mathcal{A}_Q$ such that $(...., \frac{j_i+1}{m_i}, ...) \in Q$.

Definition 10.4.3 1. We define an ideal of quasi-adjunction as an ideal coinciding with an ideal $\mathcal{A}_Q$ for some strict polytope of quasi-adjunction $Q \subset [0, 1]^r$.

[23] By a polytope we mean a set of solutions to a finite collection of inequalities. All polytopes considered here are bounded (subsets of a unit cube) and hence are the convex hulls of the sets of their vertices. Faces are subsets of the boundary of a polytope which are the convex hulls of a subset of the set of vertices of the polytope. The dimension of a polytope (including a face) is the maximal dimension of the balls in its interior (with the dimension of a vertex being zero).

2. For a polytope in $[0, 1]^r \subset \mathbb{R}^r$ as above, a *face of quasi-adjunction* $\mathcal{F}$ of dimension p is the p-dimensional intersection of the polytope of quasi-adjunction with an affine half-space in $\mathbb{R}^r$, transversal to all coordinated hyperplanes in $\mathbb{R}^r$, such that each point of this intersection is the boundary point of the polytope and the half-space.[24]

Corollary 10.4.4 *The polytopes, ideals and faces of quasi-adjunction depend only on the germ $f(x_1, ..., x_n)$ and not on a resolution. The inequalities (10.53) show that there exist firstly, the collection of polytopes $\mathcal{P}$, which are the unions of the polytopes Q, and secondly to each of these polytopes correspond the ideal $\mathcal{A}_Q$ such that for any $v \in Q$ and a germ $\phi \in O_P$, the corresponding form ω_ϕ can be extended to a holomorphic form on $\mathcal{X}_{m_1,...,m_r}$ iff $\phi \in \mathcal{A}_Q$. The boundary of such a polytope Q is a union of faces each being also a face the boundary of a polytope $\mathcal{P}$ and each such face being a close polytope in the intersection of hyperplanes given by the Eqs. (10.54) with*

$$\gamma_i = 1 - \frac{j_i + 1}{m_i} \quad i = 1, ..., r. \tag{10.55}$$

10.4.2 Ideals of Quasi-adjunction and Multiplier Ideals

For an exposition of the theory of multipler ideals we refer to [127] Part III. Here we recall the key definitions and relate them to the ideals of quasi-adjunction.

Definition 10.4.5 (*cf.* [127] *Definition 9.2.1*) Let X be a smooth complex variety and $D \in Div(X) \otimes \mathbb{Q}$ an effective $\mathbb{Q}$-divisor. Let $\mu : X' \to X$ be a log resolution, $K_{X'/X} = K_{X'} - \mu^*(K_X)$ is a relative canonical class. The multiplier ideal sheaf $\mathcal{J}(X, D)$ of D is the direct image

$$\mu_*(K_{X'/X} - \lfloor \mu^*(D) \rfloor) \tag{10.56}$$

where for a $\mathbb{Q}$-divisor $D = \sum \gamma_i D_i, a_i \in \mathbb{Q}, D_i \in Div(X)$, $\lfloor D \rfloor = \sum \lfloor \gamma_i \rfloor D_i$ and $\lfloor \gamma \rfloor \in \mathbb{Z}$ denotes the integral part of $\gamma \in \mathbb{Q}$.

For a collection of $\mathbb{Q}$-divisors $F_1, ..., F_r$ (resp. the ideals $\wp_1, ...\wp_r$), the mixed multiplier ideal $\mathcal{J}(c_1 F_1, ..., c_r F_r)$ (resp, $\mathcal{J}(\wp^{c_1} \cdot .... \cdot \wp_r^{c_r})$) (cf. [127] 9.2.8, [2]) is defined as the multipler ideal

$$\mu_*(K_{X'/X} - \lfloor c_1 F_1 + ... + c + r F_r \rfloor) \tag{10.57}$$

(in the case of mixed multiplier ideals attached to $\wp_1, .., \wp_r$, the divisors F_i are determined from $\wp_i O_{X'} = O_{X'}(-F_i)$).

Proposition 10.4.6 *The ideal of quasi-adjunction $\mathcal{A}(f|j_1, ..., j_r|, m_1, .., m_r)$ coincides with the multiplier ideal $\mathcal{J}(\sum((1 - \frac{j_i+1}{m_i})D_i)$*

[24] I.e. the set of solutions to a linear inequality.

Proof A germ ϕ is a section of the sheaf given by (10.56) where $D = \sum \gamma_i D_i$ if and only if it satisfies the inequality

$$e_k(\phi) + c_k - \sum \gamma_i a_{k,i} \geq 0 \tag{10.58}$$

This is equivalent to (10.53).

Following [145] one can define the LCT polytope, which is a multi-divisor analog of the log canonical threshold (cf. [127]).

Definition 10.4.7

$$LCT(D_1, \ldots\ldots, D_r) = \{(\lambda_1, \ldots., \lambda_r) \in \mathbb{R}^r_+ | (X, \sum_1^r \lambda_i D_i) \ is\ log\ canonical\}$$

Proposition 10.4.8 (log-canonical polytopes and polytopes of quasi-adjunction) *Let $I : [0, 1]^r \to [0, 1]^r$ be the involution of the unite cube given by $(\gamma_1, \ldots, \gamma_r) \to (1 - \gamma_1, \ldots, 1 - \gamma_r)$. Then $LCT(D_1, .., D_r)$ is the I-image of the part in the interior of the unite cube of the boundary of the polytope of quasiadjuction containing the origin.*

Proof Clearly for the vectors (10.50) sufficiently close to zero, the n-form (10.49) is extendable for any ϕ in the local ring of P i.e. the "ideal" of quasi-adjunction is not proper. The ideal of quasi-adjunction is constant for all vectors (10.50) in the polytope bounded by the faces of quasi-adjunction closest to the origin. Hence the claim follows from the characterization of log-canonical thresholds in terms of multiplier ideals (cf. [127] Sect. 9.3.B) and Proposition 10.4.6).

Remark 10.4.9 1. We often use the following correspondence between vectors (10.50) and the characters of local fundamental group of the complement to the germ of $D = \bigcup D_i$. Consider the embedding of $[0, 1)^r$ into $H^1(B_P \setminus B_P \cap \bigcup D_i, \mathbb{R})$ using the basis dual to the meridians δ_i of the divisors D_i i.e. assigning to a vector $(\gamma_1, \ldots., \gamma_r)$ the cohomology class h such that $h(\delta_i) = \gamma_i$.

This embedding induces the identification of the cube $[0, 1)^r$ with the group of characters of the local fundamental group. Indeed, any point in $H^1(B_P \setminus B_P \cap \bigcup D_i, \mathbb{R})$ via exponential map $t \to exp(2\pi i t)$ determines an element in $H^1(B_P \setminus B_P \cap \bigcup D_i, U(1))$ i.e. an unitary character of the local fundamental group.[25] Vice versa, to a character in $H^1(B_P \setminus B_P \cap \bigcup D_i, U(1))$ we can assign its unique preimage belonging to the fundamental domain of the action of $H^1(B_P \setminus B_P \cap \bigcup D_i, \mathbb{Z})$ on $H^1(B_P \setminus B_P \cap \bigcup D_i, \mathbb{R})$ which is the unit cube in the coordinates of the above basis.

2. Exact sequence of a pair $(B_P, B_P \setminus B_P \cap D)$ gives the identification:

$$H^1(B_P \setminus B_P \cap D, \mathbb{R}) = H^2(B_P, B_P \setminus D, \mathbb{R}) = H_2(B_P \cap D, \partial(B_P \cap D), \mathbb{R}) \tag{10.59}$$

[25] This is a local version of the global construction used in Proposition 10.3.9. Here the rank of the fundamental group coincides with the number of irreducible components of the divisor.

and similarly for the coefficients $\mathbb{Z}$ and $U(1)$. $B_P \cap D$ is homeomoprhic to a bouquet of disks and $\partial(B_P \cap D)$ is a disjoint union of circles, both having the cardinality coinciding with the number of branches of D at P. In particular (10.59) is a vector space with canonical direct sum decomposition with summands corresponding to the branches of D at P.

10.4.3 Local Polytopes of Quasi-adjunction and Spectrum of Singularities

Cyclic Theory

The relation between the constants of quasi-adjunction (i.e. the faces of quasi-adjunction in cyclic case) and the Hodge theory was first discussed in [152] in the case of isolated singularities. This was continued in [36, 39].

Let $f(x_0, x_1, ..x_n)$ be a germ of an isolated singularity at the origin. Recall that the cohomology of the Milnor fiber F_f support a mixed Hodge structure defined by Steenbrink and Varchenko (cf. [11, 180, 207, 216]). This mixed Hodge structure provides the cohomology $H^n(F_f, \mathbb{C})$ with a decreasing filtration $F^p \cap H_\lambda \subseteq F^{p-1} \cap H_\lambda$ on each summand of the direct sum decomposition of $H^n(F_f, \mathbb{C})$ into the eigenspaces of the monodromy operator. A rational number α is an element of the spectrum of f of multiplicity k if there is an eigenvalue of the monodromy λ such that

$$n - p - 1 < \alpha \leq n - p \ \ exp(2\pi i\alpha) = \lambda \ \ dim F^p \cap H_\lambda / F^{p+1} \cap H_\lambda = k \quad (10.60)$$

(cf. [152])

Theorem 10.4.10 *A rational number α belongs to the spectrum of $f(x_0, ...x_n) = 0$ and satisfies $0 < \alpha < 1$ if and only iff $-\alpha$ is a face of quasi-adjunction of f (i.e. is a constant of quasi-adjunction in terms of [130]; the definition given there is in terms of the adjoint ideals and is a special case of Definition 10.4.3 corresponding to cycilc covers o f$\mathbb{C}^2$.)*

There are many cases when spectrum can be easily calculated explicitly. In the case of quasi-homogeneous singularities (with weights w_0, ..., w_n i.e. when defining polynomial is a sum of monomials $ax_0^{m_0} ....x_n^{m_n}$ such that $\sum_{i=0}^{i=n} w_i m_i = 1$) the generating function i.e. $\sum_\alpha t^\alpha$ where α runs through the spectrum of the singularity is given by

$$\frac{1}{t} \prod_{i=0}^{n} \frac{t^{w_i} - t}{1 - t^{w_i}} \quad (10.61)$$

cf. [187]; we included the factor $\frac{1}{t}$ since we use the same normalization of the spectrum as in [152] where the left end of the spectrum is -1, cf. (10.60). In particular for an irreducible germ with one characteristic pair $x^p = y^q$ we obtain

$(t^{-\frac{1}{p}} + ... + t^{-\frac{p-1}{p}})(t^{\frac{1}{q}} + ... + t^{\frac{q-1}{q}})$ which for the cusp $x^2 = y^3$ gives $t^{-\frac{1}{6}} + t^{\frac{1}{6}}$. Explicite calculations of the spectrum and hence the constants of quasi-adjunction of irreducible plane curve singularities in terms of Puiseux pairs was made in [187] For related calculations see [36, 96].

Local Abelian Theory

Calculations of the polytopes of quasi-adjunction of singularities of plane curves were made in [39, 40, 138]. Curves on surfaces with rational singularities were considered in [2].

Several examples of calculation of polytopes and ideals of quasi-adjunction for isolated non-normal crossings were considered in [140, 143] which lead to expressions for characteristic varieties mentioned in Example 10.3.23. Related results are presented in [145].

10.4.4 Ideals of Quasi-adjunction and Homology of Branched Covers

One of the first applications of ideals of quasi-adjunction was a procedure that allowed to express the Hodge numbers of abelian covers in terms of dimensions of linear systems determined by the branch locus and the data of singularities. The relation between the constants and ideals of quasi-adjunction and Hodge numbers of the cyclic covers in [130] followed by numerous works (cf. for example [37, 152, 170, 171, 215]), many using the terminology of multiplier ideals.[26]

In the case of curves on surfaces one has the following result. For a high dimensional extension leading to calculations of the dimensions of the space of holomorphic forms on the resolutions of branched covers (the only Hodge numbers which are independent of resolutions) see [37, 83, 134] etc.

Proposition 10.4.11 *Let $f : \bar{X}_{\phi} \to X$ be a n abelian branched cover of a smooth projective surface X ramified over a divisor D with r irreducible components and let $f_*(O_{\bar{X}_{\phi}}) = \oplus L_{\chi}^{-1}$ be the decomposition (10.18). For a character χ, let $\mathcal{A}_{\chi} \subset O_X, \chi = exp(....2\pi i\alpha(\chi), ...), \alpha \in [0, 1)^r$ be the ideal sheaf having as stalk at $p \in Sing(D)$ the ideal of quasi-adjunction of singularity at p corresponding to the polytope of quasi-adjunction containing $\alpha(\chi)$.*

Then the dimension of the χ-eigenspace of the covering group acting on the space of holomorphic 1-forms on a resolution $\widetilde{X_{\phi}}$ of singularities of $\bar{X}_{\phi}$ is given by

$$dim H^0(\widetilde{X_{\phi}}, \Omega^1_{\widetilde{X_{\phi}}})_{\chi} = H^1(X, K_X \otimes L_{\chi} \otimes \mathcal{A}_{Sing D}(\chi)) \qquad (10.62)$$

---

[26] The term introduced by A. Nadel in 1990, cf. [169].

10.4.5 Hodge Decomposition of Characteristic Varieties

Calculation of Characteristic Varieties: Deligne Extensions

In this section we sketch a method for calculation of the variety of unitary characters with corresponding local systems having positive first Betti number and which is based on the Hodge theoretical description of the cohomology of local systems due to Deligne and Timmersheidt ([67–69, 210], related works include works of S. Zucker, M . Saito, El Zein and Illusie cf. discussion in H. Esnault review of [210] in Math. Reviews). For details we refer to [19, 37, 84, 143].

The starting point is deRham type description of the cohomology of local systems on smooth quasi-projective varieties using logarithmic forms already mentioned in Sect. 10.3.5. With notations used in this section, we assume now that Y is smooth and quasi-projective and that $\bar{Y}$ is a smooth projective compactification, such that $\bar{Y} \setminus Y$ is a normal crossing divisor $Y^{\infty} = \bigcup_{i \geq 1} Y_i$. The connection ∇ in (10.26) can be selected to be holomorphic on Y and meromorphic on $\bar{Y}$ i.e. in having a matrix given in local coordinates by meromorphic functions with poles along $\bar{Y} \setminus Y$. Moreover, this selection can be made so that the matrix of connection has as its entries the 1-forms having logarithmic poles along Y^{∞} (i.e. linear combinations: $\omega = \sum \alpha_i \frac{dz_i}{z_i}$ where z_i are local equations of irreducible components of Y^{∞} and α_i are holomorphic in a chart in $\bar{Y}$ (cf. [67], Proposition 3.2). Globally, logarithmic connection can be viewed as a $\mathbb{C}$-linear map $E \to \Omega^1(logY^{\infty}) \otimes E$ where E is a vector bundle on $\bar{Y}$satisfying Leibnitz rule. Note that the matrix of connection, in the rank one case, is just a logarithmic 1-form. One has a well defined Poincare residue map $\Omega^1(logY^{\infty}) \otimes E \to O_{Y_i} \otimes E$ along each irreducible component Y_i. Locally, residue depends on trivialization and globally on the bundle E.

Definition 10.4.12 (*Deligne's extension of a flat connection* ∇ *on* Y) is a logarithmic connection on a bundle E on $\bar{Y}$ such that its residues satisfy the inequality

$$0 \leq Res_{Y_i}(\nabla) < 1 \tag{10.63}$$

for any i.

The description of the jumping loci of local systems in terms of Deligne's extensions, is based on the degeneration of the Hodge-DeRham spectral sequence

$$E_1^{p,q} = H^p(\Omega^q(logY^{\infty}) \otimes V_\rho) \to H^{p+q}(\mathcal{V}_\rho). \tag{10.64}$$

in term E_1. Here V_ρ is a Deligne's extensions of a unitary connection corresponding to the local system $\mathcal{V}_\rho$ where $\rho : \pi_1(Y) \to U(n)$ is a unitary representation (cf. [210]). For a rank one local system $\mathcal{V}_\chi$, corresponding to a character $\chi = exp(2\pi i u)$ and the Deligne extension L_χ of the corresponding connection, this degeneration implies $rkH^1(Y, \mathcal{V}_\chi) = rkH^0(\Omega^1(logY^{\infty} \otimes L_\chi)) + rkH^1(L_\chi)$ and the following.

Proposition 10.4.13 *A subset of* $[0, 1)^{b_1(X \setminus D)} \subset H^1(X \setminus D, \mathbb{R})$ *such that the Deligne extension of the connection corresponding to a character* $\chi = exp(2\pi i u)$ *coincide with a fixed line bundle L is a polytope* Δ_L *in* $H^1(X \setminus D, \mathbb{R})$. *The rank* $rk H^1(Y, \mathcal{V}_\chi)$ *is constant when u varies within* Δ_L. *The subset of the torus* $\mathbb{C}^{*b_1(X \setminus D)} = \mathrm{Char} H_1(X \setminus D, \mathbb{Z})$ *consisting of the characters* $exp(2\pi i u)$, $u \in \Delta_L$ *is a jumping set of the Hodge numbers of unitary local systems. The Zariski closure in* $(\mathbb{C}^*)^{b_1(X \setminus D)}$ *of this subset is a translated by a finite order character a connected subgroup of* $\mathbb{C}^{*b_1(X \setminus D)}$ *which is an irreducible component of the characteristic variety of the fundamental group. Vice versa, any irreducible component of characteristic variety of* $\pi_1(X \setminus D)$, *is Zariski closure of a set* $exp(\Delta_L)$.

Indeed, after selecting a basis in $H^1(X \setminus D, \mathbb{R})$, one readily sees that inequality (10.63) for each component translates into a linear inequality on components of logarithm of χ in coordinates in this basis. It is not hard to see that there are only finitely many bundles on X which are the Deligne extensions of a connection corresponding to a character in $[0, 1)^{b_1(X \setminus D)}$ can occur (cf. [143]). This provides an explicite description of the unitary part of the components of characteristic varieties. Since by [9] all irreducible components of the characteristic variety are translated subtori of $\mathbb{C}^{*b_1(X \setminus D)}$, it follows that in this way we obtain all the components as the Zariski closures of the exponential images of the polytopes Δ_L.

Calculation of Characteristic Varieties: Quasi-adjunction

We will focus on the case of characteristic varieties of curves on surfaces. Similar results are expected for characteristic varieties associated with higher homotopy groups. We refer to [137, 141] for some of the results in this direction.

Calculation in terms of ideal of quasi-adjunction is based on comparison of topological and algebro-geometric calculation of the dimensions of eigenspaces of the action of the Galois group on the homology of abelian covers given respectively by Propositions 10.3.15 and 10.4.11. However, the Proposition 10.3.15 considers only the characters of $\pi_1(X \setminus D)$ which values on the meridians of all irreducible components of D are non-trivial. This motivates the following definition.

Definition 10.4.14 Let $D = D_1 \cup D_2$ be a decomposition of a reduced divisor on a smooth projective simply connected surface X. Let $s_{D/D_1} : \pi_1(X \setminus D) \to \pi_1(X \setminus D_1)$ be the surjection of the fundamental groups induced by inclusion $X \setminus D \subset X \setminus D_1$ and let $s^{H_1}_{D/D_1}, s^{\pi_1'}_{D/D_1}, s^{\pi_1'/\pi_1''}_{D/D_1}$ be the corresponding surjections respectively on the homology, commutator of the fundamental group and the abelianization of the commutator. Let $s^{Char_i}_{D,D_1} : Char^1_i(X \setminus D_1) \to Char^1_i(X \setminus D)$ be induced map of supports of the i-th exterior powers of the homology modules (over $\mathbb{C}[H_1(X \setminus D_1)]$ and $\mathbb{C}[H_1(X \setminus D)]$ respectively, cf. Definition 10.3.13). An irreducible component C_D of the characteristic variety of $\pi_1(X \setminus D)$ is called *non-essential* if there is a decomposition $D = D_1 \cup D_2$ and a component C_{D_1} of the characteristic variety of $\pi_1(X \setminus D_1)$ such that $s^{Char_i}_{D/D_1}(C_{D_1}) = C_D$. *An essential component* is an irreducible component of $Char_i(X \setminus D)$ which fails to be non-essential.

Below we describe a calculation of only essential components, for simplicity assuming that $H_1(X \setminus D, \mathbb{Z})$ has no torsion. The results certainly can be described without this assumption. In fact, the first examples of calculations of characteristic varieties (the roots of the Alexander polynomials) in terms of ideals of quasi-adjunction, were made in [130] in the cases when D is a an irreducible plane curve i.e. when $H_1(\mathbb{P}^2 \setminus D, \mathbb{Z})$ is a cyclic group of order $deg D$. See also Example (10.4.21) where the complement is torsion as well. Non essential components may exhibit a subtle behavior: the depth of a component may increase considered as component of C_D instead of C_{D_1}. We refer to discussion of this phenomenon to [17, 19, 21].

Recall (cf. Sect. 10.4.1 and Remark 10.4.9) that with each singular point P of the reduced divisor $D = \bigcup_1^N D_i$ on a surface X we associated a collection of polytopes of quasi-adjunction $Q_j(P)$, $j = 1, ..., n(P)$ in $U_{s(P)} = [0, 1)^{s(P)} \subset H^1(B_P \setminus D \cap B_P, \mathbb{R}) = H_2(B_P \cap D, \partial B_P \cap D, \mathbb{R})$ (recall that B_P is a small ball in X centered at P). Note that the latter locally homology groups can be endowed with the maps to the corresponding global groups for each group of coefficients $\mathbb{K} = \mathbb{Z}, \mathbb{R}, U(1)$ leading to the diagram:

$$\begin{array}{ccc} H^1(B_P \setminus B_P \cap D, \mathbb{K}) & \stackrel{\delta_P}{\to} & H_2(B_P \cap D, \partial(B_P \cap D), \mathbb{K}) \\ \uparrow i_P & & \uparrow \epsilon_P \\ H^1(X \setminus D, \mathbb{K}) & \to & H_2(D, \mathbb{K}) \end{array} \tag{10.65}$$

Here the top horizontal map δ_P is the isomorphisms (10.59), the left vertical map i_P is induced by embedding and the right vertical map ϵ_P is the homology boundary map: $H_2(D, \mathbb{K}) \to H_2(D, D \setminus B_P \cap D, \mathbb{K}) = H_2(B_P \cap D, \partial(B_P \cap D, \mathbb{K}))$.
For each polytope $Q_j(P) \subset H^1(B_P \setminus D \cap B_P, \mathbb{R})$ we consider preimages

$$Q_j^X(P) = (i_P^*)^{-1}(Q_j(P)) \subset H^1(X \setminus D, \mathbb{R}) \text{ and } Q_j^D(P) = \epsilon_P^{-1}(\delta_P(Q_j(P))) \subset H_2(D, \mathbb{R})$$

In (10.65), each group in the top row for $\mathbb{K} = \mathbb{R}$ and the group $H_2(D, \mathbb{R})$ contains the canonical fundamental domain for the action of respective lattice which one obtains taking for each group $\mathbb{K} = \mathbb{Z}$. These fundamental domains are the unit cubes in the bases corresponding to the fundamental classes of appropriate irreducible component of D. The image of each such fundamental domain in $H^1(B_P, B \setminus B_P \cap D, \mathbb{R})$, induced by embedding $H^1(B_P, B \setminus B_P \cap D, \mathbb{R}) \to \oplus_{P \in Sing(D)} H_2(B_P \cap D, \partial(B_P \cap D), \mathbb{R})$, is a face of the unit cube in the latter. The intersection of the image of $H_2(D, \mathbb{R})$ in $\oplus_P H_2(B_P \cap D, \partial(B_P \cap D), \mathbb{R})$ is either such a face, if the branches of D at P belong to different irreducible components of D, or a diagonal in such a face, if different branches at P belong to the same irreducible component of D. We denote by $U_{X,D}$ the unit cube in $H_2(D, \mathbb{R})$. From now on we will use the same nation $Q_j^X(P)$, $Q_j^D(P)$ for their intersections with the respective unite cubes: these and only parts of respective polytopes contain the information we need below.

The unit cubes in $H_2(D,\mathbb{R})$ and $H_2(D\cap B_P, \partial(D\cap B_P),\mathbb{R})$, have canonical involution corresponding to the lift of the conjugation of characters via inverse of the map induced on cohomology by $\exp : \mathbb{R}\to U(1)$. For example on each unit cube in $H^1(B_P \setminus B_P\cap D,\mathbb{R})$ this involution is given by $(u_1,..,u_r)\to(1-u_1,....,1-u_r)$ and similarly in other cases. For a subset $\mathcal{U}$ in such a cube we denote the image of this involution as$-\mathcal{U}$.

Definition 10.4.15 Let $Sing(D)$ be the set of singularities of D and let Q be the set of collections $Q=\{Q_j(P_k)|P_k\in Sing(D), j=1,...J(P_k)\}$ (here $J(P_k)$ is the cardinality of the set of local polytopes of quasi-adjunciton of singularty P_k) of (strict[27]) local polytopes of quasi-adjunction $Q_j(P_k)$, one for each singularity $P_k\in Sing(D)$.

a. *The divisorial global polytope of quasi-adjunction* is the intersection

$$\mathcal{G}_Q=\bigcap_{P_k\in Sing(D),Q_j(P_k)\in Q} pr_{P_k}{}^{-1}Q_j(P_k)\subset U_{X,D}\subset H_2(D,\mathbb{R}) \tag{10.66}$$

of preimages of polytopes of quasiadjunction, one for each singular point of D.[28]

b. A *global divisorial face of quasi-adjunction* is a face $\mathcal{F}$ of a polytope $\mathcal{G}_Q$ (cf. 10.66) corresponding to a collection Q of local polytopes of quasi-adjunction. We say that a face $\mathcal{F}$ of a global polytope of quasi-adjunction *correspond to a subset* $\mathcal{S}\subset Sing(D)$ if $\mathcal{F}$ is a face of a polytope determined already by the local polytopes of singularities only from $\mathcal{S}$: $\bigcup_{P_k\in\mathcal{S},Q_j(P_k)\in Q} pr_{P_k}{}^{-1}Q_j(P_k)$.

c. *The sheaf of quasi-adjunction* $\mathcal{A}_Q$ *(or* $\mathcal{A}_{\mathcal{G}(Q)}$*) corresponding to a choice collection* Q *of local polytopes of quasi-adjunction,* is the ideal sheaf in $\mathcal{O}_X$ having as the stalk at $P\notin\mathcal{S}$ the local ring of $P\in X$ and the ideal of quasi-adjunction $\mathcal{A}_{Q_j(P)}$ corresponding to selected local polytope of quasi-adjunction for singularity $P\in\mathcal{S}$.

d. *The homological global polytope (resp. face) of quasi-adjunction* is a polytope in $H^1(X\setminus D,\mathbb{R})$, viewed as the trivial coset of $H_2(D,\mathbb{R})/H^1(X\setminus D,\mathbb{R})$, which is the translation to this trivial coset[29] of the intersection of divisorial global polytope (resp. face) of quasiadjunction described in a. (resp. b.) of this definition with a coset in $H_2(D,\mathbb{R})/H^1(X\setminus D,\mathbb{R})$ which image in $H^2(X,\mathbb{R})$ (i.e. the image via the map $i_{\mathbb{R}}$ in (10.22)) is an integral cohomology class (i.e. the first Chern class of a line bundle).

The following Proposition shows that in the case when irreducible components of D are big and nef, only characters of the fundamental group, which after lift to $H^1(X\setminus D,\mathbb{R})$ give classes belonging to the faces of quasi-adjunction, can have non-trivial eigenspaces for the action of Galois groups on the abelian covers of X ramified along D.

[27] Strict polytopes of quasiadjunction were described just before Definition 10.4.3.

[28] The number of global polytopes of quasi-adjunction is at most $\prod_{k\in Sing(D)} n(P_k)$ where $n(P_k)$ is the number of local polytopes of quasi-adjunction at singular point P_k.

[29] cf. construction described in Proposition 10.3.9.

Proposition 10.4.16 *Assume that irreducible components of D are big and nef. Let $u \in U_{X,D}$ be such that u is in interior of all global polytopes of quasi-adjunction $\mathcal{G}_Q$ of divisor D or their images $\bar{\mathcal{P}}_Q$ for the involution $u \to \bar{u}$ sending $(x_1, .., x_k) \in U_{D,X}$ to $(1 - x_1, ..., 1 - x_k)$. If $\tilde{X}_G$ is a resolution of singularities of a cover X_G of X with abelian Galois group G and $\chi = exp(2\pi i u)$, then the eigenspace $H^1(\tilde{X}_G, \mathbb{C})_\chi = \{v \in H^1(\tilde{X}_G, \mathbb{C}) | g \cdot v = \chi(g)v, \forall g \in G\}$ is trivial.*

Proof Since the characters of $H_1(X \setminus D, \mathbb{Z})$ having a finite order are the characters $\chi = exp(2\pi i u)$ with $u \in \mathbb{Q}$, the density of those in $U_{X,D}$ implies that we can assume that χ is a character of a finite abelian group G. Let $\mathcal{L}_\chi^{-1}$ be the corresponding line bundle (cf. Proposition 10.3.9). Since the action of G is holomorphic and hence preserves the Hodge decomposition of $H^1(\tilde{X}_G, \mathbb{C})$, after possibly replacing the character χ by the conjugate $\bar{\chi}$, we can assume that χ has non-trivial eigenspace for G acting on the holomorphic forms of the cover. Then one has (cf. Proposition 10.4.11)

$$\dim H^0(\Omega^1_{\tilde{X}_G})_\chi = dim H^1(X, \Omega^2_X \otimes \mathcal{L}_\chi \otimes \mathcal{A}_{\mathcal{G}(Q)}) \tag{10.67}$$

where $\mathcal{P}(Q)$ is the global polytope of quasi-adjunction containing u, $\chi = exp(2\pi i u)$.

Since u is an interior point of $\mathcal{G}(Q)$ one can take a small perturbation of it along the intersection with $U_{X,D}$ with the coset of $H^1(X \setminus D, \mathbb{R})$ corresponding to $\mathcal{L}_\chi$ (i.e. an affine subspace) so that it remains inside $\mathcal{G}(Q)$. The corresponding line bundle L_χ is unchanged in this deformation of u. Using multiplier ideal interpretation of ideals of quasi-adjunction and Kawamata-Viehweg-Nadel vanishing (cf. [127] Sect. 9.4B) we obtain that the terms in (10.67) are zeros.

Definition 10.4.17 A global divisorial face of quasi-adjunction $\mathcal{F} \subset \mathcal{G}(\mathcal{F})$ is called *contributing* if for $u \in \mathcal{F}$ and the resolution of singularities of the cyclic cover $\pi_* X_\chi \to X$ corresponding to the surjection $\chi : H_1(X \setminus D, \mathbb{Z}) \to Im(\chi) \subset \mathbb{C}^*$ one has $H^1(X, \Omega^2_X \otimes \mathcal{L}_\chi \otimes \mathcal{A}_{\mathcal{G}(F))}) \neq 0$ (here $\mathcal{L}_\chi$ is the dual χ-eigenbundle of $\pi_*(O_{\tilde{X}_\chi})$. A homological face of quasi-adjunction is called contributing if its translation to a coset $H_2(D, \mathbb{R})/H^1(X \setminus D, \mathbb{R})$ is a contributing divisorial face.

Theorem 10.4.18 *Let X be a simply connected smooth projective surface and let $D = \sum D_i$ be a reduced divisor with irreducible components D_i which are big and nef. Assuming as above that $H_1(X \setminus D, \mathbb{Z})$ is torsion free, let $r = rkCoker$ $H_2(X, \mathbb{Z}) \to H^2(D, \mathbb{Z})$ denote its rank (cf. 10.6). For any essential component of characteristic variety $\mathcal{V}_i$ having positive dimension i.e. a coset of the r-dimensional torus $H^1(X \setminus D, U(1))$ such that $dim \mathcal{V}_i \geq 1$ there is:*

(a) a collection of singularities $\mathcal{S}$ of D

(b) a contributing face $\mathcal{F}$ of a global polytope of quasi-adjunction $\mathcal{G}(\mathcal{S})$ which is determined by the collection $\mathcal{S}$ and a collection of the polytopes of quasi-adjunction $\mathcal{Q}(P)$, one at each of singularities $P \in \mathcal{S}$

(c) a line bundle $\mathcal{L}_{\mathcal{G}(\mathcal{S})}$

such that $\mathcal{V}_i$ is the Zariski closure of $exp(\pm 2\pi i \mathcal{F})$ in the maximal compact subgroup of the r-dimensional torus $Char H_1(X \setminus D, \mathbb{Z})$ and

$$dim\mathcal{V}_i = dim\mathcal{F} = dim H^1(X, \Omega_X^2 \otimes \mathcal{L}_{\pm Q(S)}^{-1} \otimes \mathcal{A}_{\mathcal{G}(S)}) + 1$$

Moreover, $\mathcal{L}_{\pm\mathcal{G}(S)}$ is the line bundle which is part of the building data of the cyclic cover corresponding to surjection $\chi : \pi_1(X \setminus D) \to \mathbb{Z}_{ord(\chi)}$ for a character χ which is generic in the component $\mathcal{V}_i$.

Vice versa, given a maximal[30] contributing face $\mathcal{F} \subset \mathcal{G}(S)$ of a global polytope of quasiadjunction, with the ideal of quasiadjunction $A_{\mathcal{G}(S)}$, such that the line bundle corresponding to the characters $exp(2\pi i u)$, $u \in \mathcal{F}$ is a $\mathcal{L}$ (satisfying $H^1(X, \Omega_X^2 \otimes \mathcal{L} \otimes \mathcal{A}_{\mathcal{G}(S)}) \neq 0$) then the Zariski closure of the set of characters $exp(2\pi i u)$, $u \in \mathcal{F}$ is a component of characteristic variety of $\pi_1(X \setminus D)$.

Proof Let Q be a maximal contributing face of quasi-adjunction. The Zariski closure in $H^1(X \setminus D, \mathbb{C}^*)$ of the set $exp(2\pi i u)$, $u \in Q$, belongs to a component of characteristic variety, as follows from the assumptions. If this Zariski closure i s a proper subset of a component, then preimage of the unitary part (i.e. the intersection with $H^1(X, U(1)) \subset H^1(X, \mathbb{C}^*)$) of the full component must belong to the same $H^1(X \setminus D, \mathbb{R})$ coset in $H_2(D, \mathbb{R})$ as Q and, as follows from Proposition 10.4.16, its preimage in $H_2(D, \mathbb{R})$ must be a face of the same polytope as $\mathcal{F}$ i.e. coincide with $\mathcal{F}$ due to maximality assumption.

Now, let $\mathcal{V}_i$ be an irreducible component of characteristic variety. A theorem of D.Arapura implies that $\mathcal{V}_i$ is a translated subtorus of the torus $H^1(X \setminus D, \mathbb{C}^*)$. The subset $\exp^{-1}(\mathcal{V}_i \cap H^1(X \setminus D, U(1)) \subset H^1(X \setminus D, \mathbb{R})$ consist of a set of $H^1(X \setminus D, \mathbb{Z})$-translates of a linear subspace of $H^1(X \setminus D, \mathbb{R})$. The eigenbundles of the characters in $\mathcal{V}_i$, for the push forward of the structure sheaf of a cyclic cover of X corresponding to characters from $\mathcal{V}_i$, define a collection of translates of $H^1(X \setminus D, \mathbb{R}) \subset H_2(D, \mathbb{R})$ (cf. Proposition 10.3.9) which intersect the fundamental domain (i.e. the unit cube) $U_{X,D}$ for the action of $H_2(D, \mathbb{Z})$ on $H_2(D, \mathbb{R})$. Due to identification in Proposition 10.4.11 of cohomology of the local systems and the cohomology of sheaves of quasi-adjunction, one obtains that at least one of translates belongs to a contributing face of quasi-adjunction. It is maximal since otherwise $\mathcal{V}_i$ will be a proper subset of a component of larger dimension.

Corollary 10.4.19 *Let X, D be as in Theorem 10.4.18 and let C be a smooth big and nef curve intersecting all irreducible components of D at smooth points transversally. Then $H_2(D, \mathbb{R}) \subset H_2(D + C, \mathbb{R})$ has codimension one and divisorial contributing faces of quasi-adjunction of D coincide with those of $D + C$.*

Proof Since polytope quasiadjunction of ordinary node coincides with the unit square (node does not impose conditions of quasi-adjunction) it follows that the global polytopes in $H_2(D + C, \mathbb{R})$ are the cylinders over the global polytopes of $H_2(D, C)$ (preimages of projection of $H_2(D + C, \mathbb{R})$ onto the later). Kawamata-Viehweg-Nadel vanishing implies that the characters in a contributing faces of the eigenbundles $\mathcal{L}_\chi$ must have trivial ramification along C i.e. belong to $H_2(D, \mathbb{R})$ (triviality of ramification also follows from Divisibility Theorem 10.3.3).

---

[30] I.e. not contained properly in a contributing face of the same strict global polytope of quasi-adjunction.

Remark 10.4.20 The removal a line at infinity, transversal to a curve, was used extensively in [129, 137]. The main theorem in [137] follows immediately from Theorem 10.4.18 and Corollary 10.4.19.

Numerous examples to the Theorem 10.4.18 can be found in the paper [137] in the case of line arrangements is a plane and in [130] in the case of irreducible curves. The local counterpart of the Theorem 10.4.18 and many examples of calculations of multivariable Alexander polynomials of the links (i.e. the characteristic varieties, cf. discussion after Definition 10.3.13) of singularities in terms of polytopes and ideals of quasi-adjunction are given in [39]. For results on zero dimensional components of characteristic varieties we refer to [19, 20]. We will finish this section with an example of calculation on a large class of surfaces generalizing 6-cuspidal sextic of Zariski.

Example 10.4.21 Let X be a smooth projective simply connected surface and let L be a very ample line bundle on X. Let $s_2 \in H^0(X, L^2), s_3 \in H^0(X, L^3)$ be generic sections of the corresponding tensor powers of L. Let D be the zero set of $s = s_2^3 + s_3^2 \in H^0(X, L^6)$. Then the Alexander polynomial of this curve with $6L^2$ cusps, corresponding to the surjection $H_1(X \setminus D, \mathbb{Z}) \to \mathbb{Z}_6$, is $t^2 - t + 1$.

To see this, first note that the existence of the surjection follows from (10.6) since the class of D in $H_2(X, \mathbb{Z})$ is divisible by 6. Using (10.62), the eigenspace of the generator of $\mathbb{Z}_6$ acting on $H^{1,0}$ of the 6-fold cyclic can be identified with $H^1(X, K_x \otimes L^5 \otimes \mathcal{I}_{Sing})$ where $\mathcal{I}_{sing}$ is the ideal sheaf such that $\mathcal{O}_X/\mathcal{I}_{sing}$ is the reduced 0-dimensional subscheme of X with support at the set of cusps of D. One has the following Koszul resolution of $\mathcal{I}_{sing}$:

$$0 \to L^{-5} \to L^{-2} \oplus L^{-3} \to \mathcal{I}_{sing} \to 0$$

After taking the tensor product of this sequence with $K \otimes L^5$ and considering the corresponding cohomology sequence:

$$H^1(X, K_X \otimes L^2) \oplus H^1(X, K_X \otimes L^3) \to H^1(X, K_x \otimes L^5 \otimes \mathcal{I}_{Sing}) \to H^2(X, K_X) \to 0 \tag{10.68}$$

we see that Kodaira vanishing implies that $dim H^1(X, K_x \otimes L^5 \otimes \mathcal{I}_{Sing}) = 1$. This shows that $\frac{1}{6} \in [0, 1]$ is the contributing face of quasi-adjunction and now the claim about the Alexander polynomial follows from the Theorem 10.4.18. Note that this example also can be analyzed using methods of orbifold pencils discussed in [19–21].

10.4.6 Bernstein-Sato Ideals and Polytopes of Quasi-adjunction

Let $f_1, .., f_r$ be germs of holomorphic functions in n variables. The Bernstein-Sato ideal $\mathcal{B}(f_1, ..., f_r)$ is the ideal generated by polynomials $b(s_1, ..., s_r)$ such that there exist a differential operator $P \in \mathbb{C}[x_1, ..., x_n, \frac{\partial}{\partial x_1}, ..., \frac{\partial}{\partial x_1}, s_1, ..., s_r]$ satisfying the identity:

$$b(s_1, ..., s_r) f^{s_1} .... f^{s_r} = P f_1^{s_1+1} .... f_r^{s_r+1} \tag{10.69}$$

(cf. [25, 109, 160, 185], in 1-dimensional case cf. [121, 155]). In the case of plane curves singularities one has the following:

Theorem 10.4.22 *Let $f_1, .., f_r$ be the germs of holomorphic functions in two variables. Let P be the product of the linear forms $L_i(s_1 + 1, ..., s_r + 1)$ where L_i runs through linear forms vanishing on $r - 1$-dimensional faces of polytopes of quasi-adjunction corresponding to a germ with r irreducible components $f_1, ..., f_r$. Then any $b \in \mathcal{B}(f_1, ..., f_r)$ is divisible by P.*

The same argument as used in [39], provides extension to isolated non-normal crossings (cf. [140]). For a general conjecture of the structure of Bernstein ideals we refer to [38] and for a discussion of the case of arrangements, other related problems and references cf. [217].

10.5 Asymptotic of Invariants of Fundamental Groups

The problem of characterization of fundamental groups of smooth quasi-projective varieties is intractable at the moment. Nevertheless some questions about distribution of Alexander invariants can be addressed. We will see below that one can make some conclusions about distribution of dimensions of characteristic varieties of such fundamental groups. A different type of asymptotics, is suggested by the relation between the degrees of Alexander polynomials and Mordell-Weil ranks of isotrivial families of abelian varieties (cf. [46, 146]) since it allows to restate the problem of asymptotic behavior of such degrees in terms of the conjectures on distribution of Mordell Weil ranks of curves over the function fields. In this section we shall survey the results in [47] concerning distribution of the dimensions of characteristic varieties.[31]

Let X be a smooth simply connected projective variety, D a reduced divisor and let Δ be a subset of the effective cone $Eff(X) \subset NS(X)$ in the Neron Severi group of X. We shall call the set Δ *saturated* if $d_1 \in \Delta$ and $d_2 \in Eff(X)$ are such that $d_1 - d_2 \in Eff(X)$ implies that $d_2 \in \Delta$ and $d_1 - d_2 \in \Delta$. We are interested in distribution of invariants of $\pi_1(X \setminus D)$ when the class of $D \subset Eff(X)$ is a linear combination of classes in Δ with non-negative coefficients. We are specifically interested in curves D with large dimension of a component of characteristic variety of $\pi_1(X \setminus D)$ and D being a curve with all its irreducible components having classes in Δ. It follows

[31] Such circle of problems is inspired by conjectural asymptotic of number fields extensions having a given group as the Galois group or the group of its Galois closure, which are unramified outside an arbitrary subset of primes while the size of the norm of discriminant grows [156]: Malle conjectures implies a positive answer to the inverse problem of the Galois with little hope for solution in near future (as is obtaining a characterization of quasi-projective group).

from [9], that existence of a component of dimension r implies existence of surjection $\pi_1(X \setminus D) \to F_r$. Vice versa, existence of the latter implies that the characteristic variety of $\pi_1(X \setminus D)$ contains a component of dimension not smaller than r. Note right away that for the purpose of enumeration of reduced divisors D for which one has a surjection $\pi_1(X \setminus D) \to F_r$ we must impose some conditions on such surjections. For example, given any D with such property and any reduced divisor D' one has

$$\pi_1(X \setminus D \bigcup D') \to \pi_1(X \setminus D) \to F_r \tag{10.70}$$

and hence, given a curve admitting a surjection of its fundamental group onto F_r, there are enlargements of this curve with the same property parametrized by all the curves on the surface. This motivates the following.

Definition 10.5.1 (*cf.* [137]) Let $\mathcal{D}$ be a reduced divisor on a smooth projective surface X. A surjection $\pi_1(X \setminus \mathcal{D}) \to F_r$ is called *essential* if $\mathcal{D}$ does not admit split $\mathcal{D} = D \bigcup D'$ for which one has factorization (10.70).

A surjection $\pi_1(X \setminus \mathcal{D}) \to F_r$ is called *reduced* if there exist a choice of good ordered system of generators $\{x_1, ..., x_{r+1} | x_1 \cdot .... \cdot x_{r+1} = 1\}$ of F_r such that this surjection takes meridian of each irreducible component of $\mathcal{D}$ to a conjugate of a generator.

We also will say that singularities of $\mathcal{D}$ satisfy condition (*) if all singular points belonging to more than one irreducible component are ordinary singularities i.e. are intersections of smooth transversal branches.[32]

A rather detailed information about such curves was obtained in [136] in the case $X = \mathbb{P}^2$, $\Delta = \{[1]\} \in \mathbb{Z} = Pic(\mathbb{P}^2)$ i.e. the fundamental groups of the complements to arrangements of lines in a plane (see [86, 157] for related results).

Theorem 10.5.2 ([136, 179]) *Let $\mathcal{A}$ be an arrangement of lines in $\mathbb{P}^2$. If there exist an essential surjection $\pi_1(\mathbb{P}^2 \setminus \mathcal{A}) \to F_r$, $r \geq 4$ then $\mathcal{A}$ is a union of concurrent lines, in which case the last surjection is an isomorphism.*

Moreover, there is only one known example of essential surjections of the complements to an arrangement line which admits surjection onto F_3[33] and for any d there exist an arrangement of non-concurrent lines admitting essential surjection onto F_2 (e.g. $3d$ lines forming the zero set of $(x^d - y^d)(y^d - z^d)(x^d - z^d) = 0$).

Work [47] contains an extension of this theorem to reduced divisors on arbitrary simply connected surfaces. Before stating the main result, let us describe the analog of the case of concurrent lines in Theorem 10.5.2, which is a family of the curves with irreducible components in Δ and for which the fundamental group of the complements may have a free quotient of arbitrary large rank. For this family of curves,

[32] The results in this section make this assumption. It should be possible to eliminate it with essential conclusions remaining intact.

[33] I.e. the Hesse arrangement of 12 lines formed by lines containing triples of inflection points of plane smooth cubic cf. [137].

the fundamental groups of the complement form *a finite* set of groups, having cardinality depending on Δ and, moreover, a presentation of each group in this set can be described in terms of geometric data we specify. However, unlike the case of Theorem 10.5.2, the problem of characterizing which specific data is realizable by curves in this class remains open in general. Enumeration of fundamental groups of such curves for a class $\delta \in \Delta \subset Pic(X)$ can be made as follows.

Proposition 10.5.3 *For any $r \geq 1$ and a movable divisor $\delta \in Pic(X)$,*[34] *there is a divisor D with classes of components in the linear system of δ and such that $\pi_1(X \setminus D)$ admits essential surjection onto F_r. Vice versa, if D has all its irreducible components being members of a pencil of curves in complete linear system of δ (i.e. a line in $\mathbb{P}(H^0(X, \mathcal{O}_X(\delta)))$), and $\pi_1(X \setminus D)$ admits surjection onto F_r, $r \geq 2$ this group is an amalgamated product $\mathcal{G} *_{F_a} \mathcal{H}$ with $\mathcal{G}$ belonging to a finite collection of groups depending on δ, obtained by a construction below and $\mathcal{H}$ is an extension:*

$$0 \to F_a \to \mathcal{H} \to F_{r'} \to 0 \quad r' \leq r \tag{10.71}$$

defined by a homomorphism $F_{r'} \to Aut(F_a)$ coming from a finite set cardinality depending only on δ. The number of isomorphism classes of such groups $\pi_1(X \setminus D)$ with a fixed class δ, stabilises for large r.

Proof Indeed, for any pencil in the linear system containing δ, a union on its $r+1$ members yields a divisor $D \in \mathbb{P}(H^0(X, (r+1)\delta))$ with $\pi_1(X \setminus D)$ admitting a surjection onto F_r since such a pencil induces a dominant map onto the complement in $\mathbb{P}^1$ to $r+1$ points.

To enumerate all possible fundamental groups of the complements to the curves with all irreducible components belonging to a pencil let us consider the discriminant $Disc(\mathbb{P}(H^0(X, \mathcal{O}_X(\delta))))$ of the complete linear system $\mathbb{P}(H^0(X, \mathcal{O}_X(\delta))$ i.e. the subvariety consisting of the divisors having singularities worse than singularities of a generic element in $\mathbb{P}(H^0(X, \mathcal{O}_X(\delta))$. Consider also the stratification of the discriminant into connected components of equisingularity strata, adding to this stratification the complement to the discriminant as a codimension zero stratum (cf. [5] on some information about geometry of these strata).

We will use finite sets of collections of such equisingularity strata $\mathcal{S}_1, \ldots \mathcal{S}_t$ for which there exists a pencil $\mathcal{P}$ in $\mathbb{P}(H^0(X, \mathcal{O}_X(\delta)))$ with the following property: there exists a union D of members of $\mathcal{P}$ such that the curve D satisfies condition (*) (cf. Definition 10.5.1). Let $N(\delta, t)$ be the number of isotopy classes of pencils in $\mathbb{P}(H^0(X, \mathcal{O}_X(\delta))$ such that the number of the strata of this stratification intersected by the pencil is t and let T be the least upper bound for integers t for all pencils in δ. Finiteness of these numbers is a consequence of the finiteness of the number of strata of stratifications since those are an algebraic subsets of discriminant (cf. [100]).

Let D is a curve having $r+1$ irreducible components belonging to a pencil $\mathcal{P}$ in $\mathbb{P}(H^0(X, \mathcal{O}_X(\delta))$ in which the members of $\mathcal{P}$ have t (where $t \leq r+1$) equisingularity types. We claim that $\pi_1(X \setminus D)$, can have at most 2^t isomorphism types. More

[34] I.e. such that the codimension of the base locus of the linear system it defines is at least 2.

precisely for each subset $\mathcal{T}$ of the set of strata $\mathcal{S}_1, ..., \mathcal{S}_t$ there is at most one isomorphism type of the fundamental groups $\pi_1(X \setminus D)$ where the set of equisingularity starta of components of $D \in \mathbb{P}(H^0(X, \mathcal{O}_X((r+1)\delta)))$ having non-generic equisingularity type in the pencil coincides with $\mathcal{T}$. This is the case when D is a union of $|\mathcal{T}|$ curves from the strata $\mathcal{S}_1, ..., \mathcal{S}_t$ and $r' = r + 1 - |\mathcal{T}|$ curves from codimension zero stratum and none of remaining $t - |\mathcal{T}|$ singular members of the pencil are not components of D. In particular, for $r > t$ there are at most $\sum_{t=0}^{T} 2^t N(\delta(t))$ isomorphism classes of the fundamental groups and for $r > T$ the number of isomorphism classes of fundamental groups of curves with components in the linear system of δ and admitting surjections onto F_r is bounded, with bound depending only on $\delta \in \Delta$.

To describe the structure of the fundamental groups of the complement to a union D of several members of a pencil $\mathcal{P}$ of curves in δ, with the set equisingularity types of singular members of $\mathcal{P}$ consisting of equisingularity strata $\mathcal{S}_1, .., \mathcal{S}_t$, such that non-generic types of components of D are exactly those in $\mathcal{T}$, and also to enumerate such fundamental groups, consider the blow up $\tilde{X}$ of X at the base points of the pencil. We obtain a regular map $\pi : \tilde{X} \setminus \tilde{D} \to \mathbb{P}^1 \setminus S_{r+1}$ where S_{r+1} a finite subset of $\mathbb{P}^1$ with cardinality $r + 1$.

Let $\mathbb{P}^1 = B_1 \cup B_2$ be partition into union of two disks intersecting along their common boundary and having the following properties: B_1 contains all $t - |\mathcal{T}|$ fibers of π which do not have generic equisingularity type in $\mathbb{P}(H^0(X, \mathcal{O}_X(\delta))$ and are not components of D, while $B_2 = \mathbb{P}^1 \setminus B_1$ contains $|\mathcal{T}|$ non-generic fibers if π which are components of D and remaining $r' = r + 1 - (t - |T|)$ fibers of π which all are generic in the latter linear system. Over the complement in B_2 to the subset over which the fibers of π are the components of D, the map π is a locally trivial fibration which global type is determined by δ. Van Kampen Theorem 10.2.2 implies the following: if Σ is generic fiber of π, $\mathcal{G} = \pi_1(\pi^{-1}(B_1))$, $\mathcal{H} = \pi_1(\pi^{-1}(B_2))$ then

$$\pi_1(X \setminus D) = \mathcal{G} *_{\pi_1(\Sigma)} \mathcal{H}, \quad \textit{and} \quad 1 \to \pi_1(\Sigma) \to \mathcal{H} \to F_{r'} \to 1 \qquad (10.72)$$

Σ is complement in the generic fiber of the pencil to the set of base points of the pencil i.e. $\pi_1(\Sigma)$ is a free group F_a for some a. The group $\mathcal{G}$ belongs to a collection having at most 2^t elements (i.e. the number of subsets in $\mathcal{S}_1, ..., \mathcal{S}_t$). The claim follows.

Example 10.5.4 Let us enumerate the fundamental groups of the complements to conic-line arrangements which admit a surjection onto a free group of rank greater than 5. The starting point is that a conic-line arrangement (satisfying condition (*)) having such fundamental group is a union of $r + 1$ (possibly reducible) quadrics belonging to a pencil. This is content of improvement for conic-line arrangements of the general bound in Theorem 10.5.5 below (cf. Example 10.5.6 (2).) Equisingular stratification of $\mathbb{P}(H^0(\mathbb{P}^2, \mathcal{O}_{\mathbb{P}^2}(2))$ consists of 3 strata: smooth quadrics, reduced and reducible quadrics i.e. a union of two transversal lines and non-reduced quadrics i.e. the double lines. The degree of discriminant is 3. We denote these equisingular strata respectively as $\mathcal{S}_0, \mathcal{S}_1, \mathcal{S}_2$.

Any pencil of quadrics containing as generic element a smooth quadric in $\mathcal{S}_0$, has at most 3 singular fibers which are either 3 reducible quadrics or contains 2 singular fibers one reduced and one non reduced. In the latter case, the condition (*) on D fails. Moreover, there are pencils with generic element inside the stratum $\mathcal{S}_1$. For such a pencil, the divisor $D \in \mathbb{P}(H^0(\mathbb{P}^2, \mathcal{O}(2(r+1)))$ is a union of $2r+2$ concurrent lines and hence $\pi_1(\mathbb{P}^2 \setminus D) = F_{2r+1}$.

There are 4 equisingular classes of divisors $D \in \mathbb{P}(H^0(\mathbb{P}^2, \mathcal{O}((r+1)2)))$ with components formed by curves in a pencil δ, corresponding to the cases when the number of quadrics which are the singular elements of the pencil and formed by components of D, is either 0 (i.e. all components of D are smooth quadrics), or is $1, 2$ or 3. Respectively, there are 4 corresponding types of fundamental groups.

For example, let us take as D a union of $r+1$ quadrics belonging to a pencil, one of which is reducible. Let B_1 be a disk containing remaining 2 reducible fibers of the pencil and let $B_2 = \mathbb{P}^1 \setminus B_1$. Then B_2 is a disk containing the points corresponding to the fibers containing the components of the pencils comprising D. Over B_1, the map π is a fibration with generic fiber being a smooth quadrics and which has two special fibers which are the union of lines and therefore can be calculated using van Kampen Theorem 10.2.2. Over the complement in B_2 to the points corresponding to the components of D one has a locally trivial fibration with the fiber being the complement in a smooth quadric to 4 base points of the pencil. Hence $\mathcal{H} = \pi_1(B_2)$ is an extension of free group F_3 by the free group F_r with only one type of extension possible since there is only one isotopy class of generic pencils of quadrics.

Now we turn to the main result of [47] which can be stated as follows:

Theorem 10.5.5 *Given a saturated set Δ of classes in $NS(V)$ consider the following trichotomy for the distribution of the curves $\mathcal{D}$ with classes of irreducible components in Δ having a free essential reduced quotient of a fixed rank r and satisfying conditions (*)*

(1) There exist infinitely many isotopy classes of curves $\mathcal{D}$ admitting surjections $\pi_1(V \setminus \mathcal{D}) \to F_r, , r > 1$.
(2) There are finitely many isotopy classes of curves $\mathcal{D}$ admitting surjections $\pi_1(V \setminus \mathcal{D}) \to F_r, r > 1$.
(3) $\mathcal{D}$ admitting a surjection $\pi_1(V \setminus \mathcal{D}) \to F_r$, is composed of curves of a pencil. There are finitely many isotopy classes of such $\mathcal{D}$ for given Δ.

All three cases are realizable at least for some (V, Δ). Case (2) takes place for $r \geq 10$. There exists a constant $M(V, \Delta)$ such that for $r > M(V, \Delta)$ one has case (3). In the latter case, $\pi_1(V \setminus \mathcal{D})$ splits as an amalgamated product $H *_{\pi_1(\Sigma)} G$ where Σ is an open Riemann surface which is a smooth member of the pencil, H is coming from a finite set of groups associated with the linear system $H^0(V, \mathcal{O}(D))$, D is a divisor having class $\delta \in \Delta$ and G is an extension:

$$0 \to \pi_1(\Sigma) \to G \to F_r \to 0 \tag{10.73}$$

In specific cases of (X, Δ) information about the constants 10 and $M(X\Delta)$ can be improved.

Example 10.5.6 1. Above results for the arrangements of lines shows that in this case one can replace 10 by 2 and $M(\mathbb{P}^2, [1]) = 3$.

2. Again in the case $X = \mathbb{P}^2$ but $\Delta = \{[1], [2]\}$, the curves $\mathcal{D}$ for which there exist a surjection $\pi_1(\mathbb{P}^2 \setminus \mathcal{D}) \to F_r$ must have the type only as described in Example 10.5.4, provided $r > 5$. However, a generic pencil in the linear system:

$$\lambda_0 x_0(x_1^2 - x_2^2) + \lambda_1 x_1(x_2^2 - x_0^2) + \lambda_2 x_2(x_0^2 - x_1^2) = 0 \tag{10.74}$$

has 6 members which are unions of lines and quadrics. This gives a curve $\mathcal{D}$ of degree 18 for which $\pi_1(\mathbb{P}^2 \setminus \mathcal{D})$ admits a surjection onto F_5 and is not isotopic to a curve as in Example 10.5.4.

The Theorem 10.5.5 can be restated as follows: if $N(X, \Delta, r)$ denotes the number of equisingular isotopy classes of curves on X with irreducible components having numerical classes in Δ and fundamental groups admitting a surjection onto a free group F_r then for $r > M(X, \Delta)$, $N(X, \Delta, r)$ is finite and all curves have special type as in the case (C) of the trichotomy. For $10 < r \leq M(X, \Delta)$, $N(X, \Delta, r)$ is also finite but the type of the curves may vary. Finally, for $r < 10$ the number of isotopy classes $N(X, \Delta, r)$ may be infinite.

Some information on dependence of the constant $M(X, \Delta)$ on Δ and X is also available. For example if $X = \mathbb{P}^2$, $\Delta_d = \{[1], ..., [d]\}$ then $M(\mathbb{P}^2, \Delta_d) \geq 3d$. Indeed, Ruppert (cf. [184]) found a pencil of curves of degree $d + 1$ with $3d$ fibers being a union of a line an a curve of degree d. In particular a union of these $3d$ fibers yields a curve of degree $3d(d + 1)$ with irreducible components in Δ_d and having surjection on the free group of rank $3d - 1$. In particular the sequence $M(X, \Delta_d)$ is unbounded. The Ruppert pencil is a generic pencil in 2-dimension linear system of curves given by equation (which for $d = 2$ it is given in Example 10.5.6):

$$\lambda_0 x_0(x_1^d - x_2^d) + \lambda_1 x_1(x_2^d - x_0^d) + \lambda_2 x_2(x_0^d - x_1^d) = 0 \tag{10.75}$$

More precisely, the curve (10.75) is singular if and only if

$$(\lambda_0^d - \lambda_1^d)(\lambda_1^d - \lambda_2^d)(\lambda_2^d - \lambda_0^d) = 0$$

and all reducible fibers are unions of a line and a curve of degree d. Hence generic line in variables λ_i is a pencil with $3d$ reducible members as described.

We refer to [47] for examples of surjections onto free groups of the fundamental groups of the complements to curves on surfaces besides $\mathbb{P}^2$.

This discussion suggests the following problems:

Problem 10.5.7 1. Determine the rate of growth of $N(X, \Delta, r)$ for various X and Δ when $r \to \infty$, i.e. how many types of reducible curves admitting surjections onto F_r, which r large (i.e. $r > M(X, \Delta)$) exist?

2. Find a bound on $M(X, \Delta)$ in terms of invariants of X, Δ i.e. how large should be r such that there exist curves admitting surjection onto F_r and which are not the unions of the fibers of a pencil.
3. For $n \in \mathbb{N}$ let $\Delta_n = \{\sum n_i \delta_i | \delta_i \in \Delta, n_i \leq n\} \subset NS(X)$. Determine the asymptotic of the number of curves admitting surjection onto $F_r, r < 10$ with the classes in Δ_n when $n \to \infty$.
4. Determine algebraic properties of the fundamental groups described in Proposition 10.5.3.

Some partial results, mainly in the case of plane, are discussed above and in [47]: for example the curves $(x^n - y^n)(y^n - z^n)(x^n - z^n) = 0$ formed by $3n$ lines show that the growth in Problem 3 for $\mathbb{P}^2$, [1] for $r = 2$ is at least linear. The growth of $N(X, \Delta)$ appears to be related to the asymptotic of the number of strata (cf. the proof of Proposition 10.5.3) and possibly is exponential.

10.6 Special Curves

This section surveys examples of calculations of the fundamental groups and other topological information about the complements, the properties of fundamental groups and applications. An important step in each such inquiry is finding a class of curves with interesting topology of the complements. Most examples in this section are plane curves.

10.6.1 Arrangements of Lines, Hyperplanes and Plane Curves

There are many calculations of the fundamental groups of the complements to arrangements of lines. The braid monodromy can be calculated algorithmically. In the case of real arrangements finding the braid monodromy and the presentation are particularly simple: see [116, 191]. In some instances this leads to presentations allowing a more intrinsic characterization: for example in [87] conditions on arrangement were found for the fundamental groups to be products o f fee groups.

The fundamental groups and more subtle questions on the topology of the complements to arrangements formed by hyperplanes fixed by the groups generated by reflections were very actively studied in many case. In case of real reflection groups, the fundamental groups of the complements to corresponding complexified real arrangements were found [29] with presentations closely related to the Dynkin diagrams of the corresponding Coxeter groups. The topology of the complements to hyperplanes corresponding to the complex reflection groups also were actively studied with many deep results. The number of striking results is too large to survey here and we refer for example to [28, 34] for some particularly important ones and for further references.

Several calculations were made for the fundamental groups of the complements to unions of lines and quadrics. Work [8] includes the arrangements formed by unions a quadric and lines with various tangency conditions. Few example of such type of arrangements, more specifically those real arrangements of quadrics and lines which admit projections to a line with all critical points being real, were considered in [172]. Here the standard methods of calculation of the braid monodromy are almost as simple as in the case of real arrangements of lines and lead quickly to presentations in terms generators and relators.

Cardinality of the set of connected components of the equisingular families of reducible curves with fixed combinatorial type (cf. Definition 10.6.2) was investigated in several cases of plane curves of small degree. In particular the classification for curves of degree 5 was carried out in [64]. The case of arrangements of small cardinality and irreducibility of equisingular component was studied for arrangements up to 9 lines as well as arrangements of 10 and 11 lines with many different types of combinatorics with some results in the case of arrangements of 12 lines (cf. [6, 14, 89, 107, 173] the latter are in connection with Rybnikov's example of combinatorially equivalent arrangements with distinct homotopy types). Specific types of presentations of the fundamental groups of arrangements were studied in [80].

10.6.2 Generic Projections

Study of the fundamental groups of the complements to the branching curves of generic projections[35] was initiated by B. Moishezon in work [164] and continued jointly with M. Teicher and later by M. Teicher and her collaborators. Given a smooth surface $X \subset \mathbb{P}^N$, a projection from a generic $\mathbb{P}^{N-3}$ gives a generic branched cover ramified along a curve $R \subset X$. The image of R is the branching curve $B \subset \mathbb{P}^2$ of this projection. If the center of projection $\mathbb{P}^{n-3}$ is sufficiently generic, then B has nodes and cusps as the only singularities. The number of cusps and nodes can be found in terms of intersection indices of Chern classes of X and the class of hyperplane section (cf. [131]). Work [164] considers the case when X is a smooth surface in $\mathbb{P}^3$. Then the branching curve B has degree $n(n-1)$, $n(n-1)(n-2)$ cusps and $\frac{1}{2}n(n-1)(n-2)(n-3)$ nodes (for $n = 3$ one obtains sextic with six cusps). The fundamental group of the complement is isomorphic to the quotient of the braid group on n strings by its center (cf. [164]). The relation between the fundamental groups of the complements to the branching curves of generic projections and the fundamental groups of smooth models of Galois closures of these projections is discussed in [147, 166].

---

[35] Important results on geometry of such curves were obtained much earlier by italian school, notably B.Segre, Chisini and his school cf. [193].

Works [166] consider generic projections of quadrics $X = \mathbb{P}^1 \times \mathbb{P}^1$ using a family of embeddings $i_{a,b}, a, b \in \mathbb{Z}$ corresponding to various ample divisors in $NS(X)$. Interest in this class stems form the fact that Galois covers of $\mathbb{P}^2$ with branching curve of generic projections of these surfaces provide examples of simply connected surface of general type for which $c_1^2 > 2c_2$. The key step in the showing the simply connectedness is the calculation of the fundamental group of the complement to the branching curve.

Since then, the class of surfaces which generic projections produces the curves for which one has a presentation of the fundamental groups of the complements was greatly increased. Calculations produced over the span of more than 30 years include complete intersections in projective spaces [182], very ample embeddings of Hirzebruch surfaces, embeddings of K3 surfaces, very ample embeddings of ruled surfaces which are the products of $\mathbb{P}^1$ and smooth curves of positive genus and others. In many instances a quite different than in the case of surfaces in $\mathbb{P}^3$ pattern emerged for the fundamental groups (cf. [208] for references to these calculations). One has to mention that the main technical tool in such calculation is appropriate degeneration of the surface resulting in degeneration of the branching curve. Steps of calculation include calculation of the braid monodromy of degenerate curve (which may be reducible) and then applying rules of regeneration i.e. relating the braid monodromy of degenerate curve to the braid monodromy of the curve prior to degeneration. We refer to a survey article [7] which has useful references to these numerous calculations.

An interesting property of branching curves of generic projections was discovered by Chisini: (with a small number of exceptions) the cover given by generic projection is determined by the curve alone, i.e. no subgroup of the fundamental group to specify the cover (cf. Sect. 10.3.3) is needed. A proof of this result was found in [125] (cf. also, [42]).

10.6.3 Complements to Discriminants of Universal Unfoldings

With a germ of isolated hypersurface singularity $f(x_1, ..., x_n) = 0$ one associates the germ of the universal unfolding $\mathbb{C}^N$, $N = dim\mathbb{C}[x_1, ..., x_n]/(f, \frac{\partial f}{\partial x_1}, ..., \frac{\partial f}{\partial x_i})$ which comes with the germ of discriminantal hypersurface $Disc$ (corresponding to the germs having a critical point (cf. [106])).

The fundamental groups of the complements to the germs $Disc$ have appearance in a variety of questions spreading from singularity theory and topology to representation theory and beyond. An important feature of the fundamental groups of such complements (as well as complements to other discriminants) is that they come endowed with geometric monodromy i.e. the homomorphism to the mapping class group of the Milnor fiber, i.e. the group of diffeomorphisms of the Milnor fiber constant on its boundary modulo isotopy. This induces the homological monodromy via the action of the mapping class group on the homology of the Milnor fiber. For

ADE singularities one obtains the corresponding Coxeter groups (cf. [79]). Moreover, these complements to germs often can be identified with the complement to the whole affine hypersurfaces in $\mathbb{C}^N$, so these local fundamental groups are quasi-projective. In the case of simple ADE surface singularities, the fundamental groups of the complement were identified by Brieskorn (cf. [29]) with the braid groups corresponding to the respective Coxeter systems.

Calculations for several more complicated classes of singularities were made also. An important case of Brieskorn-Pham polynomials $f(x_1, ..., x_n) = x_1^d + ... + x_n^d$ was considered by Lonne (cf. [153] and references there). Generators and relations of the fundamental group of the complement to discriminant are described in terms of combinatorial data given by the graph associated to singularity, analogous to Dynkin diagram or, equivalently, in terms of the corresponding bilinear form. Vertices correspond to the integer points in the interior of the cube $I_{d,n} = \{\mathbf{i} = (i_1, ..., i_n) | 1 \leq i_k \leq d-1\}$. Edges described in terms of bilinear form on the vector space with basis $v_{\mathbf{i}}, \mathbf{i} \in I_{d,n}$ given by

$$\langle v_{\mathbf{i}}, v_{\mathbf{j}} \rangle = \begin{cases} 0\, if\ |i_\nu - j_\nu| \geq 2\ for\ some\ \nu \\ 0\, if\ (i_\nu - j_\nu)(i_\mu - j_\mu) < 1\ for\ some\ \mu, \nu \\ -2\, if\ i = j \\ -1\ otherwise \end{cases} \tag{10.76}$$

The edges of the graph connect the pairs of vertices $\mathbf{i}, \mathbf{j}$ such that $< v_{\mathbf{i}}, v_{\mathbf{j}} > \neq 0$. In terms of this bilinear form or the graph the fundamental group of the complement to discriminant has generators $t_{\mathbf{i}}$ corresponding to the vertices and the relations as follows

$$\begin{array}{rc} t_{\mathbf{i}}t_{\mathbf{j}} = t_{\mathbf{j}}t_{\mathbf{i}} & if\ < v_{\mathbf{i}}, v_{\mathbf{j}} >= 0, \\ t_{\mathbf{i}}t_{\mathbf{j}}t_{\mathbf{i}} = t_{\mathbf{j}}t_{\mathbf{i}}t_{\mathbf{j}} & < v_{\mathbf{i}}, v_{\mathbf{j}} > \neq 0 \\ t_{\mathbf{i}}t_{\mathbf{j}}t_{\mathbf{k}}t_{\mathbf{i}} = t_{\mathbf{j}}t_{\mathbf{i}}t_{\mathbf{k}}t_{\mathbf{j}} & < v_{\mathbf{i}}, v_{\mathbf{j}} >< v_{\mathbf{j}}, v_{\mathbf{k}} >< v_{\mathbf{k}}, v_{\mathbf{i}} > \neq 0 \\ & i_\nu \leq j_\nu \leq k_\nu\ for\ all\ \nu \end{array} \tag{10.77}$$

10.6.4 Complements to Discriminants of Complete Linear Systems

This class of singular curves comprised of the curves where the fundamental groups come endowed with the homomorphisms into non-abelian groups given by either geometric monodromy i.e. with values in a mapping class group or (co)homological monodromy (with values in the linear group of automorphisms of the homology). Homological monodromies often are surjective or are close to such (i.e. the fundamental group itself is non-abelian). The construction of these curve is as follows. Let X be a smooth projective variety and let $\mathcal{L}$ be a line bundle. The linear system $\mathbb{P}(H^0(X, \mathcal{L}))$ contains the discriminant consisting of the elements having singularities worse than singularities of its generic element. With rare exceptions the discriminant has codimension 1 (identifying varieties with a small dual is an interesting problem). Its intersection with a generic plane[36] in $\mathbb{P}(H^0(X, \mathcal{L}))$ produces a plane curve which fundamental group of the complement has monodromy map into the mapping class group of the generic fiber of the universal element of this linear system i.e. the group of diffeomorphisms modulo isotopy of the fiber of the incidence correspondence $I_{\mathcal{L}} \subset X \times \mathbb{P}(H^0(X, \mathcal{L})$ set theoretically consisting of pairs $\{(x, C)|x \in X, C \in \mathbb{P}(H^0(X, \mathcal{L})), x \in C\}$. In [78] was considered the case $X = \mathbb{P}^2$ (resp. $X = V_2$ the quadric in $\mathbb{P}^3$) and $\mathcal{L} = \mathcal{O}_{\mathbb{P}^2}(3)$ (resp. $\mathcal{L} = \mathcal{O}_{V_2}(2)$) when one obtains as the fundamental group of the complement to discriminant the extension of $SL(2, \mathbb{Z})$ by the Heisenberg group over the field with 3 elements (resp. the ring $\mathbb{Z}_4$). The surjection onto $SL_2(\mathbb{Z})$ is the monodromy (the mapping class of 2-dimensional torus coincides with $SL_2(\mathbb{Z})$) and the kernel is the Heisenberg group. Recently, a progress was made in understanding the kernel of the monodromy in the case $\mathcal{L} = \mathcal{O}_{\mathbb{P}^2}(4)$ cf. [112].

A much more difficult case $X = \mathbb{P}^n, \mathcal{L} = \mathcal{O}_{\mathbb{P}^n}(d)$, including the case of discriminant of the family of cubic curves just described, was addressed by Lonne [153]. It also includes apparently the only other known case of this construction i.e. $X = \mathbb{P}^1, \mathcal{L} = \mathcal{O}(d)$ considered by Zariski (and mentioned in [78]) when the corresponding fundamental group is the braid group of two dimensional sphere. The fundamental group $\pi_1(\mathbb{P}(H^0(\mathbb{P}^n, \mathcal{O}_{\mathbb{P}^n}(d)) \setminus Disc))$ is the quotient of the group with generators and relations (10.77) by the normal subgroup generated by additional relations which we now shall describe. They are defined in terms of enumeration functions: $\Upsilon_k, k = 0, ..., n : \{1, ..., (d-1)^n\} \to I_{n,d}$ or equivalently the orderings of the integral points of the cube $I_{n,d}$. Among them, Υ_0 considered as the ordering of the integral points in $I_{n,d}$ according to the reverse lexicographic order: $(i_1, ..., i_n) < (i'_1, ..., i'_n)$ iff the for the smallest subscript k for which $i_k \neq i'_k$ one has $i_k > i_{k'}$ (e.g. $(d-1, d-1, d-1) < (d-1, d-1, d-2) < (d-1, d-1, d-2) < ... < (d-1, d-1, 1) < (d-1, d-2, d-1) < (d-1, d-2, d-2) < ......$). The order $<_k$ obtained from this one as follows:

$$(i_1, ..., i_n) <_k (j_1, ...j_n) \iff i_k < j_k \; or \; i_k = j_k, (i_1, ..., i_n) <_0 (j_1, ...j_n) \quad (10.78)$$

With this notations a presentation of $\pi_1(\mathbb{P}(H^0(\mathbb{P}^n, \mathcal{O}_{\mathbb{P}^n}(d)) \setminus Disc))$ is given by generators and relators (10.77) and

$$(t_i \delta_0)^{d-1} = (\delta_0 t_i^{-1})^{d-1}, \delta_0 \cdot .... \cdot \delta_n = 1 \text{where} \delta_k = \prod_{m=1}^{(d-1)^n} t_{\Upsilon_k(m)} \; k = 0, ..., n \quad (10.79)$$

It would be interesting to understand the algebraic structure of such groups and their relation to other geometrically defined group but see [153] for discussion of the

[36] Generic choice assures that the fundamental group of the complement to the intersection with the plane inside this plane is isomorphic to the fundamental group of the complement to the discriminant of the complete linear system. Non-generic section were studies in very special cases. For a recent study cf. [85].

relation of this presentation with those in cases known earlier. For results on the monodromy representations of the groups of the complements to discriminant using presentation (10.77), (10.79) we refer to [190] and for the case of monodromy of complements to discriminants on toric surfaces to [48, 189].

10.6.5 Plane Sextics and Trigonal Curves

In the last 10–20 years, many important results were obtained in the study of equisingular families of curves of degree 6 (and less; cf. [52, 53, 55, 56, 59, 63] and references below). The number of equisingular families of plane sextics measures in thousands and hence listing of possible cases is not a reasonable approach. Several classes of sextics were identified and we will describe some of them below. The methods include the use of Alexander invariants, connection with K3 surfaces and relation with the class trigonal curves on ruled rational surfaces. An interesting study of the moduli of sextics with six cusps i.e. the locus in the moduli space $\mathcal{M}_g$ given by the curves in distinct equisingular families was done in [95]. Several good surveys of the subject are already available (cf. [57], Preface and Sect. 7.2 in [54] and [1]).

A. *Simple and non-simple sextics.* A sextic is called *simple* if its only singularities are ADE singularities. Otherwise, a sextic is called *non-simple.* For irreducible non-simple sextics the type of equisingular deformation type is determined by the combinatorial type i.e. the collection of the local types of all singularities (cf. [57] Theorem 3.2.1). The key to a classification of simple sextics is the relation with the theory of K3 surfaces. Consider a double cover X_C of $\mathbb{P}^2$ branched over a sextic C. Singularities of this surface, correspond to the singularities of the branching curve and are simple of the same ADE type as the singularity of the curve. Moreover, the minimal resolution $\tilde{X}_C$ comes with the following data associated with the intersection form on $H_2(\tilde{X}_C, \mathbb{Z})$. Recall that as a lattice with bilinear form the latter is isomorphic to $\mathbb{L} = 2E_8 \oplus U^3$ where U is the intersection form of quadric surface. The data associated with the minimal resolution $\tilde{X}_C$ consists of the sublattice of $H_2(\tilde{X}_C, \mathbb{Z})$ spanned by the classes of exceptional curves of the resolution. These curves form a root system σ in this sublattice. Let $\tilde{S}_C$ be the primitive hull in $H_2(\tilde{X}_C, \mathbb{Z})$ of these sublattices and the pull back to $\tilde{X}_C$ of the class of a line in $\mathbb{P}^2$. An abstract oriented homological type of a K3 surface is a sublattice in $H_2(\tilde{X}_C, \mathbb{Z})$, which in the case of a double cover over a ADE sextics is the image of $\tilde{S}_C$, plus the orientation of the positive definite plane in real subspace spanned by transcendental lattice given by the holomorphic 2-form on $\tilde{X}_C$ (cf. [60] p. 214).

Theorem 10.6.1 (cf.[50, 213, 218]) *There is one to one correspondence between oriented abstract homological types arising from sextics and the set of equisingular deformations of sextic curves with simple singularities. Moreover, the moduli space of sextics in each equisingular component (i.e. its quotient by the group of projective isomorphisms) is isomorphic to the moduli space of K3 surfaces with such abstract homological type.*

Particularly well understood class of such sextics is the class of maximizing ones i.e. for which the sum of Milnor numbers is 19 (i.e. the maximal possible). However, there is no classification of the fundamental groups for the curves of this type though very large number of cases was made explicite.

B. Sextics of torus type. Those are sextics given by an equation of the form $f_2^3 + f_3^2 = 0$ where f_i denotes a form of degree i.

If f_i generic than for such C, $\pi_1(\mathbb{P}^2 \setminus C)$ is the quotient of the braid group B_3 by its center [219]. The fundamental group varies when $f_i = 0$ become singular or tangent to each other and there are many explicite calculations. For curves with simple singularities, having such type, the commutativity of the fundamental group of the complement is detected by the Alexander polynomial (Oka conjecture cf. [46, 51])

C. Sextics with triple points. Blow up of the plane at a triple point of a sextic results in Hirzebruch surface F_1 and a cover of degree 3 of projective line induced by projection from the triple point. Such curves and their braid monodromy were studied extensively by Degtyarev in his book [57] in a more general framework of trigonal curves on arbitrary Hirzebruch surfaces F_d. Relation between the braid monodromy and the graphs in 2-spheres leads to enumeration of extremal irreducible trigonal curves which shows that their number grows exponentially (as a function of appropriate parameter).

10.6.6 Zariski Pairs

One of applications of the fundamental groups of the complements (as was envisioned and implemented in some cases by Zariski cf. [219, 221]) is detecting the existence of different connected components of the space of equisingular deformations of curves on the surface. Indeed, those deformations do not alter the fundamental group. In fact there are several natural topological equivalence relations of curves on surfaces interrelationship between which is a natural question.

Definition 10.6.2 Let X be an algebraic surface and let D_1, D_2 be divisors on X. Pairs (X, D_1) and (X, D_2) are equivalent if one of the following conditions is satisfied:

(A) There exist an irreducible variety T, a holomorphic map $\Phi : \mathcal{X} \to T$ with a fiber biholomorphic to X, a divisor $\mathcal{D} \subset \mathcal{X}$ such that Φ is a locally trivial fibration of the pair $(\mathcal{X}, \mathcal{D})$[37] and such that there exist a pair of points $t_1, t_2 \in T$ the fibers of $\Phi|_{\mathcal{D}}$ over t_1, t_2 are D_1, D_2 respectively.

---

[37] I.e. for any $t \in T$ there is a neighborhood $U \subset T$ such that $\Phi^{-1}(U)$ and $T \times \Phi^{-1}(t)$ are equivalent as stratified spaces.

(A′) There is a symplectic isotopy of pairs (X, D_i) i.e. $(\mathcal{X}, \mathcal{D})$ in (A) is a pair of symplectic spaces with symplectic Φ with fiber being sympletomorphic to D_1, D_2 respectively.

(B) There exists a diffeomorphism (resp. PL equivalence, reps. homeomorphism, resp. a homotopy equivalence, resp. proper homotopy equivalence of the complements) of pairs $(X, D_1) = (X, D_2)$ i.e. one selects the corresponding type of a continuous map $X \to X$ taking subcomplex D_1 to D_2.

(C) There exists an isomorphism of fundamental groups $\pi_1(X \setminus D_1) = \pi_1(X \setminus D_2)$ (or sometimes just equality of the Alexander polynomials).

(D) There exist the following:

(i) a one to one correspondence between irreducible components of D_i such that corresponding components are homeomorphic and

(ii) a one to one correspondence between singularities of D_i preserving the local type in X compatible with correspondence (i) between the components.

(E) There exist an automorphism of fields $\mathbb{C}/\mathbb{Q}$ which takes (a deformation as in (a)) of the pair (X, D_1) to the pair (X, D_2).

The names used in literature are respectively, equisingular deformation equivalence for (A), Zariski pairs[38] for (D)-equivalent but not (B)-equivalent pairs , π_1-equivalent for (C), combinatorially equivalent for (D) and conjugation equivalent for (E).

The relation between these conditions is as follows: (A) implies (B) (Thom isotopy theorem), (B) implies (D) and also (C) by topological invariance of the fundamental groups. Relation between equivalences in (B) corresponding to different types of homeomorphisms of pairs are unknown in real dimension 4 and finally (E) implies (D).

Large and continuing to increase volume of papers deals with finding examples confirming that these implications cannot be reversed, though until 80s connected components of the strata were viewed as an aberration. The conditions found in [105] delineate the range of combinatorial data for which the strata are connected but numerous examples found up to date outside of this range, suggest that disconnectedness of equisingular families is a widespread occurrence. At the same time no systematic theory of Zariski pairs as to classification or distribution did emerge. A good survey of this vast subject is given in [15]. Further non-trivial results on the relations between above equivalences are as follows.

(C) or (D) does not imply (A): Shirane [205] showed that curves in equisingular families constructed earlier by Shimada (cf. [199]) cannot be transformed by a homeomorphism of $\mathbb{P}^2$ though fundamental groups are isomorphic. Work [65] contains examples of such type in the case of sextics.

(D) does not imply (C) for arrangements of lines defined over $\mathbb{Q}$: cf. [107] and references there for other numerous examples found by those authors giving arrangements of lines for which (D) does not imply (C). (D) does not imply (C) for reducible

---

[38] The term was coined in [22] in reference to first example found by Zariski in 1930s.

curves with components being a smooth curve and a union of certain 3 tangent lines cf. [204]. k-tuples of pairwise distinct reducible curves with one component of degree 4 and several quadrics were considered in [26] (also, see there the references to the works of these two authors presenting many other examples of failure of this implication).

(E) does not imply (C): [14] gives examples of conjugate line arrangements with non-isomorphic fundamental groups. See also [13] where one has conjugacy over $\mathbb{Q}$ and isomorphism of the fundamental groups and even homeomorphism of the complements but there is no homeomorphism of pairs. Examples are the appropriately chosen unions of sextics and lines. Moreover, (E) and (C) do not imply (B) (cf. [12])

Distinct connected components often even contain curves conjugate over $\mathbb{Q}$ (arithmetic Zariski pairs cf. [201]).

The examples of Zariski pairs or multiplets[39] fall in the following groups

A. Arrangements of lines and conics [108]
B. Curves of degree 6 and trigonal curves (cf. [61])
C. Other sporadic examples such as reducible curves with components of low degree (cf. [176]).

Methods employed in these works include study of the Alexander invariants, Hurwitz equivalence classes of braid monodromy and more ad hoc invariants of the fundamental groups (e.g. existence of dihedral cover of the complement to a curve is an invariant of the fundamental group and hence can be used to distinguish classes of equisingular deformations) and other sporadic methods (cf. [202, 203, 206]). Problems here include the question of combinatorial invariance of the Alexander polynomials and more generally the characteristic varieties or existence of Zariski pairs defined over $\mathbb{Q}$.

Many examples of fundamental groups of Zariski pairs were computed in [58].

An interesting problem about Zariski multiplets is understanding the asymptotic of the number of connected components of equisingular families when the number of classes of the curves grows. Consideration of families of trigonal curves on Hirzebruch surfaces, shows that the number of equisingular components grows exponentially. One can show exponential growth of the number of connected components of equisingular families of plane curves with nodes and cusps when degree grows by considering generic projections of surfaces of general type in a families which have exponentially large growth of the number of connected components of the moduli spaces (cf. [154]). This follows from the explicite formulas in terms of Chern numbers of the surfaces for the numbers of cusps, nodes and the degree of the branching curves of generic projection (cf. [131]).

[39] Zariski k-tuples are sets of k curves in distinct classes of equivalence (B) but in the same class (D); sometimes, in a more loose usage, the reference is to sets of k curves in distinct classes for some equivalences (A)–(E) but not another.

References

1. A. Akyol, A.Degtyarev, Geography of irreducible plane sextics. Proc. Lond. Math. Soc. (3) 111 (2015), no. 6, 1307–1337.
2. M. Alberich-Carramiñana, J. Alvarez Montaner, F. Dachs-Cadefau, V. González-Alonso, Poincaré series of multiplier ideals in two-dimensional local rings with rational singularities. Adv. Math. 304 (2017), 769–792.
3. J. W. Alexander, Topological invariants of knots and links. Trans. Amer. Math. Soc. 30 (1928), no. 2, 275–306.
4. V. Alexeev, V.; R. Pardini, On the existence of ramified abelian covers. Rend. Semin. Mat. Univ. Politec. Torino 71 (2013), no. 3–4, 307–315.
5. P. Aluffi, Characteristic classes of discriminants and enumerative geometry. (English summary) Comm. Algebra 26 (1998), no. 10, 3165–3193.
6. M. Amram, M. Teicher; Fei Ye, Moduli spaces of arrangements of 10 projective lines with quadruple points. Adv. in Appl. Math. 51 (2013), no. 3, 392–418.
7. M. Amram, R. Lehman, R. Shwartz, M. Teicher, Classification of fundamental groups of Galois covers of surfaces of small degree degenerating to nice plane arrangements. Topology of algebraic varieties and singularities, 65–94, Contemp. Math., 538.
8. M. Amram, M. Teicher, M. A. Uludag, Fundamental groups of some quadric-line arrangements. Topology Appl. 130 (2003), no. 2, 159–173.
9. D. Arapura, Geometry of cohomology support loci for local systems. I. J. Algebraic Geom. 6 (1997), no. 3, 563–597.
10. D. Arapura, M. Nori, Solvable fundamental groups of algebraic varieties and Kähler manifolds. Compositio Math. 116 (1999), no. 2, 173–188.
11. V. Arnold,; S. Gusein-Zade, A. Varchenko, Singularities of differentiable maps. Vol. II. Monodromy and asymptotics of integrals. Monographs in Mathematics, 83. Birkhäuser Boston Inc, Boston, MA, 1988.
12. E. Artal Bartolo, J. I. Cogolludo-Agustín, J. Martín-Morales, Triangular curves and cyclotomic Zariski tuples. Collect. Math. 71 (2020), no. 3, 427–441.
13. E. Artal Bartolo, J. I. Cogolludo-Agustín, Some open questions on arithmetic Zariski pairs.Singularities in geometry, topology, foliations and dynamics, 31–54, Trends Math., Birkhäuser/Springer, 2017.
14. E. Artal Bartolo, J. I. Cogolludo-Agustín, B.Guerville-Ballé, M.Marco-Buzunáriz, An arithmetic Zariski pair of line arrangements with non-isomorphic fundamental group. Rev. R. Acad. Cienc. Exactas Fis. Nat. Ser. A Mat. RACSAM 111 (2017), no. 2, 377–402.
15. E. Artal Bartolo, J. I. Cogolludo, H.Tokunaga, A survey on Zariski pairs. Algebraic geometry in East Asia-Hanoi, 2005, 1–100, Adv. Stud. Pure Math., 50, Math. Soc. Japan, Tokyo, 2008.
16. E. Artal Bartolo, R. Carmona Ruber, J. I. Cogolludo-Agustín, Braid monodromy and topology of plane curves. Duke Math. J. 118 (2003), no. 2, 261–278.
17. E. Artal Bartolo, Enrique, R. Carmona Ruber, J. I. Cogolludo Agustín, Essential coordinate components of characteristic varieties. Math. Proc. Cambridge Philos. Soc. 136 (2004), no. 2, 287–299.
18. E. Artal Bartolo, Topology of arrangements and position of singularities. Ann. Fac. Sci. Toulouse Math. (6) 23 (2014), no. 2, 223–265.
19. E. Artal Bartolo, J.I.Cogolludo-Agustín, D.Matei, Characteristic varieties of quasi-projective manifolds and orbifolds. Geom. Topol. 17 (2013), no. 1, 273–309.
20. E. Artal Bartolo, J. I. Cogolludo-Agustín, Jose Ignacio, A. Libgober, Characters of fundamental groups of curve complements and orbifold pencils. Configuration spaces, 81–109, CRM Series, 14, Ed. Norm., Pisa, 2012.
21. E. Artal Bartolo, J. I. Cogolludo-Agustín, A. Libgober, Depth of cohomology support loci for quasi-projective varieties via orbifold pencils. Rev. Mat. Iberoam. 30 (2014), no. 2, 373–404.
22. E. Artal-Bartolo, Sur les couples de Zariski. J. Algebraic Geom. 3 (1994), no. 2, 223–247.
23. D. Auroux, L. Katzarkov, Branched coverings of CP^2 and invariants of symplectic 4-manifolds. Invent. Math. 142 (2000), no. 3, 631–673.

24. D. Auroux, S. Donaldson, L. Katzarkov, M. Yotov, Fundamental groups of complements of plane curves and symplectic invariants. Topology 43 (2004), no. 6, 1285–1318.
25. R. Bahloul, Demonstration constructive de l'existence de polynomes de Bernstein-Sato pour plusieurs fonctions analytiques, Compos. Math. 141 (2005), no. 1, p. 175–191.
26. S. Bannai, H. Tokunaga, Geometry of bisections of elliptic surfaces and Zariski N-plets II. Topology Appl. 231 (2017), 10–25.
27. V. Batyrev, Dual polyhedra and mirror symmetry for Calabi-Yau hypersurfaces in toric varieties. J. Algebraic Geom. 3 (1994), no. 3, 493–535.
28. D. Bessis, Finite complex reflection arrangements are $K(\pi, 1)$. Ann. of Math. (2) 181 (2015), no. 3, 809–904.
29. E. Brieskorn, Die Fundamentalgruppe des Raumes der regularen Orbits einer endlichen komplexen Spiegelungsgruppe. Invent. Math. 12 (1971), 57–61.
30. A. Beauville, Annulation du H^1 pour les fibres en droites plats, Complex algebraic varieties (Bayreuth, 1990), 1–15, Lecture Notes in Math., 1507, Springer, Berlin, 1992.
31. R. Bieri, B. Eckmann, Groups with homological duality generalizing Poincaré duality. Invent. Math. 20 (1973), 103–124.
32. I. Biswas, M. Mj, Quasiprojective three-manifold groups and complexification of three-manifolds. Int. Math. Res. Not. IMRN 2015, no. 20, 10041–10068.
33. G. Bredon, Introduction to compact transformation groups. Pure and Applied Mathematics, Vol. 46. Academic Press, New York-London, 1972.
34. M. Broué, G. Malle, R. Rouquier, Complex reflection groups, braid groups, Hecke algebras. J. Reine Angew. Math. 500 (1998), 127–190.
35. D. Buchsbaum, D. Eisenbud, What annihilates a module? J. Algebra 47 (1977), no. 2, 231–243.
36. N. Budur, On Hodge spectrum and multiplier ideals. Math. Ann. 327 (2003), no. 2, 257–270.
37. N. Budur, Unitary local systems, multiplier ideals, and polynomial periodicity of Hodge numbers. Adv. Math. 221 (2009), no. 1, 217–250.
38. N. Budur, Bernstein-Sato ideals and local systems. Ann. Inst. Fourier (Grenoble) 65 (2015), no. 2, 549–603.
39. P. Cassou-Nogués, A. Libgober, Multivariable Hodge theoretical invariants of germs of plane curves. J. Knot Theory Ramifications 20 (2011), no. 6, 787–805.
40. P. Cassou-Nogués, A. Libgober, Multivariable Hodge theoretical invariants of germs of plane curves. II. Valuation theory in interaction, 82–135, EMS Ser. Congr. Rep., Eur. Math. Soc., Zürich, 2014.
41. S. E. Cappell, J. L. Shaneson, Singular spaces, characteristic classes, and intersection homology, Ann. of Math. 134 (1991).
42. F. Catanese, On a problem of Chisini. Duke Math. J. 53 (1986), no. 1, 33–42.
43. F. Catanese, C. Ciliberto, On the irregularity of cyclic coverings of algebraic surfaces. Geometry of complex projective varieties (Cetraro, 1990), 89–115, Sem. Conf., 9, Mediterranean, Rende, 1993.
44. J. I. Cogolludo Agustín, V. Florens, Twisted Alexander polynomials of plane algebraic curves. J. Lond. Math. Soc. (2) 76 (2007), no. 1, 105–121.
45. Topology of algebraic varieties and singularities. Papers from the Conference on Topology of Algebraic Varieties, in honor of Anatoly Libgober's 60th birthday, held in Jaca, June 22–26, 2009. Edited by José Ignacio Cogolludo-Agustín and Eriko Hironaka. Contemporary Mathematics, 538. American Mathematical Society, Providence, RI; Real Sociedad Matematica Espanola, Madrid, 2011.
46. J. I. Cogolludo and A. Libgober, Mordell-Weil groups of elliptic threefolds and the Alexander module of plane curves. J. Reine Angew. Math. 697 (2014), 15–55.
47. J. I. Cogolludo, A. Libgober, Free quotients of fundamental groups of smooth quasi-projective varieties, arxiv:1904.10852.
48. R. Cretois, L. Lang, The vanishing cycles of curves in toric surfaces I. Compos. Math. 154 (2018), no. 8, 1659–1697.

49. J. Davis, F. P. Kirk, Lecture notes in algebraic topology. Graduate Studies in Mathematics, 35. American Mathematical Society, Providence, RI, 2001.
50. A. Degtyarev, On deformations of singular plane sextics. J. Algebraic Geom. 17 (2008), no. 1, 101–135.
51. A. Degtyarev, Oka's conjecture on irreducible plane sextics. J. Lond. Math. Soc. (2) 78 (2008), no. 2, 329–351.
52. A. Degtyarev, On plane sextics with double singular points. Pacific J. Math. 265 (2013), no. 2, 327–348.
53. A. Degtyarev, Dihedral coverings of trigonal curves. Indiana Univ. Math. J. 61 (2012), no. 3, 901–938.
54. A. Degtyarev, Topology of algebraic curves. An approach via dessins d'enfants. De Gruyter Studies in Mathematics, 44. Walter de Gruyter and Co., Berlin, 2012.
55. A. Degtyarev, Hurwitz equivalence of braid monodromies and extremal elliptic surfaces. Proc. Lond. Math. Soc. (3) 103 (2011), no. 6, 1083–1120.
56. A. Degtyarev, The fundamental group of a generalized trigonal curve. Osaka J. Math. 48 (2011), no. 3, 749–782.
57. A. Degtyarev, Topology of plane algebraic curves: the algebraic approach. Topology of algebraic varieties and singularities, 137–161, Contemp. Math., 538, Amer. Math. Soc., Providence, RI, 2011.
58. A. Degtyarev, Classical Zariski pairs. J. Singul. 2 (2010), 51–55.
59. A. Degtyarev Plane sextics with a type E8 singular point. Tohoku Math. J. (2) 62 (2010), no. 3, 329–355.
60. A. Degtyarev, Plane sextics via dessins d'enfants. Geom. Topol. 14 (2010), no. 1, 393–433.
61. A. Degtyarev, Zariski k-plets via dessins d'enfants. Comment. Math. Helv. 84 (2009), no. 3, 639–671.
62. A. Degtyarev, A divisibility theorem for the Alexander polynomial of a plane algebraic curve. Zap. Nauchn. Sem. S.-Peterburg. Otdel. Mat. Inst. Steklov. (POMI) 280 (2001), Geom. i Topol. 7, 146–156, 300; translation in J. Math. Sci. (N.Y.) 119 (2004), no. 2, 205–210.
63. A. Degtyarev, Quintics in CP^2 with nonabelian fundamental group. (Russian) Algebra i Analiz 11 (1999), no. 5, 130–151; translation in St. Petersburg Math. J. 11 (2000), no. 5, 809–826.
64. A. Degtyarev, I. Isotopic classification of complex plane projective curves of degree 5. Algebra i Analiz 1 (1989), no. 4, 78–101; translation in Leningrad Math. J. 1 (1990), no. 4, 881–904.
65. A. Degtyarev, On the Artal-Carmona-Cogolludo construction, J. Knot Theory Ramifications 23 (2014), no. 5.
66. P. Deligne, Le groupe fondamental du complément d'une courbe plane n'ayant que des points doubles ordinaires est abelien (d'apres W. Fulton). Bourbaki Seminar, Vol. 1979/80, pp. 1–10, Lecture Notes in Math., 842, Springer, Berlin-New York, 1981.
67. P. Deligne, Equations differentielles a points singuliers reguliers. Lecture Notes in Mathematics, Vol. 163. Springer-Verlag, Berlin-New York, 1970.
68. P. Deligne, Theorie de Hodge. II. Inst. Hautes Etudes Sci. Publ. Math. No. 40 (1971), 5–57.
69. P. Deligne, Theorie de Hodge. III. Inst. Hautes Etudes Sci. Publ. Math. No. 44 (1974), 5–77.
70. G. Denham, A.Suciu, Multinets, parallel connections, and Milnor fibrations of arrangements. Proc. Lond. Math. Soc. (3) 108 (2014), no. 6, 1435–1470.
71. G. Denham, A. Suciu, S. Yuzvinsky, Abelian duality and propagation of resonance. Selecta Math. (N.S.) 23 (2017), no. 4, 2331–2367.
72. G. Dethloff, S. Orevkov, M. Zaidenberg, Plane curves with a big fundamental group of the complement, Voronezh Winter Mathematical Schools, 63–84, Amer. Math. Soc. Transl. Ser. 2, 184, Adv. Math. Sci., 37, Amer. Math. Soc., Providence, RI, 1998.
73. A. Dimca, Singularities and topology of hypersurfaces. Universitext. Springer-Verlag, New York, 1992.
74. A. Dimca, Hyperplane arrangements. An introduction. Universitext. Springer, 2017.
75. A. Dimca, L. Maxim, Multivariable Alexander invariants of hypersurface complements. Trans. Amer. Math. Soc. 359 (2007), no. 7, 3505–3528.

76. Dimca, A., Suciu, A.: Which 3-manifold groups are Kähler groups? J. Eur. Math. Soc. 11(3), 521–528 (2009).
77. A. Dimca, S. Papadima, A. Suciu, Alexander I Quasi-Kähler groups, 3-manifold groups, and formality. Math. Z. 268 (2011), no. 1–2, 169–186.
78. I. Dolgachev, A. Libgober, On the fundamental group of the complement to a discriminant variety. Algebraic geometry (Chicago, Ill., 1980), pp. 1–25, Lecture Notes in Math., 862, Springer, Berlin-New York, 1981.
79. W. Ebeling, Distinguished bases and monodromy of complex hypersurface singularities. Handbook of Geometry and Topology of Singularities, 2020. p. 427–466.
80. M. Eliyahu, D. Garber, M. Teicher, A conjugation-free geometric presentation of fundamental groups of arrangements. Manuscripta Math. 133 (2010), no. 1–2, 247–271.
81. D. Eisenbud, Commutative algebra. With a view toward algebraic geometry. Graduate Texts in Mathematics, 150. Springer-Verlag, New York, 1995.
82. D. Eisenbud, S. Popescu, S. Yuzvinsky, Hyperplane arrangement cohomology and monomials in the exterior algebra. Trans. Am. Math. Soc. 355(11), 4365–4383 (2003).
83. H. Esnault, Fibre de Milnor d'un cone sur une courbe plane singuliére. Invent. Math. 68 (1982), no. 3, 477–496.
84. H. Esnault, E. Viehweg, Lectures on vanishing theorems. DMV Seminar, 20. Birkhäuser Verlag, Basel, 1992.
85. A. Esterov, L. Lang, Braid monodromy of univariate fewnomials, arXiv:2001.01634.
86. M. Falk, S. Yuzvinsky, Multinets, resonance varieties, and pencils of plane curves. Compos. Math. 143 (4), 1069–1088 (2007).
87. K. M. Fan, Direct product of free groups as the fundamental group of the complement of a union of lines. Michigan Math. J. 44 (1997), no. 2, 283–291.
88. B. Farb, D. Margalit, A primer on mapping class groups. Princeton Mathematical Series, 49. Princeton University Press, Princeton, NJ, 2012.
89. Fei Ye, Classification of moduli spaces of arrangements of nine projective lines. Pacific J. Math. 265 (2013), no. 1, 243–256.
90. R. H. Fox, A quick trip through knot theory. 1962 Topology of 3-manifolds and related topics (Proc. The Univ. of Georgia Institute, 1961) pp. 120–167 Prentice-Hall, Englewood Cliffs, N.J.
91. S. Friedl, A. Suciu, Kähler groups, quasi-projective groups and 3-manifold groups. J. Lond. Math. Soc. (2) 89 (2014), no. 1, 151–168.
92. T. Fujita, On the topology of noncomplete algebraic surfaces. J. Fac. Sci. Univ. Tokyo Sect. IA Math. 29 (1982), no. 3, 503–566.
93. W. Fulton, On the fundamental group of the complement of a node curve. Ann. of Math. (2) 111 (1980), no. 2, 407–409.
94. W. Fulton, On the topology of algebraic varieties. Algebraic geometry, Bowdoin, 1985 (Brunswick, Maine, 1985), 15–46, Proc. Sympos. Pure Math., 46, Part 1, Amer. Math. Soc., Providence, RI, 1987.
95. C. Galati, On the number of moduli of plane sextics with six cusps. Ann. Mat. Pura Appl. (4) 188 (2009), no. 2, 359–368.
96. C. Galindo, F. Hernando, F. Monserrat, The log-canonical threshold of a plane curve. Math. Proc. Cambridge Philos. Soc. 160 (2016), no. 3, 513–535.
97. M. Golla, L. Starkston, The symplectic isotopy problem for rational cuspidal curves. arXiv:1907.06787.
98. J. González-Meneses, Basic results on braid groups, Annales Mathématiques Blaise Pascal, Volume 18 (2011) no. 1, p. 15–59.
99. M. González-Villa, A. Libgober, L.Maxim, Motivic infinite cyclic covers. Adv. Math. 298 (2016), 413–447.
100. M. Goresky, R. MacPherson, Stratified Morse theory. Ergebnisse der Mathematik und ihrer Grenzgebiete (3) 14. Springer-Verlag, Berlin, 1988.
101. H. Grauert, R. Remmert, Coherent analytic sheaves. Grundlehren der Mathematischen Wissenschaften, 265. Springer-Verlag, Berlin, 1984.

102. D. Greb, S. Kebekus, S. J. Kovacs, T. Peternell. Differential forms on log canonical spaces. Inst. Hautes Etudes Sci. Publ. Math, 114(1), 87–169, 2011.
103. M. Green, R. Lazarsfeld, Higher obstructions to deforming cohomology groups of line bundles. J. Amer. Math. Soc. 4 (1991), no. 1, 87–103.
104. Revêtements Êtales et Groupe Fondamental, Lecture Notes in Mathematics, 224, Springer-Verlag, 1971, by Alexander Grothendieck et al. Updating remarks by Michel Raynaud.
105. G. M. Greuel, C. Lossen, E. Shustin, Singular algebraic curves. With an appendix by Oleg Viro. Springer Monographs in Mathematics. Springer, Cham, 2018.
106. G. M. Greuel, Deformations and Smoothings of Singularities, Handbook of Geometry and Topology of Singularities, 2020. p. 369–426.
107. B. Guerville-Ballé, An arithmetic Zariski 4-tuple of twelve lines. Geom. Topol. 20 (2016), no. 1, 537–553.
108. B. Guerville-Ballé, J. Viu-Sos, Configurations of points and topology of real line arrangements. Math. Ann. 374 (2019), no. 1–2, 1–35.
109. A. Gyoja, Bernstein-Sato's polynomial for several analytic functions, J. Math. Kyoto Univ. 33(2) (1993) 399–411.
110. R. Hain, Monodromy of codimension 1 subfamilies of universal curves. Duke Math. J. 161 (2012), no. 7, 1351–1378.
111. J. Harris, On the Severi problem, Invent. Math. 84 (1986), 445–461.
112. R. Harris, The kernel of the monodromy of the universal family of degree d smooth plane curves. arxiv 1904.10355.
113. R. Hartshone, Algebraic geometry. Graduate Texts in Mathematics, No. 52. Springer-Verlag, New York-Heidelberg, 1977.
114. H. Hamm, Le Dung Trang, Un theoreme du type de Lefschetz. C. R. Acad. Sci. Paris Ser. A-B 272 (1971), A946–A949.
115. A. Hatcher, Algebraic topology. Cambridge University Press, Cambridge, 2002.
116. E. Hironaka, Abelian coverings of the complex projective plane branched along configurations of real lines. Mem. Amer. Math. Soc. 105 (1993), no. 502.
117. E. Hironaka, Alexander stratifications of character varieties. Ann. Inst. Fourier (Grenoble) 47 (1997), no. 2, 555–583.
118. F. Hirzebruch, Arrangements of lines and algebraic surfaces. Arithmetic and geometry, Vol. II, 113–140, Progr. Math., 36, Birkhäuser, Boston, Mass., 1983.
119. M. Ishida, The irregularities of Hirzebruch's examples of surfaces of general type with $c_1^2 = 3c_2$. Math. Ann. 262 (1983), no. 3, 407–420.
120. M. Kapovich and J. J. Millson, Inst. Hautes Études Sci. Publ. Math. No. 88 (1998), 5–95 (1999); C. R. Acad. Sci. Paris Sér. I Math. 325(8), 1997, pp. 871–876.
121. M. Kashiwara, "B-functions and holonomic systems. Rationality of roots of B-functions", Invent. Math. 38 (1976/77), no. 1, p. 33–53.
122. A. Kawauchi, A survey of knot theory. Burkhauser, 1990.
123. P. Kirk, C. Livingston, Twisted Alexander invariants, Reidemeister torsion, and Casson-Gordon invariants, Topology 38 (1999), no. 3, 635–661.
124. D. Kotschick, Kählerian three-manifold groups, Math. Res. Lett. 20 (2013), no. 3, 521–525.
125. Vik. S. Kulikov, On Chisini's conjecture, Izv. Ross. Akad. Nauk Ser. Mat. 63:6 (1999), 83–116.
126. B. Lasell, Complex local systems and morphisms of varieties. Compositio Math. 98 (1995), no. 2, 141–166.
127. R. Lazarsfeld, Positivity in algebraic geometry. vol I and, vol. II. Classical setting: line bundles and linear series. Ergebnisse der Mathematik und ihrer Grenzgebiete. 3. Folge. A Series of Modern Surveys in Mathematics 48. Springer-Verlag, Berlin, 2004.
128. Lê Dũng Tráng, Sur les noeuds algébriques. Compositio Math. 25 (1972), 281–321.
129. A. Libgober, Alexander polynomial of plane algebraic curves and cyclic multiple planes. Duke Math. J. 49 (1982), no. 4, 833–851.
130. A. Libgober, Alexander invariants of plane algebraic curves. Singularities, Part 2 (Arcata, Calif., 1981), 135–143, Proc. Sympos. Pure Math., 40, Amer. Math. Soc., Providence, RI, 1983.

131. A. Libgober, Fundamental groups of the complements to plane singular curves. Algebraic geometry, Bowdoin, 1985 (Brunswick, Maine, 1985), 29–45, Proc. Sympos. Pure Math., 46, Part 2, Amer. Math. Soc., Providence, RI, 1987.
132. A. Libgober, On the homology of finite abelian coverings. Topology Appl. 43 (1992), no. 2, 157–166.
133. A. Libgober, Groups which cannot be realized as fundamental groups of the complements to hypersurfaces in C^N. Algebraic geometry and its applications (West Lafayette, IN, 1990), 203–207, Springer, New York, 1994.
134. A. Libgober, Homotopy groups of the complements to singular hypersurfaces. II. Ann. of Math. (2) 139 (1994), no. 1, 117–144.
135. A. Libgober, On the homotopy type of the complement to plane algebraic curves. J. Reine Angew. Math. 367 (1986), 103–114.
136. A. Libgober, S. Yuzvinsky, Cohomology of the Orlik-Solomon algebras and local systems. Compositio Math. 121 (2000), no. 3, 337–361.
137. A. Libgober, Characteristic varieties of algebraic curves. Applications of algebraic geometry to coding theory, physics and computation (Eilat, 2001), 215–254, NATO Sci. Ser. II Math. Phys. Chem., 36, Kluwer Acad. Publ., Dordrecht, 2001.
138. A. Libgober, Hodge decomposition of Alexander invariants. Manuscripta Math. 107 (2002), no. 2, 251–269.
139. A. Libgober, Eigenvalues for the monodromy of the Milnor fibers of arrangements. Trends in singularities, 141–150, Trends Math., Birkhäuser, Basel, 2002.
140. A. Libgober, Isolated non-normal crossings. Real and complex singularities, 145–160, Contemp. Math., 354, Amer. Math. Soc., Providence, RI, 2004.
141. A. Libgober, Homotopy groups of complements to ample divisors. (English summary) Singularity theory and its applications, 179–204, Adv. Stud. Pure Math., 43, Math. Soc. Japan, Tokyo, 2006.
142. A. Libgober, Lectures on topology of complements and fundamental groups. Singularity theory, 71–137, World Sci. Publ., Hackensack, NJ, 2007.
143. A. Libgober, Non vanishing loci of Hodge numbers of local systems. Manuscripta Math. 128 (2009), no. 1, 1–31.
144. A. Libgober, Development of the theory of Alexander invariants in algebraic geometry. Topology of algebraic varieties and singularities, 3–17, Contemp. Math., 538, Amer. Math. Soc., Providence, RI, 2011.
145. A. Libgober, M. Mustata, Sequences of LCT-polytopes. Math. Res. Lett. 18 (2011), no. 4, 733–746.
146. A. Libgober, On Mordell-Weil groups of isotrivial abelian varieties over function fields. Math. Ann. 357 (2013), no. 2, 605–629.
147. C. Liedtke, Fundamental groups of Galois closures of generic projections. Trans. Amer. Math. Soc. 362 (2010), no. 4, 2167–2188.
148. Y. Liu, Nearby cycles and Alexander modules of hypersurface complements. Adv. Math. 291 (2016), 330–361.
149. Y. Liu, L. Maxim, B. Wang, Mellin transformation, propagation, and abelian duality spaces. Adv. Math. 335 (2018), 231–260.
150. Y. Liu, L. Maxim, B. Wang, Perverse sheaves on semi-abelian varieties-a survey of properties and applications, arXiv:1902.05430.
151. Y. Liu and L. Maxim, Reidemeister torsion, peripheral complex and Alexander polynomials of hypersurface complements. Algebr. Geom. Topol. 15 (2015), no. 5, 2757–2787.
152. F. Loeser, M. Vaquié, Le polynôme d'Alexander d'une courbe plane projective. Topology 29 (1990), no. 2, 163–173.
153. M. Lonne, Fundamental groups of projective discriminant complements. Duke Math. J. 150 (2009), no. 2, 357–405.
154. M. Lonne, M. Penegini, On asymptotic bounds for the number of irreducible components of the moduli space of surfaces of general type II. Doc. Math. 21 (2016), 197–204.

155. B. Malgrange, Polynômes de Bernstein-Sato et cohomologie evanescente, in Analysis and topology on singular spaces, II, III (Luminy, 1981), Asterisque, vol. 101, Soc. Math. France, Paris, 1983, p. 243–267.
156. G. Malle, On the distribution of Galois groups. J. Number Theory 92 (2002), no. 2, 315–329.
157. M. A. Marco Buzunáriz, A description of the resonance variety of a line combinatorics via combinatorial pencils. Graphs Combin. 25 (2009), no. 4, 469–488.
158. L. Maxim, Intersection homology and Alexander modules of hypersurface complements. Comment. Math. Helv. 81 (2006), no. 1, 123–155.
159. L. Maxim, Intersection Homology and Perverse Sheaves with Applications to Singularity Theory. Graduate Text in Mathenmatics, 218, Springer, 2019.
160. H. Maynadier, Polynomes de Bernstein-Sato associes a une intersection complète quasi-homogene á singularite isolee. Bull. Soc. Math. France 125 (1997), no. 4, 547–571.
161. J. Milnor, Infinite cyclic coverings. 1968 Conference on the Topology of Manifolds (Michigan State Univ., E. Lansing, Mich., 1967) pp. 115–133 Prindle, Weber and Schmidt, Boston, Mass.
162. J. Milnor, A duality theorem for Reidemeister torsion. Ann. of Math. (2) 76 (1962), 137–147.
163. J. Milnor, Singular points of complex hypersurfaces. Annals of Mathematics Studies, No. 61 Princeton University Press, Princeton, N.J.; University of Tokyo Press, Tokyo 1968.
164. B. Moishezon, Stable branch curves and braid monodromies. Algebraic geometry (Chicago, Ill., 1980), pp. 107–192, Lecture Notes in Math., 862, Springer, Berlin-New York, 1981.
165. B. Moishezon. The arithmetic of braids and a statement of Chisini. Geometric topology (Haifa, 1992), 151–175, Contemp. Math., 164, Amer. Math. Soc., Providence, RI, 1994.
166. B. Moishezon, M. Teicher, Simply-connected algebraic surfaces of positive index. Invent. Math. 89 (1987), no. 3, 601–643 and also B. Moishezon, M. Teicher, Galois coverings in the theory of algebraic surfaces, Proc. Symp. Pure Math. 46 (1987), 47–65. in the theory of algebraic surfaces, Proc. Symp. Pure Math. 46 (1987), 47–65.
167. D. Mumford, The topology of normal singularities of an algebraic surface and a criterion for simplicity. Inst. Hautes Etudes Sci. Publ. Math. No. 9 (1961), 5–22.
168. D. Mumford, Appendix 2 to Chapter VIII in [221].
169. A. Nadel, Multiplier ideal sheaves and Kahler-Einstein metrics of positive scalar curvature. Ann. of Math. (2) 132 (1990), no. 3, 549–596.
170. D. Naie, The irregularity of cyclic multiple planes after Zariski. Enseign. Math. (2) 53 (2007), no. 3–4, 265–305.
171. D. Naie, Mixed multiplier ideals and the irregularity of abelian coverings of smooth projective surfaces. Expo. Math. 31 (2013), no. 1, 40–72.
172. M. Namba, H. Tsuchihashi, On the fundamental groups of Galois covering spaces of the projective plane. Geom. Dedicata 105 (2004), 85–105.
173. S. Nazir, M. Yoshinaga, On the connectivity of the realization spaces of line arrangements. Ann. Sc. Norm. Super. Pisa Cl. Sci. (5) 11 (2012), no. 4, 921–937.
174. M. Nori, Zariski's conjecture and related problems. Ann. Sci. École Norm. Sup. (4) 16 (1983), no. 2, 305–344.
175. M. Oka, A survey on Alexander polynomials of plane curves. Singularités Franco-Japonaises, 209–232, Sémin. Congr., 10, Soc. Math. France, Paris, 2005.
176. A. Ozguner, Classical Zariski Pairs with nodes, Master Thesis, Bilkent University, 2007.
177. S. Papadima, A. Suciu, Bieri-Neumann-Strebel-Renz invariants and homology jumping loci. Proc. Lond. Math. Soc. (3) 100 (2010), no. 3, 795–834.
178. R. Pardini, Abelian covers of algebraic varieties. J. Reine Angew. Math. 417 (1991), 191–213.
179. J.V. Pereira, S. Yuzvinsky, Completely reducible hypersurfaces in a pencil. Adv. Math. 219 (2008), no. 2, 672–688.
180. C. Peters, J. Steenbrink, Mixed Hodge structures. Ergebnisse der Mathematik und ihrer Grenzgebiete. 52. Springer-Verlag, Berlin, 2008.
181. H. Popp, Fundamentalgruppen algebraischer Mannigfaltigkeiten. Lecture Notes in Mathematics, Vol. 176 Springer-Verlag, Berlin-New York 1970.
182. A. Robb, M. Teicher, Applications of braid group techniques to the decomposition of moduli spaces, new examples. Special issue on braid groups and related topics (Jerusalem, 1995). Topology Appl. 78 (1997), no. 1–2, 143–151.

183. X. Roulleau, G. Urzúa, Chern slopes of simply connected complex surfaces of general type are dense in [2, 3]. Ann. of Math. (2) 182 (2015), no. 1, 287–306.
184. W. Ruppert, Reduzibilitat ebener Kurven. J. Reine Angew. Math. 369 (1986), 167–191.
185. C. Sabbah, Proximite evanescente. I. La structure polaire d'un $\mathcal{D}$-module, Compositio Math. 62 (1987), no. 3, p. 283–328.
186. M. Saito, Mixed Hodge modules. Publ. Res. Inst. Math. Sci. 26 (1990), no. 2, 221–333. '.
187. M. Saito, Exponents of an irreducible plane curve singularity, arXiv:math.0009133.
188. M. Sakuma, Homology of abelian coverings of links and spatial graphs. Canad. J. Math. 47 (1995), no. 1, 201–224.
189. N. Salter, Monodromy and vanishing cycles in toric surfaces. Invent. Math. 216 (2019), no. 1, 153–213.
190. N. Salter. On the monodromy group of the family of smooth plane curves. arXiv:1610.04920.
191. M. Salvetti, Arrangements of lines and monodromy of plane curves. Compositio Math. 68 (1988), no. 1, 103–122.
192. C. Schnell, An overview of Morihiko Saito's theory of mixed Hodge modules, arXiv:1405.3096.
193. B. Segre, Sulla Caratterizzazione delle curve di diramazione dei piani multipli generali Mem. R. Acc. d'Italia, I 4 (1930), 531.
194. F. Severi, Vorlesungen uber algebraische Geometrie. Johnson Pub. reprinted, 1968; 1st. ed., Leipzig 1921.
195. C. Simpson, Subspaces of moduli spaces of rank one local systems. Ann. Sci. Ecole Norm. Sup. (4) 26 (1993), no. 3, 361–401.
196. C. Simpson, Local systems on proper algebraic V-manifolds. Pure Appl. Math. Q. 7 (2011), no. 4, Special Issue: In memory of Eckart Viehweg, 1675–1759.
197. I. Shimada, Fundamental groups of complements to singular plane curves. Amer. J. Math. 119 (1997), no. 1, 127–157.
198. I. Shimada, On the commutativity of fundamental groups of complements to plane curves. Math. Proc. Cambridge Philos. Soc. 123 (1998), no. 1, 49–52.
199. I. Shimada, Fundamental groups of algebraic fiber spaces. Comment. Math. Helv. 78 (2003), no. 2, 335–362.
200. I. Shimada, Equisingular families of plane curves with many connected components, Vietnam J. Math. 31 (2003), no. 2.
201. I. Shimada, On arithmetic Zariski pairs in degree 6. Adv. Geom. 8 (2008), no. 2, 205–225.
202. I. Shimada, Topology of curves on a surface and lattice-theoretic invariants of coverings of the surface. Algebraic geometry in East Asia-Seoul 2008, 361–382, Adv. Stud. Pure Math., 60, Math. Soc. Japan, Tokyo, 2010.
203. I. Shimada, Lattice Zariski k-ples of plane sextic curves and Z-splitting curves for double plane sextics. Michigan Math. J. 59 (2010), no. 3, 621–665.
204. T. Shirane, Galois covers of graphs and embedded topology of plane curves. Topology Appl. 257 (2019), 122–143.
205. T. Shirane, Connected numbers and the embedded topology of plane curves. Canad. Math. Bull. 61 (2018), no. 3, 650–658.
206. T. Shirane, A note on splitting numbers for Galois covers and π_1-equivalent Zariski k-plets. Proc. Amer. Math. Soc. 145 (2017), no. 3, 1009–1017.
207. J. Steenbrink, Mixed Hodge structure on the vanishing cohomology. Real and complex singularities (Proc. Ninth Nordic Summer School/NAVF Sympos. Math., Oslo, 1976), pp. 525–563. Sijthoff and Noordhoff, Alphen aan den Rijn, 1977.
208. M. Teicher, The fundamental group of a CP2 complement of a branch curve as an extension of a solvable group by a symmetric group. Math. Ann. 314 (1999), no. 1, 19–38.
209. M. Teicher, Braid monodromy type invariants of surfaces and 4-manifolds. Trends in singularities, 215–222, Trends Math., Birkhäuser, Basel, 2002.
210. K. Timmerscheidt, Mixed Hodge theory for unitary local systems. J. Reine Angew. Math. 379 (1987), 152–171.

211. H. Tokunaga, Dihedral coverings of algebraic surfaces and their application. Trans. Amer. Math. Soc. 352 (2000), no. 9, 4007–4017.
212. H. Tokunaga, Galois covers for S4 and U4 and their applications. Osaka J. Math. 39 (2002), no. 3, 621–645.
213. T. Urabe, Combinations of rational singularities on plane sextic curves with the sum of Milnor numbers less than sixteen. Singularities (Warsaw, 1985), 429–456, Banach Center Publ., 20, PWN, Warsaw, 1988.
214. E. Van Kampen, On the Fundamental Group of an Algebraic Curve. Amer. J. Math. 55 (1933), no. 1–4, 255–260.
215. M. Vaquié, Irregularité des revêtements cycliques des surfaces projectives non singuliéres. Amer. J. Math. 114 (1992), 1187–1199.
216. A. N. Varchenko, Asymptotic Hodge structure on vanishing cohomology. Izv. Akad. Nauk SSSR Ser. Mat. 45 (1981), no. 3, 540–591.
217. U. Walther, Bernstein-Sato polynomial versus cohomology of the Milnor fiber for generic hyperplane arrangements. Compos. Math. 141 (2005), no. 1, 121–145.
218. Jin-Gen Yang, Sextic curves with simple singularities. Tohoku Math. J. (2) 48 (1996), no. 2, 203–227.
219. O. Zariski, On the Problem of Existence of Algebraic Functions of Two Variables Possessing a Given Branch Curve. Amer. J. Math. 51 (1929), no. 2, 305–328.
220. O. Zariski, A theorem on the Poincare group of an algebraic hypersurface. Ann. of Math. (2) 38 (1937), no. 1, 131–141.
221. O. Zariski, Algebraic surfaces. With appendices by S. S. Abhyankar, J. Lipman and D. Mumford. Preface to the appendices by Mumford. Classics in Mathematics. Springer-Verlag, Berlin, 1995.
222. O. Zariski, On the purity of the branch locus of algebraic functions. Proc. Nat. Acad. Sci. U.S.A. 44 (1958), 791–796.
223. O. Zariski, P. Samuel, Commutative algebra. Graduate Texts in Mathematics, Vol. 29. Springer-Verlag, New York-Heidelberg, 1975.

Index

J. L. Cisneros-Molina et al. (eds.), *Handbook of Geometry and Topology of Singularities II*, https://doi.org/10.1007/978-3-030-78024-1